60 Springer Series in Solid-State Sciences

Edited by Peter Fulde

Springer Series in Solid-State Sciences

Editors: M. Cardona   P. Fulde   K. von Klitzing   H.-J. Queisser

Volume 1 – 39 are listed on the back inside cover

Excitonic Processes in Solids

By M. Ueta, H. Kanzaki, K. Kobayashi,
Y. Toyozawa, and E. Hanamura

With 307 Figures

Springer-Verlag
Berlin Heidelberg New York Tokyo

Professor Dr. Masayasu Ueta

Department of Physics, Tohoku University, Sendai 980, Japan
Present address: Department of Physics, Kyoto Sangyo University, Kyoto 603, Japan

Professor Dr. Hiroshi Kanzaki*
Professor Dr. Koichi Kobayashi**
Professor Dr. Yutaka Toyozawa

Institute for Solid State Physics, University of Tokyo, Minato-ku, Tokyo 106, Japan
* Present address: Fuji Film Research Lab., Minami-Ashigara, Kanagawa 250-01, Japan
** Present address: College of Liberal Arts, Toyama University, Toyama 930, Japan

Professor Dr. Eiichi Hanamura

Department of Applied Physics, University of Tokyo, Bunkyo-ku, Tokyo 113, Japan

Series Editors:

Professor Dr., Dr. h. c. Manuel Cardona
Professor Dr., Dr. h. c. Peter Fulde
Professor Dr. Klaus von Klitzing
Professor Dr. Hans-Joachim Queisser

Max-Planck-Institut für Festkörperforschung, Heisenbergstrasse 1
D-7000 Stuttgart 80, Fed. Rep. of Germany

ISBN-13:978-3-642-82604-7   e-ISBN-13:978-3-642-82602-3
DOI: 10.1007/978-3-642-82602-3

Library of Congress Cataloging-in-Publication Data. Main entry under title: Excitonic processes in solids. (Springer series in solid-state sciences; 60) Bibliography: p. Includes index. 1. Exciton theory. 2. Solids. I. Ueta, M. (Masayasu), 1919-. II. Series. QC176.8.E9E94 1985   530.4'1   85-27898

© Springer-Verlag Berlin Heidelberg 1986
Softcover reprint of the hardcover 1st edition 1986

Typesetting: Schwetzinger Verlagsdruckerei, 6830 Schwetzingen

2153/3130-543210

Preface

An exciton is an electronic excitation wave consisting of an electron-hole pair which propagates in a nonmetallic solid. Since the pioneering research of Frenkel, Wannier and the Pohl group in the 1930s, a large number of experimental and theoretical studies have been made. Due to these investigations the exciton is now a well-established concept and the electronic structure has been clarified in great detail.

The next subjects for investigation are, naturally, dynamical processes of excitons such as excitation, relaxation, annihilation and molecule formation and, in fact, many interesting phenomena have been disclosed by recent works. These excitonic processes have been recognized to be quite important in solid-state physics because they involve a number of basic interactions between excitons and other elementary excitations. It is the aim of this quasi monograph to describe these excitonic processes from both theoretical and experimental points of view.

To discuss and illustrate the excitonic processes in solids, we take a few important and well-investigated insulating crystals as playgrounds for excitons on which they play in a manner characteristic of each material. The selection of the materials is made in such a way that they possess some unique properties of excitonic processes and are adequate to cover important interactions in which excitons are involved. In each material, excitonic processes are described in detail from the experimental side in order to show the whole story of excitons in a particular material. Part of this book is devoted to the theoretical description of the excitonic processes which play particularly important roles in the materials chosen in this book but are not necessarily restricted to these materials. The theory is presented in a general fashion so as to cover a variety of phenomena which have been of recent interest.

It should be remarked that, although this book has been written through the cooperation of five authors, the main contribution to Chap. 3 was made by Ueta, Chaps. 5 and 6 by Kanzaki, Chaps. 7, 8, and 9 by Kobayashi, Chaps. 1 and 4 by Toyozawa and Chap. 2 by Hanamura.

The authors wish to express their gratitude to all of their colleagues for collaboration and discussions at various stages of their researches on excitons. One of the authors (MU) would like to acknowledge the assistance of Prof. T. Itoh and Dr. Y. Nozue in completing the manuscript. The authors thank the original authors of the figures used in this book who kindly gave permission to reproduce them. Thanks are also due to the Physical Society of Japan, The American Physical Society, The Institute of Physics, Progress of Theoretical

Physics, Akademie-Verlag, International Union of Crystallography, North-Holland Publishing Company, Pergamon Press Ltd. and Plenum Press for granting them permission for the reproduction of the figures. The authors are grateful to Miss Chikako Okada, Miss Takako Tokanai, Miss Yoko Kobayashi and Dr. Atsuko Sumi for typing the manuscript. Finally the authors should like to acknowledge the constant help and encouragement furnished by their wives, Chisako Ueta, Kiyo Kanzaki, Rei Kobayashi, Asako Toyozawa and Toshiko Hanamura without whose aid the work on excitons, on which this book is based, would probably not have been done.

Tokyo and Sendai 1984   *M. Ueta, H. Kanzaki, K. Kobayashi,*
                                          Y. Toyozawa, E. Hanamura

Contents

1. Introduction

This first chapter introduces the basic concept of an exciton as an elementary excitation in a many-electron system of an insulator. Various kinds of interaction associated with translational and internal motions of this composite particle, such as electron-hole Coulomb and exchange interactions, spin-orbit interactions and interatomic or intermolecular overlap and dipole-dipole energies, are described with particular attention to their interplay and to their reflection in the optical spectra. The coupled mode of light and electronic polarization waves, whose quanta are a photon and an exciton, respectively, is described in terms of a polariton picture, whereby the optical response of the bulk is related to that of the surface, with additional boundary conditions in the case of spatial dispersion. This chapter provides the conceptual basis for more dynamical aspects of the excitons to be described in later chapters.

1.1 The Ground State of Many-Body Systems and the Modes of Excitation

Which kind of ground state is preferred by a system consisting of a great number of like particles interacting with each other has always been a matter of fundamental interest; yet a question difficult to answer in a general way. The ground state should be a perfectly ordered state in the sense of the third law of thermodynamics that the entropy of a macroscopic system approaches zero with vanishing temperature [1.1]. Two typical ways of perfect ordering are a periodic array in r-space and condensation in k-space. The former is preferred when the interparticle interaction dominates the kinetic energy – atoms (except He) and molecules form crystalline lattices while low-density electrons form a Wigner lattice. The latter is preferred in the opposite situation – outer electrons in a solid are Fermi-condensed in the Bloch band because of their large interatomic transfer energy and the small pseudopotential from atoms and other electrons, while ^{4}He atoms, the lightest closed-shell atoms, become Bose-condensed in k-space. It should be noted, however, that many-particle systems in general can take ground states with much more varied and intriguing features [1.2] than the two limiting situations mentioned above.

Once the ground state is known, the excited states are next to be considered. We have two aims in doing this. Firstly, the responses of the system to small external fields are described in terms of virtual and/or real excitations of the system. Secondly, if the assumed ground state is an approximate or an inappropriate one, one has to consider mixing of some of the excited states to get the true ground state. This state may turn out to be significantly different from the one initially conceived, perhaps eventually involving symmetry breaking such as in super-structure formation.

The modes of low-lying excitations of a many-particle system can be classified into individual and collective excitations [1.3, 4]. The individual excitation in r-space is exemplified by the formation of a Frenkel defect in the perfect crystal lattice, while that in k-space is typified by the excitation of an electron across the Fermi energy (metal) or the band gap (insulator). The collective motions can have arbitrarily small amplitude within the classical mechanics, and the potential energy for such small displacements of collective coordinates from their equilibrium points can be approximated by a quadratic form, diagonalization of which gives a set of non interacting harmonic oscillators each with angular frequency ω_j. . By quantizing them, one finds that the low-lying excited states of the collective motions can be described in terms of an integral number n_j of energy quanta $\hbar\omega_j$ for each mode j:

$$E = \sum_j n_j \hbar\omega_j \quad (n_j = 0, 1, 2, \ldots) . \tag{1.1}$$

This system behaves as though it consists of an assemblage of noninteracting fictitious particles with energy $\hbar\omega_j$ ($j = 1, 2, \ldots$), which are called the elementary excitations. For example, a collective oscillation of atoms in solids is a lattice vibration, whose energy quantum is called a phonon. The collective oscillation of charge density in a system consisting of mobile electrons and ions (e.g., a metal and an ionized gas) is known as a plasma oscillation with a plasmon for its energy quantum. The deviation of spin orientation from its ordered state in a magnetic material propagates from site to site as a wave, which is called a spin wave; its energy quantum being a magnon.

The collective motions are not completely independent of the individual motions; on the contrary, the former are superpositions of the latter. The former claim their own significance as better eigenmodes of motion in the many-particle system than the latter, when the interparticle interaction is not small. To be more exact, the interaction can be partly incorporated, as a sort of average, into the effective field (e.g., the Hartree-Fock field for the electrons in the Bloch band) which acts upon the particles themselves thus governing their individual motion. The fluctuating part of the interaction is responsible for the collective motion on the one hand and for the interparticle correlation on the other. For example, the plasma oscillation in the Fermi-degenerate electrons in a metal has a much higher frequency than the individual excitations (whose energy range starts from zero) across the Fermi energy although the former is nothing but a superposition of the latter; the long-range fluctuating part of the Coulomb repulsion gives rise to the plasma oscillation and at the same time to the interelectron correlation which amounts to a screening of the repulsion into a short-range one [1.3].

1.2 Electronic Excitation in Insulators and the Wannier-Mott Exciton

The situation is different in insulators in which the individual electronic excitations have a lower bound ε_g – the band-gap energy. Let us assume the conduction (c) and valence (v) bands to be parabolic with isotropic effective masses m_e and m_h with a minimum and maximum, respectively, at $k = 0$:

$$\varepsilon_c(k) = \varepsilon_c(0) + \frac{\hbar^2 k^2}{2m_e} ,$$

$$\varepsilon_v(k) = \varepsilon_c(0) - \varepsilon_g - \frac{\hbar^2 k^2}{2m_h} . \tag{1.2}$$

One can then write the energy for one-electron excitation: $(v, k - K) \rightarrow (c, k)$ as [1.5]

$$\varepsilon_c(k) - \varepsilon_v(k - K) = \varepsilon_g(K) + \frac{\hbar^2 k'^2}{2\mu} \equiv E(k, K) ,$$

$$\varepsilon_g(K) \equiv \varepsilon_g + \frac{\hbar^2 K^2}{2M} , \quad \text{where} \tag{1.3}$$

$$M = m_e + m_h , \quad \mu^{-1} = m_e^{-1} + m_h^{-1} , \tag{1.4}$$

$$k' \equiv k - \frac{m_e}{M} K . \tag{1.5}$$

Here K and k' represent the wave vectors for the translational (center-of-mass) and the relative motions, respectively, of the electron in the conduction band (e) and the hole in the valence band (h), as is obvious from their associated masses M and μ (reduced mass) given by (1.4).

The Coulomb potentials from all the nuclei and other electrons have been incorporated into the periodic potential with which we solve for the one electron states $\varepsilon_c(k)$, $\varepsilon_v(k)$, ... self-consistently, assuming the ground state in which electrons fill the core and the valence bands. In the excited state (1.3) where we have a pair of e and h, there should be attractive Coulomb and repulsive exchange interactions between them. The former is given by $-e^2/\epsilon r$ if the e–h distance $r \equiv |r_e - r_h|$ is large compared to the lattice constant a_0, since the crystal lattice consisting of electrons and nuclei can then be regarded quasi-macroscopically as a polarizable medium specified by a dielectric constant ϵ [1.6]. From (1.3), the e–h relative motion is subject to the Schrödinger equation $(k' \rightarrow -i\nabla_r)$ [1.7]

$$\left(-\frac{\hbar^2}{2\mu} \nabla_r^2 - \frac{e^2}{\epsilon r} \right) \psi_\lambda(r) = \varepsilon_\lambda \psi_\lambda(r) \tag{1.6}$$

of the hydrogen atom type, which has the discrete and continuous eigenvalues

$$\varepsilon_{nlm} = -\frac{E_{ex}^b}{n^2} \quad (n = 1, 2, \ldots),$$

$$\varepsilon_{klm} = +\frac{\hbar^2 k^2}{2\mu}.$$

(1.7)

The e–h binding energy E_{ex}^b and the effective Bohr radius a_B in the 1s ($n = 1$, $l = m = 0$) state are given by rescaling those of the hydrogen atom (H):

$$E_{ex}^b = Ry = \frac{\mu e^4}{2\epsilon^2 \hbar^2} = \frac{e^2}{2\epsilon a_B} = \frac{\hbar^2}{2\mu a_B^2} = \frac{1}{\epsilon^2}\left(\frac{\mu}{m_0}\right) R_H,$$

(1.8)

$$a_B = \frac{\epsilon \hbar^2}{\mu e^2} = \epsilon\left(\frac{m_0}{\mu}\right) a_H,$$

(1.9)

where m_0 is the true mass of an electron.

The excitation energy, with the e–h Coulomb attraction taken into account, is then given by

$$E_{\lambda,K} = \varepsilon_g(K) + \varepsilon_\lambda$$

(1.10)

since the translational wave vector K remains a constant of motion. The discrete state $\lambda = (n, l, m)$ represents a bound pair of e and h which is called an exciton. The whole spectrum of an e–h pair excitation is shown schematically in Fig. 1.1 as a function of K. Since the wave vector of a photon in this region of excitation energy is negligibly small compared to the reciprocal lattice vector, only those

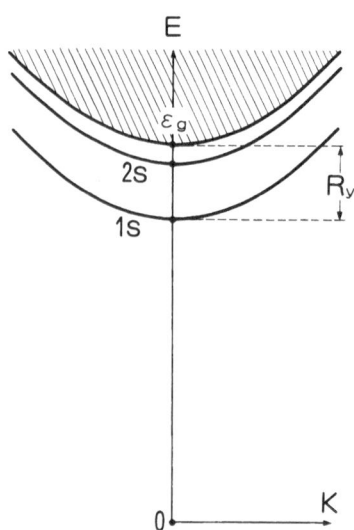

Fig. 1.1. Excitation energy of an e–h pair as a function of translational wave vector K

excited states with $K = 0$ contribute to the optical absorption spectra, which therefore consists of a series of discrete lines followed by the ionization continuum.

The above description of an exciton is based on two approximations: the effective mass approximation [the neglection of terms higher than quadratic in (1.2)] [1.7] and the dielectric continuum model for Coulomb screening (the use of macroscopic ε). They are justified only when $a_B \gg a_0$. Such an exciton is specifically called a Wannier-Mott exciton. This situation is well realized in narrow-gap semiconductors which usually have large $\epsilon(\gg 1)$ and small $\mu(\ll m_0)$.

1.3 The Frenkel Exciton

With narrow Bloch bands and hence with large effective masses m_e and m_h as in molecular crystals (small intermolecular overlapping), and/or with a smaller dielectric constant ϵ as realized in large band gap crystals, the relative motion wavefunction $\psi_\lambda(r)$ of the lowest exciton may be so localized that e and h are almost located on the same atom or molecule. One can then consider the exciton to be the intraatomic or intramolecular excitation energy which propagates through the lattice with wave vector K. This was in fact the model of an exciton as first conceived by *Frenkel* [1.8].

Let us consider, for simplicity, N identical molecules, each with one spinless electron, arrayed in a crystal lattice. We denote by Φ_n the Slater determinant of the electronic configuration in which only the nth molecule at R_n is in the excited state $a_c(r_n - R_n)$ with excitation energy ε, all other molecules (n') being in the ground state: $a_v(r_{n'} - R_{n'})$. Due to the intermolecular interactions, the excitation energy can propagate from site to site, resulting in the eigenstate:

$$\Psi_K = N^{-1/2} \sum_n [\exp(iK \cdot R_n)] \Phi_n , \tag{1.11}$$

similar in form to the tight-binding model for a one-electron Bloch state. In contrast to the latter case, however, the excitation energy can be transferred even without the intermolecular overlap: the Coulomb interaction $v_{nm} = e^2/|r_n - r_m|$ between the electrons on the nth and the mth molecules gives the matrix element

$$(\Phi_n^{cv} v_{nm} \Phi_m^{cv}) = \int dr_n \int dr_m a_c^*(r_n - R_n) a_v^*(r_m - R_m)$$

$$\times \frac{e^2}{|r_n - r_m|} a_v(r_n - R_n) a_c(r_m - R_m) \equiv w(R_{nm}) , \tag{1.12}$$

under the assumption of intermolecular orthogonality. The multiple expansion of v_{nm} gives the dipole-dipole interaction

$$H_{nm} \sim D(R_{nm}) = \frac{|\boldsymbol{\mu}|^2}{R_{nm}^3} - \frac{3|\boldsymbol{\mu} \cdot R_{nm}|^2}{R_{nm}^5} , \quad (n \neq m) \tag{1.13}$$

to be predominant at long distances $R_{nm} = |R_n - R_m|$, where μ is the intramolecular transition dipole moment:

$$\mu = \int a_c^*(r)(-er)a_v(r)\,dr \ . \tag{1.14}$$

The energy of the Frenkel exciton state (1.11) is then given by

$$E_K = \varepsilon + \sum_{n(\neq 0)} D(R_n)\exp(-iK \cdot R_n) \equiv \varepsilon + D_K \ . \tag{1.15}$$

Replacing the sum over distant molecules by integrals, as is justified for $Ka_0 \ll 1$, one finds

$$D_K \approx \text{const.} \ -\frac{4\pi}{3}N_0\left(|\mu|^2 - 3\frac{|\mu \cdot K|^2}{K^2}\right), \tag{1.16}$$

where N_0 is the number of molecules per unit volume. Equation (1.16) is singular at $K = 0$; the longitudinal exciton $(K\|\mu)$ has higher energy than the transverse one $(K\perp\mu)$ by $4\pi N_0\mu^2$. This difference originates from the depolarizing electric field due to the longitudinal component of the electronic polarization wave. The exciton is nothing but a quantum of this classical polarization wave.

1.4 The General Case

We have so far considered the simplest systems in the two limiting situations $(a \gtrless a_0)$. In the realistic case, we have first to consider the spin and orbital degeneracies of the bands which we denote by $v, v',\ldots$ and $\mu, \mu',\ldots$ for the conduction and the valence bands, respectively. Secondly, we have to introduce a discrete function $F(R_l)$ for the e–h relative motion with finite spatial extension. An exciton state with translational wave vector K can then be written as

$$\Psi_{\lambda K} = N^{-1/2} \sum_m \sum_l \sum_v \sum_\mu [\exp(iK \cdot R_m)] F_{\lambda K}^{v\mu}(R_l)\,\Phi_{m+l,m}^{v,\mu} \ , \tag{1.17}$$

$$\sum_v \sum_\mu \sum_l |F^{v\mu}(R_l)|^2 = 1 \ , \tag{1.18}$$

where $\Phi_{n,m}^{v,\mu}$ denotes the Slater determinant of the configuration in which the electron in the atomic or Wannier state $a_\mu(r - R_m)$ is excited into $a_v(r - R_n)$. The associated Bloch states

$$\psi_k(r) = N^{-1/2} \sum_m [\exp(ik \cdot R_m)]a_\mu(r - R_m) \tag{1.19}$$

constitute the one-electron energy matrix for the degenerate valence band:

$$\varepsilon_{\mu\mu'}(k) \ , \tag{1.20}$$

and the same for the conduction bands.

The eigenequation to determine the wave function $F^{\nu\mu}(R_l)$ and the excitation energy $E_{\lambda K}$ is then given, under the neglection of overlap-type energy integrals between different atoms or molecules, by [1.5, 9, 10]

$$\sum_{\nu'}\sum_{\mu'}\sum_{l'}[H_{\nu\mu l,\nu'\mu'l'}(K)-\delta_{\nu\nu'}\delta_{\mu\mu'}\delta_{ll'}E_{\lambda K}]F_{\lambda K}^{\nu'\mu'}(R_{l'})=0 . \tag{1.21}$$

The effective Hamiltonian for the exciton is given by

$$H_{\nu\mu l,\nu'\mu'l'} = [\delta_{\mu\mu'}\varepsilon_{\nu\nu'}(-i\nabla_l)-\delta_{\nu\nu'}\varepsilon_{\mu\mu'}(-i\nabla_l-K)$$
$$+ \delta_{\nu\nu'}\delta_{\mu\mu'}v(R_l)]\delta_{ll'} + \delta_{l0}\delta_{l'0}w_{K,\nu\mu,\nu'\mu'} . \tag{1.22}$$

In (1.22)

$$v(R_l) = -\int\int|a_\nu(r-R_l)|^2\frac{e^2}{|r-r'|}|a_\mu(r)|^2\,dr\,dr' \tag{1.23}$$

represents the e–h Coulomb energy, while w_K is the Fourier sum

$$w_{K,\nu\mu,\nu'\mu'} = \sum_m[\exp(-iK\cdot R_m)]w_{\nu\mu,\nu'\mu'}(R_m) \tag{1.24}$$

of the exchange-type energies

$$w_{\nu\mu,\nu'\mu'}(R_m) = \int\int dr\,dr'\,a_\nu^*(r-R_m)a_\mu(r-R_m)$$
$$\times \frac{e^2}{|r-r'|}a_{\mu'}^*(r')a_{\nu'}(r') \tag{1.25}$$

which correspond to (1.12).

With $R_l \neq 0$, (1.23) gives approximately $-e^2/R_l$. As was mentioned in Sect. 1.2, however, one has to replace it phenomenologically by a screened Coulomb potential $-e^2/\epsilon R_l$. The latter can be derived from first principles [1.11] on the basis of the electronic polaron [1.12] – an electron in the conduction band, as well as a hole in the valence band, is always accompanied by electronic polarization of the medium around itself. The electronic polarization, in turn, is equivalent to virtual excitation of e–h pairs or excitons as will be described in Sect. 1.5. Namely, the electronic configurations with multiply excited e–h pairs must also be considered in order to obtain the effective Hamiltonian for an exciton with the appropriately screened e–h interaction.

Similarly, (1.25) with $R_m \neq 0$ gives a dipole-dipole interaction of the form (1.12), which is again to be screened. The screening factor ϵ' appropriate for this dipolar term is obtained from the dielectric constant ϵ at the relevant frequency ($\omega = E_{\lambda 0}/\hbar$) by subtracting the contribution of this exciton state (λ), as will be shown in Sect. 1.6.

1.4.1 Effective Mass Approximation

With nondegenerate bands ($v = v' = c$, $\mu = \mu' = v$), we expand the one-electron excitation energy as

$$\varepsilon_v(k) - \varepsilon_\mu(k - K) = \varepsilon_g(K) + (k - k_m) \cdot \frac{\hbar^2}{2\mu} \cdot (k - k_m) + \ldots, \qquad (1.26)$$

around its minimum point $k_m(K)$ for a given K, where $\mu(K)$ is the reduced mass tensor. Then (1.21) reduces to the form similar to (1.6):

$$\left[-\boldsymbol{\nabla} \cdot \frac{\hbar^2}{2\mu} \cdot \boldsymbol{\nabla} - v(R) \right] \bar{F}_{\lambda K}(R) = [E_{\lambda K} - \varepsilon_g(K)] \bar{F}_{\lambda K}(R) , \qquad (1.27)$$

$$\bar{F}_{\lambda K}(R) = [\exp(-ik_m(K) \cdot R)] F_{\lambda K}(R) , \qquad (1.28)$$

under the neglection of the last term of (1.22) whose expectation value is given by $|F_{\lambda K}(0)|^2 w_K$. This is justified for the Wannier exciton ($a \gg a_0$) with small $|F_{\lambda K}(0)|^2 \sim O(a_0^3/a^3)$ and smooth $\bar{F}_{\lambda K}(R_l)$; then the expansion up to the quadratic term in (1.26) is also justified since the important k region is given by $(k/k_0) \sim (a_0/a) \ll 1$ (k_0 is the Debye cut off wave number).

1.4.2 The Role of Spin

Considering the spin degeneracy by explicitly rewriting $\mu \rightarrow \mu, \sigma$ and $v \rightarrow v, \tau$ with purely orbital indices μ, v and spin indices σ, τ, one obtains an extra factor $\delta_{\sigma\sigma'}\delta_{\tau\tau'}$ from the square bracket term in (1.22) representing the spin-independent one-electron energy, while the last term gives an extra factor $\delta_{\tau\sigma}\delta_{\tau'\sigma'}$ (note that r integration in (1.20) includes the summation over spin indices). The latter factor has eigenvalues 2 and 0 for spin singlet ($S = 0$) and triplet ($S = 1$) states, respectively, where $S = \tau - \sigma$ is the total spin of the e–h pair. Thus, the last term of (1.22) gives the expectation values

$$2\delta_{S,0}|F_{\lambda K}(0)|^2 \sum_m [\exp(-iK \cdot R_m)] w(R_m) , \qquad (1.29)$$

of which the term $R_m = 0$ represents the intramolecular exchange (1.25) splitting between $S = 0$ and 1. The other terms ($R_m \neq 0$) in (1.29), usually of the dipole-dipole type, contribute to the exciton transfer (through the K dependence of w_K), as the overlap-type energies of individual e and h do [through the term $\hbar^2 K^2/2M$ in (1.3)]. In molecular crystals with narrow Bloch bands, the former (dipole-dipole) interaction makes the dominant contribution to the transfer of the singlet exciton. The triplet exciton, which has to resort to the latter (overlap energies), can propagate only very slowly.

1.4.3 Interplay of Spin-Orbit and Exchange Interactions

As a typical model of band structures in cubic crystals with an optically allowed transition at the band edge ($k = 0$), one can assume a valence band (μ) consisting of triply degenerate p-like atomic orbitals ($\mu = p_x, p_y, p_z$) and an s-like conduction band ($\nu = s$). The atomic p-states ($l_v = 1$) are split into $j_v = 3/2$ and $1/2$ ($j_v \equiv l_v + \sigma$) through the spin-orbit interaction λ while the s-state ($l_c = 0$) has $j_c = 1/2$ only ($j_c \equiv l_c + \tau$). These structures are conserved under the cubic field. Neglecting the width of the valence band, one can write the square bracket of (1.22) with $K = 0$, as

$$-\frac{\hbar^2}{2m_e} \nabla_l^2 + v(R_l) + \varepsilon_g + \begin{cases} -\frac{1}{3}\lambda, \ j_v = \frac{3}{2}, \\ +\frac{2}{3}\lambda, \ j_v = \frac{1}{2}. \end{cases} \tag{1.30}$$

According to the j–j coupling scheme, with the total angular momentum of the e–h pair being given by $J = j_c - j_v$, $|j_c, j_v\rangle = |\frac{1}{2}, \frac{3}{2}\rangle$ generates $J = 1$ and 2 and $|\frac{1}{2}, \frac{1}{2}\rangle$ generates $J = 0$ and 1. With L–S coupling: $J = L + S$, one has only $L = 1$ ($l_c = 0, l_v = 1$), whence both $J = 2$ and 0 correspond to spin triplets: $S = 1$. Therefore, the exchange-dipole energy (1.29) contributes to the states with $J = 1$ only. Making use of the coefficients of transformation from L–S to j–j coupling, one finds the energy matrix of spin-orbit (λ) and exchange (Δ) interactions in the states $J = 1$:

$$\begin{array}{c} \langle\frac{1}{2}, \frac{3}{2}| \\ \\ \langle\frac{1}{2}, \frac{1}{2}| \end{array} \begin{pmatrix} -\frac{1}{3}\lambda + \frac{2}{3}\Delta & \frac{\sqrt{2}}{3}\Delta \\ \frac{\sqrt{2}}{3}\Delta & \frac{2}{3}\lambda + \frac{1}{3}\Delta \end{pmatrix} . \tag{1.31}$$

For the optically allowed transverse exciton ($\mu \perp K$), one finds Δ, from (1.29 and 16), to be

$$\Delta = 2|F_{\lambda 0}(0)|^2 \left[\iint a_s^*(r) a_x(r) \frac{e^2}{|r - r'|} a_x^*(r') a_s(r') \, dr \, dr' - \frac{4\pi}{3} N_0 \mu^2 \right]. \tag{1.32}$$

In Fig. 1.2a, we show the energy level scheme of the two $J = 1$ states as given by (1.31) and the $J = 0$ and 2 states of the spin triplet ($S = 1$), for varying λ with fixed $\Delta > 0$. The oscillator strength for optical transitions (allowed only for the $S = 0$ exciton) is shared between the former two with the intensity ratio as shown in Fig. 1.2b as a function of λ. From these two figures one can see how the L–S coupling scheme (Δ dominant) changes into the j–j coupling scheme (λ dominant).

The above argument for the simplest case [1.13] has been extended to much more general cases inclusive of crystals with lower symmetries under external fields [1.10].

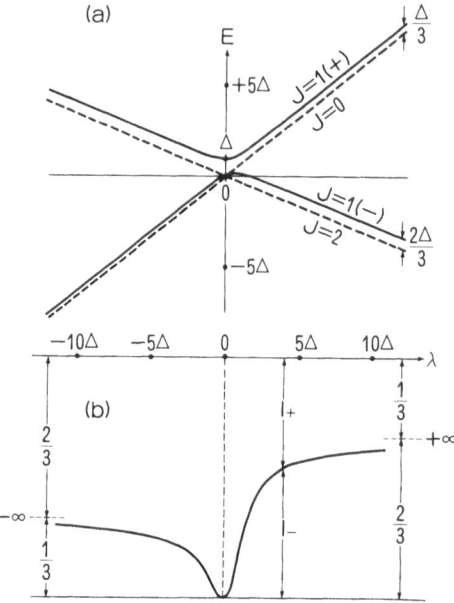

Fig. 1.2. (a) Energy levels and (b) oscillator strength ratio of a 1s exciton in a cubic crystal are shown against the spin-orbit splitting energy λ with fixed e–h exchange energy $\Delta(>0)$. They are reproduced from [ref. 1.13, Fig. 2] by changing abscissa from Δ to λ so as to facilitate comparison with the observations in $CuBr_xCl_{1-x}$ described in 3.2.6 (Fig. 3.11)

1.4.4 Davydov Splitting

If a unit cell contains $\varrho(\geq 2)$ inequivalent (e.g., differently oriented) molecules ($\nu = 1, 2, \ldots, \varrho$) of the same species, the singlet Frenkel exciton band consists of ϱ branches obtained by diagonalizing the energy matrix $2w_{K,\nu,\nu'}$ [(1.24, 25) where the suffix ν represents the intramolecular excitation of the ν-th molecule in the unit cell $(\Phi_{m,m}^{\nu,\nu})$ in (1.17)]. At $K = 0$, $w_{K,\nu,\nu'}$ is singular as it depends on the direction K/K [see (1.16) for the case of $\varrho = 1$]. For each direction of incident photon, there are two or more absorption lines with different polarizations since different eigenstates have different linear combinations of transition dipole moments: $\mu = \sum_\nu c_\nu \mu_\nu$ [1.14]. This structure in the absorption spectra is called Davydov splitting. It originates from the dipole-dipole interactions since the intramolecular exchange energy is common to all ν. The splitting can be easily related to the absorption intensities if the molecular arrangements within the unit cell are known.

1.4.5 Charge Transfer Excitons

In crystals consisting of alternately arrayed donor (D) and acceptor (A) molecules, the lowest excited state may be the charge transfer (CT) state from a donor to its neighboring acceptors rather than the intramolecular excitation within D or A. In such a case a_μ and a_ν represent the highest occupied molecular orbital (HOMO) of D and the lowest unoccupied molecular orbital (LUMO) of

A, respectively, and the e–h Coulomb and exchange terms take somewhat different forms from those given by (1.23, 25). The exchange splitting between the singlet and triplet excitons is small since a_μ and a_ν in (1.25) are always located on different molecules. In some compounds, the triplet intramolecular exciton of D or A appears below the CT exciton although the singlet intramolecular excitons are much higher [1.15].

1.5 Optical Absorption Spectra

The dielectric response of homogeneous matter to an electric field of the form $E_0 \exp(i q \cdot r - i\omega t)$ is given by the complex dielectric constant $\epsilon(\omega, q) = \epsilon_1(\omega, q) + i\epsilon_2(\omega, q)$, where ϵ_1 and ϵ_2 represent dispersive and dissipative parts, respectively. In order to calculate ϵ, we note that a system of unit volume is subject to the perturbing energy

$$H'(t) = - \int dr\, P(r) \cdot E_0 \exp(i q \cdot r - i\omega t) = -P_{-q} \cdot E_0 e^{-i\omega t}$$

where $P(r)$ is the polarization density. The polarizability $\alpha(\omega, q) \equiv [\epsilon(\omega, q) - 1]/4\pi$ is then given by calculating the response of the q component of $P(r)$: $\langle P_q(t) \rangle = \alpha(\omega, q) E_0 \exp(-i\omega t)$. Time-dependent perturbation theory gives immediately

$$\epsilon_2(\omega, q) = 4\pi^2 \sum_{e(\neq g)} (P_q)_{g,e} (P_{-q})_{e,g}\, \delta(E_e - E_g - \hbar\omega) \tag{1.33}$$

if the system is in the ground state g. The summation is to be extended over all the excited states e. We have used the dyadic expression for the tensor ϵ_2. The real part ϵ_1 of the dielectric constant is given in terms of ϵ_2 through the Kramers-Kronig relations:

$$\epsilon_1(\omega, q) - 1 = \frac{1}{\pi} \mathcal{P} \int_{-\infty}^{+\infty} \frac{d\omega'}{\omega' - \omega} \epsilon_2(\omega', q) . \tag{1.34}$$

How $\epsilon_2(\omega)$ is related to the absorption constant $A(\omega)$ will be described in Sect. 1.6.

The wave function of the excited state $e = (\lambda K)$ of our system, (1.17), can be rewritten, with the use of (1.19), as

$$\Psi_K = N^{-1/2} \sum_\nu \sum_\mu \sum_k f^{\nu\mu}_{\lambda K}(k) \Phi^{\nu,\mu}_{k,k-K} \tag{1.35}$$

in terms of the Slater determinant of the excited state with e and h in the (ν, k) and $(\mu, k - K)$ Bloch states, where

$$f^{\nu\mu}_{\lambda K}(k) \equiv N^{-1/2} \sum_l F^{\nu\mu}_{\lambda K}(R_l) e^{-i k \cdot R_l} . \tag{1.36}$$

Let us consider the current density of our electronic system,

$$\dot{P}(r) = j(r) = \frac{-e}{m_0} \sum_i \frac{1}{2} [\delta(r - r_i)p_i + p_i\delta(r - r_i)]$$

and take the matrix element $(\lambda K, g)$ of its Fourier component:

$$\frac{iE_{\lambda K}}{\hbar}(P_{-q})_{\lambda K, g} = \frac{-e}{m_0} \left[\sum_i \frac{1}{2}(e^{iq \cdot r_i}p_i + p_i e^{iq \cdot r_i}) \right]_{\lambda K, g}. \tag{1.37}$$

Making use of the fact that Φ in (1.35) is reached from the ground state by one electron excitation across the gap, one can write, in the right-hand side of (1.37),

$$[\cdots]_{\lambda K, g} = \delta_{K, q} \sum_{\nu\mu k} p_{\nu\mu}(k) f_{\lambda 0}^{\nu\mu}(k)^* \quad \text{where} \tag{1.38}$$

$$p_{\nu\mu}(k) \equiv \int \psi_{\nu, k}^*(r)(-i\hbar \nabla)\psi_{\mu, k}(r)\, dr . \tag{1.39}$$

We have neglected the light wave vector q in the summands of (1.38) since it is usually much smaller than the reciprocal lattice vector.

We will now study the Wannier-Mott exciton with large radius $(a \gg a_0)$. Since the relative motion wave function in (1.21) is slowly varying, $f_{\lambda K}(k)$ defined by (1.36) is strongly localized around $k_m(K)$ with $|\Delta k| \sim a^{-1}$. Since $p_{\nu\mu}(k)$ is much more slowly varying, we can expand the k summation in (1.38) as

$$\left[\sum_k f_{\lambda 0}(k) \right] p_{\nu\mu}(k_m(0)) + \left\{ \sum_k f_{\lambda 0}(k)[k - k_m(0)] \right\} \cdot \nabla p_{\nu\mu}(k_m(K)) + \dots \tag{1.40}$$

with convergence factor a_0/a. One has to consider the following two cases [1.16].

1.5.1 Allowed Edge Case: $p_{\nu\mu}(k_m(0)) \neq 0$

It is enough to consider the first term of (1.40). Then we obtain

$$\epsilon_2(\omega) = \frac{4\pi^2 N_0 e^2}{m_0^2 \omega^2} p_{\mu\nu}^{(m)} p_{\nu\mu}^{(m)} \sum_\lambda |\bar{F}_{\lambda 0}(0)|^2 \delta(E_{\lambda 0} - \hbar\omega) \tag{1.41}$$

where N_0 is the number of unit cells per unit volume. For an isotropic reduced mass $\mu(0)$, we have nonvanishing $\bar{F}_{\lambda 0}(0)$ only for the s-states. Thus we have a hydrogenic series of discrete lines

$$E_{n, 0} = \varepsilon_g(0) - \frac{E_{ex}^b}{n^2} \quad (n = 1, 2, \dots) \tag{1.42}$$

with respective intensities

$$|\bar{F}_{ns}(0)|^2 = \frac{v_0}{\pi a^3} \frac{1}{n^3} \tag{1.43}$$

and a continuous spectrum

$$\sum_{k'} |\bar{F}_{k'}(0)|^2 \, \delta\left(\varepsilon_g(0) + \frac{\hbar^2 k'^2}{2\mu} - \hbar\omega\right)$$

$$= \frac{\nu_0}{2\pi a^3 E_{ex}^b} \frac{1}{1 - \exp(-2\pi x)}, \qquad x \equiv \left(\frac{\hbar\omega - \varepsilon_g}{E_{ex}^b}\right)^{-1/2}. \tag{1.44}$$

The spectral density of the discrete lines, $|\bar{F}_n(0)|^2 \, |dE_n/dn|^{-1}$, tends with $n \to \infty$ to the step value $\nu_0/2\pi a^3 E_{ex}^b$ at the low-energy edge of the continuum (1.44).

1.5.2 Forbidden Edge Case: $p_{\nu\mu}(k_m(0)) = 0$

With the first nonvanishing term of (1.40), one has to replace $p_{\nu\mu}^{(m)} \bar{F}_{\lambda 0}(0)$ in (1.41) by $\nabla_k p_{\nu\mu}^{(m)} \cdot \nabla_R \bar{F}_{\lambda 0}(0)$ where the scalar product is to be taken between ∇_k and ∇_R. In the hydrogenic case, only the p-states have nonvanishing $\nabla_R \bar{F}_{\lambda 0}(0)$. Thus we obtain the discrete lines (1.42) without $n = 1$, with respective intensities

$$\left|\frac{\partial}{\partial x} F_{np_x}(0)\right|^2 = \frac{\nu_0}{\pi a^5}\left(\frac{1}{n^3} - \frac{1}{n^5}\right) \tag{1.45}$$

and a continuous spectrum

$$\frac{\nu_0}{2\pi a^5 E_{ex}^b} \frac{x^{-2} + 1}{1 - \exp(-2\pi x)}. \tag{1.46}$$

As in the allowed edge type, the spectral density behaves smoothly around $\hbar\omega = \varepsilon_g$, as

$$\frac{\nu_0}{2\pi a^5 E_{ex}^b} \frac{\hbar\omega - (\varepsilon_g - E_{ex}^b)}{E_{ex}^b},$$

whether it is the quasi continuum ($\hbar\omega < \varepsilon_g$) or the true continuum ($\hbar\omega > \varepsilon_g$). It should be noted that the forbidden edge type spectra near and below the interband edge ε_g are weaker than the allowed edge type by the order of $(a_0/a)^2$.

1.5.3 Transition from Frenkel to Wannier-Mott Exciton

In Sects. 1.2 and 1.3, we considered the excitons in the limiting situations of weak (Wannier) and strong (Frenkel) e–h binding. We will now study more systematically how the features of the absorption spectra vary between the two limits.

Let us rewrite the Schrödinger equation (1.21) for the relative motion of the $K = 0$ exciton as

$$(h + v - E_\lambda)F_\lambda = 0 \tag{1.47}$$

where h is the relative-motion kinetic energy with eigenvalues: $\varepsilon(k) \equiv \varepsilon_v(k) - \varepsilon_\mu(k)$ and eigenfunctions: $F_k(R_l) \equiv \langle R_l | k \rangle = \exp(ik \cdot R_l)$, and v is the e–h Coulomb potential. The exchange term w is neglected for simplicity. Noting that the eigenstate $|\lambda\rangle$ of $(h + v)$ contributes $|F_\lambda(0)|^2 = |\langle 0|\lambda\rangle|^2$ to the optical spectra of the allowed edge type (1.41) and that $\sum_\lambda |\langle 0|\lambda\rangle|^2 = 1$ due to the closure theorem, we find that the introduction of v displaces the average energy (the first moment) of the absorption spectra by

$$\sum_\lambda |F_\lambda(0)|^2 \, E_\lambda - \sum_k |F_k(0)|^2 \, \varepsilon_k$$

$$= \langle 0|h + v|0\rangle - \langle 0|h|0\rangle = v(0) . \tag{1.48}$$

$v(0)$ is the intramolecular (-atomic) e–h Coulomb energy given by (1.23) with $R_l = 0$. Usually $-v(0)$ is a few eV, being not much dependent on the material. In contrast, the dielectric constant ε effective in screening $v(R_l \neq 0)$, and the reduced effective mass μ related to the relative-motion band width $2B_r \equiv$ Max $[\varepsilon_v(k) - \varepsilon_\mu(k)]$ – Min $[\varepsilon_v(k) - \varepsilon_\mu(k)]$ by $B_r \sim (\hbar\pi)^2/2\mu a_0^2$, vary significantly with material.

In molecular crystals $(B_r \ll |v(0)|)$ with a large energy gap $(\to \epsilon \sim 1)$, one obtains the Frenkel exciton at $E_1 \sim v(0)$ (measured from the center of the relative-motion band, $\langle 0|h|0\rangle$) with $|F_1(0)|^2 \sim 1$ in conformity with (1.48), only a small fraction of oscillator strength being left for higher excitons and interband transitions. In semiconductors with large band widths $(B_r \gtrsim |v(0)|)$ and a small energy gap $(\to \epsilon \gg 1)$, the Coulomb-induced change of the absorption spectra given by the first moment, (1.48), is shared partly by discrete hydrogenic lines but mostly by the enhancement of the continuum over a wide range $(\gg E_{ex}^b)$ starting from the interband edge ε_g. In fact, the power series expansion of (1.44) gives the first term $\propto x^{-1} \propto (\hbar\omega - \varepsilon_g)^{1/2}$ representing the interband transition without Coulomb interaction, and the second, constant term representing the Coulomb enhancement. According to (1.48), this enhancement extends over the energy range of $\sim (4\pi a^3/v_0) \, [-v(0)/B_r]E_{ex}^b \gg E_{ex}^b$. Thus, the overall effect of the Coulomb interaction on the absorption spectra, given by the fractional shift of the first moment: $v(0)/B_r$, is much greater than is supposed from the fractional intensity of the discrete Wannier lines given by $\sim v_0/\pi a^3 \ll 1$. The fact that the Coulomb effect on the first moment of absorption spectra is always of the order of eV irrespective of the dielectric constant ε should be borne in mind in analysing the optical spectra of semiconductors in terms of the calculated band structures.

The variation, with $[-v(0)/B_r]$, of the energy level structures of an exciton and their reflection in the absorption spectra has been studied in [1.17, 18] with simplified models.

1.6 The Polariton and Spatial Dispersion

In the preceding section, we derived the dielectric constant ϵ of the electronic system by calculating the polarization P as a response to the electric field E. These two fields, as they propagate through the dielectric medium, must satisfy Maxwell's equations. Confining ourselves to the transverse part relevant to light propagation, we can rewrite them as

$$c^2 \nabla^2 E = \frac{\partial^2 D}{\partial t^2} ,$$ (1.49)

which has a plane-wave solution: $\exp(i K \cdot r - i \omega t)$ with the dispersion given by

$$\frac{c^2 K^2}{\omega^2} = n(\omega, K)^2 \equiv \epsilon(\omega, K) \equiv \frac{D}{E} ,$$ (1.50)

$n(\omega, K)$ being the refractive index.

The electronic polarization is contributed by various internal states λ of the exciton (inclusive of ionized states) as shown in previous sections. Out of them, we consider the contribution P from a particular state (say λ) explicitly, describing the contributions from all the other states ($\lambda' \neq \lambda$) phenomenologically by the residual dielectric constant ϵ' at the relevant frequency (we assume $\epsilon' > 0$):

$$D = \epsilon' E + 4\pi P .$$ (1.51)

The equation of motion for this particular mode of polarization P, with its own frequency $\omega_t \equiv E_{\lambda_0 K}/\hbar$, must be coupled to the electric field through the (static) polarizability $\alpha_0 \equiv \alpha(\omega = 0, K)$. The coupled equations now read

$$\ddot{P} + \omega_t^2 P = \omega_t^2 \alpha_0 E ,$$ (1.52)

$$\epsilon' \ddot{E} - c^2 \nabla^2 E = -4\pi \ddot{P} .$$ (1.53)

In isotropic media where we can assume P to be parallel to E for long waves ($K \ll a_0^{-1}$), the plane-wave solutions of (1.52, 53) satisfy

$$\begin{pmatrix} -\omega^2 + \omega_t^2 & -\alpha_0 \omega_t^2 \\ -4\pi\omega^2 & -\epsilon'\omega^2 + c^2 K^2 \end{pmatrix} \begin{pmatrix} P \\ E \end{pmatrix} = 0 .$$ (1.54)

By equating the determinant of coefficients to zero, one obtains the eigenfrequencies ω of the P–E coupled modes as functions of wave vector K. Or solving for K as a function of ω, one simply obtains

$$\frac{c^2 K^2}{\omega^2} = \epsilon(\omega) = \epsilon' + \frac{4\pi\alpha_0\omega_t^2}{\omega_t^2 - \omega^2} = \epsilon' \frac{\omega_l^2 - \omega^2}{\omega_t^2 - \omega^2}$$ (1.55)

where

$$\omega_l = \omega_t + \delta , \qquad \delta = \left[\left(1 + \frac{4\pi\alpha_0}{\epsilon'} \right)^{1/2} - 1 \right] \omega_t . \tag{1.56}$$

For the longitudinal polarization wave, we have, instead of (1.49), the equation $D = 0$ (note that $0 = \operatorname{div} D = i K \cdot D$ and $D \| K$) and hence $\epsilon(\omega) = 0$. Therefore, ω_l in (1.55) represents the frequency of the longitudinal mode.

It may be assumed that ϵ' is independent of ω if the other exciton states ($\lambda \neq \lambda_0$) are far enough from the region $\omega \sim \omega_t$ with which we are concerned. For the moment, we shall also neglect the K dependences of α_0, ω_t, and ϵ' since we are concerned only with long wavelengths $K \ll a_0^{-1}$. Equation (1.55) then gives the wave vector K and dielectric constant ϵ as functions of ω, which are shown schematically in Fig. 1.3a, b. Namely, the electromagnetic wave propagating through matter with dielectric dispersion given in Fig. 1.3b has $\omega(K)$ dispersion as shown in Fig. 1.3a. The quantum of this coupled wave is called the polariton. It is the stationary solution of to and fro conversion between a photon and an exciton which are the quanta of light waves and electronic polarization waves, respectively.

As is seen from (1.54), the upper polariton branch reduces to a pure E mode (with index of refraction $n = \sqrt{\epsilon'}$) when $\omega - \omega_t \gg \delta$, while the lower polariton branch reduces to a pure P mode when $0 < \omega_t - \omega \ll \delta$. In the region $\omega_t - \omega \gg \delta$, the polariton is essentially an electromagnetic wave with a constant index of refraction $n = \sqrt{\epsilon' + 4\pi\alpha_0}$, the difference from the region $\omega - \omega_t \gg \delta$ being the contribution of that particular exciton state to the polarization. In the bottom region of the upper branch such that $K \ll \sqrt{\epsilon'}\omega/c$, the polariton is subject to the depolarizing field $E = -4\pi P/\epsilon'$ since the second term on the left side of (1.53) is negligible. This is why the frequency of the upper polariton tends, with vanishing K, to the longitudinal frequency ω_l in spite of its transverse nature.

It is interesting to see how the longitudinal-transverse (L–T) splitting energy $\hbar\delta$, (1.56), obtained by the phenomenological description is related to $4\pi N_0 \mu^2$,

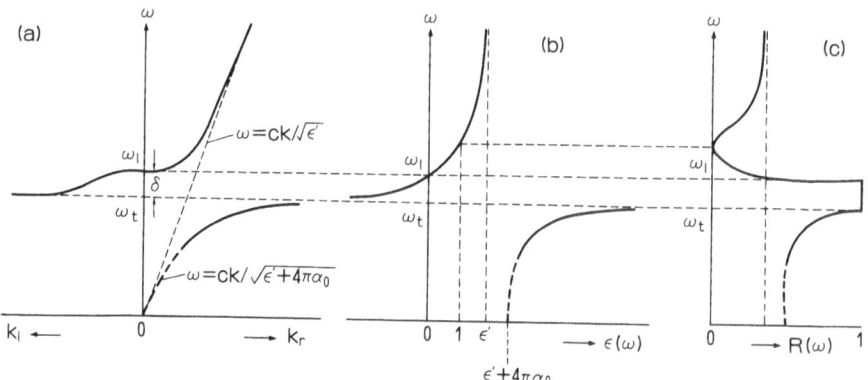

Fig. 1.3. (a) Dispersion of the polariton: $\omega(k)$. (b) Dielectric dispersion: $\epsilon(\omega)$. (c) Reflectivity

(1.16), obtained microscopically for the Frenkel exciton. Since the transition dipole moment $(P_{-K})_{g,K}$ is given by $\sqrt{N_0}\mu$, (1.11, 14), we obtain, from (1.33, 34, 56) $\hbar\delta \approx 4\pi N_0\mu^2/\varepsilon'$ when $\delta \ll \omega_t$. Namely, the dipole-dipole interaction in the relevant exciton state is screened through the residual dielectric constant ε' as it should be.

Since E and P form coupled waves inside the matter, the spectroscopic observation of the polarization waves reduces to the problem of the boundary condition through which the electromagnetic wave outside is connected to the polariton inside. An electromagnetic wave of the form: $\exp(iK_0 z - i\omega t)$ incident perpendicularly upon the surface of matter will give rise to a polariton wave of the form $\exp[iK(\omega)z - i\omega t]$ inside the matter ($z > 0$), where K_0 and $K(\omega)$ are given by ω/c and $\sqrt{\epsilon(\omega)}\omega/c$ (solid line in Fig. 1.3a), respectively. The fraction R of incident electromagnetic energy flux which is reflected by the surface is given, by the electromagnetic boundary conditions derivable from Maxwell's equations, as

$$R(\omega) = \left| \frac{n(\omega) - 1}{n(\omega) + 1} \right|^2 = \frac{(n_1 - 1)^2 + n_2^2}{(n_1 + 1)^2 + n_2^2}, \tag{1.57}$$

where $n(\omega) = n_1(\omega) + in_2(\omega) = \sqrt{\epsilon(\omega)}$ is the complex refractive index. The reflectivity $R(\omega)$ is shown in Fig. 1.3c.

In the region $\omega_t < \omega < \omega_1$, $\epsilon(\omega)$ is negative while $n(\omega)$ and $K(\omega)$ are purely imaginary, as shown in the left halves of Figs. 1.3a, b. Instead of the propagating mode, there is an evanescent mode of the form $\exp[-K_2(\omega)z - i\omega t]$ in this ω region, and hence, the energy density decreases exponentially with the attenuation constant

$$A(\omega) = 2K_2(\omega) . \tag{1.58}$$

This does not mean, however, that the incident electromagnetic energy is absorbed by matter. Instead, it is totally reflected by the surface without energy loss as shown in Fig. 1.3c. In fact, $\epsilon_2(\omega)$ relevant to dissipation is zero in this ω region.

On the other hand, the electromagnetic wave with $\omega = \omega_t$ should be subject to resonance absorption: $\epsilon(\omega)$ with real part (1.55) should have an imaginary part of $\delta(\omega - \omega_t)$ type according to (1.34). To be more realistic, let us introduce phenomenologically a damping term $-\gamma\dot{P}$ in the right hand side of (1.52). This modifies (1.55) into

$$\epsilon(\omega) = \epsilon' + \frac{4\pi\alpha_0\omega_t^2}{\omega_t^2 - \omega^2 - i\gamma\omega} . \tag{1.59}$$

The situation varies depending on the relative magnitude of the damping constant γ and the L–T splitting δ. In the "dissipative" case: $\delta \ll \gamma/2 (\ll \omega_t)$, one finds from (1.56) that $2\pi\alpha_0/\epsilon' \approx \delta/\omega_t \ll 1$, and hence, that the second term of (1.59) is smaller than the first term by a factor $\delta/(\gamma/2)$ even at resonance. One then obtains, from (1.58)

$$A(\omega) = 2K_2(\omega) \approx \frac{\omega}{c\sqrt{\epsilon'}}\epsilon_2(\omega) \approx \frac{\sqrt{\epsilon'}\omega}{c}\delta\frac{\gamma/2}{(\omega_t - \omega)^2 + (\gamma/2)^2}. \qquad (1.60)$$

Namely, the spatial attenuation of an electromagnetic wave is mainly due to the absorption of its energy by matter. $A(\omega)$ can then be called the absorption constant, and is proportional to $\epsilon_2(\omega)$ in the resonance region, being of Lorentzian shape with its width given by γ. The oscillator makes a significant contribution to $\epsilon_2(\omega)$ but little to $\epsilon_1(\omega)$.

It is in the "dispersive" case: $\gamma/2 \ll \delta$ that the polariton effect shows up clearly. In the greatest part of the L–T gap ($\omega_t < \omega < \omega_l$) excepting the very small region $|\omega - \omega_t| \lesssim \gamma/2$, the attenuation $A(\omega)$ is mainly due to total reflection and not due to absorption by matter.

The polariton picture is equivalent to the dielectric description if one neglects the spatial dispersion (K dependence) of ω_t, α_0, ϵ', and ϵ. Really new aspects of the polariton picture show up with the spatial dispersion, as the solution $K(\omega)^2$ of (1.59) and hence $n(\omega)^2 \equiv c^2 K(\omega)^2/\omega^2$ can be a multivalued function of ω. A monochromatic wave will then excite more than one (say, ν) polariton waves with different K's and n's, which phenomenon is known as birefringence. To relate the phases and amplitudes of these polariton waves and the reflected waves to those of the incident wave, ($\nu - 1$) additional boundary conditions (sometimes abbreviated as ABCs) are required besides the electromagnetic boundary condition. In contrast to the latter which is essentially macroscopic, the ABCs are related to the microscopic nature of the surface and the polarization waves. An extensive survey of the physics of spatial dispersion is provided in [1.10, 19, 20].

Finally, we should mention that the notion of the **E–P** coupled mode or polariton was introduced independently by *Born* and *Huang* [1.21] for the optical mode of lattice vibrations and by *Hopfield* [1.22] for the exciton.

1.7 Scope of the Present Book

We have given, in previous sections, a brief survey of the basic aspects of an exciton in an insulator, with particular attention to the electronic structures and the optical response. A great number of experimental and theoretical studies during recent decades have in fact revealed the finest details of the electronic structures of some typical materials.

On the basis of this accumulated knowledge, recent interest is growing in two directions. One is the extension of similar studies to a variety of rapidly increasing new (or less familiar) materials which are being discovered, explored, or synthesized. Another is to launch dynamical studies of exciton-exciton, exciton-phonon and exciton-photon interactions which are possible only on those materials with well-known electronic structures.

The present book mainly follows the second direction, except possibly for the last chapter which deals with less well understood materials. The chapters have been constructed and arranged as follows. Each of the experimental chapters, Chaps. 3, and 5–9, deals with a particular group of materials on which a variety of studies on dynamical behaviors of excitons have been made on the basis of more or less well-known electronic structures. They are arranged, together with the theoretical Chaps. 2 and 4, in such a way that the main interest of Chaps. 2 and 3 is the exciton-exciton and the exciton-photon interactions while that of later chapters is the exciton-phonon interaction. The former two interactions, which simultaneously become important under intense excitation, can cause new condensed phases of transiently created quasi particles as well as nonlinear optical effects of various types (some of which may be useful as devices). The exciton-phonon interaction causes not only scattering but also localization which may lead to catastrophes of local or bulk structural changes. We hope that the dynamical studies of interaction processes presented in this book will bear important relations with various other areas of condensed matter science and will provide a fertile basis for new technology.

The present book, which puts emphasis on dynamical aspects and concentrates on particular groups of materials, would be more useful if read together with other available books on excitons (authored or edited) by *Dexter* and *Knox* [1.23] (introductory), *Knox* [1.9] (standard), *Davydov* [1.14] (molecular crystals), *Cho* [1.10] (electronic structures and spectroscopy), *Agranovich* and *Ginzburg* [1.19] (spatial dispersion), *Rashba* and *Sturge* [1.24] (a variety of new topics), *Haken* and *Nikitine* [1.25] (high density excitons), *Silinsh* [1.26], *Kenkre* and *Reineker* [1.27], *Reineker* et al. [1.28], and *Rashba* et al. [1.29] (molecular crystals and aggregates).

2. Theoretical Aspects of Excitonic Molecules

An exciton in a pure crystal interacts with other excitons or/and lattice vibrations. The exciton is not always the most stable elementary excitation in solids and when excitons are created at such a high density that they enter each other's range of interaction before their decay, they condense into the more stable states of elementary excitations, i.e., excitonic molecules (EM) or electron-hole (e–h) metallic droplets. The type of condensation depends on the electronic band structure of the material. For example, it depends on whether the lowest band-to-band transition is direct or indirect (in many-valley structure) in the wave-vector space and whether the direct transition is dipole-allowed or -forbidden and the indirect one is phonon-allowed or -forbidden.

We can observe the EM as the characteristic luminescence spectrum and as the induced absorption due to the conversion of a single exciton into the EM. Furthermore, the EM can be coherently created by the giant two-photon absorption. The density of the EM (or its chemical potential) can be controlled by the power of the irradiating laser field. The extremely enhanced nonlinear optical response due to the giant two-photon excitation of the EM produces a variety of coherent optical phenomena, e.g., hyper-Raman scattering via the EM, four-wave mixing due to the EM and phase-conjugation by four-wave mixing. The dispersion of the polaritons and the EM has been determined precisely by hyper-Raman scattering. The anomalous dispersion of the exciton polariton which has been found under intense laser irradiation, is polariton renormalization due to the giant two-photon formation of the EM. This modification of the polariton dispersion under intense laser irradiation will be used for optical bistability. The Bose condensation of the EMs has been claimed to be observed as a sharp emission line under the high-density excitation of the EM at a low lattice temperature. This Bose condensation results from the boson nature and the large quantum effect of the EMs and will be discussed in relation to the relaxation of the EM due to mutual collisions under coherent high-density excitation of the EMs.

2.1 Fission and Fusion of Excitons vs. Chemical Reaction into Excitonic Molecules

When the electronic system of a semiconductor, or an ionic crystal, absorbs a photon with an energy larger than the band-gap energy, an electron is excited into the conduction band leaving a hole in the valence band. Bound e–h pair states, known as excitons, are the lowest energy electronic excitations in weakly excited, pure simiconductors. In a semiconductor with a small effective electron or hole mass and a large dielectric constant, the exciton wave funtion spreads over many unit cells and this weakly bound pair state is called a Wannier exciton. The relative motion of an e–h pair in a Wannier exciton is hydrogen-like if one neglects the e–h exchange interaction. This arises because the electrons in the conduction band and valence band are indistinguishable, and is relatively small for Wannier excitons due to the large mean e–h distance. Thus, Wannier excitons are in many respects analogous to hydrogen atoms. However, excitons have finite lifetimes which range from milli- and microseconds in indirect semiconductors to only nanoseconds in direct band gap materials. Under weak stationary excitation one obtains an exciton gas with a concentration n, in which the mean distance $d \equiv n^{-1/3}$ between two excitions is much larger than the exciton Bohr radius a_B. Traditional low excitation spectroscopy of excitions refers to the limit $na_B^3 \to 0$ where the interaction between the excitons is negligible in comparison with the interactions of the excitons with phonons and photons.

Laser or electron-beam excitation of semiconductors can generate e–h concentrations as high as $na_B^3 \gtrsim 1$ and many new physical phenomena occur as the exciton concentration increases from the low-density limit. The effective interaction potential between two excitons is attractive in many cases. In analogy with hydrogen, one expects the formation of molecules and indeed EMs (sometimes called biexcitons) have been observed in various semiconductors. However, the EM has three prominent characteristics in contrast to the hydrogen molecule.

The first characteristic of the EM gas is its large quantum effect as an assembly of Bose particles. The translational mass of the EM is two times the sum of electron and hole effective masses and this is very small for CuCl and CdS compared with the hydrogen molecule. As a result, a large quantum effect is expected for the gas system of EMs. This fact incidentally gives rise to the question as to whether the EM is stable in itself for any values of the electron to hole mass ratio, σ, against dissociation into two single excitons. We can answer this by evaluating the binding energy of the EM as a function of σ by a variational method. It was concluded [2.1, 2] that the EM is stable for every value of σ. Therefore, the large quantum effect due to the light translational mass and the boson-like nature of the EM made us expect Bose condensation of the EMs [2.3].

The second characteristic of the EM comes from the fact that the EM gas constitutes an open system in close contact with a radiational field. By making use of this character, we can create the EM gas at any concentration under laser irradiation and the chemical potential of the EM gas can be sustained by direct excitation of the EM through giant two-photon absorption [2.4] as will be discussed in Sect. 2.3. The characteristics of the EM gas are reflected in its emission spectrum. Coherent nonlinear optical phenomena are induced through the coherent giant two-photon excitation of the EMs. These subjects will be discussed in Sect. 2.4.

The third characteristic comes from the variety of the conduction and valence band structures. The EM gas is not always stable and a metallic droplet of electron and hole liquid is a more stable quasi-stationary state of excitons at high densities, e.g., in Ge and Si [2.5]. This metallic droplet may also be considered as a macromolecule. Which one is realized depends on the band structure of the crystal and on the degree of excitation. Figure 2.1 shows the average energy per e–h pair against the inverse concentration of e–h pairs. The crystal of CuCl, for example, has a simple band structure without valley degeneracy. In such a case, the gas state of the EM has a lower average energy than the metallic state of the e–h gas. As a result, the EM gas is stable below concentration A, and the metallic state of the electron-hole gas will be realized above concentration B. Between these concentrations the metallic e–h gas and the insulating gas of EMs may coexist spatially. On the other hand, in Ge and Si the metallic droplets of the e–h gas are more stable than the molecular gas state due to the many-valley structure of the conduction band. This point will be discussed in Sect. 2.2.

Just as the process of binding two excitons into an EM may be visualized by analogy to chemical reaction, so we have processes which can be understood as the fission and fusion of excitons. In organic molecular crystals the lowest electronic excited states are triplet exciton states and the singlet exciton has almost twice the energy of the triplet exciton. This energy difference comes from the e–h exchange effect in the Frenkel exciton. Both types of exciton can be generated with light, although the absorption coefficient leading to the generation of triplet excitons by near-infrared light is very small ($\sim 10^{-4}$ cm^{-1} in anthracene) because the transition occurs between states of different multiplicity. As two diffusing triplet excitons enter each other's range of interaction, they can undergo pairwise annihilation, giving rise to a singlet exciton. This is the fusion of two triplet

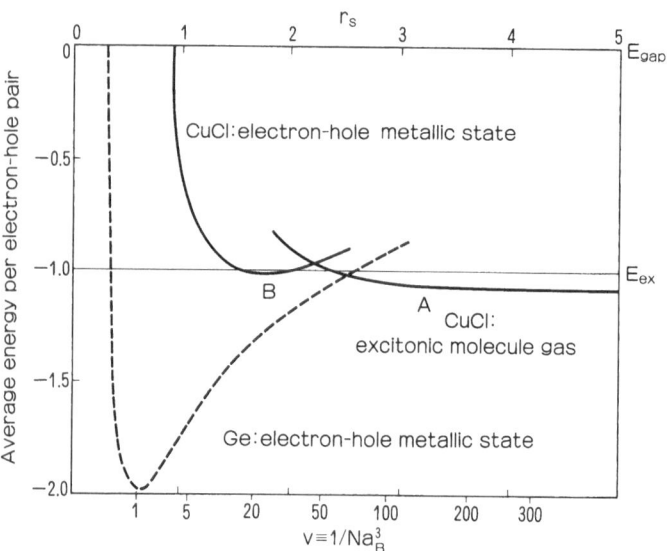

Fig. 2.1. Average energy per e–h pair as a function of pair concentration for CuCl and Ge [2.3]

excitons into a singlet exciton and the luminescence from radiative decay of these singlet excitons, resulting from triplet-triplet fusion, is called delayed fluorescence. The prompt fluorescence of singlet excitons is observed under the excitation of singlet excitons with blue or ultraviolet light. The fission or splitting of a singlet exciton into two triplet excitons (i.e., the inverse of triplet-triplet annihilation) in tetracene crystals was proposed and suggested as the dominant loss mechanism for singlets in order to explain the anomalously low quantum yield of 0.0002 at room temperature. Furthermore, the thermal quenching of the quantum yield was explained on the assumption that the singlet exciton fission reaction is endothermic, and that an activation energy $E(= 2E_T - E_S)$ is required. When the magnetic and temperature dependence of prompt and delayed fluorescence was observed [2.6], changes opposite in sign for the prompt and delayed fluorescence were found confirming the model of exciton fusion and fission satisfactorily.

In this chapter we confine ourselves to the EM in semiconductors and ionic crystals.

2.2 The Excitonic Molecule and Electron-Hole Liquid

2.2.1 Binding Energy and Electronic Structure of Excitonic Molecule in a Simple System – CuX

We can derive the envelope function of the EM in the framework of the effective mass approximation indpendently of the detailed structure of the Bloch functions. Here we neglect, in the zeroth order approximation, the e–h exchange interaction in comparison with the Coulomb interaction among the composite particles of the EM in the system of Wannier excitons. Then the EM can be described in terms of the effective masses m_e and m_h at the band extrema and the dielectric constant ε_0 only.

In many semiconductors the ratio of the electron to hole masses is much larger than the electron-proton mass ratio, which has a value of about 1/1840. This fact gives rise to the question whether the EM with an arbitrary mass ratio will be stable. To answer this, the binding energy of two excitons is evaluated by the variational method in the framework of the effective mass approximation. The Hamiltonian H_{ex-ex} for two electrons (1, 2) and two holes (a, b) can be derived in the same way as the exciton Hamiltonian in Chap. 1, if one considers a four particle state $\sum C_{k_1 k_2 k_a k_b} a_{k_1}^\dagger a_{k_2}^\dagger b_{k_a}^\dagger b_{k_b}^\dagger |0\rangle$. The resulting Hamiltonian is given by

$$H_{ex-ex} = -\frac{\hbar^2}{2\,m_e}\,(\nabla_1^2 + \nabla_2^2) - \frac{\hbar^2}{2\,m_h}\,(\nabla_a^2 + \nabla_b^2) + V \,, \qquad (2.1)$$

where the potential energy V is given by

$$V = \frac{e^2}{\epsilon_0}\left(\frac{1}{r_{12}} + \frac{1}{r_{ab}} - \frac{1}{r_{1a}} - \frac{1}{r_{1b}} - \frac{1}{r_{2b}} - \frac{1}{r_{2a}}\right) . \tag{2.2}$$

Throughout this subsection, the unit of length is the exciton Bohr radius a_B and the unit of energy is twice the exciton binding energy $A_{ex} = 2E_{ex}^b$.

Neglecting the spin-orbit interaction, we have four electronic states for the system of two electrons and two holes with an arbitrary mass ratio:

$$\Phi_{s,s} = C_1(R)[\phi_a(1)\phi_b(2) + \phi_b(1)\phi_a(2)]\chi_e(0, 0)\chi_h(0, 0) ,$$

$$\Phi_{s,t} = C_2(R)[\phi_a(1)\phi_b(2) + \phi_b(1)\phi_a(2)]\chi_e(0, 0)\chi_h(1, M_h) ,$$

$$\Phi_{t,s} = C_3(R)[\phi_a(1)\phi_b(2) - \phi_b(1)\phi_a(2)]\chi_e(1, M_e)\chi_h(0, 0),$$

$$\Phi_{t,t} = C_4(R)[\phi_a(1)\phi_b(2) - \phi_b(1)\phi_a(2)]\chi_e(1, M_e)\chi_h(1, M_h) . \tag{2.3}$$

Here $C_1(R)$ and $C_4(R)$ are symmetric and $C_2(R)$ and $C_3(R)$ are antisymmetric with respect to the exchange of two holes. In these eqations $\phi_a(1)$ denotes the envelope function of an exciton composed of the 1 electron and the a hole, and $\chi_e(S, M)$ and $\chi_h(S, M)$ are the spin functions for a pair of two electrons and two holes respectively.

The lowest bound state of two excition is described by $\Phi_{s,s}$ because these singlet structures, both for electrons and holes, minimize the intraband exchange energy of two electrons and two holes which is dominant in the binding of two excition into an EM in semiconductors. In the limit of a small electron to hole mass ratio $\sigma \equiv m_e/m_h$, the state $\Phi_{s,t}$ corresponds to the first excited rotational state.

Following the approach of [2.1], we use, as a trial function for the ground state, a slightly modified form of the function $\Phi_{s,s}$ in (2.3) which can describe a radial deformation as well as a polarization of the exciton wave functions in a molecule:

$$\Phi_{s,s} = f(R)g(\xi_1, \eta_1, \xi_2, \eta_2, R) , \tag{2.4}$$

where $f(R) = R^{\gamma/2} \exp(-\delta R/2)$, and

$$g = \exp[-\alpha(\xi_1 + \xi_2)R/2]\cosh[\beta(\eta_1 - \eta_2)R/2] ,$$

with four independent parameters α, β, γ, and δ. The relations

$$\xi_i = (r_{ia} + r_{ib})/R , \qquad \eta_i = (r_{ia} - r_{ib})/R$$

define ξ_i and η_i ($i = 1, 2$) which are two components of the vector $R_i = r_i - (r_a + r_b)/2$, expressed in terms of prolate spheroidal coordinates, the foci of which are taken at the positions of the holes. For simplicity it is assumed that there is no dependence of g on the third angular component of R_i. The function g is a symmetrized product of deformed atomic orbitals, as can be seen by rewriting it as follows:

$$g = \tfrac{1}{2}[\phi_a'(1)\phi_b'(2) + \phi_b'(1)\phi_a'(2)] \; , \quad \text{where} \tag{2.5}$$

$$\phi_a'(i) = \exp\left[-(\alpha + \beta)\left(r_{ia} + \frac{\alpha - \beta}{\alpha + \beta}\, r_{ib}\right)\right] ,$$

$$\phi_b'(i) = \exp\left[-(\alpha + \beta)\left(r_{ib} + \frac{\alpha - \beta}{\alpha + \beta}\, r_{ia}\right)\right] .$$

Here the quantities $(\alpha + \beta)$ and $(\alpha - \beta)/(\alpha + \beta)$ measure the uniform radial deformation of the $1s$-exciton orbitals in a molecule and the degree of their polarization, respectively. If one considers the limit of large values γ and δ with a fixed ratio of $\gamma/\delta = R_0$, the function (2.4) approaches the trial function of *Inui* [2.7] for a hydrogen molecule. If we put $f(R) = 1$ by setting $\gamma = \delta = 0$, $\Phi_{s,s}$ reduces to the function of *Hylleraas* and *Ore* [2.8] for a positronium molecule. In this limit, $\Phi_{s,s}$ is symmetric with respect to electrons and holes. Thus, the trial function (2.4) describes our system adequately.

Evaluating the expectation value of the Hamiltonian $H_{\mathrm{ex-ex}}$ with $\Phi_{s,s}$ and minimizing it with respect to the four variational parameters, we obtain the ground-state energy E as function of the mass ratio $\sigma \equiv m_e/m_h$. The binding energy $E_{\mathrm{mol}}^{\mathrm{b}}$ of the EM is the energy which is required to dissociate a molecule into two noninteracting excitons, i.e., $E_{\mathrm{mol}}^{\mathrm{b}} = -E - 2E_{\mathrm{ex}}^{\mathrm{b}}$. In Fig. 2.2 we plot the ratio $E_{\mathrm{mol}}^{\mathrm{b}}/E_{\mathrm{ex}}^{\mathrm{b}}$ as a function of σ in the range $0 \leq \sigma \leq 1$. For values of σ larger than unity, we have only to read the abcissa as σ^{-1}, owing to the symmetry of the system. Figure 2.2 shows that the EM should be stable for any value of the mass ratio σ [2.1]. Similar results have been found also in [2.2]. The ratio $E_{\mathrm{mol}}^{\mathrm{b}}/E_{\mathrm{ex}}^{\mathrm{b}}$ is a monotonically decreasing function of σ in the range of $0 \leq \sigma \leq 1$, as was first shown in [2.9 and 2.10].

It is interesting to know how the size of the EM depends on the parameter σ. For $\sigma = 0$ the interhole distance is fixed. With increasing σ the kinetic energy of the holes increases leading to their delocalization and to the growth of the mean interhole distance $\langle R \rangle$, which is calculated from the formula

$$\langle R \rangle = \int \psi^2 R \, d\tau_r / \int \psi^2 d\tau_r \; . \tag{2.6}$$

The evaluation of $\langle R \rangle$ as a function of the parameter σ was performed [2.1] and the result is shown in Fig. 2.3. From this it is seen that the ratio $\langle R \rangle/a_{\mathrm{B}}$ increases monotonically from 1.44 to 3.47 in the interval $0 \leq \sigma \leq 1$. Thus the EM is a looser formation than the hydrogen molecule. It is also interesting to note that $\langle R \rangle$ changes much more slowly than $E_{\mathrm{mol}}^{\mathrm{b}}/E_{\mathrm{ex}}^{\mathrm{b}}$.

Here we have described the EM in terms of the isotropic effective masses of the electron and hole and the dielectric constant only. This is applicable to the e–h system in a cubic crystal such as CuCl and CuBr. However, when we want to

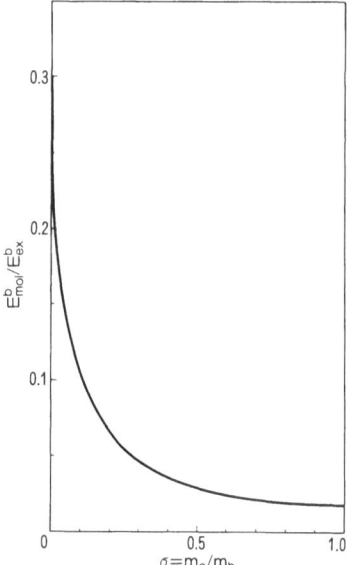

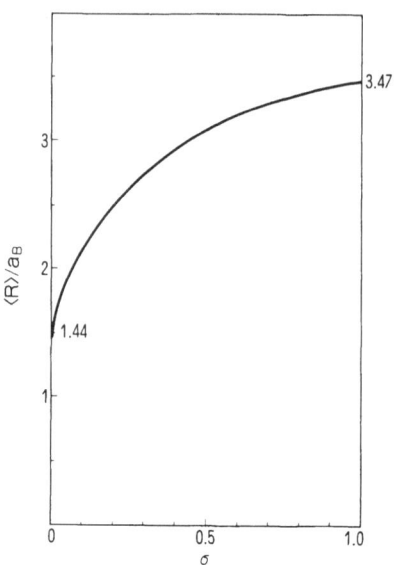

Fig. 2.2. Binding energy E_{mol}^b of the EM normalized by the exciton binding energy E_{ex}^b as a function of the electron to hole mass ratio $\sigma = m_e/m_h$ [2.1]

Fig. 2.3. Mean hole-hole distance $\langle R \rangle$ normalized by the exciton Bohr radius a_B as a function of the electron to hole mass ratio $\sigma = m_e/m_h$ [2.1]

know the effect of the e–h exchange and the optical response in the EM, we need the detailed structure of the Bloch functions at the band extrema.

We construct the wave function of an EM as a linear combination of four-particle functions obtained as the product of two-electron functions at the bottom of the conduction band and two-hole functions at the top of the valence band [2.11, 12]. These combinations are chosen to be partner functions of one of the irreducible representations of the crystal and to have definite parities under permutation of the two electrons and the two holes; the antisymmetric character of the total four-particle function is then assured by the opposite permutational properties of the envelope function. This procedure will be shown to be justified in the case of Wannier excitons. When effects due to spin-orbit interaction can be neglected, the total spin of the four particles is a good quantum number and the states can be classified according to their spin multiplets. Under the effect of spin-orbit interaction, total spin is no longer a good quantum number, and the combination of the products of four Bloch functions, in general, cannot be separated into the product of a space part and a spin part of definite total spin, but it is rather the superposition of different spin states.

In this section, we discuss the detailed electronic structure of the EM in the zinc blende crystal. This crystal has a band structure with the top of the valence band at $k = 0$ (symmetry Γ_7) and the bottom of the conduction band also at $k = 0$ (symmetry Γ_6). A lower fourfold-degererate valence band of symmetry Γ_8

is split off by the spin orbit interaction. In the case of CuCl, the energy gap $E_g(\Gamma_6 - \Gamma_7)$ is 3.428 eV and the spin-orbit splitting $\Delta(\Gamma_7 - \Gamma_8)$ is 69 meV.

The conduction Bloch functions Γ_6 can be written as

$$c_{1/2} = \psi_c|\alpha\rangle \quad \text{and} \quad c_{-1/2} = \psi_c|\beta\rangle , \tag{2.7}$$

where ψ_c can be identified with the $4s$ functions of Cu. The valence Bloch functions Γ_7 are of the form

$$v_{-1/2} = \sqrt{\tfrac{2}{3}}\psi_{-1}|\alpha\rangle - \sqrt{\tfrac{1}{3}}\psi_0|\beta\rangle \quad \text{and}$$

$$v_{1/2} = \sqrt{\tfrac{2}{3}}\psi_1|\beta\rangle + \sqrt{\tfrac{1}{3}}\psi_0|\alpha\rangle , \tag{2.8}$$

where the orbital parts may be expressed as $\psi_{\pm 1} = (p_x \pm ip_y)/\sqrt{2}$ and $\psi_0 = p_z$, and p_x, p_y, p_z are the normalized p-like functions. The mixing of a d-like orbital with this state may be neglected in the discussions of optical properties of excitions and EMs in CuCl.

The superposition of excitations of a valence electron into the conduction band gives two kinds of excitons which have Γ_2 and Γ_5 symmetry, respectively. The Γ_2 state corresponds to a state of total angular momentum $J_{ex} = 0$ and can be shown to be a superposition of pure triplet e–h spin states. The Γ_5 state corresponds to a state of $J_{ex} = 1$ and is a superposition of singlet and triplet e–h spin states. The wave functions of Γ_2 and Γ_5 excitons can be given as follows:

$$\begin{pmatrix} |\Gamma_2^{ex}(n, \boldsymbol{K})\rangle \\ |\Gamma_{5,m}^{ex}(n, \boldsymbol{K})\rangle \end{pmatrix} = \frac{1}{\sqrt{V}} \exp[i\boldsymbol{K} \cdot (\alpha\boldsymbol{r}_1 + \beta\boldsymbol{r}_a)] \cdot \phi_n(\boldsymbol{r}_1 - \boldsymbol{r}_a) \begin{pmatrix} |P_0^0(1, a)\rangle \\ |P_1^m(1, a)\rangle \end{pmatrix}$$

$$m = 1, 0, -1 \tag{2.9}$$

where the parts of Bloch functions are given by

$$|P_0^0(1, a)\rangle = \frac{1}{\sqrt{2}}[c_{1/2}(1)v_{-1/2}(a) - c_{-1/2}(1)v_{1/2}(a)]$$

$$\equiv \frac{1}{\sqrt{2}}\left[\left(\tfrac{1}{2}, -\tfrac{1}{2}\right) - \left(-\tfrac{1}{2}, \tfrac{1}{2}\right)\right] ,$$

$$|P_1^1(1, a)\rangle = c_{1/2}(1)v_{1/2}(a) \equiv \left(\tfrac{1}{2}, \tfrac{1}{2}\right) ,$$

$$|P_1^0(1, a)\rangle = \frac{1}{\sqrt{2}}[c_{1/2}(1)v_{-1/2}(a) + c_{-1/2}(1)v_{1/2}(a)] \tag{2.10}$$

$$\equiv \frac{1}{\sqrt{2}}\left[\left(\tfrac{1}{2}, -\tfrac{1}{2}\right) + \left(-\tfrac{1}{2}, \tfrac{1}{2}\right)\right] ,$$

$$|P_1^{-1}(1, a)\rangle = c_{-1/2}(1)v_{-1/2}(a) \equiv \left(-\tfrac{1}{2}, -\tfrac{1}{2}\right) ,$$

and $\alpha = m_h/(m_e + m_h)$ and $\beta = 1 - \alpha$. The envelope function $\phi(r_1 - r_a)$ which describes the relative motion of electron and hole in an exciton is obtained under the effective mass approximation as $1/\sqrt{\pi a_B^3} \exp(-r/a_B)$, where $r = |r_1 - r_a|$ and the effective Bohr radius a_B is given by $\hbar \epsilon_0/\mu e^2$ in terms of the e–h reduced mass $\mu = (1/m_e + 1/m_h)^{-1}$ and the static dielectric constant ϵ_0.

The wave functions of EMs are also superpositions of excitations of two valence electrons into the conduction band, as stated at the beginning of this section. The ground state of the molecule has singlet structures, for both electrons and holes, which minimize the intraband exchange energy of two electrons and two holes. As a result, it is a Γ_1 state with zero total angular momentum $J_t = 0$, and the envelope function $\Phi^{++}(r, r', R)$, where $r = r_1 - r_a$, $r' = r_2 - r_b$ and $R = r_a - r_b$, has even parity with respect to the permutations of two electrons and two holes. The next higher state with symmetry Γ_4 is made by sacrificing the exchange energy of two holes, which is much smaller than that of electrons due to the heavier mass, so giving the envelope function $\Phi^{+-}(r, r', R)$. This corresponds to the first rotational state of the hydrogen molecule in the limit of $\sigma \to 0$. The first and the second suffixes $\pm$ mean the parity of the envelope function with respect to the permutation of electrons and holes, respectively. Here the wave functions of the EM are represented for two low-lying states as follows:

$$|\Gamma_1^{\text{mol}}(K; 1, a, 2, b)\rangle$$

$$= \frac{1}{\sqrt{V}} [\exp(iK \cdot R_0)] \Phi^{++}(r, r', R)|\Gamma_1(1, a, 2, b)\rangle ,$$

$$|\Gamma_4^{\text{mol}}(K; 1, s, 2, b)\rangle \tag{2.11}$$

$$= \frac{1}{\sqrt{V}} [\exp(iK \cdot R_0)] \Phi^{+-}(r, r', R)|\Gamma_4(1, a, 2, b)\rangle ,$$

where

$$\left.\begin{array}{c}|\Gamma_1(1, a, 2, b)\rangle\rangle \\ |\Gamma_4(1, a, 2, b)\rangle\rangle\end{array}\right\} \equiv \left\{\begin{array}{c}|0, 0\rangle \\ |1, 0\rangle\end{array}\right\}$$

$$= \tfrac{1}{2}[c_{1/2}(1)c_{-1/2}(2) - c_{-1/2}(1)c_{1/2}(2)]$$

$$\cdot [v_{1/2}(a)v_{-1/2}(b) \mp v_{-1/2}(a)v_{1/2}(b)] , \tag{2.12}$$

and $R_0 = [\alpha(r_1 + r_2) + \beta(r_a + r_b)]/2$ describes the center-of-mass coordinate of the system. The envelope function $\Phi^{++}(r, r', R)$ in (2.11) has been solved in the effective mass approximation [2.1] and is given by

$$\Phi^{++}(r, r', R) = \Phi(r + \tfrac{1}{2}R, r' - \tfrac{1}{2}R, R) = f(R)g(\xi_1, \eta_1, \xi_2, \eta_2; R) .$$

The envelope function of the first rotational state may be expressed as

$$\Phi^{+-}(r, r', R) = \Phi'(r + \tfrac{1}{2}R, r' - \tfrac{1}{2}R, R)$$

$$= f'(R)Y_{1m}(\theta, \phi)g(\xi_1, \eta_1, \xi_2, \eta_2; R) ,$$

where $Y_{1m}(\theta, \phi)$ $(m = 1, 0, -1)$ is a spherical harmonic of the first order and (θ, ϕ) describes the orientation of the vector R.

The energy splitting between the Γ_1^{mol} and Γ_4^{mol} molecular states originates from the intraband exchange energy of two holes or two electrons in a molecule and is estimated to be of the order of the molecular binding energy, 30 meV. On the other hand, the splitting between the Γ_2^{ex} and Γ_5^{ex} exciton states comes from the effective e–h interband exchange interaction and is observed to be 6.2 meV [2.13] for CuCl. The former is much larger than the latter. Therefore we have been justified in forming the lowest state of the EM by superposing the product of the two-electron state with $J_e = 0$ and the two-hole state with $J_h = 0$ which minimize the electron-electron and hole-hole intraband exchange energies, respectively. After that we can take into account the effect of e–h interband exchange by the perturbation method as will be given in the last part of this subsection. In the opposite case of such molecular crystals as anthracene and benzene, the e–h exchange interaction is effectuated within one molecule and is therefore much larger than the electron-electron or hole-hole exchange interaction, which occurs only between neighboring molecules.

For reference, the lowest state, $J_t = 0$, of the EM, in the limit of zero e–h exchange splitting, consists of the linear combination of two $J_t = 0$ states which are made up of two $J_{ex} = 0$ excitons and two $J_{ex} = 1$ excitons, respectively, as follows:

$$|0, 0\rangle = \tfrac{1}{2}[|J_t = 0; (J_{ex} = 0)^2\rangle + \sqrt{3}|J_t = 0; (J_{ex} = 1)^2\rangle]$$

$$= \tfrac{1}{2}[|P_1^1\rangle|P_1^{-1}\rangle + |P_1^{-1}\rangle|P_1^1\rangle - |P_1^0\rangle|P_1^0\rangle + |P_0^0\rangle|P_0^0\rangle] . \tag{2.13}$$

The exciton state is hardly modified in the EM [2.1]. Under this assumption, the e–h interband exchange effect works only on the Γ_5 exciton component of an EM and it is evaluated as a perturbation in terms of the wave function (2.13), as follows:

$$\langle\Gamma_1^{mol}|H_{e-h}^{exch}|\Gamma_1^{mol}\rangle = 2 \times \tfrac{3}{4} \times \Delta E_{ex}^{exch} = 9.3 \text{ meV} , \tag{2.14}$$

where ΔE_{ex}^{exch} is the energy splitting of the Γ_2 and Γ_5 excitons. As a result, the energy of the lowest state of the EM is expressed by

$$E_m(k) = 2E_{ex}(\Gamma_2) - E_{mol}^b + \tfrac{3}{2}\Delta E_{ex}^{exch} + \hbar^2k^2/2m_{mol}$$

where the molecular binding energy E_{mol}^b is measured with respect to the lowest two-Γ_2 exciton state, which is not influenced by the e–h exchange effect, and $m_{mol} = 2(m_e + m_h)$.

In the case of CuBr, the bottom of the conduction band has the same symmetry, Γ_6, as in CuCl but the top of the valence band is composed of a fourfold-degenerate Γ_8 state:

$$v_{3/2} = -\psi_1|a\rangle, \qquad v_{1/2} = \sqrt{\tfrac{2}{3}}\psi_0|a\rangle - \sqrt{\tfrac{1}{3}}\psi_1|\beta\rangle ,$$

$$v_{-1/2} = \psi_{-1}|\beta\rangle, \qquad v_{-1/2} = \sqrt{\tfrac{1}{3}}\psi_{-1}|a\rangle + \sqrt{\tfrac{2}{3}}\psi_0|\beta\rangle . \tag{2.15}$$

Here d-like states (d_{xy}, d_{yz}, d_{zx}) of Cu are mixed with p-like states (p_x, p_y, p_z) of Br at the top of the valence band and the orbital parts ψ_0 and $\psi_{\pm1}$ can be written with the relative amplitudes a and b ($a^2 + b^2 = 1$) as follows:

$$\psi_0 = ap_z + bd_{xy} ,$$

$$\psi_{\pm1} = \frac{a}{\sqrt{2}}(p_x \pm ip_y) + \frac{b}{\sqrt{2}}(d_{yz} \pm id_{zx}) .$$

Two kinds of excitons with total angular momentum $J_{ex} = 1$ (Γ_5) and $J_{ex} = 2$ ($\Gamma_3 + \Gamma_4$) are formed from the superposition of excitations of a valence electron into the conduction band. A $J_{ex} = 2$ ($\Gamma_3 + \Gamma_4$) state of an exciton has a pure triplet e–h spin structure and is optically inactive. The wave functions of these excitons can be written as follows:

$$\begin{pmatrix} |J_{ex} = 1, m\rangle \\ |J_{ex} = 2, m'\rangle \end{pmatrix} = \frac{1}{\sqrt{V}} \exp[i\boldsymbol{K} \cdot (\alpha\boldsymbol{r}_1 + \beta\boldsymbol{r}_a)]\phi_n(\boldsymbol{r}_1 - \boldsymbol{r}_a) \begin{pmatrix} |P_1^m(1, a)\rangle \\ |P_2^{m'}(1, a)\rangle \end{pmatrix} ,$$

$$m = \pm1, 0 \quad \text{and} \quad m' = \pm2, \pm1, 0 . \tag{2.16}$$

In this material also, the electron-electron and the hole-hole intraband exchange energies which are taken into account in the effective mass calculation may be much larger than the e–h interband exchange. Therefore the $J_e = 0$ state of two electrons is accepted as a basis function of the EM, because it is antisymmetric under permutation of the two electrons and as a result it minimizes the electron-electron intra-band exchange energy. For the same reason the $J_h = 0$ and $J_h = 2$ states of two holes must be chosen from the four states $J_h = 0, 1, 2,$ and 3 which can be made from two $j = 3/2$ holes. The basis functions for the low-lying states of an EM with the total angular momentum $J_t = 0$ and 2, are given by

$$|J_t, m_t\rangle = |0, 0\rangle_e \cdot |J_h, m_h\rangle_h , \qquad (J_t = 0, 2 \quad \text{and} \quad J_h = 0, 2) . \tag{2.17}$$

In terms of these basis functions, the molecular state is written as follows:

$$|\Gamma^{mol}(J_t, m_t)\rangle = \frac{1}{\sqrt{V}} [\exp(i\boldsymbol{K} \cdot \boldsymbol{R}_0)]\Phi^{++}(\boldsymbol{r}, \boldsymbol{r}', \boldsymbol{R})|J_t, m_t\rangle . \tag{2.18}$$

In the framework of the effective mass approximation, these molecular states are degenerate in energy because they have the same envelope function in this approximation. The deviations from the effective mass approximations, i.e., the interband e–h exchange interaction and the interband scattering of two holes, must be considered to explain the shift and splitting of these molecular states. Let us discuss how these states are influenced by these deviations.

Only the diagonal matrix elements of the effective e–h exchange interaction remain non-zero and they have the same value for all the low-lying molecular states:

$$\langle \Gamma^{\text{mol}}(0, 0)|H_{\text{e–h}}^{\text{exch}}|\Gamma^{\text{mol}}(0, 0)\rangle$$

$$= \langle \Gamma^{\text{mol}}(2, m)|H_{\text{e–h}}^{\text{exch}}|\Gamma^{\text{mol}}(2, m)\rangle$$

$$= 2 \times \tfrac{3}{8} \times \Delta E_{\text{ex}}^{\text{exch}} \quad \text{for all } m ,$$

where the exciton wave function was assumed not to be modified in the EM and $\Delta E_{\text{ex}}^{\text{exch}}$ is the energy splitting of the $J_{\text{ex}} = 1$ and $J_{\text{ex}} = 2$ excitons due to the e–h exchange interaction. As a result, the e–h exchange interaction does not modify the molecular state besides reducing the molecular binding energy by $(3/4)\Delta E_{\text{ex}}^{\text{exch}}$. The intraband components of the hole-hole and electron-electron Coulomb interactions have been taken into account in the expression $\Phi^{++}(r, r', R)$. As for the interband components of the hole-hole interaction, only the following matrix elements remain non-zero:

$$\langle \Gamma^{\text{mol}}(J_{\text{t}}, m_{\text{t}})|H_{\text{h–h}}^{\text{interband}}|\Gamma^{\text{mol}}(J_{\text{t}}', m_{\text{t}}')\rangle$$

$$= \int d\mathbf{R} g^2(\mathbf{R}) V \langle J_{\text{t}}, m_{\text{t}} \left| \frac{e^2}{r_{ab}} \right| J_{\text{t}}', m_{\text{t}}' \rangle \delta(\mathbf{R}) ,$$

where we use the same assumption of a short-range interaction as in the e–h exchange and $g(\mathbf{R})$ describes the relative motion of two holes [2.12].

In evaluating the matrix elements of the hole-hole Coulomb interaction, we use the expansion:

$$\frac{1}{r_{ab}} = \sum_{n=0}^{\infty} \frac{1}{r_>} \left(\frac{r_<}{r_>}\right)^n P_n(\cos \omega_{ab}) ,$$

where $r_>$ ($r_<$) means the larger (smaller) of $|r_a|$ and $|r_b|$ and $P_n(x)$ is the nth order Legendre function. The leading term of $n = 0$ has already been included in the molecular envelope function in the effective mass approximation. The term with $n = 2$ is the next higher-order contribution. We have three contributions: A, B, and C. The A term shifts all the molecular levels by the same quantity, while B describes the inberband scattering of two holes and this gives the splitting of the low-lying molecular states into two levels with $J_{\text{t}} = 0$ and $J_{\text{t}} = 2$. The term C is finite only when the d-like state of Cu is mixed with the p-like state of Br at the top of the valence band and it is understood as the quadrupole-quadrupole interaction of two holes effective when one of them is located at Cu and the other at Br. It is this C that splits the fivefold degenerate $J_t = 2$ level into two molecular states Γ_3 and Γ_5. When we diagonalize the interband matrix elements of the hole-hole Coulomb interaction and include the energy shift due to the interband e–h exchange, the energy spectrum of an EM in CuBr is given as follows:

$$E_{mol}(\Gamma_1) = \Delta + 2\Delta' + \Delta'',$$

$$E_{mol}(\Gamma_3) = \Delta + \Delta'', \tag{2.19}$$

$$E_{mol}(\Gamma_5) = \Delta - \Delta'',$$

with the energy reference at $2E_{tr} - E^b_{mol} + (3/4)\Delta E^{exch}_{ex}$, where E_{tr} is the eigenenergy of a pure triplet exciton, ΔE^{exch}_{ex} is the e–h exchange splitting in an exciton, and $(\Delta, \Delta', \Delta'') = V|g(0)|^2(A, B, C)$. The envelope function is common to all these states and is given in (2.18). The parameters Δ, Δ', and Δ'' are determined by comparing the observed energy levels of an EM with (2.19) or the polarization characteristics of the optical response of EM as will be discussed in Sect. 2.3.4.

2.2.2 The Metallic Droplet and Excitonic Molecule in Many-Valley Structures – Ge and Si

In Ge (Si) the bottom of the conduction band is composed of four (six) equivalent valleys in the Brillouin zone and the top of the valence band is orbital degenerate. Therefore it is possibly consistent with the Pauli exclusion principle to place eight or twelve electrons in the same molecular orbital in Ge and Si, respectively. Based upon this consideration *Wang* and *Kittel* [2.14] proposed the possible stability of exciton complexes, $(exc)_8$ in Ge and $(exc)_{12}$ in Si under the assumption of infinite hole mass. In any case, the many-valley structure for the lighter particles as well as the anisotropic effective mass in each valley make the e–h metallic state more stable than the gas of excitons or biexcitons (EM). The evidence for the existence of such metallic electron-hole liquid (EHL) droplets was given by the observation of shifted recombination radiation attributed to the droplet state [2.15]. In this subsection, we will show how the metallic e–h droplet is stabilized in Si and Ge and then how the EM is observed unter pressure.

At very high excitation levels, the screening length λ becomes equal to or smaller than the exciton Bohr radius a_B. For the screened Coulomb potential with $\lambda \lesssim a_B$, no bound state exists for the e–h relative motion, so that one expects at high densities a plasma of degenerate electrons and holes. The dominant term in the high-density limit is the kinetic energy of the quantum-statistically degenerate electrons and holes. For single, parabolic conduction and valence bands the contribution to the ground state energy per e–h pair is

$$E_{kin} = \frac{3}{5}\left(\frac{k_F^2}{2\,m_e} + \frac{k_F^2}{2\,m_h}\right) = \frac{2.21}{r_s^2}\,E_{ex}^b ,\tag{2.20}$$

where k_F is the Fermi momentum of the electrons and the holes, and r_s is the radius of a sphere whose volume is equal to the inverse e–h concentration n^{-1} measured in units of a_B^3.

The first correction to (2.20) is the exchange energy which is the expectation value of the potential energy in the Hartree-Fock state

$$|0\rangle_{HF} = \prod_{|k| < k_F} a_k^\dagger b_{-k}^\dagger |0\rangle :$$

$$E_{exch} = -\frac{3e^2}{2\pi\epsilon_0}\,k_F = -\frac{1.832}{r_s}\,E_{ex}^b .\tag{2.21}$$

Note that the electrons and holes make equal contributions to the exchange energy irrespective of the mass ratio m_h/m_e.

The remaining contribution is called the correlation energy. The high-density expansion of the correlation energy is represented by the sum of the second-order exchange energy and the summation of the diverging Coulomb interaction terms to the infinite order described by the ring diagrams. The divergences are caused by the piling up of factors $4\pi e^2/\epsilon_0 q^2$ coming from the long-range nature of the Coulomb interaction. Evaluating these contributions, one finds for the case of $m_e = m_h$:

$$E_{corr} = (0.498\ln r_s - 0.260)E_{ex}^b .\tag{2.22}$$

This contribution is again negative, but only weakly dependent on the density. Writing as an approximation

$$E_0 = E_g + \frac{B}{\mu}\,n^{2/3} - 2An^{1/3} + E_{corr} ,\tag{2.23}$$

one finds a minimum of E_0 for $n_0 = \mu A/B$ with

$$E_{0,\,min} = E_g - (A/B)\mu + E_{corr} .\tag{2.24}$$

The minimum value per e–h pair is $-0.35E_{ex}^b$ at $r_s = 1.7$ [2.16].

The ring approximation described above neglected all correlations between electrons and holes and treated both spin orientations on an equal footing, which is not correct because the Pauli exclusion principle keeps particles with equal spin further apart. To overcome the latter deficiency *Brinkman* and *Rice* [2.17] extended the modified random phase approximation (RPA) proposed by Hubbard to the two-component system of electrons and holes, and obtained the minimum energy $-0.86\,E_{ex}^b$ at $r_s = 1.95$. A similar modified RPA has been given by *Combescot* and *Nozieres* [2.18].

In order to take into account the correlation of an e–h pair due to the attractive interaction as well as the electron-electron and the hole-hole correlation, a variational calculation of the system energy was done for the metallic system [2.19]. The trial function was assumed to be of the following form:

$$\Phi(r^N, R^N) = D(r^N) \, D(R^N) \prod_{i<j} f_1(r_i - r_j) \prod_{i<j} f_2(R_i - R_j)$$

$$\times \prod_{i,j} f_3(r_i - R_j) \, , \tag{2.25}$$

where $D(r^N)$ and $D(R^N)$ are Slater determinants of the N-electron and the N-hole plane-wave state, respectively. The functions f_ν are of the form $f_\nu(r) = \exp[-g_\nu(r)]$ with $\nu = 1, 2, 3$. The corresponding state vector is then represented as

$$\exp\left\{-\tfrac{1}{2}\sum_k [x_1(k)\varrho_k\varrho_k^* + x_2(k)\xi_k\xi_k^* + x_3(k)(\varrho_k\xi_k^* + \xi_k\varrho_k^*)]\right\}|0\rangle_{\text{HF}} \, ,$$

where   (2.26)

$$\varrho_k^* = \sum_{q\sigma} a_{k+q\sigma}^\dagger a_{q\sigma} \quad \text{and} \quad \xi_k^* = \sum_{q\sigma} b_{k+q\sigma}^\dagger b_{q\sigma} \, .$$

Here $x_1(k)$, $x_2(k)$, and $x_3(k)$ are variational parameters. The expectation value of the total Hamiltonian with this trial function is composed of a harmonic part (expectation value of the terms second-order in ξ, ξ^*, ϱ, and ϱ^*) and an anharmonic part (third-order terms). The parameters $x_1(k)$, $x_2(k)$, and $x_3(k)$ are determined analytically by minimizing the harmonic part. The anharmonic part was evaluated in terms of the harmonic solution. The minimum energy of the system is $-0.98E_{\text{ex}}^{\text{b}}$ at $r_s = 2.0$ for the case of equal electron and hole masses. The minimum energy still lies above the energy of the free exciton and EM. Similar results have been obtained by *Vashista* et al. [2.20]. Thus, the metallic state was shown not to be bound relative to the free excitons in the system of nondegenerate and isotropic conduction and valence bands.

The anisotropies and degeneracies in the band structure stabilize the metallic phase in Ge and Si. In polar semiconductors, such as AgBr and CdS, the interaction with the polarizable lattice also tends to stabilize the plasma state [2.21]. In Ge, the minima in the conduction band are ellipsoidal in shape and located at the L point of the Brillouin zone. Four equivalent conduction-band ellipsoids exist with $m_t = 0.082 \, m_0$ and $m_l = 1.58 \, m_0$. The valence-band structure consists of two bands which are degenerate Γ_8 levels at the center of the zone, but they split into heavy- and light-hole bands away from Γ. The detailed form of the valence band is

$$E^\pm(k) = Ak^2 \pm [B^2k^4 + C^2(k_x^2 k_y^2 + k_y^2 k_z^2 + k_z^2 k_x^2)]^{1/2} \, , \tag{2.27}$$

where $A = 13.38$, $B = 8.48$, and $C = 13.15$ are in units of $\hbar^2/m_0$.

The kinetic energy per electron is proportional to the Fermi energy E_F. Distributing the electrons among four ellipsoids lowers the Fermi energy by a factor

$4^{-2/3}$. The mass which enters the Fermi energy is the geometric mean over three principal directions of the ellipsoids. Thus, the density-of-state mass m_{de} is given by

$$m_{de} = m_t^{2/3} m_l^{1/3} = 0.22 \, m_0 \quad \text{(for Ge)} ,$$

where m_0 is the electron mass. On the other hand, the binding energy of the exciton is determined by the optical mass m_{oe} of the electron

$$m_{oe}^{-1} = \tfrac{1}{3}(2 \, m_t^{-1} + m_l^{-1}) \, , \, m_{oe} = 0.12 \, m_0 \quad \text{(for Ge)} .$$

Thus the kinetic energy has a larger mass, and therefore a lower value in the metallic state than in free excitons. A similar effect is found for the holes. From (2.27) we obtain the effective masses of the light hole $m_{lh} = 0.042 \, m_0$ and heavy hole $m_{hh} = 0.347 \, m_0$. The density of state for the valence band is given by

$$n_h(E) = \frac{(2E)^{3/2}}{3\pi^2} (m_{lh}^{3/2} + m_{hh}^{3/2}) .$$

The reciprocally averaged mass, i.e., the optical mass m_{oh} of the hole $m_{oh} = 0.07 \, m_0$ is much lighter than the heavy-hole mass m_{hh} which dominates in the kinetic energy of the metallic state. From the optical effective masses, we define a reduced mass m_r as $m_r^{-1} = m_{oe}^{-1} + m_{oh}^{-1}$. In Ge we obtain $m_r = 0.046 \, m_0$. With the dielectric constant $\epsilon_0 = 15.36$ of Ge, one obtains an exciton binding energy of $E_{ex}^b = 1.65 \, \text{meV}$: The kinetic energy per e–h pair in the metallic state is

$$E_{kin} = \frac{3}{5} (E_F^e + E_F^h)$$

$$= \frac{2.21}{r_s^2} E_{ex}^b \left[\frac{m_r}{4^{2/3} m_{de}} + \frac{m_r}{m_{hh}} \left(\frac{1}{1 + (m_{lh}/m_{hh})^{3/2}} \right)^{2/3} \right] = \frac{0.468}{r_s^2} E_{ex}^b . \tag{2.28}$$

The anisotropy of the effective mass and the degeneracy also reduce the absolute value of the exchange energy but by a much smaller amount than the kinetic energy. Firstly, the distribution of the electrons among four ellipsoids leads to a lowering of the exchange energy since only electrons in the same ellipsoid will contribute to the exchange energy. Since the exchange contribution is proportional to the Fermi vector there will be a reduction of $4^{-1/3}$. Secondly, the anisotropy of the electron effective mass and the valence-band structure also bring about a small reduction in the exchange energy. Finally, one obtains

$$E_{exch} = E_{exch}^e + E_{exch}^h = \left(-\frac{0.486}{r_s} - \frac{0.650}{r_s} \right) E_{ex}^b = -\frac{1.136}{r_s} E_{ex}^b . \tag{2.29}$$

In the calculation of the correlation energy by *Brinkman* and *Rice* [2.17], the effects of the anisotropy have been neglected and a simple band structure of four isotropic conduction bands and two isotropic valence bands has been used. The optical mass has been chosen as the characteristic mass because it determines the plasma frequency. Together with these approximations the modified RPA has been used to evaluate the correlation energy. The numerical result for the concentration dependence of the average energy per e–h pair is shown in Fig. 2.4. The energy of the metallic state of the EHL at $r_s = 0.63$ lies almost one exciton binding energy below the energy of a free exciton. We may conclude from the above energetics that the e–h metallic droplet is stable in Ge and similarly in Si.

It has been shown [2.21, 22] that the polar coupling between the EHL and the lattice vibrations increases the stability of the EHL phase even in single-conduction-band structures. According to the calculations of [2.21], the ground-state energy of the EHL in CdS lies 13 meV deeper than the lowest exciton level while in CdSe it coincides approximately with the energy of the free exciton as will be discussed in Sect. 2.2.4.

Uniaxial elastic deformation lifts the degeneracy of the bands in Ge and Si. As a result of this, the mean kinetic energy per e–h pair in EHLs increases whereas the binding energy and the equilibrium pair density in the condensed phase decrease. The binding energy of the exciton and, apparently, of the EM are not altered significantly. Let us consider first the change of the EHL binding energy when the energy spectrum is changed by lifting the band degeneracy by directional deformation in Si crystals.

Kulakovskii et al. [2.23] have investigated the recombination-radiation spectra of Si deformed along the axes $\langle 100 \rangle$, $\langle 110 \rangle$ and $\langle 111 \rangle$, i.e., Si(2–1),

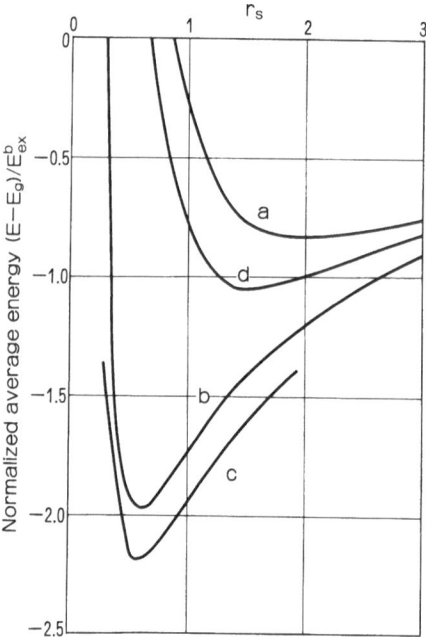

Fig. 2.4. Concentration dependence of the average energy per e–h pair. (*a*) Isotropic equal e–h masses, (*b*) Ge [2.17], (*c*) Ge [2.20] and (*d*) Ge with uniaxial stress in the [111] direction [2.17]

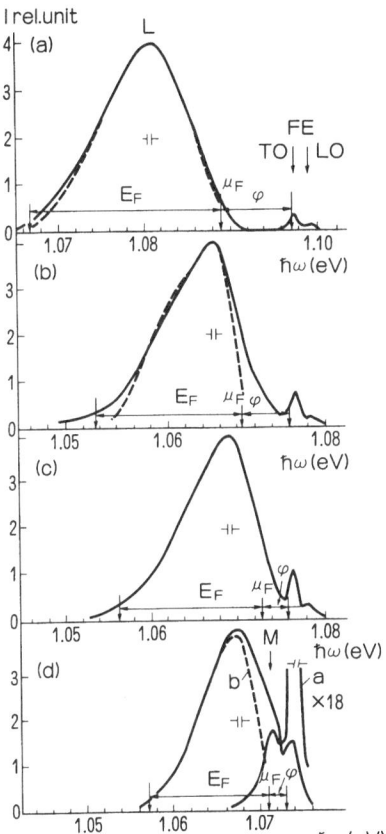

Fig. 2.5a–d. The radiative recombination spectrum. (**a**) Si (6–2), (**b**) Si (6–1), (**c**) Si (4–1) and (**d**) Si (2–1) under 30 kW/cm² pulsed laser excitation at 1.8 K. The dashed lines (**b**) show the theoretical curves. Note that the signal in (**d**) is multiplied by 18 and the M line due to the EM appears only after withdrawing the background [2.23]

Si(4–1), and Si(6–1) where the first figure denotes the number of the lowest split-off valleys at the Δ points of the conduction band and the second is the multiplicity of the valence-band degeneracy at the extremum Γ, with spin degeneracy neglected. Above the critical pressure, when the splitting of the bands Δ_c and Δ_v exceeds the corresponding values of the Fermi energies of the electron and the hole in the EHL, the recombination spectrum of the EHL becomes independent of the applied pressure.

Figure 2.5 shows the recombination spectra of Si(6–1), Si(4–1), and Si(2–1) measured at $T = 1.8$ K under pulsed laser excitation. For comparison the figure also shows the spectrum of undeformed Si(6–2). Each spectrum contains a wide band L corresponding to the EHL emission and the emission line FE connected with the gas of free excitons. Figure 2.5 demonstrates qualitatively that as the multi-valley character is decreased the EHL band becomes narrower, and its maximum and violet edge lie closer to the exciton line in accordance with curve d of Fig. 2.4. In the strained crystals the condensation threshold increases noticeably. It has been found [2.23] that at maximum lifting of degeneracy in the bands,

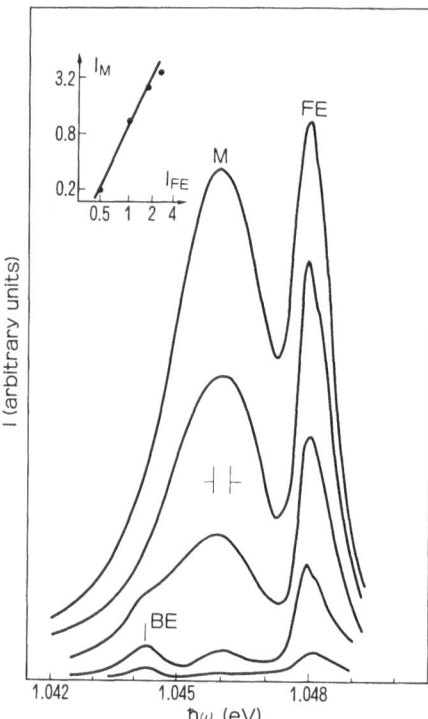

Fig. 2.6. Variation of the emission spectrum of Si (2–1) ($p \simeq 45$ kg/mm², 1.8 K) with changing exciton density as the e–h pair concentration n increases from 10^{14} to 10^{15} cm⁻³. The insert shows the relation of the emission intensity $I_M \propto (I_{FE})^2$ on a logarithmic scale [2.23]

i.e., in Si(2–1), the EHL binding energy undergoes a fourfold decrease relative to the undeformed crystal Si(6–2). Consequently in Si(2–1) the e–h gas density near the threshold of condensation into the EHL increases by almost one order of magnitude compared with Si(6–2). Qualitatively this is seen from the comparison of the Si(6–2) and Si(2–1) spectra shown in Fig. 2.5, which were measured at equal generation levels of nonequilibrium carriers. With decreasing EHL binding energy, the exciton line intensity increases.

In the spectra of Si(2–1), a separate new line M due to the emission of the EM appears between the EHL and exciton bands. The spacing between the maxima of the lines M and FE does not change with increasing pressure up to the destruction of the crystal at $p \simeq 90 \sim 100$ kg/mm² and is equal to 2 meV. This M line was attributed [2.23] to the indirect radiative decay of the EM in conformity with the conservation law,

$$E_p^{EM} = E_q^{ex} + \hbar\omega_{p-q} + \hbar\omega \;,$$

where E_p^{EM} and E_q^{ex} are the energies of the EM and the exciton, respectively, and $\hbar\omega_{p-q}$ and $\hbar\omega$ are the energies of the emitted phonon and photon, respectively. Figure 2.6 shows the evolution of the emission spectrum of Si(2–1) at $p \simeq$ 45 kg/mm² in the region of the exciton structure under stationary volume excitation ($\lambda = 1.064$ μm) at $T = 1.8$ K. At minimum excitation densities the spectrum contains only the free-exciton emission line, FE, and the line, BE, of the exciton bound to a neutral acceptor (boron atoms). With increased pumping, the intensity of the BE line saturates. The M line then appears between the BE and FE emission bands and with increasing exciton concentration its intensity increases quadratically with exciton line intensity as shown in the inset of Fig. 2.6. This quadratic dependence is one of the main confirmations of the molecular origin of the M band. The intensity ratio of the lines FE and M ceases to depend on excitation density at 1.8 K and concentrations $n_{e,h} \simeq 10^{15}$ cm^{-3}, when condensation of the exciton and EM gas into the EHL sets in. It follows from these experiments performed under the conditions of uniaxial deformation that the partial pressures of excitons and EMs are of the same order of magnitude at densities $n_{e,h}$ corresponding to the onset of condensation into EHL.

The EHL has also been confirmed in Ge crystals. The known variational estimates of the EM binding energy and the experimental values of the EHL binding energy suggest a similar situation should exist in Ge crystals. A situation favorable for the detection of the EM from emission spectra should arise in Ge crystals under conditions of uniaxial stress along the $\langle 111 \rangle$ direction, i.e., Ge(1–1). Here maximum lifting of the degeneracy in the electron and hole bands is brought about.

2.2.3 The Excitonic Molecule in Many-Valley Systems – TlX and AgX

Thallous halide crystals have CsCl structure belonging to the space group O_h^1. As will be shown in Sect. 7.1.2 they have an indirect band gap: the top of the valence band is at the X point (X_6^+), the bottom of the conduction band is at the R point (R_6^-) and the smallest direct band gap is at the X point ($X_6^+ \to X_6^-$) [2.24]. The symmetry-adapted atomic orbitals for the valence band (X_6^+) at the (001) valley and the conduction bands (R_6^-) and (X_6^-) at the (001) valley are given by (7.1, 2, 13). From this band structure, we have three kinds of excitons: the $X_6^+ \to R_6^-$ indirect exciton [2.25], the $X_6^+ \to X_6^-$ direct exciton [2.26], and the $X_6^+ \to X_6^-$ indirect exciton. Their structures are discussed in Sect. 7.1.3. From two $X_6^+ \to R_6^-$ indirect excitons which have the lowest energy, it is possible to form two kinds of EMs which have an M point wave vector (M point EM) and Γ point wave vector (Γ point EM). There are three kinds of $X_6^+ \to R_6^-$ indirect excitons with the translational momenta $M_1 = 2\pi/a$ (110), $M_2 = 2\pi/a$ (101), and $M_3 = 2\pi/a$ (011). As is discussed in detail in Sect. 7.2.5, there are three M point EMs with the wave vectors $M_3(= M_1 + M_2)$, $M_2(= M_3 + M_1)$, and $M_1(= M_2 + M_3)$. We can also make the Γ point EM from two $X_6^+ \to R_6^-$ indirect excitons in the same valley, which has zero translational momentum because $2M_i = \Gamma$ for $i = 1, 2, 3$.

There are several interesting problems concerning the stability of the EM in the thallous halides. The first problem is to investigate whether EMs or e–h droplets are preferentially formed in highly excited thallous halides. The valence band has three nonequivalent maxima at the X points and the hole mass in each valley is anisotropic. However, the anisotropy is not so large as that in Ge and Si, and the electron to hole mass ratio is almost one and the conduction-band edge at the R point is not degenerate in contrast to the case of Ge and Si. These three points work against the formation of e–h droplets, although nobody has evaluated quantitatively the stability of the EM gas against the EHL.

The second problem is the stability of the EM against the e–h exchange interaction. In the framework of the effective mass approximation [2.1], the molecular binding energy is of the order of 1 meV $[E^b_{ex}(E^b_{mol}/E_{ex})_{calculated} \doteqdot 30 \text{ meV} \times 0.03]$ because of the almost equal electron and hole masses. On the other hand, the e–h exchange energy in the indirect exciton $X^+_6 \times R^-_6$ composing the EM is also of an order of 1 meV. To answer this delicate question, we must evaluate the energy level of the EM beyond the effective mass approximation. Unfortunately this has not yet been done [2.27], therefore it is much wiser to consult the experimental result.

Nakahara and *Kobayashi* [2.28] observed the evidence of the EM in the emission spectrum of highly excited thallous halides. Prominent emission lines, which grow almost linearly with the excitation intensity under mercury lamp excitation, were assigned to the radiative decay of the forbidden indirect $X^+_6 \rightarrow R^-_6$ excitons. Thallous halides conform the space group O^1_h and the transition dipole moment has the Γ_{15} representation. In the optical emission associated with the phonon-assisted transition from R_{15} to X_1 states, only the M point phonons with even parity can participate because $X_1 \times \Gamma_{15} \times R_{15} = 2M_1 + M_2 + M_3 + M_4 + 2M_5$. However, the M point phonons of the CsCl-type crystal are composed of the odd parity M'_2, M'_4, and $2M'_5$. Therefore the indirect $X^+_6 \rightarrow R^-_6$ exciton is called forbidden.

Below these emission lines due to the forbidden indirect excitons, they found two emission lines at 3.217 eV(d) and 3.207 eV(h) in TlCl and at 2.639(d) and 2.630(h) in TlBr which grow nonlinearly with the excitation line (see Fig. 7.21). The d lines both in TlCl and TlBr have been assigned to the radiative decay of the M point and Γ point EM to the M point exciton with the emission of an M point phonon through a forbidden indirect process. The energy of the emitted photon is $E_{ex}(X^+_6 \rightarrow R^-_6) - \hbar\omega_M - E^b_{mol}$, where $\hbar\omega_M$ is the phonon energy, e.g., of the M point LA phonon in the present case. The peak energy is about $E^b_{mol} + 3k_BT/2$ lower than the peak energy of the radiative recombination of the forbidden indirect exciton. The factor 3/2 comes from the fact that it is a forbidden indirect transition.

The observed splitting between the emission lines of the exciton and the EM is 2.0 ± 0.3 meV for TlCl at 4.2 K. Therefore the binding energy of the EM is 2.0 ± 0.3 meV $- 3k_BT/2 = 1.5 \pm 0.3$ meV. The molecular binding energy is estimated as 0.9 ± 0.2 meV from the variational calculation [2.1] with $E^b_{mol}/E^b_{ex} = 0.03$ and $E^b_{ex} = 29 \pm 6$ meV [2.25]. For TlBr, the observed splitting is 1.2 ± 0.4 meV at 4.2 K and the molecular binding energy is 1.2 ± 0.4 meV $- 3k_BT/2 = 0.7 \pm 0.4$ meV. This is in good agreement with the variational calculation value of 0.7 meV obtained from $E^b_{mol}/E^b_{ex} = 0.03$ and $E^b_{ex} = 23 \pm 5$ meV [2.25].

We may conclude the new emission lines, d, in TlCl and TlBr to come from the radiative decay of the EM formed from the $X_6^+ \times R_6^-$ forbidden indirect excitons. We have three reasons for reaching this conclusion: (1) the energy position is reasonable, (2) the emission intensity shows a nonlinear dependence on the excitation power I^n ($n = 1.8$ at low excitation and $n = 1.0$ at a high excitation both in TlCl and TlBr), and (3) the observed emission spectrum agrees well with the calculated one for the forbidden indirect transition of the EMs [2.28].

The Γ point EM might decay to the $X_6^+ \rightarrow X_6^-$ direct exciton by the direct process (without phonon assistance). The emission energy is equal to $2E_{ex}(X_6^+ \rightarrow R_6^-) - E_{mol}^b - E_{ex}(X_6^+ \rightarrow X_6^-)$, i.e., 0.18 eV lower than the emission energies of the indirect forbidden $X_6^+ \rightarrow R_6^-$ exciton as will be shown in Fig. 7.20. This emission intensity may be weak because two electrons composing the EM must be scattered from the R point to the X points by the Coulomb interaction inside the EM. This radiative process was actually observed as α, β, and γ lines in Fig. 7.23. The emission lines, h, in both TlCl and TlBr cannot be assigned to this process because the emission energies are quite different from those expected theoretically. At present they cannot get the assignment of the h lines. It is noted that the effect of the electron–LO-phonon interactions on the exciton system is weak and almost cancelled out because the electron and hole masses are almost equal. This may be why the EHL is not observed in this material as it is in AgBr and CdS. The reader should refer to Sect. 7.2.5 for more detailed experimental results and further discussion on the EMs in thallous halides.

The optical absorption spectra of AgCl and AgBr differ in a fundamental way from those of the alkali halides. The alkali halides are strongly ionic materials and the fundamental absorption edge in the vicinity of 2000 Å rises steeply to above 10^5 cm^{-1} because of excitation from the ground state of the crystal to exciton states. Although AgCl and AgBr are less ionic compounds with considerable covalent bonding, they also have strong direct-exciton spectra showing the influence of the spin-orbit doublet of the halogen ion. However, in contrast to the alkali halides, AgCl and AgBr have optical absorption tails which extend towards longer wavelengths, approximately 2 eV from the direct exciton spectra [2.29].

Recent band calculations by *Kunz* [2.30] verified the old result by *Scop* and *Bassani* et al. [2.31] that while the conduction-band minimum is of standard form at the center of the Brillouin zone, the valence-band maxima are composed of four valleys at the L point in both AgCl and AgBr. Hole masses are rather anisotropic ($m_\parallel^h = 1.25\, m_0$ and $m_\perp^h = 0.52\, m_0$ in AgBr) [2.32] and the ratio of the electron to hole mass (geometrical mean = 0.64 m_0) is about 0.3 in AgBr. However the single conduction-band minimum with the lighter effective mass does not favour forming the EHL. On the other hand, *Beni* and *Rice* [2.21], and *Keldysh* and *Silin* [2.22] found independently that the electron–LO-phonon interaction increases the stability of the EHL phase relative to excitons, in particular, the EHL is strongly bound in CdS and AgBr but only weakly so in CdSe and ZnS.

The ground-state energy, E_G, of the e–h plasma including the interaction with the LO phonons is expressed in terms of the chemical potential $\zeta_e(\zeta_h)$ of the electrons (holes) as a function of normalized inverse wave number x:

$$E_G = 3r_s^3 \int_{r_s}^{\infty} dx \, x^{-4}[\zeta_e(x) + \zeta_h(x)] \,, \tag{2.30}$$

where $r_s = (3/4\pi n a_B^3)^{1/3}$ and n is the e–h pair density. Here effective atomic units are used defined by the exciton Bohr radius $a_B = \epsilon_\infty/\mu e^4$, and the energy $\mu e^4/\epsilon_\infty^2$, with $1/\mu = 1/m_e + 1/m_h$ and ϵ_∞ the high-frequency dielectric constant. The chemical potentials $\zeta_{e,h}$ are given in the random phase approximation by

$$\zeta_{e,h} = E_F^{e,h} + \Sigma_{e,h}(k_F, E_F^{e,h}) \,, \tag{2.31}$$

where $E_F^{e,h}$, k_F, and $\Sigma_{e,h}$ are, respectively, the Fermi energy, Fermi momentum, and self-energy of the electrons and holes. *Beni* and *Rice* [2.21] calculated $\Sigma_{e,h}$ for the phonon-coupled e–h plasma and then E_G using (2.30 and 31). The frequency dependent dielectric function $\epsilon_L(\omega)$ of the lattice is given by

$$\epsilon_L(\omega) = \epsilon_\infty + \frac{\epsilon_0 - \epsilon_\infty}{1 - (\omega/\omega_T)^2} \,, \tag{2.32}$$

where ϵ_0 is the static dielectric constant and ω_T is the transverse-optical-mode frequency. The dielectric function of the e–h liquid is approximated by

$$\frac{1}{\epsilon_{e-h}(q, \omega)} = \frac{1}{\epsilon_\infty}\left[1 + \frac{\omega_p^2}{2\omega_q}\left(\frac{1}{\omega - \omega_q + i\delta} - \frac{1}{\omega + \omega_q - i\delta}\right)\right] \,, \tag{2.33}$$

where the plasma frequency $\omega_p = (4\pi n e^2/\epsilon_\infty\mu)^{1/2}$ and $\omega_q^2 = \omega_p^2 + aq^2$ with $a = \omega_p^2/q_{TF}^2$ and the Thomas-Fermi wave vector q_{TF}. The dielectric function of the coupled system $\epsilon(q, \omega)$ is obtained by adding the lattice and the e–h polarizability. The self-energy $\Sigma_{e,h}$ can then be calculated by taking the lowest-order diagram with the Coulomb interaction screened by $\epsilon(q, \omega)$. Using (2.30) E_G is evaluated numerically.

This calculation can be generalized to the case of more complex band structures with anisotropy and degeneracy. The anisotropy of the hole effective mass and the four-valley valence-band structure are taken into account for AgBr. The result is shown as curve c in Fig. 2.7. This is compared with the case (curve a) of a pure Coulomb interaction with no phonons, and with the case (curve b) in which LO phonon coupling is included only by replacing ϵ_∞ by ϵ_0 and adding the polaron ground-state energy. From these comparisons, we can draw two main conclusions for AgBr:

i) At low densities (i.e., $\epsilon_F < \omega_L$ or large $r_s > 6$), the effect of the polar interaction is simply to lower the ground-state energy by the polaron energy and to further screen the Coulomb interaction by replacing ϵ_∞ with ϵ_0. However, this ϵ_0 approximation underestimates the binding energies of the EHL in the intermediate region.

ii) The polaron contribution to E_G decreases with increasing density, but depending on the band structure and the strength of the coupling, E_G may still be below the experimental binding energy of the lowest exciton. The LO interaction tends to stabilize the liquid phase versus the free-exciton gas.

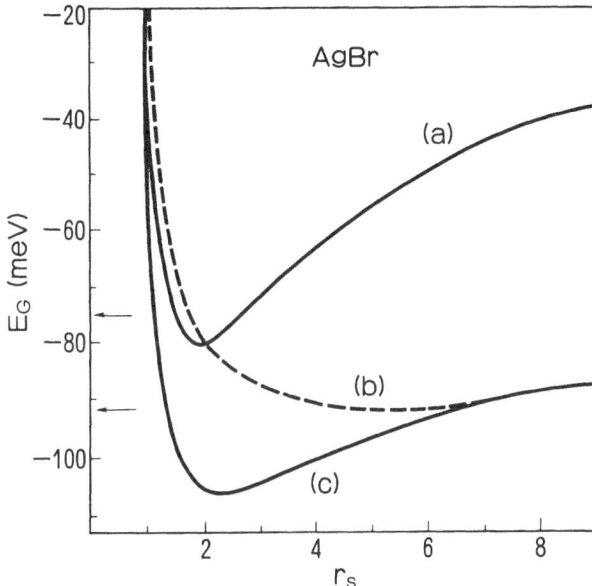

Fig. 2.7. Ground-state energy E_g plotted against interparticle distance r_s for AgBr. (*a*) pure Coulomb interaction without phonons, (*b*) LO phonon coupling included by replacing ϵ_∞ by ϵ_0 and adding the polaron ground-state energy, and (*c*) LO phonon coupling included by the method described in this section. The arrows on the left indicate the polaron ground-state energy (*upper arrow*) and the experimental binding energy of the lowest exciton (*lower arrow*) [2.21]

Pelant et al. [2.33, 34] and *Baba* and *Masumi* [2.35] observed the EM in the emission spectrum of AgBr. The stability of the EM gas against the EHL was discussed by *Hulin* et al. [2.36]. The other new emission band in AgBr was attributed to EHL recombination. They showed that EMs appear only at higher excitations and temperatures, corresponding to a limited region of the (T, n) plane called the biexciton pocket. It represents the case in which excitons, EMs, and the EHL are simultaneously present as shown in Figs. 2.8a and b. The set of spectra in Fig. 2.8a were obtained under rather low-power excitation, $I \simeq 5 \times 10^4$ W/cm². The spectra consist of the recombination of free excitons ($h\nu = 2.68$ eV) with simultaneous emission of a momentum conserving phonon *TA* (8.3 meV) at the L point and a new broad emission at $h\nu \simeq 2.6$ eV. However no EM emission is visible at $h\nu = 2.67$ eV at this low excitation. Line-shape analysis of the spectra in Fig. 2.8a were used to extract the density of carriers in the liquid, $n \sim 8 \times 10^{18}$ pairs/cm³, and the binding energy per e–h pair relative to the free exciton, $\varphi(0) \simeq 55$ meV. While the value n is close to that found theoretically by *Beni* and *Rice* [2.2] for AgBr ($n = 6 \times 10^{18}$ cm⁻³), the value of $\varphi(0)$ disagrees with the calculated energy of an EHL in AgBr, $\varphi(0)_{cal} = 14$ meV. This discrepancy is not resolved yet.

The spectra of Fig. 2.8b were recorded under identical conditions to those in Fig. 2.8a but with a much larger excitation intensity, $I \simeq 5$MW/cm². For the

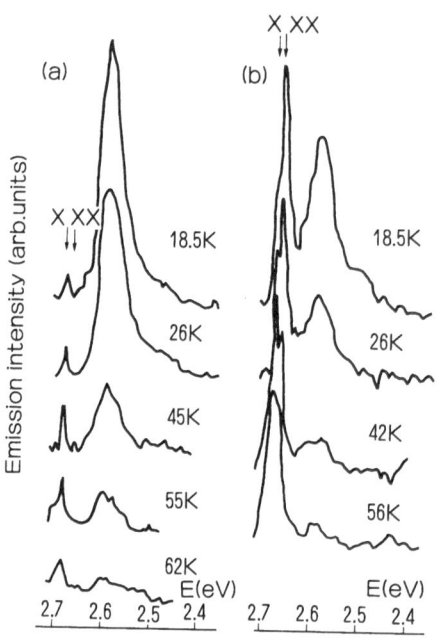

Fig. 2.8. Luminescence of AgBr at different lattice temperatures under the excitation power (**a**) 50 kW/cm² on the left and (**b**) 5 MW/cm² on the right. The two arrows indicate the positions of the free exciton X and the EM XX [2.36]

same cooling temperature, the EHL was observed to be further heated, from the broadening of the high-energy side. Another striking feature is the appearance of the EM (XX) line in addition to the exciton (X) and EHL lines, showing a change in the composition of the gas. Let E_{ex}^b, $E_{ex}^b + \Delta$ and $E_{ex}^b + \varphi$ be the binding energies of the exciton, the EM, and the EHL states, respectively. Taking into account the condition $\varphi > 2\Delta = 7$ meV, we can understand that the EHL exists with the largest binding energy at low temperature, and that EMs appear in the gas at higher temperatures and generated densities.

2.2.4 Influence of the Polarizable Lattice and the Effect of Anisotropic Effective Mass – CdS and CdSe

Many II–VI semiconductors have wurtzite structure with the space group C_{6v}^4. Wurtzite structure is obtained from the zinc blende one with T_d symmetry by deforming the latter in its [111] direction. This is called the trigonal crystal field. Usually threefold orbitally degenerate valence bands are split by the weak trigonal crystalline field and spin-orbit interaction into three double group representations Γ_9, Γ_7, and Γ_7. The direct gap semiconductors, CdS and CdSe, have the conduction-band minimum Γ_7 and the valence-band maxima Γ_9, Γ_7, and Γ_7 at the Γ point as shown in Fig. 2.9. The transition dipole moments P_x and P_y perpendicular to the c-axis and P_z parallel to the c-axis belong to the representations Γ_6 and Γ_1, respectively. From the following product representation and reduction

$$\Gamma_7 \times \Gamma_6 = \Gamma_7 + \Gamma_9, \qquad \Gamma_7 \times \Gamma_1 = \Gamma_7$$
$$\Gamma_9 \times \Gamma_6 = \Gamma_7 + \Gamma_8, \qquad \Gamma_9 \times \Gamma_1 = \Gamma_9$$

we can obtain the polarization characteristics of the $\Gamma_7 \times \Gamma_9$ and $\Gamma_7 \times \Gamma_7$ exitions as shown in Fig. 2.9.

The EMs were observed in the emission spectrum of CdS and CdSe for the first time by *Shionoya* et al. [2.37]. They assigned the new emission line (denoted M line) 5.4 meV below the free-exciton line in CdS to the emission from the EM due to (1) the line-shape analysis, taking account of the broadening due to elastic collisions and (2) the fact that the intensity I is related to the excitation power I_0 $I \propto I_0^n$, with $n \simeq 1.7$, i.e., $n > 1$. They estimated the molecular binding energy as 2.6 meV by using the geometrical mean of the hole masses $m_{h\|}$ and $m_{h\perp}$ parallel and perpendicular to the c-axis. The difference of the observed value of 5.4 from 2.6 meV was attributed mainly to the effect of the mass anisotropy and the difference of the variational result from the real value. Then the effect of the mass anisotropy on the molecular binding energy was calculated but a clear conclusion was not obtained in this respect [2.38]. The electron–LO-phonon coupling is rather strong and the electron mass is much lighter than that of the hole. As a result, we may expect enhancement of the molecular binding energy due to the particle–LO-phonon interaction. In fact the attractive potential between two excitons was calculated by taking account of this effect for CdS as well as CuCl [2.39]. However, nobody has yet calculated the molecular binding energy in the polarizable lattice.

On the other hand, *Beni* and *Rice* [2.21] found the tremendous stability of the EHL due to the electron–LO-phonon interaction in CdS as well as in AgBr. Ground-state energy, E_g, versus $r_s \equiv (3/4\pi n a_B^3)^{1/3}$ was calculated for the EHL including LO phonon coupling for CdSe, ZnS, and CdS as for AgBr. From the result of Fig. 2.7, we understand that the LO phonon interaction increases the EHL binding energy strongly enough to stabilize the EHL even for direct band gap semiconductors, especially in CdS.

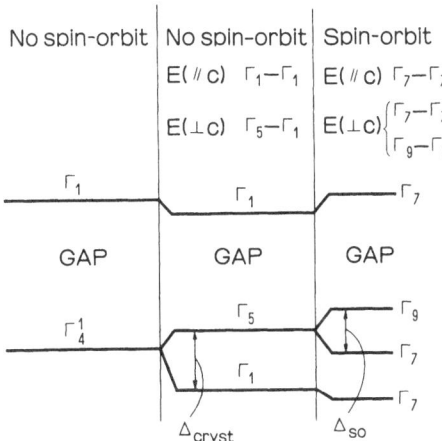

Fig. 2.9. Band structure at the Γ point in the wurtzite crystal (CdS and CdSe), and polarization dependence of the optical transition

Lysenko et al. [2.40] have investigated experimentally the radiative recombination spectra and the gain spectra of CdS and have attributed the *P* line to the EHL. They concluded that the ground state of an EHL in CdS at 9.2 K corresponds to a concentration $n \sim 10^{18}$ cm^{-3} and is 12 meV below the lowest A_T exciton. The EHL ground-state energy evaluated by including the LO phonon coupling is 13 meV lower than the exciton line at 3×10^{18} cm^{-3}. This is in good agreement with the experiments by *Lysenko* et al. [2.40]. The onset of optical gain was observed [2.41] ~ 13 meV below the exciton line while the Fermi energy is estimated as 30 meV, compared to the theoretical value of 38 meV. The onset of the optical gain in CdSe, on the other hand, is ~ 1 meV below the exciton line. This corresponds to the theoretical result of practical coincidence between the minimum of the EHL ground state and the lowest exciton line in this material. Furthermore, the calculated Fermi energy is 17.6 meV compared to the estimated experimental value of 20 meV. The e–h pair concentration at which the EM was observed under laser excitation below 40 kW cm^{-3} [2.37] is a few orders lower than when the P line is dominantly observed with the laser power between 440 kW cm^{-2} and 16 MW cm^{-2} [2.40] for both CdS and CdSe.

2.2.5 The Direct Forbidden Exciton – Cu$_2$O

When the density of excitons is increased to such a degree that interparticle interactions become important, there appear new elementary excitations of lower energy per e–h pair, as we have discussed in this chapter. In this respect, Cu$_2$O represents a special case. According to theroretical calculations, the formation of an EM is not possible. On the other hand, one expects the lowest $n = 1$ exciton state to have a very long lifetime in pure Cu$_2$O crystals because its direct radiative recombination is forbidden in the dipole approximation.

The Cu$_2$O crystal has the space group symmetry O$_h^4$. It has a direct band gap ath the Γ point, with the upper valence band and the lowest conduction band

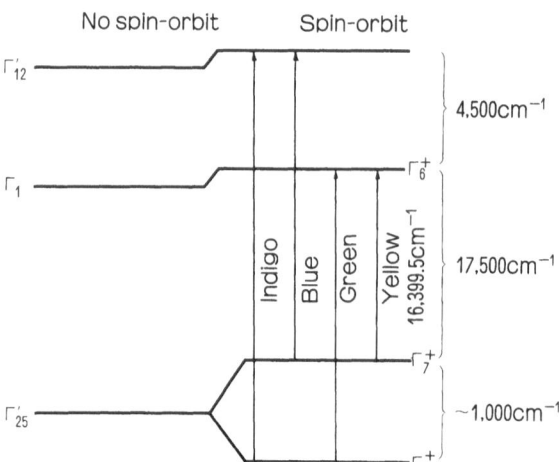

Fig. 2.10. Band structure at the Γ point in Cu$_2$O

both with positive parity, as shown in Fig. 2.10. The product representation of the exciton $\Gamma_6^+ \times \Gamma_7^+ \times \Gamma_1^+ = \Gamma_2^+ + \Gamma_5^+$ does not contain the representation Γ_4^- of the transition dipole moment. Here Γ_6^+, Γ_7^+ and Γ_1^+ are the symmetries of the conduction band, the valence band, and the s-like excitonic envelope function. Therefore the $n = 1$ exciton is optically forbidden in the dipole approximation. While the exciton binding energy in Cu_2O is 97 meV, the e–h exchange energy is also as large as 12 meV [2.42]. The e–h exchange interaction removes the degeneracy between the triply degenerate Γ_5^+ (ortho-exciton) and the nondegenerate Γ_2^+ (para-exciton) states, the para-exciton lying lower by 12 meV. *Bassani* and *Rovere* [2.43] calculated the effect of this e–h exchange on the formation of the EM. First, they performed a variational calculation of the molecular binding energy in the effective mass approximation. This gave the binding energy $(E_{mol}^b)_{var.} = 3.3$ meV for the electron to hole mass ratio $m_e/m_h = 0.73$ of Cu_2O. Then they evaluated the e–h exchange energy as a perturbation on the lowest excitonic molecular state (Γ_1, ++):

$$E_{mol}^{exch}(\Gamma_1, ++) = \tfrac{1}{2}I(++)J .$$

Here (++) means the parity under exchange of two electrons or two holes. The value of $I(++) \simeq 2.06(a_B)^{-3}$ was computed for $\sigma = m_e/m_h = 0.73$ and J is the exchange integral which appears in the exciton exchange energy $\Delta E_{ex}^{exch} = 2/3\pi(a_B)^{-3} J = 12$ meV. Then the molecular binding energy $E_{mol}^b [= (E_{mol})_{var.} - \Delta E_{mol}^{exch}]$ becomes negative. Thus they concluded that the EM is not formed but that the para-exciton gas is the elementary excitation of the lowest energy in Cu_2O.

2.3 Optical Response of an Excitonic Molecule

An EM has been experimentally and theoretically shown to be the most stable state of excitations in crystals without valley degeneracy. The EM can often be visualized by analogy to the hydrogen molecule, however, the former exists only as the transient state of excitations which decay radiatively. The EM gas therefore constitutes an open system in close contact with the radiation field and because of this, a variety of optical methods are used not only to create the gas of EMs but also to study its properties.

We have three optical processes connected with the EM. The first is the luminescence of an EM [2.44, 45], in which one of two excitons composing the EM is radiatively annihilated and the other remains in the crystal. The second is the reverse process to EM luminescence, i.e., one-photon absorption in the presence of single excitons due to the optical conversion [2.46] of an exciton into an EM. The last process is the giant two-photon absorption [2.4] due to excitation of an EM.

We have many interesting coherent optical phenomena due to the EMs as will be discussed in the following sections, but in order to study this kind of phenomena we must accumulate detailed knowledge of the EM relevant to these coherent interactions. Therefore, we confine ourselves in this section to the discussion of the three elementary optical transitions mentioned above with the purpose of understanding the EM itself.

We discuss the selection rules, the relative amplitude for the fine structure and the polarization characteristics of these three optical transitions [2.47], using the electronic structure of the EM derived in the preceding section. The luminescence, the optical conversion of an exciton into an EM, and the giant two photon absorption due to an EM are discussed in Sects. 2.3.1, 3, and 4, respectively. In Sect. 2.3.2 we calculate the relaxation of EMs due to interaction with the acoustic-phonon field and answer the question why the effective temperature observed in the luminescence of the EM is much higher than the lattice temperature. This interaction with acoustic phonons plays a role in broadening the giant two-photon absorption spectrum of the EM at high temperature, as will be discussed in Sect. 2.3.4.

2.3.1 Luminescence Spectrum

Laser light with fixed frequency can usually excite the valence electrons into the unbound states of the conduction band only. The electron in the conduction band and the hole in the valence band are bound into an exciton and then two such excitons are bound into an EM, by giving the extra energy to the lattice system. This EM can be detected by the following emission process: one e–h pair in an EM with the momentum K is radiatively annihilated and the other pair remains as a free exciton in a crystal, absorbing the momentum K-p, where p is the momentum of the emitted photon. The energy of this emitted photon is given by the energy conservation law as follows:

$$\hbar\omega = \left(2E_{1s} - E_{mol}^b + \frac{\hbar^2 K^2}{2m_{mol}}\right) - \left(E_{1s} + \frac{\hbar^2 K^2}{2m_{ex}}\right)$$

$$= E_{1s} - E_{mol}^b - \frac{\hbar^2 K^2}{2m_{mol}} ,$$

where the photon momentum $|p|(\simeq 10^4\ \text{cm}^{-1})$ was neglected in comparison with the average momentum $|K|(\simeq 10^6\ \text{cm}^{-1}$ for 4.2 K) and $m_{mol} = 2m_{ex} \simeq m_0$. This EM was first observed in the emission spectrum of highly excited CuCl by *Mysyrowicz* et al. [2.44]. The intensity of this emission line increases as the square of the incident laser power, because the EM is made up of two photons. *Souma* et al. [2.45] observed this emission line by the molecular binding energy ($E_{mol}^b \simeq 30$ meV for CuCl) below the exciton energy E_{1s} and confirmed the existence of the EM through line-shape analysis.

The emission spectrum of the EMs with distribution $f_{\mathrm{mol}}(K)$ in CuCl can be evaluated in terms of the wave functions derived in the preceding section, by a simple perturbational method, as follows:

$$W_l(\omega) = \frac{2\pi}{\hbar} \sum_{K,m} |\langle \Gamma_{5m}^{\mathrm{ex}}(K)|H'_+|\Gamma_1^{\mathrm{mol}}(K)\rangle|^2 f_{\mathrm{mol}}(K)$$

$$\times \delta(\hbar\omega - E_{\mathrm{mol}}(K) + E_{\mathrm{ex}}(K)) , \qquad (2.34)$$

where H'_+ is the photon-emission part of the electron-radiation interaction:

$$H' = \sum_i e\tilde{r}_i \cdot [E_+ \exp(-i\boldsymbol{p} \cdot r_i) + E_- \exp(i\boldsymbol{p} \cdot r_i)]$$

$$= \sqrt{\frac{8\pi\hbar\omega}{V}} \sum_i (e\tilde{r}_i \cdot e)[a_p^\dagger \exp(i\omega t - i\boldsymbol{p} \cdot r_i) + a_p \exp(-i\omega t + i\boldsymbol{p} \cdot r_i)] . \qquad (2.35)$$

Here r_i and $e\tilde{r}_i$ are the position vector and the dipole moment of the i-th electron, respectively, e is the unit polarization vector of the radiation field and a_p and $a_p^\dagger$ are the annihilation and creation operators of a photon with momentum $\boldsymbol{p}$.

It is straightforward to obtain the following expression by multiplying $|\Gamma_{5,m}^{\mathrm{ex}}\rangle$ by $1/\sqrt{2}[v_{1/2}(3)v_{-1/2}(4) - v_{-1/2}(3)v_{1/2}(4)]$ and inserting the wave function (2.9) and (2.11) of the preceeding section into (2.34);

$$W_l(\omega) \simeq \frac{4\pi^2\omega_0}{3V} [\langle \psi_c|e\tilde{x}|p_x\rangle \phi(0)]^2 \sum_K f_{\mathrm{mol}}(K)$$

$$\cdot [\int g(\boldsymbol{R}) \exp(i\boldsymbol{K} \cdot \boldsymbol{R}/2) d\boldsymbol{R}]^2 \delta(\hbar\omega - E_{\mathrm{mol}}(K) + E_{\mathrm{exc}}(K)) . \qquad (2.36)$$

Here we used the fact that $\langle \psi_c|e\tilde{x}|p_x\rangle = \langle \psi_c|e\tilde{y}|p_y\rangle = \langle \psi_c|e\tilde{z}|p_z\rangle$, approximated $\psi^{++}(r, 0, R)$ by $\phi(r)\phi(0)g(R)$, and neglected $(K_{a_B}a/2)^2$ in comparison with one. When we assume $g(R) = (\pi a_{\mathrm{mol}}^3)^{-1/2} \exp(-R/a_{\mathrm{mol}})$,

$$\int g(R) \exp(i\boldsymbol{K} \cdot \boldsymbol{R}/2) d\boldsymbol{R} = 8\sqrt{\pi a_{\mathrm{mol}}^3}/[1 + (K a_{\mathrm{mol}}/2)^2]^2 .$$

This means that the emission spectrum from $K \geq 2/a_{\mathrm{mol}}$ is very much reduced. In the case of the direct gap exitons, however, this effect is negligible and $(K a_{\mathrm{mol}}/2)^2$ is neglected in comparison with one. As we are interested only in spontaneous emission, we have replaced E^2 by the expectation value of $8\pi(\hbar\omega_0/V)a_p a_p^\dagger$ in the photon vacuum state. When the Boltzmann distribution with the effective temperature T_{eff} is assumed for the kinetic energy of the EMs, the integration in (2.36) can be done as follows:

$$W_l(\omega) = A\sqrt{E_{1s} - E_{\mathrm{mol}}^b - \hbar\omega} \exp[-(E_{1s} - E_{\mathrm{mol}}^b - \hbar\omega)/k_B T_{\mathrm{eff}}] ,$$

for $\hbar\omega \leq E_{1s} - E_{\mathrm{mol}}^b ,$ where with the EM density N_M

$$A = \left(\frac{4\pi}{k_B T_{\mathrm{eff}}}\right)^{3/2} N_M \frac{\omega_0}{3} [\langle \psi_c|e\tilde{x}|p_x\rangle \phi(0) \int g(R) dR]^2 .$$

The emission spectrum rises by the molecular binding energy below the exciton energy and it has a tail on the low energy side. *Souma* et al. [2.45] observed this characteristic spectrum in CuCl, where the intensity increased in proportion to the square of the excitation intensity, and obtained an effective temperature of 26 K in contrast to the lattice temperature of 4.2 K. This fact will be discussed in Sect. 2.3.2.

The radiative annihilation of EMs, in which the pure triplet Γ_2 free exciton remains, is forbidden in CuCl because of the group-theoretical property of the Bloch function, i.e., $\langle \Gamma_2^{ex}|H'|\Gamma_1^{mol}\rangle = 0$. This is also easily inferred from the absence of product terms of the pure triplet $\Gamma_2(|P_0^0\rangle)$ exciton and the optically active $\Gamma_5(|P_1^m\rangle)$ exciton in (2.13). The threefold degenerate state of the Γ_5 exciton is split into the longitudinal exciton and the twofold transverse exciton due to the long-range dipolar interaction. The intensity ratio of the emission lines for the processes leaving the longitudinal and the transverse exciton is one to two, corresponding to the twofold degeneracy of the transverse exciton. The energy diagram of the exciton and the EM and the selection rules are drawn in Fig. 2.11a. The splitting of the emission line was observed for the first time by *Suga* and *Koda* [2.48] for these two processes. The emission spectrum observed by *Ostertag* et al. [2.49] is shown in Fig. 2.12.

In the case of CuBr, the low-lying molecular state is composed of three levels: $\Gamma_5(J_t = 2)$, $\Gamma_3(J_t = 2)$, and $\Gamma_1(J_t = 0)$. We can selectively populate each of these molecular states by the giant two-photon absorption as will be discussed in Sect. 2.3.4. As a result, we can observe the emission spectrum from each of these lines, although other kinds of EMs are created, in a secondary process, from two excitons produced in the emission process and the emission lines of these EMs are mixed with the proper ones. If, however, we observe the emission spectrum at the initial stage when the secondary process is not effective, the proper emission lines will be predominant.

The energy diagram of the exciton and EM, and the selection rules in CuBr are drawn in Fig. 2.11b, taking into account the shift and splitting of molecular states due to hole-hole interband scattering. Two strong emission lines may be assigned to the radiative processes of the $\Gamma_1(J_t = 0, m_t = 0)$ molecule leaving longitudinal and transverse excitons with emission edges at 2.911 eV and 2.919 eV, respectively. The electronic structure of CuBr, i.e., the polariton structure and the levels of the EM will be shown to be very effectively determined by giant two-photon absorption spectroscopy, in Sect. 2.4.4, and hyper-Raman scattering via the EM, in Sect. 2.5.1.

The EM was also observed in the emission spectra of highly excited CdS and CdSe [2.37]. The binding energy of the EM is not so large in CdS and CdSe as in CuCl and CuBr, therefore the broadening effect is not negligible in the emission spectrum.

Let us discuss the spectral shape of EM luminescence and attempt to estimate the value of the molecular binding energy E_{mol}^b in CdS. With the increase of the density of EMs the number of elastic collisions between two EMs and also between an EM and a single exciton (which is produced as a result of the radia-

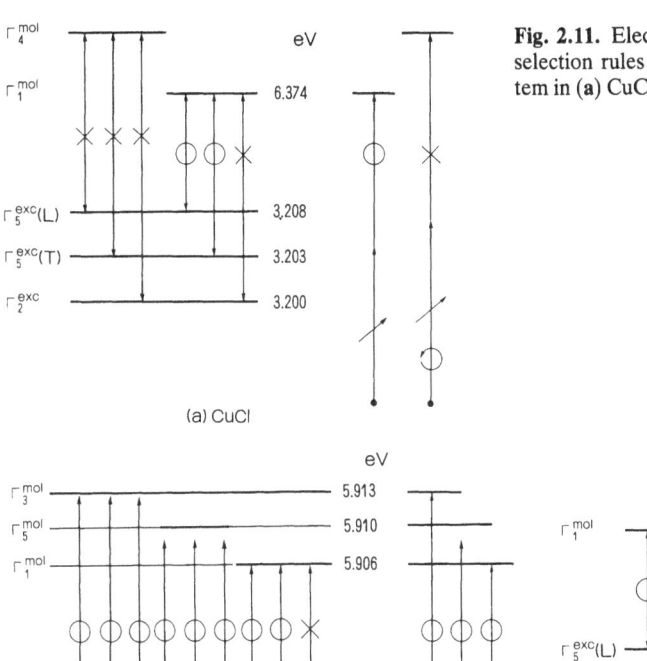

Fig. 2.11. Electronic structure and selection rules in the exciton-EM system in (**a**) CuCl, (**b**) CuBr and (**c**) CdS

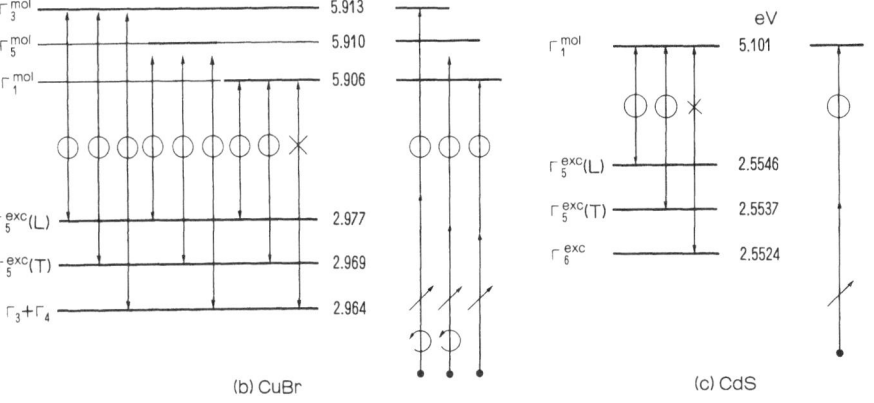

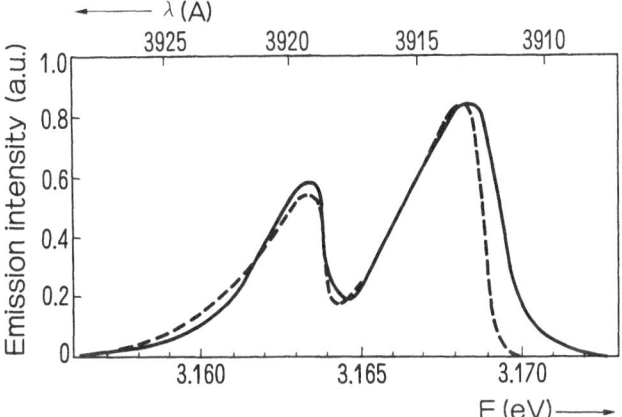

Fig. 2.12. Emission spectrum of the EM in CuCl leaving behind the transverse and longitudinal excitons. (- - -) theoretical and (——) experimental

tive annihilation of an EM), becomes considerable, so that their momentum states have relaxation times. Taking this effect into account, the emission spectrum of an EM is calculated as a function of $E \equiv 2E_{1s} - E_{mol}^b - E_{1s} - \hbar\omega$ as follows [2.37]:

$$I(E) = A \int_0^\infty \frac{n(\varepsilon)\Gamma\varepsilon^{1/2}d\varepsilon}{(E - \Sigma - \varepsilon)^2 + \Gamma^2}$$

$$= \begin{cases} A'E^{1/2}\exp(-E/k_BT) & \text{for } \Gamma \ll k_BT \\ \dfrac{A\Gamma}{(E - \Sigma)^2 + \Gamma^2} & \text{for } \Gamma \gg k_BT. \end{cases} \qquad (2.37)$$

Here, $n(\varepsilon)$ is the Boltzmann distribution function and Γ is the sum of the relaxation frequencies of a single exciton and an EM. Σ is given by $N(w_0 - v_0)$, in which N is the number of EMs, and Nw_0 and Nv_0 are the shifts of the energy levels of an EM and a single exciton, respectively, due to elastic collisions with N EMs. This can be neglected in comparison with k_BT, i.e., $\Sigma \simeq 0$. The above equation indicates that the spectral shape changes from the inverse Boltzmann distribution in the low-density limit to the Lorentzian-type in the case of high density.

The fitting of the M line shape to (2.37) has been made by taking Γ/k_BT and T as adjustable parameters. As shown in Fig. 2.13, the fit is satisfactory if one chooses $\Gamma/k_BT = 0.6$–0.9 and $T = 17$–20 K. In this way, the broadening of half-width with increasing excitation power is well interpreted. The arrows in the figure indicate the points corresponding to $E = 0$. If one considers the dissociation of an EM into two single excitons, the energy required is smallest when two triplet excitons, Γ_6 excitons in the present case, are produced. We take this energy as E_{mol}^b. Then the separation between the position of $E = 0$ in the figure and the energy of the Γ_6 exciton, 2.5524 eV, gives the value of E_{mol}^b. This is estimated to be about 5.4 meV. In the case of CdSe, detailed analysis of the M line shape is impossible, since the M line overlaps the I_2 line. Experimentally it is thought that E_{mol}^b may be roughly 4 meV.

In CdS, as described above, $m_h(\|)$ is very heavy. It might be useful to attempt the evaluation of E_{mol}^b by taking $\sigma = m_e/m_h(\|) \sim 0.037$ as the effective σ value. According to the variational calculation [2.1], $\sigma \sim 0.037$ corresponds to $E_{mol}^b/E_{ex}^b \sim 0.16$. Then E_{mol}^b is estimated to be 4.8 meV. On the other hand, when we take the geometrical mean of the hole masses $m_{h\perp} = 0.7$ and $m_{h\|} \sim 5$, E_{mol}^b becomes 2.6 meV in contrast with the value 5.4 meV obtained experimentally. This discrepancy may come from the variational error as well as the effect of the electron–LO-phonon interaction mentioned in Sect. 2.2.4.

Segawa and *Namba* [2.50] confirmed the EM from the splitting and polarization characteristics of the M emission lines in the wurtzite crystals of CdS and ZnO under uniaxial stress perpendicular to the c-axis. First, they confirmed that the peak intensity I of the M line is related to the excitation power I_0 by $I \propto I_0^n$ with $n = 1.7$ for CdS and $n = 1.3$ for ZnO, and that these M lines are polarized

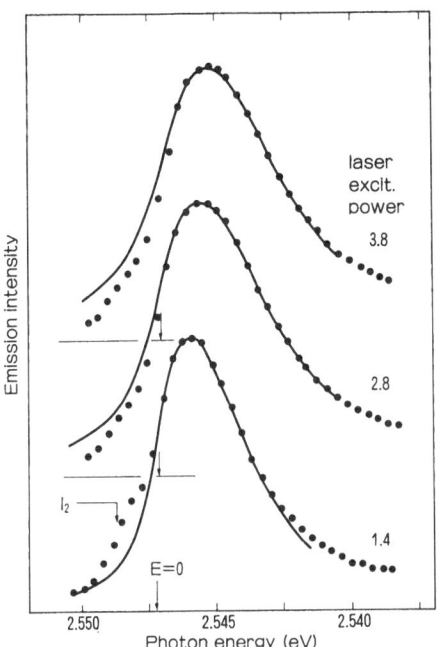

Emission intensity

laser
excit.
power

3.8

2.8

I_2

1.4

E=0

2.550   2.545   2.540

Photon energy (eV)

Fig. 2.13. Emission spectra of the EM in CdS. Dots show the observations and the solid lines are the theoretical curves including collision effects [2.37]

perpendicular to the c-axis. Second, they observed the emission and reflection spectra in the (0001) plane of the CdS crystal with the geometry: $\sigma \perp c$ and $k \| c$. It was of course confirmed that they are completely unpolarized in the stress-free crystals. Under the uniaxial stress $\sigma \perp c$, both the M line in the emission spectrum and the exciton level in the reflection spectrum were found to split into two components with the polarizations parallel and perpendicular to the direction of the stress, as shown in Fig. 2.14. The higher energy component of the exciton is polarized with $E \| \sigma$, while the polarization of the M emission line split-off on the high energy side is $E \perp \sigma$, and vice versa. These splittings and polarization characteristics can be understood when the M line is assigned to the emission of the EM.

The electronic structure was discussed in Sect. 2.2 for an EM in CdS as well as ZnO. The Bloch function part of the wave function of the EM is rewritten in terms of the basis functions of an exciton as follows:

$$|\Gamma_1(1, a, 2, b)\rangle = \tfrac{1}{2}(|\Gamma_{5x}\rangle|\Gamma_{5x}\rangle - |\Gamma_{5y}\rangle|\Gamma_{5y}\rangle$$

$$+ |\Gamma_{6-}\rangle|\Gamma_{6-}\rangle - |\Gamma_{6+}\rangle|\Gamma_{6+}\rangle) , \quad \text{where} \quad (2.38)$$

$$\begin{pmatrix} |\Gamma_{5x}\rangle \\ |\Gamma_{5y}\rangle \end{pmatrix} = \frac{1}{\sqrt{2}} (\pm |\Gamma_{5+}\rangle - |\Gamma_{5-}\rangle) = \begin{pmatrix} |\Gamma_3\rangle \\ |\Gamma_4\rangle \end{pmatrix} .$$

Therefore when we choose the x direction in the uniaxial field of the applied stress perpendicular to the c-axis, the Γ_5 exciton is split into two components Γ_{5x} and Γ_{5y} with the polarization along the x-axis and y-axis, respectively, as observed

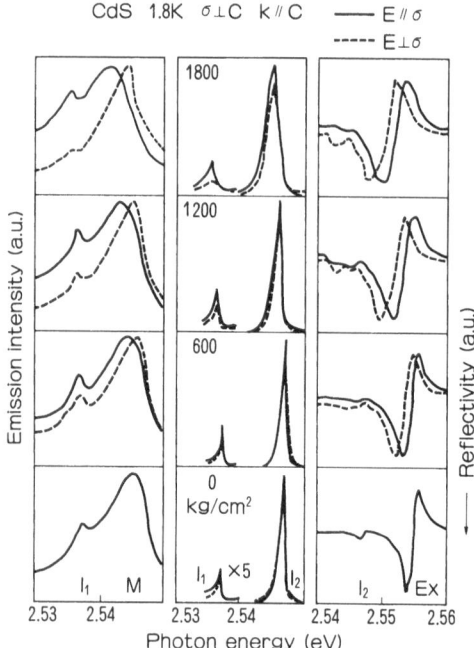

Fig. 2.14. Emission spectra from the EM (*left*) and the bound excitons (*middle*) in CdS under uniaxial stress σ perpendicular to the c-axis, and reflection spectra due to the exciton (*right*) under the stress. Solid and dotted lines describe the signal of $E\|\sigma$ and $E\perp\sigma$, respectively [2.50]

in the reflection spectra given in the right-hand parts of Fig. 2.14. The solid and dotted lines correspond to the reflection minima due to $\Gamma_{5x}(E\|\sigma)$ and $\Gamma_{5y}(E\perp\sigma)$ excitons.

On the other hand, the emission line due to decay of the EM also splits into two lines under uniaxial stress perpendicular to the c-axis, and the polarization characteristics are reversed from the case of the exciton spectrum, as shown in the left-hand part of Fig. 2.14. This can be explained as follows: In the radiative decay of the EM, when one Γ_{5x} (or Γ_{5y}) exciton is radiatively annihilated, the other Γ_{5x} (or Γ_{5y}) exciton should remain, according to (2.38). Therefore the $E\|\sigma$

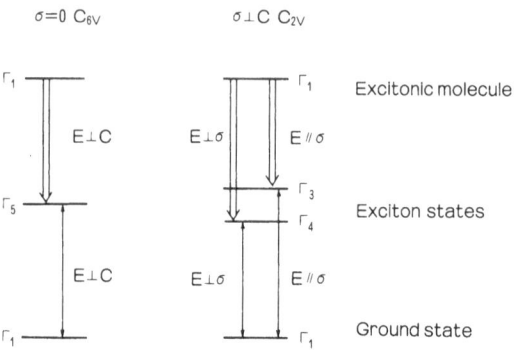

Fig. 2.15. Energy diagram of the exciton and the EM in CdS under the stress $\sigma \perp$ c-axis and the polarization characteristics of the related transitions

(stress) and $E \perp \sigma$ polarized light should be observed at $h\nu_{\parallel} = E_{mol} - E_{ex}(\Gamma_{5x})$ and $h\nu_{\perp} = E_{mol} - E_{ex}(\Gamma_{5y})$, respectively, as shown in Fig. 2.15, and the reversed characteristics of the polarization between the reflection spectrum of excitons and the emission spectrum of the M line can be explained. As a result, this experiment gave strong support to the assignment of the M line to the radiative decay of an EM in CdS and ZnO. On the other hand, the bound exciton should not split as observed in the middle part of the same figure. The EM in CdS was also confirmed by the giant two-photon absorption spectrum as will be discussed in Sect. 2.3.4.

2.3.2 Relaxation by Emission of Acoustic Phonons

In this subsection, we will discuss how the distribution of EMs over the dispersion of their kinetic energies changes in time due to their relaxation through the interaction with the acoustic-phonon field [2.51]. As a result, we can understand why the effective temperature observed in the emission spectrum of the EM gas is quite different from and much higher (30 K) than the lattice temperature (~ 4.2 K).

For the excitons and EMs with kinetic energies less than those of the optical phonons, the only remaining relaxation process which lowers the energy of the excitonic gas is the emission of acoustic phonons. Due to the smallness of the energy quanta of the acoustic phonons, the dissipation of the kinetic energy in the excitonic system is rather slow and requires many scattering events before quasi equilibrium is reached. Under these conditions the stochastic approach to the population distribution of these particles in energy space is justified as in the case of Brownian motion.

Starting with the Kolmogoroff-Chapman equation for the distribution function of an excitonic gas which interacts with acoustic phonons in momentum space, we derive a Fokker-Planck equation for the distribution function in energy space. The Fokker-Planck equation [2.52] can be derived in the same way for the exciton gas which is generated at lower excitation intensities and for the EM gas which is obtained at rather high excitation intensities. Therefore, we use the word "exciton" for the exciton as well as for the EM. Of course the parameters such as the translational mass or the "exciton"-phonon coupling constant have different values for these two cases.

The time development of the distribution function of the exciton is given by the Kolmogoroff-Chapman equation:

$$\frac{\partial n_k}{\partial t} = \sum_q 2\pi\gamma^2(q)\{n_{k+q}(1 + n_k)[(1 + f_q)\delta(\omega_q + E_k - E_{k+q})$$

$$+ f_q\delta(\omega_q - E_k + E_{k+q})] - n_k(1 + n_{k+q})[(1 + f_q)\delta(\omega_q - E_k + E_{k+q})$$

$$+ f_q\delta(\omega_q + E_k - E_{k+q})]\} , \tag{2.39}$$

where n_k and f_k are the distribution functions of the excitons and the acoustic phonons with the energy E_k and ω_k, respectively. In (2.39) $\gamma(q) = \gamma_0 \sqrt{q}$ is the matrix element of the exciton-phonon interaction. Throughout this section, we put $\hbar = 1$. Here we assume that the distribution function of the excitons n_k, depends on k only through its kinetic energy and that the heat bath of acoustic phonons remains in thermal equilibrium.

We are interested in the number density $N(E)$ in energy space, so the master equation for $N(E)$ is obtained by multiplying (2.39) by $\delta(E_k - E)$ and summing over k:

$$\frac{\partial N(E)}{\partial t} = \int_{-\infty}^{\infty} [W(E-\omega, \omega)N(E-\omega) - W(E, \omega)N(E)]\, d\omega , \qquad (2.40)$$

where the transition rate W is defined by

$$W(E, \omega) = \frac{\gamma_0^2}{2\pi s^3} \omega^2 f(\omega) \theta_E(\omega)\left(1 + \frac{N(E + \omega)}{\varrho(E + \omega)}\right),$$

and $\varrho(E)$ is the density of states of the excitons at energy E. The last enhancement factor $[1 + N(E + \omega)/\varrho(E + \omega)]$ in the transition rate is due to the boson character of the excitons, i.e., a state occupied by some number of excitons attracts other excitons to that state. The step function, $\theta_E(\omega)$, is obtained by using the dispersion relation of the excitons $E_k = k^2/2M$:

I) $\sqrt{E} > \varepsilon$

$$\theta_E(\omega) = \sqrt{\frac{M}{2E}} \begin{cases} 1 & \text{if } 0 < \omega < 4\varepsilon(\sqrt{E} + \varepsilon) \\ -1 & \text{if } -4\varepsilon(\sqrt{E} - \varepsilon) < \omega < 0 \\ 0 & \text{otherwise,} \end{cases}$$

II) $\varepsilon > \sqrt{E}$

$$\theta_E(\omega) = \sqrt{\frac{M}{2E}} \begin{cases} 1 & \text{if } 4\varepsilon(\varepsilon - \sqrt{E}) < \omega < 4\varepsilon(\sqrt{E} + \varepsilon) \\ 0 & \text{otherwise,} \end{cases}$$

where $\varepsilon = s\sqrt{M/2}$ is a small parameter which has the dimension of a square root of the energy, and s is the sound velocity.

The group velocity of the exciton coincides with the sound velocity when $E = \varepsilon^2$. The excitons with $E < \varepsilon^2$ cannot be scattered by the emission of acoustic phonons but only by absorption processes. The energy ε^2 is estimated to be 0.5 K for the exciton of CdS, and as we are interested in much larger energies, this energy can safely be treated as a small expansion parameter. Then (2.40) is written as

$$\frac{\partial N(E)}{\partial t} = \int_{-\infty}^{\infty} \left[\exp\left(-\omega \frac{\partial}{\partial E}\right) - 1 \right] W(E, \omega) N(E) \, d\omega$$

$$= -\frac{\partial}{\partial E} [\langle\omega\rangle_E N(E)] + \frac{1}{2} \frac{\partial^2}{\partial E^2} [\langle\omega^2\rangle_E N(E)] + \ldots . \tag{2.41}$$

In the following we first treat the low-density approximation, before studying the nonlinear theory which is relevant for a high-density exciton system.

In the low-density case we neglect terms quadratic in N and obtain for the nth moment:

$$\langle\omega^n\rangle_E = \frac{\gamma_0^2}{2\pi s^3} \sqrt{\frac{M}{2E}} \left\{ \frac{k_B T(4\varepsilon)^{n+2}}{n+2} [(\sqrt{E} + \varepsilon)^{n+2} + (-1)^n (\sqrt{E} - \varepsilon)^{n+2}] \right.$$

$$\left. + \frac{(-1)^n}{n+2} (4\varepsilon)^{n+3} (\sqrt{E} - \varepsilon)^{n+3} \right\} .$$

If we expand $\langle\omega^n\rangle_E$ in terms of ε, the leading term of $\langle\omega^n\rangle_E$ is of the order of ε^{n+2} for even n, and ε^{n+3} for odd n. The first and second moments are of the order of ε^4 and higher moments are negligibly small. As a result, the treatment of this system in terms of the Fokker-Planck equation is justified if the relevant energy region is much larger than ε^2. The first and the second moments are given by

$$\langle\omega\rangle_E = \frac{\gamma_0^2}{2\pi s^3} 4 \sqrt{\frac{M}{2}} (2\varepsilon)^4 (2k_B T - E) \sqrt{E} ,$$

$$\tag{2.42}$$

$$\langle\omega^2\rangle_E = \frac{\gamma_0^2}{2\pi s^3} 8 \sqrt{\frac{M}{2}} (2\varepsilon)^4 k_B T E^{3/2} .$$

We use a normalized energy $\tilde{E} = E/k_B T$ and time $\tilde{t} = t/\tau$, where τ is the characteristic relaxation time

$$\tau = \left[\frac{\gamma_0^2}{2\pi s^3} 4 \sqrt{\frac{M}{2}} (2\varepsilon)^4 \sqrt{k_B T} \right]^{-1} = \left[\frac{\gamma_0^2 s}{\pi} (2M)^{5/2} \sqrt{k_B T} \right]^{-1} .$$

Substituting (2.42) into (2.41), we obtain

$$\frac{\partial N(\tilde{E}, \tilde{t})}{\partial \tilde{t}} = \frac{\partial}{\partial \tilde{E}} \left[(\tilde{E} - 2) \sqrt{\tilde{E}} N(\tilde{E}, \tilde{t}) + \frac{\partial}{\partial \tilde{E}} \tilde{E}^{3/2} N(\tilde{E}, \tilde{t}) \right] .$$

The stationary solution of this equation is given by the Boltzmann distribution multiplied by the state density:

$$N_s(E) = \sqrt{E} \exp\left(-E/k_B T\right) .$$

In the high-density case, we include the nonlinear term $N(E + \omega)/\varrho(E + \omega)$ in $W(E, \omega)$ and expand it with respect to ω. The following Fokker-Planck equation is obtained:

$$
\frac{\partial N(\tilde{E}, \tilde{t})}{\partial \tilde{t}} = \frac{\partial}{\partial \tilde{E}} \left[\sqrt{\tilde{E}}(\tilde{E} - 2)\left(1 + \frac{N(\tilde{E}, \tilde{t})}{\varrho(\tilde{E})}\right) N(\tilde{E}, \tilde{t}) \right.
$$

$$
- 2\tilde{E}^{3/2} N(\tilde{E}, \tilde{t}) \frac{\partial}{\partial \tilde{E}} \left(\frac{N(\tilde{E}, \tilde{t})}{\varrho(\tilde{E})}\right) \Bigg]
$$

$$
+ \frac{\partial^2}{\partial \tilde{E}^2} \left[\tilde{E}^{3/2}\left(1 + \frac{N(\tilde{E}, \tilde{t})}{\varrho(\tilde{E})}\right) N(\tilde{E}, \tilde{t}) \right] . \tag{2.43}
$$

In this case, the conservation law of the total number is satisfied, (1) when the distribution function at the origin is not singular (normal case), and (2) when the distribution function is singular at the origin but in the form of the Bose distribution function with zero chemical potential (Bose-Einstein condensation). The stationary solution of (2.43) is given by the product of the Bose distribution function $f(E, \mu)$ and the state density $\varrho(E)$ as

$$N_s = \varrho(E)f(E, \mu), \quad \text{where} \quad f(E, \mu) = 1/(e^{\tilde{E}-\tilde{\mu}} - 1).$$

The quasi chemical potential μ is determined by the equation

$$
N = \sum_k \frac{1}{\exp\left[(k^2/2M - \mu)/k_B T\right] - 1}
$$

$$
= \frac{(2Mk_B T)^{3/2}}{4\pi^2} \Gamma\left(\frac{3}{2}\right) \sum_{n=1}^{\infty} \frac{1}{n^{3/2}} \exp\left(n\mu/k_B T\right) ,
$$

if N is smaller than the critical density N_c for Bose-Einstein condensation given by

$$
N_c = \frac{(2Mk_B T)^{3/2}}{4\pi^2} \Gamma\left(\frac{3}{2}\right) \zeta\left(\frac{3}{2}\right) .
$$

The numerical solution of (2.43) is plotted in Figs. 2.16 and 17 for the cases of $N < N_c$ and $N > N_c$, respectively. These figures show the exciton distribution. When the initial distribution has a rather sharp peak in energy space, it broadens quite rapidly at the initial stage and then the peak shifts towards the low-energy side. If the initial distribution peaks at a high energy it is shifted very quickly towards lower energies. The cases of broad initial distributions were also calculated. In these calculations, the initial population on the low-energy side is taken to be proportional to the state density. In this case, the distribution of "excitons" in the high-energy region shifts to and accumulates at the low-energy side as shown in Figs. 2.18 and 19. When we take into account the finite radiative

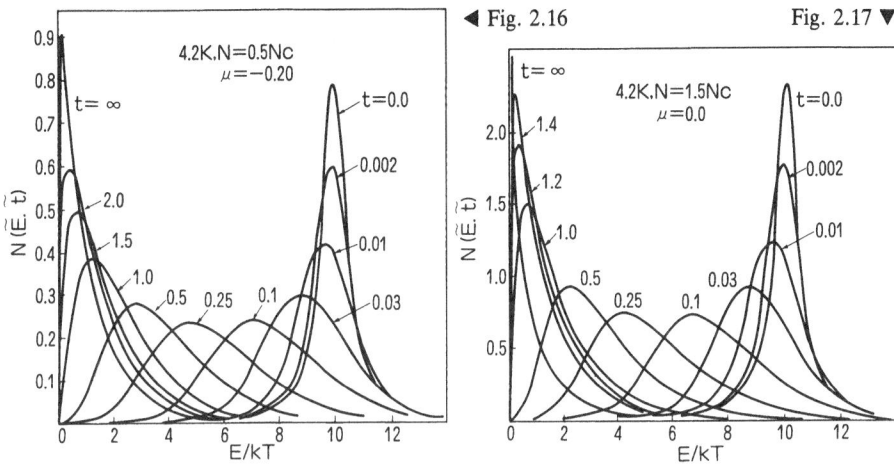

◀ Fig. 2.16   Fig. 2.17 ▼

Figs. 2.16, 17. Time development of the "exciton" distribution over normalized energy for sharp initial distributions. **Fig. 2.16** the total "exciton" concentration $N <$ the critical one N_c and **Fig. 2.17** $N > N_c$. The lattice temperature $T_L = 4.2$ K [2.51]

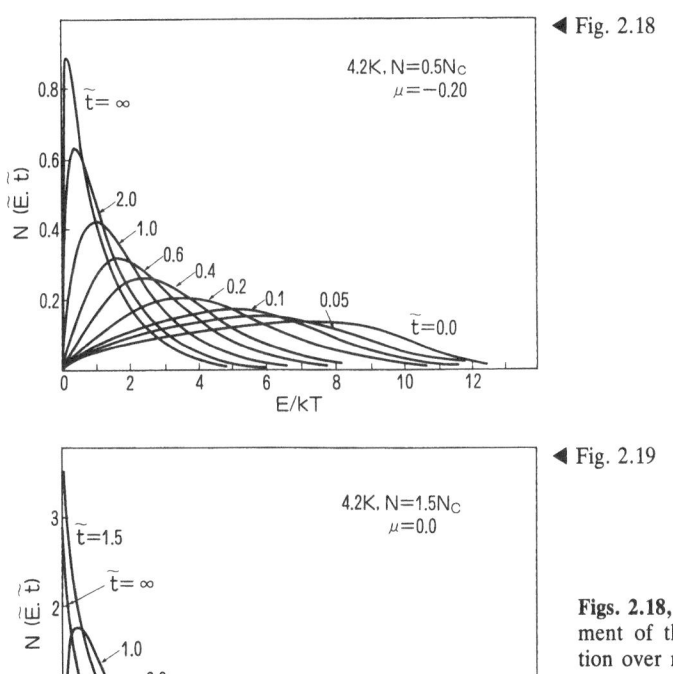

◀ Fig. 2.18

◀ Fig. 2.19

Figs. 2.18, 19. Time development of the "exciton" distribution over normalized energy for broad initial distributions modeling the case of exciton band excitation or band-to-band excitation. **Fig. 2.18** $N < N_c$ and **Fig. 2.19** $N > N_c$. The lattice temperature $T_L = 4.2$ K [2.51]

lifetime τ_R of the EM, the emission spectrum reflects the distribution at $t \fallingdotseq \tau_R$. Only in the limit t much longer than the characteristic relaxation time τ, does it have a canonical distribution with the lattice temperature. Thus we can understand that the effective temperature obtained from the luminescence spectrum is much higher than the lattice one for the case $\tau \geq \tau_R$. For such a high lattice temperature as $\tau < \tau_R$, the effective temperature of the luminescence approaches the lattice one.

2.3.3 Optical Conversion of Excitons into Excitonic Molecules

An effect of giant oscillator strength occurs in one-photon absorption in the presence of single excitons due to the conversion of an exciton into an EM, which is the reverse process to luminescence. Owing to this effect, we can expect an observable absorption coefficient for a reasonable concentration of excitons. Let us consider the initial state in which single Γ_5 excitons are accumulated in the crystal of CuCl and distributed over the exciton dispersion curve with the Boltzmann distribution $f_{ex}(K)$. Then the transition probability of one-photon absorption accompanied by the conversion of an exciton into an EM is evaluated as the reverse process of the luminescence by taking the Hermitian conjugate of the matrix element in (2.34) and replacing $f_{mol}(K)$ by the exciton distribution function $f_{ex}(K)$:

$$W_c^{(1)}(\omega) = \frac{2\pi}{\hbar} \sum_{K,m} |\langle \Gamma_1^{mol}(K+p)| \sum_i e\tilde{r}_i E |\Gamma_{5m}^{exc}(K)\rangle|^2$$

$$\times f_{ex}(K)\delta(\hbar\omega + E_{ex}(K) - E_{mol}(K+p)) . \tag{2.44}$$

When we insert the expressions of (2.9) and (2.11) into (2.44), neglect the photon momentum p in comparison with K, and assume the simplified form for the molecular envelope function: $\phi(r)\phi(r')g(R)$, the expression for $W_c^{(1)}(\omega)$ is obtained as follows:

$$W_c^{(1)}(\omega) = \frac{\pi}{6\hbar} [\langle \psi_c|e\tilde{x}|p_x\rangle \phi(0) \int dRg(R)]^2 E^2$$

$$\times \sum_K f_{ex}(K)\delta(\hbar\omega + E_{ex}(K) - E_{mol}(K)) , \tag{2.45}$$

$$= A\sqrt{E_{1s} - E_{mol}^b - \hbar\omega} \exp\left(-\frac{2(E_{1s} - E_{mol}^b - \hbar\omega)}{k_B T}\right) .$$

This last expression is also applicable to the conversion of the Γ_5 exciton into the Γ_1 EM in CdS, CdSe, and ZnO.

Here we compare the total transition probability which is obtained by integrating (2.45) over ω, with that of the single exciton Γ_5,

$$W_c^{(1)}/W_{ex}^{(1)} = [\int dRg(R)]^2 \varrho_{ex} \quad (\varrho_{ex} \text{ is the exciton density}).$$

If the relative motion of two holes, $g(R)$, is assumed to have the following form:

$$g(R) = \frac{1}{\sqrt{\pi a_{mol}^3}} \exp(-R/a_{mol}) , \quad \text{then}$$

$$W_c^{(1)}/W_{ex}^{(1)} = 64\pi a_{mol}^3 \varrho_{ex} .$$

Therefore for $\varrho_{ex} = 10^{17}/cm^3$ and $a_{mol} \simeq 2a_B \simeq 60 \text{Å}(CdS)$, we can expect the same order of absorption coefficient due to the conversion of excitons into EMs as in the exciton absorption. This strong enhancement originates from the effect of giant oscillator strength, which comes from the factor $\int dR g(R)$ in (2.45). This means that we can create an EM by exciting any valence electron into the large molecular orbital around the first exciton in the initial state.

The transition from a single exciton $\Gamma_{5m}^{ex}(m = 1, 0, -1)$ to the first rotational state Γ_4^{mol} of an EM in CuCl is forbidden. The Bloch function part of $\langle \Gamma_4^{mol} | H' | \Gamma_5^{ex} \rangle$ is non-zero but $|\Gamma_4^{mol}\rangle$ should be multiplied by the envelope function with odd parity for the exchange of two holes. As a result, the integration of this matrix element over the difference vector R between two holes vanishes. Due to the interband exchange effect, the energy level of the optically active exciton, Γ_5^{ex}, is higher than that of the pure triplet excitons Γ_2^{ex} (in CuCl) and Γ_6^{ex} (in CdS and CdSe). Therefore in the quasi equilibrium at low temperature, almost all excitons are in the pure triplet exciton state. However, these excitons cannot be converted into the EM states Γ_1^{mol} or Γ_4^{mol} by light absorption. For the transition Γ_2^{ex} (CuCl) or Γ_6^{ex} (wurtzite) $\to \Gamma_1^{mol}$, the integral over the Bloch functions of the electron-radiation interaction vanishes while for the transition Γ_2^{ex}(CuCl) or Γ_6^{ex}(wurtzite) $\to \Gamma_4^{mol}$, the symmetry properties of the envelope functions make the matrix element vanish. This can be realized from the integral $\int dR g'(R)$ in (2.45), in which $g'(R)$ shows the odd parity for the change of direction of R.

In the case of CuBr, however, the lowest pure triplet excitons of $\Gamma_3^{ex} + \Gamma_4^{ex}$ ($J_{ex} = 2$) can be converted into any states of the EM $J_t = 2$, while the transition to the molecular state $J_t = 0$ is forbidden. This can be understood from the expressions (2.17) [2.12], because $|0, 0\rangle$ is composed of only the two-exciton states $(J_{ex} = 1)^2$ and $(J_{ex} = 2)^2$ but $|2, m_t\rangle$ contains the cross term $(J_{ex} = 1)$ $(J_{ex} = 2)$. The Γ_5 excitons can be converted into any molecular states of Γ_1 and $\Gamma_3 + \Gamma_5$ by one-photon absorption. The selection rules of this transition are summarized in Fig. 2.11 for CuCl, CuBr, and CdS (CdSe and ZnO).

2.3.4 Giant Two-Photon Absorption

Originally the EM was formed from two single excitons by the relaxation process due to their interaction with the lattice system, and its existence could be detected only from the emission spectrum. If the kinetic energies of EMs relax to a thermal equilibrium in a much shorter time than the radiative lifetime, the effective temperature of the Boltzmann distribution is expected to be very close to the lattice temperature. However this is not the case as shown in Sect. 2.3.1. This was one of the obstacles to Bose condensation of EMs.

Then we had the desire to create the EM directly by an optical process and at the same time to observe its existence in the absorption spectrum. This desire was shown [2.4] theoretically to be capable of being fulfilled by using the giant two-photon absorption (GTA) due to an EM. This is extremely enhanced by the resonance effect and the effect of giant oscillator strength. *Gale* and *Mysyrowicz* [2.53] observed the EM in CuCl in this two-photon spectroscopy with the expected strong absorption coefficient. *Nozue* et al. [2.54] confirmed the EM in CdS by the giant two-photon absorption.

The transition probability is evaluated for two-photon excitation from the ground state $|g\rangle$ of the crystal to the final state $\Gamma_1^{\text{mol}}(K)$ of a created EM. This is expressed by the simple second-order perturbation with respect to the electron-radiation interaction H':

$$W^{(2)}(\Gamma_1; \omega) = \frac{2\pi}{\hbar} \left| \langle \Gamma_1^{\text{mol}}(K)|H'_- \sum_i \frac{|i\rangle \langle i|}{E_{ig} - \hbar\omega} H'_- |g\rangle \right|^2$$

$$\times \delta(2\hbar\omega - E_{\text{mol}}(K)) , \qquad (2.46)$$

where H'_- is the photon annihilation part of the electron-radiation interaction (2.35). As the intermediate state $|i\rangle$, we may take only the $1s$ exciton state $\Gamma_{5,m}^{\text{ex}}$ ($m = 1, 0, -1$) for CuCl and CuBr and Γ_5^{ex} for CdS and CdSe, because of the much larger oscillator strength and the resonant effect. By inserting into (2.46) the expressions for $\Gamma_1^{\text{mol}}(K)$ of (2.11) and $\Gamma_{5,m}^{\text{ex}}(K)$ of (2.9), the transition probability for CuCl is given as follows:

$$W^{(2)}(\Gamma_1; \omega) = \frac{2\pi}{\hbar} \left\{ \frac{1}{6(E_{1s} - \hbar\omega)} \left[\langle \psi_c|e\tilde{x}|\psi_0\rangle^2 E_\parallel E_\parallel \right. \right.$$

$$+ \langle \psi_c|e\tilde{x}_+|\psi_-\rangle \langle \psi_c|e\tilde{x}_-|\psi_+\rangle \left. (E_\perp^+ E_\perp^- + E_\perp^- E_\perp^+) \right] \phi(0)$$

$$\times \left. \int d\mathbf{r} \int d\mathbf{R} \, \phi(\mathbf{r})\psi^{++}(\mathbf{r}, 0, \mathbf{R}) \right\}^2 \delta(2\hbar\omega - E_{\text{mol}}(K)) ,$$

where $e\tilde{x}_+ = e(\tilde{x} \pm i\tilde{y})/\sqrt{2}$. When we assume the simplified envelope function: $\psi^{++}(\mathbf{r}, \mathbf{r}', \mathbf{R}) = \phi(\mathbf{r})\phi(\mathbf{r}')g(\mathbf{R})$, and use the cubic symmetry $\langle \psi_c|e\tilde{x}|p_x\rangle = \langle \phi_c|e\tilde{y}|p_y\rangle = \langle \phi_c|e\tilde{z}|p_z\rangle$ for CuCl, $W^{(2)}(\Gamma_1; \omega)$ is simplified as follows:

$$W^{(2)}(\Gamma_1^{\text{mol}}; \omega) = \frac{2\pi}{\hbar} \left(\frac{\langle \psi_c|e\tilde{x}|p_x\rangle^2}{6(E_{1s} - \hbar\omega)} \phi^2(0) \int d\mathbf{R} \, g(\mathbf{R})E^2 \right)^2$$

$$\times \delta\left(2\hbar\omega - 2E_{1s} + E_{\text{mol}}^{\text{b}} - \frac{\hbar^2(2p)^2}{2m_{\text{mol}}} \right) .$$

The transition probability of one-photon absorption due to a single exciton is

$$W^{(1)}(\Gamma_5^{\text{ex}}; \omega) = \frac{2\pi}{\hbar} \left[\frac{\langle \psi_c | e\tilde{x} | p_x \rangle}{\sqrt{3}} \phi(0) \right]^2 E^2 \delta\left(\hbar\omega - E_{1s} - \frac{\hbar^2 p^2}{2m_{\text{ex}}} \right),$$

and the ratio $W^{(2)}[\Gamma_1^{\text{mol}}; \omega = (E_{1s} - E_{\text{mol}}^{\text{b}}/2)/\hbar]/W^{(1)}(\Gamma_5^{\text{ex}}; \omega = E_{1s}/\hbar)$ is estimated to be of the order of $3 \cdot 10^{-15}$ (N/V), where (N/V) is the photon density, $p_{\text{cv}}^2/m_0 = m_0\omega^2 \langle \psi_c | \tilde{x} | p_z \rangle^2 \simeq 3$ eV, $\hbar\omega \simeq 3$ eV, $(a_{\text{mol}}/a_{\text{B}}) \simeq 2$, and the resonant energy $E_{\text{mol}}^{\text{b}}/2 = 15$ meV, corresponding to CuCl. Therefore when the dye laser is used as the excitation source with tunable frequency, above the photon density $N/V = 10^{15}/\text{cm}^3$, we can expect two-photon absorption to be stronger than one-photon absorption due to single excitons. Under this condition, it is guaranteed that the two-photon absorption coefficient due to an EM is very much greater than the absorption coefficient of an exciton in the exciton absorption tail $E_{1s} - E_{\text{mol}}^{\text{b}}/2$.

In comparison with the case of an ordinary two-photon absorption due to band-to-band transitions, we have furthermore, two strong enhancement factors in the case of the EM; the first enhancement factor $[\phi(0) \int g(R) dR]^2$ comes from the fact that our case corresponds to two-electron excitation and the second one comes from the resonance effect. The first enhancement is explained as follows: In the process of the transition from $|i\rangle$ to $\langle \Gamma_1^{\text{mol}} |$, we can choose any valence electron in the range within the large molecular radius a_{mol} around the first exciton in order to make an EM coherently. This is in contrast to the ordinary case where the electron excited to the intermediate state should interact again with the second photon. As a result, we have the factor $[\phi(0) \int g(R) dR]^2 \simeq 64(a_{\text{mol}}/a_{\text{B}})^3 \simeq 10^3$. The second enhancement factor of the resonance effect is estimated to be of an order of 10^4 at $\omega = \omega_{1s} - E_{\text{mol}}^{\text{b}}/2\hbar$ if we assume a few electron volts for the energy denominator in the band-to-band transition in contrast to the resonance denominator $E_{\text{mol}}^{\text{b}}/2 \simeq 15$ meV in the case of the EM. These two effects are combined to give the extremely strong two-photon absorption due to the EM. As a result, in two-photon spectroscopy, the EM will be confirmed as the sharp absorption peak at $E_{1s} - E_{\text{mol}}^{\text{b}}/2$ embedded in the rather weak background of the one-photon absorption tail of an exciton and the ordinary two-photon absorption due to band-to-band transitions.

Next, we discuss selection rules and polarization dependence for the direct creation of the EM by GTA. The transition probability for absorption of two photons of frequency ω_1 and ω_2 and polarization e_1 and e_2 is rewritten from (2.46) as

$$W = \frac{2\pi}{\hbar} \left(\frac{e}{m} \right)^4 \left(\frac{2\pi\hbar}{\kappa_1 V \omega_1} \right) \left(\frac{2\pi\hbar}{\kappa_2 V \omega_2} \right) N_1 N_2 \delta(E_f - E_g - \hbar\omega_1 - \hbar\omega_2)$$

$$\times |\langle f | e_1 \cdot (p\Lambda^+ p)_S \cdot e_2 + e_1 \cdot (p\Lambda^- p)_{AS} \cdot e_2 | g \rangle|^2 , \tag{2.47}$$

where

$$\Lambda^{\pm} \equiv \sum_i |i\rangle \Lambda_i^{\pm} \langle i| , \quad \Lambda_i^{\pm} = \frac{1}{E_i - E_g - \hbar\omega_1} \pm \frac{1}{E_i - E_g - \hbar\omega_2} , \tag{2.48}$$

and

$$(p\Lambda^{\pm}p)_{S,AS} \equiv \tfrac{1}{2}[(p\Lambda^{\pm}p) \pm {}'(p\Lambda^{\pm}p)] .$$

Expression (2.47) is decomposed into the geometrical factor $G_{\mu}(e_1, e_2)$ and the dynamical one as follows:

$$W(\omega_1, \omega_2) = C \sum_{f^{\mu}} G_{\mu}(e_1, e_2)\delta(E_{f^{\mu}} - E_g - \hbar\omega_1 - \hbar\omega_2)$$

$$\times \left| \sum_{\phi^{\lambda}} \Lambda_{\phi^{\lambda}}^{\pm} \langle f^{\mu}\|P^{\lambda}\|\phi^{\lambda}\rangle \langle \phi^{\lambda}\|P^{\lambda}\|g\rangle \right|^2 , \qquad (2.49)$$

where the geometrical factors, expressed in terms of Clebsch-Gordan coefficients, are

$$G_{\mu}(e_1, e_2) = \sum_{m} \left| \sum_{ll'} e_{2l}^* e_{1l'}^* (\mu m|\lambda l, \lambda l')\right|^2 , \qquad (2.50)$$

and the dynamical factors $\Lambda_{\phi^{\lambda}}^{\pm}$ are given by (2.48), the $+ (-)$ sign being appropriate for a final state which is contained in the symmetric (antisymmetric) product of the momentum representation. In the above expression, $|g\rangle$ is the crystal ground state of energy E_g with the identity representation, $\langle \phi^{\lambda}\|P^{\lambda}\|g\rangle$ is the reduced matrix element of the momentum operator $P = \sum_i p_i$ (belonging to the λ irreducible representation) between the ground state and an intermediate state $|\phi^{\lambda}\rangle$ of energy $E_{\phi^{\lambda}}$, and $\langle f^{\mu}\|P^{\lambda}\|\phi^{\lambda}\rangle$ is the reduced matrix element of the momentum operator P between an intermediate state and the final state of energy $E_{f^{\mu}}$. In (2.49) C is a constant irrelevant for the selection rules and the polarization characteristics: $C = 8\pi^3 \hbar e^4/(m^4 \kappa_1 \kappa_2 \omega_1 \omega_2 V^2)$.

We have three types of selection rules from the expression (2.49) for the two-photon absorption [2.55]:

1) Dipole selection rules. Two-photon transitions from the totally symmetric ground state are possible only to final states f^M belonging to the direct product of the irreducible representations of two transition dipole moments. We have, for CuCl and CuBr,

$$\Gamma_5 \times \Gamma_5 = \Gamma_1 + \Gamma_3 + \Gamma_4 + \Gamma_5 ,$$

and for CdS and CdSe,

$$\begin{aligned}
\Gamma_5 \times \Gamma_5 &= \Gamma_1 + \Gamma_2 + \Gamma_6 && \text{for } e_1, e_2 \perp c\text{-axis,} \\
\Gamma_1 \times \Gamma_1 &= \Gamma_1 && \text{for } e_1, e_2 \| c\text{-axis,} \\
\Gamma_1 \times \Gamma_5 &= \Gamma_5 && \text{for } e_1 \| c\text{-axis, } e_2 \perp c\text{-axis .}
\end{aligned}$$

2) Geometric selection rules. Two-photon transitions to a given final state are possible only when the corresponding geometrical factor is different from zero. Selection rules occur for particular directions of the polarization vectors with respect to one another or with respect to the crystallographic axes. The polariza-

Table 2.1. Geometrical factors for T_d and C_{6v} point groups

Point group T_d	$G_{\Gamma_1}(e_1, e_2)$	$= \frac{1}{3}\|e_1 \cdot e_2\|^2, \; G_{\Gamma_4}(e_1 \cdot e_2) = \frac{1}{2}\|e_1 \times e_2\|^2$
	$G_{\Gamma_3}(e_1, e_2)$	$= \frac{2}{3}(e_{1x}^2 e_{2x}^2 + e_{1y}^2 e_{2y}^2 + e_{1z}^2 e_{2z}^2)$
		$\quad - \frac{2}{3}(e_{1x}e_{1y}e_{2x}e_{2y} + e_{1y}e_{1z}e_{2y}e_{2z} + e_{1z}e_{1x}e_{2z}e_{2x})$
	$G_{\Gamma_5}(e_1, e_2)$	$= \frac{1}{2}[(e_{1x}e_{2y} + e_{1y}e_{2x})^2$
		$\quad + (e_{1y}e_{2z} + e_{1z}e_{2y})^2 + (e_{1z}e_{2x} + e_{1x}e_{2z})^2]$
Point group C_{6v}	$G_{\Gamma_1}(e_{1\perp}, e_{2\perp})$	$= \frac{1}{2}(e_{1x}e_{2x} + e_{1y}e_{2y})^2$
	$G_{\Gamma_2}(e_{1\perp}, e_{2\perp})$	$= \frac{1}{2}(e_{1x}e_{2y} - e_{1y}e_{2x})^2$
	$G_{\Gamma_6}(e_{1\perp}, e_{2\perp})$	$= \frac{1}{2}, \; G_{\Gamma_1}(e_{1\|}, e_{2\|}) = 1, \; G_{\Gamma_5}(e_{1\|}, e_{2\perp}) = 1$

tion dependence of the two-photon absorption intensity is listed, e.g., in [2.51b], for the two-photon allowed transition in 32 crystal point groups. The geometrical factors computed from (2.50) are listed in Table 2.1 for CuCl, CuBr, CdS, and CdSe.

3) *Dynamical selection rules.* Two-photon transitions to a final state belonging to the antisymmetric product of the irreducible representations of two transition dipole moments are forbidden when $\omega_1 = \omega_2$, as can be seen from (2.48). The dynamical selection rules come from the properties of Clebsch-Gordan coefficients: $(\mu m|\lambda l', \lambda l) = \pm (\mu m|\lambda l, \lambda l')$ for the interchange of e_1 with e_2 and ω_1 with ω_2 in the expression of second-order transition probability. Here the $+$ sign holds in the case D^μ appears in the decomposition of the symmetric product representation $[D^\lambda \times D^\lambda]$ and the $-$sign holds in the case D^μ belongs to the antisymmetric product representation $\{D^\lambda \times D^\lambda\}$. For the case of CuCl and CuBr, the symmetric and antisymmetric products are, respectively, $[\Gamma_5 \times \Gamma_5] = \Gamma_1 + \Gamma_3 + \Gamma_5$ and $\{\Gamma_5 \times \Gamma_5\} = \Gamma_4$. The strength of the transitions to the Γ_1, Γ_3, and Γ_5 final states involves the factor $\Lambda_{\phi\lambda}^+$, while that to the Γ_4 final state involves $\Lambda_{\phi\lambda}^-$, therefore the former states are strongly allowed, while the latter is weakly allowed and forbidden for $\omega_1 = \omega_2$. For the case of CdS and CdSe, the antisymmetric product is $\{\Gamma_5 \times \Gamma_5\} = \Gamma_2$, so that two-photon transitions to the final state Γ_2 are forbidden by the dynamical selection rule when $\omega_1 = \omega_2$.

The EM state of CuCl was confirmed to belong to the Γ_1 representation by measuring that the GTA intensity depends only on the relative angle between the two polarizations e_1 and e_2 as $G_{\Gamma_1} = \frac{1}{3}\|e_1 \cdot e_2\|^2$. This was confirmed experimentally by *Phach* et al. [2.56]. *Nakata* et al. [2.57] found three peaks in the excitation spectrum of CuBr due to the GTA of the EM. The assignment of these lines to the strongly allowed Γ_1, Γ_3, and Γ_5 EMs was made by observing the intensity dependence of these absorption lines on the polarization e_1 and e_2 of two incident light beams. *Phach* et al. [2.58] observed the polarization dependence of these three lines under the configuration of Fig. 2.20. They used the laser light $\hbar\omega_1(e_1, k_1)$ and the continuum light $\hbar\omega_2(e_2, k_2)$ as the first and second light

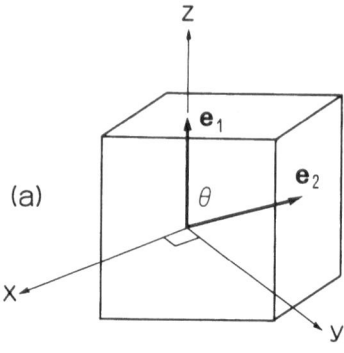

(a)

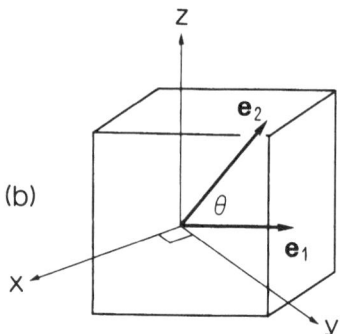

(b)

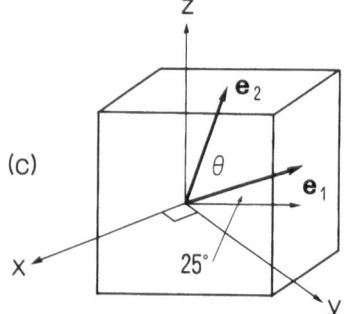

(c)

Fig. 2.20a–c. Polarization configurations of two incident laser beams. **(a)** $e_1 = [0, 0, 1]$ and $e_2 = [-(\sin\theta)/\sqrt{2},\ (\sin\theta)/\sqrt{2},\ \cos\theta]$, **(b)** $e_1 = [-1/\sqrt{2},\ 1/\sqrt{2},\ 0]$ and $e_2 = [-(\cos\theta)/\sqrt{2},\ (\cos\theta)/\sqrt{2},\ \sin\theta]$, **(c)** $e_1 = [-(\cos25°)/\sqrt{2},\ (\cos25°)/\sqrt{2},\sin25°]$ and $e_2 = [-[\cos(\theta + 25°)]/\sqrt{2},\ [\cos(\theta + 25°)]/\sqrt{2},\sin(\theta + 25°)]$

sources, respectively, and obtained the results shown in Fig. 2.21a–c. For the configurations of Fig. 2.20a–c,

$$a:\ e_1 = [0, 0, 1] \quad \text{and} \quad e_2 = \left[-\frac{1}{\sqrt{2}}\sin\theta,\ \frac{1}{\sqrt{2}}\sin\theta,\ \cos\theta\right],$$

$$b:\ e_1 = \left[-\frac{1}{\sqrt{2}},\ \frac{1}{\sqrt{2}},\ 0\right] \quad \text{and} \quad e_2 = \left[-\frac{1}{\sqrt{2}}\cos\theta,\ \frac{1}{\sqrt{2}}\cos\theta,\ \sin\theta\right],$$

$$c:\ e_1 = \left[-\frac{1}{\sqrt{2}}\cos25°,\ \frac{1}{\sqrt{2}}\cos25°,\ \sin25°\right] \quad \text{and}$$

$$e_2 = \left[-\frac{1}{\sqrt{2}}\cos(\theta + 25°),\ \frac{1}{\sqrt{2}}\cos(\theta + 25°),\ \sin(\theta + 25°)\right].$$

Here the crystallographic x-, y-, and z-axes were taken as shown in Figs. 2.20a, b and c, corresponding to the experimental configuration of *Phach* et al. [2.58] in which the two relevant faces have the indices (1 1 0). Then the geometrical factors G_Γ are given as a function of the relative angle $\theta = (e_1, e_2)$ as summarized in Table 2.2.

Table 2.2.

e_1	[001]	[$\bar{1}$10]	making an angle 25° with the [$\bar{1}$10] axis
Γ_1		$\frac{1}{3}\cos^2\theta$	
Γ_4		$\frac{1}{2}\sin^2\theta$	
Γ_3	$\frac{2}{3}\cos^2\theta,$	$\frac{1}{6}\cos^2\theta,$	$\frac{1}{3}\sin^2 25° \sin(2\theta + 50°) \times [\tan(\theta + 25°) - \cot 25°]$
Γ_5	$\frac{1}{2}\sin^2\theta,$	$\frac{1}{2},$	$\frac{1}{2}\sin^2(\theta + 50°) + \frac{1}{2}\sin^2 25° \cos^2(\theta + 25°)$

The Γ_4 level is an almost forbidden state because of the dynamical selection rule for $\omega_1 \simeq \omega_2$ so that we may assign the observed three lines to the Γ_1, Γ_3, and Γ_5 states of the EM. This fact is coincident with the following consideration. As long as we can neglect the e–h exchange interaction in comparison with the Coulomb interactions among the four composite particles, the ground state of the EM has an envelope function symmetric in the exchange of the two electrons and the two holes in analogy with the hydrogen molecule. Hence, the product of the Bloch functions for the two electrons in the conduction band and for the two holes in the valence band must be antisymmetric in the exchange of the two electrons and of the two holes, respectively,

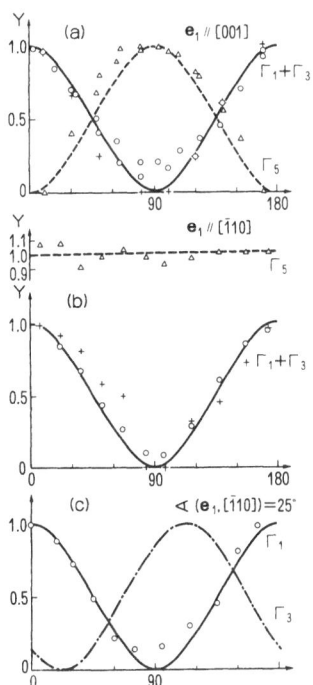

Fig. 2.21. Dependence of the giant two-photon transition intensity on the angle θ between the polarization e_1 and e_2 of two incident laser beams, corresponding to the three configurations (a), (b), and (c) in Fig. 2.20. (O): two-photon absorption at low energy ($\hbar\omega_1 + \hbar\omega_2 = 5.9061$ eV, Γ_1 symmetry), ($\triangle$) that at intermediate energy (5.9103 eV, Γ_5) and (+) that at high energy (5.9128 eV, Γ_3) [2.58]

$$\{\Gamma_6 \times \Gamma_6\}_- = \Gamma_1$$
$$\{\Gamma_8 \times \Gamma_8\}_- = \Gamma_1 + \Gamma_3 + \Gamma_5 .$$

Therefore it follows that the possible symmetries for the EM state are Γ_1, Γ_3, and Γ_5. From the polarization dependence of Figs. 2.21a and b, the intermediate (5.9103 eV) line ($\triangle$) can be assigned to the two-photon absorption due to the Γ_5 EM. From the observations of Fig. 2.21c together with those of Figs. 2.21a and b, the lowest (5.9061 eV) line ($\bigcirc$) always shows $\cos^2\theta$ dependence of the absorption intensity so that we assign this line to the Γ_1 EM and the highest (5.9128 eV) line ($+$) to the Γ_3 EM.

The GTA spectrum due to direct generation of an EM in CuCl was studied in a temperature range between 6K and 100K [2.59]. The interaction of the EM with the acoustic phonons was found to play an important role in its relaxation mechanism at high lattice temperatures. The line shape of the exciton absorption band was investigated by *Toyozawa* [2.60] by solving the self-consistent equations for the energy dependent shift and broadening with the aid of graphic calculation. The line shape in the main part of the absorption band is Lorentzian for temperatures lower than the critical temperature T_2, where the one-phonon process is dominant. For high temperatures such that $T > T_2$, the multi-phonon process prevails and the half-width is proportional to $(g_{ex}T)^2$, where g_{ex} is the exciton–accoustic-phonon coupling constant. Then the line shape is strongly asymmetric with a tail on the high-energy side.

The same discussion can be applied to the GTA spectrum due to the EM. The full width at half maximum (FWHM) W is given by

$$W = \begin{cases} \left(\dfrac{2}{\pi}\right) g_{\text{mol}} \, k_B \dfrac{T_1}{e^{T_1/T} - 1} & T < T_2 & (2.51a) \\[2ex] 0.3 \, g_{\text{mol}}^2 (k_B T)^2 / m_{\text{mol}} u^2 & T > T_2 & (2.51b) \end{cases}$$

where $T_1 = 2m_{\text{mol}}u^2/k_B$ and $T_2 = 2.6m_{\text{mol}}u^2/k_B g_{\text{mol}}$. Here, g_{mol}, m_{mol} and u are the EM–acoustic-phonon coupling constant, the translational effective mass of the EM, and the velocity of sound, respectively. This coupling constant g_{mol} is written in terms of the deformation potential E_d^{mol} of the EM and the mass density ϱ of the crystal as

$$g_{\text{mol}} = \frac{(m_{\text{mol}} E_d^{\text{mol}})^2}{\hbar^3 \varrho u} .$$

From the best fit of the experimental data to (2.51), $g_{\text{mol}} = 0.24$ and $T_2 = 53K$ are obtained, as will be shown in Fig. 3.32. Here they used $\varrho = 4.16$ g/cm^3, $u = 3.6 \times 10^5$ cm/s and $m_{\text{mol}} = 5.75 \, m_0$. The value $T_1 = 9.8K$ is also obtained from (2.51). The deformation potential difference between the conduction and valence bands $|C_c - C_v|$ is estimated as 1.2 eV, when we assume $E_d^{\text{mol}} = 2|C_c - C_v|$. We note that the half-width at low temperatures may be determined by the residual scattring of the EM by impurities or by mutual collisions. Taking into account this fact, the agreement between the experimental and theoretical results looks satisfactory. The giant two-photon absorption spectrum has also been calculated at the low temperature of 6K and at 73K in the high region in terms of Toyozawa's model. These curves match the observed characteristics very well as will be shown in Fig. 3.31.

2.4 Coherent Optical Phenomena Due to the Excitonic Molecule

2.4.1 Hyper-Raman Scattering and Luminescence

One problem which arises in the interpretation of emission spectra of EMs which are generated by GTA is that we have two emission channels; (1) the luminescence, which has been considered so far and (2) the emission due to Raman scattering. In the latter case, the EM exists only as a virtually excited state and the reemission takes place before any phase destruction sets in. In order to investigate the competition between these two channels, we examine the transition rate of a third-order optical process (absorption of two photons with the successive reemission of one photon and generation of an exciton) [2.61]. We make use of the damping theory for the interaction of the EM and the exciton in the final state with their surroundings (other EMs, excitons, and phonons).

The initial state $i = (g, N_1, N_2)$ is the crystal ground state with N_1 photons in the incoming light field of ω_1 and N_2 photons in the outgoing light field of ω_2. The interaction Hamiltonians between the light fields 1 and 2 and the electronic system are denoted by $V_1^{\pm}$ and $V_2^{\pm}$, respectively, where $-(+)$ refers to the annihilation (creation) of a photon. The intermediate state of an EM is connected with the ground state through a two-photon transition by $V_1^{\pm}$ and with the final state of a single exciton through a one-photon transition by $V_2^{\pm}$. Starting from the initial state $i = (g, N_1, N_2)$ at $t = 0$, the probability $P(t)$ of finding the final state $f = (ex, N_1 - 2, N_2 + 1)$ at time t is given by

$$
P(t) = \int_0^t dt_1 \int_0^{t_1} dt_2 \int_0^{t_2} dt_3 \int_0^t dt_1' \int_0^{t_1'} dt_2' \int_0^{t_2'} dt_3'
$$

$$
\mathrm{Tr}_R \{ \langle f | \exp[-i(t - t_1')H] V_2^+ \exp[-i(t_1' - t_2')H] V_1
$$

$$
\times \exp[-i(t_2' - t_3')H] V_1 \exp(-it_3'H) \varrho_0(H) \exp(it_3 H) V_1^+
$$

$$
\times \exp[i(t_2 - t_3)H] V_1^+ \exp[i(t_1 - t_2)H] V_2 \exp[i(t - t_1)H] | f \rangle \} .
$$

Here, V_i denotes V_i^-, Tr_R describes the trace over the reservoir and $\varrho_0(H) = |g\rangle \langle g| \varrho_R$ is the density matrix of the unperturbed system $|g\rangle \langle g|$ and reservoir ϱ_R. The process of two-photon absorption due to the EM is assumed to be completed instantaneously in comparison with the lifetime of the intermediate EM. Then we have three kinds of time orderings for the contribution to $P(t)$, as shown in Fig. 2.22.

The first contribution is expressed as follows:

$$P_i(T) = 2t \ \text{Re}\left\{ \int_0^\infty\!\!\!\int\!\!\!\int d\tau_3 d\tau_2 d\tau_1 \sum \langle\langle f\bar{m}'|\hat{U}(\tau_3)|f'''\bar{m}\rangle\rangle\right.$$

$$\times \langle\langle f''\bar{g}''|\hat{U}(\tau_2)|f'\bar{g}'\rangle\rangle \langle\langle m'\bar{g}''|\hat{U}(\tau_1)|m\bar{g}\rangle\rangle$$

$$\times \varrho_{gg}\exp\left(-i\omega_2\tau_3\right)\exp\left(i2\omega_1\tau_1\right)\exp\left[i(2\omega_1-\omega_2)\tau_2\right]$$

$$\left.\times (V_2^+)_{f'm'}(W_1)_{mg}(W_1^+)_{\bar{g}''\bar{m}}(V_2)_{\bar{m}'f}\right\}. \tag{2.52}$$

The double brackets have the following meanings

$$\langle\langle ab|U(\tau)|cd\rangle\rangle \equiv \text{Tr}_R\{\varrho_R[\langle a|\exp\left(-iH\tau\right)|c\rangle \langle d|\exp\left(iH\tau\right)|b\rangle]\}$$

$$= \delta_{ac}\delta_{bd}\exp\left(-i\omega_{ab}\tau - \Phi_{ab}\tau\right), \tag{2.53}$$

where $\Phi_{ab} = \Phi_{ba}^* = \gamma_{ab} + i\Delta_{ab}$ and $\gamma_{ab} = \gamma_{ab}^c + \frac{1}{2}(\Gamma_a + \Gamma_b)$. For the diagonal component, one has the damping constant $\Phi_{aa} = \gamma_a = \gamma_a^c + \Gamma_a$, where $(1/\Gamma_a)$ is the lifetime of level a and γ_a^c is the inelastic collision frequency of level a. The term γ_{ab}^c is composed of the elastic as well as inelastic collisions of levels a and b. The two-photon transition matrix elements $(W_1)_{mg}$ and $(W_1^+)_{\bar{g}''\bar{m}}$ are obtained after integrations with respect to $\tau' = t_2' - t_3'$ and $\tau = t_2 - t_3$. One finds $(W_1)_{mg} = i(V_1)_{gi}(V_1)_{im}/(\omega_{im} - \omega_1)$, where i is the intermediate exciton state. Here the effect of the giant oscillator strength and the resonance effect discussed in Sect. 2.4.4 are included.

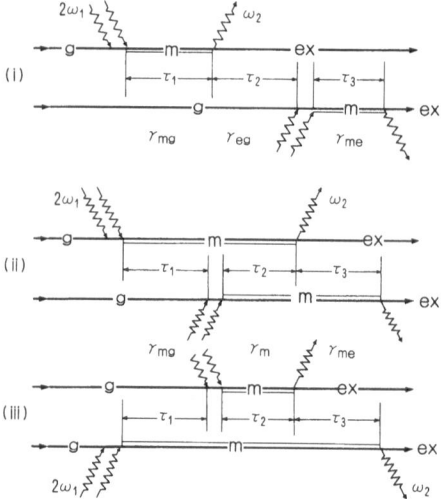

Fig. 2.22. Processes contributing to the hyper-Raman scattering and the luminescence due to the EM. (g) the initial ground state, (m) the intermediate state with an excitonic molecule and (ex) the final state with an exciton

Similarly, the second and third contributions of the diagrams (ii) and (iii) are given by

$$
\begin{aligned}
P_{\mathrm{ii}}(t) \;=\; 2t \;\; \mathrm{Re} \Bigg\{ &\iiint_0^\infty d\tau_3 d\tau_2 d\tau_1 \sum \varrho_{gg} \langle\langle f\bar{m}''|\hat{U}(\tau_3)|f'\bar{m}'\rangle\rangle \\
&\times \langle\langle m'''\bar{m}'|\hat{U}(\tau_2)|m''\bar{m}\rangle\rangle \; \langle\langle m'\bar{g}''|\hat{U}(\tau_1)|m\bar{g}'\rangle\rangle \; (V_2^+)_{f'm'''} \\
&\times (W_1)_{mg}(W_1^+)_{\bar{g}''\bar{m}}(V_2)_{\bar{m}''f} \exp(2i\omega_1\tau_1)\exp(-i\omega_2\tau_3) \Bigg\} ,
\end{aligned}
\tag{2.54}
$$

$$
\begin{aligned}
P_{\mathrm{iii}}(t) \;=\; 2t \;\; \mathrm{Re} \Bigg\{ &\iiint_0^\infty d\tau_3 d\tau_2 d\tau_1 \sum \langle\langle f\bar{m}''|\hat{U}(\tau_3)|f'\bar{m}'''\rangle\rangle \\
&\times \langle\langle m'\bar{m}'''|\hat{U}(\tau_2)|m\bar{m}'\rangle\rangle \; \langle\langle g''\bar{m}'|\hat{U}(\tau_1)|g'\bar{m}\rangle\rangle \\
&\times \varrho_{gg}(V_2^+)_{f'm'}(W_1)_{mg''}(W_1^+)_{g\bar{m}}(V_2)_{\bar{m}''f} \\
&\times \exp(-2i\omega_1\tau_1 - i\omega_2\tau_3) \Bigg\} .
\end{aligned}
\tag{2.55}
$$

Equation (2.53) is inserted into (2.52), (2.54), and (2.55) and time integrations are performed. The resulting transition probability per unit time is given by

$$
\begin{aligned}
F(2\omega_1,\, \omega_2) \;&=\; \lim_{t\to\infty}\,[P(t)/t] \\
&= A \frac{\gamma_{mg}}{(2\omega_1 - \omega_{mg})^2 + \gamma_{mg}^2} \cdot \frac{\gamma_{me}}{(\omega_2 - \omega_{me})^2 + \gamma_{me}^2} \\
&\quad \times \Bigg(\frac{2}{\gamma_m} + \frac{\gamma_{eg}[1 + (2\omega_1 - \omega_{mg})(\omega_2 - \omega_{me})/\gamma_{mg}\gamma_{me}]}{(2\omega_1 - \omega_2 - \omega_{eg})^2 + \gamma_{eg}^2} \\
&\qquad - \frac{(2\omega_1 - \omega_2 - \omega_{eg})[(2\omega_1 - \omega_{mg})/\gamma_{mg} - (\omega_2 - \omega_{me})/\gamma_{me}]}{(2\omega_1 - \omega_2 - \omega_{eg})^2 + \gamma_{eg}^2} \Bigg)
\end{aligned}
\tag{2.56}
$$

where $A \equiv N_1(N_1 - 1)(N_2 + 1)|(W_1)_{mg}|^2|(V_2)_{me}|^2$.

In the process $P_{\mathrm{i}}(t)$, the three differences of phase modulations, between the EM state and the ground state, between the final exciton state and the ground state and between the final exciton state and the EM state, work in the time intervals τ_1, τ_2, and τ_3, respectively (see Fig. 2.22). P_{i} gives rise to the second and third terms inside the large brackets of (2.56), which will be shown to be predominantly Raman scattering terms. On the other hand, in the processes P_{ii} and P_{iii}, the EM is real in the time interval τ_2 and only the radiative damping works in this time interval. Both phase modulation and radiative damping work in other time intervals. The processes P_{ii} and P_{iii} contribute to the first term inside the large brackets of (2.56), which is the luminescence term. The first term describes luminescence because the maximum emission frequency ω_2 is independent of the frequency ω_1 of the incoming laser light. The second and third terms represent mainly Raman scattering in the sence that ω_2 is correlated with ω_1 as shown in this expression.

Let us discuss the following two limiting cases: In the first case, only the EM state is strongly modulated, which is described by the inequalities: (a) $\gamma_{mg} \simeq \gamma_{me} = \bar{\gamma} \gg 2\gamma_{eg} > \gamma_m$. Then the first contribution (i.e., the second and third terms inside the large brackets) is simplified. When the spectral width of the incoming laser light is much wider than $\bar{\gamma}$, the emitted spectrum is obtained by integrating (2.56) with respect to ω_1. The second and third term of the i-th contribution cancel each other and only the first term describing the luminescence is non-zero. In the second case, only the phase of the final exciton state is strongly modulated, which is described by the following inequalities: (b) $\gamma_{eg} \simeq \gamma_{me} \equiv \bar{\gamma} \gg 2\gamma_{mg} > \gamma_m$. When the spectral width of the incoming laser light is much larger than $\bar{\gamma}$, the peaks of the Raman and the luminescence components are coincident with each other and the integrated intensity has the ratio 1 to $4\bar{\gamma}/\gamma_m$. As a result, the luminescence component is dominant also in the case (b) as long as the radiative damping γ_m is much smaller than the phase modulation of the exciton in the final state.

When the spectral width of the incident light is narrower than or of the same order as the relaxation constants, both the luminescence and Raman scattering lines can be observed distinctly. On the other hand, the Raman component disappears and only the luminescence line is observed for incident light with a much wider spectrum than any of the relaxation constants. This fact was observed in [2.62]. The Raman scattering line was observed under excitation by laser light with 0.2 meV spectral width while only the luminescence line was observed under laser radiation of 4 meV spectral width.

Mita and *Ueta* [2.63] observed the emission spectrum under the giant two-photon excitation of the EM by a narrow-band laser of 0.053 meV half-width as shown in Fig. 3.36. Around the two-photon resonance with the EM, they observed the coexistence of the Raman scattering and luminescence lines. The emission spectrum calculated in terms of (2.56) is shown in Fig. 2.23 and it can describe the observed characteristics very well (compare this with Fig. 3.36). The relevant parameters are determined as follows: γ_{me} by the half width at half maximum (HWHM) of the emission spectrum shown by the dotted line in Fig. 3.36, γ_{eg} by HWHM of the Raman line minus HWHM of the excitation light. The value of γ_{mg} can be determined by FWHM of the two-photon absorption spectrum but this depends rather sensitively on the excitation power. We fixed these parameters, therefore, by the approximate relation $\gamma_{me} \simeq \gamma_{mg} + \gamma_{eg}$, considering that the phase modulation of the ground level is negligible. The remaining parameter γ_m is determined by optimizing the agreement between the calculated and observed spectra. Thus the relevant relaxation constants are determined in units of meV as $\gamma_{mg} = 0.086$, $\gamma_{eg} = 0.066$, $\gamma_{me} = 0.152$, and $\gamma_m = 0.05$. The excitation energy of the incident light is determined by the peak position of the Raman line in Fig. 3.36 (from f to l).

Mita and *Ueta* [2.62] also observed the coexistence of the Raman and the luminescence lines under excitation by wide-band laser light [Ref. 2.62, Fig. 2]. The wide emission line at the lower energy side of the M_L line is assigned to the secondary process, i.e., the emission of hot EMs formed from two excitons left

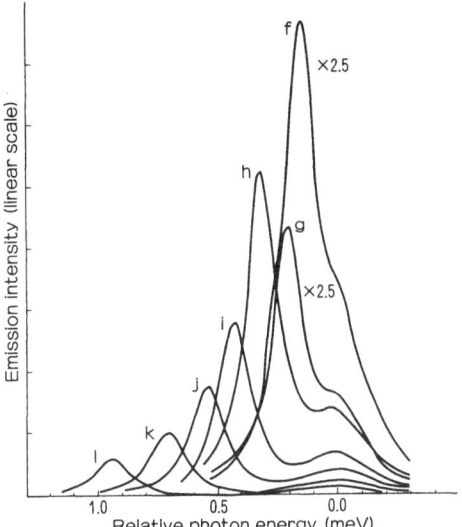

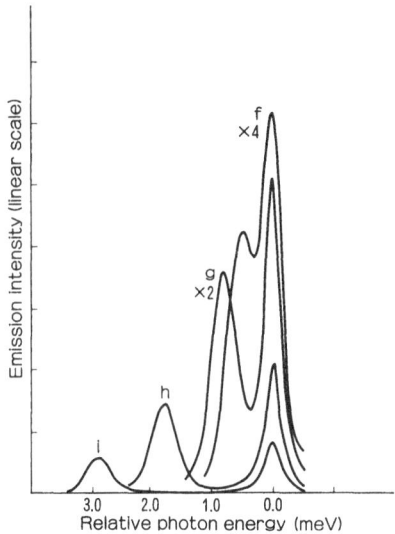

Fig. 2.23. Calculated emission spectra under the giant two-photon excitation of an EM leaving behind a longitudinal exciton under narrow band excitation (FWHM = 0.053 meV). The origin of the abscissa is M_L^0. The detuning of two-photon excitation $(2\omega_1 - \omega_{mg}) =$ 0.157 meV (f), 0.215 meV (g), 0.314 meV (h), 0.430 meV (i), 0.545 meV (j), 0.711 meV (k) and 0.942 meV (l). We fixed the relaxation constants $\gamma_{mg} = 0.086$ meV, $\gamma_{me} = 0.152$ meV, $\gamma_{eg} = 0.066$ meV, and $\gamma_m = 0.05$ meV [2.61]

Fig. 2.24. Calculated emission spectra under wide band excitation (FWHM = 0.4 meV). Parameters are the same as in Fig. 2.23. Detuning $2\omega_1 - \omega_{mg} =$ 0.632 meV (f), 0.885 meV (g), 1.839 meV (h) and 2.908 meV (i) [2.61]

behind the two-photon resonance Raman process. The emission spectra for the two-photon resonance Raman process are calculated with the same relaxation constants as in the case of Fig. 2.23. The spectral width of the excitation light was increased to 0.6 meV in the experiment but we chose the value 0.4 meV for optimum agreement between the calculated and observed spectra. The calculated spectra in terms of (2.56) are shown in Fig. 2.24 and they reproduce the observed characteristics well (compare this with [Ref. 2.63, Fig. 3]).

Segawa et al. [2.64] excited the EM by a picosecond laser pulse through the GTA and observed the emission spectrum integrated in time. Both the Raman line shifting against the change of excitation frequency and the luminescence line have been observed to coexist for both the M_L and M_T processes which leave behind a longitudinal and transverse exciton, respectively. The emission spectra corresponding to these processes are analyzed in terms of (2.56). The decay constant γ_m and other relaxation constants of the EM obtained from the comparison between the theory and experiments by *Segawa* et al. [2.64] are several times larger than those determined by experiments of *Mita* and *Ueta* [2.63]. It is specu-

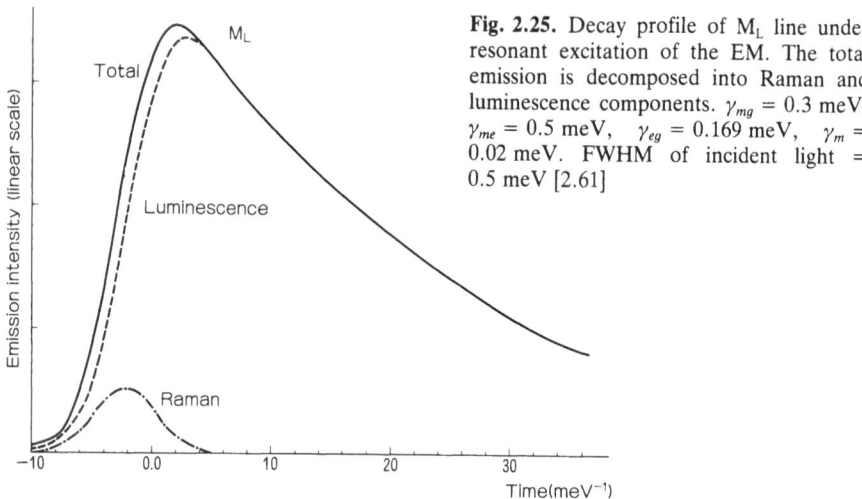

Fig. 2.25. Decay profile of M_L line under resonant excitation of the EM. The total emission is decomposed into Raman and luminescence components. $\gamma_{mg} = 0.3$ meV, $\gamma_{me} = 0.5$ meV, $\gamma_{eg} = 0.169$ meV, $\gamma_m = 0.02$ meV. FWHM of incident light = 0.5 meV [2.61]

lated that the crystals of *Segawa* et al. contained more imperfections and that collisions of the EM with these imperfections increased values of γ_{mg} and γ_{me}.

The competitive behavior of the Raman scattering and the luminescence channels is studied also in the time coordinate, i.e., in the time-resolved emission spectrum under pulse excitation. *Segawa* and his coworkers [2.64] observed the time dependence of the emission peak intensity under just-resonant excitation, and decomposed it into the Raman and luminescence components by applying population dynamics to the latter one. It was shown that the Raman intensity follows the same time dependence as the incident pulse, while the luminescence part survives for of the order of a radiative lifetime in agreement with the calculated result as shown in Fig. 2.25. In the theoretical calculation we used (2.52–56) with $2\omega_1 = \omega_{mg}$ and $\omega_2 = \omega_{me}$, i.e., under just-resonant conditions. The luminescence component is identified with $2\mathrm{Re}\{(2.54) + (2.55)\}$ and the Raman component, $2\mathrm{Re}\{(2.52)\}$. The relevant parameters for the M_L line are chosen as, $\gamma_{mg} = 0.3, \gamma_{eg} = 0.169, \gamma_{me} = 0.5, \gamma_m = 0.02$, and FWHM of the incident light = 0.5 in units of meV, the pulse duration time = 25 ps and the duration time of observation = 25 ps. The agreement between the theory and experiment is satisfactory.

Now let us list the observed features not explained by the simple damping theory presented in this section.

1) The calculated emission spectrum depends on the absolute value of the off-resonance frequency $2\omega_1 - \omega_{mg}$ so that the emission spectrum is expected to be symmetrical with respect to $2\omega_1 - \omega_{mg}$ below and above the resonance. However, *Mita* and *Ueta* [2.63] observed asymmetry below and above the resonance and saturation of the emission intensity just around the two-photon resonance with the EM. These two features may be attributed partially to the assymmetry of the two-photon absorption coefficient around the resonance frequency due to the

continuum state of the EM and to the attenuation of the excitation light intensity contributing to the Raman process under resonance conditions. If these effects cannot fully explain the observed facts, there arises the possibility of a model of population instability and optical anomalies [2.65].

2) *Segawa* et al. [2.64] observed a reduced luminescence component in the time dependence of emitted light intensity for the M_T process, while the time dependence for the M_L process could be understood by our numerical calculation as shown in Fig. 2.25. The smaller value of γ_{eg} for the M_T process can partially explain the reduction of the luminescence intensity but it cannot bring about the amount of reduction observed. *Mita* and *Ueta* [2.63] also observed the quenching of the luminescence line for the M_T process. Both these effects may be attributed mainly to reabsorption of the emitted light due to conversion of the transverse exciton into an EM. This reabsorption is of little significance for the M_L process because the concentration of longitudinal excitons is much smaller than that of transverse excitons due to relaxation into transverse excitons. The quantitative discussion of these problems is left for future studies.

2.4.2 Two-Polariton Scattering Due to the Excitonic Molecule

Two incident light waves which resonantly excite the EM may be considered as polaritons in the medium. Both the emitted light and the exciton left behind, associated with the radiative decay of the EM, also behave as polaritons in the medium. As a result, the hyper-Raman scattering can be considered as two-polariton scattering in resonance with the giant two-photon excitation of the EM.

When the two incident photons, i.e., the two polaritons, have the wave vector k_0 inside the crystal and energy $E(k_0)$, the two scattered polaritons obey the following energy and momentum conservation laws

$$2E(k_0) = E(k_s) + E(k_{s'}) ,$$

$$2k_0 = k_s + k_{s'} .$$

We have plotted the angular dependence of the energy of scattered polaritons in Fig. 2.26 [2.66]. The origin of the angle is taken in the direction of the exciting photon. The pairs of the numbers $(1,1')$, $(2,2')$, etc. given in the figure indicate that, for example, when the photon-like polariton 3 (lower polariton) is observed, the exciton-like polariton 3' (upper polariton) is left inside the crystal, and vice versa. The parameters used in the calculation are listed in Table 2.3.

For small angle scattering $\theta < \theta_m = \pi/6$ from the incident light, three branches are present as Fig. 2.26 shows. We call these the lower, upper prime, and upper branches from the low-energy side, denoting them by LEP, HEP' and HEP, respectively. The upper and lower polaritons have larger and smaller energy, respectively, than the incident one and approach the photon-like ($\theta = \pi$) and the exciton-like limit ($\theta = 0$), respectively. These features have been observed by *Itoh* and *Suzuki* [2.67]. While two emission lines M_T^R and M_L^R leaving

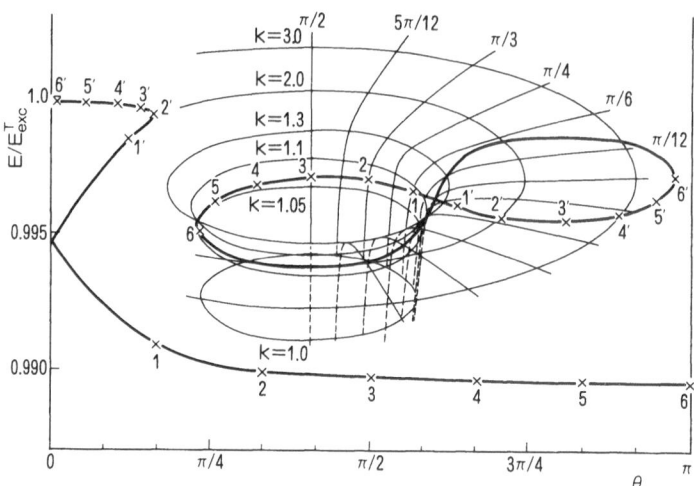

Fig. 2.26. Angular dependence of emitted polariton energy in hyper-Raman scattering due to the EM. The origin of the angle is taken in the direction of the incident photon. The pairs of the numbers (1,1'), (2,2'), etc., indicate that when the photon-like polariton n is observed, the exciton-like polariton n' is left in the crystal [2.66]

behind the transverse and longitudinal excitons are observed for backward ($\theta \simeq 180°$) scattering, they have four emission lines, i.e., LEP, HEP', and HEP in addition to the M_L line, for forward ($\theta \simeq 0°$) scattering with an incident angle of 68° under resonant two-photon excitation of the EM as will be shown in Fig. 3.20. This fact leads us inevitably to consider the hyper-Raman scattering as polariton-polariton resonant scattering (PPRS) especially for the forward scattering configuration.

Itoh and *Suzuki* [2.67] also observed the angular dependence of the emission spectrum under a nearly forward scattering configuration as shown in [Ref. 2.67, Fig. 4]. A detailed discussion will be given in Sect. 3.5. The spectral width has also been shown to be determined by the dispersion of the polariton. Figure 4 of [2.67] shows the width (FWHM) of the PPRS lines plotted as closed circles as the function of their energies. The solid line was calculated by assuming that both the incident and observed lights have an aperture angle of 2° and the incident photon has an energy width of 0.1 meV. Some observed characteristics are well reproduced. The maxima located a little below and above $\hbar\omega_i$ are mainly caused by the

Table 2.3.

$\hbar\omega_L$ = 3.2082 eV		
$\hbar\omega_T$ = 3.2027 eV		
ϵ_0 = 5.10		ϵ = 4.66
m_e = 0.4 m_0		m_h = 4.2 m_0
E_{ex}^b = 200 meV		E_{mol}^b = 34 meV

change of the inside aperture angle of the observed light. The LEP line width at the lower energy side is mainly determined by the energy width of the incident light, in agreement with the case of the M_L^R and M_T^R lines whose line width is about twice as broad as that of the incident light, while for the HEP and HEP′ lines the widths are sensitive to the aperture angles of the incident and observed light.

The EM radiatively decays into two polaritons, or a longitudinal exciton and a polariton. The polaritons are observed as emitted or hyper-Raman scattered light outside the crystal. The emitted light intensity depends on the scattering angle and the polarization of the emitted polaritons [2.68]. The PPRS process in CuCl can be separated into two independent processes as far as polarization character is concerned, i.e., two-photon absorption and the emission process, independent of the virtual or real excitation of the EM. This is because the intermediate EM state has Γ_1 symmetry. Of course the emitted light frequency is strongly correlated with that of the incident light in the case of PPRS.

Let us consider two incoming polaritons with polarization e_i and $e_{i'}$ and propagation vectors n_i and n_i'. Two incident laser beams are assumed to be normal to the surface of the crystal for simplicity, i.e., $n_i = n_i' \| y$-axis as shown in Fig. 2.27. Then the polarization dependence of the two-photon absorption intensity is given by

$$|e_i \cdot e_{i'}|^2 = \cos^2 \alpha_{ii'} .$$

This was confirmed by *Phach* et al. [2.56]. The polarization e_s of the scattered light is split into e_1 and e_2 as $e_s = e_1 \cos \alpha + e_2 \sin \alpha$. Here e_1 is chosen to be perpendicular to the propagation vector n_s and in the plane defined by n_i and n_s', and e_2 is perpendicular both to e_1 and n_s as shown in Fig. 2.27. Then the polarization vectors e_s and e_s' of the scattered polaritons are written with the Cartesian coordinates shown in Fig. 2.27 as basis:

$$e_s = \begin{pmatrix} \sin \varphi \cos \alpha \\ -\cos \varphi \cos \alpha \\ \sin \alpha \end{pmatrix} ,$$

$$e_s' = \begin{pmatrix} \cos \varphi' \\ \sin \varphi' \\ 0 \end{pmatrix} \qquad \text{for the longitudinal exciton,}$$

$$e_s' = \begin{pmatrix} \sin \varphi' \\ -\cos \varphi' \\ 0 \end{pmatrix} \text{ and } \begin{pmatrix} 0 \\ 0 \\ 1 \end{pmatrix} \quad \text{for the two polarizations of the polariton.}$$

The relevant matrix elements are evaluated from the expression

$$|\langle ik_s, jk_s'|(p \cdot e_s)(p' \cdot e_s)|m, 2k_i\rangle|^2 ,$$

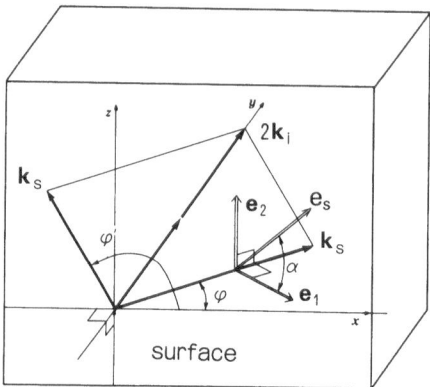

Fig. 2.27. Configuration of hyper-Raman scattering in cubic material. The incident photon k_i hits the surface normally and two scattered polaritons k_s and k_s' are in one plane

and the result is tabulated in Table 2.4. The A and B coefficients are the photon and exciton parts, respectively, of the emitted polariton, or of the remaining transverse polariton in the case of the $\Gamma_{5,T}$ branch:

$$A(E) = (1 - E^2) \left(\frac{\left(1 + \dfrac{4\pi\beta}{1 - E^2}\right)^{1/2}}{(1 - E^2)^2 + 4\pi\beta} \right)^{1/2} ,$$

$$B(E) = \left(\frac{4\pi\beta E}{(1 - E^2)^2 + 4\pi\beta} \right)^{1/2} ,$$

where the energy is measured in units of the transverse exciton energy.

The polarization character for the emission spectrum was observed by *Itoh* and *Suzuki* [2.67]. The polarization of the scattered light was measured for the forward scattering geometry of $\varphi \simeq \pi/2$. The change of φ' corresponds to the change of the Raman scattering photon energy $\hbar\omega_s$ through the polariton dispersion $E(k_s) = 2\hbar\omega_i - E(k_s')$, where $k_s^2 = (2k_i)^2 + k_{s'}^2 - 4k_i k_{s'} \sin^2 \varphi'$. Figure 3.23 shows the observed degree of polarization (D.P.) of the scattered polariton under two-photon resonant excitation by $\hbar\omega_i = 3.1861$ eV; D.P. $= (I_\perp - I_\parallel)/(I_\perp + I_\parallel)$, $I_\parallel$ and $I_\perp$ being the intensities of the scattered light having the polarization parallel and perpendicular to the scattering plane n_i and $n_{s'}$ respectively. The D.P. depends strongly on the scattered photon energy through the scattering angle $(\varphi' - \varphi)$. For the scattered light $\hbar\omega_s$ near $\hbar\omega_i$, the D.P. is very small ($\varphi' \simeq \varphi \simeq \pi/2$), while the D.P. is large both for the LEP and HEP lines of frequency far from

Table 2.4.

excitation left in the crystal	angular dependence of scattered intensity
longitudinal exciton	$\sin^2(\varphi - \varphi') \cos^2\alpha |A(\omega_s)|^2$
transverse exciton	$[1 - \sin^2(\varphi - \varphi') \cos^2\alpha] \cdot |A(\omega_s)B(\omega_{s'}) + B(\omega_s)A(\omega_{s'})|^2$

$\hbar\omega_i$ ($\varphi' \simeq 0$, $\varphi \simeq \pi/2$). The D.P. reaches a maximum (~ 1.0) around the turning point for the HEP line. The polarization character is derived from Table 2.4. Here $I_\perp = 1$ ($\alpha = 90°$) and $I_\parallel = \cos^2(\varphi - \varphi')$ ($\alpha = 0°$) so that (3.31) holds, i.e.,

$$\text{D.P.} = \left(\frac{2}{\sin^2(\varphi - \varphi')} - 1\right)^{-1}.$$

This relation is shown by the solid line in Fig. 3.23 and the observed degree of polarization can be understood very well. *Itoh* and *Suzuki* [2.67] observed the D.P. of the scattered light under the incident polarizations parallel and perpendicular to the scattering plane, denoted by the open and closed circles, respectively (see Sect. 3.5.4 for details). No apparent difference was observed between the two cases so that we can conclude that there is no correlation of the polarizations between the incident and scattered light as mentioned at the beginning of this section. The polarization dependence of the scattered light in uniaxial materials of Wurtzite type, e.g., CdS and CdSe, was also discussed by *Hennenberger* et al. [2.68].

2.4.3 Dispersion of the Exciton Polariton and Excitonic Molecule

It is shown in this subsection that the dispersion relations of the EM, the polariton and the longitudinal exciton can be determined for a wide region from the zero wave vector to a large wave vector such as one-tenth of the Brillouin zone [2.69]. This is made possible by giant two-photon absorption (GTA) and polariton-polariton scattering spectroscopy in terms of two incident beams. Then we find a large k dependence of the longitudinal and transverse splitting of the exciton which cannot be explained by the Heller-Marcus theory [2.70]. This can be understood in terms of the k-dependent Bloch functions obtained by k–p perturbation [2.71].

The absorption peak due to the giant two-photon excitation of the EM is $\hbar\Omega_0 \equiv 3.1861$ eV for a CuCl crystal at 1.6 K. With laser excitation of the crystal in the frequency region around Ω_0 ($k_0 = 4.44 \times 10^5$ cm^{-1}), two-photon resonance Raman scattering has been observed in which the EM with $k \simeq 2k_0$ and the exciton are the intermediate and final states, respectively. Corresponding to the longitudinal and transverse excitons as the final state, two Raman lines denoted M_L^R and M_T^R, respectively, are observed. From these measurements, the spatial dispersion of the exciton polariton has been determined for the small wave vector region between k_0 and $3k_0$. The two-photon excitation of the EM with a large wave vector such as $15k_0$ has been found with the simultaneous excitation of two dye laser beams.

The first laser, with energy $\hbar\omega_1$ in the L–T gap between the longitudinal and transverse excitons, excites the transverse exciton with a large wave vector, as shown in Fig. 2.28. The transmission is measured under irradiation from the first laser, with the second dye laser beam ω_2 as a probe light. First, the transmission

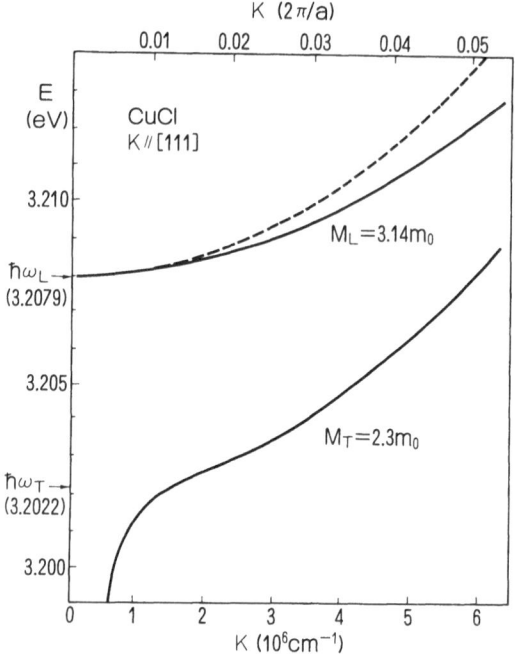

Fig. 2.28. Dispersion relation of longitudinal exciton and transverse exciton polariton in CuCl determined by hyper-Raman scattering due to the EM (——). The dashed line represents the energy of the longitudinal exciton obtained by adding the L–T splitting by Heller-Marcus to the transverse exciton energy

minimum $\tilde{\omega}_2$ is measured. Second, the two-photon resonant Raman scattering M_L^R and M_T^R under the first ω_1 laser irradiation are observed in a narrow region around $\tilde{\omega}_2$ (see Sect. 3.10 for the details). From these measurements, the dispersion relations of the EM, the longitudinal exciton and the polariton are self-consistently determined and are found to be parabolic for a large wave-number region extending to $15 K_0 = 6.6 \times 10^6$ cm^{-1}, i.e., about one-tenth of the Brillouin zone boundary. The dispersion relation of the exciton polariton $E_{ex}(k)$ is obtained by the solution of the equation

$$E(k)^4 - AE(k)^2 + B = 0 \;. \tag{2.57}$$

Here

$$A = \frac{(\hbar ck)^2}{\epsilon_\infty} + E_{ex}^T(k)^2 (1 + 4\pi\beta^0) \;, \tag{2.58a}$$

$$B = \frac{1}{\epsilon_\infty} [\hbar ck \, E_{ex}^T(k)]^2 \;, \tag{2.58b}$$

where $2\pi\beta^0 = [E_{ex}^L(0) - E_{ex}^T(0)]/E_{ex}^T(0)$ and $E_{ex}^T(k) = E_{ex}(0) + (\hbar k)^2/2 \, m_{ex}^T$. Here $E_{ex}^L(0) = 3.2079$ eV, $E_{ex}^T(0) = 3.2022$ eV, and $\epsilon_\infty = 5.59$.

From the experimental results, which will be discussed in Sect. 3.10, we can determine the dispersion $E_{ex}^{L \; or \; T}(k)$, as shown in Fig. 2.28, and $m_{ex}^T = 2.3 \, m_0$ and

$m_{ex}^L = 3.1\ m_0$. The dispersion of the EM is also determined as $E_m(k) = E_m(0) + \hbar^2 k^2/2\ m_{mol}$ with $m_{mol} = 2.3\ m_{ex}^T = 5.3\ m_0$.

Two-beam Raman scattering opened a way to determine the exciton dispersion relation in a wider range of k vectors. This dispersion relation disclosed by the new technique has a strongly k-dependent longitudinal-transverse splitting (L–T splitting) for large k. The observed exciton L–T splitting is expressed for large k values in the form

$$\Delta_{LT}(k) = \Delta_{LT}(0) + \alpha\ \frac{\hbar k^2}{2\ m_0}\ , \quad \text{where} \tag{2.59}$$

$$\Delta_{LT}(0) = 5.7\ \text{meV}\ , \tag{2.60}$$

$$\alpha = \frac{m_0}{M_L} - \frac{m_0}{M_T} = -0.12\ , \tag{2.61}$$

and m_0 stands for the free-electron mass. The L–T splitting of excitons arises from the exchange interaction between electron and hole, as noted by *Heller* and *Marcus* [2.70]. The dipole-dipole interaction that results from multipole expansion of the exchange interaction becomes

$$J_D(k) = \sum_n{}' \frac{u^2 R_n^2 - 3(u \cdot R_n)^2}{R_n^5} \exp(i k \cdot R_n)\ , \tag{2.62}$$

where u denotes the transition dipole moment, and the summation is taken over nonvanishing lattice vectors. The above expression is nonanalytic in the limit $k \to 0$, and the L–T splitting arises from (2.62).

The dipole-dipole interaction is shown to be irrelevant to the observed k dependence of $\Delta_{LT}(k)$. The dashed curve in Fig. 2.28 represents the transverse exciton energy plus the L–T splitting computed from (2.62) using the lattice constant $a = 10.24$ a.u. for CuCl and the transition dipole moment μ determined so as to fit the splitting at $k = 0$. Disagreement with experiment is apparent. In fact, the value of α obtained from the dipole sum is $\alpha = -0.00049$. It agrees with the experiment in sign, but disagreement in magnitude is significant. Consequently, the observed k dependence of L–T splitting is much larger than expected from the dipole-dipole interaction. The k dependence of the exciton L–T splitting can be traced back to the k dependence of the Bloch functions of the conduction and valence bands. This is shown by evaluating the exchange interaction with the k-dependent Bloch functions obtained by the k–p perturbation theory. The calculated value $\alpha = -0.29$ is in fair agreement with the above experimental value with regard to the sign and magnitude [2.71].

We take the Bloch functions at $k = 0$ as Γ_6 for the conduction band and Γ_7 and Γ_8 for the valence band discussed in Sect. 2.2. Bloch functions with $k \neq 0$ are obtained through the $k \cdot P$ perturbation $H' = p \cdot k/m_0$. We confine ourselves to the three-band model shown in Fig. 2.29 and take only the perturbation process-

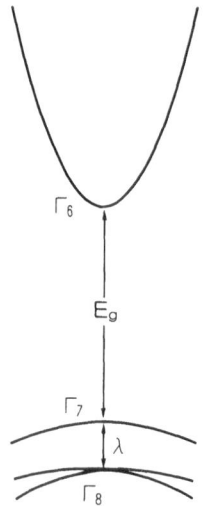

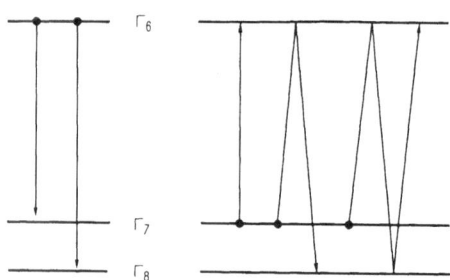

Fig. 2.30. Important $k \cdot p$ perturbation processes contributing to the present effect [2.71]

◀ **Fig. 2.29.** Electron band structure of CuCl around the Γ point. The band gap $E_g = 3.3$ eV and the spin-orbit splitting $\lambda = 71$ meV in CuCl

es shown in Fig. 2.30. Then the Bloch functions at $k \neq 0$ are presented in terms of the matrix element $P = i(\hbar/m_0) \langle s|p_z|z \rangle$, the energy gap E_g, and the spin-orbit splitting λ. The exciton state with wave vector K can be written as the linear combination of the excited states $|\mu k', \nu k\rangle$ as

$$|iK\rangle = \sum_k f(k)|\mu k - K, \nu k\rangle .$$

Here $|\mu k', \nu k\rangle$ denotes the excited state where the electron in the valence-band state $\mu k'$ is excited to the conduction-band state νk. The Fourier transform of the coefficient $f(k)$ is called the envelope function and is obtained from the effective mass equation as

$$f(k) = 8 \sqrt{\frac{\pi a_B^3}{N\Omega}} \left(1 + \left|k - \frac{m_e}{m_{ex}}K\right|^2 a_B^2\right)^{-2} ,$$

where N stands for the number of unit cells, $m_{ex} = m_e + m_h$ and a_B denotes the exciton Bohr radius. As long as the discussion is confined to the lowest exciton state, $1s$, for the relative motion, i stands for the polarization of the exciton x, y, or z.

The matrix element of the e–h exchange interaction between the exciton states i and j can be written in the form

$$J_{ij}(K) = \frac{4\pi e^2}{\Omega} \sum_n \frac{N_i(K, G_n)^* N_j(K, G_n)}{|K + G_n|^2} , \qquad (2.63)$$

utilizing the Fourier expansion of the Coulomb interaction. Ω is the unit cell volume and G_n is the reciprocal lattice vector. The $G_n = 0$ term (non-analytic part) is dominant and $N_i(K, 0)$ is expressed in the form $N_i(K, 0) =$

$N^{-1/2} \sum_k f(k) \langle vk| \exp{(\mathrm{i}K \cdot r)} |\mu k - K \rangle$. After tedious calculation, the L–T splitting is expressed as

$$\Delta_{LT}(0) = \frac{4\pi e^2}{\Omega} |\phi(0)|^2 \frac{2P^2}{3E_g^2},\tag{2.64}$$

$$\alpha = -\Delta_{LT}(0) \frac{8m_0 P^2}{3\hbar^2 \lambda E_g} \left(\frac{m_h}{m_{ex}} \right)^2.\tag{2.65}$$

Here λ is the spin-orbit splitting in the valence band and the effective mass m_h of the Γ_7 valence band is mainly determined by the Γ_6–Γ_7 transition matrix element, so that $-\dfrac{1}{m_h} = \dfrac{1}{m_0} - \dfrac{2P^2}{3\hbar^2 E_g}$.

Using this expression, (2.65) is rewritten as

$$\alpha = -4 \left(\frac{m_0}{m_h} + 1 \right) \left(\frac{m_h}{m_{ex}} \right)^2 \frac{\Delta_{LT}(0)}{\lambda}.$$

We substitute the following values: $m_h = 1.77\, m_0$, $m_e = 0.53\, m_0$ (from $\mu = 0.406\, m_0$), $m_{ex} = 2.3\, m_0$, $\Delta_{LT}(0) = 5.7$ meV, and $\lambda = 71$ meV. Then $\alpha = -0.29$. This compares rather well with the experimental value $\alpha = -0.12$. As for the value of $\Delta_{LT}(0)$, when we use $E_g = 3.3$ eV, $a_B = 13.3$ a.u., and $|\phi(0)|^2 = \Omega/\pi a_B^3$, then $\Delta_{LT}(0) = 19$ meV. Since the result depends sensitively on the choice of the envelope function and the screening constant ($\varepsilon_\infty = 5.59$), the order of magnitude agreement with experiment is satisfactory. From this comparison, the present theory accounts for the observed k dependence of the exciton L–T splitting.

2.4.4 Four-Wave Mixing Due to the Excitonic Molecule

Nonlinear polarization $P(t, r)$ of the medium is induced by the incident radiation field $E(t, r)$. The polarization may be expanded in terms of E as

$$\begin{aligned}
P_i(t, r) = &\sum_j \chi_{ij}^{(1)} E_j(t, r) \\
&+ \sum_{j,k} \chi_{ijk}^{(2)} E_j(t, r) E_k(t, r) \\
&+ \sum_{j,k,l} \chi_{ijkl}^{(3)} E_j(t, r) E_k(t, r) E_l(t, r) \\
&+ \dots \, .
\end{aligned}\tag{2.66}$$

The first term represents the linear polarization P_L and the other terms compose the nonlinear one P_{NL}. The latter terms are enhanced under some conditions and induce significant effects in the optical phenomena of the exciton and the EM system. In this section we discuss several interesting phenomena induced by this nonlinear optical response.

As we wish to discuss the stationary nonlinear optical response in this section, we take the Fourier transformation of (2.66) in time

$$
\begin{aligned}
P_i(\omega_1, r) = & \sum_j \chi_{ij}^{(1)}(-\omega_1, \omega_1) E_j(\omega_1, r) \exp(i k_1 \cdot r) \\
& + \sum_{j,k} \int_{\omega_2 + \omega_3 = \omega_1} d\omega_2 \chi_{ijk}^{(2)}(-\omega_1, \omega_2, \omega_3) E_j(\omega_2, r) E_k(\omega_3, r) \\
& \times \exp[i(k_2 + k_3) \cdot r] \\
& + \sum_{j,k,l} \int_{\omega_2 + \omega_3 + \omega_4 = \omega_1} d\omega_2 d\omega_3 \\
& \times \chi_{ijkl}^{(3)}(-\omega_1, \omega_2, \omega_3, \omega_4) E_j(\omega_2, r) E_k(\omega_3, r) E_l(\omega_4, r) \\
& \times \exp[i(k_1 + k_2 + k_3) \cdot r] .
\end{aligned}
$$

Here the Fourier transform of $E(t, r)$ is represented by $E(\omega, r) \exp(i k \cdot r)$. When the external field is composed of the three components

$$
E(t, r) = \sum_{i=2}^{4} E(\omega, r) \exp[i(k_i \cdot r - \omega_i t)] ,
$$

the third-order nonlinear polarization which plays an important role in the following is represented as

$$
\begin{aligned}
P_i^{(3)}(\omega_2 + \omega_3 + \omega_4, r) = & \sum_{j,k,l} 6 \chi_{ijkl}^{(3)}(-(\omega_2 + \omega_3 + \omega_4), \omega_2, \omega_3, \omega_4) \\
& \times E_j(\omega_2, r) E_k(\omega_3, r) E_l(\omega_4, r) \exp[i(k_2 + k_3 + k_4) \cdot r] .
\end{aligned} \tag{2.67}
$$

Here the factor 6 appears because we have six terms which are proportional to $E(\omega_2, r) E(\omega_3, r) E(\omega_4, r)$, and we have also the nonlinear polarizations which are obtained by replacing $E(\omega, r)$ by the phase conjugate $E^*(\omega, r)$.

The signal of the nonlinear optical polarization $E(t, r)$ is obtained by solving self-consistently the following Maxwell equations:

$$
\left\{
\begin{aligned}
& \operatorname{curl} E + \frac{1}{c} \frac{\partial B}{\partial t} = 0 , && (2.68a) \\[2mm]
& \operatorname{curl} H - \frac{1}{c} \frac{\partial D}{\partial t} = \frac{4\pi}{c} j , && (2.68b) \\[2mm]
& \operatorname{div} D = 4\pi \varrho , && (2.68c) \\[2mm]
& \operatorname{div} B = 0 , \quad \text{where} && (2.68d)
\end{aligned}
\right.
$$

$$
D = E + 4\pi P = E + 4\pi(P_L + P_{NL}) . \tag{2.69}
$$

We may put $j = \varrho = 0$, $\mu = 1$. We eliminate H from (2.68a) and (2.68b) and take the Fourier transfom in time of these equations. Then we obtain

$$\text{curl}\,\text{curl}\,E(\omega,\,r) - \frac{\epsilon(\omega)\omega^2}{c^2}E(\omega,\,r) = \frac{4\pi\omega^2}{c^2}P_{NL}(\omega,r) \; . \tag{2.70}$$

Here the effect of the linear polarization $P_L(\omega, r)$ is taken into account in $\epsilon_{ij}(\omega) = \delta_{ij} + 4\pi\chi_{ij}^{(1)}(-\omega, \omega)$.

When the nonlinear polarization P_{NL} is negligible, the left-hand side of (2.70) describes the propagation of light in the medium with the dielectric tensor $\epsilon(\omega)$, e.g., the exciton polariton. On the other hand, in a nonlinear optical phenomenon, the new signal field is induced with the nonlinear polarization P_{NL} as the source term on the right-hand side of (2.70). The field $E(\omega, r)$ is written as the sum of the homogeneous solution E_H and the signal field $E_S(\omega, r)$. Usually we may approximate E_H by the external field E_0 and express P_{NL} in terms of E_0 by the perturbational method. In addition, we may factor out the rapidly oscillating part of E_S as

$$E_S(\omega, r) = E_{S0}(r)\exp{(ik_s \cdot r)} \; ,$$

with $E_{S0} \perp k_s$ and $E_{S0}(r)$ assumed to vary slowly in the range of $2\pi|k_s|^{-1}$. Then we can approximate (2.70) to the following form:

$$-2i(k_s \cdot \nabla)E_{S0}(r)\exp{(ik_s \cdot r)} = \frac{4\pi\omega^2}{c^2}P_{NL}(\omega, r) \; . \tag{2.71}$$

Using this equation, we can describe the four-wave mixing; in particular the two-photon resonant, stimulated Raman scattering due to the EM (in this subsection) and the phase conjugation due to the exciton and EM system (in the next subsection).

Here let us start from the clearest optical phenomenon: the stimulated hyper-Raman scattering via the EM. The two dye laser beams (ω_1, k_1) and (ω_2, k_2) are made to hit a surface of the CuCl sample simultaneously, with incident angles $\theta/2$, as shown in Fig. 2.31 [2.72]. Then the third-order light is found to emerge from the sample as a directional beam at an angle $3\theta/2$, according to the relation $k_3 \simeq 2k_1 - k_2$ or $k_4 \simeq 2k_2 - k_1$, in addition to the undeflected beams with k_1 and k_2. The nonlinear polarization is written as

$$P_{NL}(\omega, r) = P_{NL}^0 \exp{(ik_t \cdot r)} \; ,$$

where $k_t = 2k_1 - k_2$ or $2k_2 - k_1$, and $\omega_3 = 2\omega_1 - \omega_2 = \omega$ for the degenerate case $\omega_1 = \omega_2 = \omega$. Then the signal field $E_{S0}(r)$ is obtained from (2.71) as follows:

$$\frac{dE_{S0}}{dz} = i\frac{2\pi\omega^2}{c^2 k_s}P_{NL}^0 \exp{(i\Delta k \cdot r)} \; , \quad \text{where}$$

$$\Delta k = k_t - k_s \; , \quad \text{and} \quad z \| k_s \; .$$

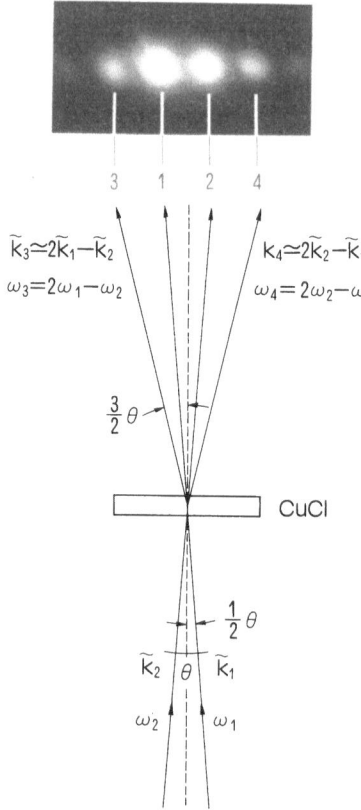

Fig. 2.31. Geometrical configuration of four-wave mixing. The third-order nonlinear light beams 3 and 4 are observed with the angular frequencies $\omega_3 = 2\omega_1 - \omega_2$ and $\omega_4 = 2\omega_2 - \omega_1$ and the wave vectors $k_3 = 2k_1 - k_2$ and $k_4 = 2k_2 - k_1$, respectively

$$|E_{s0}(z)|^2 = \left(\frac{2\pi\omega^2}{c^2 k_s}\right)^2 |\chi^{(3)}(\omega)|^2 |E_1^2 E_2^*|^2 \frac{\sin^2\left(\frac{\Delta kz}{2}\right)}{\left(\frac{\Delta k}{2}\right)^2} \tag{2.72}$$

and

$$\chi^{(3)}(\omega) = \frac{-8\pi N |\mu_{eg}|^2 |\mu_{me}|^2}{\hbar^3 [(\omega_{mg} - 2\omega) - i\gamma_{mg\perp}][(\omega_{eg} - \omega) - i\gamma_{eg\perp}]^2} . \tag{2.73}$$

The incident photon with energy $\hbar\omega_1 = \hbar\omega_2 \equiv \hbar\omega$ is converted into an excitonic polariton inside the crystal. When the energy $\hbar\omega$ approaches the two-photon resonance of the EM $\hbar\omega_m/2$ and the one-photon resonance of the single exciton $\hbar\omega_{ex}$, the signal field intensity is enhanced by (2.72, 73) and is observed as shown in Fig. 2.32 [2.72–74]. While the effect of the two-photon absorption due to the EM is slight, the single-exciton absorption is significant around $\hbar\omega_{ex}$. The phase-matching condition is represented by the factor $\sin^2(\Delta k_t l/2)/(\Delta k_t d/2)^2$ in (2.72), where Δk_t is the z component of $\Delta k = 2k_1 - k_2 - k_s$, d is the sample thickness, and l is the propagation length.

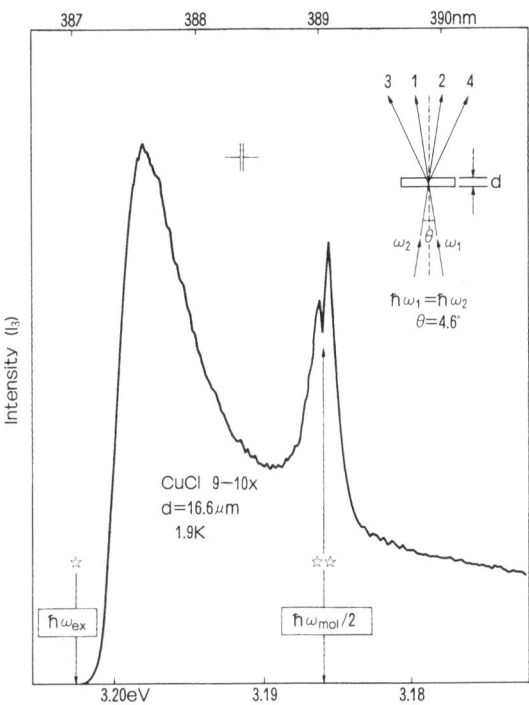

Fig. 2.32. Intensity of the third-order nonlinear light beam I_3 as a function of the incident light frequency $\omega_1 = \omega_2$

Finally let us discuss the case of nondegenerate four-wave mixing. Taking into account the exciton-polariton dispersion, Δk_t is derived from k_1, k_2, and θ:

$$n_1 = \frac{ck_1}{\omega_1}, \quad n_2 = \frac{ck_2}{\omega_2}, \quad \frac{\sin(\theta/2)}{\sin\alpha} = n_1, \quad \frac{\sin(\theta/2)}{\sin\beta} = n_2, \quad \text{and}$$

$$\Delta k_t = k_s - \sqrt{4k_1^2 + k_2^2 - 4k_1k_2\cos(\alpha + \beta)}.$$

Here n_i's are refractive indexes, and α and β are refracted angles inside the crystal. This effect is clearly demonstrated as shown in Fig. 2.33 in the third-order beam intensity I_3 as a function of incident photon energy $\hbar\omega_2$ (with $\hbar\omega_1$ fixed) and the incident angle θ for the nondegenerate case.

2.4.5 Phase-Conjugation by Four-Wave Mixing

The two-photon resonance of the Γ_1 excitonic molecule and the one-photon resonance of the Γ_5 exciton in CuCl have been found to enhance the generation of the phase-conjugated wave at 1.6 K. Each contribution has been separated

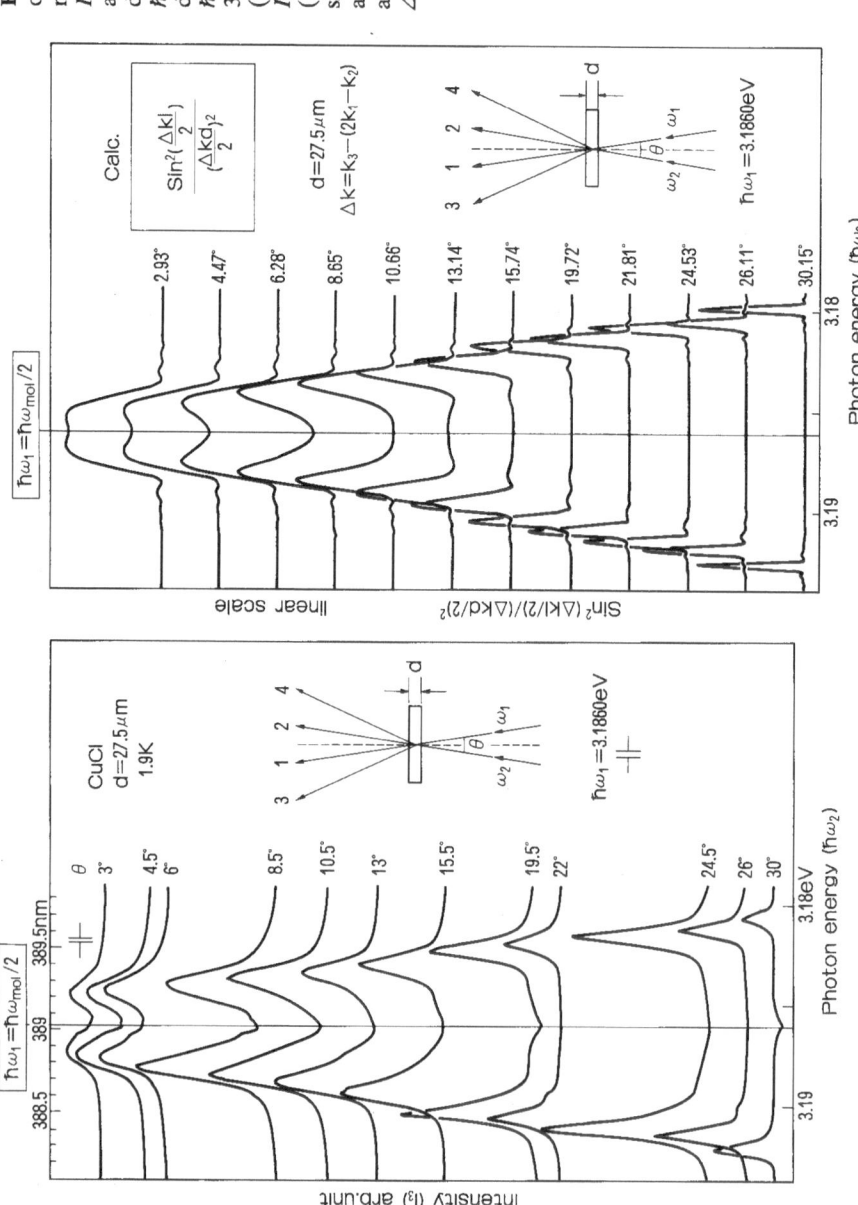

Fig. 2.33. Intensity of the third-order nonlinear light beam I_3 as functions of the angle θ and the incident photon energy $\hbar\omega_2$. The other incident photon energy $\hbar\omega_1$ is fixed at 3.1860 eV. (**a**) Experiment with $I = 1$ MW/cm^2 and (**b**) calculation of $\sin^2(\Delta kl/2)/(\Delta kd/2)^2$ as functions of $\hbar\omega_2$ and θ. $\Delta \mathbf{k} = \mathbf{k}_3 - (2\mathbf{k}_1 - \mathbf{k}_2)$

spectroscopically by adjusting the polarization configuration of the incident fields adequately. The maximum efficiency of 1% was obtained for the phase-conjugated reflection at around the two-photon resonance of the EM [2.75].

Phase-conjugation by degenerate four-wave mixing (DFWM) is especially attractive because the phase-matching requirement is automatically fulfilled and it is possible under Doppler-free conditions. This DFWM originates from the third-order nonlinear polarization: $P^{(3)}(\omega, r)$ of (2.67). Here $P_i^{(3)}(\omega, r)$ and $E_{n,i}(\omega, r) \exp(i k_n \cdot r)$ with $n = f, b, p$ and $i = x, y, z$ are the ith Cartesian components of the nonlinear polarization and the incident fields. They have the common time factor $\exp(-i\omega t)$ for the DFWM and f, b, and p represent the incident light fields, namely, two counter-propagating excitation fields E_f (forward) and E_b (backward) and a weak probe light field E_p.

When the medium is isotropic, as is CuCl, the third-order nonlinear polarization can be rewritten in the following form:

$$P^{(3)} = \alpha(\omega)[(E_p^* \cdot E_f)E_b + (E_p^* \cdot E_b)E_f] + \beta(\omega)(E_f \cdot E_b)E_p^* , \qquad (2.74)$$

where $\alpha(\omega)$ and $\beta(\omega)$ are derived microscopically for the exciton and the EM system. The first term of (2.74) describes the process in which E_b is diffracted by the population grating of the exciton excited by the E_p and E_f fields. The second term gives the diffraction of E_f by a similar grating produced by the E_p and E_b fields. The third term represents the process in which E_p is scattered by the excitation with frequency 2ω formed coherently by the E_f and E_b fields. These three contributions are observable, independently of each other, by choosing the polarizations of the three incident fields as shown in Fig. 2.34. On the basis of the three-level model of the exciton and EM shown in Fig. 2.35, the coefficients $\alpha(\omega)$ and $\beta(\omega)$ of the nonlinear polarization are derived in terms of the density matrix ϱ as $P^{(3)} = \text{Tr}\{\mu\varrho(t)\}$:

$$\alpha(\omega) = \frac{3N|\mu_{eg}|^4 \gamma_{eg\perp}}{\hbar^3 \gamma_{eg\|}[(\omega_{eg} - \omega) - i\gamma_{eg\perp}][(\omega_{eg} - \omega)^2 + \gamma_{eg\perp}^2]} , \qquad (2.75)$$

$$\beta(\omega) = -\frac{-2N|\mu_{eg}|^2|\mu_{me}|^2}{\hbar^3[(\omega_{mg} - 2\omega) - i\gamma_{mg\perp}][(\omega_{eg} - \omega) - i\gamma_{eg\perp}]^2} . \qquad (2.76)$$

Here ω_{ij}, μ_{ij}, $\gamma_{ij\|}$, and $\gamma_{ij\perp}$ are the energy difference, the transition dipole moment, and the longitudinal and the transverse relaxation constants between the ith and jth levels, respectively.

The intensity spectra of the phase-conjugated wave are shown in Fig. 2.36 for different polarization configurations of the incident beams. For $E_f\|E_b\|E_p$, all terms of (2.74) contribute to the spectrum, and it consists of two peaks at 3.186 and 3.195 eV as Fig. 2.36a shows. For $E_f\|E_p\perp E_b$ (Fig. 2.36b) and $E_f\perp E_p\|E_b$ (Fig. 2.36c), $\alpha(\omega)$ terms contribute to the spectrum and the higher energy peak is dominant as expected from (2.75). This is due to the spatial modulation of the exciton population by one-photon exciton resonance. For $E_f\|E_b\perp E_p$

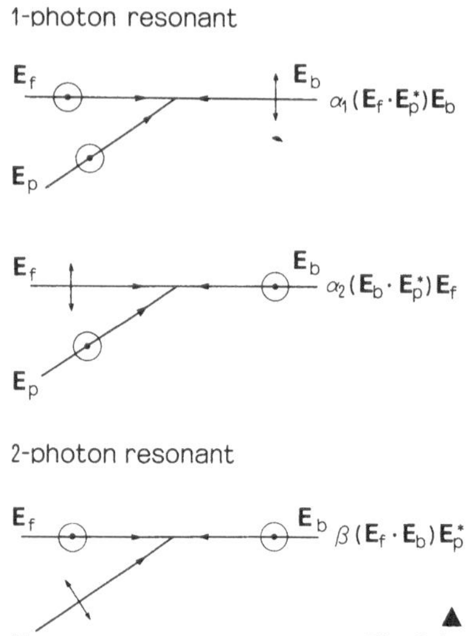

1-photon resonant

E_f E_b $\alpha_1 (E_f \cdot E_p^*) E_b$

E_p

E_f E_b $\alpha_2 (E_b \cdot E_p^*) E_f$

E_p

2-photon resonant

E_f E_b $\beta (E_f \cdot E_b) E_p^*$

E_p ▲ Fig. 2.34

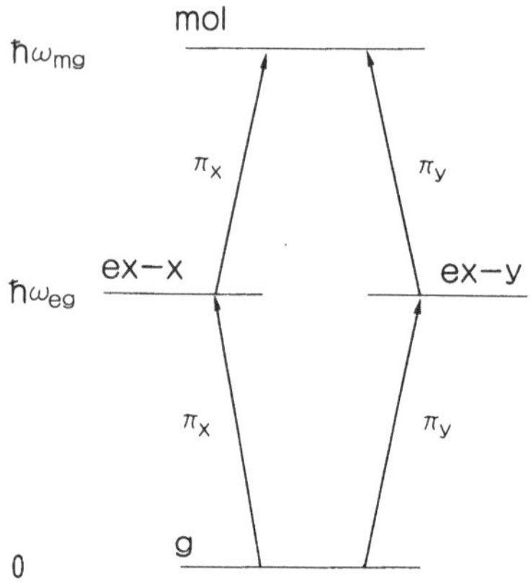

mol $\hbar\omega_{mg}$

π_x π_y

$\hbar\omega_{eg}$ ex−x   ex−y

π_x π_y

0   g

▲ Fig. 2.35

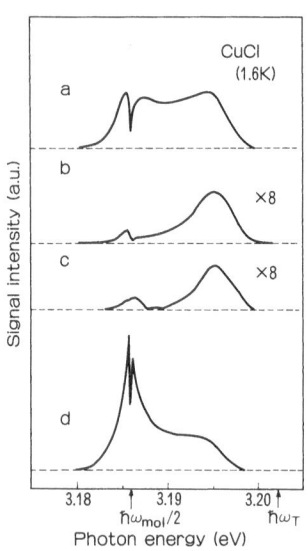

Fig. 2.34. Three configurations to obtain the phase conjugated wave

Fig. 2.35. Model of the exciton-EM system used to evaluate the nonlinear optical susceptibility

CuCl
(1.6K)

Signal intensity (a.u.)

a

b   ×8

c   ×8

d

3.18   3.19   3.20
$\hbar\omega_{mol}/2$ $\hbar\omega_T$
Photon energy (eV)

◀ **Fig. 2.36.** Intensity spectra of the phase-conjugated wave under (a) $E_f \| E_b \| E_p$, (b) $E_f \| E_p \perp E_b$, (c) $E_f \perp E_p \| E_b$ and (d) $E_f \| E_b \perp E_p$ [2.75]

(Fig. 2.36d), a peak and a hump appear at 3.186 eV, and this is due to the two-photon coherent excitation of the EM through the third term of (2.74).

For nondegenerate four-wave mixing, the phase-conjugated light of $2\hbar\omega_1 - \hbar\omega_2$ is observed for two excitation beams of $\hbar\omega_1$ and $\hbar\omega_2$, as shown in Fig. 2.37. The phase-conjugated wave is the time-reversed propagation of the original probe light and shows an inverse image of the pattern of the linearly reflected probe beam.

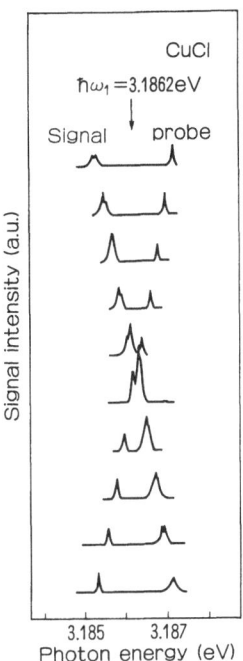

CuCl

$\hbar\omega_1 = 3.1862\,\text{eV}$

Signal   probe

Signal intensity (a.u.)

3.185   3.187
Photon energy (eV)

Fig. 2.37. Intensity spectrum of phase-conjugated wave at $2\omega_1 - \omega_2$ due to nondegenerate four-wave mixing. The energy $\hbar\omega_1$ of E_f and E_b is fixed at 3.1862 eV denoted by the arrow and $\hbar\omega_2$ of E_p is changed [2.75]

2.5 The Excitonic Molecule at High Densities

2.5.1 Renormalization of the Exciton Polariton Due to the Excitonic Molecule Giant Two-Photon Absorption

The dispersion relation of the exciton polariton has been determined very precisely for CuCl [2.69, 76], CuBr [2.77, 78], and CdS [2.79] by polariton-polariton resonant scattering (PPRS). *Itoh* and *Suzuki* [2.67] have found some interesting behavior when the incident power is increased near the giant two-photon absorption (GTA) peak due to the EM. For low excitation densities, the scattered lower and higher exciton polaritons (LEP and HEP) show a smooth shift against the incident photon energy near the EM GTA. The shift of the scattered LEP against the incident photon energy changes from $\Delta\omega_s = \Delta\omega_i$ to $\Delta\omega_s = 2\Delta\omega_i$ when the scattering angle α_s increases from 0° to 180°, while that of the HEP changes from $\Delta\omega_s = \Delta\omega_i$ to $\Delta\omega_s = 0$. As the excitation power was increased, the deviation of the dispersion from a straight line was found to be pronounced near the frequency of the EM GTA in CuCl. The sign of this singularity is opposite for the LEP and HEP and it is enhanced in forward scattering, i.e., for small α_s. These features are shown in Fig. 3.24. Furthermore the intensity of the scattered light increases more than linearly with increasing excitation power at excitation energies a little away from the GTA peak energy. On the other hand, a dip at the

GTA peak energy indicating a tendency to saturate is observed on the scattering-yield curve [2.67]. The line width is almost constant for low excitation power, while it increases at higher excitation power except near the resonance energy.

This peculiar behavior is closely connected with the EM GTA. The GTA induces a nonlinear correction to the refractive index, and consequently a modification of the exciton-polariton dispersion. The modification of the dispersion is localized near the GTA in momentum and energy space. *Itoh* and *Suzuki* [2.67] connected the GTA coefficient with the nonlinear refractive index through the Kramers-Kronig relation as will be discussed in Sect. 3.5.5. The ratio of the nonlinear to the linear refractive index is estimated to be of the order of 2×10^{-3}. However, the nonlinear correction of the polariton dispersion is very clearly observed as the energy change of the scattered polariton $\hbar\omega_s$ for small internal scattering angles. The nonlinear effect changes the magnitude of the wave vector k_i of the incident polaritons near the GTA band region and, therefore, changes that of the total wave vector $k_t = 2k_i$. As a result, a small change in a wave vector k_t induces an enhanced shift in the energy of the scattered polariton in the sharp dispersion around the GTA energy, i.e., in the forward scattering regime.

These experimental results and this analysis have been reconfirmed quantitatively by *März* et al. [2.79], and *May* et al. [2.80]. Let us consider that a large concentration of exciton-polaritons are created by irradiation of a crystal with incident light at half the frequency of the EM. The real and virtual formation of EMs from two polaritons causes a renormalization of the polariton spectrum. The polarization due to the conversion of a polariton into an EM associated with the annihilation of another polariton is coupled with the polarization of the incident polariton. As a result, a new mode is introduced. The full dielectric function is calculated by including the contribution of the new mode and the finite lifetime of the EM. The damping removes the splitting of the lower polariton branch due to the new mode and leaves behind only anomalous dispersion features, as shown in Fig. 2.38.

The dispersion of an exciton polariton is written in terms of the dielectric function $\epsilon(k, \omega)$ of the exciton polarization

$$\epsilon(k, \omega) = \left(\frac{kc}{\omega}\right)^2, \quad \text{where}$$

$$\epsilon(k, \omega) = \epsilon_\infty - (\epsilon_0 - \epsilon_\infty)\frac{\omega_{ex}^2(k)}{\omega^2 - \omega_{ex}^2(k)}$$

with the exciton dispersion $\omega_{ex}(k)$. We investigate a situation in which a strong pump beam has created a nonequilibrium polariton distribution $n_p(k) = n_p \delta_{k,k_0}$ at high density. These polaritons interact with a weak test beam and produce virtual EMs. The self-renormalization can be obtained by identifying both beams as will be done later. Then the dielectric function $\epsilon(k, \omega)$ can be expressed by

$$\epsilon(k, \omega) = \epsilon_\infty - |g_k|^2 D_{ex}(k, \omega) , \tag{2.77}$$

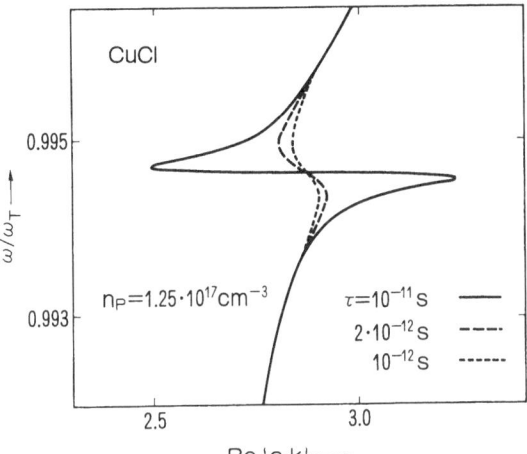

Fig. 2.38. Exciton-polariton dispersion close to the giant two-photon excitation $\omega = \omega_{mol}/2$ for CuCl with a polariton concentration $n_p = 1.25 \times 10^{17}$ cm^{-3} and various EM lifetimes [2.79]

where $D_{ex}(k, \omega)$ is the exciton Green's function and g_k is the exciton-photon interaction defined by $|g_k|^2 = (\epsilon_0 - \epsilon_\infty)\omega_{ex}(k)/2$.

The exciton Green's function is obtained from the Dyson equation in terms of the self-energy $\Sigma(k, \omega)$ as

$$D_{ex}(k, \omega) = [D_{ex}^0(k, \omega)^{-1} - \Sigma(k, \omega)]^{-1} ,$$

where $D_{ex}^0(k, \omega)$ is the Green's function for a noninteracting exciton:

$$D_{ex}^0(k, \omega) = \frac{2\omega_{ex}(k)}{\omega^2 - \omega_{ex}^2(k)} - 2\pi i n_p(k)V\delta(\omega \pm \omega_{ex}(k)) ,$$

and the self-energy $\Sigma(k, \omega)$ is evaluated by taking the simplest contribution from the virtual formation of the EM:

$$\Sigma(k, \omega) = n_p|M(k, k_0)|^2 \frac{2\Delta\omega}{\omega^2 - (\Delta\omega)^2} . \tag{2.78}$$

Here $\Delta\omega = \omega_{mol}(k + k_0) - i\gamma_{mol}(k + k_0) - \omega_p(k_0)$ and the vertex $M(k, k_0)$ is the matrix element for the transition from a single exciton into an EM. An effective relaxation rate γ_{mol} of the EM due to all scattering processes was introduced in the Green's function as

$$D_{mol}(k, \omega) = \frac{1}{\omega - \omega_{mol}(k) + i\gamma_{mol}} - \frac{1}{\omega + \omega_{mol}(k) - i\gamma_{mol}} .$$

Thus the dielectric function (2.77) is obtained, neglecting the k dependence of the matrix element $M(k, k_0)$ and of $\omega_{ex}(k)$ and $\omega_{mol}(k)$, as

$$\epsilon(\omega) \;=\; \epsilon_\infty \;+\; \frac{(\epsilon_0 - \epsilon_\infty)\omega_{ex}^2(\Delta\omega^2 - \omega^2)}{(\Delta\omega^2 - \omega^2)(\omega_{ex}^2 - \omega^2) - \omega_{ex}\Delta\omega\Omega^2} \tag{2.79}$$

where $\Omega^2 = 4n_p|M|^2$.

When the pump and the laser beam are identical, the dielectric function is obtained by putting $\omega_p(k) = \omega$ in (2.78). This describes the self-renormalization of a strong laser beam due to the formation of the EM. Then the dispersion relation is obtained by $\epsilon(\omega) = (kc/\omega)^2$ with (2.79). The numerical result is shown in Fig. 2.38 [2.79].

The renormalization of ω versus $\mathrm{Re}\{k\}$ is more pronounced at a higher polariton density and a smaller relaxation constant. Taking into account this renormalized dispersion, the frequency ω_s of the scattered polariton is calculated as a function of the incident polariton frequency ω_i for various angles α_s for CuCl. These results catch the observed characteristics of Fig. 3.24 very well as shown in Figs. 2.39a and b. A similar anomalous hyper-Raman shift due to EM formation has also been observed for CdS by *Kurtze* et al. [2.81].

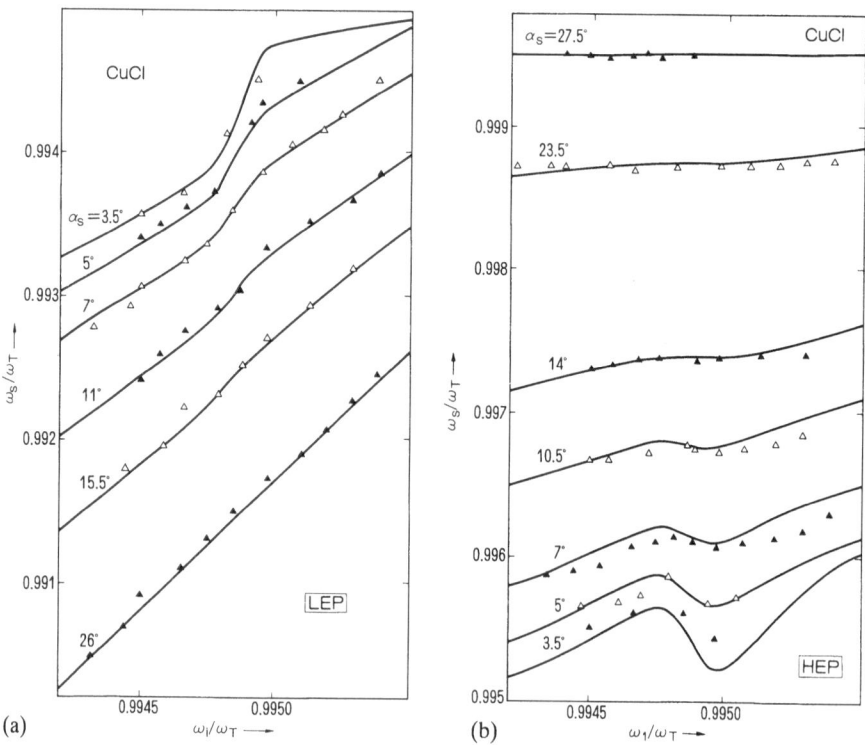

Fig. 2.39. (a) Scattered polariton frequency ω_s (LEP) as a function of incident frequency ω_i and scattering angle α_s for CuCl. ($\triangle$, $\blacktriangle$) experimental points [2.67] (——) theory [2.79]. (b) The same dependence of HEP

2.5.2 Polarization Rotation Effects Due to Two-Photon Exitation of the Excitonic Molecule

Let us consider that the polaritons are higly populated at the incident laser frequency Ω_2 corresponding to around half the EM energy $E_{mol}/2 = 3.1861$ eV. Then the refractive index, and consequently the dispersion, of an exciton polariton is modified around $E_{mol} - \hbar\Omega_2$. This is the induced renormalization of the polariton. A similar modulation is induced at $E_{mol}/2$ by the incident laser itself. This is the self-renormalization discussed in the previous subsection.

A CuCl sample is illuminated by two almost collinear laser beams: one is a strongly left circularly polarized beam σ^- of frequency Ω_2 (excitation beam) and the other is a weak linearly polarized one of frequency Ω_1 (probe beam). This probe beam is the coherent superposition of σ^+ and σ^- components. Through the giant two-photon excitation of the Γ_1 EM, the σ^+ component of the probe beam can couple with the σ^- excitation beam while the σ^- component does not couple with the excitation beam.

Consequently, dispersion renormalization of the probe beam is induced at $\hbar\Omega_1 = E_{mol} - \hbar\Omega_2$ for the σ^+ component and not for the σ^- component. Then a phase difference between these components is brought about and the polarization plane of the probe beam rotates after passing through the sample. This polarization rotation is measured as a function of the incident frequency by the analyzer set behind the sample. The analyzer is rotated by the angle θ from the position where it is perfectly crossed against the polarization of the incident probe beam. The intensity $I(\theta, \Omega_1)$ of the transmitted probe beam after the passage through the analyzer is expressed [2.82] as

$$I(\theta, \Omega_1) = I_0 \exp\left[-\alpha_0(\Omega_1) \cdot l\right]$$

$$\cdot \left\{\frac{1}{4}\left[1 + \exp\left(-\Delta\alpha \cdot l\right)\right] - \frac{1}{2}\exp\left(-\frac{\Delta\alpha \cdot l}{2}\right)\cos\left(2\theta + \delta\right)\right\},$$

$$(2.80)$$

where $\delta = \Omega_1\Delta n \cdot l/c$ and $\Delta\alpha$ are the phase difference and the difference of the absorption coefficients between the σ^+ and σ^- components, respectively. Also l is the sample thickness, α_0 the background absorption coefficient, and I_0 the incident intensity of the probe beam.

When we measure a set of spectra $I(\theta, \Omega_1)$ for two angles, e.g., θ and $-\theta$, we can determine the spectra $\Delta n(\Omega_1)$ and $\Delta\alpha(\Omega_1)$ for a fixed frequency Ω_2 of the excitation beam. The modulation spectra $\Delta n(\Omega_1)$ of the refractive index have been obtained by *Kuwata* et al. [2.82] under maximum and 49% excitation power with $\hbar\Omega_2 = 3.1796$ eV, as shown in Fig. 2.40. The remarkable dispersive anomaly found at $\hbar\Omega_1 = E_{mol} - \hbar\Omega_2 = 3.1926$ eV is due to the induced renormalization. It is noted that the maximum refractive index change reaches 3×10^{-4}. The change $\Delta n(\Omega_1)$ does not tend to zero with increasing energy up to the exciton resonant energy at 3.202 eV due to the contribution of the exciton resonance under the σ^- excitation beam.

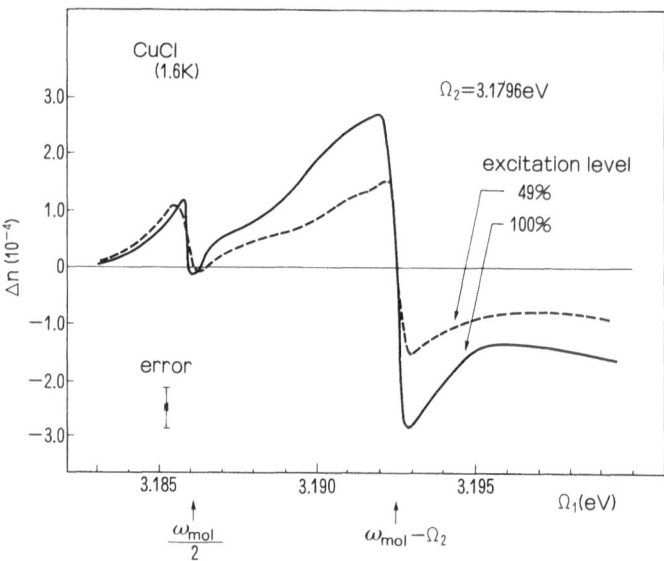

Fig. 2.40. Refractive index change due to the circularly polarized laser field Ω_2 for two excitation levels, probed by the linearly polarized probe field Ω_1 [2.82]

The other dispersive anomaly is observed at $\Omega_1 = E_{mol}/2$ independently of the excitation beam. This is due to the self-renormalization of the probe beam. An isotropic (or cubic) medium has nonlinear polarization in the following form:

$$P_i^{NL} = [AE_i(\boldsymbol{E}^* \cdot \boldsymbol{E}) + \tfrac{1}{2}BE_i^*(\boldsymbol{E} \cdot \boldsymbol{E})]\exp(-i\omega t + ikz) , \qquad (2.81)$$

where A and B are constants which are derived microscopically in terms of the electronic states of the medium. Let us assume the incident laser beam to be elliptically polarized. Then the intensities I^+ and I^- of the left (σ^+) and right (σ^-) circularly polarized components, respectively, are expressed in terms of the ellipticity, tan ψ, as

$$I_0^\pm = \tfrac{1}{2}I_0(1 \pm \sin 2\psi) .$$

The nonlinear difference Δn of the refractive index for σ^+ and σ^- is then induced through the nonlinear polarization of (2.81) as

$$\Delta n = \frac{2\pi}{n_0}\,\mathrm{Re}\{\chi^+ - \chi^-\} = \frac{2\pi}{n_0}\,B(I_0^- - I_0^+) = -\frac{2\pi}{n_0}\,BI_0\sin 2\psi .$$

As for induced renormalization, we have self-induced renormalization of the polariton as long as the probe beam is elliptically polarized and B is finite.

We consider the exciton and EM system shown in Fig. 2.41. The exciton $2(e^+)$ level is connected to the ground state $1(g)$ by the σ^+ component and to the

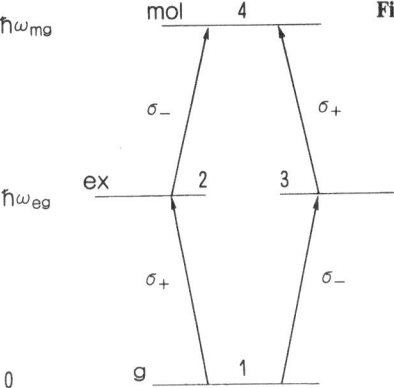

Fig. 2.41. Model for the exciton-EM system in CuCl

EM state $4(m)$ by the σ^- component of the probe beam. The $3(e^-)$ exciton level is connected oppositely. Then the susceptibility difference $\Delta\chi \equiv \chi^+ - \chi^-$ is evaluated from the nonlinear polarization of this system:

$$\Delta\chi = \frac{\mu^2}{8\hbar^3}\left[\frac{2\mu'^2}{\Delta\Sigma}\left(\frac{1}{\Delta} - \frac{1}{\Lambda}\right) + i\left(\frac{1}{\gamma_{ee}} - \frac{1}{\gamma_e}\right)\left(\frac{\mu^2}{\Delta} - \frac{\mu'^2}{\Lambda}\right)\left(\frac{1}{\Delta} - \frac{1}{\Delta^*}\right)\right]$$

$$\times (I_0^+ - I_0^-),\tag{2.82}$$

where μ and μ' are the transition dipole moments of, respectively, the exciton and the excitation from the exciton to the EM.

$$\Delta = \hbar\Omega - E_{ex} - i\gamma_{ge},$$
$$\Lambda = \hbar\Omega - E_{mol} + E_{ex} - i\gamma_{em},$$
$$\Sigma = 2\hbar\Omega - E_{mol} - i\gamma_{gm},$$

where γ_{ee} and γ_e are, respectively, the cross relaxation rate between e^+ and e^- exciton states and the decay rate of the exciton. The 1st and 2nd terms in the square bracket of (2.82) describe, respectively, the nonlinear coherent and incoherent processes. The latter is due to the real excitation of excitons and is negligible in the interesting frequency region $\hbar\Omega \simeq E_{mol}/2$ in CuCl, compared with the first term. As a result, inserting (2.82) into the expression of Δn gives

$$B = -\frac{\mu^2\mu'^2}{4\hbar^3}\,\mathrm{Re}\left\{\frac{1}{\Delta\Sigma}\left(\frac{1}{\Delta} - \frac{1}{\Lambda}\right)\right\}.$$

Kuwata and *Nagasawa* [2.83] observed the self-induced polarization rotation in CuCl with elliptically polarized light ($\psi = 20°$) and obtained a dispersive anomaly at $\Omega = E_{mol}/2$ for the phase difference $\delta \equiv \Delta n\Omega l/c$ as observed in Fig. 2.42.

Polarization-rotation spectroscopy has been shown to be an effective tool to study delicate changes of the refractive index, especially those due to the excita-

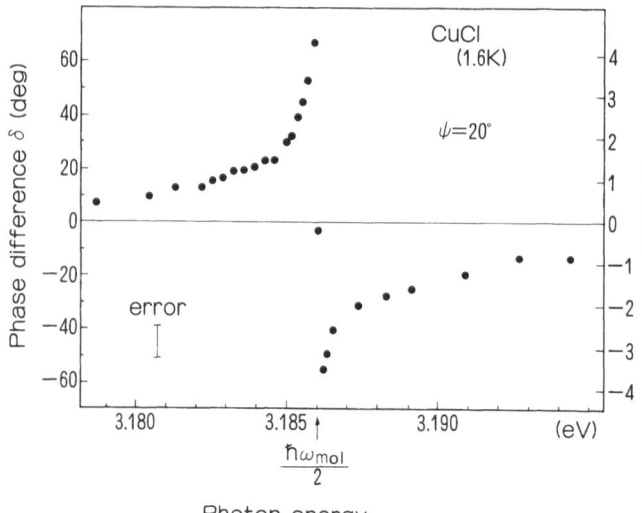

Fig. 2.42. Refractive index change due to the self-induced polarization rotation effect [2.83]

Photon energy

tion of the EM. As a consequence, this spectroscopy has been shown to be effective in studying the giant two-photon resonance of the EM and the exciton polaritons even in the strongly absorbing region.

It was through two-photon polarization (TPP) spectroscopy due to the EM, that *Kuwata* and *Nagasawa* [2.84] found a new aspect of exciton polaritons, probably connected with the additional boundary condition (ABC). An incident laser beam excites the exciton polariton inside the crystal. For excitation energies higher than the longitudinal exciton E_{ex}^L, there coexist lower- and upper-branch polaritons (LBP and UBP) as shown in Fig. 2.43. This is due to the spatial dispersion of the exciton with a finite translational mass.

A CuCl single crystal is excited by two laser beams. The first is a linearly polarized probe beam Ω_1 and the second is a circularly polarized excitation beam Ω_2. The intensity of both beams is kept weak to avoid higher-order nonlinear effects. The excitation beam Ω_2 excites two kinds of polaritons, A and B, with the same energy $\hbar\Omega_2$ but different wave vectors k_L and k_U. The ABC and the Maxwell boundary condition determine how they are populated. The 15 meV, broad spectrum, probe beam Ω_1 is incident on the crystal which is also irradiated by the excitation beam Ω_2. The transmitted light of the probe beam is analyzed, after passing through a cross polarizer, by a monochromater with spectral width 0.03 meV. Here the TPP ($\Omega_1 + \Omega_2$) spectrum due to the Γ_1 EM is observed as shown in Fig. 2.44 and the relative populations are obtained for the upper (U) and lower (L and L') polaritons.

Now the third beam Ω_3 is used simultaneously to depopulate these upper and lower polaritons by exciting them into the EM. The TPP spectrum of Ω_1 and Ω_2 (Ω_2 is fixed at 3.2104 eV) was observed under the third strong pump laser with $\Omega_3 = 3.1656$ eV (Fig. 2.45a), 3.1617 eV (Fig. 2.45b), and 3.1678 eV (Fig. 2.45c). In all these cases, three sharp lines appear at $\hbar\Omega_1 = 3.1617$ (U,

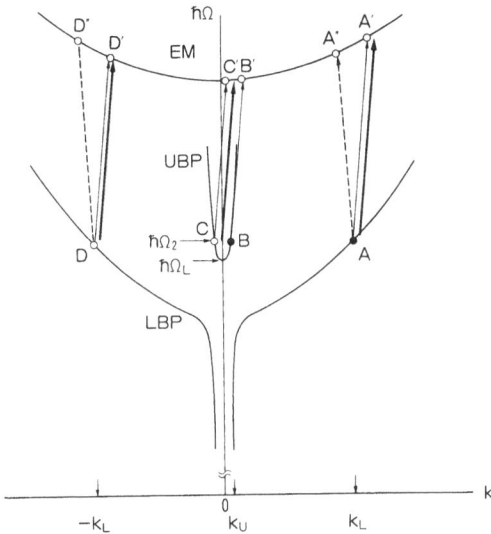

Fig. 2.43. Diagram to illustrate pumping the two exciton polaritons A and B and probing those polariton densities by excitation into EMs (A → A' and B → B'). Here C and D are polaritons reflected at the back surface

B → B' and C → C'), 3.1656 eV (L, A → A'), and 3.1649 eV (L', D → D').
The transitions of B → B' and C → C' are almost degenerate and cannot be resolved. It is noted, however, that the line intensity decreases remarkably when the polaritons involved in each spectrum are selectively depopulated by the third beam Ω_3 as shown in Fig. 2.45a, b in comparison with Fig. 2.45c. In the last case, the third beam is not resonant with any transition.

In the case of Fig. 2.45a, the lower polariton A is depopulated and in that of Fig. 2.45b, the upper polaritons B and C are depopulated by the third pump beam Ω_3. Furthermore the signal of the TPP rotation at $\hbar\Omega_1 = 3.1662$ eV (due to the A polariton, Fig. 2.46a), 3.1655 eV (D polariton, Fig. 2.46b), and 3.1629 eV (B and C polaritons, Fig. 2.46c) under the fixed $\Omega_2 = 3.2092$ eV have been observed as a function of the third pump frequency Ω_3. Actually the signals

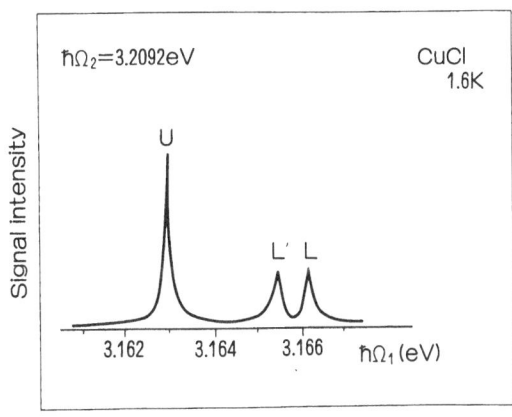

Fig. 2.44. Two-photon polarization (TPP) spectroscopy of the probe (Ω_1) beam under the excitation beam at $\hbar\Omega_2 = 3.2092$ eV [2.84]

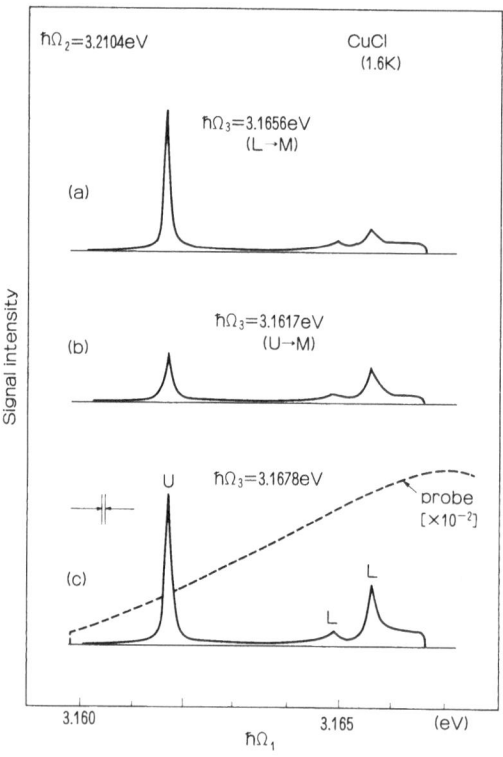

Fig. 2.45. TPP spectroscopy of the probe (Ω_1) beam under the excitation ($\hbar\Omega_2 = 3.2104$ eV) beam and the pump beam (**a**) $\hbar\Omega_3 = 3.1656$ eV (A → A′), (**b**) $\hbar\Omega_3 = 3.1617$ eV (B → B′) and (**c**) $\hbar\Omega_3 = 3.1678$ eV [2.84]

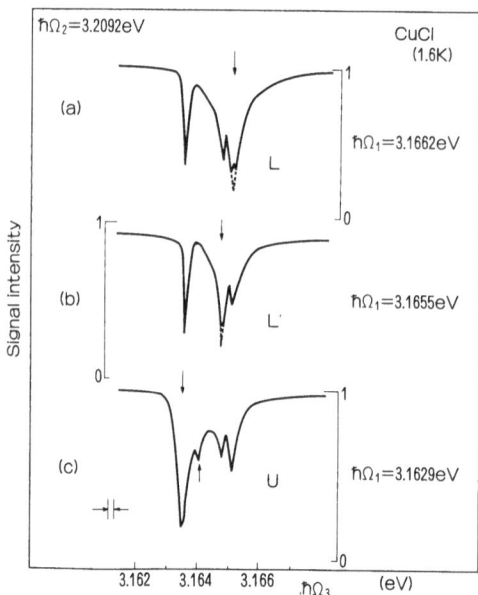

Fig. 2.46. Change in the TPP signal at (**a**) A → A′ ($\hbar\Omega_1 = 3.1662$ eV), (**b**) D → D′ ($\hbar\Omega_1 = 3.1655$ eV), (**c**) B → B′ and C → C′ ($\hbar\Omega_1 = 3.1629$ eV) under the excitation beam at $\hbar\Omega_2 = 3.2092$ eV as a function of the pump frequency Ω_3. Arrows denote the depression of A (**a**), D (**b**), B and C polaritons (**c**) [2.84]

corresponding to A → A′ (in Fig. 2.46a), D → D′ (in Fig. 2.46b) and B → B′ and C → C′ (in Fig. 2.46c) show the deepest dips, but other signals also show dips, e.g., in Fig. 2.46a, the signal due to A → A′ is also reduced under the depopulation of the B(C) and D polaritons.

We may consider a few possible mechanisms responsible for the correlation between the upper k_U and lower k_L polaritons with the same energy.

1) *Nakayama* [2.85] considered that the upper- and lower-branch polaritons are not themselves good quanta but linear combinations of these polaritons and the incident and reflected fields should be considered as quantum states in a semi-infinite medium. In this view, when a polariton, e.g., the lower-branch polariton A is excited into an EM by the pump beam Ω_3, one should consider that a coupled quantum mode is annihilated. This explains why the pumping A → A′ results in the suppression of the signal not only from the transition A → A′ but also that from B → B′, C → C′ and D → D′. There are four eigenmodes written as linear combinations of the incident and reflected electromagnetic fields and the upper- and lower-branch polaritons, which are degenerate in energy. Although only one of these may be excited by the excitation beam Ω_2, the radiation of the strong pump beam Ω_3 and the probe beam Ω_1 will mix up the four degenerate eigenmodes. The scattering of the eigenmode by imperfections in the medium will also bring about a mixing effect. Therefore we must check whether Nakayama's eigenmode is a good quantum state or not even under such conditions.

2) Under the strong pump beam Ω_3, the upper- and lower-branch polaritons will correlate with each other through nonlinear optical processes [2.86]. However, the experiments were done under as weak excitation and pump beams as possible and this possibility was ruled out by observing the signals to be proportional to the incident laser powers as far as can be observed.

3) Let us take into account the experimental conditions more quantitatively, i.e., that the sample is a 20-μm thick thin film and that the excitation pump laser has a spectral width $\Delta\nu$ corresponding to 0.08 meV. The polarization field is dephased out in a time of the order of $(\Delta\nu)^{-1}$ due to destructive interference of the polaritons composing the macroscopic polarization over $\Delta\nu$. This decays also, due to the effect of frequency modulations, in a time of the order of $[(\delta\omega)^2\tau_c]^{-1}$, where $\delta\omega$ and τ_c are the amplitude and correlation time of the frequency modulation, respectively. On the other hand, the polariton density over $\Delta\nu$ is almost free from the dephasing both due to the frequency distribution of the incident laser light and due to the frequency modulation as long as the inelastic collision and large-angle scattering of polaritons are not predominant. Under such a situation, the polariton dynamics can be discussed in terms of the polariton densities. The A or B polaritons are converted into the upper C and lower D polaritons with the wave vector opposite in sign and into the transmitting photons at the rear surface. The population ratios among them are determined by the ABC at the surfaces. The polariton dynamics were solved by taking into account these reflection effects at both the front and rear surfaces, as well as the effects of the excitation of A and B polaritons at the front surface and the pumping of the specified

polaritons into the EMs. Then the correlation effects among the polaritons were qualitatively understood [2.87].

2.5.3 Multi-Polariton Scattering Via Excitonic Molecules

New emission lines denoted by X in Fig. 3.43 [2.88] and L in Fig. 3.29 [2.89], were observed under strong excitation. They are located slightly to the lower-energy side of the two-photon Raman scattering lines in CuCl. X and L lines are assigned to the multi-polariton scatterings associated with the coherently (virtually) and incoherently (really) created excitonic molecules, respectively [2.90].

The emission spectrum for two-photon excitation of an EM is composed of two channels, i.e., a hyper-Raman scattering and a luminescence, as discussed in Sect. 2.4.1. These can be understood as, respectively, a coherent two-polariton scattering via the virtual creation of an EM and an incoherent emission from a real, excited EM. On increasing the excitation power, higher-order nonlinear optical processes are enhanced by an effect of giant oscillator strength associated with the EM resulting in the new emission lines called X and L bands. Let us list a few characteristic features of these two lines:

1) They are observed only under strong excitation slightly to the higher-energy side of the two-photon resonance of an EM.

2) The peak of the L band is located at the point midway between the two-photon Raman scattering and luminescence lines. That is, the peak frequency shifts by almost the same amount as the change of excitation frequency. On the other hand, the X band shifts in the opposite sense.

3) The peak energy and line shape of the X band do not depend on the scattering angle, while those of the L band do.

Let us consider irradiating the crystal with a strong monochromatic laser field $E_0 \exp(-i\omega_0 t) + c.c.$ near the two-photon resonance of an EM. Then the polariton with this frequency, ω_0, and wave vector, K_0, and the virtual excitation of an EM with $2\omega_0$ and $2K_0$ are induced in the crystal. At the same time, real EMs are formed on the dispersion surface $\omega_m(2K_0)$. They compose the macroscopically condensed states. With the macroscopic occupation of these three kinds of excitations as the initial states, other coherent and incoherent excitations are formed through multi-polariton scatterings. The formation rate of the polariton K is written as

$$W_K = \lim_{t \to \infty} \frac{d}{dt} \left(\langle \langle A_K^\dagger A_K \rangle \rangle_t \right) ,$$ (2.83)

where $\langle \langle A_K^\dagger A_K \rangle \rangle_t \equiv \langle \psi(t) | A_K^\dagger A_K | \psi(t) \rangle$ with the annihilation, A_K, and creation, $A_K^\dagger$, operators of the polariton and the wave function $\psi(t)$ of the total system. The emission spectrum is given by W_K times the transmittivity of the polariton at the surface, which is considered to be almost constant in the frequency region under consideration. Therefore we may consider W_K approximately as the emission spectrum.

First, the energies of a polariton and an EM are measured with respect to $\hbar\omega_0$ and $2\hbar\omega_0$, respectively, as $\Delta_{pK} \equiv \omega_p(K) - \omega_0$ and $\Delta_{mK} \equiv \omega_m(K) - 2\omega_0$ under transformation into the rotating frame. Second, we diagonalize $H_0 + V_1$:

$$H_0 = \sum_K [\hbar\omega_p(K)A_K^\dagger A_K + \hbar\omega_m(K)B_K^\dagger B_K] \, ,$$

$$V_1 = \sum_K 2\hbar g(K, K_0)(E_0 e^{-i\omega t} B_{K+K_0}^\dagger A_K + h.c.) \, ,$$

under the unitary transformation

$$\begin{pmatrix} \alpha_K \\ \beta_K \end{pmatrix} = \begin{pmatrix} C_{\alpha K}, & C_{\beta K} \\ -C_{\beta K}, & C_{\alpha K} \end{pmatrix} \begin{pmatrix} A_K \\ B_{K+K_0} \end{pmatrix} \, ,$$

into the following form:

$$H_0 + V_1 \Rightarrow \sum_K \hbar\omega_\alpha(K)\alpha_K^\dagger \alpha_K + \hbar\omega_\beta(K)\beta_K^\dagger \beta_K \, . \tag{2.84}$$

Here,

$$2\omega_{\alpha,\beta}(K) = (\Delta_{pK} + \Delta_{mK+K_0}) \pm [(\Delta_{pK} - \Delta_{mK+K_0})^2 + 4|g_K|^2]^{1/2} \tag{2.84a}$$

and

$$\begin{pmatrix} C_{\alpha K} \\ C_{\beta K} \end{pmatrix} = \begin{pmatrix} \Delta_{mK+K_0} - \omega_{\alpha K} \\ -g_K \end{pmatrix} [(\Delta_{mK+K_0} - \omega_{\alpha K})^2 + |g_K|^2]^{-1/2} \, , \tag{2.84b}$$

with $g_K \equiv g(K, K_0)E_0$. Finally, other interaction terms $V \equiv V_2^{coh} + V_2^{inc} + V_3$ are taken into account by the perturbational method. V_2^{coh} describes the process in which the coherently excited EM(F_0) at $(2K_0, 2\omega_0)$ is dissociated into two polaritons K and $2K_0 - K$ and the inverse process. V_2^{inc} represents the dissociation process of the real EM (F_Q') at $(Q, \omega_M(Q))$ into two polaritons K and $Q-K$, and vice versa. In the V_3 process the non-condensed EM is resolved into or formed from two non-condensed polaritons.

With the ground state $|0\rangle$ of $H_0 + V_1$ as the initial state at $t = -\infty$, we may rewrite (2.83) as follows:

$$W_K = 2 \text{Im} \{\langle 0|T[A_K^\dagger A_K \tilde{V}(0_-)S(\infty, -\infty)]|0\rangle\} \, . \tag{2.85}$$

Here T is a time-ordering operator, $\tilde{V}(t)$ is the interaction representation of V, and $S(\infty, -\infty) \equiv \exp\left[-i \int_{-\infty}^{\infty} V(\tau)d\tau\right]$. The hyper-Raman scattering and the luminescence from an EM are obtained as the lowest-order contribution to (2.85).

As realized from Fig. 2.47a with the coherent external lines F_0, a coherent polarization due to an EM at $2K_0$ and $2\omega_0$ is converted into two polaritons, and one of these is observed as a Raman line. We have four possible combinations of the two renormalized polaritons α and β, which are defined in (2.84). A few

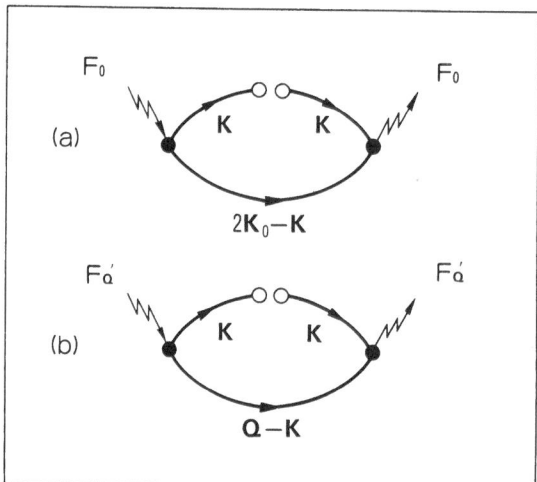

Fig. 2.47. Diagram contributing to (**a**) the Raman scattering with the coherent external lines F_0 and (**b**) the luminescence with the external lines F'_ϱ [2.90]

Raman lines were observed [2.91], depending on the frequency ω_0 and the intensity of the incident laser field E_0 as shown in (2.84a).

The luminescence line is observed when the real excited EM on the EM dispersion surface $\omega_M(2K_0)$ is converted into two polaritons. This process is represented by the diagram of Fig. 2.47b with the external lines F'_ϱ. In the next higher order processes, two EMs or four polaritons are involved. The X and L bands are shown to correspond to these emission processes:

a) As shown in Fig. 2.48a, two polaritons K and $2K_0 - K$ are formed from a coherent EM $(2K_0, 2\omega_0)$ and then both polaritons interact individually with incident polaritons (K_0, ω_0) to form two EMs. One of these EMs is resolved into two polaritons and one of these polaritons is observed as an X band. When both of the EMs are resolved into two polaritons, as shown in Fig. 2.48b with the external lines F_0, these contribute only to the correction to the Raman lines given by Fig. 2.47a.

b) We have an L band when a real EM created on the dispersion surface $\omega_m(2K_0)$ is annihilated in the diagram of Fig. 2.48b with the external lines F'_ϱ. When an EM remains in the final state in Fig. 2.48a with the external lines F'_ϱ, it gives only a modification of the luminescence line.

First we perform the calculation of the emission spectrum (2.85) corresponding to the process (a), which reveals the following facts: This process has a large contribution only for $\omega_0 \geq \frac{1}{2}\omega_m(K_0)$, in agreement with the first characteristics cited at the beginning. The numerical calculation of the emission spectrum shows double peaks, which matches the characteristics of the X_T band. This is understood as an emission from the EM on a dispersion sphere in wave-vector space with its center at $2K_0$ that satisfies the relation $\omega_m(4K_0 - K - q) + \omega_m(K + q) = 4\omega_0$. This emission frequency decreases as the excitation frequency ω_0 increases. This dependence agrees with the one observed for the X band as shown in Fig. 2.49. The energy shift can be understood as follows: As the

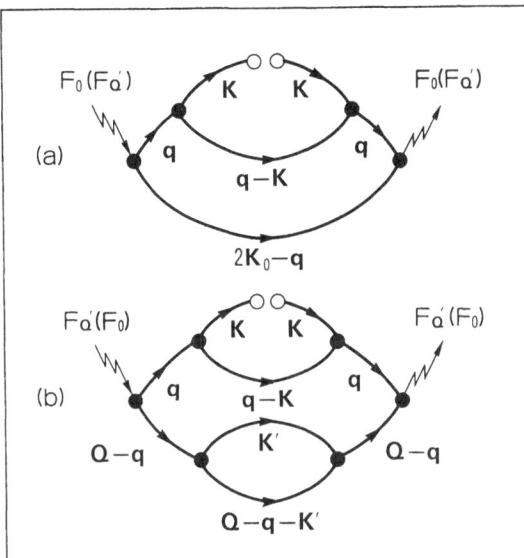

Fig. 2.48. (a) Diagram inducing the X_T and X_L bands when the external lines are F_0. (b) Diagram inducing the L_T and L_L bands when the external lines are F'_ϱ [2.90]

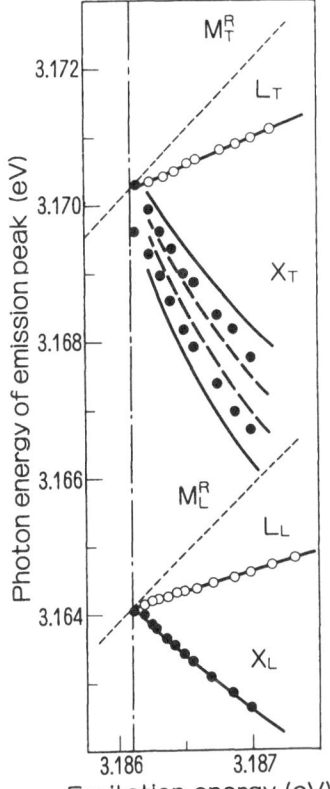

Fig. 2.49. Peak frequencies of the emission spectra as a function of the incident laser frequency ω_0. The open and closed circles denote the observations [2.88] and the solid lines are calculated in [2.90] with the material constants for CuCl [2.88]. The heavy and light broken lines show, respectively, the M^R bands and the calculated X_T bands for the forward scattering configuration [2.90]

laser frequency, ω_0, increases, the energy of the created EM increases so that the energy separation between the EM and the polariton, i.e., the emission frequency, decreases due to the dispersion relationships.

The second process of Fig. 2.48 b consists of two V_2^{inc} and four V_3. The emission spectrum due to this process consists of the product of the luminescence and the Raman spectra. From this fact, we realize that this second process has an intermediate character. Actually for $|2\omega_0 - \omega_m(2K_0)| < \gamma_m$, this emission spectrum has a single peak at $\omega_p(K) = \omega_0 - \omega_p(2K_0 - K) + \frac{1}{2}\omega_m(2K_0)$. This peak frequency increases in proportion to the increase of ω_0, and is coincident with the peak of the L band as shown in Fig. 2.49. It is noted that the L band is observed even under the off-resonant excitation of $\omega_0 - \frac{1}{2}\omega_m(2K_0) \sim 1$ meV. This is because under such a strong excitation real EMs are created at $\omega_m(2K_0)$ by this off-resonant laser light due to the power-dependent γ_m as Fig. 3.40 shows.

We have discussed two new emission bands which are denoted by X_T and L_T bands and observed on the low frequency side of the Raman line M_T^R in which a transverse polariton is left inside the crystal. Just as they observed another Raman line M_L^R in which a longitudinal exciton is left in the final state, *Itoh* et al. [2.88, 89] also observed two new emission lines on the low frequency side of the M_L^R line. When we repeat similar calculations for the process leaving behind a longitudinal exciton, two emission peaks below the Raman line M_L^R can be calculated as a function of ω_0. These agree with the observed X_L and L_L bands as shown in Fig. 2.49.

We conclude that the two emission lines observed below both of the M_T^R and M_L^R Raman lines for $\omega_0 > \frac{1}{2}\omega_m(2K_0)$ under rather strong excitation can be assigned to the four-polariton scatterings associated with the virtual and real excitations of an EM at $(2\omega_0, 2K_0)$ and $(\omega_m(2K_0), 2K_0)$, respectively. The ω_0 dependence of these four lines can be perfectly explained by these processes as shown in Fig. 2.49. The only discrepancy is in the magnitude of the splitting of the two peaks in the X_T band. The observed splitting is almost independent of K, i.e., the direction of the observation, as similar emission spectra were observed in the backward and the forward scattering [2.88]. In this calculation, the splitting in the forward configuration is almost twice as large as that in the backward one, as shown in Fig. 2.49. This may be mainly due to neglection of the relaxation and higher order effects. In reality the relaxation works toward reducing the difference between the backward and forward scattering.

2.5.4 Optical Bistability Due to the Excitonic Molecule

In this subsection the optical bistability due to the coherent, dispersive, nonlinear response of the exciton-EM system of CuCl is discussed [2.92, 93]. Here the dispersion anomaly of the polariton due to EM formation, discussed in Sect. 2.5.1, is used.

Optical bistability results from the combined effect of nonlinear optical response and the feedback provided by the Fabry-Perot or Ring cavity to the beams. The power-dependent dielectric function (2.79) is rewritten in the vicinity of the giant two-photon excitation $2\hbar\omega \simeq E_{mol}$ as

$$\epsilon(\omega) = \epsilon_\infty + (\epsilon_0 - \epsilon_\infty)\omega_{ex}^2 \left((\omega_{ex}^2 - \omega^2) - \frac{4n_p|M|^2\omega_{ex}(\omega_{mol} - \omega)}{\omega_{mol}(\omega_{mol} - 2\omega - i\gamma_{mol})}\right)^{-1} \quad (2.86)$$

where the effect of the giant oscillator strength is taken into account in the expression $M \equiv \mu'$ [2.4]. The corresponding reflection coefficient for normal incidence $R = [(n' - 1)^2 + n''^2]/[(n' + 1)^2 + n''^2]$ with $(n' + in'')^2 = \epsilon(\omega)$ is plotted for two polariton concentrations in Fig. 2.50. The intensity which is transmitted through a Fabry-Perot resonator filled with an active medium is given by

$$\frac{I_t}{I_0} = \frac{(1 - R)^2}{(e^{\delta''} - Re^{-\delta''})^2 + 4R\sin^2\delta'} , \quad (2.87)$$

where I_0 and I_t are the intensities of the normally incident and transmitted beams, respectively. The phase shift for a single path through a resonator of thickness l is $\delta = \delta' + i\delta'' = \sqrt{\epsilon(\omega)}\omega l/c$. The dependence of δ' on the polariton concentration n_p is essential to the optical bistability. The transmitted beam intensity I_t is related to the internal intensity I_i, which in turn can be expessed by the polariton concentration n_p:

$$I_t \simeq (1 - R)I_i/2 , \quad I_i \simeq n_p\hbar\omega v_g , \quad (2.88)$$

where $v_g(\omega)$ is the polariton group velocity. Inserting (2.88) into (2.87), one obtains a nonlinear equation for the polariton density n_p (or I_t). The resulting polariton density n_p is plotted in Fig. 2.51 as a function of I_0 and ω for a plate of

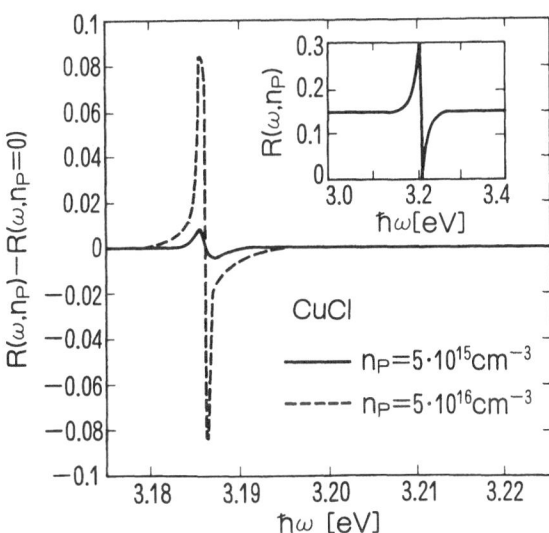

Fig. 2.50. Spectrum of the reflection coefficient change $R(\omega, n_p) - R(\omega, 0)$ of CuCl shows an anomaly due to two polaritons scattering resonantly into an EM for two polariton concentrations. The insert shows the exciton resonance in the reflection spectrum and the two-polariton resonant anomaly occurs on the low-energy side of this [2.93]

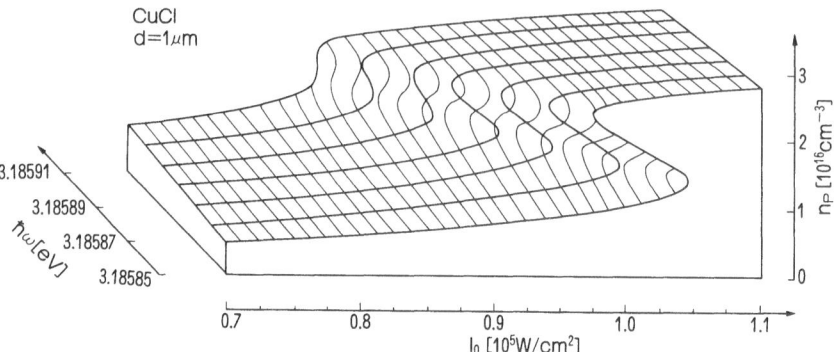

Fig. 2.51. Polariton concentration n_p in a CuCl resonator of 1 μm thickness as a function of the intensity and the frequency of the incident beam [2.93]

CuCl 1 μm thick. In the fold, n_p has multiple solutions. If the intensity I_0 of the incident beam is increased at a constant frequency, the polariton density jumps discontinuously at the edge of the fold from the lower to the upper branch. The critical intensities of the incident beam are $I_0 \simeq 0.1$ MW/cm^2 which can be easily obtained with available dye lasers. Figure 2.51 shows that the bistability vanishes due to increasing absorption when $\hbar\omega$ approaches the two-photon resonance at $E_{mol}/2$. The bistability becomes more pronounced with decreasing ω but at the same time the necessary intensities become larger.

It is considered that this optical bistability will be used in the future in the essential parts of an optical integrated circuit as well as an optical computer. Here light plays the role of electrons in large-scale integrated (LSI) circuits in modern electronic computers. In order to get logic and memory devices more effective than very large scale integrated circuit (VLSI) or Josephson devices, the system is required to be switched between the off and on states in times of the order of picoseconds by a light pulse with power of the order of picojoules. An obstacle to reaching these requirements is that the switching-off time from the high to the low transmission state is too long, e.g., 40 ns for the system of GaAs [2.94] and of the order of 100 ns for that of InSb [2.95]. The measurements used the dispersive nonlinear response which was induced by the presence of real, excited, elementary excitations. Therefore it takes the decay time of an exciton, in the case of GaAs, and that of an e–h pair, in the case of InSb, for the switching-off to be completed.

We wish to investigate the possibility of the optical bistability responding in picoseconds to the picojoule pulse. For this purpose, we study equations of motion for polarizations and populations of the exciton-EM system under strong, coherent, laser irradiation. Then we solve the Maxwell equations for the internal field, consistently with the equations of the medium. There are two kinds of contributions to the nonlinear dispersive response, i.e., the coherent two-photon process described by the off-diagonal density matrix $\varrho_{mg} \equiv \langle \text{mol} | \varrho(t) | g \rangle$ and the incoherent one described by the diagonal density matrix $\varrho_{ee} \equiv \langle \text{exc}|\varrho(t)|\text{exc}\rangle$ and $\varrho_{mm} \equiv \langle \text{mol}|\varrho(t)|\text{mol}\rangle$. The dynamics due to the incoherent nonlinear process are governed by the longitudinal relaxation time T_1, while those due to the coherent one are determined by the transverse relaxation time T_2. For elementary excitations in solids, T_1 is usually of the order of nanoseconds while T_2 is of the order of picoseconds. The optical bistability observed by *Gibbs* et al. [2.94] for a GaAs wafer is due to this incoherent process and it takes 40 ns to switch off.

As expected, we can show that not only switching-on but also switching-off can be completed in times of the order of picoseconds for the case where the coherent process is dominant. When the incident laser light is nearly in resonance with the exciton, the incoherent nonlinear process, which is accompanied by the creation of excitons, prevails. This, however, decreases exponentially as the excitation energy shifts downward from the exciton peak, obeying the Urbach-Martienssen rule. Consequently, the contribution from the incoherent process almost vanishes while that due to the coherent process decreases only very gradually as the inverse of the difference from the resonant energy. As a result, under suitable off-resonant excitation, e.g., at the excitation frequency 16 meV lower than the exciton peak in the CuCl system, the coherent process is shown to be dominant.

Usually higher exciting power is required to fulfill the condition of optical bistability under off-resonant excitation, but, very fortunately, the transition dipole moment to excite a single exciton into the EM is about ten times larger than that of the single exciton, through the effect of the giant oscillator strength, as discussed in Sect. 2.3.2. As a result, the required incident power can be reduced by 2 orders of magnitude so that $1\,\text{MW/cm}^2$ incident laser power is enough to fulfill the condition under the 16 meV off-resonance excitation for the CuCl sample with 1-µm thickness reflectivity of 0.9 at both ends.

Optical limiting behavior and bistability were actually observed by *Peyghambarian* et al. [2.96] and *Hönerlage* et al. [2.97] with a 15–30 MW/cm² incident laser for a 10-µm thick CuCl etalon with mirrors with reflectivity of 0.9. Switching time is estimated to be shorter than 100 psec limited by the response time of the detection system. This optical bistability may be considered to be due to the incoherent nonlinear optical process associated with the resonant excitation of the EMs for two reasons. First, the incident laser is nearly resonant with the GTA by the EM although it is quite off-resonant with the one-photon resonance with the exciton. The detuning from the two-photon resonance is only of the order of a fraction of a milli-electron volt. Second, a large holding power such as 15 to $30\,\text{MW/cm}^2$ is required in contrast with the theoretically predicted values of $0.1\,\text{MW/cm}^2$ [2.93] and $1\,\text{MW/cm}^2$ [2.92]. The optical bistability due to the coherent nonlinear optical process is possible at incident power a few orders of magnitude lower than in the case of that due to the incoherent nonlinear optical process. Therefore we may consider that the coherent optical bistability has not been realized yet.

2.5.5 Relaxation and Bose Condensation of Excitonic Molecules

The relaxation process of EMs is studied from the EM giant two-photon absorption (GTA) spectrum and from the competition between the hyper-Raman scattering and the luminescence due to the EM [2.98]. The collision broadening of the GTA implies a collision rate $\sim 10^{-12}$ s for the EMs. This is sufficiently rapid to establish a quasi thermal equilibrium for the particles on a time scale that is short in comparison with the lifetime of the EMs and the duration of the laser pump pulse. The luminescence spectrum of the highly excited EMs has been fitted using a Bose-Einstein thermal distribution. The chemical potential obtained from the fit goes to zero as the EM density is increased at a fixed temperature or if the lattice temperature is decreased at a fixed density. The sharp luminescence features that are observed at high densities and low temperatures are interpreted as resulting from a Bose-condensed state of the EM [2.99].

Under the GTA of the EM, the EM is created at twice the wave vector of the incident light inside the crystal. When any relaxation processes work on the EMs, they spread over a certain wave-vector region on the parabolic dispersion surface and the relaxation also causes broadening in energy of the EM state. This leads to broadening of the GTA band. The broadening was actually observed under intense two-photon resonant excitation and with increasing lattice temperature. The relaxation mechanism at high lattice temperatures has been mainly ascribed to the interaction of the EMs with acoustic phonons as discussed in Sect. 2.2.3. The detailed mechanism of the relaxation under intense excitation has been made clear by *Chase* et al. [2.100], *Peyghambarian* et al. [2.101], and *Itoh* et al. [2.98].

The absorption spectra due to the giant two-photon excitation of the EM have been observed under various excitation powers as shown in Fig. 3.40. The ordinate indicates the magnitude of the absorption $A \equiv (I_0 - I_t)/I_0$, where I_0 and I_t are the incident and the transmitted light intensities, respectively. Dashed lines indicate the background absorption A_b ascribed to the one-photon absorption due to the Z_3 exciton band tail which is about 0.1 and almost independent of the pump power. The absolute power at the peak $I_{max} \equiv (A - A_b)I_0/(1 - A_b)$ due to the two-photon process and the FWHM, $\gamma(N)$ obey the following relations [2.98]:

$$I_{max} \propto I_0^{1.0} , \tag{2.89a}$$

$$\gamma \propto I_0^{0.8} . \tag{2.89b}$$

These are in contrast with the results of *Chase* et al. [2.100]:

$$I_{max} \propto I_0^{1.5} , \tag{2.90a}$$

$$\gamma \propto I_0^{0.5} . \tag{2.90b}$$

Let us describe the absorbed power by

$$I(\omega) = \frac{C[\gamma(N) + \Gamma]I_0^2}{(2\omega - \omega_{mol})^2 + [\gamma(N) + \Gamma]^2} , \tag{2.91}$$

where C is a constant involving the transition matrix element, Γ is the intrinsic width independent of the density N of the EM, and $\gamma(N) \equiv N\sigma \langle v \rangle$ is the collision rate. When we take the model of binary collisions in the impact approximation, $\gamma(N) \propto I_0$ and $I_{max} \propto I_0$ are rather near the result of (2.89a, b) obtained by *Itoh* et al. [2.98]. On the other hand, when we assume that $\gamma(N) = K_a N^\alpha$, the number N of EMs created at the resonance peak is proportional to $CI_0^2/(K_a N^\alpha)$ in the limit $\gamma(N) \gg \Gamma$. Therefore N is proportional to $I_0^{2/(1+\alpha)}$, and $\gamma(N) \propto I^{2\alpha/(1+\alpha)}$ and $I_{max} \propto I_0^{2/(1+\alpha)}$. These results, with $\alpha = 1/3$, are in agreement with the experimental observations of *Chase* et al. [2.100]. This corresponds to a relaxation proportional to the mean distance between the EMs. A similar assumption is made in the cage model appropriate in dense media such as liquids.

As discussed in Sect. 2.3.1, the hyper-Raman scattering and luminescence are the secondary emissions from the EM, before and after relaxation, respectively [2.59]. Therefore, information about the relaxation is available from the observation of the secondary emissions. Excitation-power dependences of the integrated intensities of the Raman line R and the luminescence L were observed [2.88] as shown in Fig. 3.44a: under just-resonant excitation at 3.18603 eV, and b: under off-resonant excitation at 3.18622 eV. Under just-resonant excitation,

$$R \propto I_0^{0.5} , \tag{2.92a}$$

$$L \propto I_0^{1.5} . \tag{2.92b}$$

Under the off-resonant excitation,

$$R \propto I_0^{2.3} , \tag{2.93a}$$

$$L \propto I_0^3 . \tag{2.93b}$$

Theoretically the integrated Raman and luminescence intensities are given [2.96] as

$$R \sim B_0 \frac{\pi I_0^2}{(2\omega - \omega_{mol})^2 + [\gamma(N) + \Gamma]^2} , \tag{2.94a}$$

$$L \sim B_0 \frac{2\tau[\gamma(N) + \Gamma]I_0^2}{(2\omega - \omega_{mol})^2 + [\gamma(N) + \Gamma]^2} . \tag{2.94b}$$

Here the relaxation constants γ_{mg} and γ_{me} of the EM relative to the ground and exciton states, respectively, are assumed to have the same I_0 dependence at high pump power. When we take $\gamma(N) \propto I_0^{0.8}$ from (2.89b), i.e., FWHM of the two-photon absorption spectrum, for the resonant excitation,

$$R \sim B_0 \frac{\pi}{\gamma(N)^2} I_0^2 \propto I_0^{0.4} ,$$

$$L \sim B_0 \frac{2\pi}{\gamma(N)} I_0^2 \propto I_0^{1.2} .$$

and for the off-resonant excitation such as $|2\omega - \omega_{\text{mol}}| > \gamma(N)$,

$$R \sim B_0 \frac{\pi}{(2\omega - \omega_{\text{mol}})^2} I_0^2 \propto I_0^2 ,$$

$$L \sim B_0 \frac{2\tau\gamma(N)}{(2\omega - \omega_{\text{mol}})^2} I_0^2 \propto I_0^{2.8} .$$

On the other hand, when we accept $\gamma \propto I_0^{0.5}$ from the data of *Chase* et al. [2.100], for the just-resonant excitation,

$$R \propto I_0^{1.0} ,$$

$$L \propto I_0^{1.5} ,$$

and for the off-resonant excitation,

$$R \propto I_0^2 ,$$

$$L \propto I_0^{2.5} .$$

The calculated excitation-power dependences reproduce fairly well the observed results of both groups. *Itoh* et al. [2.98] estimated the collision time $\pi/\gamma(N) \simeq 2\,\text{psec}$, and the concentration of EMs $N_{\text{mol}} \simeq 10^{17}/\text{cm}^3$ assuming the decay time of the EM $\tau \sim 300\,\text{psec}$ [2.102]. Sufficiently frequent collisions of the EMs, before radiative decay, establish a quasi thermal equilibrium for the particles. Under just-resonant excitation, the emission spectrum is composed of the sharp lines M_T^0 and M_L^0 and the broad bands M_T and M_L. Under off-resonant excitation, it has only two broad bands M_T and M_L, as shown in Figs. 3.42a and b. The effective temperatures of the EMs, $T_{\text{eff}} = 18$ K for Fig. 3.42a and 24 K for Fig. 3.42b, are obtained by fitting the emission curve under the assumption of the Maxwell-Boltzman distribution. These temperatures are higher than the lattice temperature of 4.2 K.

The emission spectrum from the Bose-condensed EMs has been calculated [2.65]. It consists of a sharp line with a broad band on the low-energy side as shown in Fig. 2.52. The sharp line comes from the emission of the condensed EM and the broad one from the emission with recoil of the other EMs and the emission of the EM thermally activated in the finite momentum state. *Nagasawa* et al. [2.103] claimed the first observation of the Bose condensation of EMs created by the GTA in CuCl. Later the sharp emission line, shifting in frequency as the exciting laser frequency was changed, was attributed to a resonant hyper-Raman scattering process in which two laser photons excite a virtual EM and one photon is reemitted leaving an exciton-polariton in the crystal [2.104]. These sharp luminescence features have also been assigned, by *Grun* and his coworkers [2.104–106], to a cold gas of EMs that radiatively decay before undergoing thermalizing collisions with phonons or thermalized excitonic particles. On the other

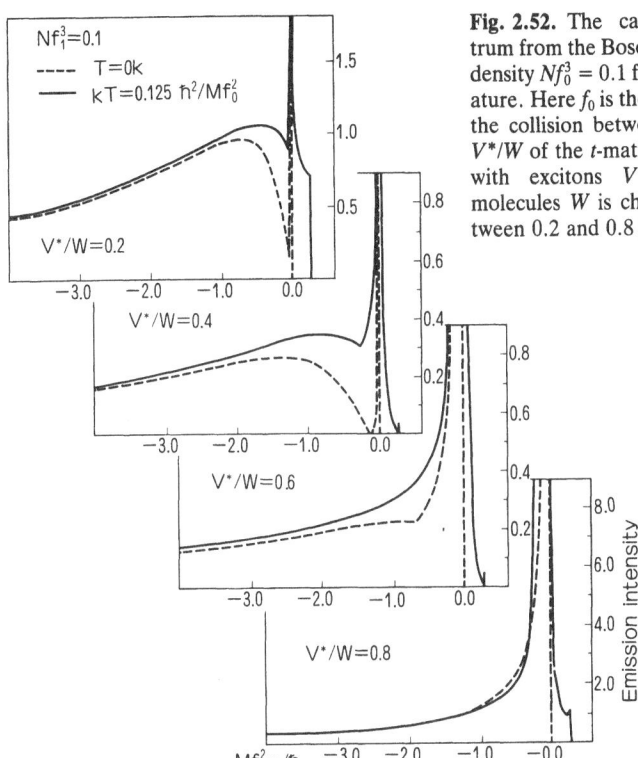

Fig. 2.52. The calculated emission spectrum from the Bose-condensed EM gas for a density $Nf_0^3 = 0.1$ for zero and finite temperature. Here f_0 is the scattering amplitude for the collision between two EMs. The ratio V^*/W of the t-matrix of the EM interaction with excitons V^* and other excitonic molecules W is chosen as a parameter between 0.2 and 0.8 [2.66]

hand, they were also suggested to come from stimulated emission [2.107]. Recently *Peyghambarian* and coworkers [2.108, 109, 110] confirmed the Bose condensation of EMs by eliminating the above three possibilities.

First, the intensity of the sharp emission line diminishes rapidly as the laser is detuned from resonance, and it is unobservable when the excitation laser is detuned by one to two half-widths. At higher pump intensities, a weak component with a hyper-Raman-type shift, equal to twice the shift of the pump-laser wavelength, is observed but the sharp line shifts by $\delta v = m\Delta\omega$ (laser) with $m = 0.8$ and 0.9 in contrast with the hyper-Raman shift $\Delta v = 2\Delta\omega$. This shift was attributed to the inhomogeneous broadening of the exciton level as well as the EM level. Therefore they assigned the sharp line not to the hyper-Raman scattering but to the emission from the Bose-condensed EMs.

Second, at moderate excitation levels, the emission data show that most of the emitting particles are in the thermalized distribution, which would coexist with a so-called cold gas in the steady state if we accept a cold gas model, regardless of how the thermalized EMs are produced by the resonant pumping. In the absence of the quantum attraction due to the Bose nature of the particles, any collisions of newly created EMs should scatter them predominantly into the thermal distribution, not into some separate cold distribution. Note, however, that the collision frequency γ is a few orders larger than the radiative decay constant τ^{-1}. It is a characteristic of Bose particles that they are scattered more densely into the highly populated states $k = 2k_p$ or $k = 0$, respectively, in the single-beam excitation and when pumped by the counter-propagating beams. This is a microscopic origin of Bose-condensation.

Third, stimulated hyper-Raman scattering and Bose condensation at the pump wave vector result from the presence of a coherent EM field at $2k_p$ or 0. There is the possibility that stimulated hyper-Raman scattering might account for the increase in the relative intensities of the sharp component with increasing excitation level. At higher excitation intensities well above $10\,\text{MW/cm}^2$, a very dominant, sharp, and strongly stimulated hyper-Raman line M_L^R is observed. Near the threshold, the M_L^R intensity increases by nearly 2 orders of magnitude when the pump intensity is doubled. All of the results on Bose condensation were obtained at much lower excitation levels where the total luminescence intensity increased smoothly with pump power I_0 as $I_0^{1.5}$.

Another experiment to confirm the Bose condensation was done at the same time by *Peyghambarian* et al. [2.108, 110]. One of the distinctive properties of a condensed system of bosons is that, if additional particles are added to the system, they will be preferentially attracted into the condensed state. This would normally have negligible statistical weight in the classical limit. An intense pump beam with a frequency ω_2 tuned to the center of the GTA spectrum creates the EMs at $2k_p$. Two, much weaker, counter-propagating probe beams with frequencies $\omega_1 = \omega_2 + \Delta\omega$ and $\omega_3 = \omega_2 - \Delta\omega$ create EMs only near $k = 0$ by the simultaneous absorption of one photon from each beam. Here the detuning $\hbar\Delta\omega$ of the probe beams is 1.64 meV and is too large compared with the GTA line width for any measurable excitation of EMs at $2k_p$ by either probe beam alone. Experiments are designed to seach for such an effect.

The probe-induced luminescence in the absence of pumping at $2k_p$ is due to the EMs scattered from $k = 0$ into a thermal distribution. In order to detect the probe-induced luminescence in the presence of the much stronger emission due to the pump beam, the probe beam at frequency ω_3 is mechanically chopped, and the luminescence is synchronously detected. When the pump beam is not present, the probe luminescence line shape indicates the expected thermal distribution of EMs. In the presence of the pump, the probe emission is strongly redistributed; sharp lines appear at the positions of the M_T^0 and M_L^0 lines. The synchronously detected spectrum contains no emission features of the EMs whenever either probe beam is blocked. Spectra with sharp M_T^0 and M_L^0 lines were obtained at lattice temperatures up to 40 K. This is the evidence of the boson attraction to the condensed state. The integrated probe emission intensity decreases as the lattice temperature is raised and is a few percent or less of the integrated pump intensity. The effect of the stimulated emission of the EMs was carefully eliminated. In addition to this, evidence of stimulated emission is found at laser intensities above 5–$10\,\text{MW/cm}^2$, and most of the experiments for Bose condensation have been done under much weaker pump and probe laser beams. As a result, we may conclude that the sharp emission lines come from these condensed EMs coherently excited by the laser pump.

Concluding this chapter, we will list the references of review articles mainly on EMs and point out a few characteristics of this chapter in contrast with those articles. The book *Excitons* edited very recently by *Rashba* and *Sturge* contains two related articles on EMs [2.111, 112]. The discussion in this chapter has concentrated on the coherent nonlinear optical phenomena due to the EM which are missing in [2.111, 112]. These have been found and extensively studied very recently. Therefore these subjects were not discussed in the old review articles published in 1976 [2.113, 114], 1977 [2.115], and 1981 [2.116]. Due to recent developments in laser spectroscopy and the effect of the giant oscillator strength in the EM, we have found the fascinating nonlinear and coherent optical phenomena related to the EM, as shown mainly in Sects 2.4 and 5. We hope to be able to continue to enjoy studies related to the EM in the years to come.

3. The Exciton and Excitonic Molecule in Cuprous Halides

The exciton in cuprous halides has a large binding energy and this makes it possible to realize the high-density exciton state in wide temperature regions. The CuCl crystal is the first case where the excitonic molecule (EM) has been found to be created through exciton-exciton collisions. The EM is annihilated radiatively leaving an exciton behind. It can also be created directly through two-photon excitation with giant oscillator strength. Raman scattering to the exciton states occurs simultaneously with the generation of a real EM. In the case of resonant excitation the Raman scattering and luminescence processes coexist. The redistribution of EMs, after being created with a definite wave vector by the two-photon excitation, has been clarified through the line-shape analysis of the luminescence bands in the backward and forward scattering configurations. A technique called stepwise two-photon excitation is found to be effective for the creation of EMs of large wave vector and for the determination of the spatial dispersions of the exciton and EM.

A nonlinear anomaly of the exciton-polariton dispersion has been found, which is due to the large oscillator strength of the two-photon absorption, and this causes various nonlinear effects in the secondary emission. The relaxation of the EM and the line shape of the two-photon absorption band are discussed in terms of collisions between EMs.

3.1 Band Structure and Excitonic States

Cuprous halides have zinc blende structure below the transition temperature of about 660 K and wurtzite structure above it. The exciton is of the Wannier type and shows two kinds of hydrogen-like series, $n = 1, 2, 3$, and $n' = 1, 2, 3$ [3.1]. The two excitons arise from the spin-orbit splitting of the hole in the $(p)^5$ electron configuration of the halogen atom left after the $(p)^6 \rightarrow s(p)^5$ transition. The excitons consisting of holes of $j_h = 3/2$ and $1/2$ are called respectively the $Z_{1,2}$ and Z_3 excitons [3.2]. The symmetry of the cuprous halides belong to T_d and the exciton absorption is of the direct allowed type at the Γ point. The exciton absorption bands in CuCl and CuBr shift to the *high* energy side with a temperature rise. The energy separation between the $\Gamma_7(j_h = 1/2)$ and $\Gamma_8(j_h = 3/2)$ valence bands, split from Γ_{15} by the spin-orbit interaction, is much smaller than those in alkali halides and further the Γ_7 band in CuCl is located above the Γ_8 band, contrary to the cases of general zinc blende type crystals. These characteristic properties of the band structure in cuprous halides arise from the fact that the upper part of the valence band consists mainly of Cu3d, the spin-orbit splitting of which has the opposite sign to that of the halogen p-state [3.2, 3]. The mixing of the Cu3d state amounts to 75, 64, and 50% in CuCl, CuBr, and CuI, respectively [3.2, 4].

Evidence of Cu3d mixing is revealed in photoelectron emission studies. A photoelectron emission measurement with excitation by a monochromatic x-ray showed that the valence band Γ_{15} consisted of two bands, each of which had a small structure [3.5]. Since the efficiencies of the photoelectron emission from d- and p-states show different dependences on the excitation energy further work has been carried out varying the photon energy of the exciting ultraviolet light. The results show that the relative intensity of the two bands changes greatly with increasing excitation photon energy and thus the upper part of the valence band is confirmed to arise mainly from the Cu3d state [3.6].

The $Z_{1,2}$ exciton made up of a $\Gamma_8 (j_h = 3/2)$ hole and Γ_6 conduction electron $(j_e = 1/2)$ and the Z_3 exciton consisting of a Γ_7 hole $(j_h = 1/2)$ and Γ_6 electron have the following states of different total angular momentum J:

$$Z_{1,2} : \Gamma_8 \times \Gamma_6 = \Gamma_3(J = 2), \Gamma_4(J = 2), \Gamma_5(J = 1) ,$$

$$Z_3 \; : \Gamma_7 \times \Gamma_6 = \Gamma_2(J = 0), \Gamma_5(J = 1) .$$

The $\Gamma_5(J = 1)$ state splits into a longitudinal exciton(L) and a doubly degenerate transverse exciton(T). The Γ_3 and Γ_4 states are doubly and triply degenerate, respectively. Exciton states having $J = 2$ or 0 are observable under an external perturbation such as a magnetic or stress field.

The fundamental absorption of CuCl, CuBr, and CuI has been measured by *Cardona* [3.2] in the vacuum ultraviolet region down to 30 eV. The correspondence of the absorption to the band structure derived from the x-ray excited photoelectron emission and from a calculation is shown in Fig. 3.1 [3.5]. Sharp doublet peaks, Z, are the first members of the two series of the S-state excitons.

3.2 Exciton Absorption, Reflection, and Emission Spectra

3.2.1 Absorption and Reflection Spectra

In Fig. 3.2, a detailed absorption spectrum of the $Z_{1,2}$ and Z_3 band regions is shown for an evaporated CuCl film at 4.2 K. The absorption curve duplicates well the first result of the microphotometer curve of a photographic plate by *Nikitine* [3.1]. The LO phonon structures associated with the $n = 2 Z_{1,2}$ band are clearly seen.

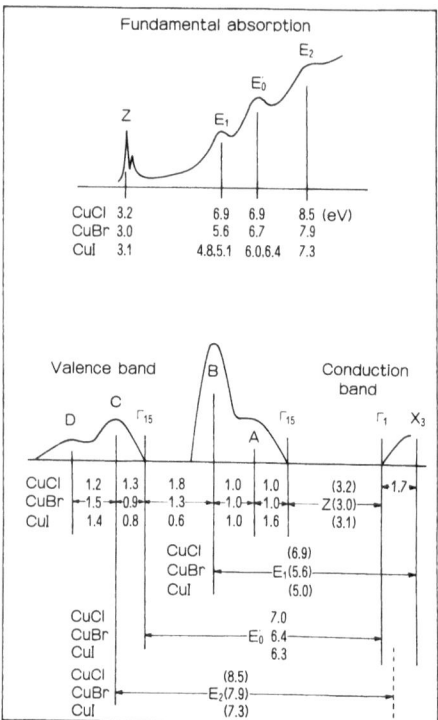

Fig. 3.1. Fundamental absorption and structure of valence bands of cuprous halides [3.2, 5]

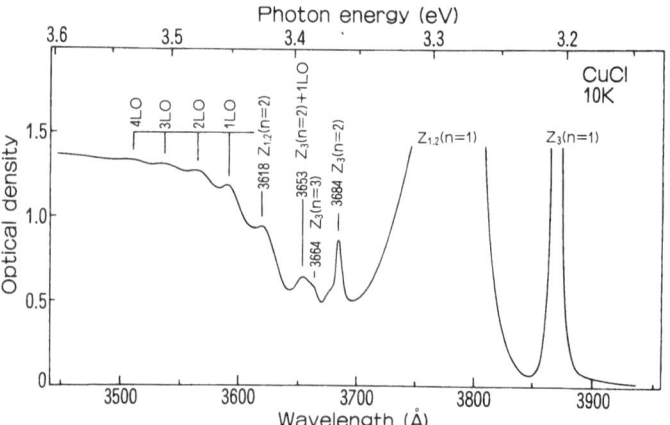

Fig. 3.2. Exciton absorption spectrum of an evaporated CuCl film at 10 K [3.7]

Table 3.1. Exciton parameters and phonon energies etc. in cuprous halides at liquid helium temperature

	CuCl	CuBr	CuI
Exciton energy [eV]			
E_{ex}^T	3.2022[a]	2.9645[b]	3.0586[c]
E_{ex}^L	3.2079[a]	2.9770[b]	3.0647[c]
Exciton binding energy [meV]	213[d] 190[e]	108	62[d]
Exciton reduced mass [m_0]	0.39[d]	0.23[d] 0.26[f]	0.27[d]
Exciton translational mass [m_0]	$m_{ex}^T = 2.3$[a] $m_{ex}^L = 3.1$[a]	$m_{ex}^T \begin{cases} \text{heavy: } 2.6^b \\ \text{light: } 0.9^b \end{cases}$ $m_{ex}^L = 1.0$	
Band-gap energy [eV]	3.416[d]	3.0726[g]	3.122[d]
Phonon energy [meV]			
TO	20.2[h]	16.8[h]	16.6[h]
LO	25.9[h]	20.7[h]	18.7[h]
Spin-orbit splitting of valence band [meV]	60[i]	150[i]	640[j]
ϵ	5.59[a]	5.7[f]	6.2[c]

[a] [3.8]   [b] [3.9]   [c] [3.10]   [d] [3.1]   [e] [3.11]   [f] [3.12]   [g] [3.13]   [h] [3.14]   [i] [3.15]   [j] [3.2]

The absorption line series in CuCl and CuI are expressed by the following hydrogenic series formulas [3.1]:

$$\left. \begin{array}{l} \text{CuCl } Z_3: \ \nu_n = 27560 - 1720/n^2 \ \text{cm}^{-1} \\ \text{CuI } Z_{1,2}: \ \nu_n = 25189 - 504/n^2 \ \text{cm}^{-1} \end{array} \right\} \ n = 1, 2, 3 \ .$$

Exciton absorption energy, binding energy, reduced mass, etc. are listed in Table 3.1 with the phonon energy measured by *Nanba* et al. [3.14].

In CuBr, the S-state exciton absorption lines are not fitted with a hydrogenic series, however, the P-state exciton lines of $n = 2$, 3 and 4, measured by *Mattaush* and *Uihlein* [3.13] through the two-photon absorption method, are found to form a series. Thus, the band-gap energy was precisely determined to be 3.0726 eV. The binding energy in Table 3.1 is given by the difference between the gap energy and the $n = 1 Z_{1,2}$ exciton energy.

3.2.2 Splitting of Exciton Bands by Perturbations

a) **Magnetic Field**. In a magnetic field, a sharp spike of a reflection maximum in the Z_3 band of CuCl was found by *Certier* et al. [3.16] inside the low-energy side of the Γ_5 band. The spike is assigned to $\Gamma_2(J = 0)$ which is allowed due to mixing

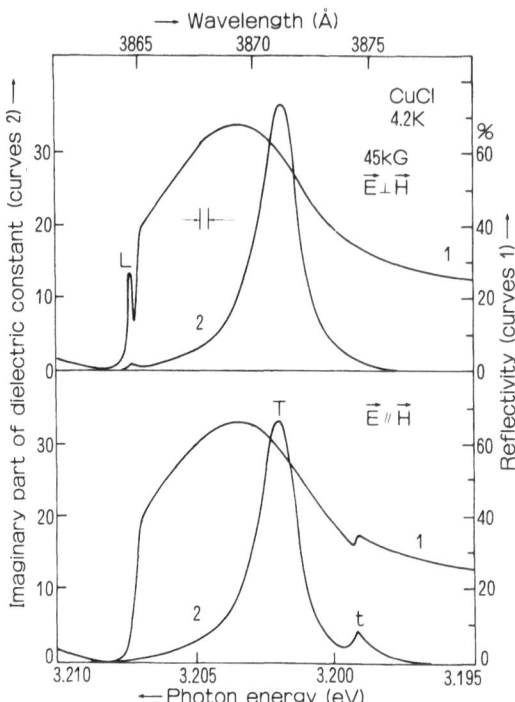

Fig. 3.3. Magneto-reflectivity and the imaginary part of the dielectric constant of CuCl, measured for $E\|H$ and $E\perp H$ configurations [3.17]

with $\Gamma_5(J = 1)$. As shown in Fig. 3.3, *Staude* [3.17] has examined the polarization character of the reflection spectrum of CuCl by measuring it with the E vector of the incident light parallel and perpendicular to the applied field. The imaginary part of the dielectric constant, which is obtained through the Kramers-Kronig transformation, is also shown. A small peak on the low-energy side is due to the Γ_2 exciton and another spike on the high energy side in the reflection spectrum is concluded to be due to the longitudinal exciton because of its appearance in the $E\perp H$ configuration. The L–T splitting of the exciton is obtained as 5.4 meV. The energy difference between states of $J = 0$ and $J = 1$ gives the exchange interaction energy of the electron and hole in the exciton and it is obtained as $\Delta = 8.7$ meV.

b) Stress Effect. *Koda* et al. [3.18] have examined reflection spectra of CuCl under the application of stress for two configurations, which are with the E vector of the incident light parallel and perpendicular to the stress field. The Z_3 band splits into three; the transverse, longitudinal, and Γ_2 states. The $Z_{1,2}$ band shows splitting due to the lifting of the orbital degeneracy of the Γ_8 valence band. A remarkable property in this case is that the longitudinal exciton band is observed under the stress and its intensity increases with increasing stress until it becomes more intense than the Γ_5 allowed state band. These stress effects are explained by *Sakoda* and *Onodera* [3.19] by introducing higher-order perturbations which show that the exciton energy has a term linear in k. The theory expects the $Z_{1,2}$ band to split into six components.

c) **Electric Field.** *Mohler* [3.20] has examined the reflection spectrum in an electric field with the expectation that the first-order Stark effect of the exciton band in CuCl would be observed since the Γ_7 valence band is a mixed state of Cu d- and Cl p-states. A sinusoidal electric field of frequency f is applied and the reflection change is measured with a lock-in amplifier synchronized to frequency f or $2f$. Thus the first- and second-order effects can be measured separately. The first-order effect is related to the splitting due to the exchange interaction and it is determined as $\Delta = 7.4$ meV. The second-order effect has shown no anisotropy which is reasonable for the S-state exciton.

3.2.3 Emission Spectra

a) **CuCl.** Figure 3.4 shows emission spectra of CuCl and CuBr obtained by *Goto* et al. [3.21]. Generally the spectrum at 4.2 K consists of two narrow lines at 3871

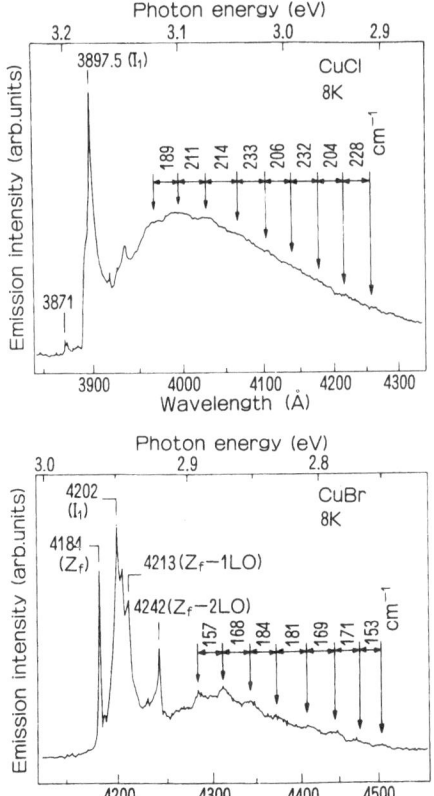

Fig. 3.4. Emission spectra of CuCl and CuBr [3.21]

and 3898 Å and a broad band having LO phonon structure in the case of excitation into the band-to-band transition region or into the Z_3 exciton band. The emission at 3871 Å is due to the radiative decay of the Z_3 exciton and its line shape is modified by self-absorption, so that the emission peak is apparently shifted somewhat to the low-energy side of the exciton absorption peak. This emission is called hereafter the free-exciton emission. Another narrow line at 3898 Å is very intense at low temperatures. It is due to an exciton bound at a neutral acceptor which is called the I_1 exciton. Both the emissions of the free exciton and the I_1 bound exciton show the same temperature shift of 16.8 meV to higher energies when the temperature increases from 4 K to 80 K.

The I_1 emission decreases in intensity with rising temperature and disappears above ~ 70 K. On the other hand the intensity of the free-exciton emission is almost constant up to 150 K but decreases rapidly above it. Temperature variations of these emission intensities are shown in Fig. 3.5 [3.21]. Curve a shows the total intensity of the two narrow lines, free and I_1 bound excitons, measured together. The curve consists of two portions, the low-temperature part is for the I_1 emission and the higher-temperature part is for the free-exciton emission. The separation into two curves, shown by dotted lines, was made using the observation that both lines had the same intensity at 55 K.

The rapid decrease of the free-exciton emission above 150 K is expressed by a well-known formula of thermal quenching:

$$\eta = \eta_0 [1 + A \exp(-E/k_B T)]^{-1},$$

where E is an activation energy of the thermal quenching. From the slope of the curve, E is calculated to be 0.15 eV which is nearly in agreement with the exciton binding energy.

The thermal quenching of the free-exciton emission above 150 K is also verified from the rapid increase of the photoconductivity above 150 K shown by curve b in Fig. 3.5, measured by the excitation into the exciton absorption band region. Thus the exciton is considered to decompose into a free electron and hole above 150 K. In a similar way the activation energy of the I_1 emission quenching is obtained as 0.07 eV, much larger than the binding energy of the I_1, 0.02 eV. Furthermore the thermal quenching of the I_1 does not cause the increase of the free-exciton emission, hence the decrease of the I_1 emission with temperature does not mean the thermal release of free excitons.

The broad emission is found to be of the pair recombination type and its intensity variation with temperature shown by curve c in Fig. 3.5 resembles that of the I_1 emission. From these facts the decrease of the I_1 emission with a temperature rise is ascribed to the thermal instability of the neutral acceptor responsible for the I_1 bound exciton.

In an evaporated film at 8 K the broad band ranging from 3950 to 4600 Å is found to consist of two bands. One of these is intense with a peak at 4030 Å and an asymmetric band shape tailing off to the low-energy side, while the other is weak with a symmetric line shape having a peak at 4250 Å [3.22]. The first band

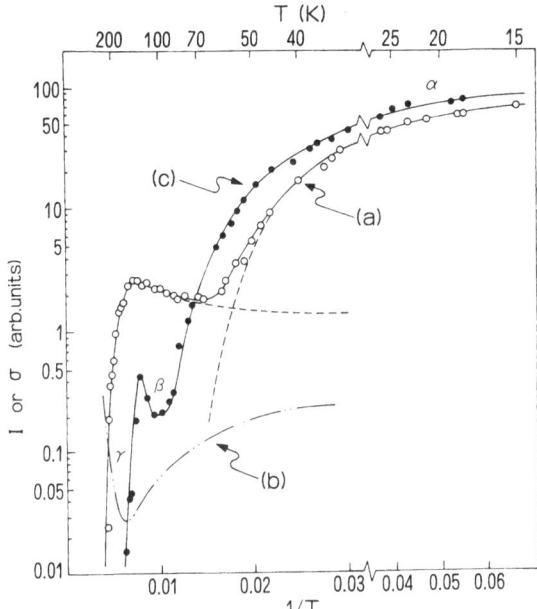

Fig. 3.5. Temperature variations of emission intensity and photoconductivity in CuCl [3.21]

decreases in intensity rapidly and disappears above 55 K, as shown in the low-temperature part (α to β) of curve c in Fig. 3.5, and its temperature variation is quite similar to that of the I_1 emission. The 4250 Å band remains up to 150 K as shown in the higher-temperature part (β to γ), and thermally quenches above it just like the free-exciton emission.

Regarding the free-exciton emission, the Z_3 exciton emission is clearly observed but the case for the $Z_{1,2}$ exciton emission is not so clear. In an evaporated sample the $Z_{1,2}$ exciton emission seemed to be found at 4 K by excitation into the band-to-band region below 3600 Å [3.22]. There was some question as to whether the observed emission might be due to stray light of the excitation source being reflected in the exciton band region owing to the reflection maximum at the $Z_{1,2}$ band peak. A careful search for this effect of stray light seemed to confirm that the $Z_{1,2}$ exciton emission occurred. There is no reason why the $Z_{1,2}$ exciton does not radiatively decay, although nonradiative scattering into the low-lying Z_3 exciton state should be more probable as inferred from its wider band width. In crystalline samples the $Z_{1,2}$ emission is not observed because of the high rate of self-absorption.

b) CuBr. The emission consists of the free-exciton emission at 4184 Å, an intense I_1 bound-exciton emission at 4202 Å and a broad band of pair recombination type. The 4184 Å line is located at the same energy as the Z_1 absorption, which is split from the $Z_{1,2}$ by strain due to the contraction difference between the film and the fused quartz substrate, so the emission has previously been assigned as the Z_1 resonance emission [3.23]. The emission has finally been assigned as the resonance emission of the lowest-energy triplet exciton. This exciton emission is usually forbidden, so it is named the Z_f emission. The broad band extends from 4250 to 5000 Å and consists of two bands, one of which is intense and has an asymmetric shape with a tail to the low energy side with a peak at 4280 Å. This band decreases in intensity with rising temperature. The other smaller intensity band, which has a symmetric line shape with a peak at 4460 Å, remains stable up to 130 K. The temperature variation of the 4280 Å band is similar to that of the I_1 emission. These properties of the broad band resemble those in CuCl.

The broad band emissions in CuCl and CuBr have structures separated by 210 and 169 cm^{-1} which are the LO phonon energies. As seen in Fig. 3.4 a few additional narrow lines are found at 4213 and 4242 Å, being separated by 164 and 326 cm^{-1} respectively from the Z_f emission. Thus, the lines are ascribed to the one- and two-phonon side bands of the Z_f exciton emission.

Figure 3.6 shows the temperature variation of emission spectra in a region covering the Z_f and its phonon side bands and a reflection spectrum at 8 K [3.21]. A small but sharp spike is seen in the reflection spectrum on the low-energy shoulder of the $Z_{1,2}$ exciton band. The energy of the spike is exactly the same as the Z_f emission line. The line shape of the two-phonon side band, which is asymmetric, with a tail on the high-energy side, and becomes broader with temperature, is similar to that found in CdS [3.24] and can be expressed as

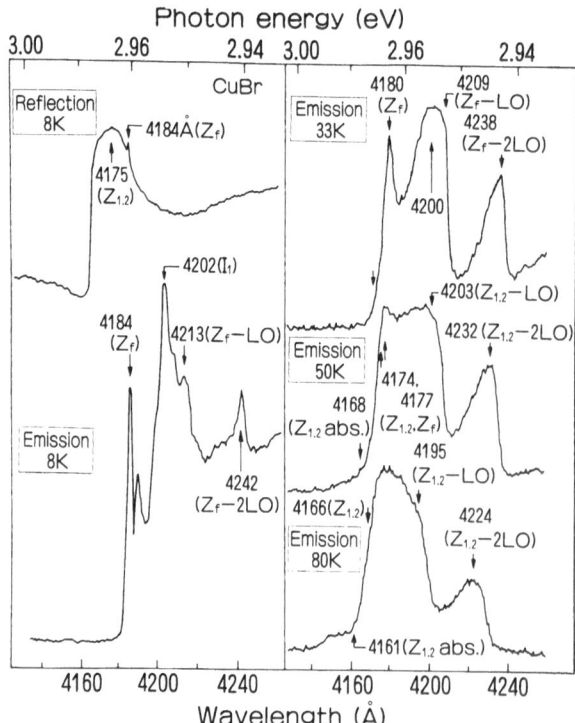

Fig. 3.6. Reflection and emission spectra of a CuBr crystal [3.21]

$$I(E) \propto E^{1/2} \exp(- E/k_B T) ,$$

where E is measured from the low-energy edge. Thus the line shape shows that the Z_f exciton is in thermal equilibrium with the lattice. The Z_f exciton state is due to the splitting of the $Z_{1,2}$ exciton and is ascribed to a state belonging to Γ_3 or Γ_4.

A comparison of the two spectra at 33 and 50 K shows that the Z_f exciton almost vanishes and the radiative decay of the allowed exciton (Γ_5) and its phonon side bands appear near 50 K. From these facts it is concluded that the exciton emission in CuBr arises from the lowest-energy triplet state at temperatures below 50 K and above that the triplet state is thermally raised to the allowed Γ_5 state and the Γ_5 state resonance emission dominates.

3.2.4 Phonon Structure in the Excitation Spectra of Free-Exciton Emission

The LO phonon structure is also found in the excitation spectra of the free-exciton emission. Figure 3.7 shows excitation spectra of emissions in CuBr for (a)

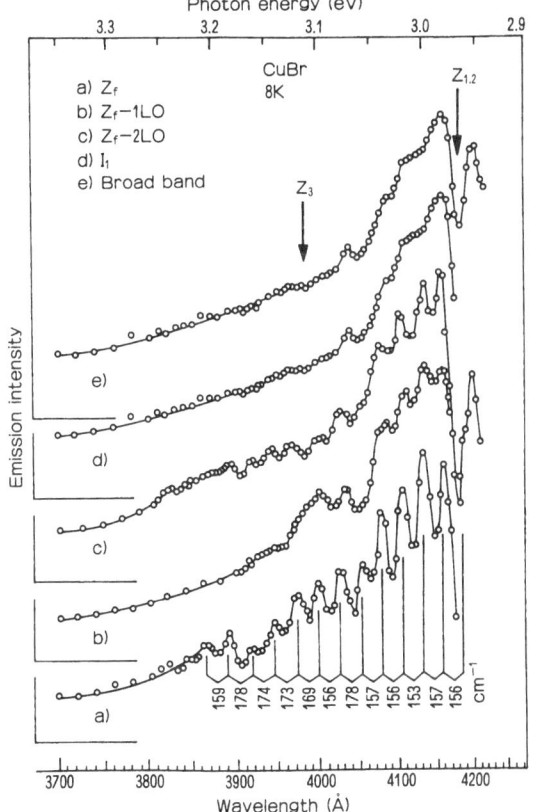

Fig. 3.7. Excitation spectra of free- and bound-exciton emissions and a broad band of pair recombination type in CuBr [3.21]

free exciton Z_f, (b) Z_f – LO, (c) Z_f – 2LO, (d) I_1 bound exciton, and (e) broad band [3.21]. In case (a) the LO phonon structure is easily seen, with a 164 ± 8 cm^{-1} interval. In the cases of the bound-exciton and broad band emissions shown in (d) and (e) the phonon structure is not observed. The fact that structures in the cases of the phonon side bands, curves (b) and (c), are not so remarkable compared to (a) is due to the overlap of the I_1 and broad emissions with the Z_f – LO and Z_f – 2LO lines, respectively. The temperature variation of the excitation spectrum of the free-exciton emission indicates that the phonon structure becomes obscure with rising temperature and it vanishes at 50 K [3.25].

The explanation of why the LO phonon structure is seen only for the free-exciton emission is given in terms of the theory of the direct radiative annihilation of the exciton by *Toyozawa* [3.26]. The incident radiation in the exciton band region is converted to the so-called exciton polariton inside the crystal and it has two modes, the upper and lower branches. The energy of the polariton is shown in Fig. 3.8 as a function of wave vector k. Here E_0 is the energy of a polarization wave of the exciton in the absence of the interaction between the crystal medium and the radiation field. The broken line represents the relation between the energy and wave vector of the photon in vacuum. The absorption of a photon, the energy of which is designated by a_0, creates an upper-branch polariton at a, which is then scattered to state b of the lower-branch polariton. The polariton thus formed is thermalized from b to c through phonon and point-defect scattering. The thermalized polariton can be annihilated as a photon at d (direct radiative annihilation) or move in the crystal to be ionized or annihilated nonradia-

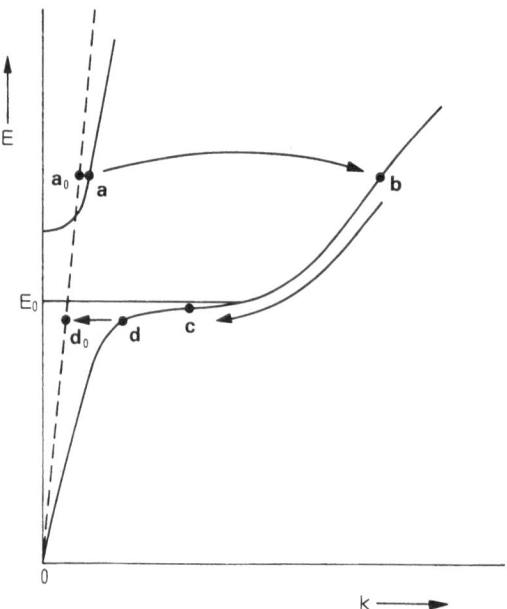

Fig. 3.8. Energy diagram of a system consisting of a crystal and radiation

tively at imperfections, since the thermalized state has a longer lifetime. If the incident radiation energy is below the bottom of the upper-branch polariton, it generates the lower polariton directly.

Here two assumptions are necessary. First, most inelastic scattering is caused by the LO phonon. Second, the probability for a thermalized exciton to be ionized or annihilated at imperfections is larger than that for it to be annihilated as a photon at d after being scattered from b to c. Under these assumptions if the energy difference between the upper polariton state a and the lower polariton state d is an integral multiple of the LO phonon energy, the cascade scattering of the polariton into state d occurs very efficiently and it is annihilated as a photon d_0. If this energy difference is not an integral multiple of the phonon energy, the polariton is scattered and thermalized in the state near c in the lower-branch polariton and then it is ionized or trapped at imperfections before reaching state d. Thus the LO phonon structure is expected in the excitation spectrum of the free-exciton emission.

Since the cascade scattering by the LO phonon into state d enhances the resonance emission, it reduces the thermalization of the exciton. Therefore, the phonon structure would also be expected in the excitation spectrum of the I_1 and broad emissions and in the photoconductivity as well, in the opposite phase to that for the resonance emission. However the structures are not observed. If the efficiency for creating the thermalized exciton is very high, the relative magnitude of reduction due to the resonance scattering will be too small to be detected. In such a case the phonon structure can be observed only in the resonance emission.

As mentioned before, the braod band emission is ascribed to the pair recombination of a trapped electron and hole. An excitation spectrum of the emission shows that excitation into the exciton band is effective for enhancing the emission with nearly the same yield as excitation into the band-to-band region. This fact shows that the exciton can be decomposed into a free electron and hole, other than by radiative decay, through the interaction with impurities or defects and this contributes to the photoconduction and pair emission.

3.2.5 Bound Excitons

As mentioned in Sect. 3.2.3, CuCl and CuBr crystals have an intense bound-exciton emission at 3898 and 4202 Å, respectively, at 4.2 K. It is observed even in the thinnest evaporated film and is also seen as an absorption line in single crystals with exactly the same energy as for emission. In crystals it is most intense and insensitive to the sample preparation and its absorption coefficient amounts to 10^3 cm^{-1}. Thus, the origin of the bound exciton is considered to be intrinsic and it is named the I_1 exciton. The designation of I_1 for the 3898 Å line in CuCl comes from the Zeeman splitting of its emission or absorption line, measured by *Certier* et al. [3.27]. The line splits into four components, the outer two of which are π components and inner two are σ components. This fact shows that the line

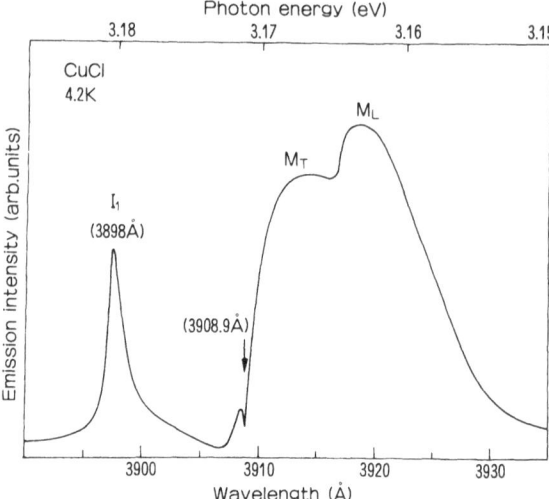

Fig. 3.9. Emission spectrum of CuCl with nitrogen laser excitation at 4.2 K, showing a narrow dip at 3908.9 Å due to a bound exciton [3.28]

is ascribed to an exciton bound to a neutral acceptor. The g factors of the ground and excited states are determined as

$$g_0 = 1.1 \quad \text{and} \quad g_{exc} = 2.05 .$$

The former is the g factor of a hole and the latter is that of an electron.

Another remarkable bound exciton was found by *Anzai* et al. [3.28] when CuCl crystals were highly excited by a laser beam at a low temperature. As described later in Sect. 3.3.3, broad emissions, M_L and M_T, due to the radiative decay of the EM are caused by laser illumination. At the high-energy edge of the M_T band, a sharp intensity minimum or dip is observed at 3908.9 Å at 2 K as shown in Fig. 3.9. The Zeeman splitting of this intensity minimum or dip shows four components, the two outer component lines are π and the inner two are σ as in the case of the I_1 line, but the intensity change of the split components with changing magnetic field strength and temperature indicates that the ground- and excited-state splittings in this case originate from the electron and hole, respectively, which is opposite to the I_1 exciton. Thus the intensity minimum or dip is concluded to be caused by the creation of a bound exciton at a neutral donor and is named the I_2 exciton in analogy with the case in CdS. The g factors of the electron and hole in this case are the same as those in I_1.

The Cu^+ and Cl^- ion vacancies are considered to be the most probable defects for the neutral acceptor and donor which bind excitons in forms of the I_1 and I_2, respectively. Since the I_2 bound exciton is observed only when the crystal is under laser irradiation, the neutral donor, which is analogous to the F center in alkali halides, might be observed through its absorption band under intense laser excitation.

In a previous report a bound exciton of a little broader width was observed at 3890 Å in CuCl and it has been tentatively named as I_2 [3.29]. Its nature is not yet clear.

3.2.6 CuCl–CuBr Solid Solutions

The electron and hole exchange interaction is found to play an important role on the relative intensity of $Z_{1,2}$ and Z_3 exciton bands in solid solutions of CuCl–CuBr. The relative position of the $Z_{1,2}$ and Z_3 exciton bands in CuCl is opposite to that in CuBr. As the solid solution of the CuCl–CuBr system has been found to belong to the amalgamation type [3.2], the exciton band consists of two bands throughout the whole range of composition and their absorption peaks shift continuously as the Br ion concentration is changed. Curves of shifts of the $Z_{1,2}$ and Z_3 band peaks with respect to the Br concentration, x, seem to cross each other at $x = 0.23$ and hence here the spin-orbit splitting becomes zero.

Takahashi and *Goto* [3.30] have found that the relative intensity of the two bands, measured photographically, varies a great deal and in a solid solution at $x = 0.23$, the intensity of the exciton band on the low-energy side becomes very weak. *Kato* et al. [3.15] investigated absorption bands in evaporated films of $Cu(Br_xCl_{1-x})$ solid solutions in detail, photoelectrically, as shown in Fig. 3.10 for some examples. In CuCl the absorption area of the higher-energy band, $Z_{1,2}$, is larger than that of Z_3 and with increasing x the difference in absorption area of two bands becomes much larger. In CuBr the area of the lower-energy band, $Z_{1,2}$, is larger than that of the Z_3 band but with increasing Cl concentration the area of Z_3 becomes larger than that of $Z_{1,2}$. That is to say with increasing separa-

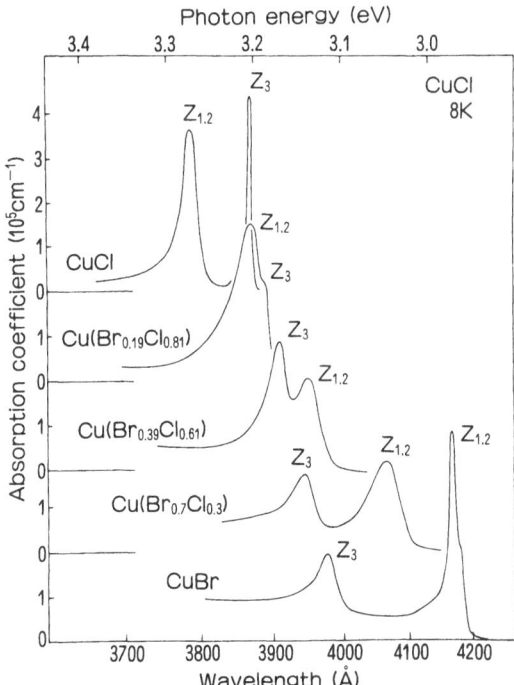

Fig. 3.10. Absorption spectra of the $Z_{1,2}$ and Z_3 exciton bands in CuCl–CuBr solid solutions, measured at 8 K [3.15]

tion between two bands or with increasing spin-orbit splitting, the area of the higher-energy band becomes greater than that of the lower-energy one regardless of whether it is $Z_{1,2}$ or Z_3. In CuCl and CuBr the exciton band shapes are asymmetric Lorentzian as in the case of alkali halides. In the solid solutions of cuprous halides in which two exciton bands overlap, each band is also assumed to have the same asymmetric Lorentzian shape and the overall absorption curve is analysed into two bands. The absorption area of each band is obtained in the following way [3.31].

The absorption coefficient, k_λ, is expressed by the following formula when the exciton bands coexist in the same spectral range and if the refractive index is assumed not to change in both band regions:

$$k_\lambda = \sum_j k_{0j}(\Gamma_j/2) \cdot \frac{\Gamma_j/2 + 2A_j(E_\lambda - E_{0j})}{(E_\lambda - E_{0j})^2 + (\Gamma_j/2)^2} ,$$

where Γ_j is the band width, A_j the order of asymmetry, E_{0j} the energy gap of the optical transition taking account of the self-energy of the exciton in the phonon field, and k_{0j} is the absorption coefficient due to the j-th exciton band at E_{0j}. The summation is taken over the two $Z_{1,2}$ and Z_3 bands in this case. The absorption curve can be separated into two bands by parameter fitting. The absorption area of each band thus resolved is obtained as [3.32]

$$\int_{E_1}^{E_2} k_0(\Gamma/2) \cdot \frac{\Gamma/2 + 2A(E_\lambda - E_0)}{(E_\lambda - E_0)^2 + (\Gamma/2)^2} \, dE_\lambda .$$

Here, E_1 is the photon energy at which the absorption coefficient of a given band is zero, i.e., $E_1 = E_0 - \Gamma/4A$, and E_2 is the photon energy corresponding to the absorption band of the first excited state of the exciton. The values E_0, the energy gaps for the optical transition plus the self-energy of the phonon, are plotted in Fig. 3.11 a as a function of x. The two solid curves do not cross but repel each other at $x = 0.23$. The repulsion of two curves is in accordance with a theoretical expectation studied by *Knox* and *Inchauspe* [3.33]. The energy separation between two exciton bands, α, which depends on both the spin-orbit splitting of the hole, λ, and the electron-hole (e–h) exchange energy, Δ, is given by

$$\alpha = \sqrt{(\Delta - \lambda/3)^2 + (8/9)\lambda^2} . \qquad (3.1)$$

In the solid solution $Cu(Br_{0.23}Cl_{0.77})$, in which λ becomes zero, α is equal to the exchange energy and it does not approach zero. The exchange energy in this case is found to be 10 meV.

The relative absorption area of two exciton bands in solid solutions are plotted in Fig. 3.11 b by open circles as a function of x. The intensity ratio of absorption bands is found to deviate a great deal from 2:1 around $x = 0.23$ which is expected from the multiplicities of Γ_8 and Γ_7 holes. The variation of the relative intensity in Fig. 3.11 b is explained well by a theory of *Onodera* and *Toyozawa*

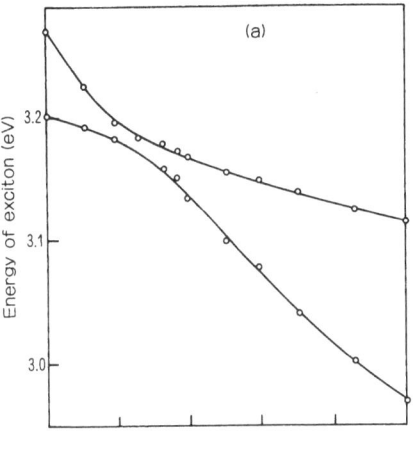

Fig. 3.11. Energy shifts and the variation of relative intensity of exciton bands in CuCl–CuBr solid solutions with changes of Br concentration [3.15]

[3.34]. According to the theory the relative intensity β can be expressed in terms of Δ and λ as follows:

$$\beta \equiv \frac{I(Z_{1,2})}{I(Z_{1,2}) + I(Z_3)} = \frac{1}{2}\left(1 \pm \sqrt{\frac{(3\Delta - \lambda)^2}{8\lambda^2 + (3\Delta - \lambda)^2}}\right),\qquad(3.2)$$

where the $+$ and $-$ signs apply to $\lambda/(3\Delta - \lambda) < 0$ and $\lambda/(3\Delta - \lambda) > 0$, respectively. For $\Delta = 0$, β is equal to 2/3 which is independent of λ.

For the case where $\Delta \neq 0$, β changes as a function of λ and approaches zero or unity in the region of $|\lambda| \ll \Delta$. This is shown in the figure. Values of λ and Δ are obtained as functions of x using α and β obtained experimentally, and λ is found to change smoothly with x from -60 meV in CuCl to $+150$ meV in CuBr but Δ does not change much. Values of β, calculated from (3.2) using λ and Δ thus obtained, are shown by a solid line in Fig. 3.11 b which is in good agreement with the experimental result. Thus, the system of the CuCl–CuBr solid solution is the unique example in which the exchange effect reveals itself remarkably because, at a certain composition, the spin-orbit splitting becomes zero.

3.3 High-Density Excitation Effects

Nonlinear emissions of ionic semiconductors under high-density excitation by laser or electron beam irradiation have been studied in many types of crystals since 1965. The dependence of the emission intensity upon the excitation density, and the energy separation between the emission peak and the exciton band, have led to proposals of varieties of interaction involving the exciton. In ZnO, CdS, and CdSe crystals, nonlinear emissions have been observed which are considered to arise from exciton-exciton, exciton–trapped-electron or hole, exciton–free-charge-carrier, exciton–bound-exiton interactions etc. In CuCl and CuBr crystals, a superlinear luminescence band called the M band, which is attributed to the radiative decay of EMs, has been found for the first time. As the binding energy of the EM is relatively large it is stable up to 50 K. As mentioned before, free excitons become unstable above 150 K and decompose to electrons and holes. Thus, the exciton-exciton interaction leads to the formation of EMs at low temperatures while at higher temperatures exciton-free charge carrier interactions occur and a nonlinear emission called the H band is enhanced.

3.3.1 Exciton-Electron Interaction

a) CuCl. With the excitation of a N_2 laser beam at 150 K, CuCl crystals showed the free-exciton emission at 3840 Å and a new broad band at 3900 Å together with an unknown emission named N_3 [3.35]. On raising the temperature from 150 K the free-exciton emission and the N_3 band shift to the high-energy side but the newly found emission, called the H band, shifts to the low-energy side. The H band has the following properties:

i) The band appears only when the crystal is highly excited by a laser beam and the band intensity changes nonlinearly with the excitation intensity.
ii) The temperature shift of the band is opposite to that of the exciton band.
iii) The band appears at temperatures above 150 K.
iv) The band shape is asymmetric with a tail to the low-energy side.

The energy separation of the H band from the exciton band, $E_{ex} - \hbar\omega_H$, is found to increase linearly with temperature as expressed by

$$\hbar\omega_H = E_{ex} - 5.4\, k_B T .\tag{3.3}$$

Emissions similar to the H band have been reported in CdS and CdSe [3.36] and the exciton-electron inelastic collision is proposed to explain the linear temperature shift of the emission peak. In CuCl the effective mass ratio of electron and hole, m_e/m_h, was considered to be very small, e.g., 0.02 [3.37] or 0.019 [3.38]. Hence, the electron is assumed to participate in the collision. When the free exciton and electron collide, the momentum of the exciton is given to the electron and it becomes hot. Simultaneously the exciton is annihilated radiatively. In this inelastic collision the energy and momentum conservation laws yield the following relations:

$$\left(E_{ex} + \frac{\hbar^2 K^2}{2\,m_{ex}}\right) + \left(E_{ex} + R + \frac{\hbar^2 k^2}{2\,m_e}\right)$$

$$= \left(E_{ex} + R + \frac{\hbar^2 k^{*2}}{2\,m_e}\right) + \hbar\omega_H \tag{3.4}$$

$$K + k = k^* + \omega/c \tag{3.5}$$

where m_{ex} is the total mass of the exciton, E_{ex} the exciton energy at $K = 0$, R the binding energy of the exciton, K the momentum of exciton, and k and k^* are the electron momenta in the initial and final states, respectively. Also ω and c are the angular frequency of the emission and the velocity of light, respectively.

As the wave vector of light is much smaller than K or k, (3.5) can be written as

$$K + k = k^* \, . \tag{3.6}$$

At the time of the previous report m_e was considered to be very small compared to m_h, so that the approximation $k^* = K$ was made together with the assumptions that the kinetic energy of the hot electron was much larger than that of the exciton and electron in the initial state, and in the thermal equilibrium of the initial state the kinetic energy of the exciton and electron were of the same order. By these rough approximations (3.4) can be simplified as

$$\hbar\omega_H = E_{ex} - \frac{\hbar^2 k^{*2}}{2\,m_e} = E_{ex} - \frac{m_{ex}}{m_e}\frac{\hbar^2 K^2}{2\,m_{ex}} \, . \tag{3.7}$$

The line shape of the H band, I_H, is determined by the dependence of the exciton density upon its kinetic energy $E_K = \hbar^2 K^2/2m_{ex}$ which follows the Maxwell distribution in thermal equilibrium and is expressed by

$$I_H = \alpha\sqrt{E_K}\exp\left(-E_K/k_B T\right) \, , \tag{3.8}$$

where α is a constant independent of E_K. The maximum value of I_H occurs at

$$E_K = k_B T/2 \, . \tag{3.9}$$

Thus, the following equation is derived from (3.7) for the emission peak energy:

$$\hbar\omega_H(\text{peak}) = E_{ex} - (m_{ex}/2m_e)k_B T \, . \tag{3.10}$$

In CuCl the experimental relation (3.3) is in agreement with (3.10) when

$$(m_{ex}/2m_e) = 5.4 \, . \tag{3.11}$$

The linear temperature shift of the H band is thus explained by exciton-electron inelastic collisions.

From the hydrogenic series of the exciton bands, the reduced mass of the Z_3 exciton is obtained as

$$\mu = 0.39 \, m_0 \tag{3.12}$$

With (3.11 and 12) the effective masses of the electron and hole are determined as

$$m_e = 0.43 \, m_0 \quad \text{and} \quad m_h = 4.2 \, m_0 \,.$$

The m_e value is almost equal to 0.4 m_0 obtained by *Nikitine* et al. [3.37] but is larger than 0.2 m_0 obtained by *Song* [3.38]. The m_h is considerably smaller than 20 m_0 [3.37] or 13 m_0 [3.38].

Apart from the linear temperature shift of the H band there are several other characteristic properties as mentioned before. These properties are well explained by the exciton-electron collision. First, the asymmetric shape is explained qualitatively by (3.7) and (3.8). The kinetic energy of the exciton which follows the Maxwell distribution is involved as a negative term in (3.7), therefore the expression for $\hbar\omega_H$ has a tail to the low-energy side. Second, the fact that the H band is emitted only when the crystals are excited at high density is easily understood by the exciton-electron collision. Third, the photoconductivity shows the rapid increase of the conduction-electron density above 150 K. The nature of the H band, which appears at temperatures above 150 K, is understood by the increase of the electron density.

b) CuBr. In CuBr crystals an emission ascribed to the H band is observed at 4265 Å at 194 K. The band appears above 120 K with N_2 laser excitation. The extrapolation to 0 K of a straight line showing the relation between $E_{ex} - \hbar\omega_H$ and temperature gives 37 meV, hence the linear shift of the H band is expressed by

$$\hbar\omega_H = E_{ex} - 37 - 3.0 \, k_B T \, [\text{meV}] \,. \tag{3.13}$$

c) CuI. The H band is found at 4137 Å at 84 K and it shifts to 4173 Å at 167 K. In this temperature region CuI crystals show no emission other than the H band and the $Z_{1,2}$ exciton emission. The peak energy of the H band is given by

$$\hbar\omega_H = E_{ex} - 45 - 2.6 \, k_B T \, [\text{meV}] \,. \tag{3.14}$$

In our previous report, 37 meV and 45 meV in (3.13) and (3.14) were assumed to be the binding energies of donor electrons E_D, and the H bands in CuBr and CuI were ascribed to exciton–trapped-electron inelastic collisions. The interaction process yields the next relation for the peak energy of the H band:

$$\hbar\omega_H(\text{peak}) = E_{ex} - E_D - (m_{ex}/2m_e)k_B T \,. \tag{3.15}$$

Comparing it with the experimental expression, (3.13), we get $(m_{ex}/2m_e) = 3.0$. In CuBr the reduced mass is not well known but if we take $\mu = 0.23\ m_0$ [3.37] the effective masses are obtained as $m_e = 0.28\ m_0$ and $m_h = 1.4\ m_0$. In CuI, with $\mu = 0.27\ m_0$ [3.37] and $(m_{ex}/2m_e) = 2.6$, m_e and m_h are obtained as $0.33\ m_0$ and $1.4\ m_0$, respectively.

Hönerlage et al. [3.39] investigated the temperature dependent emission due to free-exciton–electron inelastic scattering in CuBr and CuI and rhe results were fit to calculated curves taking into account the polariton effect and the momentum dependent scattering matrix element. The results show that the temperature shift of the H band is not linear in the whole range and the shift depends on the electron density. The analysis of their data gave $m_e = 0.28\ m_0$, $m_h = 1.4\ m_0$ for CuBr and $m_e = 0.30\ m_0$, $m_h = 2.4\ m_0$ for CuI. These values are in very good agreement with those mentioned above; in particular, the values in CuBr obtained by the two groups are exactly the same. In CuI, m_e is very close to $\mu = 0.28\ m_0$, so that the small difference of m_e causes a large difference in m_h. In CuCl, *Hönerlage* et al. [3.39] used data from [3.35] and obtained $m_e = 0.50\ m_0$ and $m_h = 2.0\ m_0$. These values are reasonable because the exciton mass (T exciton) has been precisely determined to be $2.3\ m_0$ as noted later in Sect. 3.10.1. With $\mu = 0.39\ m_0$, m_e and m_h are determined to be $0.50\ m_0$ and $1.8\ m_0$, respectively. In CuBr, the exciton has a heavy mass of $2.6\ m_0$ (Sect. 3.10.2) and μ is estimated to be $0.26\ m_0$ [3.12]. Using these values, m_e and m_h are determined as $0.29\ m_0$ and $2.3\ m_0$, respectively. These values are listed in Table 3.2.

In the above treatment k was neglected in (3.6) because m_e was considered to be much smaller than m_h. Since the difference between m_e and $m_{ex} = m_e + m_h$ has now been found not to be large, k is not negligible. If one uses the relation (3.6), $(m_{ex}/2m_e)$ in (3.11) is slightly modified and a simple calculation gives $m_e = 0.48\ m_0$ and $m_h = 2.01\ m_0$ in agreement with the values obtained by *Hönerlage* et al. [3.39]. It is remarkable that in spite of the very rough approximations made the resulting values of effective mass are in excellent agreement with those obtained through detailed calculation.

Table 3.2. Effective masses of electrons and holes in units of the free electron mass

		Yu et al.[a]	*Hönerlage* et al.[b]	*Kato* et al.[c]	From exciton reduced and effective masses
CuCl	m_e	0.48	0.50	0.44	0.50 $\}$ $m_{ex} = 2.3$[d]
	m_h	2.01	2.00	3.6	1.8 $\}$ $\mu = 0.39$[e]
CuBr	m_e	0.28	0.28	0.28	0.29 $\}$ $m_{ex} = 2.6$[f]
	m_h	1.4	1.4	1.3	2.3 $\}$ $\mu = 0.26$[g]
CuI	m_e	0.33	0.3		
	m_h	1.4	2.4		

[a] [3.35] [b] [3.39] [c] [3.7] [d] [3.8] [e] [3.1] [f] [3.9] [g] [3.12]

In CuBr and CuI the H band had been ascribed to the exciton–donor-electron interaction. However there is little possibility of the stable existence of a donor at such a high temperature. It has been pointed out (Sect. 3.10.2) that in CuBr the exciton dispersion has a term linear in K and the energy of the H band extrapolated to 0 K is expected to lie below the exciton energy. *Hönerlage* et al. [3.39] have also pointed out that the H band shift is not linear over the whole temperature region, which makes the extrapolation of H band peak deviate from the exciton energy.

3.3.2 Effect on Exciton Absorption Bands

Keldysh and *Kozlov* [3.40] and *Hanamura* [3.41] have shown that the repulsive force between two excitons due to the Pauli exclusion principle and Coulomb interaction is more prominent than the attractive force due to the exchange interaction, so that the energy necessary for adding one more exciton to the majority of excitons increases as the exciton density does. This means that the exciton absorption peak shifts to the high-energy side with increasing exciton density. Conversely, the band gap is suggested to become smaller because of a strong attractive Coulomb interaction between an exciton and an electron.

The experimental results in this section, obtained by *Kato* et al. [3.7], concern a quantitative study of the change in peak energy of an exciton band with respect to the laser excitation power and the comparison with Hanamura's theoretical works. The samples used were CuCl crystals which were evaporated onto fused quartz plates. A pulsed nitrogen laser beam was divided into two beams, one of which was used to excite the samples to generate excitons and the other was focused on a quartz cell filled with AVCO-17/66 dye solution. The spontaneous emission from the dye solution, which has no structure and has a spectral distribution covering the exciton band region, was made to be incident on the samples in the area irradiated with the nitrogen laser beam by adjusting the spatial and temporal coincidence of two beams. The emission from the dye solution was focused on the entrance slit of a monochromator after being passed through the sample. The spontaneous emission from the dye solution had so rapid a rise and decay that one could measure the transmitted light intensity only while the sample was being excited by the laser light. The laser light intensity on the crystal surface was ~ 400 KWcm^{-2}.

Since the spontaneous emission transmitted through the sample is much stronger than the emission from the sample due to exciton annihilation, the emission from the crystal does not influence the absorption measurements.

Figure 3.12 shows exciton absorption spectra of a CuCl film of 0.05-μm thickness at 10 K during laser excitation with various powers. With increasing laser power both exciton bands, Z_3 and $Z_{1,2}$, shift to the high-energy side and their band widths become larger. The energy shifts of both bands have the same laser power dependence and they are almost linearly proportional to the laser power but the peak shift of the Z_3 band, $\Delta E(Z_3)$, is larger than that $\Delta E(Z_{1,2})$, of $Z_{1,2}$.

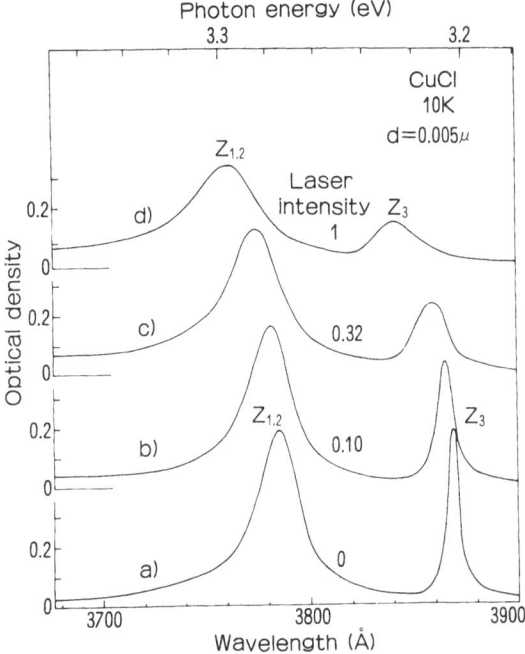

Photon energy (eV)

CuCl
10K
d=0.005μ

Fig. 3.12. Exciton bands of an evaporated CuCl film at 10 K during irradiation with a nitrogen laser light of various powers [3.7]

The peak shift of the Z_3 band amounts to 24 meV with maximum power. The ratio of peak shifts of both bands is given by

$$\Delta E(Z_{1,2})/\Delta E(Z_3) = 0.89 .$$

According to Hanamura's theory the energy necessary to create one more exciton when N excitons exist is given as

$$E = E_0 + 42\pi E_{ex}^b a_B^3 N - (100/3)\pi E_{ex}^b a_B^3 N \qquad (3.16)$$

where E_0, E_{ex}^b, and a_B are the formation energy, binding energy, and Bohr radius of the exciton, respectively. The second and third terms represent the energy shift owing to the repulsive interaction due to the Pauli exclusion principle and the attractive exchange interaction between two excitons, respectively. If the spins of the electron and hole in an exciton and the multiple scattering between two excitons are taken into account, the energy shift is written as

$$\Delta E = E - E_0 = 8\pi \frac{\mu}{m_{ex}} E_{ex}^b a_B^2 f_0 N . \qquad (3.17)$$

The factor f_0 is the amplitude of a zero momentum scattering and is nearly equal to $3.3a_B$ in CuCl. The above relation is derived with the assumption that an additional exciton is created by being associated with the same conduction and valence bands as those of the thermalized N excitons and that there is no binding between excitons. The number of EMs formed is found to be small compared to the exciton density, in the present case of the film, even at the highest laser power, as confirmed through the observation of the EM emission. Moreover, almost all the excitons are thought to be thermalized into the Z_3 state at low temperatures as it is 72 meV lower energy than the $Z_{1,2}$ state. As the sample is very thin, the thermalized excitons are considered to be distributed uniformly inside the sample. Under these conditions the density of the Z_3 exciton, N, is proportional to the laser power. Thus the experimental results are in good agreement with (3.17).

The difference of the peak shift between Z_3 and $Z_{1,2}$ excitons is treated by Hanamura as follows. The wave function of the 1s exciton with momentum K is written as

$$|1s, K\rangle = \sum_{P,p} \sqrt{V} \delta_{K, P-p} F_{1s}(P, p) a_{cP}^+ a_{vp} |g\rangle \quad \text{where,} \tag{3.18}$$

$$F_{1s}(P, p) = \frac{8\sqrt{\pi a_B^3}}{[1 + \hbar^{-2} a_B^2 (\alpha P + \beta p)^2]^2}, \quad \alpha = m_h/(m_e + m_h), \; \beta = 1 - \alpha. \tag{3.19}$$

Here P and p are the momenta of a conduction electron and a missing electron in the valence band, respectively, V is the volume of the crystal, a_{cP} and a_{vp} are the electron annihilation operators in the conduction and valence bands, respectively, and $|g\rangle$ is the wave function of the ground state. The Fourier transformation of the envelope function, $F_{1s}(P, p)$, represents the e–h relative motion of an exciton. For the $K = 0$ exciton, $P = p$ and (3.19) is written as

$$F_{1s}(P, p) = \frac{8\sqrt{\pi a_B^3}}{(1 + \hbar^{-2} a_B^2 P^2)^2}.$$

This equation shows that $F_{1s}(P, p)$ has a large value when

$$|P| = |p| < \hbar/a_B.$$

This means that the energy surface of the Γ_6 conduction electron having kinetic energy below $\Delta E_c = (\hbar^2/2m_e)(1/a_B)^2$ and that of the Γ_7 valence band having kinetic energy below $\Delta E_v = (\hbar^2/2m_h)(1/a_B)^2$ are cut down, so that the Z_3 exciton is formed with binding energy

$$E_{ex}^b = \Delta E_c + \Delta E_v.$$

This effect is thought to give the repulsive force between Z_3 excitons and thus the Z_3 exciton energy increases by

$$\Delta E(Z_3) = 8\pi \frac{\mu}{m_{ex}} (\Delta E_c + \Delta E_v)(3.3 \, a_B) a_B^2 N.$$

In the case of the $Z_{1,2}$ exciton, only the Γ_6 electron comprising the $Z_{1,2}$ exciton is predicted to feel the repulsive interaction. The Γ_8 valence band is not affected by the existence of the Z_3 excitons. Therefore the shift of the formation energy of the $Z_{1,2}$ exciton is given by

$$\Delta E(Z_{1,2}) = 8\pi \frac{\mu}{m_{ex}} (\Delta E_c)(3.3\ a_B)a_B^2 N\ .$$

The theoretical ratio of the shifts of both exciton-band peaks is thus obtained as

$$\Delta E(Z_{1,2})/\Delta E(Z_3) = \Delta E_c/(\Delta E_c + \Delta E_v) = \mu/m_e\ . \tag{3.20}$$

From the experimental side the ratio is 0.89. With $\mu = 0.39\ m_0$ the effective masses are calculated to be $m_e = 0.44\ m_0$ and $m_h = 3.6\ m_0$.

In the case of CuBr the shifts of both exciton bands are also linearly proportional to the laser intensity and the shift of the low-lying $Z_{1,2}$ band is larger than that of Z_3. The ratio of shifts is given by $\Delta E(Z_3)/\Delta E(Z_{1,2}) = 0.82 = \mu/m_e$. Using $\mu = 0.23\ m_0$, the effective masses are determined as $m_e = 0.28\ m_0$ and $m_h = 1.3\ m_0$. These masses in CuCl and CuBr are listed in Table 3.2 and are in close agreement with $m_e = 0.48\ m_0$, $m_h = 2.01\ m_0$ for CuCl and $m_e = 0.28\ m_0$, $m_h = 1.4\ m_0$ for CuBr obtained in Sect. 3.3.1.

In CuCl crystals the exciton bands also shift to the high-energy side with a temperature rise. Therefore the observed shift with laser excitation might be attributed to the heating effect of laser irradiation. In the emission spectrum the I_1 exciton emission was always observable and the position of its peak did not change even at the highest laser power. The I_1 peak is known to shift by 0.43 nm with a temperature rise from 10 to 25 K. Further, shifts with temperature of the peaks of both exciton bands were examined through the analysis of the reflection spectrum of a single crystal, and both bands showed the same temperature shift of 0.284 meVK^{-1}. From these facts it can be confirmed that the observed exciton-band shift under laser excitation can be attributed to the existence of the high-density excitons and not to the heating effect.

The exciton-band shift has been discussed with the assumption that the EM is not formed. The assumption is confirmed, as mentioned before, by the emission study showing the small intensity of the EM emission compared with that of the free-exciton emission. A large concentration of defects is involved in a thin evaporated sample which inhibits excitons from diffusing and recombining with each other to form the EM.

In the present experiment the sample excitation was into the band-to-band region. This means that the first step of the excitation causes the creation of free e–h pairs and they recombine to form excitons. In the relaxation processes leading to exciton formation many phonons are emitted. The existence of the high-density free charge carriers and phonons should affect the line shape of the exciton band, in fact the exciton band becomes broader with increasing laser power and the increase in the width amounts to 15 meV in both Z_3 and $Z_{1,2}$

bands, which is too large to be attributed to collisions between excitons. The absorption area of both bands remains almost unchanged. A theoretical treatment of line shape under the existence of excitons, free charge carriers and phonons is required while experimental studies, time resolved within the lifetimes of free charge carriers and excitons and also the excitation into the exciton band, are expected to make the analysis simpler.

3.3.3 Creation of the Excitonic Molecule by Exciton-Exciton Collision

Since 1958 when *Lampert* [3.42] and *Moskalenko* [3.43] independently pointed out the possibility of the existence of the EM, which consists of two electrons and two holes, experimental and theoretical work concerning the molecular complexes consisting of an exciton and a neutral donor or acceptor has been reported [3.44]. From theoretical work dealing with the binding energy of the EM [3.45], it was considered that the EM is unstable in crystals in which the e–h mass ratio $\sigma = m_e/m_h$ lies between 0.2 and 0.4. Therefore experimental work was directed toward crystals such as CuCl having $\sigma = 0.02$ (now it is 0.3) and Si ($\sigma = 0.45$) or Ge ($\sigma = 0.67$). A new theory by *Akimoto* and *Hanamura* [3.46] shows that any crystal can have EMs. Nonlinear emissions were found at 1.08 and 0.708 eV in Si and Ge, respectively, and they were assigned to the radiative decay of EMs leaving a free electron and hole pair behind. However, they are now attributed to e–h drops. In 1969 the Sendai and Strasbourg groups independently found a new emission band in CuCl at 392 nm, the intensity of which varies as the fourth power of the ruby laser intensity. The laser light is absorbed in the crystal through the two-photon process [3.47], hence the new emission intensity changes as the square of the excitation density. The energy difference between the new emission peak from the Z_3 exciton, was nearly in agreement with the binding energy of the EM calculated with an incorrect old theory. Here an exciton is considered to be left behind in the radiative annihilation of the EM.

In highly excited states, a number of elementary interactions involving excitons and free or trapped charge carriers can lead to nonlinear emissions which are located at the low-energy side of the exciton band. Thus, the square law dependence of the emission intensity upon the excitation density and its emission energy, only, are not enough to ascribe the emission to a certain definite interaction.

The first reliable proof of the radiative annihilation of the EM was given by *Souma* et al. [3.48] in CuCl through the line-shape analysis of a new band at 3919 Å, which they named the M band, which appeared when the crystal was excited with a giant ruby laser at 4.2 K as shown in Fig. 3.13.

The M band has an asymmetric line shape with a tail to the low-energy side and the spectral distribution of the emission intensity $J(E)$ is fitted to the following Maxwell distribution:

$$J(E) = j_0 E^{1/2} \exp(-E/k_B T) \tag{3.21}$$

with $T = 26$ K. Here, E is measured from the high-energy edge, $\hbar\omega_0$, towards

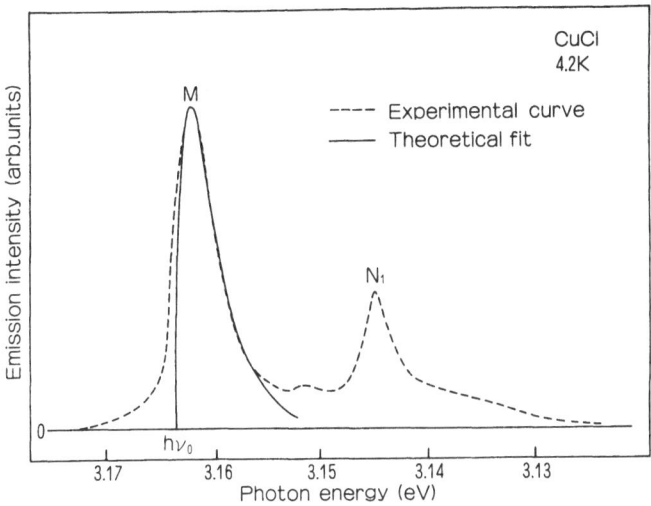

Fig. 3.13. Line shape of the M emission band due to the radiative decay of EMs in CuCl excited by a giant ruby laser at 4.2 K, showing the Maxwell distribution of the EM [3.48]

the low-energy side. For the exciton or its LO phonon side band, the emission line shape is also expressed by the Maxwell distribution, but in this case the band has a tail to the high-energy side.

Considering that the M band is emitted by the radiative annihilation of the EM leaving an exciton behind, the energy scheme for the transition between EM and exciton states is as shown in Fig. 3.14. Assuming that the dependence of the kinetic energy on the momentum k is parabolic and the total mass of the EM is twice that of an exciton, the kinetic energies of exciton and EM are expressed by

$$E_{ex}^k = \hbar^2 k^2 / 2m_{ex} \quad \text{and} \quad E_m^k = \hbar^2 k^2 / 4m_{ex} .$$

Neglecting the momentum of an emitted photon, the transition occurs vertically due to the momentum conservation law.

The emitted photon energy due to the transition is

$$\hbar\omega_m = (E_m^0 - E_{ex}^0) - \hbar^2 k^2 / 4m_{ex} . \tag{3.22}$$

The superscript 0 indicates the energy at $k = 0$. If the EMs are in thermal equilibrium they follow the Maxwell distribution. The density of the EMs as a function of their kinetic energy, E, is proportional to $E^{1/2} \exp(-E/k_B T)$, where $E = \hbar^2 k^2 / 4m_{ex}$. If the vertical transition probability is independent of k, the emission intensity $J(\hbar\omega_m)$ will be proportional to the density of the thermalized EMs in the initial state, i.e.,

$$J(\hbar\omega_m) \propto E^{1/2} \exp(-E/k_B T) .$$

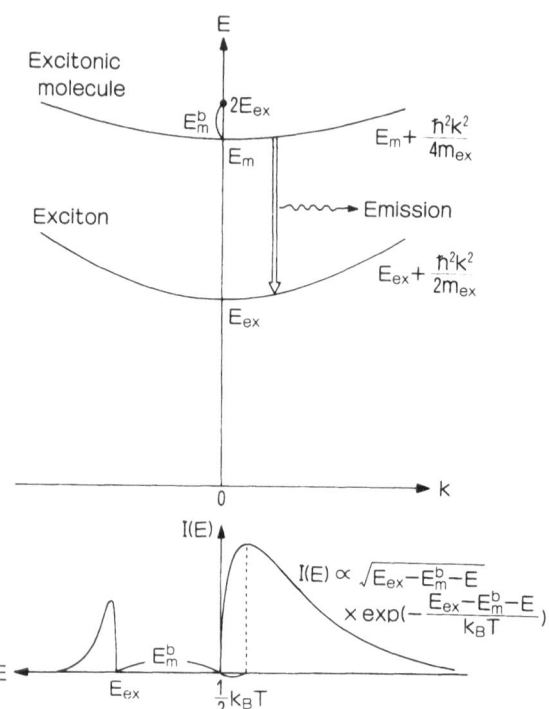

Fig. 3.14. Schematic energy diagram (*upper figure*) of an exciton and EM showing the radiative decay process of the EM, leaving an exciton behind. Line shape (*lower figure*) of the EM emission band

Using (3.22) the emission intensity is expressed as

$$J(\hbar\omega_m) \propto [(E_m^0 - E_{ex}^0) - \hbar\omega_m]^{1/2}$$

$$\times \exp\{-[(E_m^0 - E_{ex}^0) - \hbar\omega_m]/k_BT\} . \tag{3.23}$$

According to this equation the M band has a peak at $E_m^0 - E_{ex}^0 - k_BT/2$ with a tail to the low-energy side. Thus the observed line shape of the M emission band is explained by the radiative decay of EMs in a process shown in Fig. 3.14. From the peak position of the M band, the effective temperature is found to be 26 K indicating that the EM system is in thermal equilibrium at a higher temperature than that of the lattice. As the EM mass has been determined recently to be 2.3 m_{ex}, as noted in Sect. 3.10.1, the denominator 4 m_{ex} in (3.22) must be replaced by 3.5 m_{ex}. However the correction is not essential for the discussion of the line shape showing the Maxwell distribution.

The binding energy of the EM is obtained as 34 meV from the energy difference between the exciton energy and the high-energy edge of the M band.

Wehner [3.45] has derived a formula giving the ratio of the binding energies of the exciton and EM as follows:

$$E_{mol}^b/E_{ex}^b = (0.346 - 0.764\sqrt{\sigma})(1 + \sigma) .$$

According to the formula $E_{mol}^b > 0$ with $\sigma < 0.22$. *Souma* et al. reached the conclusion that the experimentally obtained value of E_{mol}^b is in near agreement with the calculated one by taking $E_{ex}^b = 190$ meV and the old value of $\sigma = 0.02$ [3.37], however, the most reliable values at present are $m_e = 0.5\, m_0$ and $m_h = 1.8\, m_0$ and thus $\sigma = 0.28$. In this case E_{mol}^b becomes negative and the EM is not formed. According to theories by *Akimoto* and *Hanamura* [3.46] and *Huang* [3.49], E_{mol}^b is 11 meV or 30 meV, respectively.

In the case of relatively weak excitation, the EM emission is found to consist of two bands, M_L and M_T [3.50], which are ascribed to the two kinds of exciton states, longitudinal and transverse, which are left behind in the EM radiative decay. The energy difference between the two bands is in fact equal to that of the L–T splitting of the exciton, being 5.4 meV. In CuBr the EM emission has another band called M_f corresponding to the recoil of the spin triplet Γ_4 exciton [3.51].

By fitting the formula (3.23) to the measured line shapes of the M_L and M_T bands, the emitted photon energies corresponding to $E_{mol}^0 - E_{ex}^0$ are found to be

$$3.164 \text{ eV} (M_L) \quad \text{and} \quad 3.169 \text{ eV} (M_T) \quad \text{for CuCl},$$

$$2.929 \text{ eV} (M_L) \quad \text{and} \quad 2.940 \text{ eV} (M_T) \quad \text{for CuBr}. \tag{3.24}$$

For the M_f band the determination was not possible because of the overlap of bound-exciton emissions. The binding energy of the EM in CuBr is determined to

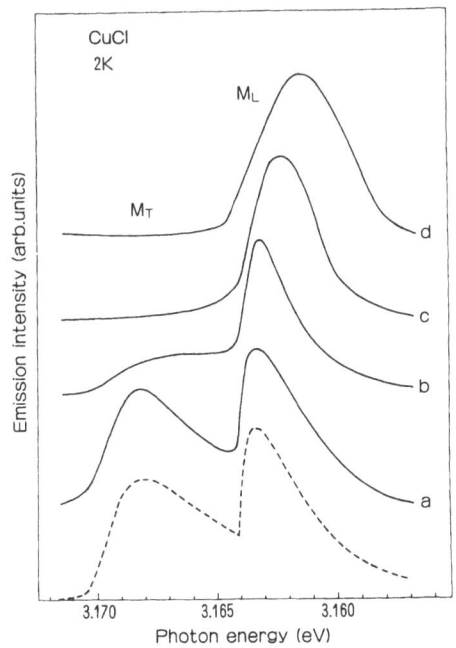

Fig. 3.15. Variation of the emission spectrum of the EM with increasing excitation intensity (*a*) to (*d*), measured at 2 K. The dashed curve shows the calculated line shape for a weak excitation with $T_{eff} = 24$ K

be 25 meV from the difference between $E_{ex}(T) = 2.965$ eV and the above 2.940 eV (M_T).

The dependence of the EM emission spectrum upon the excitation power is shown in Fig. 3.15 for CuCl. With weak excitations the M_T intensity is twice that of M_L because of the double degeneracy of the transverse exciton state. Since the L exciton relaxes rapidly to the T exciton state, the population of the T exciton becomes large with increasing excitation power, the optical transition from the T exciton to the EM state occurs [3.52], and the M_T emission intensity saturates. On the other hand the M_L band intensity continues to increase. With the highest excitation power the M_L band has a symmetric Lorentzian shape caused by mutual collisions among the EMs [3.53].

A dashed curve shows a calculated M band shape at T = 24 K using the correct values of EM and exciton masses; $m_{mol} = 5.29\ m_0$, $m_{ex}^L = 3.14\ m_0$, and $m_{ex}^T = 2.3\ m_0$ as obtained in Sect. 3.10.1.

3.4 Giant Two-Photon Excitation of the Excitonic Molecule

3.4.1 Evidence of Giant Two-Photon Creation

Hanamura [3.54] has shown that EMs can be generated directly by the giant two-photon absorption (GTA) using photons with energy

$$\hbar\omega = E_{ex} - \tfrac{1}{2}E_{mol}^b ,\tag{3.25}$$

(for details, see Sect. 2.3.4.). The absorption coefficient for the two-photon generation of EM depends on the photon density, and it amounts to of the order of $\sim 10^5$ cm^{-1} using a photon density of 10^{15} photons/cm^2. *Gale* and *Mysyrowicz* [3.55] have confirmed the two-photon generation of EMs in CuCl by observing that the absorption coefficient for the two-photon absorption increases rapidly when the incident photon energy approaches that given by (3.25).

Nagasawa et al. [3.51] confirmed the efficient generation of the EM by the two-photon excitation in CuCl as well as in CuBr through the investigation of excitation spectra of the EM emission as follows. The excitation spectra for the M_L band are shown in the upper curves of Fig. 3.16. Sharp excitation peaks are observed at 3.186 eV and at 2.953 eV in CuCl and CuBr, respectively. These photon energies are almost the same as those expected from (3.25) by adopting the known values; $E_{ex} = 3.203$ eV and $E_{mol}^b = 34$ meV for CuCl and $E_{ex} = 2.965$ eV and $E_{mol}^b = 25$ meV for CuBr. Thus, the EM was found to be generated effectively by the GTA.

When CuCl and CuBr crystals are excited into the narrow peaks in the excitation spectra, sharp emission bands are observed in each of the M_L, M_T, and M_f

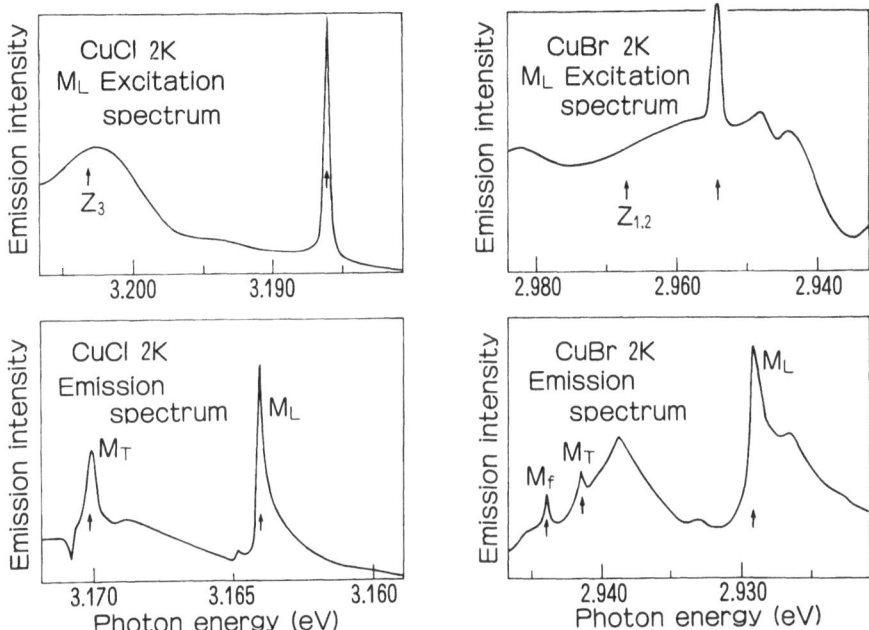

Fig. 3.16. (*Upper curves*) Excitation spectra of the M_L emission due to the EM in CuCl and CuBr at 2 K. (*Lower curves*) Emission spectra with the crystal excited into narrow peaks in the upper curves [3.51]

band regions as shown in the lower curves of Fig. 3.16. The photon energies of these bands are:

M_L: 3.1640 eV   and M_T: 3.1701 eV   for CuCl ,
M_L: 2.9291 eV, M_T: 2.941 eV(M_T), M_f: 2.9437 eV   for CuBr.

Comparing the above results with (3.24), one can see that the sharp emissions of CuCl and CuBr in the M_L band region exactly coincide in energy with the high-energy edge of the M_L band with Maxwell distribution, but the one at the M_T band region is located about 1 meV to the high-energy side of the edge of the M_T band. The energy discrepancy was later found to be due to the polariton effect mentioned in Sect. 3.5.2.

Since the EM has a boson nature, and the directly generated EM has small kinetic energy, the Bose condensation of it is expected to be realized as mentioned in Sect. 2.5.5. Hence these sharp emission lines were once ascribed to the Bose condensation of the EM at $k \sim 0$.

When one uses a dye laser having a wide spectral width, such as 2 meV, a side band appears, with increasing excitation density at the low-energy side of the sharp emission, as shown by curve a in Fig. 3.17 for the case of CuCl. For comparison, the emission due to excitation into the Z_3 exciton band is shown by

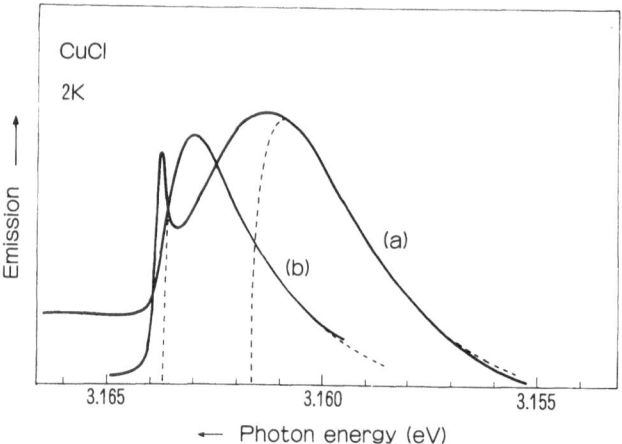

Fig. 3.17. Line shapes of M_L emission bands in CuCl excited (*a*) at exciton absorption band; (*b*) at two-photon absorption peak with high excitation level, at 2 K. The dashed curve shows the fitting of (3.21) [3.51]

curve b, the line shape of which is expressed well by (3.21). Moreover, the sharp line does not shift with changes of excitation energy. From these facts, along with the temperature and excitation density dependences of the intensity of the sharp emission, it was previously considered that the majority of the EMs generated by the two-photon absorption might have condensed at a state $k \sim 0$ and decayed radiatively from this state, and that the side band of curve a might indicate the collective motion of condensed molecules [3.56].

There have been stimulating discussions concerning the sharp emission line, as to whether it is to be ascribed to two-photon Raman scattering (Sect. 3.5.1) or to be assigned to resonant luminescence (Sect. 3.7). The sharp line has now been found to contain both the Raman and luminescence components when the crystal is resonantly excited into the EM state. However, there still remains some possibility of Bose-Einstein condensation as is discussed by *Peyghambarian* et al. [3.57] and this problem has still to be solved.

3.4.2 Giant Two-Photon Absorption

Transmission measurements were performed in the energy region given by (3.25) using an intense tunable laser. In Fig. 3.18 logarithmic absorbances of CuCl at 2 K are shown as a function of the incident one-photon energy, ω_1, for three pump powers: (a) 100%, (b) 45%, and (c) 15% of the maximum power (~ 1 MW cm^{-2}) [3.58]. A narrow peak is found at 3.1861 eV on a continuous background due to the exciton-band tail. The peak energy coincides exactly with that of the expected GTA. Since the absorption occurs as the result of a two-photon process, the absorption area increases with the increase of the excitation level.

The line shape is symmetric when the laser intensity is weak but the line width is broadened, with increasing intensity, especially to the high-energy side. The half-width varies from 0.15 meV to ~ 1 meV. The energy difference between the GTA and transverse exciton (3.203 eV) is 17 meV, thus the binding energy of the EM is finally determined precisely as 34 meV. The absorption coefficient at the peak was estimated to be $500 \sim 1000$ cm^{-1} with a dye laser of maximum power. Detailed analysis and the origin of change in the line shape dependent on the pump power will be discussed later in Sect. 3.9.

Confirmation that the absorption is really caused by the two-photon process can be obtained using two dye laser beams having broad and narrow bandwidths, both of which are pumped by the same N_2 pulsed laser. The broad band laser was used as the probe light for measuring the transmission of the crystal. The energy of the EM is obtained as 6.3722 eV from the sum of the energies of the narrow-band laser and the photon corresponding to the transmission minimum in the broad-band laser. The energy is just twice the energy of the GTA for the single-beam excitation method [3.59].

According to theory (Sec. 2.3.1) the lowest state of the EM in CuCl has Γ_1 symmetry. This state may be obtained by superposing the two-electron and two-hole states with total spin angular momentum of zero. Thus the GTA of the EM is expected to be allowed for excitation by a single beam of linearly polarized light. For excitation with two linearly polarized beams, their polarization should be parallel. Moreover it is also expected that the two-photon transition should be

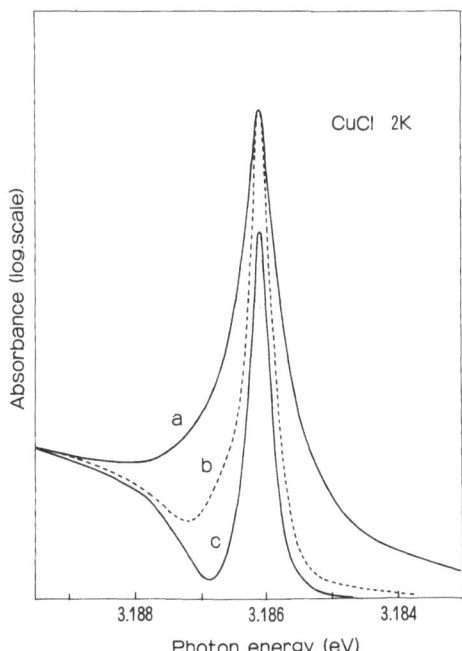

Fig. 3.18. GTA band for the direct creation of the EM in CuCl, measured with a dye laser of various powers at 2 K [3.58]

forbidden for a single beam of circularly polarized light, while, if two light beams have the same (or opposite) circular polarization but almost opposite (or the same) wave-vector direction, the transition should be allowed. These expectations have been verified experimentally [3.60, 61].

By the single-beam excitation of the crystal into the GTA peak, the energy of which is designated as $\hbar\Omega_0$, the EM is directly generated with the definite wave vector $2 k_0$, where k_0 is the wave vector of the incident light Ω_0 inside the crystal. When two laser beams are used having slightly larger and smaller energies than $\hbar\Omega_0$, which collide inside the crystal from opposite directions, the EM is created with a very small wave vector, $0 < k \ll 2k_0$. The EM having larger k extending to $k = 15 k_0$ is realized by stepwise two-photon excitation into the EM via the intermediate polariton state of large k as described in Sect. 3.10.

In the case of CuCl, the EM has $J = 0$ and $J = 1$ states, but only the former is allowed to be attained by two-photon absorption. On the other hand in CuBr, the EM is considered to have three levels $\Gamma_1(J = 0)$, $\Gamma_5(J = 2)$, and $\Gamma_3(J = 2)$ with antiparallel electron spins. These levels are all allowed for the two-photon transition from the ground state. Three transmission minima were found at 2.953, 2.955, and 2.957 eV on the low-energy tail of the exciton band only when the laser light is used for the transmission measurement. The lowest-energy minimum at 2.953 eV coincides exactly with the excitation peak for the emission of the EM. Further, under excitation into each of three minima the crystal shows the same emission of Maxwell-type M bands and three two-photon resonant Raman lines M_L^R, M_T^R, and M_f^R, due to scattering into the longitudinal, transverse, and triplet exciton states [3.62]. Therefore, the three absorption bands were ascribed to the giant two-photon excitation into the three levels of the EM mentioned above. The Γ_1 and Γ_3 states are the lowest- and highest-energy states, respectively [3.63]. The M_T^R line is seen to be a doublet. These M_f^R and doublet M_T^R lines are named later as M_I, M_{II}, and M_{IV} (Sect. 3.10.2).

3.5 Two-photon Resonant Raman Scattering Via the Excitonic Molecule

3.5.1 Backward Scattering

In order to investigate whether the sharp emission lines, which were observed under the GTA excitation and considered to be due to the Bose-condensed EM, might be due to two-photon resonance Raman scattering (TRRS), the secondary emission spectra were measured in a backward scattering configuration for CuCl. The energy of the dye laser having a spectral width of 0.2 meV, much narrower than that used in Sect. 3.4.1, was varied. As shown in Fig. 3.19, the emission consists of the Maxwell-type M_L and M_T bands and sharp lines called M_L^R and M_T^R [3.59]. The sharp lines are found to arise from TRRS in which the EM with energy $\hbar\omega_{mol}$ and exciton with energy $\hbar\omega_f$ are the intermediate and final states,

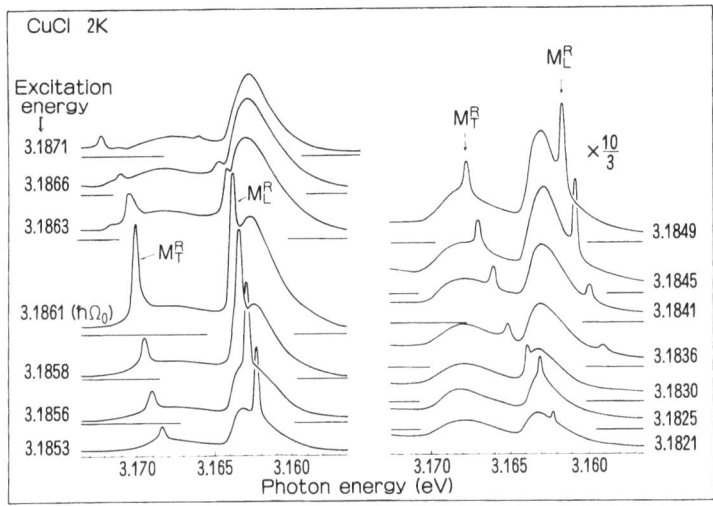

Fig. 3.19. TRRS spectra showing M_L^R and M_T^R formed via the EM state in CuCl for the backward scattering configuration at 2 K [3.59]

respectively, as verified by the fact that the peak energy of the Raman line $\hbar\omega_R$ shifts by twice as much as the change of the excitation energy $\hbar\omega_{inc}$, satisfying the following relation

$$\hbar\omega_R = 2\hbar\omega_{inc} - \hbar\omega_f .$$

The Raman lines, M_L^R and M_T^R, correspond to $\hbar\omega_L$ and $\hbar\omega_T$ for $\hbar\omega_f$, respectively. The peak intensities of these sharp lines are much enhanced around the GTA region ($2\omega_{inc} \sim \omega_{mol}$). Thus the sharp lines in Fig. 3.16 were found to have the characteristics of resonant Raman scattering.

The intensity of the Raman line has a maximum for excitation at the two-photon absorption peak, $\hbar\omega_{mol}/2$, for weak excitation but with higher power excitation it peaks for an excitation energy slightly less than $\hbar\omega_{mol}/2$, as will be mentioned in Sect. 3.9.

When one uses an intense laser beam, another emission line, called L, appears only in the case of high-energy off-resonant excitation and its peak energy shifts linearly with the shift of the excitation laser energy as seen in [Ref. 3.58, Fig. 2]. The emission has been suggested to arise from the scattering of two incident photons into one photon and one longitudinal exciton with a recoil of an EM at $k \sim 0$ to larger k. A more probable explanation is given in Sect. 3.5.5.

Generally speaking the enhancement of the two-photon resonant Raman scattering by light having energy just below that of the exciton is definite evidence of the existence of an EM state having twice that energy. The presence of the EM state has been confirmed by the observation of the TRRS in other materials such as CdS [3.64], ZnSe [3.65], ZnO [3.66], HgI$_2$ [3.67], and PbI$_2$ [3.68].

3.5.2 Forward Scattering

When a CuCl crystal is excited by light at the GTA peak, the incident light propagates inside the crystal with a wave vector $k_0 = 4.44 \times 10^5$ cm^{-1}. In the two-photon Raman scattering via the EM having $k = 2\,k_0$, the scattered light has nearly the same wave vector as k_0, and the final state exiton has $k = \sim 3\,k_0$ for the backward scattering configuration. In this wave-vector region the dispersion of the exciton is small, so that the energy shift of the scattered light with scattering angle is small. For the forward scattering configuration the exciton is left with $k = \sim k_0$. The longitudinal exciton has a negligibly small energy change in a region $0 < k < 2k_0$. In fact the difference of kinetic energy between $k = 2k_0$ and $k = k_0$ is ~ 0.07 meV with the translational mass $m_{ex}^L = 3.14\,m_0$.

On the other hand, the transverse exciton has the character of the polariton expressed by [3.69]

$$\frac{c^2 k^2}{\omega^2} = \epsilon_\infty \frac{\omega_L^2 - \omega^2}{\omega_T^2 - \omega^2} \,, \tag{3.26}$$

where ϵ_∞ is the background dielectric constant. The k dependence of the polariton energy is large in a small k region around k_0. Thus the M_T^R line shifts a great deal with scattering angle α. *Inoue* and *Hanamura* [3.56] calculated the scattering-angle dependence of the energy of M_T^R as shown in Fig. 2.26, where the energies of the two scattered polaritons (1, 1′) etc. were obtained by the following energy and momentum conservation laws, together with (3.26):

$$2\omega_{inc} = \omega_R + \omega_f \quad \text{and} \tag{3.27}$$

$$2k_{inc} = k_R + k_f \,. \tag{3.28}$$

In this case the excitonic polariton left behind as the final state is also photon-like and it comes out of the crystal as a photon. Therefore, in the forward scattering configuration, the M_T^R scattering process is described as a four-polariton process as two polaritons having the same energy and wave vector are scattered into two polaritons having different energies and wave vectors via an intermediate EM state [3.70].

Experimental results are as follows [3.59]. Figure 3.20 shows the scattering configurations with resonant excitation. The laser beams a and b are incident on a crystal with almost the same angle $\theta = 65°$. The emission is measured from the direction normal to the illuminated surface (angle of observation, $\varphi = 0$). When one of the two beams a is cut the spectrum corresponds to that for backward scattering and shows strong M_L^R and M_T^R and weak V lines as shown by curve b. On the other hand the forward scattering spectrum obtained by cutting beam b

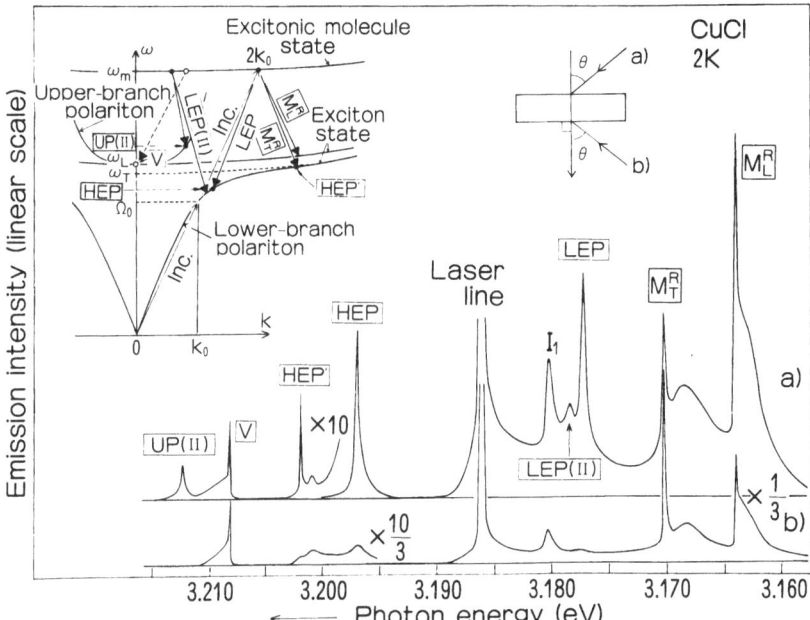

Fig. 3.20. Raman spectra in CuCl resonantly excited into the GTA peak for (*a*) forward and (*b*) backward scattering configurations measured at 2 K. Various scattering processes into the lower and upper branches of the excitonic polariton are shown schematically in the inset [3.59]

shows additional lines named as LEP, HEP, HEP′, LEP(II), and UP(II) in curve a. The M_L^R and R_T^R lines in the forward scattering spectra show no detectable shift from those in the backward scattering ones. The LEP(II), UP(II), and V lines increase in intensity when the crystal is excited with two beams.

The LEP and HEP as a pair form one solution in the forward scattering configuration and the HEP′ and M_T^R are another pair in the backward scattering configuration. However the LEP and HEP observed in a given direction are not true partners because true partners are scattered in different directions, to satisfy momentum conservation.

LEP and HEP lines change their energies with θ and φ. The energies of these lines are plotted in Fig. 3.21 against the internal scattering angle α. Both lines coincide with the excitation laser frequency when $\alpha = 0$, but with increasing α the HEP line shifts toward the high energy side while the LEP line decreases in energy and approaches the M_T^R line for $\alpha = 180°$.

From the energy dependences of the LEP and HEP lines on scattering angle, the disperison of the excitonic polariton has been determined in a small k region and the parameters in (3.26) are obtained as follows taking into account the spatial dispersion of the exciton [3.71] mentioned in Sect. 3.10.1:

$$\omega_L = 3.2079 \text{ eV}, \quad \omega_T = 3.2022 \text{ eV}, \quad \text{and} \quad \epsilon_\infty = 5.59 .$$

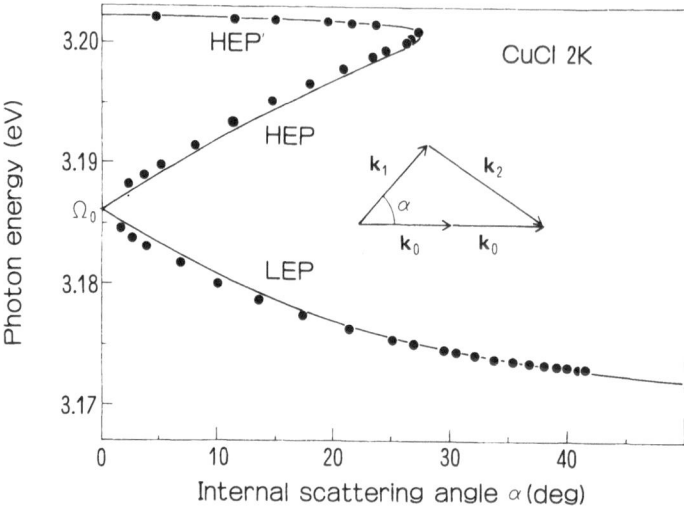

Fig. 3.21. Raman line energies versus internal scattering angle [3.59]

3.5.3 Scattering with Recoil of the Upper-Branch Polariton

When the EM is created with $k < 2k_0$ a new TRRS process occurs into the upper-branch polariton [3.72] as shown in the inset of Fig. 3.20. For the creation of the EM having a small k, the crystal is illuminated from both sides (front and rear surfaces) by two laser beams a) and b) having the same energy $\hbar\omega_{inc}$ and nearly the same incident angle θ. In this configuration of two-beam excitation, the direction of the wave vector k_{mol} of the created EM is along the crystal surface and its magnitude is simply given by

$$k_{mol} = (2\omega_{inc}/c)\sin\theta .$$

The magnitude k_{mol} can be changed from 0 to 0.7 k_0 by varying θ.

As has been mentioned in the previous section, the UP(II) and LEP(II) lines in Fig. 3.20 were much enhanced with two-beam excitation. The LEP(II) is due to a scattering process similar to the one producing LEP: from the EM to the lower-branch polariton state. In the case of LEP(II) the EM is generated by two laser beams with a combined wave vector $k_{mol} < 2k_0$. The UP(II) line is located in the energy region of the upper-branch polariton. With increasing θ, the UP(II) line approaches the longitudinal-exciton energy $\hbar\omega_L = 3.208$ eV and increases its peak intensity. Thus, the UP(II) is due to scattering into the upper-branch polariton from EM of $k_{mol} < 2k_0$. The lower-branch polariton which forms a pair with the LEP(II) or UP(II) is not observed in this scattering configuration.

Figure 3.22 shows the dependences of the UP(II) and LEP(II) lines upon k_{mol}/k_0 for $\varphi = 0°$, by dots, together with theoretically predicted curves. Since, for the UP(II) process, the energies of the related four exciton polaritons are widely different and one must take the contribution from the higher $Z_{1,2}$ exciton state into account, the following modified dispersion relation is used for the calculation:

$$k(\omega) = \frac{\omega}{c}\sqrt{\epsilon'_\infty} \left(\frac{\omega_L^2 - \omega^2}{\omega_T^2 - \omega^2}\right)^{1/2} \cdot \left(\frac{\omega_{L'}^2 - \omega^2}{\omega_{T'}^2 - \omega^2}\right)^{1/2}. \tag{3.29}$$

The relation between the energy of the scattered polariton and k_{mol}/k_0 is calculated as shown by solid lines in the figure. In (3.29) $\hbar\omega_{L'}$ and $\hbar\omega_{T'}$ are the energies of the L- and T-$Z_{1,2}$ excitons, and are given to be 3.287 eV and 3.269 eV, respectively [3.72]. Here ϵ'_∞ is taken to be 4.66. Both of the peak energies decrease as k_{mol} increases. The discrepancy between the theoretical curve and the experimental points for the UP(II) line is mainly brought about by the fact that the value of ϵ'_∞ is still not a constant but a slowy varying function of $\hbar\omega$. The V line is assigned to an emission of an upper-branch polariton from its bottom region as discussed in Sect. 3.8. The TRRS process could later be extended into the more powerful idea of the stepwise two-photon excitation in order to study the spatial dispersion of the exciton polariton and EM which is discussed in Sect. 3.10.

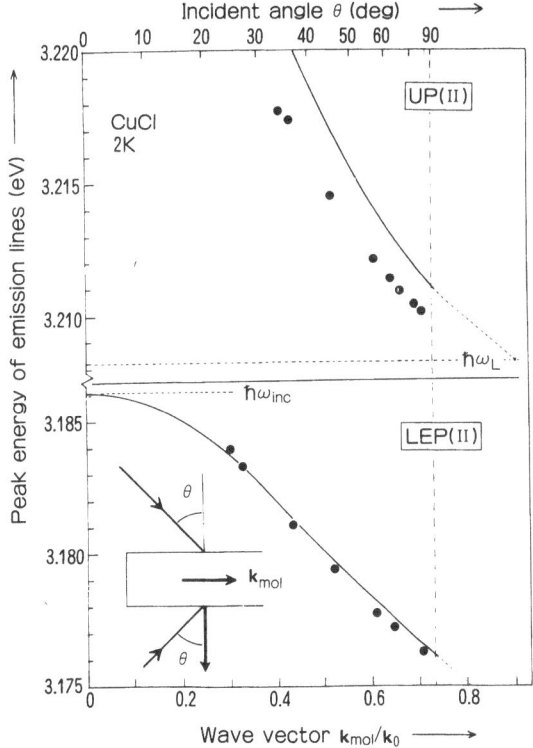

Fig. 3.22 Energies of (a) UP(II) and (b) LEP(II) lines versus k_{mol} for $\varphi = 0°$ (●) and theoretical relations (——) [3.72]

3.5.4 Polarization Character – Geometrical Selection Rules

According to *Henneberger* et al. [3.73], the polarization character of the TRRS process in CuCl can be separated into two individual parts, one for the "virtual absorption" step and the other for the "virtual emission" step, on account of the Γ_1-symmetry character of the intermediate EM state. No correlation is expected between the two steps so far as the polarization character is concerned. Transition probablilities for the emission steps of the TRRS line obey the geometrical selection rules which are proportional to the following factors:

$$1 - |e_R \cdot k_f / |k_f||^2 \quad \text{for the } M_T^R \text{ process}$$

and

$$|e_R \cdot k_f / |k_f||^2 \quad \text{for the } M_L^R \text{ process .} \tag{3.30}$$

Here e_R is the unit vector of polarization of the polariton (light) which appears as the Raman line under observation and k_f is the wave vector of the final state polariton (exciton).

The experiment on the energy dependence of the polarization of the scattered light has been performed under forward scattering geometry with the incident angle θ being changed for the M_T^R process [3.72]. Figure 3.23 shows the degree of polarization (D.P.) of the scattered polariton with respect to the scattering plane versus its energy for $\hbar\omega_{inc} = 3.1861$ eV. The polarization character can be easily analyzed as follows:

$$\text{D.P.} = \left(\frac{2}{\sin^2 \beta} - 1 \right)^{-1}, \tag{3.31}$$

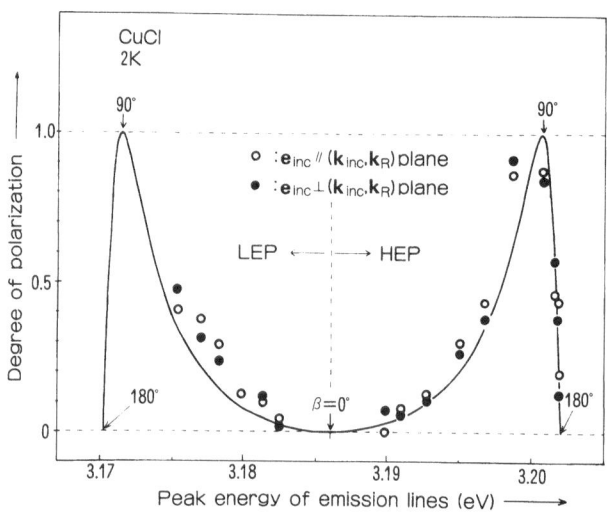

Fig. 3.23. Degree of polarization of scattered light versus its peak energy under resonant excitation in CuCl. Open circles and dots are the cases for incident light parallel and perpendicular to the scattering plane, respectively [3.72]

where $\beta = \sphericalangle (k_R, k_f)$. The relation (3.31) is shown by a solid line in the figure. Since there is no apparent difference with respect to the change in the incident polarization e_{inc} indicated by the open circles and dots it can be said that there is no correlation of polarization between the incident and scattered light.

3.5.5 Nonlinear Change of Exciton-Polariton Dispersion Associated with the GTA

Since the GTA occurs under high-intensity laser excitation, the corresponding nonlinear change of the refractive index and thus of the excitonic polariton dispersion should occur at the same energy region. Three different methods have been adopted for the study of this nonlinear change of dispersion.

a) **Anomalous Raman Shift in TRRS.** Figure 3.24 shows the relations between the incident photon energy and the peak energies of the LEP and HEP lines obtained by *Itoh* and *Suzuki* [3.72], at various internal scattering angles α in the forward scattering configuration. The anomalous behavior around the resonance energy region is clearly seen, especially for small scattering angles. The expected energy shifts of these lines without the nonlinear effect due to the GTA are shown by dashed lines. When α increases from 0° to 180°, the tangent of the slope is

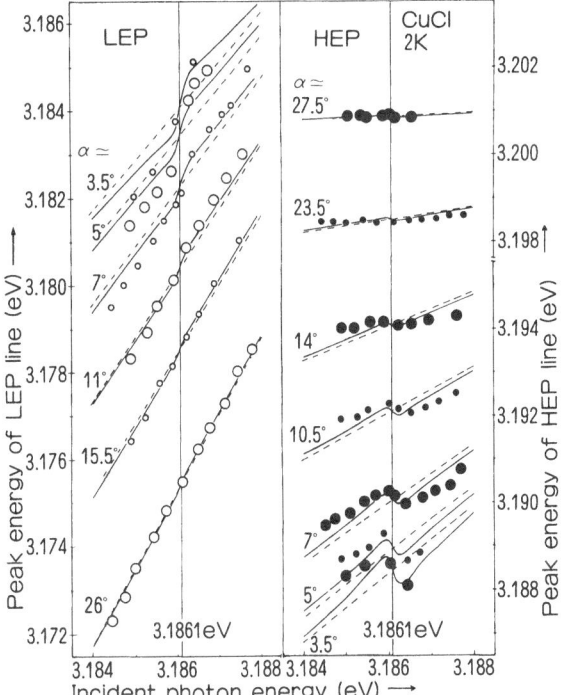

Fig. 3.24. Energy relations between LEP and HEP peaks versus incident light for various scattering angles. Solid and dashed lines are the expected relations with and without taking the nonlinear effect into account [3.72]

expected to change from one to two for the LEP line and from one to zero for the HEP line. The prediction is qualitatively in agreement with the experiment except for the anomalous behavior around the resonance energy. The anomalous behavior was found to become more pronounced with increasing excitation power.

In the first approximation, the nonlinear change is assumed to occur in the dispersion of the incident polaritons of high density, but not in the scattered polaritons, because the latter polaritons are scattered throughout the whole solid angle and are not in high density. Evaluation of the dispersion for the incident polaritons has been carried out using energy and momentum conservation throughout the Raman process.

$$2\omega_{inc} = \omega_R + \omega_f \quad \text{and}$$

$$\cos\alpha = \frac{(2k_{inc})^2 + k_R^2 - k_f^2}{2(2k_{inc})k_R}. \tag{3.32}$$

Using refractive indices obtained from the interference spectra under a conventional weak probe light, k_R, k_f, and α (for $\varphi = 20°$) are calculated, then k_{inc} is obtained from (3.32). The relation between $\hbar\omega_{inc}$ and k_{inc} thus calculated, taking $\hbar\omega_R$ as the energy of the observed LEP, is plotted in Fig. 3.25 with open circles.

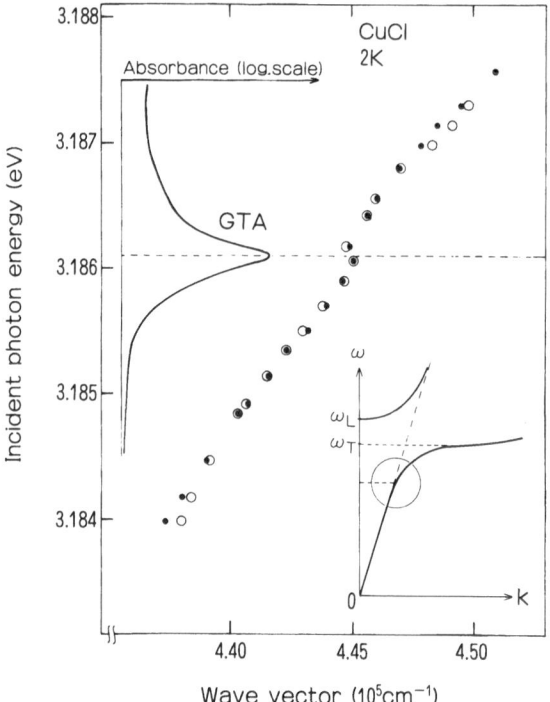

Fig. 3.25. Relations between wave vector and energy of incident polariton. Those derived from peak energies of LEP and HEP are shown with open circles and dots, respectively

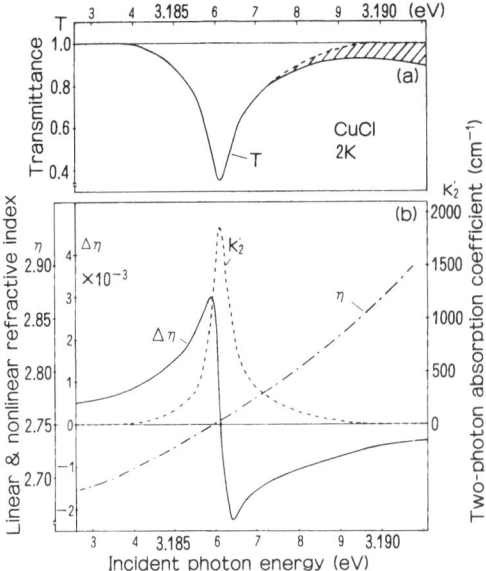

Fig. 3.26. (a): Two-photon transmission spectrum at 2 K. (b): Two-photon absorption coefficient K_2', nonlinear part of the refractive index $\Delta\eta$, and linear refractive index η [3.72]

The similar relation for the HEP is shown by dots. An anomaly in the dispersion is clearly seen just around the peak energy of the GTA.

Figure 3.26 a shows the transmission spectrum around the GTA band obtained with the dye laser beam in the same excitation condition as in Fig. 3.24. The nonlinear part of the transmittance, T, is extracted from the linear part as shown by a dashed curve. According to the procedure described by *Loudon* [3.74] the GTA coefficient for a single beam $K_2'(2\omega)$, which is shown in Fig. 3.26 b, can be given as

$$K_2'(2\omega) = \frac{2}{g^{(2)}d}\left(\frac{1}{T} - 1\right), \tag{3.33}$$

where d is the sample thickness and the second-order correlation function $g^{(2)}$ is put equal to unity for coherent light. Since the coefficient $K_2'(2\omega)$ and the associated nonlinear change of refractive index $\Delta\eta(\omega)$ are directly related to the imaginary and real parts of the third-order susceptibility $\chi^{(3)}$, respectively, the relation between them can be given approximately, using the Kramers-Kronig relation for $\chi^{(3)}$, as

$$\Delta\eta(\omega) = \frac{c}{\pi}\mathscr{P}\int_0^\infty \frac{K_2'(2\omega')}{(2\omega')^2 - (2\omega)^2}\,d(2\omega'). \tag{3.34}$$

Here, it is assumed that $|\Delta\eta(\omega)| \ll \eta(\omega)$ and $\eta(\omega)$ changes little in the GTA region. In Fig. 3.26 b $\Delta\eta(\omega)$ is drawn as a solid line, which shows a dispersive change around the two-photon resonance energy. The normal (linear) refractive index $\eta(\omega)$ in the same energy region, obtained from the interference spectra, is also shown. Since $\Delta\eta = (c/\omega)\Delta k$ the nonlinear change of the polariton dispersion is also obtained, and it amounts to the same order of magnitude as shown in Fig. 3.25. Although the ratio of the nonlinear part to the linear one is estimated to be $\sim 2 \times 10^{-3}$, the nonlinear part of the excitonic-polariton dispersion has an important effect upon the behavior of the energy shifts of the TRRS lines for forward scattering as shown by solid curves in Fig. 3.24.

Since the GTA is also caused by the combination of two photons having different energy, the nonlinear change in the polariton dispersion is also expected around the energy region of $(\omega_{mol} - \omega_1)$ under strong excitation at ω_1. For forward scattering with excitation at ω_1, the energy of the LEP or HEP line can eventually fall on $(\omega_{mol} - \omega_1)$, where the anomalous shift and the splitting of the Raman line due to the nonlinear change in the scattered polariton have in fact been observed [3.75].

b) **Induced Circular Dichroism.** The following induced circular dichroism (ICD) spectra are more useful for the precise derivation of small $\Delta\eta$ which are closely connected to the GTA. Since the EM state in CuCl has Γ_1 symmetry, the GTA is allowed only for the combination of right and left circularly polarized (CP) components of the incident light as mentioned in Sect. 3.4.2. Therefore, the strong excitation by a left CP pump light having energy near the GTA band causes an ICD. Namely, when a linearly polarized (LP) probe light is incident on the crystal from almost the same direction as the pump light, only the right CP component of it suffers a phase shift and damping during its transmission through the crystal. Thus the rotation of the polarization axis of the probe light and the change of a linear polarization into an elliptical one take place [3.76]. The angle of rotation of

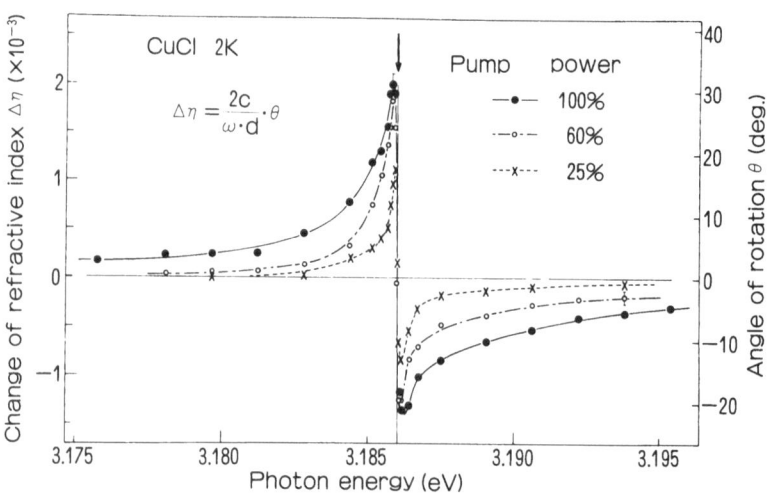

Fig. 3.27. Change of refractive index $\Delta\eta$ around the GTA region in CuCl at various pump powers [3.77]

the axis, θ, is directly related to the nonlinear change of refractive index $\Delta\eta$ of the right CP component of the probe beam and is given by $\theta = \Delta\eta\omega \cdot d/2c$, where ω is the frequency of the incident light and d is the thickness of the sample.

In the ICD experiment, the intensity ratio of the pump beam to the probe one was about $100:1$ at maximum pump power (~ 1 MW cm^{-2}). The polarization of the probe beam passed through the crystal was analyzed with the rotation of a linear polarizer. Figure 3.27 shows $\Delta\eta$ around the GTA band obtained for a single crystal ($d = 32$ μm) at 2 K for three different pump powers [3.77]. The change $\Delta\eta$ is of the order of 10^{-3} which agrees quantitatively with that obtained from the anomalous shift of the TRRS lines. As the pump power increases, $\Delta\eta$ becomes more pronounced and shows an asymmetry around the GTA peak, which corresponds to the asymmetric broadening of the GTA band.

c) **Effect of Spatial Dispersion Due to the EM.** By analogy with the spatial dispersion effect upon the exciton polariton, *Itoh* and *Kirihara* [3.78] have proposed taking the spatial dispersion of the EM into account for the nonlinear change in the polariton dispersion. Under the existence of the pump beam having energy, $\hbar\omega_1$, the dielectric function for the probe beam having energy, $\hbar\omega_2$, which is nearly equal to $\hbar(\omega_{mol} - \omega_1)$, is given approximately by

$$\epsilon(\omega_2, k_2) \cong \epsilon_0(\omega_2) + \frac{A(n_1)}{[\omega_{mol}(k_{mol}) - \omega_1(k_1)]^2 - \omega_2^2(k_2)} \tag{3.35}$$

neglecting the damping factor, where $\epsilon_0(\omega_2)$ is the background dielectric function independent of the density n_1 of the pump polariton and $\omega_{mol}(k_{mol})$, $\omega_1(k_1)$, and $\omega_2(k_2)$ are the dispersion relations of the EM, the pump polariton and the probe polariton, respectively. Therefore, the solution of $\omega_2(k_2)$ for $k_2 \gg k_0$ which gives the new spatial dispersion is written as

$$\omega_{mol}(k_{mol}) = \omega_1(k_1) + \omega_2(k_2) , \qquad k_{mol} = k_1 + k_2 . \tag{3.36}$$

There are two kinds of typical solutions for $k_1 \| k_2$: (1) if $\omega_1(k_1)$ is little affected by the nonlinear change, then $k_2 \simeq k_{mol}$ and $\omega_2(k_2) = (\omega_{mol}(0) - \omega_1) + (\hbar k_2^2/2m_{mol})$, and (2) if $\omega_1(k_1) = \omega_2(k_2)$, then $k_2 = k_{mol}/2$ and $\omega_2(k_2) = \omega_{mol}(0)/2 + \hbar k_2^2/m_{mol}$. These solutions, (1 and 2), are drawn schematically as "new branches" in Fig. 3.28 as solid and broken curves, respectively. The ordinary exciton-polariton and EM dispersion curves are also shown. From the figure, one can notice that there exist three kinds of exciton polaritons, a, b, and c, for *intense* one-beam excitation with energy a little larger than $\hbar\omega_{mol}(0)/2$, whereas for $\omega_1 < \omega_{mol}(0)/2$ one has only one kind of exciton polariton. Therefore, for $\omega_1 > \omega_{mol}(0)/2$ one has several kinds of TRRS processes typical ones of which are drawn as (1), (2), and (3).

Process (1) corresponds to the ordinary TRRS treated so far. Process (2) should be the new type of TRRS whose peak shifts by approximately the same amount as the energy shift of the incident light. This TRRS corresponds to the

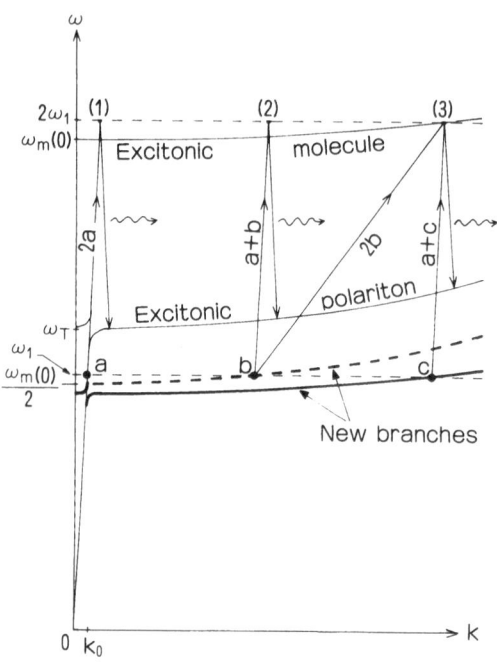

Fig. 3.28. Schematic representation of the new branches of the exciton polariton together with the ordinary exciton and EM dispersions. Several kinds of TRRS are shown

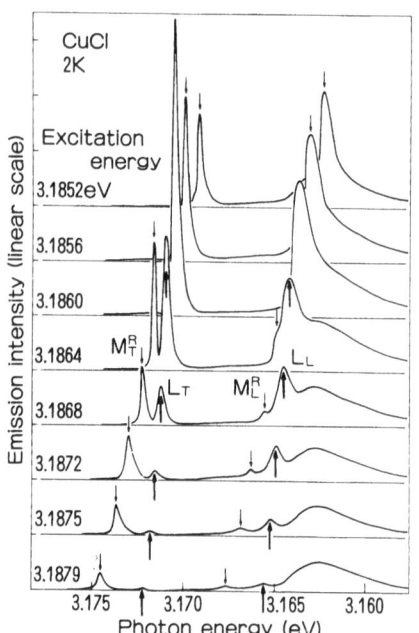

Fig. 3.29. Two-photon resonant Raman lines called L_L and L_T in CuCl which appear with intense high-energy off-resonant excitation of GTA in CuCl, measured at 2 K [3.78]

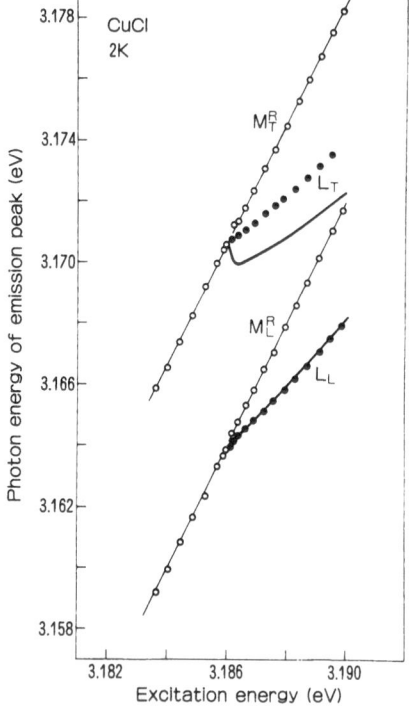

Fig. 3.30. Energy shifts of L_L and L_T lines versus excitation energy change [3.78]

Raman line called L mentioned in Sect. 3.5.1. As shown by upward arrows in Fig. 3.29 a pair of L lines, L_T and L_L, appear in the M_T and M_L band regions under intense off-resonant excitation slightly to the higher energy side of the GTA peak. The relation between the energies of the L lines and that of the incident light are shown in Fig. 3.30 by dots. The open circles correspond to those of the ordinary TRRS. The thick solid lines are the calculated energy relations derived from process (2) mentioned above, indicating fairly good agreement with the experimental data for the L_L process. The discrepancy for the L_T process is not understood.

Process (3) is a special one because it is always resonant with the EM state. Therefore, the secondary emission may have the character of luminescence on account of the strong damping at the EM state. The X bands which will be mentioned in Sect. 3.9.2 are thought to be caused by process (3).

3.6 Acoustic-Phonon Interaction of the Excitonic Molecule

The temperature dependences of the GTA line shape and the secondary emission spectra have been investigated under weak pump power [3.79] in order to prevent the occurrence of the pump power dependent broadening of the GTA spectra as explained in Sect. 3.4.2. With increasing temperature, the line shape of the GTA band becomes asymmetrically broadened with a longer tail toward the higher energy side, as shown in Fig. 3.31. In Fig. 3.32 the width (FWHM) of the GTA coefficient, $K_2'(2\omega)$, is plotted as a function of the lattice temperature.

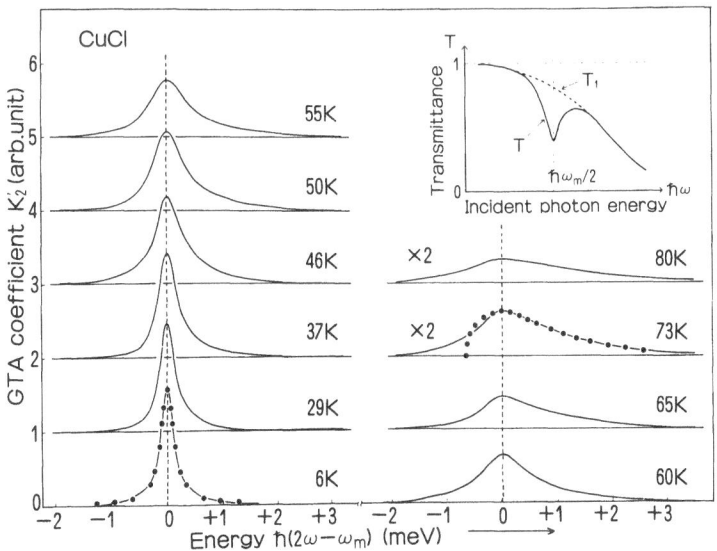

Fig. 3.31. Temperature variations of the GTA line shape in CuCl [3.79]

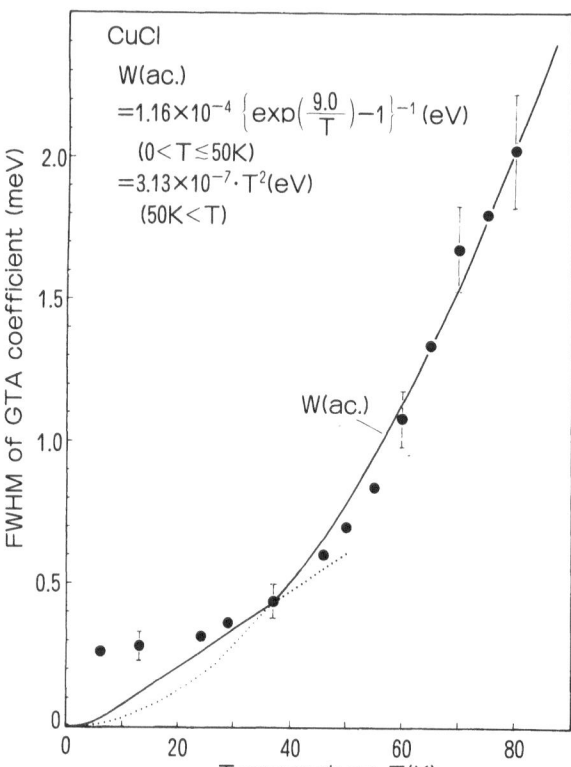

CuCl

W(ac.)

$= 1.16 \times 10^{-4} \left\{ \exp\left(\frac{9.0}{T}\right) - 1 \right\}^{-1}$ (eV)

$(0 < T \leq 50K)$

$= 3.13 \times 10^{-7} \cdot T^2$ (eV)

$(50K < T)$

W(ac.)

Fig. 3.32. Band width of the GTA versus temperature in CuCl and theoretical fit for the acoustic-phonon scattering [3.79]

The width of the GTA band can be thought to be brought about by the interaction between EMs and phonons. The interaction originates from either acoustic phonons (ap) or optical phonons (op). The phonon interaction can be treated analogously to that of the exciton [3.80] taking different values of parameters such as the translational mass and the exciton-phonon coupling constant. The intraband scattering of the EM by the LO phonon is unlikely, because under GTA band excitation there is no EM state capable of LO phonon emission, while the temperature is still too low for phonon absorption.

For the interaction of the EM with an acoustic phonon, it should be taken into consideration that the EM state under consideration is energetically the lowest one in the two-exciton quasi-particle system and has an energy minimum at the Γ point.

According to the calculation by *Toyozawa* [3.81], the line shape in the main part of the absorption band at high temperature is determined by the interactions of the exciton not only with one acoustic phonon but also with several phonons. He found the half-width to be proportional to $(g_{ex} T)^2$ where g_{ex} is the exciton-phonon coupling constant ($g_{ex} < 1$), and the line shape to be strongly asymmetric (degree of asymmetry $A = 0.33$) with a tail on the high-energy side. In the low-temperature region, the one-phonon scattering process is predominant which gives a Lorentzian line shape [3.82].

The quantitative description of the half-width (FWHM) of the GTA band W (acoustic) for two-photon processes is given as follows [3.81]:

$$W = \left(\frac{2}{\pi}\right) g_{\text{mol}} k_B \frac{T_1}{e^{T_1/T} - 1} \quad (0 < T \leq T_2) ,$$

$$= 0.3 \, g_{\text{mol}}^2 (k_B T)^2 / m_{\text{mol}} u^2 \quad (T_2 < T) , \quad \text{where} \tag{3.37}$$

$$T_1 = 2 \, m_{\text{mol}} u^2 / k_B \quad \text{and}$$

$$T_2 = 2.6 \, m_{\text{mol}} u^2 / k_B g_{\text{mol}} .$$

Here, g_{mol}, m_{mol}, and u are the EM-phonon coupling constant, the translational effective mass of the EM and the velocity of sound, respectively. With the use of the deformation potential for the EM, E_d^{mol}, and the density ϱ of the crystal, the coupling constant g_{mol} is given by

$$g_{\text{mol}} = \frac{(m_{\text{mol}} E_d^{\text{mol}})^2}{\hbar^3 \varrho u} . \tag{3.38}$$

Adopting $\varrho = 4.16$ g cm^{-3}, $u = 3.6 \times 10^5$ cm s^{-1}, and $m_{\text{mol}} = 5.29 \, m_0$, the calculated half-width shown by the solid curve in Fig. 3.32 can be best fitted to the experimental one by assuming $g_{\text{mol}} = 0.23$ which gives $E_d^{\text{mol}} = 2.6$ eV and $T_2 = 50$ K ($T_1 = 9.0$ K). Taking into consideration the fact that the experimentally obtained half-width at low temperatures is partially restricted by that of the incident dye laser (~ 0.08 meV), the agreement between the experimental value and the calculated one is considered to be fairly good.

The GTA line shape given by Toyozawa's model at high temperature is also compared with the experimentally obtained curve at 73 K as shown by a dotted curve in Fig. 3.31. Although the fit is not satisfactory for the low-energy tail part, the line shape of the high-energy tail part and the degree of asymmetry are well reproduced.

It is well understood from the simple theoretical point of view that at high temperatures the interaction with acoustic phonons plays a more important role for the GTA band than for the exciton band, since the effective coupling constant of EM with acoustic phonons, which is proportional to $g_{\text{mol}}^2 / m_{\text{mol}}$ according to (3.37), is larger than that of the exciton by two orders of magnitude. From these facts, it is concluded that scattering by acoustic phonons is an important relaxation process for the EM. With the use of the parameters derived above, the inverse of the scattering rate at 20 K is estimated to be

$$\tau_{\text{mol}-\text{ap}} = \frac{\varrho u \pi \hbar^4 (e^{T_1/T} - 1)}{2 (E_d^{\text{mol}} m_{\text{mol}})^2 k_B T_1} \sim 3.3 \text{ ps} . \tag{3.39}$$

At temperatures higher than 50 K the multi-phonon scattering process is expected to reduce τ_{mol-ap} drastically.

It is interesting to examine whether or not the EM system generated by the GTA band excitation at high lattice temperatures reaches thermal equilibrium with the lattice through collisions with acoustic phonons. For this purpose, the effective temperature T_{eff} of the EM at various lattice temperatures was obtained from the line shape of the broad M_L and M_T emission bands calculated by the formula described in Sect. 3.8.1.

Since the T_{eff} obtained is almost the same as the lattice temperature in cases above 30 K, the EM system generated by the GTA is considered to be in thermal equilibrium with the lattice through interaction with acoustic phonons before radiative decay. As the line shape of the M broad bands depends strongly on the excitation power at temperatures lower than ~ 20 K, the EM system in this temperature range is considered not to reach thermal equilibrium with the lattice on account of mutual collisions of EMs being more frequent than scattering by accoustic phonons, as will be shown in Sect. 3.9.1.

The broadening of the M_T^R line at the resonant excitation region is observed at higher temperatures and it may be considered to be brought about mainly by some kind of relaxation process in the intermediate state (the EM state) and not by broadening of the final state (the T exciton state). Near the resonant condition there is a chance that EM–acoustic-phonon scattering may be realized in the intermediate state, and emission lines due to scattering with the recoil of both the exciton and long-wavelength acoustic phonon may be observed around the main part of the Raman peak which brings about the broadening of the Raman line.

In the temperature region where multi-phonon scattering is not important, it is expected that the GTA associated with one-acoustic phonon emission may be oservable. In evaporated films of CuCl at 2 K, such a process has been observed through the secondary emission.

In evaporated samples a new band, called S, appears on the background M_T broad band as shown in Fig. 3.33 [3.83]. The S band is found only in evaporated samples and its peak shifts to lower energies with increasing excitation photon energy, in contrast to the Raman lines, as seen in Fig. 3.34. Further the S band is emitted only when the sample is excited at the higher-energy side of the GTA peak. The S emission is assumed to result from the radiative annihilation of the EM. In the case of the higher-energy off-resonant excitation, the EM is attained by two-photon absorption and one-acoustic phonon emission as shown schematically in Fig. 3.35, where q is the wave vector of an emitted phonon and $q = |q|$. This EM has a wave vector $k_{mol} = 2\,k_{inc} - q$.

With $2\Delta E$ being the excess energy in the excitation of the EM the following equation holds:

$$2\Delta E + \frac{\hbar^2 (2\,k_{inc})^2}{2\,m_{mol}} = \hbar u q + \frac{\hbar^2 (2\,k_{inc} - q)^2}{2\,m_{mol}}, \qquad (3.40)$$

where u is the velocity of sound. This equation gives $k_{mol} = 2\,k_{inc} - q$ as a locus like a shell in k-space for a given value of ΔE. Such an EM is dissociated into a

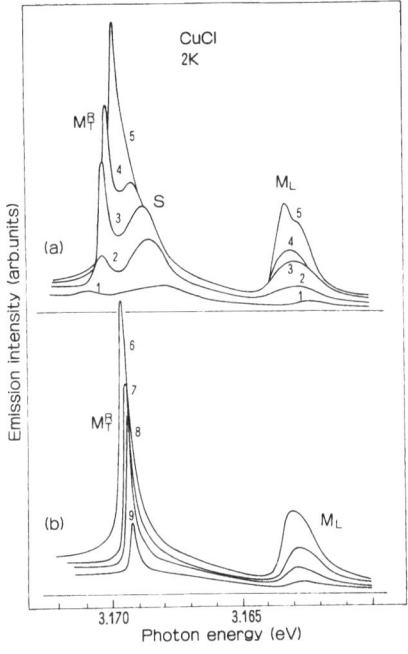

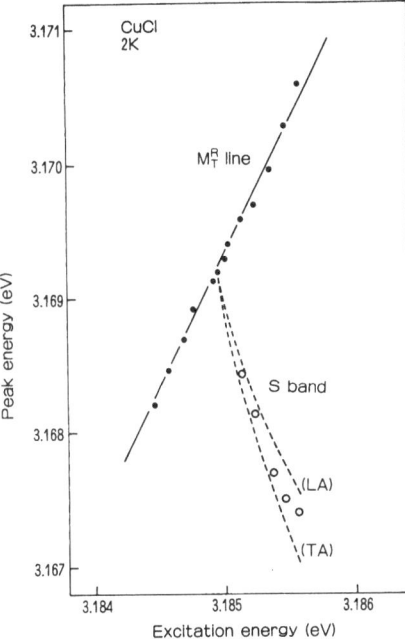

Fig. 3.33

Fig. 3.34

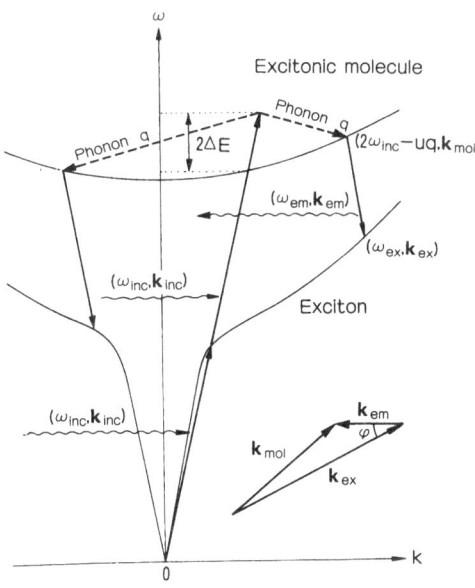

Fig. 3.33a, b. Emission spectra of evaporated CuCl film under two-photon excitation of the EM at 2 K. (**a**): excitation energies are above the resonance and (**b**): below the resonance. The spectra are numbered in order of decreasing excitation energy [3.83]

Fig. 3.34. Excitation energy dependences of M_T^R line and S band, [3.83]

◀ **Fig. 3.35.** Schematic diagram of emission processes of an EM generated by two-photon excitation followed by acoustic phonon emission, [3.83]

transverse exciton (frequency, wave vector) = (ω_{ex}, K_{ex}) and an exciton polariton (ω_{em}, k_{em}), the latter of which is observed as emitted light. The wave vectors and energies satisfy

$$k_{mol} = 2\,k_{inc} - q = k_{ex} + k_{em}\,, \quad \text{and}$$

$$2\omega_{inc} - uq = \omega_{ex} + \omega_{em}\,. \tag{3.41}$$

The relation between k_{ex} and ω_{ex} of the T exciton is obtained from the following formula involving the polariton effect:

$$\hbar\omega_{ex} = \hbar\omega_L - \frac{\Delta_{LT}}{1 - (\epsilon_\infty \omega_{ex}^2 / c^2 k_{ex}^2)} + \frac{\hbar^2 k_{ex}^2}{2\,m_{ex}^T}\,, \tag{3.42}$$

where Δ_{LT} is the L–T splitting. The relation (3.42) can be applied also to that between k_{em} and ω_{em} by replacing k_{ex} and ω_{ex} by k_{em} and ω_{em}, respectively. Taking φ as the angle between k_{em} and k_{ex} as shown in Fig. 3.35, ω_{em} observed in a direction along k_{inc} is given by (3.41) and (3.42) with intensity

$$I(\Delta E, \omega_{em})d\omega_{em} \propto A(\Delta E)\varrho(\Delta E, \omega_{em})\,(1 + \cos^2\varphi)d\omega_{em}$$

where $A(\Delta E)$ is the resonance factor of excitation into the EM level. The factor $\varrho(\Delta E, \omega_{em})$ is the density of the EM state generated by phonon emission and is nearly proportional to ΔE. The geometrical selection rule gives the factor $1 + \cos^2\varphi$. The calculated line shape shows two peaks at the higher and lower edges corresponding to the two downward arrows in Fig. 3.35. With increasing ΔE the emission band is easily seen to shift to the low-energy side. The peak energy (the lower-energy one) of the calculated spectrum is plotted as dashed lines in Fig. 3.34 for the cases of LA ($u = 3.5 \times 10^5\ \mathrm{cm\,s^{-1}}$) and TA ($u = 1.3 \times 10^5\ \mathrm{cm\,s^{-1}}$) phonons. The parameters adopted in the calculation are $m_{ex}\,(T) = 2.3\,m_0$, $m_{mol} = 5.29\,m_0$, $\epsilon_\infty = 5.59$, and $\Delta_{LT} = 5.7$ meV. The resonance factor $A(\Delta E)$ was neglected and the intensity was assumed to be proportional to the state density of the EM level obtained from phonon emission. The correspondence between the spectral line shape of the S band and the calculated one is not good because of the simplifying assumptions but approximate agreement is seen for the peak shift.

It is noted that the S band is not found in single crystals but only in evaporated films. The large strain in the latter case may be considered to cause the strong coupling between the EM and acoustic phonon.

3.7 Coexistence of Luminescence and Raman Components in the Resonant Excitation

With resonant excitation the EM is selectively created with $k = 2\,k_0$ and sharp lines are found at 3.164 and 3.170 eV; i.e., at the higher-energy edges of the broad Maxwellian M_L and M_T bands. The peak energies of these lines do not change with the shift of the excitation energy $\hbar\omega_{inc}$ when the bandwidth of the laser $\Delta\omega_{inc}$ is much broader than that of the GTA band, $\Delta\Omega$. Therefore they are considered to be due to luminescence and we call them M_L^0 and M_T^0. On the other hand, when $\Delta\omega_{inc}$ is narrower than $\Delta\Omega$ the lines shift as mentioned in Sect. 3.5.1 showing a property of two-photon Raman scattering, and they are called M_L^R and M_T^R.

The energies of the luminescence lines M_L^0 and M_T^0 are the same as those of the M_L^R and M_T^R lines with resonant excitation. Due to their narrow line width they were previously assigned as luminescence from the Bose-Einstein condensed phase at $k \sim 0$. *Levy* et al. [3.84] found that the line width increased with increasing excitation intensity and concluded that all the EMs did not occupy exactly the same state, in contradiction with Bose condensation, and the emission was due to cold EMs not in thermal equilibrium with the lattice. If they were annihilated immediately, infinitely sharp lines would be expected. Thus, the EMs are somewhat spread on their energy band. However, in general it might be expected that the Raman scattering line and resonant luminescence would coexist under resonant excitation, irrespective of the incident laser width $\Delta\omega_{inc}$.

The properties of the sharp lines under resonant excitation have been investigated using a picosecond tunable laser by *Segawa* et al. [3.85] and *Masumoto* et al. [3.86]. They found that the sharp lines involved two components of decay: the faster component has the same rise and decay times as those of the exciting laser light and is ascribed to the Raman process, whereas the slower component is a radiative decay process of the EM. *Segawa* et al [3.85] also found that the transition rates of the luminescence and Raman processes for the M_L region under resonant excitation depends on the bandwidth of the exciting laser. The luminescence apparently increases with the increase of the laser width. The above mentioned properties of the secondary emission depending on $\Delta\omega_{inc}$ have been partially explained by a damping theory by *Hanamura* and *Takagahara* [3.87].

Mita and *Ueta* [3.88] have studied the Raman scattering and luminescence in the M_L band region using a very narrow dye laser having $\hbar\Delta\omega_{inc} = 0.05$ meV and a monochromator with spectral resolution of 0.04 meV. Figure 3.36 shows the Raman lines when $\hbar\omega_{inc}$ was shifted from 3.1858 eV to 3.1865 eV around the resonance at $\hbar\Omega_0 = 3.1861$ eV with an interval of ~ 0.05 meV. With excitation on both sides of the resonant energy, the Raman lines have a similar line shape to the excitation light but with 0.15-meV half-width. On the other hand, when the excitation is very close to the resonance, curves e and f, a broader emission line of ~ 0.3 meV half-width appears and overlaps the Raman line. Its peak is located at 3.164 eV as shown by the dotted curve. It is found that the Raman yield drops at resonant excitation and is replaced by luminescence. From the peak energy the luminescence is attributed to M_L^0.

In a case of weaker excitation with a broader laser of $\Delta\omega_{inc} = 0.5$ meV, the Raman lines are observed with ~ 1 meV half-width in the cases of both low- and high-energy off-resonant excitations and M_L^0 luminescence is also observed, with

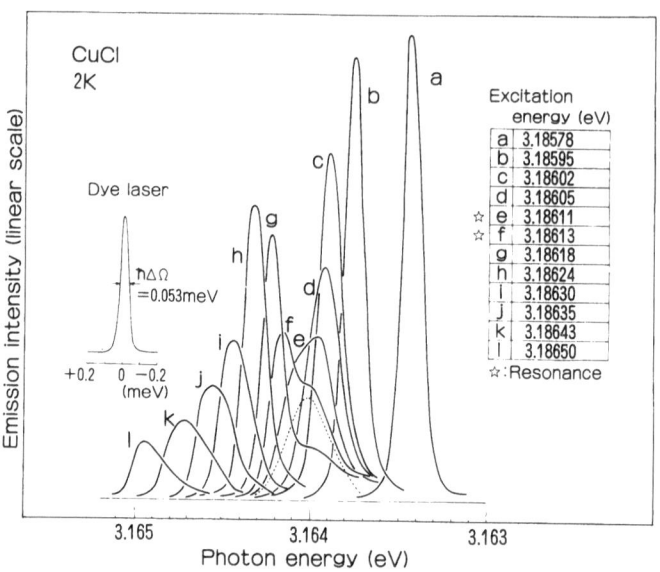

Fig. 3.36. Two-photon resonant Raman spectral line M_L^R with the recoil of a L exciton in CuCl, measured at 2 K with a dye laser of 0.05 meV half-width [3.88]

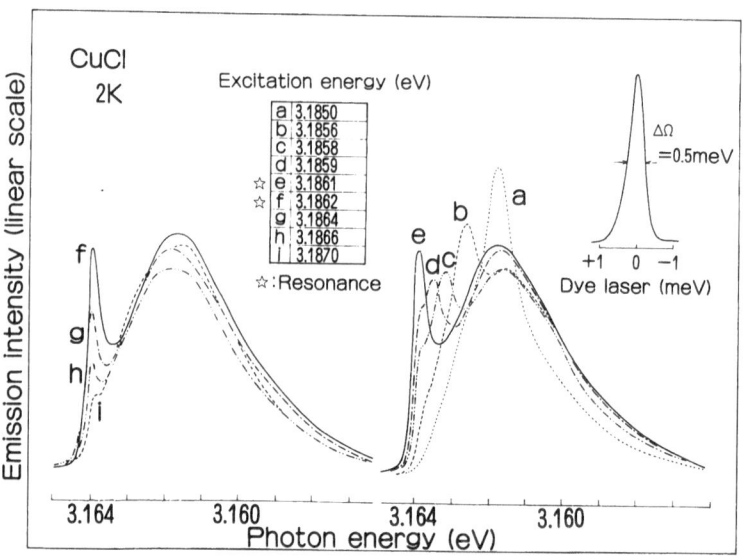

Fig. 3.37. M_L^O luminescence and M_L^R spectra of CuCl in the resonant-excitation region, excited by an intense dye laser of 0.5-meV half-width, [3.88]

~ 0.3 meV half-width, in the case of resonant excitation, together with the M_L Maxwellian band [Ref. 3.88, Fig. 2]. However, when the laser intensity is increased by a factor of ten, the Raman line is not observed in the high-energy off-resonant excitation, as shown in Fig. 3.37, and a broad band similar to that in Fig. 3.17 is found at 3.162 eV. The sharp peak at 3.164 eV is the M_L^0 luminescence.

With high-energy off-resonant excitation new emissions appear such as the L and S bands mentioned in Sects. 3.5.1 and 3.6 and another one called X which is described later in Sect. 3.9.2. All these emissions are enhanced only for high-energy off-resonant excitation. With intense laser excitation the channel to these emission processes should become pronounced and the Raman yield should decrease. It is suggested that the EMs created under resonant excitation have two kinds of distribution; one is the thermalized Maxwellian distribution causing the M broad emission and the other is the non-equilibrium localized distribution responsible for the M^0 or X band luminescence. It is possible that the latter distribution of EMs is attained as the initial step of the successive relaxation process and leads to hot luminescence, while the former leads to ordinary luminescence. However, there exists another explanation proposed by *Toyozawa* [3.89] for the coexistence of the two kinds of distribution. The population instability realizable in a high-density EM system forms two domains (hot and cold), inside the excited volume, which are responsible for the broad and sharp emission bands. The model also explains qualitatively the asymmetric line shape of the Raman yield and the GTA band. Therefore, whether the coexistence of the hot and cold distributions is spatial or temporal has still to be made clear.

Since the M_L^R line was strongly observed in Fig. 3.36, the emission was considered not to be taken in a near-backward scattering configuration but taken along the surface of excitation, in which unfavorable effects such as stimulation and reabsorption often distort the emission spectra. By shutting off emission from crystal edges and using weaker excitation the above-mentioned unfavorable effects are almost avoided [3.59].

Figure 3.38 shows emission spectra thus obtained in a near-backward scattering configuration with a dye laser of 0.07 meV spectral width. The excitation energy was gradually changed across the resonant energy in curves a to d. In this case the GTA band had 0.2 meV half-width as shown in the figure. The sharp lines of 0.13 meV half-width which shift with the change of excitation energy are the M_T^R and M_L^R lines. In pure backward scattering the M_L^R line should not appear because of the geometrical selection rule mentioned in Sect. 3.5.4. In the figure the M_L^R line is much weaker than the M_T^R one, indicating that the observation is made in a nearly pure backward scattering configuration.

The wider bands which do not change their peak energy are ascribed to the M_L^0 and M_T^0 (broken curve) luminescence. The half-width of the M_L^0 line, 0.3 meV, is the same as before and M_T^0 has ~ 1 meV half-width. The total intensity (area) of the M_T^0 line is larger than that of the M_L^0 line.

These line shapes provide information about the distribution of the EM in k-space.

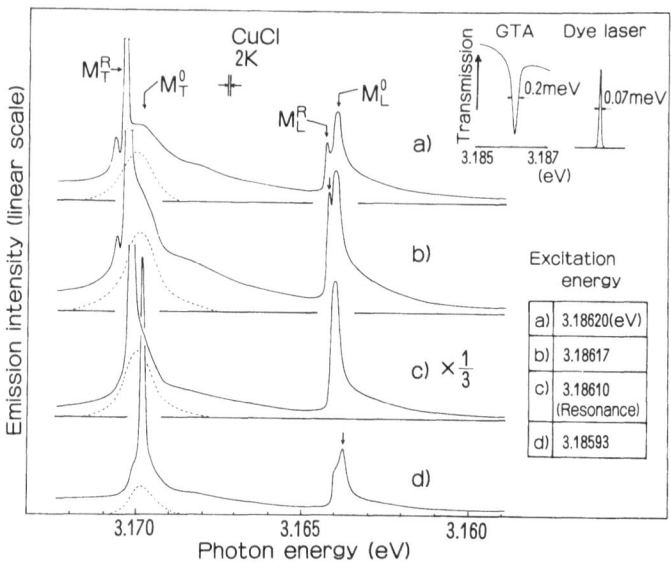

Fig. 3.38. Coexistence of luminescence (M_L^O and M_T^O) (---) and Raman (M_L^R and M_T^R) lines at resonant excitation [3.59]

3.8 Redistribution of Excitonic Molecules Resonantly Generated by Two-Photon Excitation

In a further study of the luminescence M_T^0 and M_L^0 a difference in line shape is found between the forward and the backward scattering configurations which makes it possible to determine the distribution of the EMs [3.90]. Figure 3.39 shows secondary emission spectra of the $k = 2\,k_0$ excitation for (a) the backward and (b) the forward scattering configurations, together with (c) the emission spectrum of the $k = 0$ excitation under resonant excitation.

In the forward emission spectra one can recognize some differences from the backward spectra. The most distinct change is in the width of the M_L^0 line. Although the M_L^R Raman line is found only weakly, there is a very sharp luminescence line which seems not to change its position and shape, and this is the M_L^0 line. Comparing the spectra with the backward ones, the spectral width of M_L^0 is narrower than one-half of that in the backward spectra. This fact suggests an anisotropic distribution of EMs in k-space. At the skirt of the M_T^R line a broad weak emission line, M_T^0, can be seen, the line shape of which is different from that in the backward spectrum.

In the case of the $k = 0$ excitation, with two counter-propagating circularly polarized beams having energy around 3.1861 eV, the M_L^0 luminescence line can be recognized to have a FWHM as small as 0.15 meV. The M_T^R line must be

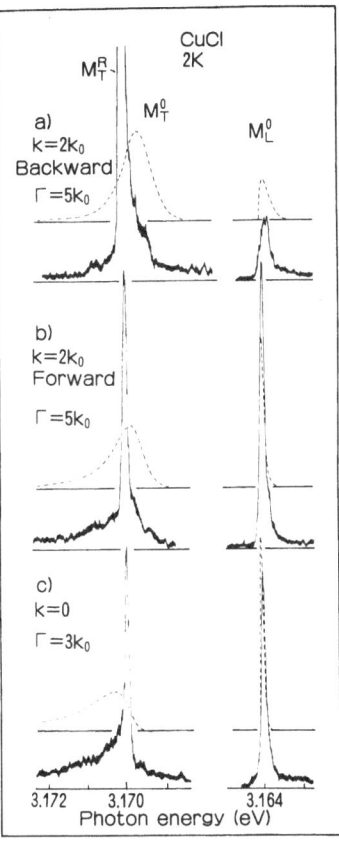

Fig. 3.39. Comparisons of secondary emission spectra in CuCl with $k = k_0$ excitation under the (**a**) backward and (**b**) forward scattering geometries and (**c**) with $k = 0$ excitation. Dotted curves are calculated luminescence line shapes [3.90]

located at 3.186 eV, at just the same energy as the incident laser light. Since the sharp line around 3.170 eV is the same as the M_T^R line in the backward spectrum under the $k = 2\,k_0$ excitation, it is caused by the linearly polarized component of the incompletely circularly polarized light. It is named the $2\,k_0 - M_T^R$ line.

As for the emission in the M_T region, a rather broad band is seen tailing off to the higher-energy side from the $2\,k_0 - M_T^R$ line, which we can assign as due to the luminescence M_T^0 line. Comparing these spectra with $k = 2\,k_0$ backward spectra, differences in the position and the shape are seen. In particular, under resonant excitation the peak of the M_T^0 line appears at the high-energy side of the $2\,k_0 - M_T^R$ line for $k = 0$ excitation, but for $k = 2\,k_0$ excitation it appears at the low-energy side.

From these luminescence line shapes the distribution of the EMs can be discussed with the aid of model calculations. The EMs are tentatively assumed to be in the following Gaussian distribution before the radiative annihilation:

$$\varrho(k_{\mathrm{mol}}) \propto \exp\left(-\frac{4(k_{\mathrm{mol}} - k_c)^2}{\Gamma^2}\right),$$

where k_{mol} is the wave vector of the EM, k_c the center of the distribution brought about by the excitation wave vector ($2 k_0$ or 0), and Γ the parameter showing the spread around k_c. Details of the calculation of the line shapes are shown in Sect. 3.8.1. The best fits to the observed spectra are illustrated in Fig. 3.39 by broken lines. The observation is qualitatively reproduced by the calculation taking $\Gamma = 5 k_0$, $k_c = 2 k_0$ and $\Gamma = 3 k_0$, $k_c = 0$ in the cases of $k = 2 k_0$ and $k = 0$ excitation, respectively.

The redistribution of EMs created at larger k values is also investigated by the stepwise two-photon excitation method described in Sect. 3.10.1. The width of the M_L^0 line is found to increase with k from ~ 0.3 meV for $k = 2.3 k_0$ to ~ 1.7 meV for $k = 12.2 k_0$, and for $k_c = 4 k_0$ and $9 k_0$ the M_L^0 line shapes in the backward and forward configurations are almost the same showing that EMs created with large k are redistributed almost isotropically in k space in contrast to $k = 0$ or $2 k_0$ excitation.

Further evidence of the redistribution of the EMs resonantly generated could be obtained as follows [3.91]. At the resonant excitation of the EM at $2 k_0$, a sharp emission line followed by a broad band called V is found, as shown in Fig. 3.20. The peak energy of the V line, 3.208 eV, is equal to that of the bottom of the upper-branch polariton. The line shape has a close relationship with M_L^R and its broad side band. Taking into account these facts together with momentum conservation, the V line is considered to be an emission of the upper-branch polariton from its bottom region. The upper polariton is created through the decomposition of EMs with $k \sim 0.9 k_0$. The lower polariton paired with it has almost the same energy as that of the M_L emission therefore the V emission process is induced by the M_L light. These facts suggest that EMs created resonantly with $k = 2 k_0$ relax in k-space and a relatively high population is attained even for $k \sim 0.9 k_0$.

3.8.1 Calculation of Line-Shapes of the Excitonic Molecule Luminescence

For convenience of calculation, the distribution of EMs in k-space is assumed to be

$$\varrho(k_{mol}) = \left(\frac{2}{\sqrt{\pi}\,\Gamma}\right)^3 \exp\left(-\frac{4(k_{mol} - k_c)^2}{\Gamma^2}\right), \qquad (3.43)$$

where, $(2/\sqrt{\pi}\,\Gamma)^3$ is the normalization factor.

In the case of the calculation of emission spectra of the EM in quasi thermal equilibrium at effective temperature T_{eff}, the wave vector k_c would be zero and the value Γ would read as

$$\Gamma = \frac{2\sqrt{2}\,m_{mol}\,k_B\,T_{eff}}{\hbar}. \qquad (3.44)$$

The emission process is the spontaneous decay of the EM into L or T excitons conserving wave vector and energy. The geometrical selection rule and state density are taken into account. The k dependence of the transition probability is neglected.

In the case of decay into a L exciton, energy and momentum conservation are given by

$$E_{mol} = E_L + E_p , \qquad k_{mol} = k_L + k_p , \tag{3.45}$$

where the suffixes L and p stand for the longitudinal exciton and emitted photon, respectively. Writing the effective masses of the EM and L exciton as m_{mol} and m_L, respectively, E_p is expressed as

$$
\begin{aligned}
E_p &= E_{mol}(k_{mol} = 0) + \frac{\hbar^2 k_{mol}^2}{2\, m_{mol}} - E_L(k_L = 0) - \frac{\hbar^2 k_L^2}{2\, m_L} \\[2mm]
&= E_{M_L}(0) + \frac{\hbar^2 k_p^2}{2\,(m_{mol} - m_L)} - \frac{(m_{mol} - m_L)\hbar^2}{2\, m_{mol} \cdot m_L} \\[2mm]
&\quad \times \left| k_{mol} - \frac{m_{mol}}{m_{mol} - m_L} k_p \right|^2 , \qquad \text{where}
\end{aligned}
\tag{3.46}
$$

$$E_{M_L}(0) = E_{mol}(k_{mol} = 0) - E_L(k_L = 0) .$$

Then, if we put $B = k_{mol} - \eta k_p$ where $\eta = m_{mol}/(m_{mol} - m_L)$, E_p is a function of $|B|$, and the distribution function (3.43) is rewritten as

$$\varrho(B) = \left(\frac{2}{\sqrt{\pi}\,\Gamma}\right)^3 \exp\left(-\frac{4\,(B + \eta k_p - k_c)^2}{\Gamma^2}\right) . \tag{3.47}$$

From the geometrical selection rule, the emission probability A is proportional to $\cos^2 \alpha$, where α is the angle between the polarization vector of the L exciton and the electric field vector of the emitted photon. It can be expressed as

$$A \propto \cos^2 \alpha = \frac{B^2 (1 - \cos^2 \theta)}{B^2 + (\eta - 1)^2 k_p^2 + 2(\eta - 1)B \cdot k_p \cos \theta} , \tag{3.48}$$

where, θ is the angle between B and k_p. Since EMs of $|B| = $ constant give photons of the same energy, the emission spectrum $I(E_p)$ is expressed as

$$I(E_p) \propto \left[\iint_{B = const.} A \cdot \varrho(B) d^2 k_{mol} \right] \cdot \left| \frac{dB}{dE_p} \right| . \tag{3.49}$$

If we treat only the case of $k_p \| k_c$, that is, the case in which the observation is made from the forward or backward direction relative to the incident light, we can express $I(E_p)$ in polar coordinates (B, θ, φ) centered at ηk_p as

$$I(E_p) \propto \int_0^\pi \int_0^{2\pi} A \cdot \varrho(B) B^2 \sin \theta d\theta d\varphi \cdot \left| \frac{dB}{dE_p} \right| . \tag{3.50}$$

The T exciton energy is approximated by the polariton dispersion

$$E_T = E_L(k_L = 0) + \frac{\hbar^2 k^2}{2\,m_T} - \frac{\Delta_{LT}}{1 - \epsilon_b \omega_0^2/c^2 k^2} \, , \tag{3.51}$$

where m_T is the effective mass of the T exciton, 2.3 m_0, Δ_{LT} the L–T splitting of the exciton, 5.7 meV, ϵ_b the background dielectric constant, 5.59, and $\hbar\omega_0$ is taken to be 3.202 eV.

Calculation of the line shape of the EM in CuCl was carried out using the following wave numbers and effective masses [3.8]:

$$
\begin{aligned}
k_0 &= 4.44 \times 10^5 \text{ cm}^{-1} \, , \\
k_p &= 4.12 \times 10^5 \text{ cm}^{-1} \, , \\
m_L &= 3.14 \, m_0 \, , \quad \text{and} \\
m_{mol} &= 5.29 \, m_0 \, .
\end{aligned}
$$

Since the variation of wave number of emitted photons in the relevant spectral range is small, k_p is set as a constant and the value at 3.170 eV is adopted. For the M_T^0 luminescence, the geometrical selection rule requires that $A \propto (1 + \cos^2\theta)$. For the cases of $k = 2\,k_0$ or $k = 0$ excitation, k_c is assumed to be equal to $2\,k_0$ or 0, respectively.

3.9 Relaxation of the Excitonic Molecule Due to Intermolecular Collisions: Influence on the GTA and Secondary Emissions

The EM resonantly created by two-photon excitation is known to redistribute over a certain wave-vector region along its energy band before its radiative decay. The relaxation leading to redistribution is considered to cause the broadening of the EM state, which makes the GTA band broad. The broadening has actually been observed under intense two-photon resonant (TPR) excitation (Sect. 3.4.2) or at high lattice temperatures (Sect. 3.6). The detailed mechanism of broadening under intense TPR excitation at low temperatures is not well known, although possibilities other than the relaxation were also proposed, such as super broadening of the incident light [3.92] or the renormalization effect of the incident polariton [3.93, 94].

In this section, the pump power dependence of the intensity and line shape of the GTA spectrum and the secondary emission of the resonantly generated EMs are described in detail with the use of an improved high-resolution spectroscopic method, in order to clarify the role of the mutual collision of EMs [3.77, 95] in their relaxation.

3.9.1 Effect on the GTA Spectra

Figure 3.40 shows precise measurements of the GTA band for various pump powers using a dye laser having a spectral width of 0.03 meV [3.96], where the ordinate indicates the magnitude of the absorption, A, defined by $(I_0 - I_t)/I_0$. Here, I_0 and I_t are the intensities of the incident light (corrected for reflection) and the transmitted light, respectively. The maximum pump power is ~ 1 MW cm^{-2}. Dashed lines indicate the background one-photon absorption ascribed to the exciton band tail. It is remarkable that the peak height of the GTA does not change much with the increase of the pump power, in spite of the two-photon absorption, while the band width gradually increases asymmetrically with a more pronounced broadening toward the higher-energy side as mentioned in Sect. 3.4.2. The peak position is found to shift slightly toward the lower-energy side. Except for at low pump power the following relations hold for the absorbed power at the GTA peak, P, and the FWHM of the GTA band, W, for two-photon excitation

$$P \simeq AI_0 \propto I_0^{1.0} \, , \quad \text{and} \quad W \propto I_0^{0.8} \, . \tag{3.52}$$

These pump power dependences are different from those obtained by *Chase* et al. [3.95], but the present results are more reliable because of the better quality of

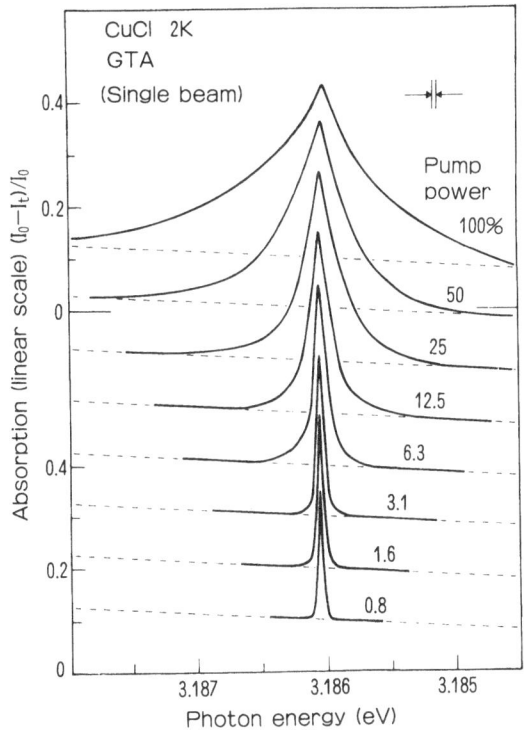

Fig. 3.40. Pump power dependence of the absorption spectrum around the GTA band at 2 K. Dashed lines indicate the background due to one-photon absorption [3.96]

samples and the higher resolution of the spectra. Taking the finite width of the laser into account, the GTA width at the lowest pump power is estimated to be less than 0.06 meV.

It is obvious from the spectra in Fig. 3.40 that the two-photon absorption coefficient K_2' is dependent not only on the two-photon energy 2ω but also on the intensity of light, through the intensity dependent damping γ_d. Therefore, K_2' can be written as

$$K_2'(2\omega, I) = B_0 \frac{1}{\gamma_d} f\left(\frac{2\omega - \omega_{mol}}{\gamma_d}\right) I_0 ,$$

where, B_0 is a constant including the two-photon transition probability, and $f(x)$ defines the GTA line shape $\left[\int_{-\infty}^{\infty} f(x)dx = 1\right]$. Then the change in the coherent light intensity I ascribed to the GTA during the passage of a small distance ΔZ is given by [3.74]

$$\Delta I = -\frac{1}{2}K_2'\frac{I^2}{I_0}\Delta Z = -\frac{1}{2}B_0I^2\frac{1}{\gamma_d}f\left(\frac{2\omega - \omega_{mol}}{\gamma_d}\right)\Delta Z .$$

In order to satisfy the relations (3.52) at $2\omega = \omega_{mol}$ it is found that the damping γ_d should be proportional to the pump power I.

On the other hand, the density of the EM, N_{mol}, generated at Z under TPR excitation with energy $\hbar\omega$ and intenstiy I is written as

$$N_{mol}(2\omega, I) = \frac{1}{2}\left(-\frac{\Delta I}{\Delta Z}\right)\tau/\hbar\omega = \frac{B_0\tau}{4\hbar\omega\gamma_d}f\left(\frac{2\omega - \omega_{mol}}{\gamma_d}\right)I^2 , \qquad (3.53)$$

where τ is the lifetime of the EM. From (3.53) at $2\omega = \omega_{mol}$, the intenstiy dependent damping γ_d is derived to be proportional to N_{mol}, which is the preferred support for the mutual collision of EMs as the main origin of the damping of the EM state.

In order to obtain the pump power dependence of the width of the GTA band, it is reasonable to take $2\hbar\gamma_d$ as the effective width W_{eff} of the GTA band. Then at the front surface of excitation, W_{eff} is proportional to the pump power I_0. The slight discrepancy of the pump power dependence of the width between W_{eff} and W in (3.52) may be brought about by oversimplification.

The following typical values are obtained from the spectrum in Fig. 3.40 for the maximum pump power (~ 1 MW cm^{-2}) by taking $\tau \sim 300$ ps [3.97]: $K_2'(\omega_{mol}I_0) = 500$ cm^{-1}, $N_{mol}(\omega_{mol}, I_0) = 8 \times 10^{16}$ cm^{-3} and $1/\gamma_d = 0.7$ ps. Since the time interval between successive collisions, $1/\gamma_d$, is much less than the lifetime of the EM, τ, sufficiently frequent collisions between EMs before their radiative decay have been demonstrated.

3.9.2 Effect on Secondary Emissions

a) Two-photon Resonant Raman Scattering. The effect of collision damping of the EMs is expected to appear in the secondary emission under TPR excitation as

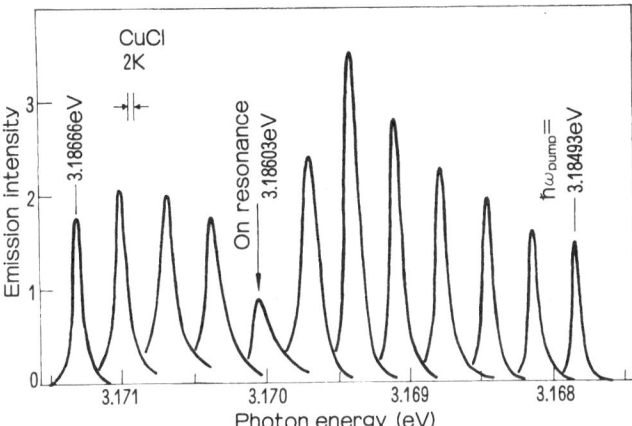

Fig. 3.41. Change in the spectrum of the M_T^R line with the variation of pump photon energy across the GTA band. Pump photon energies are also shown [3.96]

well. Figure 3.41 shows the change in the M_T^R spectrum with the variation of the pump photon energy across the GTA band, step by step, with an energy interval of ~ 0.16 meV. The vertical arrow indicates the energy position of the M_T^R peak under resonant excitation. Two distinct facts are shown by the spectra: (i) its intensity decreases significantly just around the GTA peak. (ii) the line width of the M_T^R line gradually broadens as the pump photon energy approaches the GTA peak and the line broadening becomes more pronounced with increasing pump power. Similar behavior of the M_T^R line width was observed when the lattice temperature was raised (Sect. 3.6) and it was ascribed to acoustic-phonon damping at the EM state [3.79]. The same explanation is valid for the pump power dependent broadening, replacing acoustic-phonon damping with mutual collision damping.

b) Effect on the Luminescence of the EM. Figure 3.42 shows the pump power dependence of the whole secondary emission spectrum of the EM (a) under the GTA band excitation at 3.18603 eV and (b) under the exciton band excitation at 3.201 eV with the same experimental conditions. In (a) two broad bands M_T and M_L and two narrow bands M_T^0 and M_L^0 are observed. The peak position of the intense M_T^R line is indicated by a vertical dashed line.

In (b) the main luminescence is rather simple and consists of the M_T and M_L bands only. Fine structures observed at 3.168 eV and 3.171 eV are thought to be some kind of bound-exciton luminescence.

The line shape of the M bands was calculated, under the assumption of a Maxwell-Boltzman distribution of EMs and was fitted to the experimental data, from which the effective temperature of the EM was obtained. Typical examples of line-shape fits are drawn as dots in Fig. 3.42 a, with $T_{eff} = 18$ K, and in Fig. 3.42 b with $T_{eff} = 24$ K.

From the above-mentioned analysis the following facts are known: The effective temperature in the case of (a) falls as the pump power decreases and is below 10 K at the lowest pump power, but in (b) the temperature does not decrease

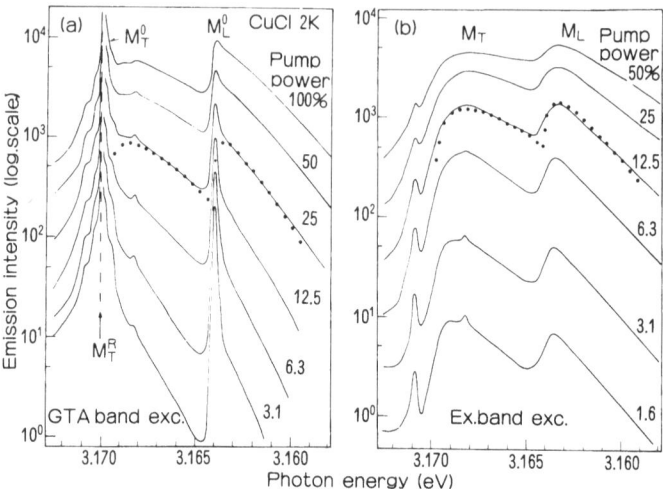

Fig. 3.42. Pump power dependence of the secondary emission spectrum of the EM (**a**) under resonant GTA band excitation and (**b**) under exciton band excitation [3.96]

below 20 K and is almost pump power independent at the lower pump powers. This fact clearly demonstrates that, under TPR excitation, the EMs giving the M broad bands are not mainly generated through the inelastic collision of the excitons produced in the TRRS or luminescence process (the secondary processes), but are generated directly through the GTA process (the primary process).

As will be shown in Fig. 3.44a the integrated intensity of the M broad bands at high pump power is larger than those of the TRRS lines and the M^0 bands. This fact is further evidence for the direct generation of EMs distributed over a large k giving the M broad bands through the primary process.

Two collision processes are effective for raising the temperature of the EM system at high pump power: the intra-band collision and an Auger-type one in which two EMs generated at $2 k_0$ collide with each other and one EM is scattered into a larger wave vector, whereas the other one is decomposed into an exciton and a photon.

Secondary emission was further investigated under high-energy off-resonant excitation. New emission bands designated X_T and X_L are observed [3.98] and as shown in Fig. 3.43 the X_T band consists of two peaks, while the X_L band has only peak. With decreasing pump photon energy, the M_T^R and M_L^R lines shift toward the lower-energy side, whereas the X_T and X_L bands shift toward the higher-energy side. Their intensities gradually increase and at resonant excitation the X bands merge into the M^0 bands so that these bands are observable only with high-energy off-resonant excitation. The origin of these bands has been investigated with the aid of polarization and pump power dependences of the spectra under backward and forward scattering configurations. Based on the above results, together with a model calculation, the X bands are assigned to a hot luminesc-

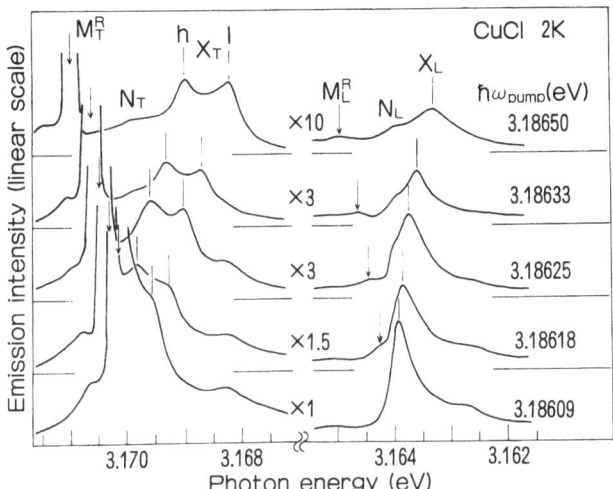

Fig. 3.43. Emission spectra for the backward scattering configuration around the M_T and M_L broad band regions for various pump photon energies, higher than the GTA peak, measured at 2 K. Peak positions of the X bands are indicated by vertical bars [3.98]

ence of the EMs isotropically distributed along their energy band through a certain kind of transverse relaxation process. The M_L^0 bands are found to be decomposed into two components: one is isotropic having the nature of the X bands and the other is strongly anisotropic as mentioned in Sect. 3.8. The appearance of X bands suggests that even under high-energy off-resonant excitation the nonlinear change in the polariton dispersion mentioned in Sect. 3.5.5 makes it possible for the EM to be directly excited with a large wave vector.

The transverse relaxation producing the X bands and the isotropic part of the M^0 band is found to be closely related to the mutual collision of EMs. However, the relaxation process causing the anisotropic part of the M^0 band is not yet clarified.

In evaporated films an emission called S was found (Sect. 3.6). There is some resemblance between the X_T and S bands with respect to the pump photon energy dependence of their peak energies. However, the complete emission spectra of the two are quite different. The S band has only one peak and the ratio of its peak intenstiy to that of the M_T^R line is much larger than that of the X_T band. A strain effect was also suggested for the enhancement of the acoustic-phonon scattering in evaporated films [3.83].

c) Pump Power Dependence of the Secondary Emissions. The comparison of the pump power dependences of the integrated intensities of the M_T^R line (indicated by ×) and the total luminescence (○) are shown in Fig. 3.44a under resonant excitation at 3.18603 eV and in Fig. 3.44b under higher-energy off-resonant excitation at 3.18622 eV. The behavior of the M broad band (▲) is also shown for reference. In (a) the M_T^R intensity increases as $I_0^{0.5}$ with respect to the pump power I_0, while the total intensity of the luminescence of the M broad bands and the M^0 bands increases as $I_0^{1.5}$. On the other hand, in (b) the M_T^R line increases as $I_0^{2.3}$ whereas the total luminescence of the M broad bands and the X bands

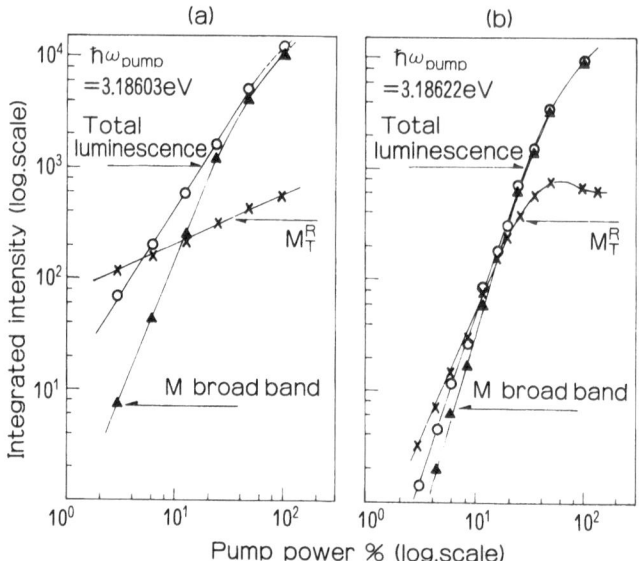

Fig. 3.44. Integrated intensities of the M_T^R line, M broad band, and total luminescence versus the pump power (**a**) under resonant excitation and (**b**) under high-energy off-resonant excitation [3.98]

increases as I_0^3, except for at much higher pump power where their intensities saturate.

From the calculation by *Hanamura* and *Takagahara* [3.87] (Sect. 2.5.5) the integrated intensities of the TRRS line, R, and the luminescence, L, under TPR excitation are obtained approximately, apart from an unimportant factor, as

$$L \sim B_0 \frac{2\tau\gamma_d}{(2\omega - \omega_{mol})^2 + \gamma_d^2} I_0^2, \quad \text{and}$$

$$R \sim B_0 \frac{\pi}{(2\omega - \omega_{mol})^2 + \gamma_d^2} I_0^2.$$

$$(3.54)$$

The strong collision damping in the EM state reduces the Raman yield, and instead, increases the luminescence yield as shown below.

For resonant excitation, where $2\omega = \omega_{mol}$ and γ_d is experimentally proportional to $I_0^{0.8}$ the following relations are obtained from (3.54)

$$L \sim B_0 \frac{2\tau}{\gamma_d} I_0^2 \propto I_0^{1.2}, \quad \text{and}$$

$$R \sim B_0 \frac{\pi}{\gamma_d^2} I_0^2 \propto I_0^{0.4}.$$

$$(3.55)$$

On the other hand, for off-resonant excitation

$$L \sim B_0 \frac{2\tau\gamma_d}{(2\omega - \omega_{\mathrm{mol}})^2} I_0^2 \propto I_0^{2.8} , \quad \text{and}$$

$$R \sim B_0 \frac{\pi}{(2\omega - \omega_{\mathrm{mol}})^2} I_0^2 \propto I_0^2 ,$$

(3.56)

as long as γ_d is smaller than $|2\omega - \omega_{\mathrm{mol}}|$, whereas for $\gamma_d > |2\omega - \omega_{\mathrm{mol}}|$ the results are approximately given by those for resonant excitation. Under the assumption of strong collision damping in the EM state, the calculated pump power dependences reproduce fairly well those obtained experimentally including the saturation tendency.

3.10 Spatial Dispersion of the Exciton and Excitonic Molecule

The selective excitation technique afforded by a tunable laser has developed studies of spatial dispersion. Single-beam excitations at the GTA, however, enable us to determine the dispersion of the exciton only in narrow regions of k around k_0 or $3k_0$. For the determination of it in a wider k region, a new technique

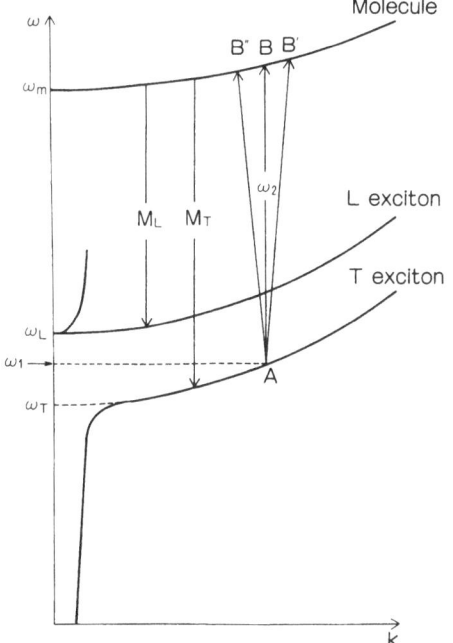

Fig. 3.45. Stepwise two-photon excitation into the EM state with $\hbar\omega_1$ and $\hbar\omega_2$ via the excitonic polariton at a large wave vector. The luminescence transitions are indicated by M_L and M_T

of stepwise two-photon excitation has been developed. The principle is as follows. When one incident photon has an energy $\hbar\omega_1(k_1)$ in a region between the L and T exciton energies at $k = 0$, $\hbar\omega_L$ and $\hbar\omega_T$, the photon inside the crystal becomes a polariton at A having a large wave vector as shown in Fig. 3.45. Simultaneous excitation with another laser beam of energy $\hbar\omega_2(k_2)$ allows the EM to be created at B. When the directions of the two laser beams are the same or opposite, the EM is attained at B' or B". At the same time as the excitation into the EM state TRRS occurs to the exciton states causing the emission of Raman light which is also a polariton $\hbar\omega_R(k_R)$. The energy and wave vector, ω_{ex} and k_{ex}, of the final state exciton in the scattering process are given by

$$\omega_{ex} = \omega_1 + \omega_2 - \omega_R \quad \text{and} \quad k_{ex} = k_1 + k_2 - k_R .$$

Under the resonant condition of TRRS, the relation between $\omega_1 + \omega_2$ and $k_1 + k_2$ gives the dispersion of the EM.

3.10.1 CuCl

Mita et al. [3.71] could observe two-photon excitation into the EM by finding an intensity minimum in the transmission spectrum measured with the second laser

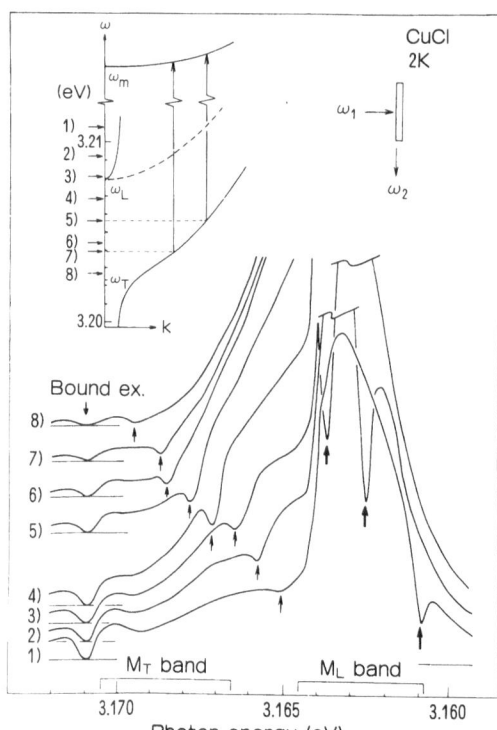

Fig. 3.46. Intensity minima of frequencies ω_2, shown by arrows, in the M_T and M_L emission bands. The minima are due to two-photon absorption into the EM state in cooperation with excitation of frequencies ω_1, the corresponding energies of which are shown in the inset [3.71]

beam as a probe light or else by finding an intensity minimum inside the M emission band because the transition from A to B is the reverse process to the M emission.

Figure 3.46 shows emission spectra of EM under normal incident excitation of frequency ω_1 for various energies in the region around the exciton band. Emission is observed in a direction perpendicular to k_1. The excitation energies for spectra (1)–(8) are shown inside the figure. Shallow minima are found in the M_T band region as shown by the thin arrows. The energy of this minimum varies with the change of excitation energy ω_1. When ω_1 exceeds ω_L, cases (1)–(3), other deep minima are found in the M_L band region as shown by the thick arrows.

As will be clarified later, the intensity minimum in the M_T band region corresponds to the energy of the photon which causes the stepwise two-photon excitation into the EM cooperatively with ω_1. When ω_1 exceeds ω_L the upper-branch polariton is created and with subsequent absorption of the second photon the EM is formed at $k \sim 2k_0$. The energy of the second photon lies in the M_L band region, thus, dips are observed in the M_L band.

In Fig. 3.47 the relation between ω_1 and the peak energy of the dip, ω_2, is plotted with open circles and dots for those in the M_L and M_T band regions, respectively. The dips in the M_L band region lie on a straight line, a, representing

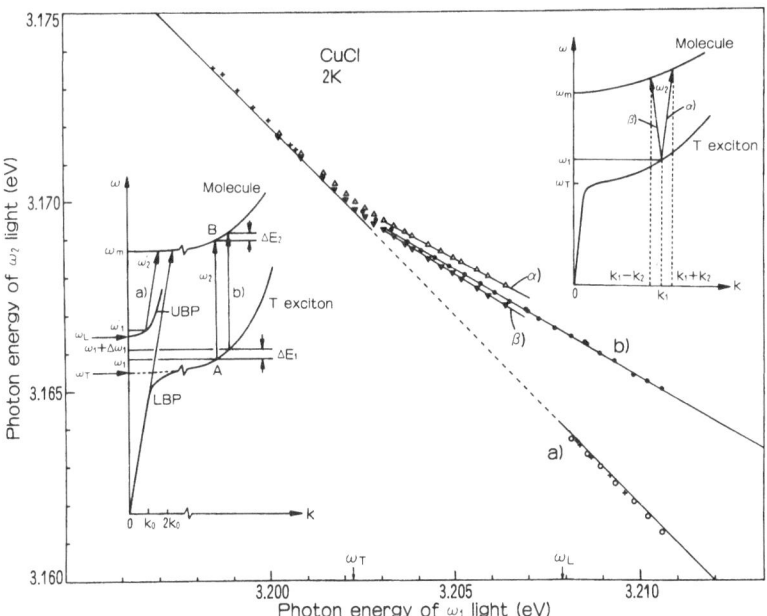

Fig. 3.47. Plot of ω_2 in Fig. 3.46 against ω_1. Curves (a) and (b) are the cases for ω_2 in the M_L and M_T band regions, respectively. Excitation processes in (a) and (b) are shown in the lower inset. Curves (α) and (β) are the relations of ω_2 and ω_1 in the cases of A → B' and A → B" transitions in Fig. 3.45, respectively as shown in the upper inset [3.71]

the relation, $\omega_1 + \omega_2 = 6.372$ eV. The energy 6.372 eV is just twice that of the GTA peak, $\hbar\Omega_0 = 3.1861$ eV, for the direct generation of the EM at $k = 2k_0$. Considering that the spatial dispersion of the EM around $k = 2k_0$ is very small, the dip in the M_L band region is explained to be due to stepwise two-photon excitation into the EM via the upper-branch polariton created with $\omega_1 > \omega_L$.

The energy of the dip in the M_T band region is also found to lie on a straight line, b, of gradient -0.57. This is due to the excitation into the EM as shown below. We assume parabolic energy dispersion of the transverse exciton with mass m_{ex}^T and the EM with mass m_{mol}. With excitation with $\omega_1 > \omega_T$, excitons are created with wave vector k_1, which is much larger than k_0. With the help of $\omega_2(k_2 \fallingdotseq k_0)$ corresponding to the dip in the M_T band, the EM is created with almost the same wave vector as that of the exciton, $\sqrt{(k_1^2 + k_2^2)} \fallingdotseq k_1$, because k_1 and k_2 are perpendicular to each other. Thus, the transition from the exciton to the EM state is considered to occur vertically (perpendicular to the k-axis), $A \rightarrow B$, when an exciton is created by ω_1 at $k \gg k_0$. By these assumptions, the gradient of the straight line b is given by $(\Delta E_2 - \Delta E_1)/\Delta E_1$, where ΔE_1 is the energy difference between the exciton states attained by the ω_1 and $\omega_1 + \Delta\omega_1$ excitations and ΔE_2 is that between the EM states, as shown in the inset. From the relation $\Delta E_2/\Delta E_1 = m_{ex}^T/m_{mol}$, we obtain the relation: [gradient of line b] $= m_{ex}^T/m_{mol} - 1 = -0.57$. Thus the EM mass is determined to be 2.3 times the transverse exciton.

Under the excitation of the crystal with ω_1 the transmission was measured by the second dye laser beam being scanned in a region of 3.160–3.175 eV. The transmission minimum was observed at ω_2 which varied with changes in ω_1. The energy of the transmission minimum ω_2 is plotted by crosses in Fig. 3.47. The data points lie on the straight line a. Thus, ω_2 in the 3.161–3.164 eV region cooperates with ω_1 to cause the excitation into the EM at $k \sim 2k_0$ via the upper-branch-polariton state, while ω_2 in the 3.171–3.174 eV region does so via the lower-branch polariton.

In the case where ω_1 falls in the region between ω_L and ω_T, ω_2 appears at different energies in two experimental configuration, α and β, where k_1 and k_2 are in the same and opposite directions, respectively, as shown by $A \rightarrow B'$ $(k = k_1 + k_2)$ and $A \rightarrow B''$ $(k = k_1 - k_2)$ in Fig. 3.45.

The relations between ω_2 and ω_1 in the two configurations are shown by curves α and β in Fig. 3.47. The energy difference of ω_2 in the two configurations arises directly from the spatial dispersion of the EM. Relations between $\hbar\omega_2 = E_{mol}(k_1 \pm k_2) - E_{ex}(k_1)$ and $\hbar\omega_1 = E_{ex}(k_1)$ were calculated for various values of the transverse exciton mass, m_{ex}^T, in a range 2–4 m_0 with the assumption of parabolic dispersions and with the mass of the EM 2.3 times that of the transverse exciton. The best fit was found with $m_{ex}^T \cong 2.5\, m_0$.

At the same time as the stepwise two-photon absorption of the EM, resonant Raman scattering into the L and T exciton states was also found. These lines originate from the same processes as the M_L^R and M_T^R lines but in the region of large wave vectors. From these Raman scattering measurements the relation between the L and T exciton dispersions is obtained and the translational masses

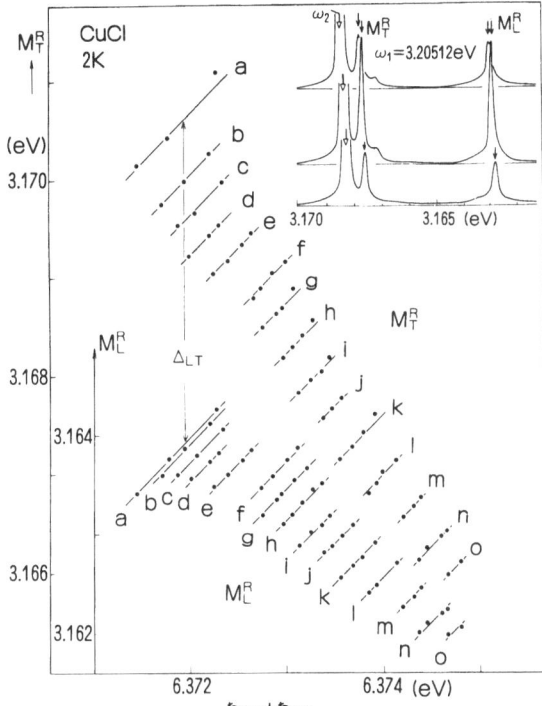

Fig. 3.48. Energy plot of the M_L^R and M_T^R Raman lines versus $\hbar(\omega_1 + \omega_2)$ with ω_1 fixed and ω_2 varied, in CuCl. Raman spectra are shown for the case of (j) [3.8]

of the L and T excitons are determined as $(3.1 \pm 0.1)\,m_0$ and $(2.3 \pm 0.1)\,m_0$, respectively, as follows [3.8].

Examples of the Raman spectra are shown at the top of Fig. 3.48 where ω_1 is fixed at 3.20512 eV and ω_2 is varied. The Raman scattering is found in a very narrow excitation region and it is resonantly enhanced at $\hbar\omega_2 = 3.16835$ eV with M_L^R and M_T^R lines at 3.16295 and 3.16772 eV, respectively. The energy difference between the two Raman lines gives the L–T splitting Δ_{LT} of the exciton. The Raman spectra were measured with ω_1 at several different energies in a region around the L–T gap.

The energies of the M_L^R and M_T^R lines are plotted by dots in Fig. 3.48 against $\omega_1 + \omega_2$. The paired lines of (j) correspond to the spectra shown at the top of the figure. The energy shift of the Raman lines is the same as that of ω_2 as seen by the solid lines of slope one. The vertical separation in energy between the paired solid lines gives Δ_{LT} and it is clearly seen to decrease with the increase of $\omega_1 + \omega_2$, i.e., with the increase of k of the intermediate EM. The final state exciton energies, E_{ex}^L and E_{ex}^T, were obtained from the differences between $\hbar(\omega_1 + \omega_2)$ and the Raman line energies.

The dispersion curve for $m_{ex}^T = 2.3m_0$ is drawn in Fig. 3.49 adopting the values [3.71] $\omega_L = 3.2079$ eV, $\omega_T = 3.2022$ eV, and $\epsilon_\infty = 5.59$. For example, from the solid line (d) in Fig. 3.48 the final state T exciton energy is obtained as

$$E_{ex}^T = \hbar(\omega_1 + \omega_2) - (M_T^R\text{-line energy}) = 3.20276 \text{ eV} .$$

In this case, the first laser photon ω_1 was 3.20233 eV. The wave vector k_1 of the exciton created by ω_1 is read out of the dispersion curve to be 1.655×10^6 cm^{-1} and k_{ex} is calculated as 2.107×10^6 cm^{-1}. Thus, one pair of E_{ex}^T and k_{ex} obtained from the solid line (d), is plotted by a dot designated (d) in Fig. 3.49. Pairs of E_{ex}^T and k_{ex} obtained similarly from solid lines (a), (b), (c), ..., (o) in Fig. 3.48 are similarly plotted in Fig. 3.49. The $E_{ex}^T : k_{ex}$ relation thus obtained is nothing but the dispersion, and it is found to be best fit to the dispersion curve when it is initially drawn with $m_{ex}^T = 2.3 \, m_0$. Therefore, the effective mass of the T exciton is finally determined to be $(2.3 \pm 0.1) \, m_0$. The L exciton dispersion is obtained with $E_{ex}^T + \Delta_{LT}$ as shown by dots in the figure, and it is expressed by the parabolic k dependence shown by the solid curve drawn with the effective mass $m_{ex}^L = 3.14 \, m_0$. Thus the L exciton mass is found to be $(3.1 \pm 0.1) \, m_0$ which is fairly heavy compared to the T exciton mass.

The L–T splitting of the exciton arises from the e–h exchange interaction. Based on the Frenkel or Wannier exciton formalism, the calculation of the dipole-dipole interaction which results from the multipole expansion of the exchange interaction gives very small k dependence of the L–T splitting; only four-hundredths of that observed [3.99]. However, the k dependence calculated

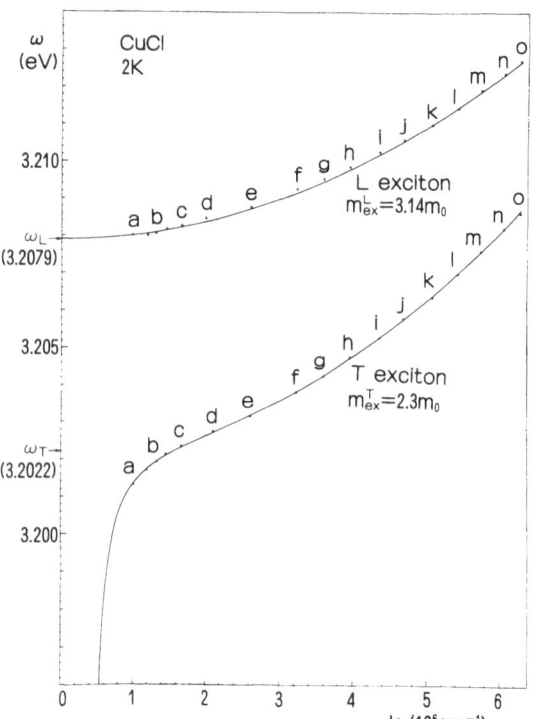

Fig. 3.49 (——) spatial dispersions of the T and L excitons in CuCl calculated with $m_{ex}^T = 2.3 m_0$ and $m_{ex}^L = 3.14 m_0$. Small dots show the dispersions obtained from Raman scattering [3.8]

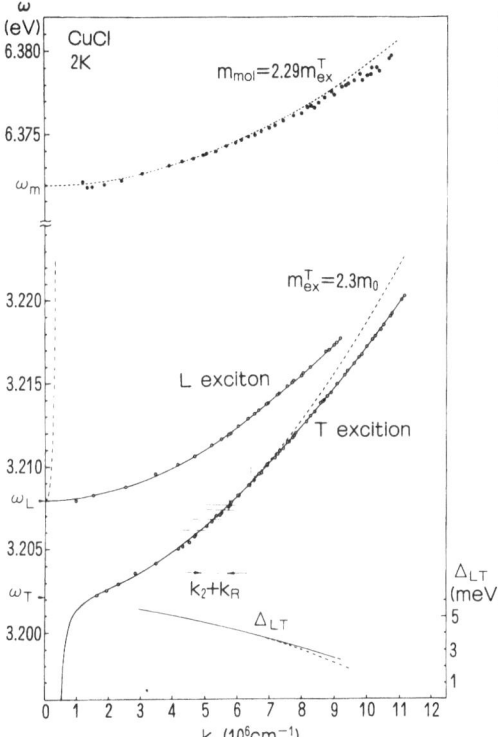

Fig. 3.50. Spatial dispersions of the exciton and EM determined in wider k regions in CuCl. Broken lines show the parabolic k dependences. The k dependence of the L–T splitting Δ_{LT} is also shown [3.101]

by k–p perturbation using the exciton wave function constructed with the k-dependent Bloch functions of the conduction and valence bands has been found to be in fair agreement with that determined experimentally [3.100] (Sect. 2.4.3).

In Fig. 3.50, the dispersions are given for an extended large k region of $k > 6 \times 10^6$ cm^{-1} [3.101]. Since the Raman scattering was measured for the backward scattering configuration in this case, k_{ex} is given by $k_{ex} = k_1 + k_2 + k_R$. At first, ω_1 was chosen at several energies a little below ω_L, and k_1 was obtained from the exciton dispersion curve, shown by a broken line, which is a reproduction of that in Fig. 3.49. Next, ω_1 was chosen in an extended region and a similar procedure enabled us to determine the dispersion in a further extended region. These processes were repeated with increasing ω_1. In the figure, small horizontal bars indicate some examples of the chosen ω_1 positions and the vertical ones show the wave vectors of the exciton left behind. The plot of $\omega_{mol} = \omega_1 + \omega_2$ with respect to $k_{mol} = k_1 + k_2$ gives the EM dispersion as shown in the figure. Parabolic dispersion shows the effective mass of the EM to be $m_{mol} = 5.3\, m_0$, however, deviation from parabolic k dependence is found for $k > 7 \times 10^6$ cm^{-1}.

We found that the upper- and lower-branch polaritons are simultaneously created by ω_1 excitation when $\omega_1 > \omega_L$. It is remarkable that in such a large k

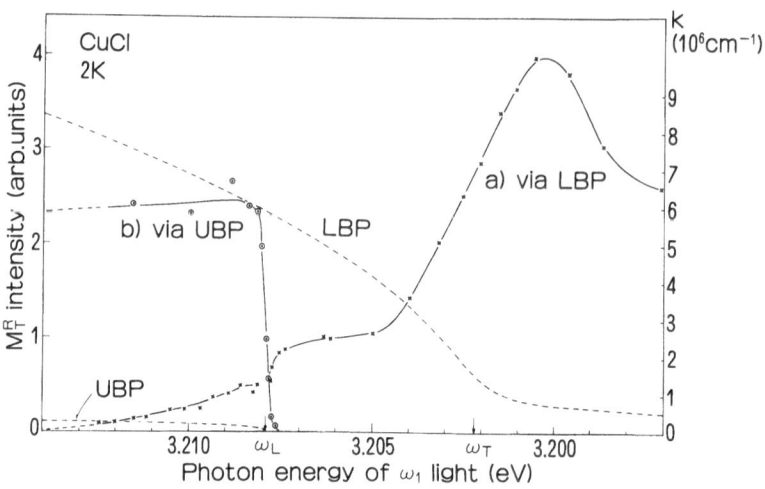

Fig. 3.51. ω_1 dependence of the M_T^R Raman line intensities via upper- and lower-branch polaritons [3.101]

excitation, the created polaritons remember the k vector and make possible excitation into an EM state with a sharply defined wave vector.

The M_T^R Raman intensity from the EM states attained via upper- and lower-branch polaritons are plotted in Fig. 3.51 against ω_1 [3.101]. Assuming that the two-photon excitation cross sections into the EM via the two branches are the same, the figure indicates the allotment of the incident polariton into the two modes of the upper and lower branches, namely, it might provide information for the so-called additional boundary condition (ABC) [3.102].

3.10.2 CuBr

The exciton of CuBr is split into Γ_5 and $\Gamma_3 + \Gamma_4$ states by the spin exchange interaction and has eightfold degeneracy. At a finite wave vector, K, the K dependent mixing among the eight states provide the multiple branch and anisotropic dispersion [3.103]. In the zinc blende structure, a term linear in k is allowed in the band structure due to the lack of inversion symmetry [3.104]. The term linear in K in the exciton arises from that in the Γ_8 valence band and it plays an essential role in the reflectance anomaly at the energy of the spin-triplet exciton of CuBr [3.20], [3.105], CuI [3.10], and CdTe [3.106]. Terms quadratic in K lead to the so-called heavy- and light-mass excitons.

a) Theoretical Foundation. The electron and hole Hamiltonians, $H^{(e)}$ and $H^{(h)}$, in the effective mass approximation are expressed by the wave vector k and the angular momentum operator $J(J = 3/2)$ for the hole as

$$H^{(e)} = \frac{\hbar^2}{2\,m_e} k^2 ,$$

$$
\begin{aligned}
H^{(h)} = \frac{\hbar^2}{m_0} &\left[\left(\gamma_1 + \frac{5}{2}\gamma_2 \right) \frac{k^2}{2} - \gamma_2 (k_x^2 J_x^2 + k_y^2 J_y^2 + k_z^2 J_z^2) \right. \\
&\left. -2\gamma_3 (\{k_x k_y\}\{J_x J_y\} + \text{c.p.}) \right] + \hbar K_1 [k_x\{(J_y^2 - J_z^2)J_x\} + \text{c.p.}] , \quad (3.57)
\end{aligned}
$$

where $\{k_x k_y\} = (k_x k_y + k_y k_x)/2$ etc., and $(\gamma_1, \gamma_2, \gamma_3)$ are the Luttinger parameters [3.107], $\hbar K_1$ is the coefficient of the term linear in k, and c.p. means the cyclic permutation of the preceding term.

The exciton-polariton dispersions are analyzed by various interaction mechanisms which act among the eight components [3.103, 108]. In the present case, where any external perturbation such as a magnetic field, electric field or stress is absent, the relevant mechanisms are (i) e–h exchange interaction and (ii) energy terms linear and quadratic in K. The terms of (ii) include not only the usual spatial dispersion effect, but also the effect of heavy- and light-masses and the warping effect. Finally, the polariton effects on the corresponding oscillator strengths are considered.

The exciton Hamiltonian of the exchange interaction is represented by

$$H_{\text{exch}} = \Delta_1 \left(j \cdot \sigma - \frac{3}{4} \right) + \Delta_{\text{LT}} \left(\cos^2 \theta - \frac{1}{3} \right) , \qquad (3.58)$$

where σ is the angular momentum operator for the electron, Δ_1 the isotropic spin-exchange splitting, Δ_{LT} the longitudinal and transverse splitting owing to the dipole-dipole interaction and θ the angle between the dipole and K. The anisotropic spin-exchange splitting Δ_2 between the Γ_3 and Γ_4 states is expected to be very small [3.105] and is neglected here.

The K dependent part of the exciton Hamiltonian is characterized by the hole state and is given by replacing γ_1, γ_2, γ_3, and K_1 by γ_1^{ex}, γ_2^{ex}, γ_3^{ex}, and K_1^{ex}, respectively, in (3.57).

$$
\begin{aligned}
H(K) = \frac{\hbar^2}{m_0} &\left[\left(\gamma_1^{\text{ex}} + \frac{5}{2}\gamma_2^{\text{ex}} \right) \frac{K^2}{2} - \gamma_2^{\text{ex}} (K_x^2 J_x^2 + K_y^2 J_y^2 + K_z^2 J_z^2) \right. \\
&\left. -2\gamma_3^{\text{ex}} (\{K_x K_y\}\{J_x J_y\} + \text{c.p.}) \right] + \hbar K_1^{\text{ex}} [K_x\{(J_y^2 - J_z^2)J_x\} + \text{c.p.}] ,
\end{aligned}
$$

$$(3.59)$$

where γ_1^{ex}, γ_2^{ex}, and γ_3^{ex} are, so to speak, the Luttinger parameters of excitons and $\hbar K_1^{\text{ex}}$ is the coefficient of the term linear in K for the exciton. Relations between them are as follows:

$$
\begin{aligned}
&\gamma_1^{\text{ex}} = \beta_h \gamma_1 , \qquad \gamma_2^{\text{ex}} = \beta_h^2 \gamma_2 , \\
&\gamma_3^{\text{ex}} = \beta_h^2 \gamma_3 , \qquad K_1^{\text{ex}} = \beta_h K_1 ,
\end{aligned}
$$

$$(3.60)$$

where $\beta_h = m_0/(m_0 + m_e\gamma_1)$. According to theoretical results by *Kane* [3.109], γ_1^{ex} is

$$\gamma_1^{ex} = \frac{m_0}{M_a} , \tag{3.61}$$

where M_a is given in [3.109].

Matrix elements of the exciton Hamiltonian are represented by the eight bases [3.103]

$$|2+\rangle, |20\rangle, |1+\rangle, |1-\rangle, |2-\rangle, |x\rangle, |y\rangle, |z\rangle ,$$

in the absence of any symmetry-breaking effects due to finite K and the dipole-dipole interaction. Among them, $|2+\rangle, |20\rangle, |1+\rangle, |1-\rangle, |2-\rangle$ are the bases for the Γ_3 and Γ_4 irreducible representations (the total angular momentum $J = 2$), and have energy E_{ex}^t, while $|x\rangle, |y\rangle, |z\rangle$ are the bases for the Γ_5 representation ($J = 1$), and correspond to x, y and z polarizations, respectively. They are split into one longitudinal state (energy E_{ex}^L) and two transverse states (E_{ex}^T), for any direction of K, by the dipole-dipole interaction.

K dependent matrix elements for above bases are given in Table 3.3 (exactly the same as Tables VIII and IX of [3.103],) where complex conjugated terms of the upper-right (lower-left) corner have to be added to terms of the lower-left (upper-right) corner. The symbols are defined as follows:

$$W = \gamma_1^{ex} Q(K_x^2 + K_y^2 + K_z^2) ,$$

$$U = \frac{\sqrt{3}}{2} \gamma_2^{ex} Q(K_x^2 - K_y^2) ,$$

$$V = \gamma_2^{ex} Q[K_z^2 - (K_x^2 + K_y^2)/2] ,$$

$$X = \sqrt{3} \gamma_3^{ex} Q K_y K_z ,$$

$$Y = \sqrt{3} \gamma_3^{ex} Q K_z K_x ,$$

$$Z = \sqrt{3} \gamma_3^{ex} Q K_x K_y ,$$

$$L_x = -\frac{\sqrt{3}}{4} \hbar K_1^{ex} K_x ,$$

$$L_y = -\frac{\sqrt{3}}{4} \hbar K_1^{ex} K_y ,$$

$$L_z = -\frac{\sqrt{3}}{4} \hbar K_1^{ex} K_z , \tag{3.62}$$

where $Q = \hbar^2/2m_0$. (W, U, V, X, Y, Z) and (L_x, L_y, L_z) are quadratic and linear in K, respectively.

Table 3.3. K-dependent matrix elements. Complex conjugated terms of the upper-right (lower-left) corner have to be added to terms of the lower-left (upper-right) corner. For the notation, see text

	$|2+\rangle$	$|20\rangle$	$|1+\rangle$	$|1-\rangle$	$|2-\rangle$	$|x\rangle$	$|y\rangle$	$|z\rangle$
$|2+\rangle$	$W-2V$	$-2U$	$-i\sqrt{3}X$	$i\sqrt{3}Y$	0	$-iX$	$-iY$	$2iZ$
$|20\rangle$		$W+2V$	iX	iY	$-2iZ$	$-i\sqrt{3}X$	$i\sqrt{3}Y$	0
$|1+\rangle$	$-\sqrt{3}L_x$	L_x	$W+V-\sqrt{3}U$	$\sqrt{3}Z$	$\sqrt{3}Y$	$\sqrt{3}V+U$	Z	$-Y$
$|1-\rangle$	$\sqrt{3}L_y$	L_y		$W+V+\sqrt{3}U$	$\sqrt{3}X$	$-Z$	$-\sqrt{3}V+U$	X
$|2-\rangle$	0	$-2L_z$			$W-2V$	Y	$-X$	$-2U$
$|x\rangle$	L_x	$\sqrt{3}L_x$		$-2iL_z$	$2iL_y$	$W-V+\sqrt{3}U$	$\sqrt{3}Z$	$\sqrt{3}Y$
$|y\rangle$	L_y	$-\sqrt{3}L_y$	$2iL_z$		$-2iL_x$		$W-V-\sqrt{3}U$	$\sqrt{3}X$
$|z\rangle$	$-2L_z$	0	$-2iL_y$	$2iL_x$				$W+2V$

The dispersions quadratic in K representing the heavy and light masses are considerably changed by the terms linear in K, but are given formally by assuming $K_1^{ex} = 0$, as follows:

$$E(K^2) = \begin{cases} (\gamma_1^{ex} \pm 2\gamma_2^{ex}) \, QK^2 \,, & (K \| [001]) \\ (\gamma_1^{ex} \pm 2\gamma_3^{ex}) \, QK^2 \,, & (K \| [111]) \\ [\gamma_1^{ex} \pm \sqrt{(\gamma_2^{ex})^2 + 3\,(\gamma_3^{ex})^2}] \, QK^2 \,, & (K \| [110]) \end{cases} \tag{3.63}$$

The light and heavy masses of the exciton in units of m_0 are the reciprocals of the coefficients of QK^2 in (3.63), and have the same forms for the heavy and light holes with γ_i^{ex} replaced by γ_i.

Next, the exciton Hamiltonian is represented for the specific direction of K by introducing new bases.

i) $K \| [111]$. The quantization axis is parallel to the [111] direction. The bases $|2, 2\rangle, |2, 1\rangle, |2, 0\rangle, |2, -1\rangle, |2, -2\rangle, |1, 1\rangle, |1, 0\rangle, |1, -1\rangle$ are defined as follows.

$$|2, 2\rangle \;\; = -\frac{i}{3} \, (i\sqrt{3}|20\rangle + 2|2-\rangle - |1+\rangle - |1-\rangle) \,,$$

$$|2, 1\rangle \;\; = \frac{1}{\sqrt{6}} \, (|1+\rangle - |1-\rangle - 2i|2+\rangle) \,,$$

$$|2, 0\rangle \;\; = \frac{1}{\sqrt{3}} \, (|1+\rangle + |1-\rangle + |2-\rangle) \,,$$

$$|2, -1\rangle = \frac{1}{3\sqrt{2}} \, (-2i\sqrt{3}|20\rangle + 2|2-\rangle - |1+\rangle - |1-\rangle) \,,$$

$$|2, -2\rangle = \frac{1}{\sqrt{3}}(|2+\rangle - i|1+\rangle + i|1-\rangle) \,,$$

$$|1, 1\rangle \;\; = \frac{1}{\sqrt{6}} \, (2|z\rangle - |x\rangle - |y\rangle) \,,$$

$$|1, 0\rangle \;\; = \frac{1}{\sqrt{3}} \, (|x\rangle + |y\rangle + |z\rangle) \,,$$

$$|1, -1\rangle = -\frac{1}{\sqrt{2}} \, (|x\rangle - |y\rangle) \,, \tag{3.64}$$

where $|1, \pm1\rangle$ are the transverse states and $|1, 0\rangle$ is the longitudinal one. The exciton Hamiltonian is decomposed into two equivalent 3×3 blocks and two 1×1 blocks, as shown in Table 3.4.

Table 3.4. Block-diagonal forms of the exciton Hamiltonian for $K \parallel [111]$

	$\|1, 0\rangle$	
	$[\Delta_{LT} + E_{Tt} + (\gamma_1^{ex} + 2\gamma_3^{ex})G^2]$	

	$\|2, 0\rangle$	
	$[(\gamma_1^{ex} + 2\gamma_3^{ex})G^2]$	

$\|2, \mp2\rangle$	$\|2, \pm1\rangle$	$\|1, \pm1\rangle$
$(\gamma_1^{ex} - 2\gamma_3^{ex})G^2$	T	$\sqrt{3}\,T$
T	$(\gamma_1^{ex} + \gamma_3^{ex})G^2$	$-\sqrt{3}\,\gamma_3^{ex}G^2$
$\sqrt{3}\,T$	$-\sqrt{3}\,\gamma_3^{ex}G^2$	$E_{Tt} + (\gamma_1^{ex} - \gamma_3^{ex})G^2$

The $\|1, 0\rangle$ and $\|2, 0\rangle$ states are pure longitudinal and pure spin-triplet, respectively, even at a finite K. The origin of the energy is chosen to be the spin-triplet state energy E_{ex}^t. The short-range exchange energy E_{Tt} is defined as

$$E_{Tt} = E_{ex}^T - E_{ex}^t = -(2\Delta_1 + \tfrac{1}{3}\Delta_{LT}) , \tag{3.65}$$

and $G^2 = \hbar^2 K^2/2m_0 = QK^2$. The term linear in K, T, is given by

$$T = \frac{\sqrt{3}}{2\sqrt{2}} \hbar K_1^{ex} K . \tag{3.66}$$

Blocks for the bases $\|2, \mp2\rangle$, $\|2, \pm1\rangle$, $\|1, \pm1\rangle$ mean the twofold-degenerate polaritons. Linear in K dispersions of the spin-triplet excitons near the Γ point are $\pm T$ (twofold degenerate) and 0 (horizontal).

ii) $K \parallel [110]$. The quantization axis is parallel to the $[1\bar{1}0]$ direction which is perpendicular to the direction of K. The new bases $i\|2-\rangle$, $\|20\rangle$, $\|x''\rangle$, $i\|2+\rangle$, $\|y''\rangle$, $\|x'\rangle$, $\|y'\rangle$, $\|z\rangle$ are introduced, where $\|x''\rangle$, $\|y''\rangle$, $\|x'\rangle$ and $\|y'\rangle$ are defined as follows.

$$|x''\rangle = \frac{1}{\sqrt{2}} (|1+\rangle + |1-\rangle) ,$$

$$|y''\rangle = \frac{i}{\sqrt{2}} (|1+\rangle - |1-\rangle) ,$$

$$|x'\rangle = \frac{i}{\sqrt{2}} (|x\rangle + |y\rangle) ,$$

$$|y'\rangle = \frac{1}{\sqrt{2}} (|x\rangle - |y\rangle) . \tag{3.67}$$

Table 3.5. Block-diagonal forms of the exciton Hamiltonian for $\mathbf{K}\parallel[110]$

	$\mathrm{i}\lvert 2+\rangle$	$\lvert y''\rangle$	$\lvert x'\rangle$	$\lvert z\rangle$
$\mathrm{i}\langle 2+\rvert$	$(\gamma_1^{\mathrm{ex}}+\gamma_2^{\mathrm{ex}})G^2$	$\sqrt{\tfrac{3}{2}}\,T$	$-\dfrac{1}{\sqrt{2}}T$	$\sqrt{3}\,\gamma_3^{\mathrm{ex}}G^2$
$\langle y''\rvert$	$\sqrt{\tfrac{3}{2}}\,T$	$(\gamma_1^{\mathrm{ex}}-\tfrac{1}{2}\gamma_2^{\mathrm{ex}}-\tfrac{3}{2}\gamma_3^{\mathrm{ex}})G^2$	$-\dfrac{\sqrt{3}}{2}(\gamma_2^{\mathrm{ex}}-\gamma_3^{\mathrm{ex}})G^2$	$-\sqrt{2}\,T$
$\langle x'\rvert$	$-\dfrac{1}{\sqrt{2}}T$	$-\dfrac{\sqrt{3}}{2}(\gamma_2^{\mathrm{ex}}-\gamma_3^{\mathrm{ex}})G^2$	$\Delta_{\mathrm{LT}}+E_{\mathrm{Tt}}+(\gamma_1^{\mathrm{ex}}+\tfrac{1}{2}\gamma_2^{\mathrm{ex}}+\tfrac{3}{2}\gamma_3^{\mathrm{ex}})G^2$	0
$\langle z\rvert$	$\sqrt{3}\,\gamma_3^{\mathrm{ex}}G^2$	$-\sqrt{2}\,T$	0	$E_{\mathrm{Tt}}+(\gamma_1^{\mathrm{ex}}-\gamma_2^{\mathrm{ex}})G^2$

	$\mathrm{i}\lvert 2-\rangle$	$\lvert 20\rangle$	$\lvert x''\rangle$	$\lvert y'\rangle$
$\mathrm{i}\langle 2-\rvert$	$(\gamma_1^{\mathrm{ex}}+\gamma_2^{\mathrm{ex}})G^2$	$\sqrt{3}\,\gamma_3^{\mathrm{ex}}G^2$	0	$\sqrt{2}\,T$
$\langle 20\rvert$	$\sqrt{3}\,\gamma_3^{\mathrm{ex}}G^2$	$(\gamma_1^{\mathrm{ex}}-\gamma_2^{\mathrm{ex}})G^2$	$-\dfrac{1}{\sqrt{2}}T$	$-\sqrt{\tfrac{3}{2}}\,T$
$\langle x''\rvert$	0	$-\dfrac{1}{\sqrt{2}}T$	$(\gamma_1^{\mathrm{ex}}-\tfrac{1}{2}\gamma_2^{\mathrm{ex}}+\tfrac{3}{2}\gamma_3^{\mathrm{ex}})G^2$	$-\dfrac{\sqrt{3}}{2}(\gamma_2^{\mathrm{ex}}+\gamma_3^{\mathrm{ex}})G^2$
$\langle y'\rvert$	$\sqrt{2}\,T$	$-\sqrt{\tfrac{3}{2}}\,T$	$-\dfrac{\sqrt{3}}{2}(\gamma_2^{\mathrm{ex}}+\gamma_3^{\mathrm{ex}})G^2$	$E_{\mathrm{Tt}}+(\gamma_1^{\mathrm{ex}}-\tfrac{1}{2}\gamma_2^{\mathrm{ex}}-\tfrac{3}{2}\gamma_3^{\mathrm{ex}})G^2$

The bases $|y'\rangle$ and $|z\rangle$ correspond to the transverse states whose dipoles are parallel to the [1$\bar{1}$0] and [001] directions, respectively, and $|x'\rangle$ the longitudinal state whose dipole is parallel to the [110] direction. As shown in Table 3.5 the exciton Hamiltonian has the block-diagonal form of two independent 4×4 matrices, which means all of the degeneracies are removed at a finite K. It is noted that the longitudinal mode $|x'\rangle$ should mix with the transverse mode $|z\rangle$ at a finite K, and have the polariton effect. The linear in K dispersions of the spin-triplet exciton are $\pm \sqrt{3/2}\,T$, $\pm T/\sqrt{2}$, and 0 near the Γ point.

Finally, dispersion curves of the multicomponent polariton are described by

$$\frac{c^2 K^2}{\omega^2} = \epsilon_b + \sum_i \frac{4\pi\beta_i(K)}{1 - [\omega/\omega_i(K)]^2} \ ,$$

where $4\pi\beta_i(K)$ is the K-dependent oscillator strength of the i-th eigenstate having the eigenenergy $\hbar\omega_i(K)$.

b) **Experimental Results.** Figures 3.52 and 3.53 show the dispersion curves for $K\|[111]$ and $K\|[110]$ obtained by *Nozue* [3.9]. The large dots indicate the experimental results analyzed by assuming $\epsilon_b = 5.7$ [3.110], the solid lines are calculated curves fitted to the experimental results and the dotted lines are those without the polariton effect. The fitting parameters are E_{ex}^L, E_{ex}^T, E_{ex}^t, γ_1^{ex}, γ_3^{ex}, and $\hbar K_1^{ex}$. Of these E_{ex}^L is determined by extrapolating the parabolic dispersion

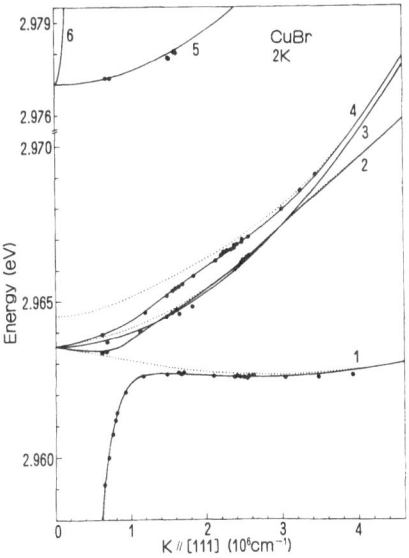

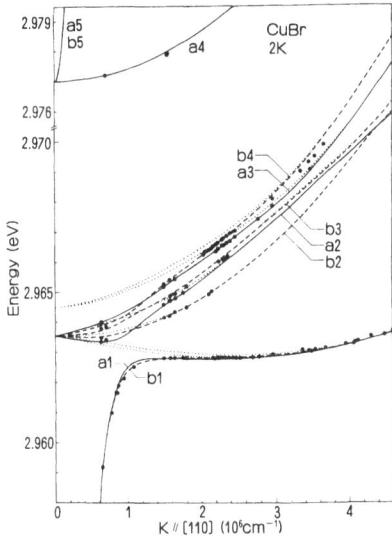

Fig. 3.52. Spatial dispersions of the $Z_{1,2}$ exciton polariton for $K\|[111]$ in CuBr. Dots indicate experimental results. Solid lines are the calculated fits [3.9]

Fig. 3.53. Spatial dispersions of the $Z_{1,2}$ exciton polariton for $K\|[110]$ [3.9]

Table 3.6. Parameters of the $Z_{1,2}$ exciton

Parameters Units	γ_1^{ex} —	γ_2^{ex} —	γ_3^{ex} —	$\hbar K_1^{ex}$ 10^{-10} eV cm	Δ_{LT} meV	E_{T_1} meV	δ_1 $\hbar^2/2m_0$	δ_2 $\hbar^2/2m_0$	δ_3 $\hbar^2/2m_0$
Present work[a]	0.75	0.07	0.18	6.9	12.5	1.0	0	0	0
TRRS[b]	0.69	0	0	7.3 ± 0.4	12.2	1.7	0.24 ± 0.02	0	0
Resonant Brillouin scattering[c]	0.65 ± 0.01	0.050 ± 0.015	0.040 ± 0.005	5.0 ± 0.6	–	–	0.10 ± 0.05	0	0
Two-photon absorption[d]	~ 0.7	~ 0.1	~ 0.4	–	–	–	–	–	–

[a] The parameters depend on the value used for the dielectric constant, ϵ_b. In the present work, ϵ_b is assumed to be 5.7. The parameters (γ_1^{ex}, γ_2^{ex}, γ_3^{ex}) and K_1^{ex} are inversely proportional to ϵ_b and $\sqrt{\epsilon_b}$, respectively while Δ_{LT} and E_{T_1} are unchanged. Here δ_i are the coefficients of the K-dependent exchange-energy terms.

[b] [3.111]
[c] [3.112]
[d] [3.13]

curve of the longitudinal branch 5 to $K = 0$, and then the light mass of the exciton, $m_0/(\gamma_1^{ex} + 2\gamma_3^{ex})$, is also determined. The value of E_{ex}^t is about the average value of the branches 2, 3, and 4, which are split linearly in K, at small K. The curvature of branch 1 is greatly affected by $\hbar K_1^{ex}$ and the heavy mass of the exciton $m_0/(\gamma_1^{ex} - 2\gamma_3^{ex})$. The dispersion of branch 4 determines E_{ex}^T. Finally, all the parameters are adjusted to the best fit. The parameters used for the solid lines are listed in Table 3.6. Heavy and light exciton masses are given as 2.6 m_0 and 0.9 m_0, respectively. The longitudinal exciton mass is determined to be (1.0 ± 0.1) m_0.

Branch 1, which has mainly the character of the spin-triplet exciton, contains a considerable amount of the transverse exciton component due to the term linear in K. Therefore, it is strongly dipole active and has a considerable polariton effect. Branch 1 has a shallow minimum at $K \sim 2.6 \times 10^6$ cm^{-1}. Branch 2 is the upper-polariton branch of branch 1 at small K, and is almost spin-triplet exciton like. Branch 4 is the upper-polariton branch of branch 2 at small K and is mostly transverse exciton like. Branches 1, 2, and 4 are twofold degenerate. Branches 2 and 4 cross each other at $K \sim 3 \times 10^6$ cm^{-1} because they have the light and heavy masses respectively, for rather small K. Branch 3 is the pure spin-triplet state $|2, 0\rangle$, and has the light mass.

Branches 3 and 5 have the light mass in theory, but the experimental results show somewhat heavier masses. Such small discrepancies are neglected in the fitting procedure. These discrepancies are discussed later.

The experimental data around $K \sim 2.3 \times 10^6$ cm^{-1} in Fig. 3.53 are obtained by using the rotational and/or vibrational states of the EM as the two-photon resonant state as described in the next section.

The solid and broken lines in Fig. 3.53 are calculated curves fitted to the experimental results, and the large dots and dotted lines have the same meanings as in Fig. 3.52. The only fitting parameter to be added is γ_2^{ex}, as the other parameters, γ_1^{ex}, γ_3^{ex}, $\hbar K_1^{ex}$, E_{ex}^L, E_{ex}^T, and E_{ex}^t have the values already given. The value of γ_2^{ex} is determined as 0.07, which shows some warping effect. The branches a1–a5 and b1–b5 have electric dipoles parallel to the [001] and [1$\bar{1}$0] directions, respectively. The spin-triplet exciton like branches a1 and b1 are strongly dipole active due to the terms linear in K. The shallow minimum which appeared for $K \| [111]$ disappears here. Branches a2, a3, b2, b3, and b4 at a small K are the upper-polariton branches of a1, a2, b1, b2, and b3, respectively. Branches b2, a2, and b3 are spin-triplet like and weakly dipole active for fairly small K. Branches a3 and b4 are mostly transverse exciton like. Branches a2 and a3 cross each other at $K \sim 3 \times 10^6$ cm^{-1} and branches b2 and b3 and b3 and b4 do so at $K \sim 4.5 \times 10^6$ and $K \sim 3 \times 10^6$ cm^{-1}, respectively. All of the above branches become considerably dipole active at large K due to strong mixing with the transverse state. The longitudinal branch a4, however, becomes only slightly dipole active even at large K.

Thus, the dispersion relations of the $Z_{1,2}$ exciton polaritons for $K \| [111]$ and $K \| [110]$ are analyzed self-consistently by the theory of the symmetry-breaking effect due to a finite wave vector.

In Table 3.6 parameters obtained by various experiments are listed. From the multiplet structure of P excitons observed in the two-photon absorption spectrum [3.13], the Luttinger parameters are evaluated. Their Luttinger parameters correspond well to the present values, and also the warping effect, which is reflected in the difference between γ_2^{ex} and γ_3^{ex}, appears in some degree.

The parameters obtained by two-photon resonant Raman scattering (TRRS) [3.111] give nearly the same values for γ_1^{ex} and $\hbar K_1^{\mathrm{ex}}$ as the present work. However, γ_2^{ex} and γ_3^{ex} are missing, which is probably due to the restriction of the measured K-space ($K \sim 0.6 \times 10^6$ and $\sim 1.5 \times 10^6$ cm^{-1}).

The parameters determined by resonant Brillouin scattering [3.112], however, differ considerably from those in the present case. A very small γ_3^{ex} is found and the warping effect is scarcely recognized. The coefficient of the term linear in K is too small compared with the present value. The TRRS spectra cannot be explained at all by the dispersion curves expected from the resonant Brillouin scattering. On the other hand, the Brillouin shifts expected from the present experiments are found to agree well with the measured Brillouin shifts for $K\|[111]$ and $K\|[110]$, but the assignments of Brillouin lines do not correspond to each other. For example, the scattering from branch 1 to branch 1' (negative K) in Fig. 3.52 is not found in the measured Brillouin spectrum. These facts make us expect that if the assignments of Brillouin lines were changed into suitable ones, the Brillouin spectra would give the same parameters as those obtained in the present work.

According to calculations of the intensity of resonant Brillouin scattering [3.113], the intra-band Brillouin scattering inside a branch split K linearly, for example, scattering $1 \rightarrow 1'$ in Fig. 3.52, becomes very weak at large K. These expectations support the above discussions.

The fitted curves 3 and 5 in Fig. 3.52 and b3, b4, and a4 in Fig. 3.53 deviate slightly from the experimental results. In CuCl, it is significant that Δ_{LT} of the Z_3 excitons decreases in proportion to K^2, which leads to the different masses, 3.14 m_0 and 2.3 m_0 for the longitudinal and transverse excitons [3.8]. The large K dependence of the Δ_{LT} is ascribed to the K-dependent oscillator strength of the exciton which arises from the $k \cdot p$ mixing of the Γ_8 valence band into the Γ_7 band [3.100]. In CuBr, the same condition would be present due to the $k \cdot p$ mixing of the Γ_7 valence band into the Γ_8 band. If Δ_{LT} depends on K^2 as

$$\Delta_{\mathrm{LT}}(K) = \Delta_{\mathrm{LT}}(0) + \alpha Q K^2 \,,$$

the longitudinal branch in Figs. 3.52 and 3.53 can be fitted further by putting $\alpha = -0.1$ which corresponds to the theoretically expected value

$$\alpha = -\Delta_{\mathrm{LT}}(0)/\lambda \,,$$

where λ is the spin-orbit splitting energy which is 150 meV [3.15]. The $k \cdot p$ mixing between the Γ_7 and Γ_8 valence bands would cause the compensation for the dispersions of other branches [3.103]. In addition to them, the K-dependent exchange term may play a supplemental role, but too many parameters are impractical for the present accuracy.

By the way, in CuCl, CuBr, and CuI, the nonlinear luminescence band, called H, has been found as mentioned in Sect. 3.3.1. The H band in CuCl is extrapolated into the exciton band at $T = 0K$ and is interpreted as exciton-free electron scattering. In CuBr and CuI, however, the peak energies of the H band extrapolated to $T = 0K$ lie 37 and 45 meV below the exciton, respectively. This discrepancy can be explained by the K-linear effect as follows. At $T = 0K$, the excitons are populated at a finite wave vector K_{ex} due to the K-linear effect. The emission band due to the scattering between these excitons and electrons, which are at the Γ point, shifts away from the exciton band toward lower energies by an amount equal to the kinetic energy of the electron with $k = K_{ex}$. For example, if $K_{ex} = 3 \times 10^6$ cm^{-1} and $m_e = 0.2 \, m_0$, the shift is ~ 30 meV. The term linear in K, however, is absent in CuCl. From this point of view, it is not necessary to ascribe the H band in CuBr and CuI to exciton-bound electron scattering, and a common mechanism such as exciton-free electron scattering becomes valid.

Figure 3.54 shows the dispersion relation of the lowest EM state having Γ_1 symmetry. The open circles and dots indicate experimental results for $K \| [111]$ and $K \| [110]$, respectively. The solid line is the parabolic dispersion curve fitted to the experimental results. There is no significant difference between the dispersions for $K \| [111]$ and $K \| [110]$, indicating that the effective mass of the Γ_1 state is isotropic. The effective mass is determined as $3.1 \, m_0$.

The dispersions of the EM in CuBr have already been measured for rather small K ($\lesssim 1.7 \times 10^6$ cm^{-1}), and the dispersion linear K and the small dispersions quadratic in K have been observed [3.114]. Neglecting the heavy- and light-mass effect and the warping effect, the effective mass is given by $2m_0/\gamma_1^{ex}$, which has the value $2.7 \, m_0$.

As-grown crystals from the vapor phase have a (111) surface and the (110) surface can be prepared by cleaving, hence the measurements for $K \| [111]$ or [110] are easy. For the case of $K \| [100]$, the dispersion curves are estimated as shown in [Ref. 3.9, Fig. 6].

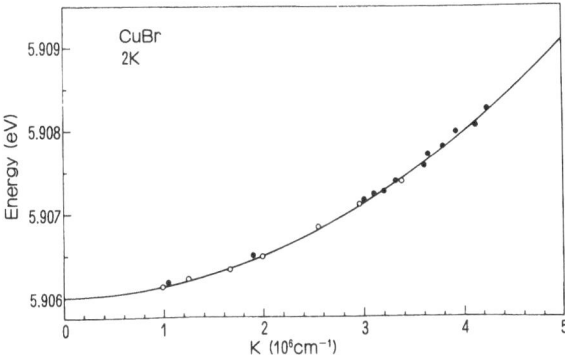

Fig. 3.54. Dispersion of the lowest-energy EM in CuBr. Open circles and dots indicate the experimental results for $K \| [111]$ and $K \| [110]$, respectively. The solid line is the parabolic dispersion curve fitted to the experimental results [3.9]

For ZnSe having zinc blende structure, the exciton-polariton dispersion curves were determined for $K \parallel [111]$ in a similar way [3.115] with the following fitting parameters:

$$E_{ex}^L = 2.8039 \text{ eV}, \ E_{ex}^T = E_{ex}^t = 2.8023 \text{ eV} ,$$

$$\gamma_1^{ex} = 1.7, \ \gamma_3^{ex} = 0.54 \quad \text{and} \quad \hbar K_l^{ex} = 0.12 \times 10^{-9} \text{ eV cm} .$$

The heavy and light exciton masses are 1.7 m_0 and 0.36 m_0, respectively.

3.11 Higher Excited States of the Excitonic Molecule

Many theoretical works have suggested the existence of rotational and vibrational states of the EM by analogy with the hydrogen molecule. Calculations of rotational and vibrational states by *Karp* and *Moskalenko* [3.116] are based on the system of two interacting excitons having the Morse potential which is generalized to an arbitrary e–h mass ratio σ. Finite binding energies of the states are expected for fairly small σ. The calculation shows that the first rotational state $J = 1$ can exist for $\sigma < 0.11$. The rotational quantum energy $E_J(\sigma)$ is given by

$$\frac{E_J(\sigma)}{E_{ex}^b(\sigma)} = \left[\frac{\sigma(1 + \sigma^2)^2}{(1 + \sigma)^6} J(J + 1) \right] \left[1 - 1.37 \frac{\sigma(1 + \sigma^2)}{(1 + \sigma)^4} J(J + 1) \right] \quad (3.68)$$

where, $E_{ex}^b(\sigma)$ is the exciton binding energy. The binding energy of the rotational state is smaller than that of the ground state of the EM by this energy. Experimental evidence of such an excited state has been revealed in CuBr through the discovery of new resonance states in two-photon Raman scattering and also in the excitation spectrum of the EM emission.

As mentioned in Sect. 3.4.2 the GTA bands for the direct creation of EM are located at 2.953, 2.955, and 2.957 eV in CuBr. These bands are here called the A, B, and C bands, respectively. In detailed measurement of the excitation spectrum of the EM emission, another peak, D, was found at 2.9617 eV [3.12]. When an excitation laser is tuned around 2.962 eV, the resonant Raman scattering appears as well [3.12]. Raman lines M_1, M_2, and M_4 are found which are due to scattering into exciton branches 1, 2, and 4 at $k \sim 2.3 \times 10^6$ cm^{-1} in Fig. 3.52. The Raman lines are the same as M_f^R and the doublet M_T^R mentioned in Sect. 3.4.2 or M_{II}, M_{III}, and M_{IV} in [3.12].

Figure 3.55 shows the Raman line intensities as a function of the incident laser energy, where the open circles, dots, and open squares indicate the results for the M_1, M_2, and M_4 Raman lines, respectively [3.12]. Four resonant structures, marked D, E, F, and G, can be seen, indicating that the excited state consists of at least four levels which can be excited by the two-photon process. The energy position of D is exactly the same as the excitation peak at 2.9617 eV for the EM

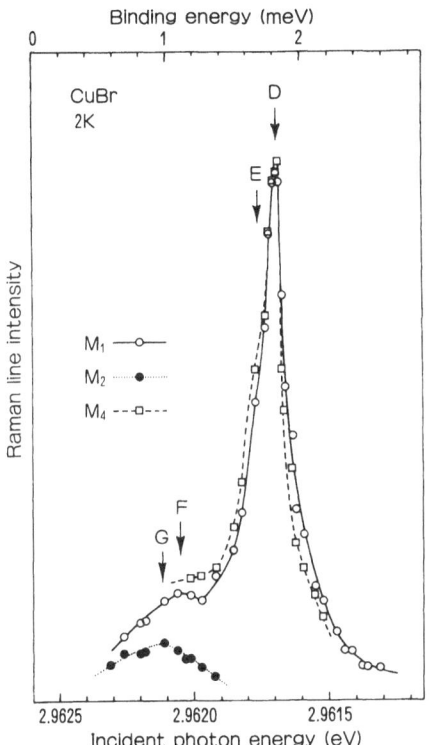

Fig. 3.55. Raman line intensity versus excitation energy associated with resonant excitation into the higher excited state of the EM in CuBr [3.12]

mission. From these facts it is confirmed that these new levels are those of the EM.

The exciton of CuBr has a remarkable term linear in K and the energy minimum is located at a finite wave vector. The binding energy of the EM, defined as the energy difference between the EM and a pair of the lowest excitons, is indicated in the upper scale of Fig. 3.55 where the origin at 2.9626 eV is the minimum energy of the exciton without the polariton effect. Levels D, E, F, and G have binding energies of 1.8, 1.7, 1.1, and 1.0 meV, respectively. The ground states of the EM have binding energies of 19.1, 15.3, and 12.1 meV for the A, B, and C levels, respectively.

For the consideration of the symmetries of the excited states of EM, it is convenient [3.117] to start from the direct products of two electrons $\Gamma_6 \times \Gamma_6 = \Gamma_1 + \Gamma_4$, and of two holes $\Gamma_8 \times \Gamma_8 = \Gamma_1 + \Gamma_2 + \Gamma_3 + 2\Gamma_4 + 2\Gamma_5$. To obtain a bound state, an odd combination of two electrons is required, hence, the Γ_4 symmetry of two electrons is excluded. The total angular momentum of the two holes can be taken as $J_{hh} = 0, 1, 2$, and 3. The ground states are required to be $J_{hh} = 0, 2$ (even), because these states are antisymmetric for the exchange of two holes. The corresponding two-hole states are $\Gamma_1(J_{hh} = 0)$, $\Gamma_3(2)$, and $\Gamma_5(2)$.

The total symmetry of the molecular state, Γ_{tot}, is expressed by the direct product of the symmetries of three wave functions; that of two electrons Γ_{ee}, two holes Γ_{hh} and the envelope Γ_{env}:

$$\Gamma_{\text{tot}} = \Gamma_{\text{ee}} \times \Gamma_{\text{hh}} \times \Gamma_{\text{env}} .$$

For the ground states it is given by

$$\Gamma_1 \times (\Gamma_1 + \Gamma_3 + \Gamma_5) \times \Gamma_1 = \Gamma_1 + \Gamma_3 + \Gamma_5 .$$

All of the ground states Γ_1, Γ_3, and Γ_5 can be excited by two photons, and the levels A, B, and C correspond to them.

The first rotational state ($J = 1$) is required to be $J_{\text{hh}} = 1, 3$ (odd), because these states are symmetric for the exchange of the two holes. Corresponding two-hole states are $\Gamma_2(J_{\text{hh}} = 3)$, $\Gamma_4(1)$, $\Gamma_4(3)$, and $\Gamma_5(3)$. The total symmetries are expressed by the direct product with the Γ_5 symmetry of the envelope function of the first rotational state;

$$\Gamma_1 \times (\Gamma_2 + 2\Gamma_4 + \Gamma_5) \times \Gamma_5 = \Gamma_1 + 2\Gamma_2 + 3\Gamma_3 + 4\Gamma_4 + 3\Gamma_5 .$$

Among them, the Γ_1, $3\Gamma_3$, and $3\Gamma_5$ states are two-photon allowed. The exciton binding energy is given by $E_{\text{ex}}^{\text{b}} = 13.6 \, \mu/\epsilon_{\infty}^2$ [eV]. Assuming $\epsilon_{\infty} = 5.7$, μ is obtained as $0.26 \, m_0$. If we adopt the heavy exciton mass, $2.6 \, m_0$ (Sect. 3.10.2), σ becomes 0.126. The first rotational energy is calculated by (3.68) to be 10.7 meV. This value is nearly in agreement with the energy difference between the (A, B, C) and (D, E, F) levels. Thus, one of the possible interpretations of the D–G levels is the first rotational state ($J = 1$, $v = 0$) split by the hole-hole exchange interaction. However, for the F and G levels two alternative assignments are possible because their levels have smaller binding energies than the D and E levels. One is the second rotational state ($J = 2$, $v = 0$), since the energy of which, calculated as 13.9 meV, is still smaller than the averaged binding energies of the ground state. Another is the first vibrational state ($J = 0$, $v = 1$), although σ exceeds the theoretical limit $\sigma < 0.11$, but taking into account the ambiguities in σ and the theoretical limit, this possibility may not be excluded.

In the case of ZnSe, five molecular levels have been observed [3.115]. Among them, the lower three levels A, B, and C are considered to be ground states. However the higher two levels, D and E, cannot be interpreted as ground states. The mass ratio σ in ZnSe is estimated as 0.11 [3.115], then the rotational energy for $J = 1$ state becomes 1.7 meV. This value is nearly the same as the difference between the (A, B, C) and (D, E) levels.

4. Theory of Excitons in Phonon Fields

In Chaps. 1 and 2, we have mainly considered an exciton or an excitonic molecule (EM) as an elementary electronic excitation in the static and spatially periodic field of a rigid crystal, but the real crystal lattice is deformable and vibrating. The lattice vibrations cause a spatially and temporally fluctuating potential for the exciton. The *spatial* fluctuation gives rise to scattering and momentary localization of the exciton, as reflected in the broadening and the low-energy tail, respectively, of the excitonic absorption spectra. The *temporal* fluctuation is responsible for the fine structure (phonon sidebands) in the spectra. The *deformability* of the lattice causes the reverse action: the exciton induces a lattice distortion around itself, namely, the exciton is always dressed with phonons. This has the effect of stabilizing – in some cases, even immobilizing – the exciton, as reflected in the energy shift of excitonic absorption and luminescence spectra. Not only the *translational*, but also the *relative* motion of the electron and the hole in the exciton is subject to the effect of self-induced phonons; in particular, optical phonons give rise to dynamical screening of the electron-hole (e–h) Coulomb attraction, thus reducing the effective binding energy significantly.

When the exciton-phonon coupling exceeds a critical value, the exciton becomes immobilized, being trapped by the self-induced lattice distortion. This *self-trapping* is realized in a variety of ways, depending on the material, because of different electron-phonon and hole-phonon interactions; sometimes it leads to defect formation with e–h decomposition or annihilation. The hot and ordinary excitonic luminescence tell us a great deal about the exciton in its various stages of relaxation in the deformable lattice.

It may happen, with very strong exciton-phonon coupling, that the energy of the lattice relaxation induced by the exciton is greater than the excitonic energy in the rigid lattice, namely, that the self-trapped exciton has negative energy referred to the ground state of the rigid crystal. In such a situation, self-trapped excitons should be spontaneously created everywhere in the crystal; one should rather say, the *true ground state* of the whole crystal is *electronically and structurally different* from the fictitious ground state from which we started. Thus, it is not only instructive but also useful to describe various possible forms of the ground state in the context of exciton-phonon interactions.

4.1 Electron-Phonon Interactions

4.1.1 Types and Ranges of Electron-Phonon Interactions

Since an exciton is a composite particle consisting of an electron in the conduction band and a hole in the valence (or inner-core) band, its interaction with the phonon field can be expressed in terms of electron-phonon and hole-phonon interactions. We will start with a general but simplified prescription for deriving the electron(hole)-phonon interaction and then present several examples of the interaction with different force ranges and different phonon dispersions.

Let us consider a *generalized displacement* $u(R_n)$, within the nth unit cell at R_n, which can be Fourier-expanded, in terms of the normal coordinates u_q of a particular mode of lattice vibrations, as

$$u(R_n) = N^{-1/2} \sum_q u_q e^{iq \cdot R_n}, \quad (u_q = u_{-q}^\dagger) . \tag{4.1}$$

Here q is the wave vector and N is the total number of unit cells contained in the crystal. The generalized displacement $u(R_n)$ gives rise to a *generalized distortion* $v(R_n)$ of the lattice which may be $u(R_n)$ itself or its spatial derivative depending upon the particular case, as will be considered later on. The distortion-induced potential $H_I(R_n)$ for the electron at R_n (the position of an electron within a particular band can be specified only by discrete lattice points in the Wannier sense, in contrast to an electron in vacuum) can be written, within the linear approximation, as

$$
\begin{aligned}
H_I(R_n) &= \sum_{n'} \psi(R_n - R_{n'})\, v(R_{n'}) \\
&= \sum_q \psi_q v_q e^{iq \cdot R_n} \tag{4.2}
\end{aligned}
$$

where $\psi(R_n - R_{n'})$ represents the *interaction kernel* relating the source field $v(R_{n'})$ to the resulting potential, and ψ_q are defined as in (4.1).

The energy of the harmonic lattice vibrations can be written as

$$H_L = K_L + U_L = \frac{M}{2} \sum_n [\dot{u}(R_n)]^2 + \frac{C}{2} \sum_n [v(R_n)]^2$$

$$= \sum_q \left(\frac{\omega_q^2}{2C} p_q p_q^\dagger + \frac{C}{2} v_q v_q^\dagger \right) \tag{4.3}$$

where $\dot{u}$ denotes the time derivative of u, M is the generalized mass defined appropriately, C the force constant, ω_q the frequency given by

$$\omega_q = (C/M)^{1/2} |f_q| , \tag{4.4}$$

$$f_q = v_q/u_q \quad \text{(assumed to be constant in time)} \tag{4.5}$$

and p_q is the momentum conjugate to v_q as defined by

$$p_q \equiv \partial H_\mathrm{L}/\partial \dot{v}_q^\dagger = (C/\omega_q^2)\dot{v}_q \ . \tag{4.6}$$

Let us now quantize the lattice vibrations by introducing the phonon operators

$$b_q \ = \ \sqrt{\frac{C}{2\hbar\omega_q}} v_q \ + \ i\sqrt{\frac{\omega_q}{2\hbar C}} p_{-q} \ ,$$

$$b_{-q}^\dagger \ = \ \sqrt{\frac{C}{2\hbar\omega_q}} v_q \ - \ i\sqrt{\frac{\omega_q}{2\hbar C}} p_{-q} \ , \tag{4.7}$$

with commutation relations

$$[b_q, b_{q'}^\dagger] = \delta_{q,q'} \ , \quad [v_q, p_{q'}] = i\hbar\,\delta_{q,q'} \ . \tag{4.8}$$

Then Eqs. (4.3) and (4.2) are written as

$$H_\mathrm{L} \ \ = \sum_q \frac{\hbar\omega_q}{2} \left(b_q b_q^\dagger + b_q^\dagger b_q\right) \ , \tag{4.9}$$

$$H_\mathrm{I}(\boldsymbol{R}_n) = \sum_q (\gamma_q b_q \mathrm{e}^{i q \cdot R_n} + \gamma_q^* b_q^\dagger \mathrm{e}^{-i q \cdot R_n}) \ , \tag{4.10}$$

$$\gamma_q \ \ \ \equiv \left(\frac{\hbar\omega_q}{2C}\right)^{1/2} \psi_q = \gamma_{-q}^* \ . \tag{4.11}$$

The electron-phonon interaction (4.10) is therefore characterized by the q dependences of ψ_q (force range) and ω_q (phonon dispersion), typical cases of which will be described below.

a) **Deformation Potential.** The long-wavelength acoustic modes of lattice vibrations can be described in terms of the strain field represented by the tensor

$$e_{ij}(\boldsymbol{R}) \equiv (\partial u_i/\partial R_j + \partial u_j/\partial R_i)/2$$

The electron is then subject to the potential, which in the linear approximation can be written as [4.1]

$$H_\mathrm{I}(\boldsymbol{R}_n) \ = \sum_{ij} \Xi_{ij} e_{ij}(\boldsymbol{R}_n) \ . \tag{4.12}$$

For an isotropic crystal with the conduction band bottom at the Γ point one has $\Xi_{ij} = \Xi_{ji} = 0 \ (i \neq j)$ and $\Xi_{11} = \Xi_{22} = \Xi_{33} \equiv \Xi$ because of symmetry. Then one obtains

$$H_\mathrm{I}(\boldsymbol{R}_n) = \Xi\Delta(\boldsymbol{R}_n) \ , \quad \Delta(\boldsymbol{R}_n) \equiv \operatorname{div} \boldsymbol{u}(\boldsymbol{R}_n) \ . \tag{4.13}$$

If v is defined by

$$v(\boldsymbol{R}_n) \equiv v_0^{1/2} \varDelta(\boldsymbol{R}_n) , \tag{4.14}$$

with the unit cell volume v_0, C in eq. (4.3) turns out to be the elastic constant for dilation $\varDelta$:

$$U_L = \frac{C}{2} \int [\varDelta(\boldsymbol{R})]^2 d\boldsymbol{R} . \tag{4.15}$$

Since the electron interacts only with the longitudinal modes, according to (4.13), both sides of eq. (4.1) should be replaced by vectors with $\boldsymbol{u}_q = (\boldsymbol{q}/q)u_q$. From (4.1, 4, 5, 11, 14), one obtains

$$\omega_q = sq , \quad s \equiv (Cv_0/M)^{1/2} , \tag{4.16}$$

$$\psi(\boldsymbol{R}_n - \boldsymbol{R}_{n'}) = v_0^{-1/2} \varXi \delta_{n,n'} , \tag{4.17}$$

$$\gamma_q = \left(\frac{\hbar q}{2NMs}\right)^{1/2} \varXi , \tag{4.18}$$

where s represents the velocity of sound and M the mass of a unit cell. Equation (4.12 or 13) is called the deformation potential; it is a short-range potential as is obvious from (4.17).

b) Fröhlich Interaction. An optical mode in an ionic crystal is accompanied by an electric dipole $v(\boldsymbol{R}_{n'}) \equiv u(\boldsymbol{R}_{n'})$ ($f_q = 1$) at $\boldsymbol{R}_{n'}$ which gives rise to a dipolar potential for an electron at $\boldsymbol{R}_n$:

$$\psi(\boldsymbol{R}_n - \boldsymbol{R}_{n'}) \cdot v(\boldsymbol{R}_{n'}) \equiv e\eta |\boldsymbol{R}_n - \boldsymbol{R}_{n'}|^{-3} (\boldsymbol{R}_n - \boldsymbol{R}_{n'}) \cdot v(\boldsymbol{R}_{n'}) . \tag{4.19}$$

where η is a screening factor. In (4.2) $\psi_q v_q$ should then be replaced by the scalar product $\psi_q \cdot v_q$ where

$$\psi_q = [4\pi e\eta/(N^{1/2} v_0)] i\boldsymbol{q}/q^2 \equiv \psi_q \boldsymbol{q}/q , \tag{4.20}$$

$$v_q = v_q \boldsymbol{q}/q . \tag{4.21}$$

Only the longitudinal optical (LO) mode interacts with the electron according to (4.21).

The sceening factor η can be related to the dielectric constant ε as follows. Put a classical point charge e at $\boldsymbol{R}_n$. The lattice is then subject to the force derived from the potential (4.10), and is displaced by $b_q = -(\gamma_q/\hbar\omega_q)\exp(i\boldsymbol{q} \cdot \boldsymbol{R}_n)$ as is seen by minimizing $H_L + H_I$ with respect to b_q. Putting this into (4.10) with n replaced by n' and making use of (4.11) and (4.20), one obtains the self-induced potential at $\boldsymbol{R}_{n'}$:

$$\overline{H}_I(\boldsymbol{R}_{n'}) = -4\pi\eta^2 e^2/(Cv_0|\boldsymbol{R}_n - \boldsymbol{R}_{n'}|) .$$

This should be equal to $-(1 - \epsilon^{-1})e^2/|\boldsymbol{R}_n - \boldsymbol{R}_{n'}|$ according to electrostatics. However, we are concerned here with the displacement polarization which is the difference between the total and the electronic polarizations. Thus, we have to replace $(1 - \epsilon^{-1})$ by $(1 - \epsilon_0^{-1}) - (1 - \epsilon_\infty^{-1})$ where ϵ_0 and ϵ_∞ are the static and high-frequency dielectric constants. Making use of (4.11, 20) and neglecting the dispersion of the longitudinal optical mode ($\omega_q - \omega_0$), we obtain

$$\gamma_q = i \left[\frac{2\pi e^2 \hbar \omega_0}{N v_0} \left(\frac{1}{\epsilon_\infty} - \frac{1}{\epsilon_0} \right) \right]^{1/2} \frac{1}{q} \tag{4.22}$$

as was first derived by *Fröhlich* [4.2].

It should be noted that the point charge-dipole interaction (4.19) is a long-range one in contrast to the short-range deformation potential (4.17). This difference in force range is reflected in the different q dependence of γ_q.

c) **Piezoelectric Interaction.** In piezoelectric crystals, elastic strain e_{ij} is accompanied by electric polarization which in turn gives rise to an electrostatic potential as described in b). Thus, we obtain the long-range piezoelectric interaction [4.3] between the electron and the acoustic modes with $\gamma_q \propto q^{-1/2}$ with an imaginary coefficient – being out of phase with the deformation potential (4.18) and hence free of interference with it.

d) **Interaction with Nonpolar Optical Modes.** The optical modes in nonpolar crystals and the intramolecular vibrations in molecular crystals cause a short-range potential (proportional to the displacement) which acts on the electron, with $\gamma_q = $ real constant [4.4].

The electron-phonon interaction (4.10) gives rise to electron scattering from $\boldsymbol{K}$ to $\boldsymbol{K}+\boldsymbol{q}$ ($\boldsymbol{K} - \boldsymbol{q}$) states by absorbing (emitting) a phonon with momentum $\boldsymbol{q}$ since the nonvanishing matrix element of b_q is given by $\langle n_q - 1|b_q|n_q \rangle = \sqrt{n_q}$ [$n_q (= 0, 1, 2, \ldots)$ is the number of phonons]. The velocity and temperature dependences of electron scattering probability and hence, the temperature dependence of electron mobility [4.5], are thus governed by the q dependences of γ_q, ω_q, and thermal phonon number $\bar{n}_q = [\exp(\hbar\omega_q/k_B T) - 1]^{-1}$. In addition to this real process which conserves energy, the electron is always dressed with phonons through the virtual emission and absorption of phonons. The latter effect – renormalization of an electron in the phonon field – has been studied in great detail for the Fröhlich interaction which was thought to be of predominant importance for an electron in ionic crystals [4.2, 6, 7]. The electron dressed with optical phonons, namely, accompanied by the displacement polarization, is called a "polaron". It has an effective mass greater than the band effective mass in a rigid lattice, as will be shown in Sect. 4.1.2. In some cases, however, the short-range deformation-potential interaction gives rise to a more drastic effect, triggering the self-trapping of an electron. We will see, in Sect. 4.4, how the nature of renormalization is governed by the force range of the electron-phonon interaction and the dimensionality of the lattice.

4.1.2 The Polaron

Let us consider in more detail the problems mentioned in the last paragraph of Sect. 4.1.1 taking the Fröhlich type interaction as an example. We start with a noninteracting state: a conduction electron with wave number K and the phonon vacuum state $|0\rangle$ ($n_q = 0$ for all q's). The interaction (4.10) has the effect of mixing into this state the one-phonon states $b_q^\dagger|0\rangle$ when the electron recoils to $K - q$. The wave function of the electron-phonon system can be written, up to the first-order perturbation, as

$$|\Psi_K\rangle = \left[1 - \sum_q \gamma_q^* \left(\frac{\hbar^2(K-q)^2}{2\,m_e} - \frac{\hbar^2 K^2}{2\,m_e} + \hbar\omega_q\right)^{-1} e^{-iq\cdot r}b_q\right] V^{-1/2}e^{-iK\cdot r}|0\rangle$$

$$(4.23)$$

where we have replaced the discrete electron coordinate (R_n) by the continuous one (r) and assumed a parabolic conduction band with effective mass m_e. The total volume of the crystal is $V \equiv Nv_0$.

The wave function (4.23) represents the *phonon-dressed* electron – the polaron. One can probe its structure by putting another point charge e at position r_1. The latter is subject to the polarization-induced potential $H_{eL}(r_1)$ of (4.10), the expectation value of which in state (4.23) is calculated, with the use of (4.22) and the neglection of phonon dispersion ($\omega_q \sim \omega_0$), as [4.2]

$$\langle \Psi_K|H_I(r_1)|\Psi_K\rangle$$

$$= \frac{1}{V}\int dr \sum_q \frac{-|\gamma_q|^2[e^{iq\cdot(r_1-r)} + \text{c.c.}]}{(\hbar^2/2\,m_e)\,[(K-q)^2 - K^2] + \hbar\omega_0}$$

$$\rightarrow \frac{1}{V}\int dr \frac{-e^2}{\bar{\epsilon}|r-r_1|}\,[1-\exp(-q_p|r-r_1|)]\,, \quad (K\rightarrow 0) \qquad (4.24)$$

where

$$q_p \equiv \left(\frac{2\,m_e\omega_0}{\hbar}\right)^{1/2} \equiv a_p^{-1}\,, \qquad (4.25)$$

$$\frac{1}{\bar{\epsilon}} \equiv \frac{1}{\epsilon_\infty} - \frac{1}{\epsilon_0}\,. \qquad (4.26)$$

The integrand in (4.24) represents the phonon-mediated potential between the electron at r and the probe (classical) at r_1. At large distances, $r \gg a_p$, it reduces to $-e^2/\bar{\epsilon}|r-r_1|$ which, together with the direct Coulomb repulsion $e^2/\epsilon_\infty|r-r_1|$, gives the (statically) screened Coulomb potential $e^2/\epsilon_0|r-r_1|$ as it should. The spatial smearing of the electron, $\delta r \sim a_p$, as seen by the probe, is related, through the uncertainty principle, to the recoil kinetic energy of the electron due to phonon emission, $\hbar^2q_p^2/2\,m_e = \hbar\omega_0$. The quantity a_p is called the *polaron radius*. It is several to ten times as great as the interatomic distance a_0 in typical ionic crystals. This justifies a posteriori the neglection of phonon dispersion, the effective mass approximation and the dielectric continuum model; in fact, those q's which make important contributions in (4.23 and 24) are far smaller than the reciprocal lattice vector.

The second-order energy of the total system turns out to be

$$\Sigma_K = \sum_q \frac{-|\gamma_q|^2}{(\hbar^2/2\,m)\,[(K-q)^2 - K^2] + \hbar\omega_q}$$

$$= -\alpha\hbar\omega_0 - \frac{\alpha}{6}\frac{\hbar^2 K^2}{2\,m} + O(K^4)\,, \quad \text{where} \tag{4.27}$$

$$\alpha \equiv \frac{1}{2}\frac{e^2/\bar{\epsilon}a_p}{\hbar\omega_0} = \frac{1}{2}\left(\frac{1}{\epsilon_\infty} - \frac{1}{\epsilon_0}\right)\frac{e^2}{\hbar\omega_0}\left(\frac{2\,m\omega_0}{\hbar}\right)^{1/2} \tag{4.28}$$

is called the polaron coupling constant. It is interesting to note that the first term of (4.27) is equal to half of the expectation value of the self-induced potential $H_{eL}(r)$ as is seen by putting $r_1 \to r$ in the integand of (4.24). Their difference comes half from the electron kinetic energy and half from the phonon energy as is easily confirmed. Adding the second-order energy to the nonperturbed energy $\hbar^2 K^2/2\,m$ one finds that not only is the total energy lowered by $\alpha\hbar\omega_0$ but also the effective mass of the electron is enhanced to $m^* = (1 - \alpha/6)^{-1}\,m$.

The intermediate coupling theory [4.8] – essentially a variational method expected to be valid up to greater values of α than the perturbation theory is – gives the same energy shift $-\alpha\hbar\omega_0$ (as an upper bound this time) but the nondivergent effective mass $m^* = (1 + \alpha/6)\,m$, both agreeing, in their lowest order, with the perturbation theory. This improvement is not trivial since α is of the order of 3 to 8 in typical ionic crystals.

With greater values of α, the electron binding energy ($> \alpha\hbar\omega_0$ according to the above) in the self-induced phonon field becomes so great that the latter can see only the time average of the rapid orbital motion of the former. The variational calculation with a Gaussian orbital function gives the energy shift $-(\alpha^2/3\pi)\hbar\omega_0$ [4.9], which is lower than that of the intermediate coupling theory when $\alpha > 3\pi$. The orbital radius becomes smaller than a_p defined in (4.25) in the same region of α. In this strong coupling regime, the lattice displacement is essentially static as is seen from the fact that the energy shift is independent of ω_0 [$\alpha \propto \omega_0^{-1/2}$ according to (4.28)]. The effective mass, on the other hand, is given by $m^* = (16/81\pi^2)\alpha^4 m \propto \omega_0^{-2}$, namely, the polaron can have translational motion because of nonvanishing ω_0.

According to the path integral method [4.10] which is considered to be the most powerful tool for treating the polaron over the whole range of α, the effective mass as well as the binding energy of a polaron increase continuously and smoothly with α. This is in contrast to the short-range electron-phonon interaction, as will be discussed in Sect. 4.4.

4.1.3 Exciton-Phonon Interactions and the Form Factor

The exciton is an elementary excitation in a many-electron system, and each electron ($i = 1, 2, \ldots$) in the crystal is subject to the phonon interactions $H_I(r_i)$ as described in Sect. 4.1.1. As long as we confine ourselves to a single exciton subspace and consider the scattering of an exciton between the states λK and $\lambda' K'$ of the type (1.17), we have only to calculate the matrix element:

$$\int \ldots \int \Phi_{n,m}^* \left[\sum_{i=1}^{N} H_I(r_i) \right] \Phi_{n',m'}\, dr_1 \ldots dr_N$$

$$= \delta_{mm'} \int a_c^*(r - R_n) H_I(r) a_c(r - R_{n'})\, dr$$

$$- \delta_{nn'} \int a_v(r - R_m) H_I(r) a_v^*(r - R_{m'})\, dr$$

$$= \delta_{mm'} \delta_{nn'} [H_I^{(c)}(R_n) - H_I^{(v)}(R_m)] \tag{4.29}$$

where we have made use of the properties of Φ as a Slater determinant, $H_I(r)$ as a one-electron operator and $a(r)$ as a Wannier function. It should be noted that the electron-phonon coupling coefficient γ_q (4.11) is different for the conduction ($\gamma_q^{(c)}$) and valence ($\gamma_q^{(v)}$) bands as is distinguished by the superscripts (c) and (v) to H_I in (4.29). Making use of (1.17, 28, and 4.10), and the effective mass approximation (1.2) of the standard form, we obtain the matrix element for scattering of an exciton [4.11, 12]

$$\int \ldots \int \Psi_{\lambda K}^* \left[\sum_{i=1}^{N} H_I(r_i) \right] \Psi_{\lambda' K'}\, dr_1 \ldots dr_N$$

$$= \gamma_{\lambda\lambda'}(K - K') b_{K - K'} + \gamma_{\lambda\lambda'}^*(-K + K') b_{-K + K'}^\dagger \tag{4.30}$$

where

$$\gamma_{\lambda\lambda'}(q) \equiv \gamma_q^{(c)} \eta_{\lambda\lambda'}(p_e q) - \gamma_q^{(v)} \eta_{\lambda\lambda'}(-p_h q) , \tag{4.31}$$

$$\eta_{\lambda\lambda'}(q) \equiv \int \bar{F}_\lambda(R) \bar{F}_{\lambda'}(R) \exp(iq \cdot R)\, dR , \tag{4.32}$$

$$p_{e,h} \equiv m_{h,e}/(m_e + m_h) . \tag{4.33}$$

The form factors $\pm \eta_{\lambda\lambda'}(\pm p_{e,h} q)$ represent the effective charges of the electron and the hole in the exciton seen by the phonon q when the exciton is scattered from λ' to λ of the internal motion. Note that $\pm p_{e,h} R$ represent the electron and the hole coordinates referred to the center of mass.

For the scattering of the $1s$ exciton within the same exciton band ($\lambda = \lambda' = 1s$), one obtains

$$\eta(q) = [1 + (a_B q/2)]^{-2} \tag{4.34}$$

since $F_{1s}(R) \propto \exp(-R/a_B)$. This means that only those phonons with wavelengths of the same order or greater than the exciton radius a_B are effective in scattering the exciton.

4.1.4 Polaron Effects of an Exciton

If we consider the Fröhlich type electron(hole)-optical phonon interaction, we have $\gamma_q^{(c)} = -\gamma_q^{(v)} = \gamma_q$ of (4.22) since this electrostatic interaction is common to all of the bands in contrast to the deformation potential which differs from band to band. Thus we have the *cancellation effect* of $1s \rightarrow 1s$ (more generally for $\lambda \rightarrow \lambda$) scattering

$$\gamma_{1s,1s}(q) = \gamma_q\{[1 + (p_e a_B q/2)^2]^{-2} - [1 + (p_h a_B q/2)^2]^{-2}\} \tag{4.35}$$

for small q's. Now, a slow ($K \sim 0$) $1s$ exciton can be scattered only by absorbing an optical phonon with $q \sim q_0 \equiv [2(m_e + m_h)\omega_0/\hbar]^{1/2}$ due to energy and momentum conservation. Then, if $p_e a_B q_0 \gg 1 \gg p_h a_B q_0$, namely, if

$$\frac{m_h}{m_e} \gg \frac{\varepsilon_B}{\hbar\omega_0} \gg \frac{m_e}{m_h} \tag{4.36}$$

is valid, see (1.8), the cancellation effect is insignificant for this real scattering, the lighter particle (usually the electron) making little contribution. Eq. (4.36) is well satisfied in those crystals with $\varepsilon_B \sim \hbar\omega_0$ and $m_e \ll m_h$. In fact, in II–VI compounds [4.13, 14], silver halides (Chap. 5) and thallous halides (Chap. 7) which satisfy this condition, the optical phonon scattering is significant in spite of the neutrality of the exciton, as seen in the optical-phonon structures in the absorption spectra and the resonance Raman scattering spectra [4.15]. In alkali halides [4.16] and cuprous halides (Chap. 3) with $\varepsilon_B \gg \hbar\omega_0$, the cancellation effect is significant for the $1s$ exciton but much less significant for the $2s$ exciton with its smaller binding energy. This is reflected in the effective mass of the phonon-dressed exciton and in the intensity ratio of the LO phonon sideband to the zero-phonon line of the exciton spectra, both of which are greater in the $2s$ than in the $1s$ exciton [4.17].

The matrix elements $\gamma_{\lambda\lambda'}(q)$ with $\lambda \neq \lambda'$ give rise to mixing of different states of relative motion. This results, for instance, in LO phonon mediated screening of the Coulomb interaction between an electron and a hole with the force changing from $-e^2/\epsilon_\infty|r_e - r_h|$ to $-e^2/\epsilon_0|r_e - r_h|$ when their distance $r = |r_e - r_h|$ is greater than the sum of their polaron radii a_{ep} and a_{hp} [4.18], as was described in Sect. 4.1.2. Therefore, the effective dielectric constant ϵ for the screening of the e–h interaction in an exciton is given by ϵ_∞ or ϵ_0 according as $\varepsilon_B \gg \hbar\omega_0$ or $\varepsilon_B \ll \hbar\omega_0$ since the latter condition is equivalent to $a_{ex}^2 \gg a_{ep}^2 + a_{hp}^2$, see (1.8) and (4.25). The marginal situation $\varepsilon_B \sim \hbar\omega_0$ is complicated also because of strong intraband ($\lambda \leftrightarrow \lambda$) scattering as mentioned above and because of strong resonance between the e–h relative motion and the LO phonon as will be discussed in Sect. 4.3.3.

In addition to the matrix elements (4.30) which conserve the exciton number, we have of course those which do not; for instance, those corresponding to creation or annihilation of an exciton. The latter, though as large as the former, have no significant effect on the exciton as long as the exciton energy ($\sim$ band gap) is large enough compared with the exciton-phonon interaction energy. The situation in which the latter play an important role will be considered in Sect. 4.7.

4.2 The Exciton in Spatially Fluctuating Fields

The phonon fields act on an exciton as potential sources with spatial and temporal fluctuations. While these two types of fluctuations are in general inseparably reflected in the exciton dynamics and the optical spectra, it is sometimes useful and instructive to describe them separately. In fact, the lattice vibrations are usually much slower than the excitonic motions, because of the great difference between atomic and electronic masses, and the phonon energies manifest themselves as subtle quantal effects in exciton dynamics and as fine structures superposed on usually broader optical spectra. For this reason, we will first discuss the effect of spatial fluctuation, neglecting the temporal fluctuation for the moment. That is in accordance with the Franck-Condon principle according to which the atomic positions are fixed during the optical absorption process. This is legitimate as long as the overall width w of the absorption band is much greater than the phonon energies $\hbar\omega$ since measuring the overall line shape requires the duration of the absorption process to be of the order of $\hbar w^{-1}$ (because of the uncertainty principle) which is still shorter than ω^{-1}.

In order to extract the most important aspect of the spatial fluctuation due to lattice vibrations, we make further simplifications by using the following model Hamiltonian for the exciton-phonon system:

$$H = H_e + H_I + U_L + K_L , \quad \text{where} \tag{4.37}$$

$$H_e = \sum_n |n\rangle E_a \langle n| + \sum_{n \neq m} |n\rangle t_{nm} \langle m| , \tag{4.38}$$

$$H_I = -\sum_n |n\rangle cQ_n \langle n| , \tag{4.39}$$

$$U_L = \sum_n Q_n^2/2 , \quad K_L = \sum_n \sum_{n'} (\Omega^2)_{nn'} P_n P_{n'}/2 . \tag{4.40}$$

We are considering a Frenkel exciton (or a particular internal state of a Wannier exciton) in the tight-binding picture, with atomic excitation energy E_a and intersite transfer energy t_{nm}. The lattice energy (4.40) is obtained from (4.3) by rewriting $\sqrt{C}v(\boldsymbol{R}_n)$ as Q_n and introducing the momenta P_n conjugate to Q_n. As long as ω_q depends on $\boldsymbol{q}$, K_L is not diagonal in P_n's, and hence the local modes, Q_n's, are not the normal modes. The exciton-phonon interaction (4.39) is obtained from (4.2) by putting $\psi(\boldsymbol{R}_n - \boldsymbol{R}_{n'}) = -\delta_{nn'} c\sqrt{C}$ (short-range force).

4.2.1 Localization Versus Delocalization

In the absence of the exciton-phonon interaction ($c = 0$), the eigenstates of an exciton are given by the Bloch states

$$|K\rangle = N^{-1/2} \sum_n \exp(iK \cdot R_n)|n\rangle \tag{4.41}$$

with energies

$$E_K = E_a + t_K , \qquad t_K \equiv \sum_{n(\neq 0)} t_{0n} \exp(iK \cdot R_n) \tag{4.42}$$

as shown in Figs. 4.1a and a' for small and large $|t|$, respectively. The intersite transfer thus allows the excitation energy to be lowered by

$$B = E_a - (E_K)_{\text{min.}} = (-t_K)_{\text{max.}} , \qquad (> 0) . \tag{4.43}$$

The optical transition from the ground state is allowed only to the state $|K = 0\rangle$ of the exciton due to the K-selection rule (Sect. 1.2). The normalized absorption spectra is then given by

$$F_a(E) = \delta(E_0 - E) . \tag{4.44}$$

In the absence of transfer ($t_{nm} = 0$), the localized excitation (say, $|n\rangle$) gives an eigenstate, with adiabatic potential

$$W_{en}(Q) = \langle n|H_e + H_I|n\rangle + U_L(Q) = E_a - cQ_n + \tfrac{1}{2} \sum_{n'} Q_{n'}^2$$

$$= E_a - E_{LR} + (Q_n - c)^2/2 + \tfrac{1}{2} \sum_{n'(\neq n)} Q_{n'}^2 , \tag{4.45}$$

$$E_{LR} \equiv c^2/2 . \tag{4.46}$$

The adiabatic potential of the ground state is simply given by

$$W_g(Q) = U_L(Q) . \tag{4.47}$$

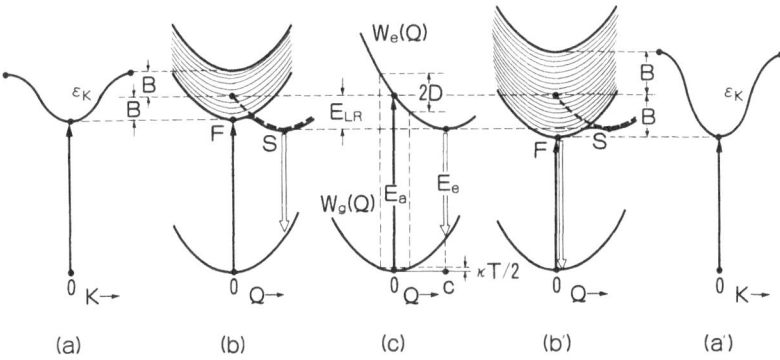

Fig. 4.1. Configuration coordinate models for localized excitation (c) and the exciton [(b) or (b')], and the dispersion of the exciton band in a rigid lattice [(a) or (a')]. The strong and weak coupling cases ($B \lessgtr E_{LR}$) are represented by (a), (b) and (a'), (b'), respectively

After Franck-Condon excitation (vertical transition in Fig. 4.1c) at $Q = 0$, the lattice will relax to a new equilibrium at $Q_n = c$, $Q_{n'} = 0$ $(n' \neq 0)$, thereby releasing the *lattice relaxation energy* E_{LR} to other sites through K_L of (4.40).

Let us take into account the effect of lattice vibrations on the absorption spectra by assuming the Boltzmann distribution $\exp[-W_g(Q)/k_BT[$ in the initial state. Applying the Franck-Condon principle we can calculate the normalized absorption spectra as

$$F_a(E) = \int \ldots \int dQ_1 \ldots dQ_N \delta[W_{en}(Q) - W_g(Q) - E] \exp[-W_g(Q)/k_BT]$$

$$\div \int \ldots \int dQ_N \exp[-W_g(Q)/k_BT]$$

$$= (2\pi D^2)^{1/2} \exp[-(E - E_a)^2/2D^2] , \qquad (4.48)$$

$$D^2 \equiv \langle v_n^2 \rangle = 2E_{LR}k_BT \qquad (4.49)$$

under the assumption of Q independence of the transition dipole moment. The amplitude of the local energy fluctuation $v_n = -cQ_n$ is reflected in the Gaussian width D. The emission spectra can be calculated in the same way if thermal equilibrium within the excited state *en* is reached before the emission process; the spectral shape is the same as (4.48) with E_a replaced by $E_e = E_a - 2E_{LR}$.

In the general case with $c \neq 0$ and $t \neq 0$, the configuration coordinate model corresponding to Fig. 4.1c is obtained by incorporating the continuum of the exciton band of Fig. 4.1a or a', as shown schematically in Fig. 4.1b or b' [4.19, 20]. As we switch on the local distortion Q_n and hence the local potential $-cQ_n$ at the cost of lattice energy $\frac{1}{2}Q_n^2$, the extended exciton states are subject to an energy change of $O(N^{-1})$ only (so their levels are parallel to the ground-state parabola) while a bound state splits off out of this continuum, possibly forming another minimum S. The state S is nothing but the self-trapped state – the exciton is stabilized by being trapped by the lattice distortion it has induced. With an increase of Q_n, this discrete level asymptotically approaches the broken line which is a reproduction of the $W_e(Q)$ line (completely localized excitation) in Fig. 4.1c with the same value of c. Neglecting the small energy difference between the solid and broken lines at S, one finds that S has lower or higher energy than F (the lowest free-exciton state without lattice distortion) according as $B \lessgtr E_{LR}$, corresponding to the case b or b'. It will be shown in Sect. 4.4 that there always exists a potential barrier between F and S (if the latter minimum exists). Therefore, the stable exciton state in the phonon field changes abruptly from the F- to S-state as the exciton-phonon coupling constant, defined by

$$g \equiv E_{LR}/B , \qquad (4.50)$$

exceeds unity. The exciton will stabilize itself in the phonon field long before it emits a photon, the emission band usually appears at $E_a - B = E_0$ (F-type: sharp line resonant to the absorption band; Fig. 4.1b') or $E_a - 2E_{LR}$ (S-type: broad, Stokes-shifted band; Fig. 4.1b) according as $g \lessgtr 1$, as also shown in the lower half of Fig. 4.2 [4.20].

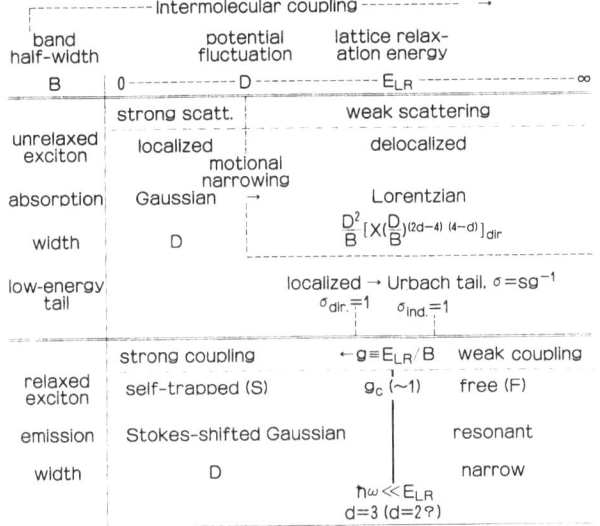

In contrast to the relaxed excited states, the configuration coordinates (Q_n) relevant to the absorption spectra are confined to much smaller thermal vibrations in the ground state. The normalized line-shape of the absorption spectra is given by

$$F_a(E) = \overline{\sum_i \langle 0|i(Q)\rangle \, \delta(E - E_i(Q)) \, \langle i(Q)|0\rangle} \qquad (4.51)$$

where $|0\rangle$ means the free-exciton state with $K = 0$, $|i(Q)\rangle$ and $E_i(Q)$ are eigenstates and eigenenergies of $H_e + H_1(Q)$ for a configuration $Q \equiv (Q_1, Q_2, \ldots Q_N)$ and —— means to take the average over the Boltzmann distribution of Q as was done in (4.48). Thus, we have to solve a typical eigenvalue problem of a random lattice where the electron with transfer t_{nm} is subject to a spatially random potential $v_n = -cQ_n$ obeying the Gaussian distribution $\exp(-v_n^2/2D^2)$ without mutual correlation: $\overline{v_n v_m} = 0$ $(n \neq m)$. It would be interesting to see how the line-shape varies from the delocalization limit (4.44) to the localization limit (4.48) as we increase c/t from 0 to ∞.

4.2.2 Overall Line-Shape of the Absorption Spectra

To study the line-shape, we will first apply perturbation theory starting from the delocalization limit. It is convenient to rewrite (4.51) as

$$F_a(E) = \pi^{-1} \operatorname{Im} \{\langle 0|\overline{R(z, Q)}|0\rangle\} , \qquad (4.52)$$

with the use of the resolvent operator (the sign being reversed from the conventional definition)

$$R(z, Q) \equiv \frac{1}{z - H_e - H_I(Q)} , \quad z = E - i\eta , \quad (\eta \to +0) . \tag{4.53}$$

Expanding the operator $R(z)$ in powers of H_I, rearranging terms appropriately and taking the average over Q (note that $\overline{Q^n} = 0$ for odd n), one can derive an exact formula [4.21, 22]

$$\overline{[R(z, Q)]_d} \equiv D(z) = \frac{1}{z - H_e - \Sigma(z)} , \tag{4.54}$$

$$\Sigma(z) = [\overline{H_I(Q)D(z)H_I(Q)} + \overline{H_I(Q)D(z)H_I(Q)D(z)H_I(Q)} + \dots]_{id} . \tag{4.55}$$

Here, R_d with subscript d means the diagonal operator obtained by taking that part of the operator R which is diagonal in the representation in which H_e is diagonal, namely, $[R_d]_{KK'} = \delta_{KK'} R_{KK}$. As a trivial example, $[H_e]_d = H_e$ since $H_{e,KK'} = \delta_{KK'} E_K$. The subscript id is an abbreviation of "irreducible diagonal", it means, in addition to taking the diagonal part, to restrict the intermediate states in such a way that none of them are identical to each other nor to the initial (= final) state. Thus, the self-energy Σ is explicitly written as

$$\Sigma_K(z) = \sum_{K' (\neq K)} \frac{\overline{H_I(Q)_{KK'} H_I(Q)_{K'K}}}{z - E_{K'} - \Sigma_{K'}(z)}$$

$$+ \text{(higher-order terms)} . \tag{4.56}$$

The real and imaginary parts of $\Sigma_K(z) \equiv \Delta_K(E) + i\Gamma_K(E)$ are the level shift and the level broadening of the exciton state $|K\rangle$ due to its phonon scattering, as is also obvious from the line-shape formula which can be rewritten exactly as, see (4.52, 54),

$$F_a(E) = \frac{1}{\pi} \frac{\Gamma_0(E)}{[E - E_0 - \Delta_0(E)]^2 + \Gamma_0(E)^2} . \tag{4.57}$$

If E_0 is at the band bottom (or top), the line-shape deviates significantly from a Lorentzian due to the peculiar E dependence of $\Gamma_0(E)$ and $\Delta_0(E)$ at $E \sim E_0$, as will be shown below.

Let us calculate $\Sigma_K(z)$ of (4.56), neglecting higher-order terms. Noting that $(H_I)_{KK'} = -cN^{-1} \sum_n \exp[i(K' - K) \cdot R_n]Q_n$, one obtains $\overline{H_I(Q)_{KK'} H_I(Q)_{K'K}} = N^{-1}D^2$, and hence the self-energy turns out to be K independent:

$$\Sigma_K(z) = \Sigma_0(z) = D^2 N^{-1} \sum_{K'} \frac{1}{z - E_{K'} - \Sigma_0(z)} . \tag{4.58}$$

Define a function

$$G^{(0)}(z) \equiv N^{-1} \sum_{K'} \frac{1}{z - E_{K'}} \tag{4.59}$$

which is analytic in the lower half of the complex z-plane and whose imaginary part gives π times the normalized density of exciton states $\varrho^{(0)}(E)$ in the rigid lattice. Then (4.58) reduces to

$$\Sigma_0(z) = D^2 G^{(0)}(z - \Sigma_0(z)) \tag{4.60}$$

which must be solved for $\Sigma_0(z)$.

As a solvable example of (4.60), we set [4.23]

$$G^{(0)}(z) \equiv 2B^{-2}(z - \sqrt{z^2 - B^2}) , \tag{4.61}$$

where for convenience we take the center of the band as the origin of energy. The corresponding density of states (DOS) $\varrho^{(0)}(E) = 2\pi^{-1}B^{-2}\sqrt{B^2 - E^2}$ rises from the band edges as $\sqrt{B - |E|}$ as is characteristic of a three-dimensional lattice. Then (4.60) is solved to give

$$\Sigma_0(z) = \frac{2D}{B^2 + 4D^2}[z - \sqrt{z^2 - (B^2 + 4D^2)}] = \Delta_0(E) + i\Gamma_0(E) . \tag{4.62}$$

Putting this into (4.57) and assuming that $K = 0$ is at the bottom of the band $(E_0 = -B)$, one obtains

$$F_a(E) = \frac{1}{\pi}\frac{2D^2}{B^2}\frac{\sqrt{(B^2 + 4D^2) - E^2}}{[E + B + (2D^2/B)]^2} \tag{4.63a}$$

which can be approximated by

$$F_a(E) = \frac{2}{\pi}\frac{\sqrt{W}\sqrt{E - \tilde{E}_0}}{(E - \tilde{E}_0 + W)^2} , \tag{4.63b}$$

$$\tilde{E}_0 \equiv -(B^2 + 4D^2)^{1/2} , \qquad W = \frac{2D^4}{B^3} \propto T^2 \tag{4.64}$$

if $D \ll B$. The full width at half maximum (FWHM) of (4.63b) is given by $\sim 1.9W$. However, this numerical coefficient must be multiplied by an order of magnitude if higher-order terms are taken into account [4.22]. This is characteristic of a three-dimensional lattice where $\varrho^{(0)}(E)$ rises rapidly from the band bottom; the higher-order scattering is more favorable since it can take advantage of a greater number of intermediate states with higher energy. This situation is different in two- or one-dimensional lattices where $\varrho^{(0)}(E)$ stays constant or decreases as E increases from the band bottom.

The temperature dependence of the line-width for different dimensionalities d in the case that $K = 0$ is at the band edge can be obtained by the following argument. According to (4.58) $\Gamma_0(E)$ is given by $\sim D^2 \varrho^{(0)}(E)$ when $\varrho^{(0)}(E) \sim B^{-d/2}|E - E_0|^{d/2-1}$ near the band edge. The width w of the quasi-Lorentzian line shape (4.57) is given by setting $w \sim |E - E_0| \sim \Gamma_0(E)$ in the denominator, whence we obtain $w \propto D(D/B)^{d/(4-d)} \propto T^{2/(4-d)}$ [4.20].

The line-shape will be nearly Lorentzian with the width proportional to T if the DOS is finite and gently varying around E_0. This is realized when E_0 is at neither the bottom nor the top of the band, or when there are other exciton bands into which the optically created exciton can be scattered [4.12].

The width obtained above is smaller than the Gaussian width D of localized excitation as long as $D \ll B$. This can be explained as follows. In order for the exciton to sufficiently perceive the local fluctuation of the potential with amplitude D, it must stay at the same site for at least a time $\sim \hbar D^{-1}$ because of the uncertainty principle. If the time of transfer to neighboring sites, $\hbar B^{-1}$, is shorter than this, the exciton will feel the averaged potential of many sites. The line-width is then *motionally narrowed* [4.12], by a d-dependent power of the narrowing factor D/B. As D increases beyond B, the line-shape tends to a Gaussian, characteristic of localized excitation.

Thus the competition between localization and delocalization of the *unrelaxed exciton* as reflected in the absorption spectra is governed by D/B, in contrast to that of the *relaxed exciton* which is governed by $g \equiv E_{LR}/B$ (Fig. 4.2). Moreover, the transition from one limit to the other is *gradual* in the former case but *abrupt* in the latter. The cases $D \lessgtr B$ will be called *weak* and *strong scattering*, respectively [4.24], and $E_{LR} \lessgtr B$ *weak* and *strong coupling*, respectively. Since B and E_{LR} are of the order of 1 eV in typical insulators and much greater than the thermal energy $k_B T$, we usually have $D \ll E_{LR}$. Thus we have an intermediate region of weak scattering but strong coupling, namely, the exciton is delocalized when optically created but becomes localized after relaxation. As a unifying method of describing the whole range of weak through strong scattering, we will introduce, in Sect. 4.2.3, the coherent potential approximation and give its applications to the present system as well as to the exciton in a mixed crystal.

Here we present some results of numerical calculation of the line shape of exciton absorption based on the Franck-Condon type formula (4.51) as applied to the model Hamiltonian (4.37–40) [4.25]. The calculation has been made for a linear chain ($d = 1$) of 30 sites, a square lattice ($d = 2$) of 13×12 sites and a simple cubic lattice ($d = 3$) of $10 \times 9 \times 8$ sites, with only the nearest neighbor transfer t. It is assumed that t is negative, hence the band bottom is at $K = 0$. The results at typical temperatures for each dimensionality are shown in Fig. 4.3, where energy is measured with E_a as the origin, and g and B ($= 2d|t|$) were chosen to be 1 and 0.5 eV, respectively (note that we are well within the weak scattering regime: $D^2/B^2 \equiv 2gk_B T/B \ll 1$).

Most remarkable is the asymmetry of the line shape, with longer tails on the high-energy side which can be fitted to Lorentzian curves: $F(E) \propto \Gamma/[(E - \tilde{E}_0)^2 + \Gamma^2]$ as shwon by the broken lines. The deviations at higher ener-

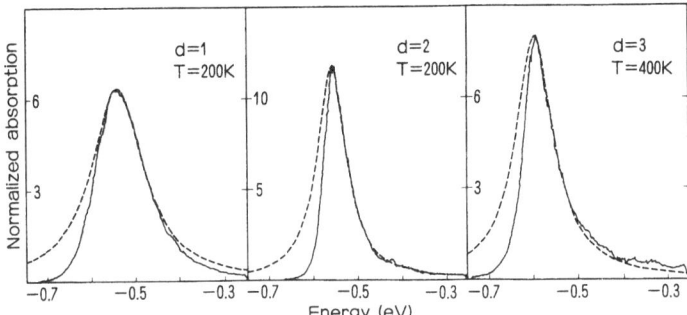

Fig. 4.3. Calculated exciton absorption spectra (——), compared with Lorentzian (---) best fitted to the high-energy side of the peaks

gies – downward for $d = 1$, none for $d = 2$ and upward for $d = 3$ – can be explained as follows. The broadening $\Gamma(E)$ is an energy dependent quantity approximately proportional to the density $\varrho(E)$ of the final states into which the photo-created exciton is scattered [see (4.56, 57) and susequent statements, and also a model calculation (4.63b) for $d = 3$]. In fact, if one calculates the normalized density of states

$$\varrho(E) = N^{-1} \overline{\sum_i \delta(E - E_i(Q))} \tag{4.65}$$

and the average oscillator strength per state (AOSPS) defined by

$$f(E) \equiv F_a(E)/\varrho(E)$$
$$= N \overline{[|\langle 0|i(Q)\rangle|^2]}_{E_i(Q) = E}, \tag{4.66}$$

one finds that the high-energy side of $f(E)$ fits almost exactly to a Lorentzian curve in the weak scattering regime in accordance with our perturbation theory interpretation of the line-width as (scattering-limited) lifetime broadening. This behavior of $F_a(E)$ and $f(E)$ will be analysed in more detail in Sect. 4.2.4, with particular attention to the different origins of high- and low-energy tails.

The widths of the calculated line-shapes are plotted against temperature in Fig. 4.4, which are also in good agreement with the perturbation theory calculations described before, especially in that the temperature dependence is given by $T^{2/(4-d)}$. The deviations from this in the $d = 3$ case at low and high temperatures are due, respectively, to the number of lattice sites considered – too few to calculate this small width – and to the onset of the intermediate scattering regime. It should also be noted that inclusion of higher-order scattering terms in the perturbation theory is essential to obtain the right order of magnitude of the line width.

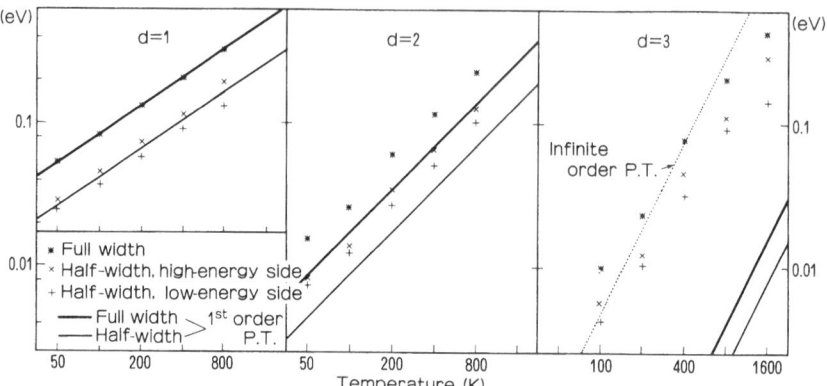

Fig. 4.4. Calculated line widths are plotted against temperature and compared with perturbation theory (*lines*)

Quantitative analysis of the observed line-shapes of the exciton absorption bands has been done for a variety of materials. Among them, we mention the works by *Tomiki* and his collaborators on alkali halides [4.26] and those by *Burland* et al. on some organic molecular crystals [4.27]. See also Sect. 3.6 for excitonic molecules. While the general features of the line-shapes are in accordance with the above-mentioned theory, quantitative explanation of their details seems to need more elaborate models for the scatterers (various modes of phonons and impurities) and the exciton band structures.

4.2.3 Coherent Potential Approximation for an Exciton in a Mixed Crystal and in a Phonon Field

As was mentioned in the preceding sections, it is desirable to have a unifying theory covering the whole region from weak to strong scattering, namely, from delocalization to localization limits. A kind of interpolation theory introduced to describe the corresponding two limits of an exciton in binary mixed crystals – amalgamation and persistence limits [4.28] – turned out to be equivalent [4.29] to the coherent potential approximation (CPA) devised to treat electrons and phonons in alloys [4.30]. The CPA was then applied to an exciton in a phonon field [4.23], not only to see the gradual transition from weak to strong scattering but also to analyse the low-energy tail of the absorption spectra which will be described in Sect. 4.2.4. In the present section we will give a brief account of the CPA [4.31] and then apply it to the excitons in a binary mixed crystal and in a phonon field, with particular attention to the transition from weak to strong scattering.

Let us consider an exciton, moving through the lattice with a translationally symmetric Hamiltonian H_e and spatially fluctuating potential V. There is an arbitrariness in the decomposition of the total Hamiltonian H into H_e and V. We will choose decomposition such that the ensemble (or spatial) average of the potential V vanishes

$$\overline{V} = 0 . \tag{4.67}$$

As a model Hamiltonian, we consider a Frenkel exciton in a binary mixed crystal consisting of atoms A and atoms B with respective concentrations $c_A \equiv c$ and $c_B = 1 - c$ randomly distributed over the regular lattice sites. We assume different values E_A and E_B for the intraatomic excitation energy E_n, but a common t_{nm} for the interatomic $(n \to m)$ transfer energy irrespective of the atomic species occupying the relevant sites. Then we can write the total Hamiltonian and then decompose it according to the condition (4.67) as follows:

$$H = \sum_n |n\rangle E_n \langle n| + \sum_n \sum_m |n\rangle t_{nm} \langle m| \quad (t_{nn} = 0) \tag{4.68a}$$

$$= H_e + V , \tag{4.68b}$$

$$H_e = \sum_n |n\rangle \overline{E_n} \langle n| + \sum_n \sum_m |n\rangle t_{nm} \langle m| \tag{4.69a}$$

$$= \sum_K |K\rangle E_K \langle K| , \tag{4.69b}$$

$$V = \sum_n |n\rangle v_n \langle n| \equiv \sum_n V_n , \quad v_n \equiv E_n - \overline{E_n} , \tag{4.70}$$

$$\left.\begin{array}{l} \overline{E_n} = c_A E_A + c_B E_B , \quad E_K = \overline{E_n} + t_K \\[4pt] t_K \equiv \sum_n t_{0n} \exp(i\mathbf{K} \cdot \mathbf{R}_n) . \end{array}\right\} \tag{4.71}$$

For an exciton in a phonon field the decomposition $H = H_e + H_I$ $(H_I \leftrightarrow V)$ as prescribed in (4.37–39) satisfies (4.67). The random potential $v_n = -cQ_n$ due to lattice vibrations obeys a continuous and symmetric Gaussian distribution, $\exp(-v_n^2/2D^2)$, in contrast to the mixed crystal with v_n obeying a discrete and asymmetric (unless $c_A = c_B = 1/2$) distribution. In both systems, v_n's at different sites have no correlation:

$$\overline{v_n v_m} = \overline{v_n}\,\overline{v_m} = 0 \quad (n \neq m) , \tag{4.72}$$

The spirit of the CPA is to further incorporate into H_e the main effect of exciton scattering by V, in the form of an *average effective potential* u, in such a way that the effect of scattering by the remaining potential

$$V' \equiv V - u\underline{1} = \sum_n |n\rangle v'_n \langle n| \equiv \sum_n V'_n, \quad v'_n \equiv v_n - u \tag{4.73}$$

is minimized. Thus, we expand the resolvent as

$$R(z) = (z\underline{1} - H_e - V)^{-1} = [z\underline{1} - (H_e + u\underline{1}) - (V - u\underline{1})]^{-1}$$

$$= D + D \sum_n V'_n D + D \sum_n V'_n D \sum_m V'_m D + \ldots \tag{4.74}$$

where D is an operator diagonal in K:

$$D(z) \equiv [z\underline{1} - (H_e + u\underline{1})]^{-1} = \sum_K |K\rangle (z - E_K - u)^{-1} \langle K| , \qquad (4.75)$$

but will be distinguished for the moment from $D(z)$ of (4.54).

The expansion (4.74) can be rearranged in such a way that successive scatterings at the same site are collected into a t-matrix:

$$V'_n + V'_n D V'_n + V'_n D V'_n D V'_n + \ldots = V'_n (1 - G V'_n)^{-1}$$

$$\equiv T_n = |n\rangle t_n \langle n| \quad \text{where} \qquad (4.76)$$

$$G(z) \equiv D(z)_{nn} = N^{-1} \sum_K (z - E_K - u)^{-1} = G^{(0)}(z - u) , \qquad (4.77)$$

$$G^{(0)}(z) \equiv D^{(0)}(z)_{nn} = N^{-1} \sum_K (z - E_K)^{-1} . \qquad (4.78)$$

Note that the imaginary part of $G^{(0)}(z)$ represents π times the density of states $\varrho^{(0)}(E)$ in the unperturbed lattice. Equation (4.74) can then be rewritten as

$$R(z) = D + D \sum_n T_n D + D \sum_n T_n D \sum_{m(\neq n)} T_m D$$

$$+ D \sum_n T_n D \sum_{m(\neq n)} T_m D \sum_{l(\neq m)} T_l D + \ldots \qquad (4.79)$$

where none of successive pairs of t scattering take place at the same site. Then, in view of (4.72), it would be a fairly good approximation to replace the ensemble average of the product of t's by the product of the averages of t:

$$\overline{R(z)} \approx D + D \sum_n \overline{T_n} D + D \sum_n \overline{T_n} D \sum_{m(\neq n)} \overline{T_m} D$$

$$+ D \sum_n \overline{T_n} D \sum_{m(\neq n)} \overline{T_m} D \sum_{l(\neq m)} \overline{T_l} D + \ldots . \qquad (4.80)$$

In fact, the lowest-order discrepancy appears in the third-order terms:

$$\overline{t_n t_m t_l} \neq \overline{t_n}\, \overline{t_m}\, \overline{t_l} \quad \text{when} \quad l = n .$$

Let us choose the effective potential u in such a way that the ensemble average of the t-matrix vanishes:

$$\overline{t_n} \equiv \overline{v'_n (1 - G v'_n)^{-1}} = 0 , \quad \text{namely,} \qquad (4.81a)$$

$$G(z) = \overline{[G(z)^{-1} - v_n + u]^{-1}} . \qquad (4.81b)$$

The "coherent potential" $u(z)$ as well as the individual potential $v'_n(z)$ thus determined are energy (z) dependent complex quantities. Equation (4.80) is then written, with the use of (4.75), as

$$\overline{R(z)} \approx D(z) = \{z\underline{1} - [H_e + u(z)\underline{1}]\}^{-1}$$

$$= \sum_K |K\rangle [z - E_K - u(z)]^{-1} \langle K| . \tag{4.82}$$

The approximation (4.82) with $u(z)$ determined by (4.81a) or (4.81b) is the CPA. Comparison of this with (4.54) indicates that our $D(z)$ and $u(z)\underline{1}$ correspond to $D(z)$ and $\Sigma(z)$ of Sect. 4.2.2, and that our self-energy $u(z)\underline{1}$ is K independent as in the approximation (4.58).

The merit of the CPA is that it is exact in various limiting situations. In the weak scattering regime, it agrees with the perturbation theory up to the second order as is obvious from the statement which follows (4.80). Remembering (4.70) and (4.73), one obtains from (4.81)

$$u(z) = \langle v_n^2 \rangle G(z) \quad (+ \text{ higher-order terms in } v) \tag{4.83}$$

which reduces, in its lowest order, to the renormalized second-order perturbation theory (4.58) for the exciton-phonon system. In the case of a mixed crystal, the amplitude D of the potential fluctuation is given by

$$D^2 \equiv \langle v_n^2 \rangle = c(1-c)\Delta^2 , \tag{4.84}$$

$$\Delta \equiv E_B - E_A . \tag{4.85}$$

The CPA becomes exact again in the strong scattering limit: $D \gg B$. Neglecting the band width, one can put $E_K \approx \overline{E}_n$ and $G^{(0)}(z) \approx (z - \overline{E}_n)^{-1}$. Because of (4.77), (4.81b) can be written as

$$[z - \overline{E}_n - u(z)]^{-1} \approx \overline{(z - \overline{E}_n - V_n)^{-1}} . \tag{4.86}$$

Hence, the density of states and the absorption line-shape reduce to

$$\varrho(E) \approx F_a(E) = \pi^{-1} \operatorname{Im} \{[z - \overline{E}_n - u(z)]^{-1}\} \tag{4.87}$$

$$= \begin{cases} c_A \delta(E - E_A) + c_B \delta(E - E_B) \\ \\ \dfrac{1}{\sqrt{2\pi D^2}} \exp[-(E - E_a)^2/2D^2] \end{cases} \tag{4.88}$$

for a mixed crystal and phonon field, respectively.

The CPA turns out to be exact also in the dilution limit ($c \equiv c_A \to 0$ or $c_B \to 0$) of a mixed crystal, as is obvious from the fact that single site scattering is considered up to infinite order (4.76). Rewrite (4.81b) for a mixed crystal as

$$u(z) - v_B = \frac{-c\Delta}{1 + G(z) \{\Delta + [u(z) - v_B]\}} \tag{4.89}$$

and then neglect $[u(z) - v_B]$ in the denominator since it is $O(c^2)$ in the dilution limit ($c \to 0$). Then the $K = 0$ element of (4.82) can be expanded as

$$[\overline{R(z)}]_{K=0} = \{z - (E_0 + v_B) - [u(z) - v_B]\}^{-1}$$

$$= \frac{1}{z - E_0^B} - \frac{c\Delta}{(z - E_0^B)^2 \,[1 + \Delta G_B(z)]} + O(c^2) \tag{4.90}$$

where

$$G_B(z) \equiv D_B(z)_{nn} = N^{-1} \sum_K (z - E_K^B)^{-1} , \tag{4.91}$$

$$D_B(z) \equiv (z\underline{1} - H_e^B)^{-1} \quad \text{with} \tag{4.92}$$

$$E_K^B = E_B + t_K = v_B + E_K \tag{4.93}$$

are defined for a pure B crystal. If there is a solution $E = E_b$, below the band edge E_0^B of the B crystal, of the equation

$$1 - (-\Delta) G_B(E) = 0 , \quad (E < E_0^B) \tag{4.94}$$

the term linear in c in (4.90) contributes

$$F_a(E)_{E < E_0^B} = \frac{c}{[-G_B'(E_b)] \,(E_0^B - E_b)^2} \delta(E - E_b) \tag{4.95}$$

to the absorption spectrum. The solution E_b represents a bound-exciton state due to a guest atom A substituted for a host atom B, say at the site $n = 0$. In fact, the eigenequation for this system:

$$(H_e^B + V)\psi = E\psi , \quad V_{nm} = -\delta_{n0}\delta_{m0} \tag{4.96}$$

can be solved, with the use of (4.92), as

$$\psi_b(n) = -\Delta [D_B(E_b)]_{n0} \psi_b(0) . \tag{4.97}$$

Putting $n = 0$, one obtains (4.94), see (4.91).

The bound state wave function $\psi_b(n)$ can be normalized with the use of the identity:

$$-[D_B'(z)]_{00} = \sum_n |[D_B(z)]_{n0}|^2$$

which is proved by differentiating (4.92). Noting that

$$\sum_n [D_B(z)]_{n0} = [D_B(z)]_{K=0} ,$$

one obtains

$$f_b \equiv \left| \sum_n \psi_b(n) \right|^2 = \frac{1}{[-G_B'(E_b)] \,(E_0^B - E_b)^2} . \tag{4.98}$$

The quantity f_b represents the oscillator strength of the bound state in units of the atomic oscillator strength, as is obvious from comparison with (4.95) and also from the fact that

$$\left| \sum_n \psi_b(n) \right|^2 = N |\langle \psi_{K=0} | \psi_b \rangle|^2 . \tag{4.99}$$

The right-hand side represents the oscillator strength borrowed from $\psi_{K=0}$ into which the total oscillator strength of N atoms is concentrated in a pure crystal, see also (4.66). The left-hand side of (4.99) represents the number M of those sites over which the bound-state wave function extends with significant amplitude, since each $\psi(n)$ is of the order of $M^{-1/2}$. The binding energy $\varepsilon_b \equiv E_0^B - E_b$ and the orbital radius a_b of the bound exciton are correlated by $\varepsilon_b \propto a_b^{-2}$ due to the uncertainty principle. Thus we find

$$f_b \sim M \propto (a_b)^d \propto \varepsilon_b^{-d/2} \tag{4.100}$$

where d is the dimensionality. Thus, a shallowly bound exciton bears *giant oscillator strength*, corresponding to the number of atoms it covers [4.32].

The ε_b dependence of f_b in (4.100) can also be shown directly from the right-hand side of (4.98) with the use of the dispersion formula (Kramers-Kronig relation)

$$\text{Re} \{G_B(E)\} = \mathscr{P} \int \frac{dE'}{E - E'} [\pi^{-1} \text{Im} \{G_B(E')\}] \tag{4.101}$$

which follows from the analyticity of $G_B(z)$ in the lower half of the complex z-plane. Note that the square bracket in the integrand is the density of states $\varrho_B(E')$ which behaves around the band edge E_0^B as $(E' - E_0^B)^{(d/2)-1}$ $(E' > E_0^B)$ and 0 (otherwise). By differentiating (4.101), one immediately finds that $-G_B'(E_b) \propto \varepsilon_b^{(d/2)-2}$ provided $\varepsilon_b \ll B$.

Model calculations have been made for excitons in mixed crystals [4.28] and in a phonon field [4.23], assuming the semielliptic form for the state density of the unperturbed band as specified by (4.61). Figure 4.5 shows how the density of states (dashed line), the absorption spectrum (solid line) and the imaginary part of the self-energy (dot-dash line) of an exciton in a mixed crystal vary with $\Delta/2B$ for fixed $c_A : c_B = 1 : 1$. The A and B atomic states are *amalgamated* into a single band or *persist* as split bands according as $\Delta/2B$ is small or large. As shown by the dot-dash lines, the perturbation is most remarkable near the (pseudo-)gap of the bands where mixing of the two atomic states is greatest. The transition from the amalgamation type to the persistence type depends of course on the composition, as shown in Fig. 4.6. The amalgamated band and the split bands are shown against concentration c in Figs. 4.7, 8. The lower edge of the amalgamated band, which is close to the absorption peak, shows downward quadratic bending against c from the linearly shifting averaged band edge, $E_{K=0} = c_A E_A + c_B E_B - B$, due to the self-energy shift given by

$$\text{Re} \{u(E_0)\} \sim c(1-c)\Delta^2 G^{(0)}(E_0) , \quad [G^{(0)}(E_0) < 0] . \tag{4.102}$$

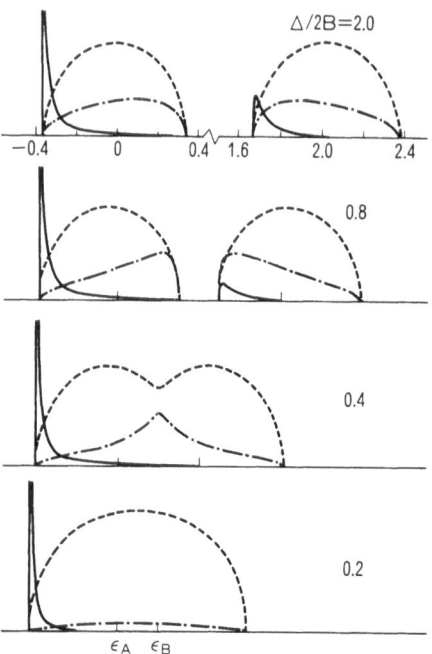

Fig. 4.5. Density of states (– – –), optical absorption spectrum (——), and imaginary part of self-energy (– · –) of an exciton in 50 : 50 mixed crystals with various values of $\Delta/2B$

In Fig. 4.8 one sees how the impurity–bound-exciton states form the impurity band which finally grows into the host band as c increases. In the intermediate case ($0.25 < \Delta/2B < 0.5$, see Fig. 4.6), the shallow bound exciton states form the *impurity band* which then merges into the host band, as shown in Fig. 4.9.

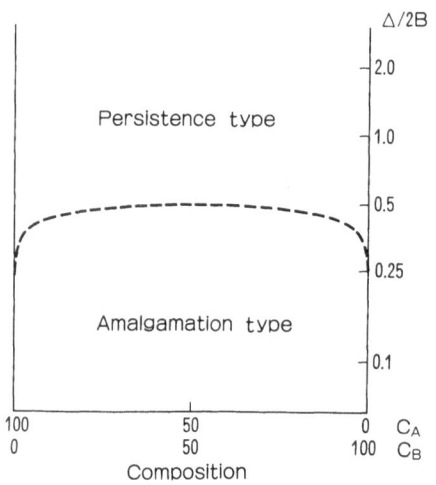

Fig. 4.6. Amalgamation and persistence regions in the coordinates space of composition and $\Delta/2B$

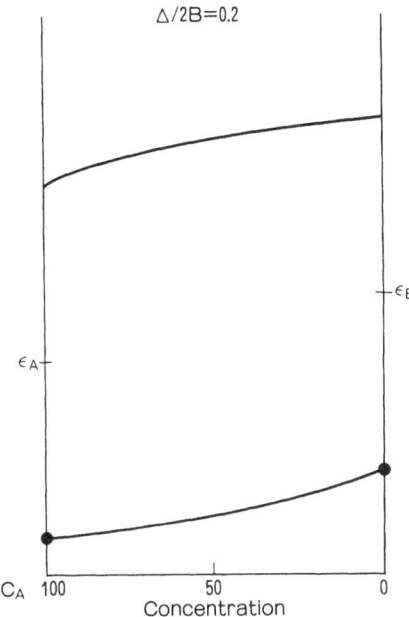

Fig. 4.7. Amalgamated band against concentration

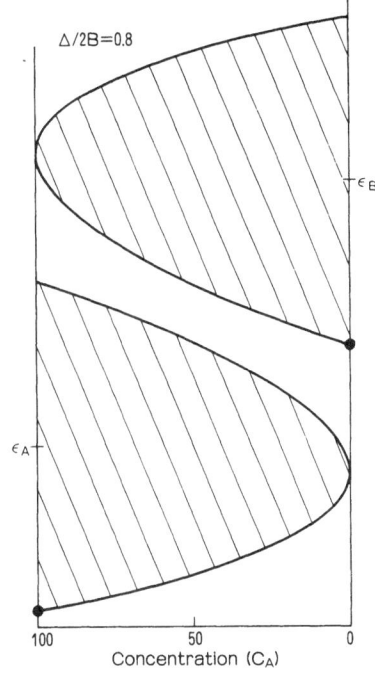

Fig. 4.8. Split bands against concentration

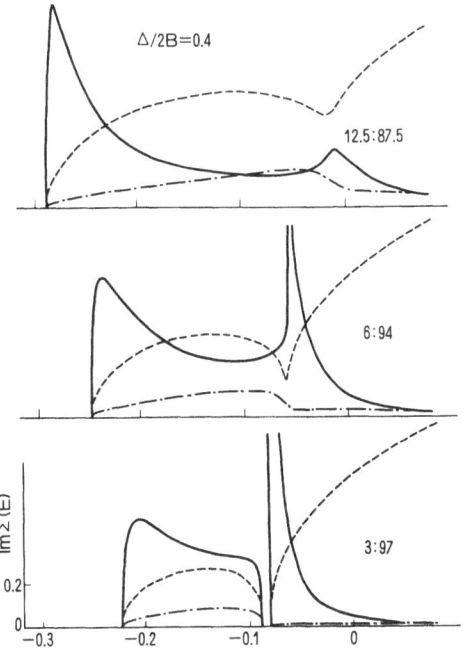

Fig. 4.9. Impurity band merges into host band with increase in concentration for an intermediate value of $\Delta/2B = 0.4$. Energy is measured from E_A in units of $2B$. Line designations are the same as in Fig. 4.5

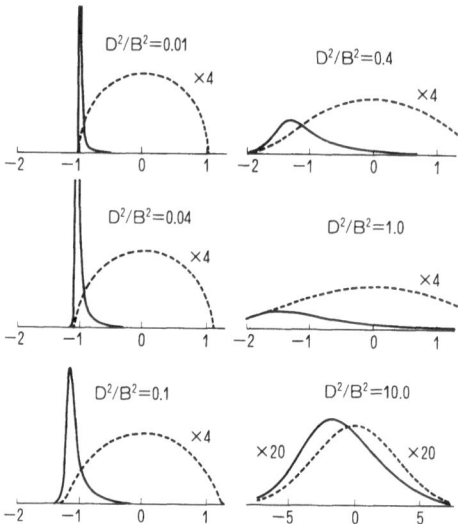

Fig. 4.10. Optical absorption spectra (——) and density of states (–––) of an exciton for various values of $(D/B)^2$ which is proportional to temperature. Energy is measured from the band center in units of B

One can clearly notice the giant oscillator strength characteristic of a shallow bound exciton; the ratio of the oscillator strengths of the impurity band and the host band is much greater than the concentration ratio.

Typical examples of amalgamation-type excitons are found in $CuBr_{1-x}Cl_x$ (Chap. 3), $TlBr_{1-x}Cl_x$ (direct and indirect excitons, Chap. 7) and mixed-alkali halides [4.33], and the persistence type in alkali mixed-halides [4.34] and mixed rare gas solids [4.35]. The indirect exciton in $AgBr_xCl_{1-x}$ is almost amalgamated but slightly intermediate: the Br impurity band shows up only for very small x (Chap. 6).

The transition from weak to strong scattering of an exciton in the phonon field [4.23] is shown in Fig. 4.10. By comparing the absorption spectra (solid lines) and the density of states (dashed lines), one can clearly see how the K-conserving delocalized exciton gradually transforms to the localized exciton as D/B increases. Starting from the opposite limit, one sees how the Gaussian band due to the site-energy fluctuation is motionally narrowed into the (asymmetric) Lorentzian band of lifetime-broadened type.

4.2.4 The Urbach Rule and Exciton Localization

Since the discovery by Urbach [4.36], it has been established experimentally for a variety of insulators – organic as well as inorganic [4.37] – that the low-energy tail of the exciton absorption spectrum depends exponentially on energy, with a decay constant proportional to the reciprocal temperature at high temperatures:

$$F_a(E) = A \exp\left(-\sigma \frac{E_0' - E}{k_B T}\right).$$

(4.103)

The convergence point, E_0', of the semilogarithmic plots of the absorption spectra at various temperatures is usually situated near the absorption peak at the lowest temperature. The dimensionless "steepness coefficient" σ is a material constant of the order of unity. The empirical rule (4.103) holds over several decades in $F_a(E)$.

The universality, simplicity and wide validity range of this empirical rule – now called the Urbach rule – have evoked a number of theoretical and experimental studies to explain it on a microscopic basis. It is extremely difficult to review these studies, which are so varied and diverse, within a resonable space. We will trace here only a few of them from our viewpoint, referring to [4.37, 25] for more complete references.

In view of its temperature dependence, the exponential tail is undoubtedly an effect of lattice vibrations on the electronic state, as was first discussed by *Dexter* [4.38]. Secondly, it is almost obvious, in view of the absence of phonon structures in this tail part (except possibly at very low temperatures where this rule no longer holds), that the adiabatic and classical treatment of lattice vibrations is enough for its explanation.

An ad hoc model to derive (4.103) analytically was then proposed [4.39], in which an exciton was supposed to couple linearly with one mode of lattice vibrations but only quadratically with another mode. The derivation was not well-founded, however, because of its neglection of the exciton recoil effect. This two-mode model was revived by *Mahr* [4.40] to explain his data on the absorption band due to an impurity-bound exciton which was Gaussian in its central part (ascribed to linear coupling with, presumably, even parity modes) and Urbach-like in the low-energy tail (ascribed to quadratic coupling with, presumably, odd modes). His argument is legitimate since a bound exciton does not recoil.

In view of the apparent similarity between the intrinsic and extrinsic Urbach tails another attempt was made to apply the two-mode model to an intrinsic exciton, this time not directly to its free states but to the "momentarily localized states" (which vary from time to time, adiabatically following the lattice vibrations) [4.41]. This notion was introduced, not only for the ad hoc reason of being recoilless, but also due to the realization [4.22] that successive perturbation theory starting from free-exciton states is an extremely roundabout approach to describing the exponential tail. The existence of localized states below the mobile states in a randomly perturbed lattice was also becoming widely realized after the famous paper by *Anderson* [4.42]. However, it seemed extremely difficult to give a quantitative study of the configurational statistics for momentarily localized states. Although a relation between the steepness coefficient σ and the exciton-phonon coupling constant g (which will be mentioned later) was noted in the qualitative derivation of the Urbach rule, it was desired to find a theoretical framework to treat quantitatively the localized states induced by the randomness of the lattice perturbations, together with their energy dependence.

Substantial progress was made by the use of the average t-matrix approximation (ATA) [4.24] and the CPA [4.23] which were being developed to study random lattice problems, as already mentioned in Sect. 4.2.3. In these approximations, the localized states are directly and properly incorporated in the self-energy. As a result, the low-energy tail of the Gaussian distribution of site energies manifests itself as a pseudoexponential tail of the absorption spectrum resembling (4.103) in its energy and temperature dependence. Quantitatively speaking, however, there is a nontrivial discrepancy. In particular, the fact that the calculated line shape is convex just below the peak seems to originate from the well-known drawback of the ATA and CPA that coherent scattering by more than two sites is not properly taken into account. Any correction to this underestimation would result in significantly improving the agreement of the line shape with (4.103), because the shallowly localized states due to coherently negative potentials would contribute giant oscillator strength to this spectral region.

A completely different mechanism was proposed by Dow and Redfield [4.43] who ascribed the exponential tail to the Franz-Keldysh type broadening due to exciton ionization under the fluctuating electric field caused by lattice vibrations. What is concerned here is the electron–hole relative motion, instead of the translational motion considered so far. An experimental study of the effect of a static electric field on the exponential tail [4.44] seemed to favor this theory. A difficulty with the theory is the unrealistic assumption of a uniform electric field in calculating the ionization rate of a large-radius Wannier exciton although the statistical distribution of the electric fields is considered carefully. Moreover, the Urbach rule has been confirmed in nonpolar crystals in which the lattice vibrations are not accompanied by strong electric fields.

Coming back to the translational motion, one has to overcome the aforementioned difficulty inherent in one-site approximations such as the ATA and CPA. We will present here the results of a direct numerical calculation [4.25] which is free of such a difficulty. We have already presented a part of the numerical results – the overall line-shape with its temperature dependence – in Sect.4.2.2 (Figs. 4.3, 4). Here we will concentrate on the low-energy tails of the absorption spectra.

If the Urbach rule, (4.103), is derivable from our model Hamiltonian (4.37–40) with nearest neighbor transfer only, the steepness coefficient σ in the former should be related to physical parameters appearing in the latter in the following way. Let us scale various energies appearing in the Hamiltonian in units of B: $(E - E_0)/B$, cQ_n/B, etc. The temperature comes in only through the Boltzmann distribution, that is, in the form $\overline{(cQ_n/B)^2} = 2E_{LR}k_BT/B^2 = D^2/B^2$, see (4.46, 49). Therefore, the exponent in (4.103) with its energy and temperature dependence can be obtained only in the form of

$$\text{const.} \frac{(E_0' - E)/B}{E_{LR}k_BT/B^2} .$$

This means that the steepness coefficient σ is related to the exciton-phonon coupling constant $g = E_{LR}/B$, (4.50), by

$$\sigma = sg^{-1} . \tag{4.104}$$

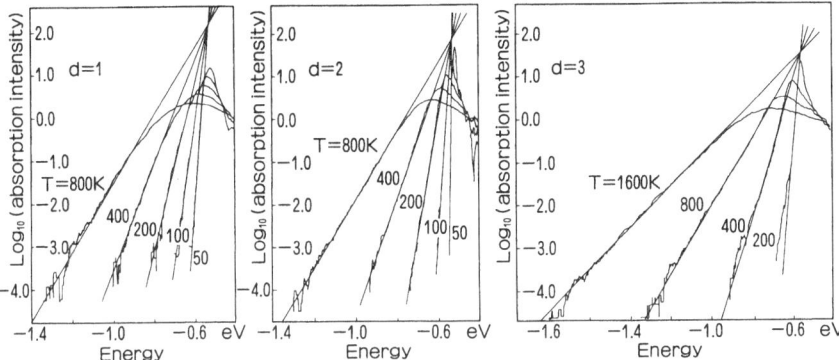

Fig. 4.11. Semilogarithmic plots of calculated absorption spectra of the direct exciton at various temperatures, with $g = 1$, $E_a = 0$ and $B = 0.5$ eV

The dimensionless constant s no longer depends on the material constants (such as g) but only on the geometrical structure of the lattice, in particular on its dimensionality.

Figure 4.11 is the semilogarithmic plot of the calculated line-shapes for one- to three-dimensional lattices, where $g = 1$, $B = 0.5$ eV, and $E_a = 0$ were chosen, as before. For all dimensionalities one finds the exponential dependence on energy over a range of three to five decades of intensity, at various temperatures. As for the temperature dependence of the decay constant, however, one finds proportionality to T^{-1} for $d = 2$ and 3 but to $T^{-2/3}$ for $d = 1$. This is shown by broken lines in Fig. 4.12, where the σ value obtained by best fitting of (4.103) to

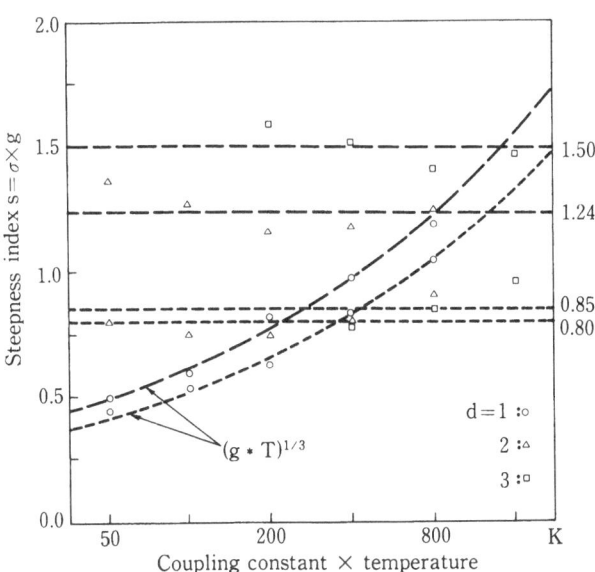

Fig. 4.12. Calculated steepness index s (plotted by ○, △, and □ for $d = 1, 2$, and 3, respectively) of the Urbach tail for direct and indirect excitons fitted by longer- and shorter-dashed lines, respectively. The one-dimensional results are best fitted by $T^{1/3}$ lines

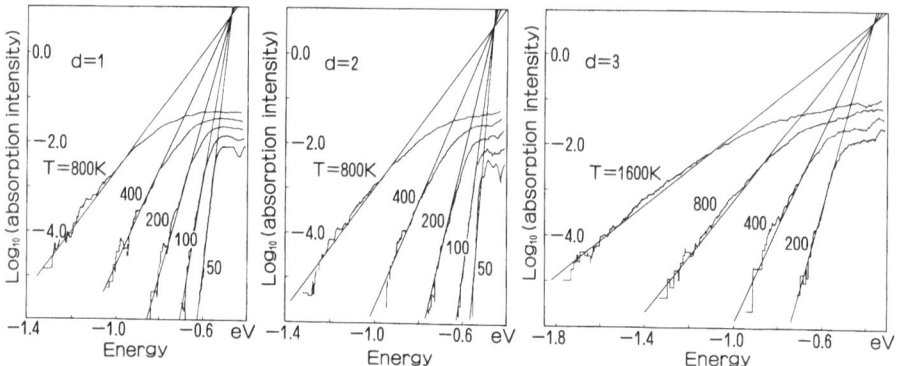

Fig. 4.13. Semilogarithmic plots of calculated absorption spectra of the indirect exciton at various temperatures, with $g = 1$, $E_a = 0$ and $B = 0.5$ eV

the calculated line-shape at each temperature is plotted by the symbols ○, △ and □ for $d = 1$, 2, and 3.

By inverting the sign of the transfer energy t, one obtains the band structure with the direct exciton ($K = 0$) at the top and the indirect exciton at the bottom of the band. Figure 4.13 is the semilogarithmic plot of the thus calculated indirect absorption edge, with the same parameter values as before. One finds again the exponential dependence on energy over a range of 2.5–4 decades of intensity for various temperatures. The temperature dependence of the decay constant is the same as before as shown by dotted lines in Fig. 4.12. The slope is more gentle in the indirect edge than in the direct edge for each dimensionality.

In order to see the physical origins of the low- and high-energy tails of the absorption spectra, we show in Fig. 4.14 the calculated AOSPS spectra $f(E)$ defined by (4.66), for the direct edge only. (The calculated DOS curves are not shown here.) What is most remarkable here is the *energy* and *temperature*

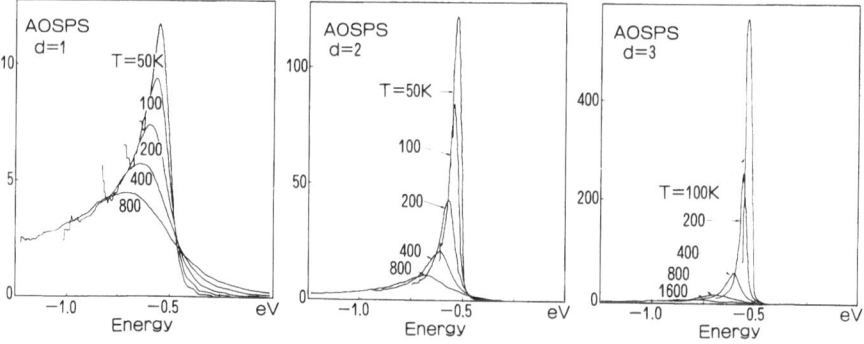

Fig. 4.14. Calculated AOSPS spectra, with the same parameter values as Figs. 4.11 and 13

dependence which is completely different for the high- and low-energy sides of the peak. For all dimensionalities, the high-energy side of $f(E)$ fits to a Lorentzian (with temperature dependent width as shown in Fig. 4.4) much better than $F_a(E)$ does. This indicates that the high-energy tail is due to the scattering-limited lifetime broadening of a delocalized exciton, as was already discussed in Sect. 4.2.2. In contrast, the low-energy tail is rather insensitive to temperature and decays with decreasing energy more rapidly for a higher dimensionality. These features clearly indicate that the low-energy tail is due to the localized exciton states. As was shown by (4.98) for the case of single site attractive potential and by subsequent qualitative analysis applicable to more general cases, the oscillator strength of a localized state is essentially a function of its binding energy only (measured from the effective band edge E_0 which is close to the absorption peak E_p), being insensitive to the individual site energies responsible for that localized state. In fact, one finds easily that the low-energy tail of $f(E)$ in Fig. 4.14 decays as $(E_p - E)^{-d/2}$ in accordance with (4.100) if one takes for E_p the peak of the calculated $F_a(E)$ at each temperature.

Starting from the model Hamiltonian (4.37–40), we have derived numerically the empirical rule (4.103) without resorting to any physical picture. Through the analysis of the calculated spectra of $f(E)$, however, we have been able to unambiguously ascribe the exponential tail of $F_a(E)$ to the localized exciton states. Thus, the introduction of the ATA and CPA into the present problem turns out to have been a step towards reality; the mathematical difficulty inherent in these one-site approximations, however, had to be resolved by resorting to direct numerical calculation.

One may well ask if the simplified model Hamiltonian (4.37–40) can represent realistic situations. For instance, can the interaction between a large-radius exciton and lattice vibrations be described by the short-range and on-site interaction (4.39)? This question is of a similar nature to that which was asked about the field ionization model above, and is rather difficult to answer directly. Another question of a different nature is: Can a theoretician be really satisfied with a numerical derivation of an analytically very simple empirical rule? Useful hints concerning this question, if not an answer, are found in the scaling argument given for this problem by *Ihm* and *Phillips* [4.45]. Finally, has the two-mode model nothing to do with the present numerical theory? The former gave an analytical explanation of the Urbach rule for an impurity bound exciton, and one is tempted to explain the apparently same rule for intrinsic and extrinsic excitons by the same mechanism. It should be mentioned, in this connection, that the numerical calculation of line-shapes for the crystal with an impurity was also performed using the same Hamiltonian as above (except for the inclusion of an attractive potential at the impurity site), and that the Urbach tails were thereby obtained for extrinsic [4.25] as well as for intrinsic edges. It is easy to show that the numerical theory is equivalent to the two-mode model as far as the extrinsic absorption band (discrete level) is concerned. It should then be a serious question, how and where the two-mode mechanism is concealed in the numerical theory for the intrinsic exciton.

Table 4.1. Correlation between the observed steepness coefficient σ and the observed nature (F or S) of the relaxed exciton. See Sect. 4.4.7 and [4.37] for experimental references

predicted:					
direct: $\sigma_C = 1.64$			indirect: $\sigma_C = 0.93$		
observed:					
Se	0.6	?			
NaCl	0.76	S			
KBr	0.79	S			
KI	0.82	S			
α-perylene	0.93	S			
β-perylene	1.38	S			
pyrene	1.38	S			
GeS	1.45	?	AgCl	0.78	S
PbI$_2$	1.48	?	AgCl$_{.45}$Br$_{.55}$	0.89←S=F	
anthracene	1.7~1.5	F	AgBr	0.98	F
CdS	2.2	F	TlCl	1.16	F
ZeSe	2.4	F			
CdSe	2.5	F			
ZnTe	2.8	F			
CdTe	3.0	F			
Te	3.1	?			

The relation (4.104) between σ and g, together with the values of s obtained by the numerical calculation, serves to *correlate two independent groups of experiments*. While σ is obtained from the absorption spectra, g characterizes the emission spectra which are predicted to be F-type or S-type according as $g \lessgtr g_c \sim 1$ (Sect. 4.2.1. and Fig. 4.2). A more elaborate theory in Sect. 4.4 will give $g_c = 1 - (2\nu)^{-1}$ where ν is the number of nearest neighbors in the crystal lattice. Considering the simple cubic lattice for which $g_c = 0.917$ and $s = 1.50$ (direct) or 0.85 (indirect), we find the critical values $\sigma_c = 1.64$ (direct) or 0.93 (indirect) below which the exciton is expected to be self-trapped. Table 4.1 shows some examples of materials for which the Urbach rule has been established for the direct or indirect edge, in increasing order of σ. In the third column we indicate the nature of relaxed excitons found from luminescence studies. Thus, the two groups of experimental data can be correlated with each other in satisfactory consistency with the theory. Of particular interest in Table 4.1 is the indirect exciton in $AgBr_{1-x}Cl_x$ (Sect. 6.2.2) where the gradual decrease of σ and the abrupt *F–S* transition of the luminescence are in excellent agreement with the theory.

4.3 Phonon Structures in Exciton Spectra

We have so far studied the line shapes of optical spectra within the classical Franck-Condon principle on the assumption that each atom is fixed in its instantaneous position during the electronic transition. In fact, the atoms are vibrating with finite velocities, and the vibrational frequencies, namely, the phonon energies, will show up as fine structures in the optical spectra, especially clearly when the temperature is low or the exciton-phonon coupling is small. Instead of considering a vertical transition in the configuration coordinate model, one considers an electronic transition with the simultaneous absorption and/or emission of phonons according to this quantal picture. These phonon structures in the optical spectra provide us with more detailed information about the exciton-phonon dynamics as will be seen.

4.3.1 Motional Reduction of Phonon Sidebands

As the simplest example of phonon structures, let us quantize the one-dimensional harmonic motion of the lattice in the ground and excited states of a localized excitation as depicted in Fig. 4.1a, under the assumption of $t_{nm} = 0$ ($n \neq m$) and $(\Omega^2)_{nm} = \delta_{n,m}\omega^2$ in (4.38, 40). In fact, such a model applies to a localized electron at an impurity atom or lattice defect whose excitation energy is not resonant with that of the host atom. Each electronic state has equally spaced vibrational levels with energies $(v + 1/2)\hbar\omega$ ($v = 0, 1, 2, \ldots$) measured from the minimum of its adiabatic potential (Fig. 4.15), with vibrational wave functions $\chi_v(Q_n)$ and $\chi_v(Q_n - c)$ for the ground (g) and the excited (e) states, respectively. At the absolute zero of temperature, the absorption spectra consist of equally spaced lines at $E^{(0)} + v\hbar\omega$ ($v = 0, 1, 2, \ldots$, $E^{(0)} = E_a - E_{LR}$) corresponding to the vibronic transition: $(g, 0) \rightarrow (e, v)$, with relative intensities obeying the Poisson distribution:

$$I_v = |\int \chi_v(Q - c)\chi_0(Q)dQ|^2 = \frac{S^v}{v!}e^{-S} , \tag{4.105}$$

$$S \equiv E_{LR}/\hbar\omega , \quad (E_{LR} = c^2/2) . \tag{4.106}$$

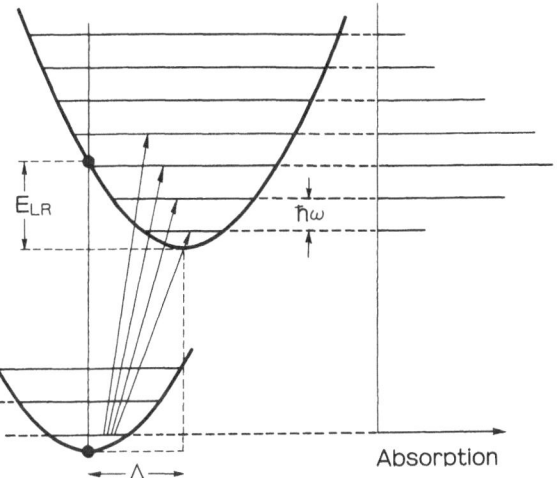

Fig. 4.15. Adiabatic potentials with vibrational levels (*left*) and an absorption spectrum with phonon structures (*right*) of a localized excitation

The line $\nu = 0$ is called the zero-phonon line, while the lines $\nu \neq 0$ corresponding to the simultaneous emission of ν phonons are called phonon sidebands of order ν [4.46]. Hereafter, S, representing the average number of phonons emitted, will be called the coupling strength.

How will these phonon structures be changed as we introduce intersite transfer (t_{nm}) of excitation energy? In the weak scattering (large transfer) regime, we can start from the exciton band (Fig. 4.1c) and study successive phonon sidebands by perturbation theory. The exciton phonon coupling coefficient γ_q, introduced in (4.11), turns out to be

$$\gamma_q = \sqrt{\hbar\omega/2N}\, c \tag{4.107}$$

in our case of short-range coupling (4.39) with dispersionless phonons ($\omega_q = \omega$), as is obvious from the fact that the lattice relaxation energy E_{LR} is given by $c^2/2 = \sum_q (\gamma_q)^2/\hbar\omega_q$. Normalizing the intensity I_0 of the zero-phonon exciton line at $E_0 \,(\equiv E_{K=0})$ to unity, we find that the one-phonon sideband is given by

$$F_1(E) = \sum_K \frac{\hbar\omega c^2/2N}{(E - E_0)^2}\, \delta\!\left(E_0 + \frac{\hbar^2 K^2}{2M} + \hbar\omega - E\right)$$

$$= S\left(\frac{\hbar\omega}{E - E_0}\right)^2 \varrho^{(0)}(E - \hbar\omega) \tag{4.108}$$

where M is the effective mass of the exciton and $\varrho^{(0)}(E)$ is the normalized density of states of the exciton band. $F_1(E)$ starts to rise at $E = E_0 + \hbar\omega$ as $[E - (E_0 + \hbar\omega)]^{(d/2)-1}\, (E - E_0)^{-2}$ (d is the dimensionality) and its integrated intensity is of the order of

$$I_1 \sim S\left(\frac{\hbar\omega}{B}\right)^{d/2} = g\left(\frac{\hbar\omega}{B}\right)^{(d/2)-1} \tag{4.109}$$

if $\hbar\omega \ll B$. The intensity of the one-phonon sideband relative to that of the zero-phonon line is thus reduced by $(\hbar\omega/B)^{d/2}$ (4.105) due to the translational motion of the exciton. The same holds for the intensity ratios of successive phonon sidebands. This "motional reduction" [4.47] of the phonon sidebands is the quantal version of the motional narrowing of the overall lines shapes discussed in Sect. 4.2.2. The moving exciton feels much less phonon field than the localized exciton does.

With the more general form of exciton-phonon interaction as given by (4.10), the one-phonon sideband of the localized excitation is a continuous band given by

$$F_1(E) = s(E - E^{(0)})\exp(-S)\,, \tag{4.110}$$

$$s(\varepsilon) \equiv \sum_q \frac{|\gamma_q|^2}{(\hbar\omega_q)^2}\, \delta(\hbar\omega_q - \varepsilon)\,, \tag{4.111}$$

$$S \equiv \int d\varepsilon\, s(\varepsilon) \tag{4.112}$$

and the ν-phonon sideband is given by its ν-fold convolution:

$$F_\nu(E) = [\exp(-S)] \, (\nu!)^{-1} \int d\varepsilon_1 \dots \int d\varepsilon_\nu$$

$$\times s(\varepsilon_1) \dots s(\varepsilon_\nu) \delta(E^{(0)} + \varepsilon_1 + \dots + \varepsilon_\nu - E) \; . \qquad (4.113)$$

With intersite transfer being considered, the one phonon sideband $\hat{s}(\varepsilon)$ with ε being measured from E_0 (the zero-phonon exciton line) is given by replacing $\hbar\omega_q$ in (4.111) by

$$\hbar\Omega_q \equiv \hbar\omega_q + \frac{\hbar^2 q^2}{2M} \qquad (4.114)$$

as was already shown for a simpler case, compare (4.108 and 111). By changing the variable from $\varepsilon = \hbar\Omega_q$ to $\hbar\omega_q = \eta(\varepsilon)$ through the relation (4.114) under the assumption of one-to-one correspondence, we obtain [4.47]

$$\hat{s}(\varepsilon) \equiv \sum_q \frac{|\gamma_q|^2}{\omega^2} \delta(\hbar\Omega_q - \varepsilon)$$

$$= \left(\frac{\eta}{\varepsilon}\right)^2 \frac{d\eta(\varepsilon)}{d\varepsilon} s(\eta(\varepsilon)) \; . \qquad (4.115)$$

As a typical example, the acoustic- and optical-phonon sidebands of the first order are shown schematically in Fig. 4.16 for localized (dotted line) and moving

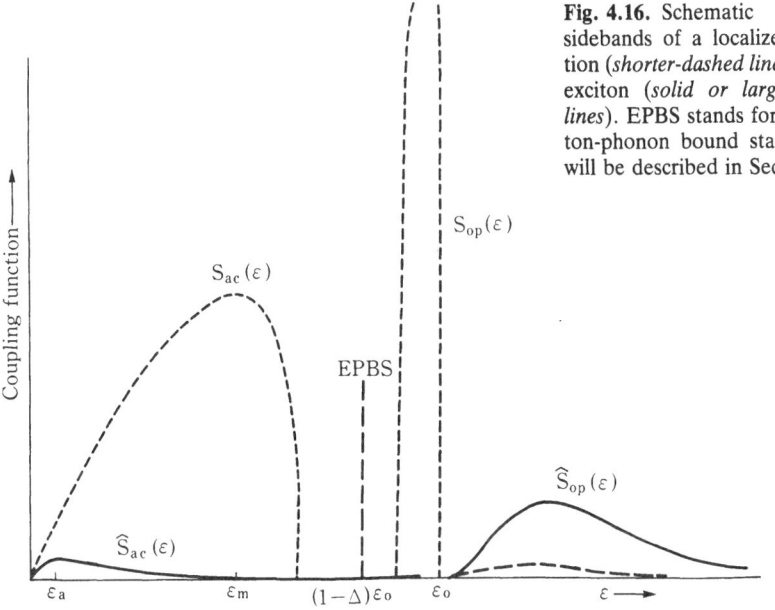

Fig. 4.16. Schematic phonon sidebands of a localized excitation (*shorter-dashed lines*) and an exciton (*solid or larger-dashed lines*). EPBS stands for the exciton-phonon bound state, which will be described in Sect. 4.3.3

(solid line) excitons. The motional reduction, which is governed by the factor $(\eta/\varepsilon)^2 = [\hbar\omega_q/(\hbar\omega_q + \hbar^2q^2/2M)]^2$ in (4.115) [see also (4.109) for its order estimation], is more noticeable for the acoustic phonon sideband because these phonon energies are smaller than those of the optical phonons. The S value for localized excitation of a tightly bound electron (e.g., the color centers in alkali halides [4.46]) is typically as large as a few to several tens since E_{LR} is also of the order of eV, and it is usually difficult even at $T = 0$ to observe the zero-phonon line with intensity $\exp(-S)$ much smaller than those of the phonon sidebands. In the case of host excitation, the relative intensity of the phonon sidebands is significantly reduced, due to the motional effect of the exciton, being smaller than that of the zero-phonon line in the three-dimensional lattice in the usual situation with $g = O(1)$ and $\hbar\omega \ll B$ (4.109). However, the motional effect also makes the acoustic-phonon sideband so close to the zero-phonon line [4.48] (the separation is given by $Ms^2/2$, being of the order of 10^{-4} eV) that they are usually observed as an inseparable band.

It is noted, in passing, that the S value for excitation of a localized electron (or hole) depends sensitively on its orbital radius r, and decreases as r^{-n} ($n = 1$ and 2 for optical and acoustic modes, respectively) when r is much greater than the lattice constant [4.49]. For this reason, the shallow bound electrons, holes, and excitons in semiconductors have small S values and weak phonon sidebands. This can also be considered as a sort of motional reduction.

Another problem of importance is how the phonon structure changes when g exceeds g_c, namely, when the self-trapped state becomes more stable. This will be described in Sect. 4.4.4.

4.3.2 Multicomponent Line-Shape Formula

In the adiabatic description of the overall line-shape in Sect. 4.2.2 and in the perturbation theory for the phonon sideband in Sect 4.3.1, we have already seen that the tail part (on the high-energy side if the bottom of the exciton band is at $K = 0$) of the absorption line originates from the phonon-assisted creation of an exciton with $K \neq 0$. As it is away from that particular absorption peak, this tail part is also contributed to by other exciton bands (different states, λ, of e–h relative motion) as possible intermediate states of this second-order process (Fig. 4.17). One may well expect an interference effect, especially when the tail overlaps with other exciton peaks. It is then desirable to derive a line-shape formula taking account of all the states λ of e–h relative motion (inclusive of ionized states, of course).

Instead of a direct extension of the method described in Sect. 4.2.2, we will start here with the second-order perturbation theory for phonon-assisted exciton creation with all λ being considered as intermediate states, which is a heuristic way to obtain an exact line-shape formula for this multicomponent system [4.50]. Denoting by $M_{\lambda g}$ the transition dipole moment for creating the λ exciton with $K = 0$, one can write the probability that a photon with energy E is absorbed through the second-order process as

$$F(E) = \sum_{\lambda K\mu, \pm} \left| \sum_{\lambda'} \frac{(H_I)^{(\mu\pm)}_{\lambda K, \lambda'0} M_{\lambda'g}}{E - E_{\lambda'0}} \right|^2 \delta(E_{\lambda K} \pm \hbar\omega_{\mu, \mp K} - E) \qquad (4.116)$$

apart from unimportant factors. Here

$$(H_I)^{(\mu\pm)}_{\lambda K, \lambda'0} = \begin{cases} \gamma^{(\mu)*}_{\lambda\lambda'}(-K)\sqrt{n_{\mu,-K} + 1} & (+) \\ \gamma^{(\mu)}_{\lambda\lambda'}(K)\sqrt{n_{\mu,K}} & (-) \end{cases} \qquad (4.117)$$

represents the exciton scattering $(\lambda'0 \to \lambda K)$ by emission $(+)$ or absorption $(-)$ of a phonon of the μth branch. The thermal average over the phonon numbers, $n_{\mu,K}$, is to be taken on the right-hand side of (4.116).

It is convenient to introduce the second-order self-energy matrix $\Sigma_{K'}(E)$ defined by

$$\Sigma_{\lambda'\lambda''K'}(E) \equiv \Delta_{\lambda'\lambda''K'}(E) + i\Gamma_{\lambda'\lambda''K'}(E)$$

$$= \sum_{\lambda K\mu, \pm} (H_I)^{(\mu\pm)}_{\lambda'K', \lambda K} (H_I)^{(\mu\pm)}_{\lambda K, \lambda''K'}$$

$$\times \left[\frac{\mathcal{P}}{E - E_{\lambda K} \mp \hbar\omega_{\mu, \mp (K-K')}} + i\pi\delta(E - E_{\lambda K} \mp \hbar\omega_{\mu, \mp (K-K')}) \right] \qquad (4.118)$$

whose diagonal elements $(\lambda' = \lambda'')$ are similar to the lowest-order terms in (4.56). It is easy to show that the matrices Δ and Γ are Hermitian and that Γ is not negative. Similarly we define the diagonal matrix H_{eK} whose $\lambda\lambda$ element is the

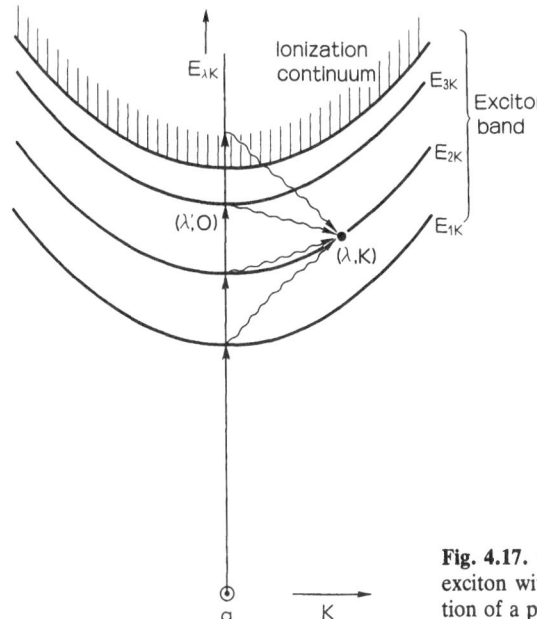

Fig. 4.17. Optical creation (*solid arrows*) of an exciton with simultaneous emission or absorption of a phonon (*wavy arrows*)

exciton energy $E_{\lambda K}$ and the one-column matrix M whose λth element is given by $M_{\lambda g}$. We can then rewrite (4.116) as

$$F(E) = \frac{1}{\pi} \sum_{\lambda'\lambda''} M_{g\lambda'} \frac{1}{E - E_{\lambda'0}} \Gamma_{\lambda'\lambda''0}(E) \frac{1}{E - E_{\lambda''0}} M_{\lambda''g} \tag{4.119a}$$

$$= \frac{1}{\pi} M^\dagger \frac{1}{E - H_{e0}} \Gamma_0(E) \frac{1}{E - H_{e0}} M . \tag{4.119b}$$

Decomposing the cross terms ($\lambda' \neq \lambda''$) in (4.119a) as

$$(E - E_{\lambda'0})^{-1} (E - E_{\lambda''0})^{-1} = (E_{\lambda'0} - E_{\lambda''0})^{-1} [(E - E_{\lambda'0})^{-1} - (E - E_{\lambda''0})^{-1}] ,$$

we can rearrange the double summation in it into a single summation:

$$F(E) = \frac{1}{\pi} \sum_\lambda f_\lambda \frac{\Gamma_{\lambda\lambda0}(E) + A_\lambda(E)(E - E_{\lambda0})}{(E - E_{\lambda0})^2} \quad \text{where} \tag{4.120}$$

$$f_\lambda \equiv |M_{\lambda g}|^2 , \tag{4.121}$$

$$f_\lambda A_\lambda(E) \equiv \sum_{\lambda'(\neq\lambda)} \frac{2\,\mathrm{Re}\,\{M_{g\lambda}\Gamma_{\lambda\lambda'0}(E)M_{\lambda'g}\}}{E_{\lambda0} - E_{\lambda'0}} . \tag{4.122}$$

The divergence of each term in (4.120) at $E = E_{\lambda0}$ is obviously an artifact of the second-order perturbation theory. Since the intermediate state $\lambda0$ is subject to shift and broadening as given by (4.118), the denominator in (4.120) is to be replaced by $[E - E_{\lambda0} - \Delta_{\lambda\lambda0}(E)]^2 + [\Gamma_{\lambda\lambda0}(E)]^2$ in higher-order theory. Thus we obtain a Lorentzian-type line-shape, as studied in Sect. 4.2.2, but for the asymmetry term with A_λ which is due to the phonon-mediated interference with other exciton bands ($\lambda' \neq \lambda$) as described by the nondiagonal element $\Gamma_{\lambda\lambda'0}(E)$ in (4.122). For formal consistency, however, the diagonal and nondiagonal elements of the self-energy should together be incorporated into (4.119b) in matrix form:

$$F(E) = \frac{1}{\pi} M^\dagger \frac{1}{E - H_{e0} - \Delta_0(E) - i\Gamma_0(E)} \Gamma_0(E) \frac{1}{E - H_{e0} - \Delta_0(E) + i\Gamma_0(E)} M$$

$$= \frac{1}{\pi i} M^\dagger \left(\frac{1}{E - H_{e0} - \Delta_0(E) - i\Gamma_0(E)} - \mathrm{h.c.} \right) M . \tag{4.123}$$

In fact, it can be shown that (4.123) is the exact formula for the line-shape of a multicomponent exciton provided the exact formula for the self-energy is used:

$$\Sigma_{K'}(E) = \sum_{K\mu,\pm} (H_{\mathrm{I}})_{K'K}^{(\mu\pm)} \frac{1}{E - H_{eK} \mp \hbar\omega_{\mu,\mp(K-K')} - \Sigma_K(E \mp \hbar\omega_{\mu,\mp(K-K')})} (H_{\mathrm{I}})_{KK'}^{(\mu\pm)}$$

$$+ \text{(higher-order terms)} . \tag{4.124}$$

The matrix form of H_I [with respect to (λ, λ')] is used in (4.124). See [4.50] for an exact derivation of (4.123).

Let us diagonalize the renormalized energy matrix by a transformation $T_K(E)$ (not unitary),

$$T_K(E)[H_{eK} + \Delta_K + i\Gamma_K(E)]T_K(E)^{-1} = \tilde{H}_K(E) - i\tilde{\Gamma}_K(E) . \tag{4.125}$$

$\tilde{H}_K(E)$ and $\tilde{\Gamma}_K(E)$ are real, diagonal matrices whose λ-th element will be denoted by $\tilde{E}_{\lambda K}(E)$ and $\tilde{\Gamma}_{\lambda K}(E)$, respectively. Using (4.125), we can rewrite (4.123) as

$$
\begin{aligned}
F(E) &= \frac{1}{2\pi i} \sum_{\lambda} \left(\frac{(M^\dagger T_0^{-1})_\lambda (T_0 M)_\lambda}{E - \tilde{E}_{\lambda 0} - i\tilde{\Gamma}_{\lambda 0}} - \text{c.c.} \right) \\
&= \sum_{\lambda} f_\lambda \frac{1}{\pi} \frac{\tilde{\Gamma}_{\lambda 0} + \tilde{A}_{\lambda 0}(E - \tilde{E}_{\lambda 0})}{(E - \tilde{E}_{\lambda 0})^2 + \tilde{\Gamma}_{\lambda 0}^2} \quad \text{where}
\end{aligned}
\tag{4.126}
$$

$$(M^\dagger T_0^{-1})_\lambda (T_0 M)_\lambda \equiv \tilde{f}_\lambda + i\tilde{f}_\lambda \tilde{A}_\lambda . \tag{4.127}$$

Note that the renormalized quantities with $\tilde{\ }$ on them are energy dependent.

Equation (4.126) with (4.124) is an extension of (4.57) with (4.56) in two respects. Firstly, the effect of multicomponents is the appearance of the asymmetry term in (4.126), see (4.122) for the unrenormalized A. As already mentioned, it results from the *interference* of the two waves coming through different intermediate states: $|\lambda 0\rangle$ and $|\lambda' 0\rangle$. The asymmetric line-shape (4.126), with a dip on one side, is well known in atomic physics as the Fano effect [4.51]. A typical example in the exciton spectra is found in Cu_2O [4.52]. Secondly, the lattice vibrations are treated quantum mechanically so that the phonon structures appear in the spectra as was described in Sect. 4.3.1. for the simplest situation. The phonon structures are implicitly incorporated into (4.126) through the energy dependence of $\tilde{\Gamma}_{\lambda 0}(E)$ and other renormalized quantities as is seen from (4.124). Such an analysis was made on the first exciton band in NaI where the Lorentzian is modulated by the LO phonon sideband [4.53].

If the bottom of the lowest exciton band λ is not at $K = 0$, but at K_i (in general, there are two or more inequivalent points in the Brillouin zone), the absorption spectra at $T = 0$ starts as $\sqrt{E - (E_{\lambda K_i} + \hbar\omega_{\mu, -K_i})}$ for each phonon mode μ reflecting the density of exciton states which rises as $\sqrt{E - E_{\lambda K_i}}$, see $\Gamma(E)$ given in (4.118). With a nonzero but not too high temperature, the absorption edge consists of several such steps with phonon absorption as well as emission [4.54],

$$
\begin{aligned}
F(E) &= \sum_{\mu} C_\mu [n_{\mu, K_i} \sqrt{E - (E_{\lambda K_i} - \hbar\omega_{\mu, K_i})} \\
&\quad + (n_{\mu, K_i} + 1) \sqrt{E - (E_{\lambda K_i} + \hbar\omega_{\mu, K_i})}]
\end{aligned}
\tag{4.128}
$$

where $\sqrt{\ }$ is to be taken as zero when the argument is negative. If symmetry forbids any of the optically allowed excitons $\lambda' 0$ $(M_{\lambda' g} \neq 0)$ to be scattered into

λK_i by the (μ, K_i) phonon, the rising form of the μth absorption edge is of three-halves power instead of one-half since the $(\lambda'0 \to \lambda K)$ scattering matrix element increases proportionally to $|K - K_i|$.

The indirect absorption edge of the form (4.128) has been observed in a variety of crystals such as Si, Ge [4.54], and silver halides [4.55] (see also Chap. 6). In thallous halides with the exciton band bottom being at the corner point (R) of the first Brillouin cube, all phonon modes are forbidden due to the high symmetry of the R point, and all edges rise with three-halves power of energy (Chap. 7).

4.3.3 The Electron-Hole Relative Motion and the Phonon Sideband of an Exciton

We have seen in the preceding section how the phonon-mediated interaction between different internal states of the exciton influences the energies of these states and the features of the associated optical spectra. The effect is of particular importance for those excitons in ionic crystals with binding energy as small as the LO phonon energy. The orbital radius of the e–h relative motion is then as large as the sum of the radii of the electron and hole polarons (Sect. 4.1.4), and the dielectric screening of the e–h Coulomb attraction causes a reduction of the binding energy, which is more significantly for the internal states with higher quantum numbers.

This effect and the resulting optical spectra were studied in great detail by *Matsuura* and *Büttner* [4.17, 56] using the variational method developed by *Pollmann* and *Büttner* [4.57]. The method is an extension of the intermediate coupling theory for a polaron, mentioned in Sect. 4.1.2, in that it determines virtual phonons and e–h relative (as well as translational) motion in a self-consistent way. In effect, it takes into account, up to higher-order perturbation, the phonon-mediated interaction between different internal states of an exciton discussed in Sect. 4.3.2. The calculated binding energy of an exciton as a function of polaron coupling constant α is shown in Fig. 4.18 for various internal states, all being normalized at $\alpha = 0$. The average virtual phonon numbers, the effective mass enhancement of the exciton (Fig. 4.19) and the intensity ratio of the one-phonon sideband to the zero-phonon line (Fig. 4.20) are significant for excitons with small binding energy ($\lesssim \hbar\omega$) and small e–h effective mass ratio ($m_e \ll m_h$), due to the less effective cancellation of electron and hole charge distributions (Sect. 4.1.4). The phonon-induced reduction of the oscillator strength of the zero-phonon line (as referred to its value at $\alpha = 0$) is more significant for internal states with higher quantum numbers partly due to more effective dielectric screening (greater enhancement in the average e–h distance) and partly due to greater removal of intensity to the phonon sidebands.

As shown schematically in Fig. 4.16, the separation of the peak of the optical-phonon sideband from the zero-phonon line is greater than the optical-phonon energy itself, due to the recoil kinetic energy, $\hbar^2 K^2/2M$, of the exciton. The LO

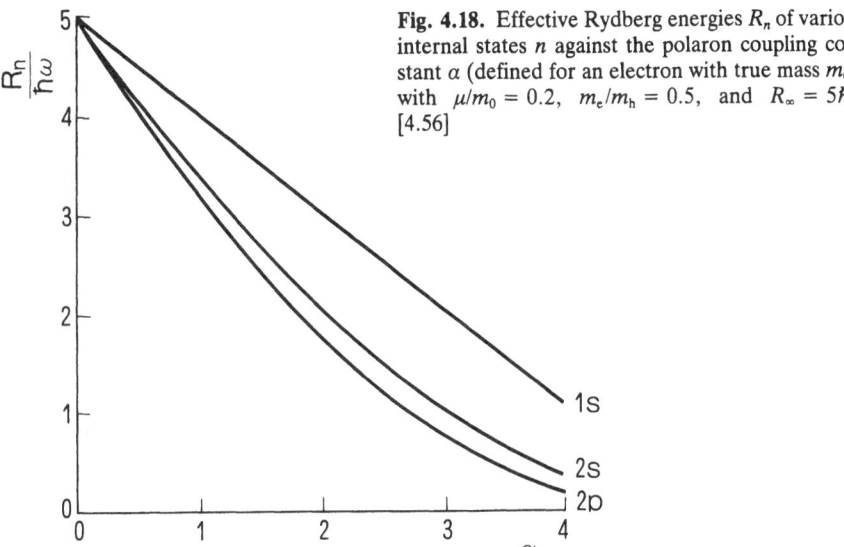

Fig. 4.18. Effective Rydberg energies R_n of various internal states n against the polaron coupling constant α (defined for an electron with true mass m_0), with $\mu/m_0 = 0.2$, $m_e/m_h = 0.5$, and $R_\infty = 5\hbar\omega$ [4.56]

phonons which scatter the Wannier exciton most effectively have wave number K of the order of the reciprocal radius of the exciton (Sect. 4.1.4). In fact, the calculated phonon sideband is sharper and closer to the one LO phonon energy above the zero-phonon line for higher quantum number excitons. This tendency will be strengthened even more if the mass renormalization effect (Fig. 4.19) is

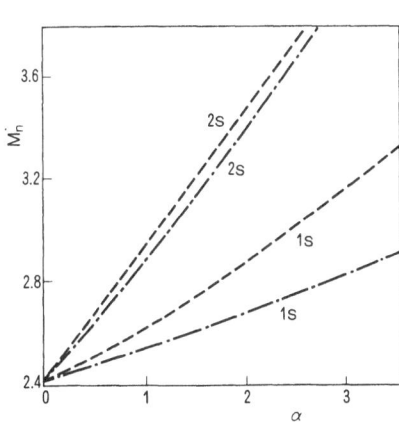

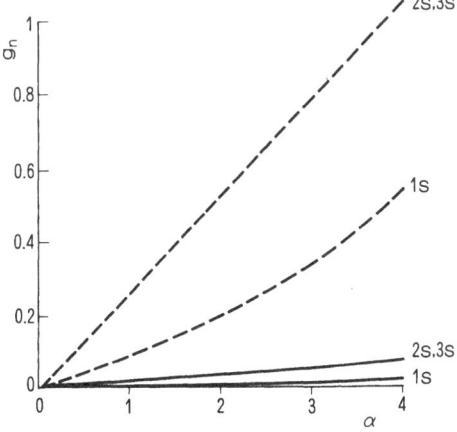

Fig. 4.19. Effective masses M_n^* (in units of true electron mass m_0) of phonon-dressed $1s$ and $2s$ excitons versus α with $\mu/m_0 = 0.2$, $m_e/m_h = 0.1$, $R_\infty/\hbar\omega = 5$ (---) and 10 (·-·-) [4.17]

Fig. 4.20. Intensity ratio of the one-phonon sideband to the zero-phonon line of excitons of various internal states versus α, with $\mu/m_0 = 0.2$, $R_\infty = 5\hbar\omega$, $m_e/m_h = 0.5$ (—) and 0.1 (---) [4.56]

taken into account. The observations in alkali halides [4.16, 53, 58] and cuprous halides [4.59] (see also Chap. 3) are fully consistent with the above results.

In some ionic crystals such as ZnO, MgO, BeO [4.14], TlCl, and TlBr [4.60, 61] (see also Chap. 7), however, the separation of the LO phonon sideband of the $1s$ exciton from its zero-phonon line is found to be several to 20% smaller than the LO phonon energy. This means the exciton recoiling from the emitted LO phonon has *negative* kinetic energy, namely, the exciton and the LO phonon form a bound state in the final state of optical absorption in which they are created simultaneously. Where does this final state interaction between these neutral particles come from? A clue is in the observations themselves: the binding energy of the $1s$ exciton is as small as the LO phonon energy in the above-mentioned crystals. Then, the LO phonon energy is enough or nearly enough to excite the $1s$ exciton into higher quantum states. The strong mixing of excited states due to near resonance will serve to lower the total energy of an exciton plus one LO phonon compared to the situation when they are far apart. This attractive potential can be formulated as follows [4.62].

In the final state of the optical transition, the exciton and the LO phonon (if any) have vanishing total momentum $\hbar K_{tot}$ which is a constant of motion. The resonant mixing mentioned above is expressed by the wave function

$$\Psi = \sum_K u_K b^{\dagger}_{-K}|1K, \ 0\rangle + \sum_{\lambda > 1} v_\lambda |\lambda 0, 0\rangle \tag{4.129}$$

where $|\lambda K, 0\rangle$ denotes the λK exciton in the phonon vacuum. Putting this into the Schrödinger equation $(H - E)\Psi = 0$ for the exciton-phonon system with $H = H_e + H_I + H_L$, and multiplying by $\langle 1K, \ 0|b_{-K}$ and $\langle 0\lambda, \ 0|$ from the left of this equation, one obtains the equations for u_K and v_λ. After elimination of v_λ, one finally obtains the equation

$$(E_{1K} + \hbar\omega_{-K})u_K + \sum_{K'} \left(\sum_{\lambda > 1} \frac{\gamma_{1\lambda}(K)\gamma^*_{1\lambda}(K')}{E - E_{\lambda 0}} \right) u_{K'} = Eu_K \ . \tag{4.130}$$

The first term represents the energy of independent quasi particles subject to $K_{tot} = 0$ and the second term their interaction originating from the phonon-induced changes of the internal states of the exciton as mentioned above. If $E < E_{\lambda 0}$ for $\lambda > 1$, this interaction is certainly attractive. If the attractive potential is deep enough, there will appear an exciton phonon bound state (EPBS) below the continuum of the free states: $E_{1K} + \hbar\omega_{-K}$.

The interaction term turns out to be small with large binding energy because of a large denominator. For an exciton with $1s$ binding energy comparable to the LO phonon energy, there are a number of excited internal states which are nearly resonant with the one-phonon state and hence make large contributions to the interaction. Because of the distribution of $E_{\lambda 0}$ towards the higher-energy side, the resultant interaction is expected to be attractive. This explains why the EPBS has been observed only in those crystals with exciton binding energy comparable to the LO phonon energy.

When the 1s binding energy is significantly smaller than the LO phonon energy, the EPBS may be situated above the ionization threshold – band gap ε_g. It is then a resonance or quasi-bound state embedded in the ionization continuum. The LO phonon sideband is then expected to be a lifetime-broadened Lorentzian of the form (4.57) due to the disintegration of this exciton–LO-phonon quasi-bound state (EPQBS) into an ionized e–h pair. The LO phonon sidebands observed in TlCl and TlBr [4.61], Chap. 7, belong to this case. Its strange lineshape – the Lorentzian with *sharp truncation below the ionization continuum* – is due to the energy dependence of the width $\Gamma_0(E)$ reflecting the density of final states (ionization continuum) into which the EPQBS disintegrates [4.63].

Equation (4.130) can also be interpreted as the equation for the eigenvibration, especially when the exciton is immobile: $E_{\lambda K} = E_\lambda$. The localized modes will appear much more readily than for the free exciton since the dispersion of the LO phonon is small. The LO phonon sideband of the bound exciton in AgBr: I [4.64] (see also Chap. 6) is a typical example; the 30% reduction of LO phonon energy due to the existence of a bound exciton is greater than any of the intrinsic EPBSs so far observed.

The EPBS in molecular crystals was discussed by Rashba [4.65], who assumed a finite reduction in the intramolecular vibrational frequency in the excited state and derived a criterion for the appearance of the bound state. The second term on the left side of (4.130) is the Wannier version of this reduction derived ab initio from the individual electron-phonon interactions.

4.4 Self-Trapping

We have already given a brief account of exciton self-trapping in Sect. 4.2.1. This section will be devoted to a more detailed and systematic study of self-trapping of an electron, a hole, and a Frenkel or Wannier exciton in phonon fields of various types, with particular attention to the roles of *force range* and *dimensionality* for the stability and distinguishability of self-trapped state versus free state. The concept of extrinsic self-trapping will be introduced as an extension. Negative U effect and the self-decomposition of two particles with the same or opposite charges will also be discussed along the same lines. The main problem here is the competition between the electron (hole)-phonon and electron-electron (hole) interactions, the many particle version of which will be the subject of Sect. 4.6.

4.4.1 Local Stabilities of Free and Self-Trapped States

We have seen in Sect. 4.2.1 that the free (F) and self-trapped (S) states are stabilized in the limits of c (electron-phonon coupling coefficient) $\to 0$ and t (intersite transfer energy) $\to 0$, respectively. We will now study the *local stabilities* of these states against lattice distortions, by introducing c or t as a perturbation. The kinetic energy of the lattice will be neglected in this adiabatic argument.

Starting with the eigenstates (4.41) of H_e, let us calculate the energy change of the lowest state (assumed to be $K = 0$) to second order in H_I. Introducing the Fourier transform: $Q_n = N^{-1/2} \sum_K \exp(i\boldsymbol{K} \cdot \boldsymbol{R}_n) Q_K$, one obtains

$$\delta_2 E(0) = - \sum_K \frac{c^2/N}{E(\boldsymbol{K}) - E(0)} \, Q_K^\dagger Q_K \, . \tag{4.131}$$

Adding this to $U_L = \sum_K Q_K^\dagger Q_K / 2$, one finds that the first candidate modes $\boldsymbol{K}$ to become unstable are those with the lowest excitation energy $E(\boldsymbol{K}) - E(0)$. With the cyclic boundary condition for the d-dimensional cube with $N^{1/d}$ sites for each edge, we obtain $K_x = 2\pi/(N^{1/d} a_0)$ and hence the condition for instability [4.66]:

$$\frac{2d}{\pi^2} N^{(2/d) - 1} g > 1 \, . \tag{4.132}$$

By letting $N \to \infty$, one finds that the F-state is always unstable or "limp" against lattice distortion for $d = 1$ but is always stable (at least locally in the Q_n space) or "stiff" for $d = 3$. In the "marginal" case of $d = 2$, the instability sets in when g exceeds $g_1 = \pi^2/4$ (this number depends on the boundary shape of the macroscopic crystal).

The above argument for short-range coupling can be extended to the general case of the force range index l defined by $\psi_K \propto K^{-l}$ ($K \to 0$), see (4.2) for the definition of ψ_K. The local stability criterion for the F-state is then given by the sign of the stability index s (not to be confused with the steepness index introduced in Sect. 4.2.4):

$$s \equiv d - 2(1 + l) \quad \begin{cases} < 0 & \text{limp}, \\ = 0 & \text{marginal}, \\ > 0 & \text{stiff}. \end{cases} \tag{4.133}$$

Long-range electron-phonon coupling, such as the Fröhlich and piezoelectric interactions [$l = 1$, (4.20)], in a three-dimensional lattice belongs to the limp case, being similar to short-range coupling ($l = 0$) in a one-dimensional lattice. The long-range interaction cannot be defined uniquely in a low-dimensional space.

Let us now turn to the opposite limit of a localized state $|n\rangle$ with lattice distortion $Q_n = c$, and introduce the nearest neighbor transfer t as a perturbation. Then the electron will be virtually transferred to the nearest neighboring sites which have site energy $-cQ_{n'} = 0$ in contrast to $-cQ_n = -c^2$ at the original site. The second-order energy turns out to be $\delta_2 E = -2dt^2/c^2 = -(4dg^2)^{-1} E_{LR}$. The total energy of this optimally localized state – the self-trapped state (S) – can thus be calculated in the form of a strong coupling expansion

$$E_{(S)} = E_a - E_{LR}[1 + (4dg^2)^{-1} + O(4dg^2)^{-2}] \tag{4.134}$$

where the electron-induced lattice distortions at other sites ($n' \neq n$) contribute to higher-order terms. The local stability test of the lattice around this optimal distortion can be performed by expanding the force constant tensor ($N \times N$) in the same way as (4.134) and examining its nonnegativity [4.67].

The S-state with energy (4.134) becomes more stable than the F-state with energy $E_a - B$ when

$$g > g_c = 1 - \frac{1}{4d} + O\left(\frac{1}{4d}\right)^2, \quad (d = 2, 3) . \tag{4.135}$$

Within this strong coupling regime, the rapid convergence of the inverse expansion (4.134) as well as the nonnegativity of the force constant can be confirmed for $d = 3$. Higher-order calculation gives $g_c = 0.903$ for $d = 3$.

Such a comparison of F and S is meaningless for $d = 1$ since F is locally unstable. We have to study the only minimum (4.134) over the whole range, $\infty > g > 0$. As g decreases, the convergence of (4.134) becomes poorer and the wave function as well as the lattice distortion become less localized.

We have so far considered the electron-phonon interaction (4.39) originating from the variation (with lattice distortion) of the site-diagonal matrix element E_a of the electronic energy (4.38). In addition to this "site-diagonal" electron-phonon interaction, we have also to consider the "site-off-diagonal" interaction originating from the variation of transfer energy t_{nm} with the interatomic distance $|R_n - R_m|$. If this interaction is large enough the electron becomes self-trapped not around a single atom but around a pair of neighboring atoms coming closer than their regular distance so as to stabilize the electron-phonon system. This problem was studied by *Song* [4.68], *Umehara* [4.69], *Iida* et al. [4.70], and *Sumi* [4.71], with particular attention to positive holes and excitons in alkali halides and rare gas solids. The deformation potential Ξ as well as coupling constant g must be redefined [4.19] when the site-off-diagonal interaction is taken into account.

Another elaboration is needed when the atomic state generating the relevant band is degenerate as in p and d bands. One has then to introduce the interactions with lower-symmetry modes $Q'_n, Q''_n, \cdots$ even into the site-diagonal part (4.39). If this Jahn-Teller type site-diagonal interaction predominates, the electron is trapped at an atom with a lower-symmetry lattice distortion around it.

4.4.2 Continuum Model for Self-Trapping

We introduce here the continuum model [4.72, 19] for the lattice which allows us to study the *global* features of the adiabatic potential including the states with intermediate localization – this problem is intractable with the discrete lattice model. With the continuum model, one can readily include the long-range electron-phonon interaction as well in order to analyse the role of force range.

Consider an electron in the conduction band (or a positive hole in the valence band), interacting with acoustic and optical phonon fields in a three-dimensional lattice. Replacing these fields by an *elastic continuum* with dilation $\Delta(r)$ and a *dielectric continuum* with electrostatic potential $\Phi(r)$ due to the ionic displacement polarization, one can write the adiabatic energy $\langle H_e + H_I \rangle + U_L$ of the coupled electron-phonon system as

$$W[\psi, \Delta, \Phi] = \frac{\hbar^2}{2m} \int (\nabla\psi)^2 \, dr$$

$$+ \, \Xi \int \psi(r)^2 \Delta(r) \, dr + e \int \psi(r)^2 \Phi(r) \, dr$$

$$+ \, \frac{C}{2} \int \Delta(r)^2 \, dr + \frac{\bar\varepsilon}{8\pi} \int \{\nabla\Phi(r)\}^2 \, dr \qquad (4.136)$$

where m and e are the band effective mass and the charge, respectively, of the electron (or the hole), Ξ the deformation potential constant (4.13) and $\bar\varepsilon$ is defined by (4.26). The electronic energy H_e is measured from the band bottom.

By minimizing (4.136) with respect to ψ, one would obtain the adiabatic potential $W[\Delta, \Phi]$ for the lowest electronic state in the ∞-dimensional configuration coordinate space $(\Delta(r), \Phi(r))$. This will be done in a simplified form in Sect. 4.4.3. If one is interested only in the extrema of the adiabatic potential $W[\Delta, \Phi]$, one can take an alternative but simpler way by inverting the order in which the extrema are taken. By first minimizing (4.136) with respect to Δ and Φ, one obtains $\Delta(r) = -(\Xi/C)\psi(r)^2$ and $\nabla^2\Phi(r) = (4\pi e/\bar\varepsilon)\psi(r)^2$. Putting them back into (4.136) gives

$$W[\psi] = \frac{\hbar^2}{2m} \int (\nabla\psi)^2 \, dr$$

$$- \frac{1}{2} \int\!\int dr \, dr' \psi(r)^2 \left[\frac{\Xi^2}{C}\delta(r - r') + \frac{e^2}{\bar\varepsilon|r - r'|} \right] \psi(r)^2 \qquad (4.137)$$

as a function of ψ. The second line represents the self-interaction, namely, the electron-induced lattice distortions acting back upon the electron. It consists of -1 from H_I and $+1/2$ from U_L [the second and third lines, respectively, of (4.136)], due to the virial theorem. The different forms of the kernels of the self-interaction originate from the different force range – the short-range electron-dilation interaction and the long-range electron-polarization interaction.

For the second step, we have to find the extrema of (4.137) with respect to ψ. Choosing the trial function $\psi(r) = (2/a^2)^{3/4} \exp[-\pi(r/a)^2]$ with variational parameter a, one obtains

$$W[\psi] = W(a) = B(a_0/a)^2 - E_{LR}^{(ac)}(a_0/a)^3 - E_{LR}^{(op)}(a_0/a) \qquad (4.138)$$

where $B \equiv 3\pi\hbar^2/2 \, ma_0^2$, $E_{LR}^{(ac)} \equiv \Xi^2/2Ca_0^3$ and $E_{LR}^{(op)} \equiv e^2/\varepsilon a_0$ with an as yet arbitrary constant a_0.

With varying dimensionality d, the exponent of (a_0/a) in the second term (the short-range part of the lattice relaxation energy) of (4.138) turns out to be d while that of the first term (kinetic energy) is always 2. Dropping the last term, we find the only minimum with finite a and W for $d = 1$ as was pointed out by *Emin* and *Holstein* in their scaling argument [4.73]. We also find the only minimum $F(a = \infty, W = 0)$ or the only minimum S $(a = 0, W = -\infty)$ depending upon the coupling constant for $d = 2$, and always two minima, F and S, for $d = 3$.

The collapse $(W \to -\infty)$ of the S-state, originating from allowing a to vary down to 0 (continuum model), should in fact be avoided by letting a vary down to the lattice constant a_0. Then the three coefficients in (4.138) represent the kinetic energy and the short- and long-range parts of the lattice relaxation energy for complete localization $(a = a_0)$. With this cut-off in a, the above mentioned features of the minima in the adiabatic potential for different dimensionalities are fully consistent with what we conjectured in Sect. 4.4.1 on the basis of the discrete lattice model.

As a function of $\lambda \equiv a_0/a$ $(0 < \lambda < 1)$, the degree of localization, (4.138) for $d = 3$ has two minima in a typical situation. The first minimum F at $\lambda_F < 1$ represents a relatively delocalized state accompanied by moderate polarization but little dilation (contraction), while the second minimum S at $\lambda_S = 1$ corresponds to a completely localized state (self-trapped state) with strong dilation or contraction $(\varDelta \gtrless 0$ according as $\varXi \lessgtr 0)$ as well as strong polarization. The maximum M between them represents the energy barrier (saddle point of the adiabatic potential) separating the states F and S. The activation energy for the F $\to$ S transition amounts to

$$\varepsilon^* = W_M - W_F = \frac{4}{27} B g_s^{-2} (1 - 3 g_s g_l)^{3/2} \tag{4.139}$$

where we have defined the short- and long-range electron-phonon coupling constants by

$$g_{s \, or \, l} = E_{LR}^{(ac) \, or \, (op)}/B . \tag{4.140}$$

Figure 4.21 is the phase diagram for the *stable* and *metastable* (shown in parentheses) states on the (g_s, g_l)-plane. The thick solid line, across which the stable state changes discontinuously from F to S, starts from the g_s-axis (at $g_s = 1$) and terminates at a critical point without reaching the g_l-axis; the only trigger for the F–S discontinuity is the short range interaction with acoustic phonons while the long-range interaction with optical phonons dominates in the F-state.

It should be noted that the polaron state of the weak $(\alpha \lesssim 3\pi)$ through strong $(\alpha \gtrsim 3\pi)$ coupling regime described in Sect. 4.1.2 corresponds to the F-state of the above phase diagram along the g_l-axis $[\alpha \propto g_l(B/\hbar\omega)^{1/2}]$. The fact that our solution for the F state apparently corresponds to the strong coupling polaron of the Pekar-type is an artifact of our adiabatic description $(\omega \to 0$ so that $\alpha \to \infty$ in spite of finite $g_l)$. Not only the weak coupling but also the strong coupling polaron should be well distinguished from the S-state. The latter wants to be as

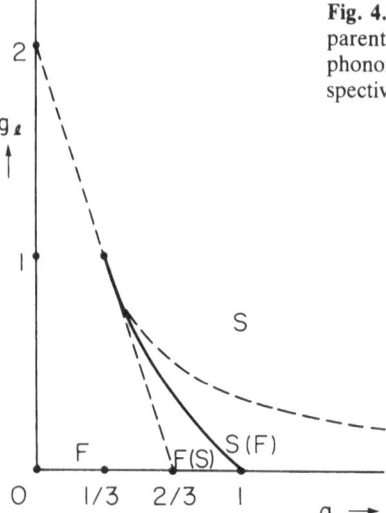

small as possible ($\sim a_0$) while the former has an optimum radius which is usually much greater than a_0. For this reason, F- and S-states are sometimes called "large" and "small" polarons, respectively. This nomenclature, although convenient, should be used with some caution since it may give the impression that the latter is simply the strong coupling limit of the electron-polarization (optical phonon) interaction. While, the real S-state, which is well distinguished from, and can coexist with, the F-state, is triggered by the short-range interaction with acoustic phonons, although the optical phonons also make a comparable contribution to this state. The importance of the short-range interaction for self-trapping was first pointed out by *Rashba* [4.74] but remained largely unnoticed for some time. See also his review article on self-trapping of excitons [4.75].

4.4.3 Adiabatic Potentials for Self-Trapping

We will now study the adiabatic potential $W[\varDelta(r)]$, 4.4.2, as a functional of the dilation field $\varDelta(r)$ for one-, two-, and three-dimensional lattices, in the absence of a long-range interaction. While it is impossible to directly treat the infinite-dimensional configuration coordinate (C.C.) space [$\varDelta(R_n)$ with $n = 1, 2, \ldots$], it is enough, for our purpose of characterizing the C.C. path to self-trapping, to consider the two-dimensional C.C. space $(\varDelta, a)$, the magnitude and the spatial extension of the dilation $\varDelta(r)$ [4.76]. For simplicity, we put $\varDelta(r) = \varDelta$ for $r < a$ and $\varDelta(r) = 0$ for $r > a$. Then the Schrödinger equation for the electron can be written, with suitable scaling, as

$$[- \lambda^{-1}\nabla_\varrho^2 - u(\varrho)]\psi(\varrho) = \frac{E(\varDelta, a)}{-\varXi\varDelta} \, \psi(\varrho) \tag{4.141}$$

where $\varrho \equiv r/a$, $u(\varrho) = 1$ or 0 according as $\varrho \lessgtr 1$,

$$\lambda \equiv \frac{-\Xi\varDelta}{K} , \qquad K \equiv \frac{\hbar^2}{2ma^2} , \tag{4.142}$$

and $E(\varDelta, a)$ denotes the lowest state of the electron. In terms of the depth, $-\Xi\varDelta$, of the square well potential, one can write the binding energy as

$$-E(\varDelta, a) = (-\Xi\varDelta)f_d(\lambda) . \tag{4.143}$$

The functions $f_d(\lambda)$ calculated for dimensionalities $d = 1, 2,$ and 3 are shown in Fig. 4.22, the features being essentially the same as in the discrete lattices.

By adding U_L, one obtains the adiabatic energy

$$W(\varDelta, a) = \gamma_d \left(\frac{C}{2}\right) a^d \varDelta^2 - (-\Xi\varDelta)f_d(\lambda) \tag{4.144a}$$

where γ_d is the volume of the d-dimensional sphere of unit radius. For one- and three-dimensional cases, one can bring (4.144) into the parameterless form

$$w(x, y) = \lambda^{d/2}x^{2-(d/2)} - f_d(\lambda)x , \qquad (\lambda = x/y) \tag{4.144b}$$

by rescaling the energy in units of W_0:

$$-\Xi\varDelta = W_0 x, \; K = W_0 y, \; W = W_0 w, \tag{4.145}$$

$$\left(\frac{W_0}{\hbar^2/2m}\right)^{(d/2)-1} \equiv \frac{\gamma_d}{2} \frac{\hbar^2 C}{2m\Xi^2} , \qquad (d = 1, 3) . \tag{4.146}$$

This is impossible with $d = 2$, the marginal dimensionality; for any choice of W one obtains

$$w(x, y) = G^{-1}\lambda x - f_2(\lambda)x , \qquad (d = 2) \tag{4.144c}$$

with irremovable coupling constant

$$G \equiv \frac{4}{\pi} \frac{m\Xi^2}{\hbar^2 C} , \qquad (d = 2) . \tag{4.147}$$

The adiabatic potentials for $d = 1, 2$ [4.20] and 3 [4.76] systems calculated with the use of (4.144b, c) and $f_d(\lambda)$ (Fig. 4.22) are shown on the x(lattice distortion)-y(inverse square of the radius of distorted region) plane in Fig. 4.23a, b and b', and c, respectively, where the negative energy region is shown by hatching. Noting that the volume $\gamma_d a^d$ of the distorted region should not be smaller than the atomic volume a_0^d of the discrete lattice, one finds the upper bound

$$y_d^{(u)} = \gamma_d^{4/(2d-d^2)}(8dg)^{-2/(2-d)} , \qquad (d = 1, 3) \tag{4.148}$$

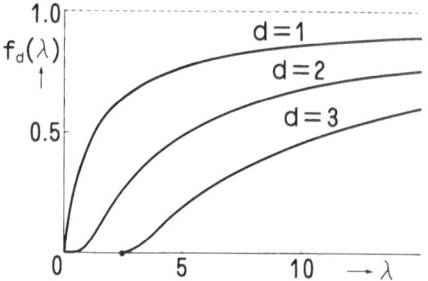

Fig. 4.22. The ratio of binding energy to square well potential depth as a function of the latter, in 1-, 2-, and 3-dimensional lattices

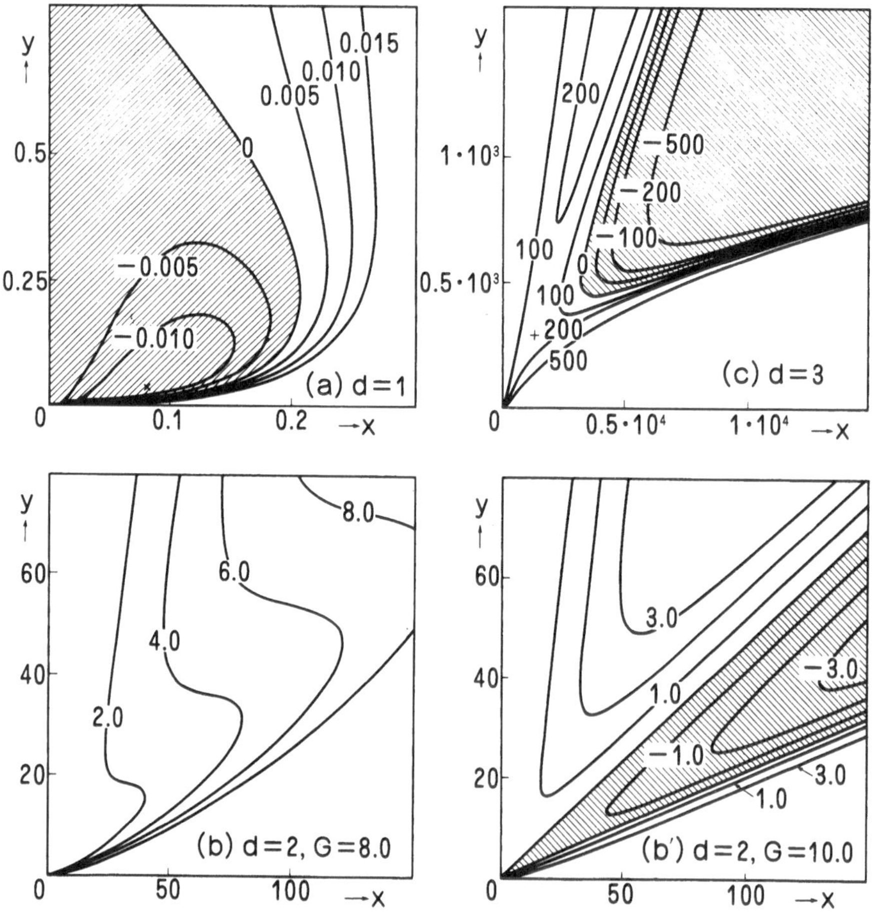

Fig. 4.23a–c. Adiabatic potentials for an electron in an elastic medium. The values of x and $y^{-1/2}$ are proportional to the strength of distortion and the radius of the distorted region, respectively. The energy is zero along the ordinate and negative in the hatched region. (**b**) and (**b'**) represent the weak and strong coupling cases, respectively, of 2-dimensional lattice

for the physically meaningful y range where use has been made of (4.145, 146) and the relations $E_{LR} = \Xi^2/2Ca_0^d$ and

$$\frac{\hbar^2}{2ma_0^2} = |t| = \frac{B}{2d} . \tag{4.149}$$

With increasing g, the critical line $y = y_d^{(u)}$ moves downward for $d = 1$ and upward for $d = 3$.

Note in Fig. 4.23 a that the free state ($\Delta = 0$) in a one-dimensional lattice is always unstable against lattice distortion Δ and that there is only the minimum indicated by $\times$. This means that as g increases the optimum radius of the distorted region and the electron orbital decreases as g^{-1} while the optimum distortion $|\Delta_m|$ and the stabilization energy $-W_m$ increase as $g^{3/2}$ and g^2, respectively (4.142, 145, 146), always continuously. Even after $y = y_1^{(u)}$ traverses this minimum point $|\Delta_m|$ and $-W_m$ still increase continuously.

In constrast, the stable state in a three-dimensional system changes discontinuously from the free state F ($\Delta = 0$) to the self-trapped state S $[\Delta \neq 0, a = (\gamma_3)^{1/3}a_0$ being the minimum possible value] when g exceeds $g_c = 2.3$ (namely, when $y = y_3^{(u)}$ reaches the negative energy region in Fig. 23 c). The height of the potential barrier from F- to S-state (the saddle point marked with $+$ in Fig. 23 c) is given by $\varepsilon_a = 0.79\,g^{-2}B$. It should also be noted that the F-state remains metastable for $g_c < g < \infty$ and that the S-state is metastable for $g_c' < g < g_c$ where $g_c' = 1.86$ corresponds to $y_3^{(u)} = y$ (saddle point).

The situation is varied in a two-dimensional lattice: as $g = (\pi/16)G$ exceeds $g_c = 1.83$, the negative energy region sets in like an opening fan. The stable state changes discontinuously from F- to S-state at $g = g_c$; however, there is neither a potential barrier nor the F–S coexistence range of g_c, in contrast to the three-dimensional system.

The dimensionality dependent features of the adiabatic potential mentioned above are qualitatively in good agreement with the results of Sects. 4.4.1, 2. Quantitatively, however, the following points should be made. The critical value g_c for self-trapping estimated here by the continuum model is at least twice as great as that of the dicrete lattice model, for two- and three-dimensional systems. One origin for this is the step function *Ansatz* for the dilation field $u(\varrho)$ which may be an unfavorable constraint on the S-state. Another origin is the use of a parabolic band with local effective mass m fitted at $k = 0$ to that of the $[1 - \cos{(k_x a_0)}]$ type band of the discrete lattice by (4.149), which results in significant overestimation (~ 1.5 times) of the electron kinetic energy of strongly localized states, thus tending to suppress self-trapping. A less trivial corollary of this consideration is that the adiabatic potentials in Fig. 4.23 based on the continuum model with a parabolic band will be lowered towards increasing y in the case of a discrete lattice, possibly resulting in the appearance of the potential barrier as well as the F–S coexistence region even in the case of $d = 2$. Conversely, if the local effective mass at $k = 0$ is greater than the average effective mass of the whole band, the F–S transition of the stable state for $d = 2$ will be of the second order: the minimum appears at $x = y = 0$ and proceeds toward the upper right corner as g exceeds a critical value. In this way, the situation in the marginal dimensionality $d = 2$ is quite varied, depending on the details of band structures as well as on other ingredients.

It is also of interest to see how the situation depends on the anisotropy. *Pertzsch* and *Rössler* [4.77] studied the effect of anisotropy in a three-dimensional system, with particular attention to the limiting situations of nearly one- and nearly two-dimensional systems.

4.4.4 Effective Mass Change in the F–S Transition

We have so far confined ourselves to the adiabatic limit where the lattice is vibrating infinitely slowly. If we treat the lattice vibrations quantum mechanically considering their finite frequencies, the localized state obtained in Sects 4.4.1 or 2 cannot be a true eigenstate since it makes a resonant transfer to a neighboring site with a finite rate. The true eigenstates, Ψ_K, of the exciton-phonon system with translational symmetry should be the plane-wave-like linear combination of the localized electronic state with localized lattice distortion.

Let us consider the standard Hamiltonian (4.37–40) with nondispersive phonons: $(\Omega^2)_{nn'} = \delta_{nn'}\omega^2$ as we did in Sect. 4.3.1, and use the following form of the trial wave function with variational parameter δ,

$$\Psi_K(\delta) = N^{-1/2} \sum_n \exp(i K \cdot R_n) \Phi_n(\delta) , \tag{4.150}$$

$$\Phi_n(\delta) = \chi_0(Q_1) \ldots \chi_0(Q_n - \delta) \ldots \chi_0(Q_N)|n\rangle . \tag{4.151}$$

In (4.151), the origin of the zero-point vibrational state $\chi_0(Q)$ is displaced by δ at the site where the exciton is sitting. With $\delta = c$, (4.151) with $n = 1, 2, \ldots, N$ gives N-fold degenerate eigenstates of the Hamiltonian with $t = 0$ as was mentioned in Sect. 4.3.1. Introduction of t will remove this degeneracy, the new eigenstates being of the form (4.150) with Φ_n being different from $\Phi_n(c)$. Our variational Ansatz is to use the same form of Φ_n but with different δ.

The expectation value of our Hamiltonian in the state (4.150) turns out, for $K = 0$, to be

$$E_{K=0}(\delta) = E_a - B \exp\left[-S(\delta/c)^2\right] - E_{LR}[2(\delta/c) - (\delta/c)^2] , \tag{4.152}$$

where the zero-point vibrational energy, $N\hbar\omega/2$, of the whole lattice has been subtracted. The exponential factor in the second term of (4.152) is nothing but the square of the overlap integral $\int dQ \, \chi(Q - \delta)\chi(Q)$ between distorted and undistorted states of the harmonic oscillator.

Equation (4.152) has two minima (Fig. 4.24):

$$\text{F:} \quad (\delta/c)_F \sim \gamma, \quad E_F \sim E_a - B(1 + g\gamma) , \tag{4.153}$$

$$\text{S:} \quad (\delta/c)_S = 1 - O(Se^{-S}), \quad E_S = E_a - E_{LR}[1 + O(e^{-S})] \tag{4.154}$$

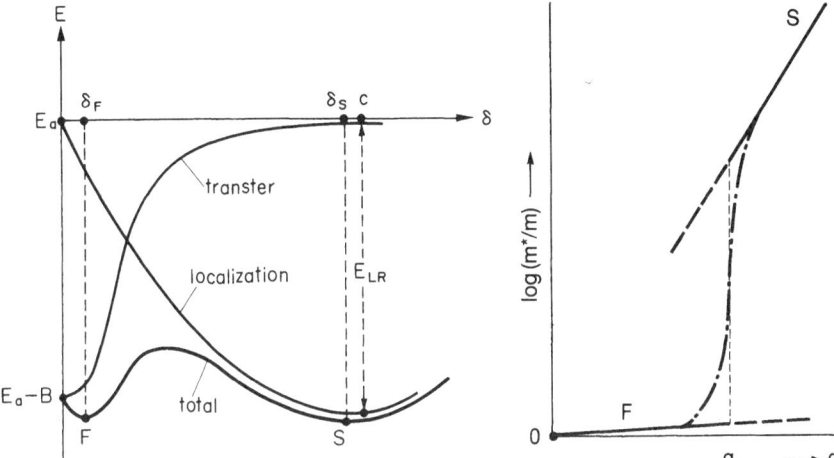

Fig. 4.24. Energy of an exciton moving (with wave vector $K = 0$) in the phonon field, as a function of accompanying distortion δ chosen as variational parameter

Fig. 4.25. Effective mass m^* of the phonon-dressed exciton as a function of coupling constant g

if the nonadiabaticity parameter defined by

$$\gamma \equiv \frac{\hbar\omega}{B} \tag{4.155}$$

is small enough compared to unity. Local distortion δ of the stable state changes discontinuously from (4.153) to (4.154) as $g \equiv E_{\text{LR}}/B$ exceeds $g_c \sim 1$, and correspondingly, the effective mass m^* of the phonon-dressed exciton, which is proportional to $(d^2E/dK^2)^{-1} \propto \exp[S(\delta/c)^2]$, increases discontinuously from $m^* \sim m(1 + g\gamma)$ to an enormous value $\sim m\exp(S)$ (note that $S = g\gamma^{-1}$ is much greater than unity as long as $\gamma \ll 1$), as shown schematically in Fig. 4.25 by the solid line.

The abrupt change in the effective mass of the phonon-dressed electron was first derived with the deformation-potential interaction with acoustic phonons, with the use of the trial wave function more general than (4.151) such that sites other than the n-th (where the electron is situated) are also allowed to be displaced [4.76]. (The same is true for a Frenkel exciton.) In contrast to these "displacement-type" wave functions, another extension by *Cho* and *Toyozawa* [4.78], who replaced $\chi_0(Q_n - \delta)$ in (4.151) by an arbitrary wave function $\chi(Q_n)$ (but the exciton-induced change in vibrational motion was confined to the nth site), led to the result that the change in the effective mass m^* is rapid but continuous around $g = g_c$ although it tends to discontinuity in the adiabatic limit $\gamma \to 0$ (see Fig. 4.25 where the dash-dot line corresponds to finite but small γ). This is also expected from the discontinuous solutions (4.153, 154) since their

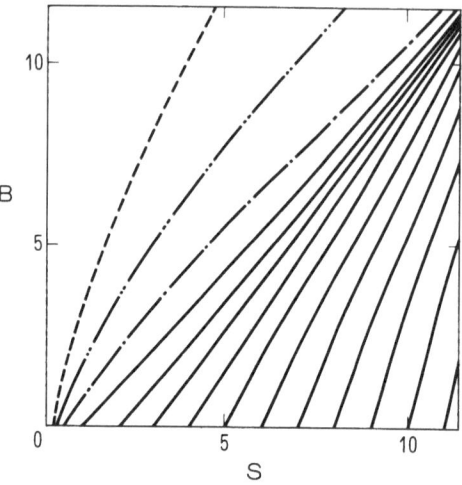

Fig. 4.26. Calculated equi-mass contours of an exciton on the S–B' plane for $\ln(m^*/m) = 0.125$ (---), $0.25, 0.5,$ $1, 2, 3, \ldots, 11$ (*rightmost solid line*) [4.79]

linear combination, giving a better solution than either one alone, recovers such continuity.

Further extension was made by *H. Sumi* [4.79] with the use of a *dynamical coherent potential approximation* (CPA), where the changes in vibrational state at other sites are taken into account in the form of an effective potential to be determined self-consistently as was described in Sect. 4.2.3. The equi-mass contours for $\ln(m^*/m)$ on the $S(\equiv E_{\mathrm{LR}}/\hbar\omega) - B'(\equiv B/\hbar\omega)$ plane thus calculated are shown in Fig. 4.26. One can see how the effective mass change as $g \equiv E_{\mathrm{LR}}/B$ exceeds $g_{\mathrm{c}} \sim 1$ becomes steeper with the increase in γ^{-1} representing the adiabaticity. With this method, H. Sumi further discussed the dispersions of low-lying excited states of the exciton-phonon system inclusive of phonon scattered and bound states. The latter state is similar to the exciton-phonon bound state described in Sect. 4.3.3, but different in its origin since the present bound state appears without consideration of the internal motion of the exciton.

Intimately related to this abrupt change in the exciton effective mass is the change in the phonon structures of the absorption spectra. In fact, one expects an abrupt reduction in the intensity of the zero-phonon line as g exceeds g_{c} under $\gamma \ll 1$ [4.78], since the vibrational wave function in the S-state of the exciton has a very small overlap integral with that of the ground state, of the order of $\exp(-S/2)$. The absorption spectra calculated by *H. Sumi* [4.80] with the use of the dynamical CPA mentioned above are shown in Fig. 4.27, where the exciton band width measured in unit of phonon energy, $B' \equiv B/\hbar\omega$, is varied for fixed $S \equiv E_{\mathrm{LR}}/\hbar\omega$. As $g \equiv E_{\mathrm{LR}}/B$ exceeds $g_{\mathrm{c}} \sim 1$, the zero-phonon line is reduced in intensity and replaced by its phonon sidebands, abruptly or gradually according as $\gamma \equiv \hbar\omega/B = g/S$ is much smaller than (left figure) or comparable to (right figure) unity.

So far we have confined ourselves to the on-site interaction with dispersionless phonons. What about the electron effective mass change with the Fröhlich-

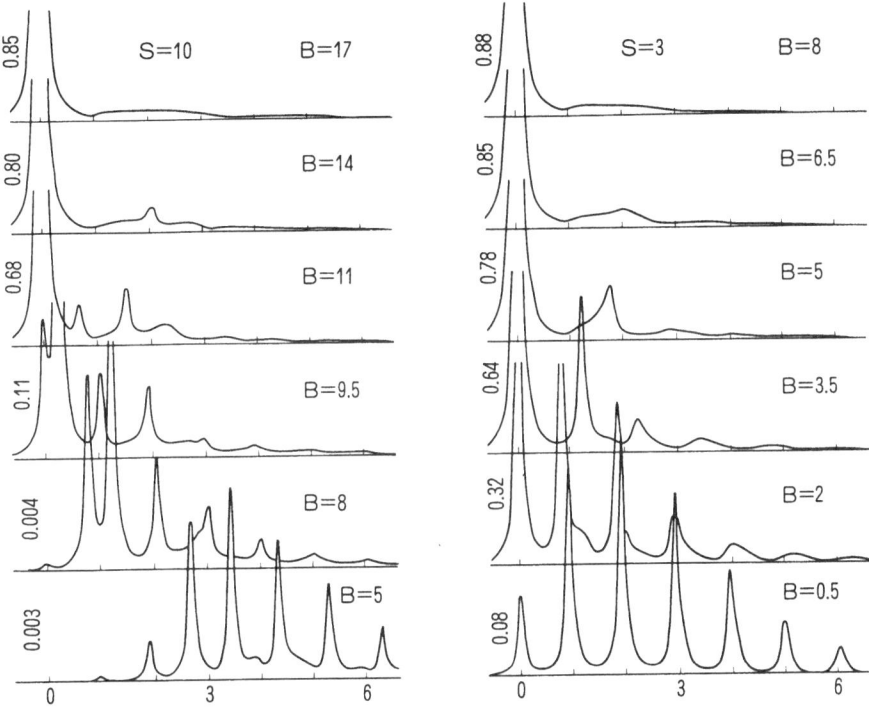

Fig. 4.27. Calculated exciton absorption spectra for various values of B with $S = 10$ and $S = 3$. The zero of energy is taken at the lowest exciton-polaron energy [4.80]

type interaction with optical modes coexistent with the deformation-potential type interaction with acoustic phonons? The "large" polaron situation, realized in the F region of Fig. 4.21, where the electron bound by the self-induced lattice distortion is fairly extended in space, is not appropriately described by wave functions of the type (4.151) or its extensions mentioned above in which the electron is confined to a single site. The only method capable of describing the whole F region – weak through strong coupling (referred to α and not to g_l) – is the Feynman path integral method as mentioned in Sect. 4.1.2. While this method assumes continuous space for the electron as well as for phonons, the discrete structure of the lattice is essential to appropriately describe the S-state and the F–S transition triggered by the short-range interaction. A. *Sumi* and *Toyozawa* [4.81] calculated the effective mass of the phonon-dressed electron under the coexistent Fröhlich and deformation-potential interactions using the Feynman path integral method with the lattice discreteness being considered through the Debye cutoff in the phonon wave number but not in that of the electron. In spite of the fact that the effective mass is underestimated in the S region according to this indirect (in k-space) and partial consideration of the discreteness, the calculated mass shows discontinuity consistent with the phase

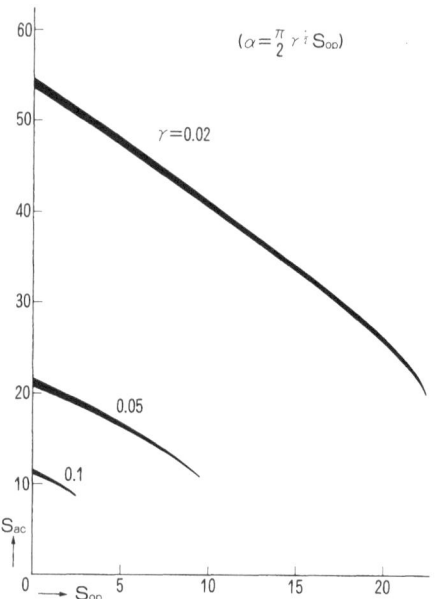

Fig. 4.28. Phase diagram for the polaron ground state due to the path integral method. The thickness of the line represents the magnitude of the discontinuity

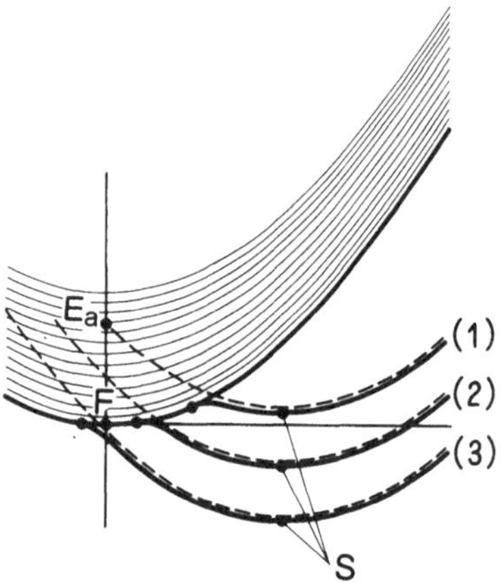

Fig. 4.29. Adiabatic potentials for the impurity bound state and the extended states in an elastic medium for various depths v of a short-range attractive potential [v increases as (1) → (3)]

diagram Fig. 4.21, as shown in Fig. 4.28 where the abscissa and the ordinate are inverted and multiplied by $B/\hbar\omega$ compared with the former.

Although the mathematical discontinuity in the effective mass is an artifact of the trial wave function or trial action adopted in the variational calculation of energy, the physically abrupt change from F- to S-states is thus established in the nearly adiabatic regime with $\gamma \equiv \hbar\omega/B$ smaller than one-fifth. This condition is usually well satisfied in inorganic insulators where the widths of conduction, valence, and exciton bands are of the order of eV.

4.4.5 Extrinsic Self-Trapping and Shallow-Deep Instability

Let us consider the lattice in which a host atom, say at the n-th site, is replaced by a guest (impurity) atom, causing a local potential $V_n = -v$, $V_{n'} = 0$ $(n' \neq n)$. In a one-dimensional configuration coordinate space Q_n, the total local potential amounts to $V_n = -(v + cQ_n)$, and the electron binding energy is obtained by shifting the abscissa in Fig. 4.22 by an amount proportional to v. The adiabatic potential for an electron in a three-dimensional lattice corresponding to Fig. 4.1b is shown schematically in Fig. 4.29 for different values of v with a fixed c. One can think of various possibilities: (0) F is the only minimum (not shown in the

Figure), (1) F is stable and S metastable, (2) F is metastable and S stable, (3) S is the only minimum (the bound state exists already for an undistorted lattice), as discussed by *Shinozuka* and *Toyozawa* [4.82].

As seen from the phase diagram of Fig. 4.30, the relative stability of the F- and S-states is determined approximately by $g_s' \equiv g_s + (v/B)$ while the potential barrier between them disappears with vanishing g_s. In the S(F) region which corresponds to case (2) of Fig. 4.29, the electron can be bound by the impurity only with the assistance of self-induced lattice distortion and hence only after overcoming a potential barrier in the adiabatic potential. This situation is called impurity-assisted self-trapping [4.82] or extrinsic self-trapping [4.83]. In particular, in the hatched region, the cooperation of the impurity potential and the electron phonon coupling is indispensable for the electron localization to take place at all, neither of them alone being sufficient. In the remaining region of S(F), we have both intrinsic (away from the impurity) and extrinsic self-trapped states which are again separated by potential barriers.

If the impurity has a positive (negative) charge, the electron (hole) always has a shallow Coulombic state (sh) and in addition a deep state minimum stabilized by strong lattice distortion [4.72, 83]. With an increase in the Coulomb attraction, the stable configuration may change discontinuously from the shallow state minimum to the deep state minimum as shown schematically in Fig. 4.31. With reservations (1) and (2) mentioned below, the criterion for this shallow-deep

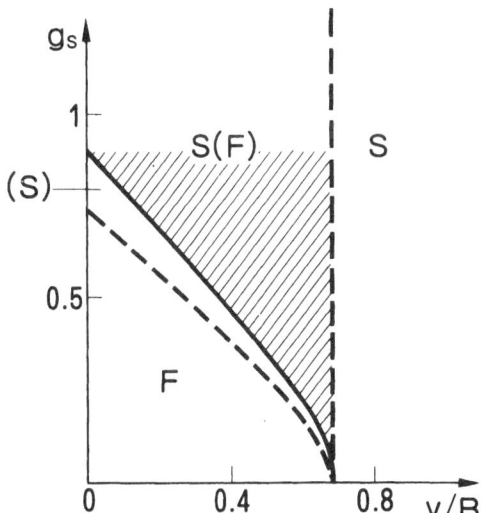

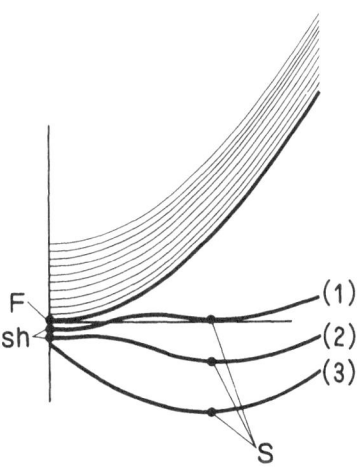

Fig. 4.30. Phase diagram for the stable and metastable (in parentheses) states of an electron under a short-range attractive impurity potential $-v$ and a short-range electron-phonon coupling constant g_s

Fig. 4.31. Adiabatic potentials for the charged-impurity bound state and the extended states of an electron in an elastic medium for varied strengths of the Coulomb attraction [*increasing as (1)* → *(3)*]

instability [4.84] is given by the phase diagram Fig. 4.21, where the abscissa and the ordinates are to be reinterpreted as $g_s' \equiv g_s + v/B$ and $g_l' = g_l + u/B$ with $-u$ being the on-site value of the attractive Coulomb potential [4.83]. In this phase diagram, (1) the F state is to be understood as a shallow Coulombic state (sh) mentioned above when $u \neq 0$ (charged impurity), and as a large polaron when $u = 0$ (neutral impurity). (2) Another point, physically obvious, is to be added: the discontinuity line for F–S transition, together with the F–F(S) and S–S(F) borderlines, recede toward the point ($g_s' = 1$, $g_l' = 0$) as the share of g_s in the total g_s' decreases; in the vanishing limit, the apparent discontinuity is a mathematical artifact of the Gaussian trial function assumed for $\psi(r)$ in (4.137).

4.4.6 Instabilities in the Relative Motion of a Pair of Charged Particles

We have so far considered the instabilities in the translational motion of a single particle in phonon fields. In the case of an exciton, however, we have matrix elements of exciton-phonon interaction between different internal states of the exciton which may have a significant effect on the electron-hole relative motion (see Sect. 4.3.3 for the Fröhlich interaction). Hence, we have also to study whether and how the relative motion is destabilized in the phonon field. It is difficult, however, to imagine the latter destabilization to take place without the former; rapid translational motion has the effect of averaging out the local distortion potentials, thereby reducing the effective exciton-phonon coupling. So we have to directly study the system of two charged particles in the phonon field, without separating the center-of-mass motion in advance. We will start with a simpler system of two electrons (a), then study an electron-hole system (b), and finally an electron-impurity system (c) as a corollary. A new aspect in these two charged particles systems is the *competition between the Coulomb interaction and the electron-phonon interaction*. We shall mainly consider the short-range electron-phonon coupling which is presumably the main trigger for the instability in the case of relative motion, too.

a) Two Electrons in the Phonon Field. *Anderson* [4.85] pointed out that the phonon-mediated attractive force between electrons may favor the occupation by two electrons (rather than by one) of localized states in amorphous semiconductors. This mechanism – the competition of electron-phonon interaction against the electron-electron interaction – turns out to underlie a much broader class of materials and phenomena. We will describe this mechanism in the context of intrinsic and extrinsic self-trapping.

Let us consider two electrons, each of which has Hamiltonian (4.37–38) with $E_a = 0$ and interacts through (4.39) with the phonon field (4.40), and which mutually interact through the on-site Coulomb repulsion U of the Hubbard type. In the zeroth-order approximation, as was taken as the starting point in Sect. 4.2.1, we can consider the following energy states as candidates for the most stable state:

2F (two free electrons): $-2B$
2S (a separated pair of self-trapped electrons): $-2E_{LR}$
S$_2$ (spin singlet of two electrons self-trapped at the same site): $U - 4E_{LR}$

Two electrons on the same site induce twice as large a lattice distortion and hence four times as large a lattice relaxation energy as one electron does. Comparing the energies of the three states, we obtain the phase diagram for the stable state as shown in Fig. 4.32 with the use of triangular coordinates ($B, U, 2E_{LR}$). In the lower-right region S$_2$ is more stable than 2S because of the negativity of the effective electron-electron interaction

$$U_{\text{eff}} = U - 2E_{LR} \ . \tag{4.156}$$

Within the hatched region, an electron does not self-trap by itself but can cooperate with another electron to be self-trapped together. In the remaining (unhatched) region of S$_2$, isolated self-trapped electrons can exist as a metastable state although they can be further stabilized by forming S$_2$ centers.

The energies of 2S and S$_2$ are lowered if virtual transfer of electrons to neighboring sites is considered, whereby the 2S–2F and S$_2$–2F border lines in Fig. 4.32 move upward and the 2S–S$_2$ border line rightward, although they continue to correspond to a first-order transition because of the finite difference in lattice distortion on their different sides (metastable states are not shown in Fig. 4.32).

The state S$_2$ obtained above is, so to speak, the small bipolaron, and is to be distinguished from the Coulomb bound state of two large polarons which is difficult to form. According to the path integral study of two Coulomb-repulsive electrons in the coexistent phonon fields of optical and acoustic modes by *Hiramoto* and *Toyozawa* [4.86], bipolaron formation is mainly triggered by the acoustic mode, and it is only in the extreme case of $\epsilon_0 \gg \epsilon_\infty$ and large α that the bipolaron can be formed with the optical mode alone. The required value of α is so large that the bipolaron as well as the polaron is well classified as small. From the dielectric viewpoint that the optical-phonon–mediated electron-electron attraction never overscreens the direct Coulomb repulsion, the main mechanism stabilizing the bipolaron in the optical-phonon field seems to be the electron correlation effect.

b) Electron-Hole Pair in the Phonon Field. The electron in the conduction band and the positive hole in the valence band have different effective masses and different deformation potential constants. For the sake of simplicity and comparison with the two-electron system, however, we assume, for the moment, the same values of $|t|$ (intersite transfer energy between nonnearest neighbors is neglected) and $|c|$ for the two particles, and the on-site attractive potential $-U$ ($U > 0$) between them.

The candidates for the most stable state are given in the lowest-order approximation as follows, for the cases of the same and opposite signs of c_e and c_h:

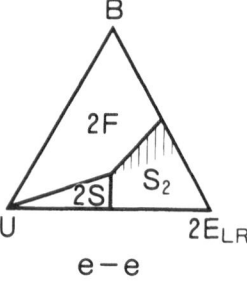

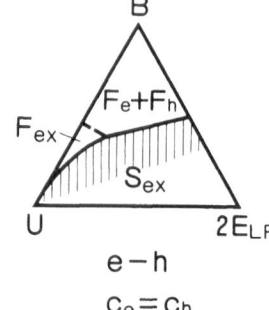

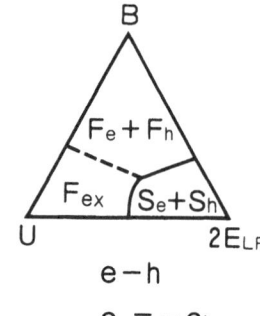

Fig. 4.32. Phase diagram for the stable state of a pair of electrons in an elastic medium, with on-site repulsive potential

Fig. 4.33. Phase diagram for the stable state of an e–h pair with on-site attractive potential $-U$, for the same sign of electron and hole deformation potentials

Fig. 4.34. The same as Fig. 4.33 except for the opposite sign of electron and hole deformation potentials

$F_e + F_h$ (a free electron and a free hole): $-2B$

F_{ex} (a free exciton): $-U - O(B^2/U)$

S_{ex} (a self-trapped exciton): $-U - 4E_{LR}$

only for the case of $c_e = c_h$

$S_e + S_h$ (a separated pair of a self-trapped electron and hole): $-2E_{LR}$

only for the case of $c_e = -c_h$

The resulting phase diagrams for the two cases are shown schematically in Figs. 4.33 and 34. It is to be noted that a tightly bound pair with energy $-U$ can be further stabilized by successive virtual transfers of the electron and the hole to neighboring sites which allow translational as well as relative motions. It is this second-order energy which can stabilize the F_{ex} state against S_{ex} in Fig. 4.33. The border lines between the F_{ex} and $F_e + F_h$ regions represent a second-order transition (shown by broken lines) in the sense that neither of the two states are accompanied by lattice distortion.

Replacing the short-range attractive potential $-U$ by a long-range Coulomb potential $-e^2/\epsilon_\infty r$ has the effect of modifying the phase diagrams of Figs. 4.33 and 34 along their B–$2E_{LR}$ edges such that the free unbound pair $F_e + F_h$ is replaced by a free exciton F_{ex} and the separated pair $S_e + S_h$ by a Coulomb bound pair $S_e : S_h$, as was shown by A. Sumi [4.87] with the use of the variational method for more general cases with effective mass m and deformation potential Ξ different for the electron and the hole. Figure 4.35 is the phase diagram of the stable state on the (Ξ_e, Ξ_h)-plane for the case of $m_h/m_e = 4$ and a relatively weak Coulomb attraction. Note that $\Xi_e \propto c_e$ and $\Xi_h \propto c_h$ correspond respectively to $\gamma_q^{(c)}$ and $-\gamma_q^{(v)}$ of (4.31) since the positive hole is an electron deficit in the valence band. As

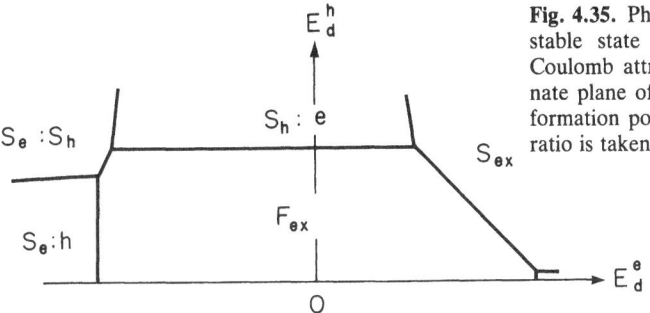

Fig. 4.35. Phase diagram for the stable state of an e–h pair with Coulomb attraction, on the coordinate plane of electron and hole deformation potentials. The e–h mass ratio is taken to be 1:4 [4.87]

we increase Ξ_h starting from the free-exciton region F_{ex}, the hole (h) becomes self-trapped around which the electron (e) is loosely bound by the Coulomb potential – the region $S_h : e$. Then, as we increase Ξ_e beyond a critical value, e becomes abruptly more deeply bound with a smaller radius (S_{ex} region) due to the self-induced distortion. This is a modification of the shallow-deep instability described in Sect. 4.4.5 (the impurity in the latter is to be replaced by the self-trapped hole in the former). Fig. 4.35 indicates that an exciton can self-trap even if neither e nor h can (in the region of S_{ex} near the F_{ex} border line) because of Ξ_e and Ξ_h acting constructively.

The situation is completely different in the second quadrant of the phase diagram where Ξ_e and Ξ_h act destructively. As we increase $-\Xi_e$ starting from the $S_h : e$ region, e also becomes self-trapped, this time not on the same site as h but on one of those neighboring sites where the phonon-mediated short-range repulsion ($\Xi_e\Xi_h < 0$) counterbalances the long-range Coulomb attraction: e wants to expand the surrounding lattice while h wants to contract it, thus preventing themselves self-trapping on the same site. The transition to $S_e : S_h$ will be called self-decomposition.

Among the various S-type states in Figs. 4.32–35, an important difference between S_2, S_{ex} on one hand and 2S, $S_e + S_h$, $S_e : S_h$ on the other is that *the parity of the relative motion is broken* in the latter while the parity conserving shallow-deep instability takes place in the former. It is readily seen from these examples that the short-range repulsive force (direct or phonon-mediated) is an important impetus for parity breaking.

c) Parity Breaking of a Defect-Bound Electron. By replacing the self-trapped hole (or electron) in the e–h system considered in b) with a point defect in the lattice, one can expect the parity breaking instability also for a defect-bound electron (or hole) – sh → S as shown in Fig. 4.36a – if the defect potential consists of the long-range Coulomb attraction and short-range strong repulsion (direct or phonon-mediated). This is to be contrasted with the shallow(sh)-deep(S) instability shown in Fig. 4.36b which has already been described in Sect. 4.4.5. It should be noted that this parity breaking of a defect-bound electron can arise even if the electron does not self-trap in the host lattice – the long-range attraction assists

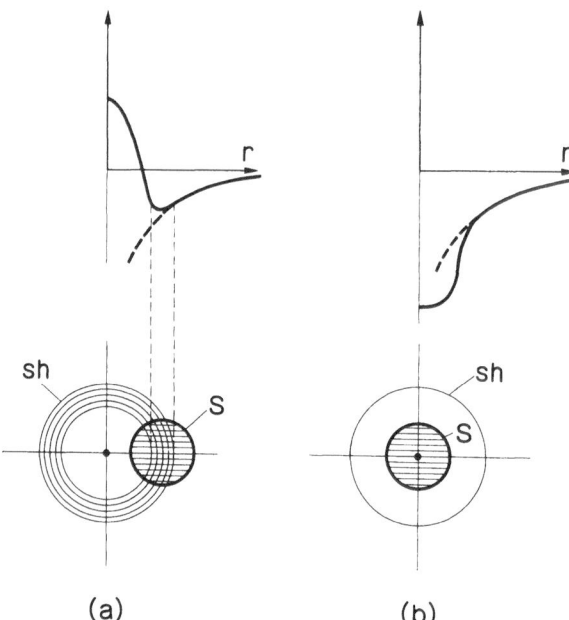

Fig. 4.36. Possible states of an electron bound by a positively charged defect with (**a**) repulsive or (**b**) attractive core potential (shown in the upper part)

(a)   (b)

self-trapping (parity breaking extrinsic self-trapping [4.84]). Another way of looking at this state is as two-dimensional self-trapping on the spherical surface (sh → S in Fig. 4.36a) on which the electron is forced to move as the result of quantization of radial motion, confined by the short-range repulsive and long-range attractive forces. Two-dimensional self-trapping takes place with smaller E_{LR} than in the three-dimensional case because of smaller $B = 2d|t|$ (Sect. 4.2.1).

4.4.7 Survey of Experimental Studies of Self-Trapping and Related Instabilities

It has been revealed, through a variety of experimental studies, that electrons, holes, and excitons in different crystals can be *classified into two distinct categories* – mobile type (F) and well-localized type (S) – with very few intermediate types. This remarkable fact can be explained by the phase diagram, Fig. 4.21, as follows. In the first place, electrons and holes in nonpolar crystals, as well as excitons in any type of crystal, have no long-range interaction with phonons ($g_1 = 0$), and should be located along the abscissa of Fig. 4.21. Secondly, most electrons and holes in typical polar crystals have $g_1 \lesssim 1$, as it turns out from available experimental data and material constants. These particles should then behave quite differently depending on which side of the discontinuity line of Fig. 4.21 they fall.

In Fig. 4.37 we allocate the electrons, holes, and excitons in various crystals into appropriate regions of the phase diagram, on the basis of a variety of experimental facts as described below. Typical inorganic semiconductors – group IV elements and III-V or II-VI compounds – have electrons and holes with high mobility and sharp cyclotron resonance, and excitons with resonant emission (Fig. 4.1b'), both being characteristic of the F-state. In contrast, holes [4.88] and excitons [4.89] in alkali halides, and excitons in rare gas solids [4.90, 91] and some molecular crystals [4.92] are relaxed into the S-state soon after they are produced in the F-state by optical excitation, as has been established by electron spin resonance (ESR) with its hyperfine structure or by the optical absorption and emission spectra therefrom (Sect. 4.2.1 and Fig. 4.1b).

The S-type emission from excitons is often accompanied by a weaker F-type emission as is expected from the phase diagram (along the abscissa, F is locally always stable, with a finite potential barrier to cross to get to S). The temperature dependence of the intensity ratio of the two types of emission in pyrene and α-perylene, as studied by *Matsui* et al. [4.93], indicates that the populations of the F(metastable)- and S(stable, excimer)-states are in thermal equilibrium, wherefrom they estimate the energy difference between the two states. In contrast, the intensity of the F (metastable) emission is a decreasing function of temperature in alkali halides according to *Lushchik* et al. [4.94], *Hayashi* et al. [4.95], *Nishimura* et al. [4.96], and *Khiem* and *Nouhailhat* [4.97]. This indicates that the popula-

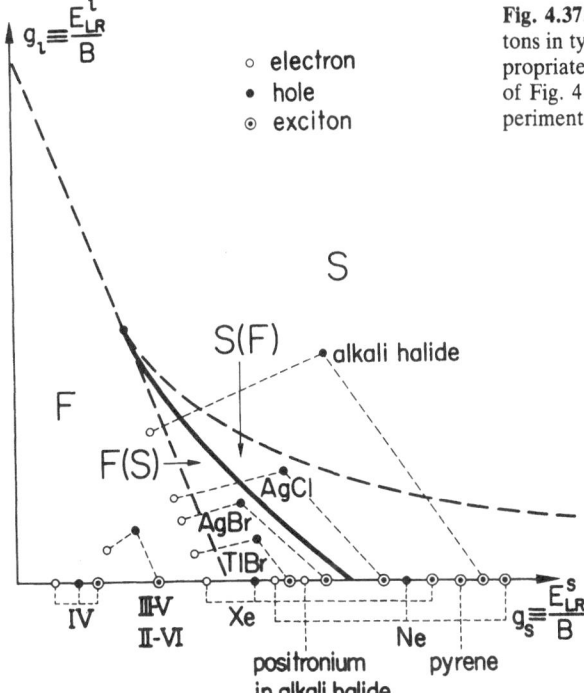

Fig. 4.37. Electrons, holes, and excitons in typical insulators allocated in appropriate regions of the phase diagram of Fig. 4.21 on the basis of various experimental studies

tions are not in thermal equilibrium but are governed by the rate of F → S transition (only the F excitons are produced initially in the optical excitation). From this temperature dependence they estimate the height of the F → S potential barrier (instead of the F–S energy difference). It is quite reasonable that thermal equilibrium of F–S populations is reached much more rapidly in molecular crystals where the F → S potential barrier is lower because of smaller B (4.139).

Silver halides are in a *marginal* situation: holes and excitons in AgBr are in the F-state while those in AgCl are in the S-state, in spite of significant similarities between the two materials in other respects. *Kanzaki* et al. [4.98] observed an abrupt change from the F- to the S-type of low-temperature excitonic luminescence in the alloy system $AgBr_{1-x}Cl_x$ at $x_c \sim 0.45$, which provides direct evidence for the existence of the F–S discontinuity line. Moreover, the observed σ value, Sect. 4.2.4 of the indirect absorption edge at this critical concentration is in excellent agreement with the theoretically calculated σ_c value (Table 4.1).

The mobility measurements in AgBr by *Hanson* [4.99] and TlBr by *Kawai* et al. [4.100] allocate the positive holes in these materials to the F(S) region and not the F region: the anomalous decrease in their mobilities with rising temperaturue can be explained only by assuming a metastable S-state a fraction of an electron volt above the F-state [4.72]. According to *Kobayashi* et al. [4.101], excitons in thallous halides are also near the marginal situation, as although they are F-type in both TlBr and TlCl, they are subject to extrinsic self-trapping in TlCl : Br; the impurity-bound exciton appears only in the luminescence (Stokes-shifted broad band, indicating a significant lattice relaxation) and not in the absorption spectra [case (2) of Fig. 4.29].

As an exotic example, the positronium in alkali halides has been located in the F(S) region of Fig. 4.35 by *Hyodo* et al. [4.102] on the basis of the anomalous temperature dependence of the momentum distribution of positronium analysed from their γ–γ angular correlation study of positron annihilation. The effective mass of the positronium in the F-state obtained at low temperatures, which is more than twice the free electron mass but significantly less than that of a free exciton, is consistent with the existence of a metastable S-state.

Positive holes in alkali halides are allocated to the S region (Fig. 4.37) according to the picosecond spectroscopy of optically produced e–h pairs by *Suzuki* et al. [4.103] which indicates the absence of a potential barrier for hole self-trapping. Positive holes in rare gas solids are self-trapped, possibly except for Xe, in view of very low mobility and its temperature dependence as described in Chap. 5.

Let us now consider the roles of electron-phonon and hole-phonon interactions in the behavior of an exciton. The fact that electrons are mobile but holes and excitons are self-trapped in most rare gas solids, alkali halides, and AgCl seems to indicate that in these materials self-trapping of an exciton is mainly due to the hole-phonon interaction, and that the electron is loosely bound around the self-trapped hole. Although this is not far from the truth, we have also to note the important role of the electron-phonon interaction. The deformation potential Ξ_h of a positive hole in large gap materials is usually positive, namely, a hole tends to contract the surrounding lattice ($\Xi_h \Delta$ becomes negative for stability, so does dilation Δ) [4.19]. This is partly because of induced electronic polarizations of surrounding atoms which in turn cause ion(hole)-dipole attractive forces and partly because of the removal of an electron from the "top" of the valence band where antibonding nature dominates (the latter applies only to large gap materials and not to narrow gap valence band semiconductors where bonding nature dominates throughout the valence band). In contrast, the deformation potential Ξ_e of a conduction electron is negative in solid He and solid Ne though positive in heavier rare gas solids. The tightly closed shell in the former results in a repulsive pseudopotential for an extra electron, namely, a negative electron affinity even in solids. Thus the extra electron forced (from outside) into the conduction band tends to stabilize itself by forming a cavity (bubble) around it ($\Xi_e \Delta$ wants to be negative with positive Δ – expansion), as is realized in solid and liquid ^{4}He and liquid Ne.

The self-trapped exciton in solid Ne is also accompanied by a bubble due to its electron, while the positive hole is shared by one or two Ne atoms (Chap. 5). Therefore, Ξ_e plays the main role in exciton self-trapping in solid Ne, especially in the case of one-center (atomic) type states. In heavier rare gas solids, Ξ_e and Ξ_h, both being positive, cooperate constructively to favor exciton self-trapping. The fact that self-trapping takes place with an exciton but with neither an electron nor a hole in solid Xe can be explained in terms of the phase diagram of Fig. 4.35 (see also Sect. 4.4.6).

The ESR study by *Känzig* and *Woodruff* [4.88], the transient optical spectroscopy by *Kabler* [4.89] and the time-resolved optical spectroscopy by *Suemoto* and *Kanzaki* [4.104] clarified that the self-trapped holes in alkali halides and the holes in the most stable self-trapped excitons in alkali halides and rare gas solids (except He) are of the two-center type. This is presumably because of the predominance of the site-off-diagonal hole-phonon interaction (see the end of Sect. 4.4.1) which in fact is important due to the bonding nature of the p orbital. *Nakai* and his collaborators [4.105] have established that alkali halides containing heavier halide impurities have two types of relaxed configurations of an impurity-bound exciton: a metastable one-center type (exciton localized on an impurity) and a stable two-center type (exciton localized on an impurity and one of its neighboring host halides). This indicates that the stabilization energy difference between two- and one-center types is greater than the electron affinity difference between the host and the guest halides.

In contrast, the positive hole in solid and liquid ^{4}He, which originates from the 1s orbital, is known to form an iceberg around a central ion (Chap. 5). The self-trapped hole in AgCl is located on a Ag$^+$ ion (to form Ag^{++}) with tetragonal lattice distortion according to the ESR studies [4.106], presumably due to the strong mixing of Ag 4d orbitals into the valence band (see the end of Sect. 4.4.1). The lattice configuration is essentially the same in the self-trapped exciton according to *Hayes* et al. [4.107].

It is of interest to see what happens if free electrons and holes are produced, by an energetic electron beam or x-ray, in materials with a strong hole-phonon interaction. In alkali halides which have been studied most extensively [4.108–110], there are two important channels for the relaxation of an e–h pair: i) radiative annihilation of the self-trapped exciton and (ii) nonradiative self-decomposition into an F center (an electron captured by an anion vacancy) and an H center (a hole captured by a doubly occupied anion site) through simultaneous creation of the relevant lattice defects as shown in Fig. 4.38.

As created initially, the free holes have much smaller velocities than the free electrons because of the narrow valence band, and will immediately self-trap to form X_2^- molecules (X denotes a halogen atom). The next step will be the cascading capture of an electron by the attractive Coulomb potential of the self-trapped hole. The experiments indicate that channel (ii) opens at excited Coulombic states of the electron, namely, while the electron is loosely bound around the X_2^- center (Fig. 4.38a). From this configuration, a slight displacement of X_2^- along the [110] direction will lead to the creation of a neighboring pair of F and H centers (Fig. 4.38b).

The process $(a) \to (b)$ can be considered [4.84] as a modification of the self-decomposition of an e–h pair described in Sect. 4.4.6b ($S_e : S_h$ in the phase diagram of Fig. 4.35), although there remains some controversy about the details of the relaxation channels [4.110, 111]. While the electron does not self-trap in a perfect lattice, self-decomposition is realized with the assistance of the self-created defect pair. The condition of opposite signs of Ξ_e and Ξ_h, which is necessary for the decomposition, is in fact satisfied if we consider only the anion sublattice which plays a major role here: the electron wants to expand (repel) and the hole to contract (attract) the anion sublattice. The fact that channel (ii) opens only from loosely bound states of the electron is consistent with the theoretical prediction [4.87] that the state $S_e : S_h$ is stable only when the e–h Coulomb binding energy is small enough to be overcome by the phonon-mediated e–h repulsion.

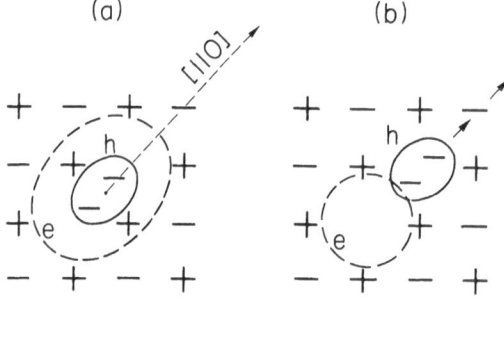

Fig. 4.38. A model for the formation of (b) F and H centers from (a) a self-trapped exciton in an alkali halide

Under direct optical excitation of $1s$ excitons in alkali iodides, another luminescence (E_x band) with a smaller Stokes-shift appears [4.112] before they relax into the molecular-type exciton with the well-known luminescence. This new center has recently been identified by *Itoh* [4.113] as an atomic Jahn-Teller type self-trapped exciton with two neighboring iodides along the [110] direction coming closer to the central iodide. Because of the tightly bound electron, the hole may not be as readily relaxed into the two-center type as in the cases of a loosely bound (higher exciton states) or ionized electron.

Finally, we give a brief survey of experimental studies revealing large lattice relaxation induced by the capture of an electron or hole by a defect center in semiconductors. The configuration coordinate (C.C.) model proposed for CdF_2 : In on the basis of optical, electrical and thermal experiments [4.114] is a typical example of case (2) of Fig. 4.31, namely, the shallow Coulomb bound state (In^{+++} replacing Cd^{++} behaves as singly charged) coexists with the deep state with large lattice distortion. The electron-phonon interaction is considered to be weaker in more covalent semiconductors such as III-V compounds and group IV elements, resulting in high mobility of free carriers and small lattice distortion for shallow capture. On the other hand, it has also been pointed out that the captured carrier sometimes causes a strong distortion of the surrounding lattice, stabilizing itself into a deep level [4.115–117]. The significant difference between the two situations realized in the same host lattice is reasonably understood in terms of the shallow-deep instability discussed in Sect. 4.4.5. The C.C. model for the DX center in a III-V semiconductor, which was introduced [4.118] to explain the persistent photoconductivity, is a typical example of extrinsic self-trapping, corresponding to case (2) of Fig. 4.29 and the hatched region in Fig. 4.30; at low temperatures, the recapture of a photoelectron by this neutral center is prevented by the potential barrier – the energy to set up enough lattice distortion to capture the electron at the center.

The negative U effect described in Sect. 4.4.6a has been demonstrated for some of the deep defect states in semiconductors – Si vacancy [4.119] and Si : B (interstitial) [4.120] in addition to the identified and unidentified defects in glassy semiconductors [4.121].

The negative U effect in the host crystalline lattice (S_2 state of Fig. 4.32) has been reported for some transition metal oxides, in particular, Ti_4O_7 [4.122] and WO_{3-x} [4.123]. In the latter, the thermally activated conduction (characteristic of S_2 rather than of the large bipolaron) and the optical dissociation of the paired electrons, have been observed by ESR monitoring of W^{5+}.

Parity breaking of a defect-bound hole, as described in Sect. 4.4.6c, has been studied in detail in $MgO : V^-$, where the hole is localized on one of the six neighbors of the cation vacancy [4.124]. It is not known whether the hole self-traps in the host MgO lattice. Parity breaking of holes trapped at cation vacancies are well-known in alkali halides where holes self-trap already in the host lattice. It is suggested [4.84] that the holes in AgBr and TlBr, being close to the discontinuity line, are candidates for parity breaking extrinsic self-trapping if appropriate defects with negative charge and short-range repulsive potential exist.

4.5 Electron-Hole Recombination

In insulators and semiconductors, electrons, holes, and/or excitons produced by excitation or injection can return to their thermal equilibrium populations only through the recombination process in which energy of the order of the band gap ε_g, per an e–h pair, must be disposed of at one coup. It is usually assumed that long before the recombination the energy distribution of each species of carriers reaches quasi thermal equilibrium within its own band, typically in picoseconds. In fact, various channels of recombination in semiconductors with moderate ε_g (of the order of 1 eV) have relatively small rates for the following reasons: The radiative recombination rate, which is proportional to ε_g^2 according to Einstein's formula for spontaneous emission, is of the order of 10^8 s^{-1} in semiconductors with a moderate gap. Of nonradiative recombination channels, the Auger processes [4.125] have significant rates only when the number of carriers is great enough. Nonradiative multiphonon recombination rates are usually much smaller than ω when ε_g is large compared with the average phonon energy $\hbar\omega$ since it is the perturbation process of higher order $\sim (\varepsilon_g/\hbar\omega)$ in the electron-phonon interaction.

In this way, each recombination channel has its rate-determinig *bottleneck* across which the population is far from thermal equilibrium or even inverted. It is in this region that *an e–h pair spends most of its time, awaiting something athermal, dynamic, or even catastrophic to happen.* The purpose of the present section is to briefly describe the typical channels of recombination with particular attention to their respective bottlenecks.

In semiconductors without self-trapping, free electrons and holes are subject to Coulombic capture to form free excitons, or to form donors and acceptors if they can find ionized impurities of opposite charge. Those processes have giant cross sections [4.126] since the carriers are initially captured to very large orbital states and can then immediately cascade down the Coulombic levels emitting one phonon each time, except possibly for the last few steps which might need multiphonon emission. The main decay channel of a distant pair of a shallow donor and acceptor or a free or impurity-bound Wannier-Mott exciton is radiative recombination since a large orbital radius greatly reduces the electron-phonon interactions [4.47], suppressing the nonradiative multiphonon processes. In contrast, the deep impurity states as well as the self-trapped states can act as efficient centers of nonradiative multiphonon processes leading to recombination or defect reaction [4.83, 84, 117]. An extensive review of nonradiative transitions in semiconductors was presented by *Stoneham* [4.127].

4.5.1 Polariton Bottleneck

Radiative recombination of a free direct exciton has a peculiar feature. As long as we treat the exciton-photon interaction as a perturbation, only an exciton with $K \sim 0$ can be converted into a photon because of momentum-energy conservation. However, the nonvanishing matrix element of this interaction is of the order of $N^{1/2}$ since the oscillator strengths of all the atoms (N) in the crystal are concentrated to this particular exciton state. This singular structure of the exciton-photon interaction is in fact removed by invoking the polariton picture as was described in Sect. 1.6. According to this picture, a hot exciton created in some way will lose its kinetic energy by successively emitting phonons, along the lower polariton branch in the final stage (see the arrows in Fig. 4.39) whereby the exciton gradually changes its nature into a photon. The decrease of the exciton component in the polariton results in the decrease of the polariton-phonon interaction, and hence in the decrease of its energy relaxation rate. On the other hand, the increase of the photon component (which amounts to the decrease of refractive index n according to Fig. 1.3) results in the increased rate of conversion of the polariton into a photon outside the crystal (the transmittance of light from inside to outside is given by $2n^{-1}/(1 + n^{-1})^2$ which is an increasing function of n^{-1} when $n^{-1} < 1$). As the result of these two effects, the polaritons are populated only down to the bottleneck region b across which the population is inverted [4.128, 129]. The earliest experimental study relevant to this is due to *Goto* et al. [4.130] who ascribed the equally spaced LO phonon structures in the excitation spectrum of free-exciton luminescence in CuBr to the bottleneck effect (see Sect. 3.2.4). Relaxation kinetics and dynamics of exciton polaritons in the bottleneck region have been studied with nanosecond-resolved luminescence spectroscopy by *Wiesner* and *Heim* [4.131] and with picosecond-resolved absorption spectroscopy (transition into the excitonic molecule) by *Segawa* et al. [4.132] and by *Masumoto* and *Shionoya* [4.133].

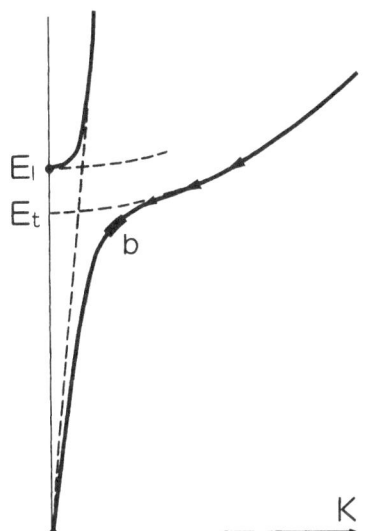

Fig. 4.39. Phonon scattering of a polariton towards the bottleneck (*b*)

4.5.2 Resonant Secondary Radiation

In some of the above-mentioned processes, we are already confronted with that controversial problem of the resonant secondary radiation process. Can the photo-excited luminescence be distinguished from the resonant Raman scattering (RRS)? (The generally accepted notion is of course that the two represent the absence and the presence of correlation between the incident and emitted light.) The answer to this question depends on the nature of the system and on the way of excitation and observation [4.134]. The two-level system with dephasing relaxation in the rapid modulation limit of the transition energy is a typical case in which the luminescence and the resonant scattering can be spectrally (under monochromatic excitation) or temporally (under pulse excitation) distinguished [4.135, 136]. With the two-level system without dephasing relaxation (as in an isolated atom), monochromatic light can be only elastically scattered while white light is subject to real absorption followed by luminescence with natural width [4.137]. The secondary radiation from a localized electron strongly coupled with phonons [4.138] cannot simply be decomposed into the two components; on the contrary, we even have the hot luminescence component of intermediate nature.

Secondary radiation under optical excitation in and above the exciton absorption peak of an ionic crystal has a feature which is simultaneously RRS-like and luminescence-like. It consists of equally spaced lines Stokes shifted from the incident photon energy by integral multiples of LO-phonon energy, which implies the higher-order Raman processes. In the corresponding higher-order matrix element, however, one can always find sets of intermediate exciton states which consecutively conserve energy (as long as the emitted photon has energy higher than the exciton bottom minus the LO phonon energy), and it is in fact such sets which make the predominant contribution [4.139]. In other words, the higher-order resonant Raman process in the exciton region is almost equivalent to a cascade of *real* processes – the absorption of an incident photon to create an exciton with simultaneous emission of a LO phonon, followed by successive emission of LO phonons, and completed by simultaneous emission of a photon and a LO phonon. In this sense, it can also be considered as hot luminescence. The cascade picture is useful in explaining the fact that the intensities of the equidistant lines in the secondary radiation are all of the same order as long as they are within the exciton band [4.15].

Such a coalescence of two notions – RRS and luminescence – becomes even more thorough if one resorts to the polariton picture. Here RRS is simply the real scattering of a polariton (in or between the upper and lower branches) by phonons.

Besides the LO phonon lines mentioned above, one finds the acoustic-phonon satellite with a much smaller Stokes shift in the secondary radiation. The latter – resonant Brillouin scattering as it is called – serves as a useful tool for determining the polariton dispersion because of the dispersion of acoustic phonons with finite slope at $q = 0$ [4.140]. In general, the resonant secondary radiation consists of all possible combinations of multiple emission and absorption of various phonons. The multiple absorption and emission of acoustic phonons contribute to the wings of each LO phonon line and also to the ordinary luminescence due to thermalized excitons around the band bottom – the bottleneck region according to the polariton picture – where the LO phonon emission is negligible because of the e–h cancellation effect of the Fröhlich interaction. The simple juxtaposition of luminescence and RRS turns out to be inadequate to describe these complications.

Another interesting aspect of RRS in the exciton region is that the divergence of the unrenormalized perturbation theory due to a vanishing denominator for the intermediate state can be taken advantage of in directly studying the energy dependent dephasing rate $\Gamma(E)$ (the imaginary part of the self-energy as introduced in Sect. 4.2.2) to which the RRS cross section is inversely proportional [4.141, 142].

4.5.3 Capture, Recombination, and Enhanced Defect Reaction Via a Deep Impurity Level in a Semiconductor

Electron(hole)-phonon interactions in II–VI, III–V and group IV semiconductors (except diamond) are not strong enough to cause intrinsic self-trapping but enough to cause extrinsic self-trapping (see Sect. 4.4.5 and the end of Sect. 4.4.7) whereby the electron is deeply bound at an impurity stabilizing itself by inducing strong lattice distortion. Conversely, any deep level impurity must be accompanied by significant lattice distortion as it captures an electron since the well-localized electron has strong phonon coupling. Such an impurity level, deep in the gap and with large lattice distortion, is expected to act as an efficient multi-phonon recombination center, as is seen from Fig. 4.40 where the C.C. model for electron capture followed by hole capture is presented [4.83, 84, 143]. Here the initial state (top) is a double continuum with a free e–h pair, the intermediate

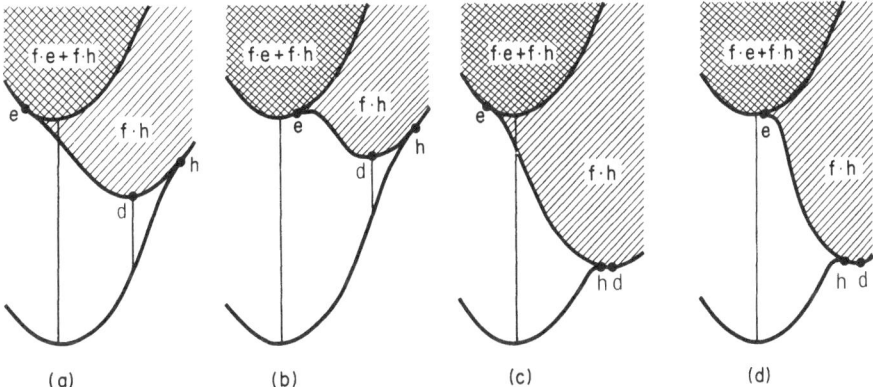

Fig. 4.40a–d. C.C. models for nonradiative e–h recombination via a deep impurity state

(middle) a single continuum with a free hole, and the final (bottom) a discrete ground state. Four possible cases are shown such that the electron [hole] capture needs finite lattice distortion in cases (b) and (d) [(c) and (d)]. The Coulomb part of the impurity potential will introduce some modifications. The impurity may also be doubly occupied, in which case we have to combine another C.C. model with the present one. The M center in n-type InP studied by *Stavola* et al. [4.144] is a facinating example with extrinsic self-trapping, negative U and recombination in it.

Looking at (a) and (b) of Fig. 4.40, one may well suppose that after the electron capture (at around point e) the lattice relaxes, overshoots the minimum point d (deep impurity state) and climbs up to point h where a free hole can be captured (whereby the recombination is completed) if the hole density n_h is large enough. This explains the mechanism of the well-known but unidentified defects acting as efficient nonradiative recombination centers without forming deep levels on the way. *H. Sumi* [4.145] evaluated the critical density of the majority carriers above which recombination dominates deep level formation.

The above statement contains a more general problem of nonradiative electronic transition during lattice relaxation. The conventional theory of this process assumes an initial state with the lattice configuration thermally distributed around the minimum m of the adiabatic potential, whereas in the present problem one has to start from a different point i (Fig. 4.41). The nonradiative transition is enhanced in the latter situation due to overshooting with surplus kinetic energy of the lattice. Point i is the Franck-Condon state in the case of optical excitation, and the transition $1 \rightarrow 2$ may be the de-excitation as in the luminescence quenching in the F center discussed by *Dexter* et al. [4.146] or the energy transfer to another center as in the hot transfer process considered by *Hizhnyakov* and *Tehver* [4.147]. It has been found in the study of degradation in III–V semiconductor lasers that recombination enhances the defect reaction around the recombination center; the activation energy E_A of the same reaction process in thermal equilibrium is reduced in recombination by energy which can be almost

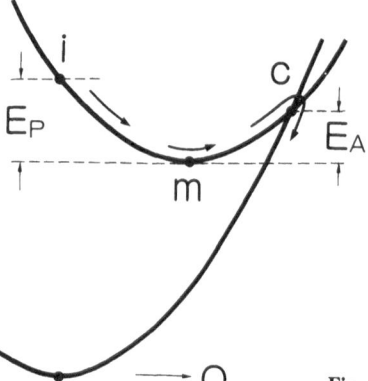

Fig. 4.41. C.C. model for a "hot" nonradiative transition

as large as the gap energy [4.148]. This means that the lattice energy released during the recombination is very efficiently and promptly utilized for the local defect reaction before it diffuses away. The C.C. used in diagrams such as Fig. 4.41 is in fact a sort of interaction mode [4.149] which is a linear combination of a great number of normal coordinates with dispersed frequencies, and the starting point Q_i and the reacting point Q_c are not always collinear with the relaxed point Q_m in the multi-dimensional C.C. According to H. Sumi [4.150], the reduction of activation energy can be equal to the lattice relaxation energy E_p but is in general smaller, being a decreasing function of $\sin^2 \theta$ where θ is the angle between $\overline{Q_i Q_m}$ and $\overline{Q_m Q_c}$ (the collinearity enhances the efficiency!).

4.5.4 Self-Trapping and Recombination of an Exciton as a Multiphonon Process

If an exciton or a hole is self-trapped, it can be an efficient center of multi-phonon recombination since the localization enhances the electron-phonon interaction. Self-trapping itself is already a multiphonon process, and in the case of an exciton, the lattice C.C. must tunnel through (at low temperature) or climb up (at high temperature) the adiabatic potential barrier B between the free (F) and the self-trapped (S) states (see Fig. 4.42). The crucial point in the evaluation of the self-trapping rate is how to find the most favorable C.C. path connecting the two states which are quite different in localization and in symmetry. For this purpose, *Iordanskii* and *Rashba* [4.151] use the continuum model of the lattice which is adequate in the strong coupling limit, while *Nasu* and *Toyozawa* [4.152] use the discrete lattice model applicable to intermediate coupling. *H. Sumi* [4.71] studied the self-trapping rates into one- and two-center type states (see the end of Sect. 4.2.1) with varying site-diagonal and site-off-diagonal exciton-phonon coupling constants. These theories are compared with the observations by *Masumoto* et al. [4.153], *Roick* et al. [4.154], and *Nishimura* et al. [4.96].

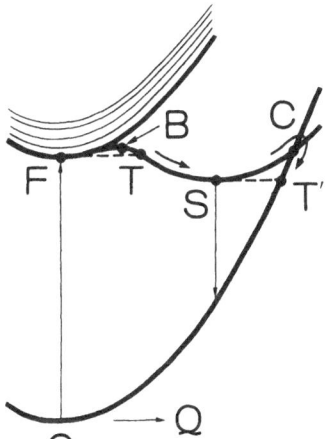

Fig. 4.42. C.C. for an exciton with tunneling self-trapping (FTS), tunneling recombination (ST'G), hot non-radiative recombination (TSCG) and radiative recombination from the self-trapped state (*downward arrow*)

It is assumed in all of the above theories that from the end point T of the tunneling path the lattice relaxes immediately towards the stable configuration S (Fig. 4.42), namely, that the tunneling is the rate-determining bottleneck of self-trapping at low temperatures. This is quite reasonable since the lattice relaxation from T to S means the irreversible conversion of free energy $E_T - E_S$ into entropy of thermal phonons extended throughout the crystal [4.149], as was the case in the i $\rightarrow$ m relaxation in Fig. 4.41. Here again, the lattice configuration over-shoots S and may climb up to the crossing point C, where the system can make the transition to the electronic ground state and relax to its minimum point G. The bottleneck in the exciton recombination via the self-trapped state at low temperature may be (i) FT tunneling, (ii) luminescence from S or (iii) ST' tunneling depending on the relative height of T versus C, the tunneling distances of FT and ST', and the radiative lifetime. The population anomaly arises between F and S states if the recombination bottleneck is at the F–S barrier. This is born out by A. Sumi's calculation of resonant secondary radiation [4.155]; because of overpopulation of F against S compared to thermal equilibrium, the emission from the F state is "cold" or "hot" luminescence according as $E_F \lessgtr E_S$.

We mentioned previously in Sect. 4.4.7 that the self-trapped exciton (in its higher internal states) in alkali halides triggers the formation of defects. It is more catastrophic than the recombination-enhanced defect reaction described in Sect. 4.5.3 in two respects: it needs no impurity, and it creates defects rather than enhances defect reactions.

4.6 Excitonic Instability and Phase Changes

We considered in previous sections the symmetry-breaking instabilities of the translational and relative motions of an exciton in the deformable lattice – self-trapping plus self-decomposition. If the total energy of these instabilities – lattice relaxation energy – were to exceed the energy of the unrelaxed exciton, the relaxed e–h pairs (each with negative energy) would be spontaneously generated (without getting energy from optical excitation) all over the lattice (Fig. 4.43). This should be the true ground state, rather than the one from which we started, the two states being different in lattice structure as well as in electronic charge distribution (or in valency).

A trivial example is an ionic crystal, e.g., NaCl. Starting from the assemblage of neutral Na and Cl atoms forming a NaCl-type lattice with a very large lattice constant a, one can transfer an electron from a Na atom to a nearest Cl atom. The energy needed to produce this "charge transfer" exciton is given by I (ionization energy of Na) – A (electron affinity of Cl) minus the attractive Coulomb energy e^2/a between Na^+ and Cl^-. This energy becomes negative as the two particular ions come closer (the excitation energy is negative as the result of lattice relaxation). If the lattice constant a of the whole crystal were allowed to change so as to stabilize the total energy, such charge transferred pairs would be produced spon-

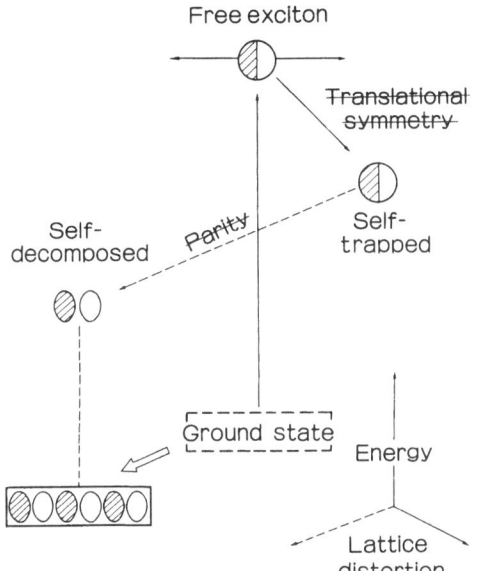

Fig. 4.43. Symmetry-breaking excitonic instabilities in a deformable lattice

taneously everywhere with simultaneous reduction of a, finally forming the ionic crystal we know. Starting from this real ground state, one obtains an exciton by transferring an electron in the opposite direction – from Cl^- to Na^+. This real exciton produced optically or by an energetic particle beam is again lattice-relaxed into (i) the self-trapped states from which it radiatively annihilates or (ii) the self-decomposed states – a pair of F and H centers (Sect. 4.4.7). The channel (ii) amounts, so to speak, to producing a pair of Na^0 and Cl^0. The existence of a similar but more efficient channel qualifies silver halides as good photographic materials: exposure to light results finally in the separation of Ag^0 atoms and halogen gas, although the intermediary electronic and ionic processes are less well-known. In any case, the "oxidized" state (ionic state) is the ground state while the "reduction" is realized in the relaxed excited states (metastable).

In contrast, many of the organic charge transfer complexes [4.156], consisting of alternate stacking of donor (D) and acceptor (A) molecules, have only a slight charge transfer in the ground state but almost complete charge transfer ($D \rightarrow A$) in the lowest excited state. In some other compounds such as TTF-chloranil, however, the neutral phase and the ionic phase have free energies so close to each other that the phase transition takes place with applied pressure [4.157] or change of temperature [4.158].

The most basic quantities governing the relative stabilities of the various states described above and in preceding sections are: $B \, (= 2d|t|)$, the energy gain due to intersite transfer; E_{LR}, the lattice relaxation energy; U, the e–h attractive potential or e–e repulsive potential, and the electron affinity difference between different constituents. Systematic study of competition among these quantities would be extremely interesting and useful.

4.6.1 t-U-S Problem

For simplicity, we shall confine ourselves to the system of a single constituent where the fourth quantity mentioned at the end of the previous section vanishes, and consider the Hubbard Hamiltonian plus the phonon field:

$$H = H_e + H_{ep} + H_p$$

$$= -t \sum_l \sum_{l'} \sum_\sigma a_{l\sigma}^\dagger a_{l'\sigma} + U \sum_l n_{l\alpha} n_{l\beta}$$

$$- \sqrt{S} \sum_l \sum_\sigma Q_l n_{l\sigma} + \sum_l \frac{1}{2} (Q_l^2 + \omega_l^2 P_l^e) \tag{4.157}$$

where we have used $S \equiv c^2 = 2E_{LR}$ instead of c or E_{LR}. In (4.157) $a_{l\sigma}$ denotes the annihilation operator for an electron with spin σ ($\uparrow$ or $\downarrow$) at the lth site, and $n_{l\sigma} \equiv a_{l\sigma}^\dagger a_{l\sigma}$ the corresponding number operator. This system comprises quite a number of well-known problems of basic importance in condensed matter theory. Superconductivity, for instance, results from the phonon-mediated attraction between electrons in k-space, provided it overcomes the Coulomb repulsion. Although the finite phonon energy $\hbar\omega$ plays an essential role there, we shall confine ourselves to the adiabatic limit ($\gamma = \hbar\omega/B \to 0$) and study the "$t$-$U$-$S$" problem in the following.

A t predominant system behaves as a metal, U brings in the correlation effect, possibly causing spin density waves of various types, while S may cause charge density waves accompanied by structural change (Q's). The instability of the metallic state against U or S depends very much upon the dimensionality of the lattice. The one-dimensional metal is unstable and becomes an insulator as soon as we switch on U or S (Peierls instability [4.159] in the case of S), in contrast to other dimensionalities which have finite thresholds for the instabilities. For one electron in the phonon field (t-S problem) we have already seen different features of instability in different dimensionalities in Sects. 4.4.1–3. The negative U problem is the competition between U and S (Sect. 4.4.6), with t playing a subsidiary role there.

It is instructive to study in some detail the small systems which are exactly solvable, and then to proceed to larger scale systems. We will confine ourselves to the ring system of N sites with N electrons.

4.6.2 Two-Site Two-Electron System

Let us introduce the new variables defined by

$$\left.\begin{matrix} Q \\ q \end{matrix}\right\} \equiv \frac{1}{\sqrt{2}} (Q_1 \pm Q_2), \qquad \left.\begin{matrix} n \\ \nu \end{matrix}\right\} \equiv \sum_\sigma (n_{1\sigma} \pm n_{2\sigma}). \tag{4.158}$$

Substituting (4.158) into (4.157), one immediately finds that Q and n ($= 2$) appear only in the form: $(Q - cn/\sqrt{2})^2/2 - c^2 n^2/4$ and hence, that the minimum point is given by $Q = \sqrt{2}c$ independent of the electronic states. The remaining part of (4.157) is given by

$$\left(H_e - \frac{c}{\sqrt{2}} vq \right) + \frac{1}{2}q^2 \equiv H(q) + \frac{1}{2}q^2 . \qquad (4.159)$$

Namely, the charge transfer v interacts only with the antisymmetric mode q as it should.

As for the spin, there are one triplet (t) and three singlets (s). The triplet state has $v = 0$ and $H^{(t)}(q) = H^{(t)} = 0$ since each site is occupied by one electron. The singlets consist of homopolar (h) and ionized states (i):

$$|h\rangle = \frac{1}{\sqrt{2}} (a_{1\uparrow}^\dagger a_{2\downarrow}^\dagger + a_{2\uparrow}^\dagger a_{1\downarrow}^\dagger)|0\rangle ,$$

$$|i\rangle = a_{i\uparrow}^\dagger a_{i\downarrow}^\dagger |0\rangle , \quad (i = 1, 2) \qquad (4.160)$$

with the corresponding energy matrix:

$$H^{(s)}(q): \quad \begin{matrix} & |h\rangle & |1\rangle & |2\rangle \\ & \overline{\begin{pmatrix} 0 & -\sqrt{2}t & -\sqrt{2}t \\ -\sqrt{2}t & U - \sqrt{2}cq & 0 \\ -\sqrt{2}t & 0 & U + \sqrt{2}cq \end{pmatrix}} \end{matrix} . \qquad (4.161)$$

The adiabatic energies (4.159) of three eigenstates of (4.161) with vanishing and nonvanishing t are shown in Fig. 4.44 by solid and dotted lines, respectively. With very small t, subsidiary minima (metastable configurations) of the lowest branch, which originate from the ionized states, appear for intermediate electron-phonon coupling: $U/2 \lesssim S \lesssim U$ and become most stable for strong coupling: $S \gtrsim U$. With finite t and U but vanishing c or q, the system is nothing but a hydrogen molecule: (4.161) has eigenvalues $E_{g,e'} = (U \mp \sqrt{U^2 + 16t^2})/2$ and

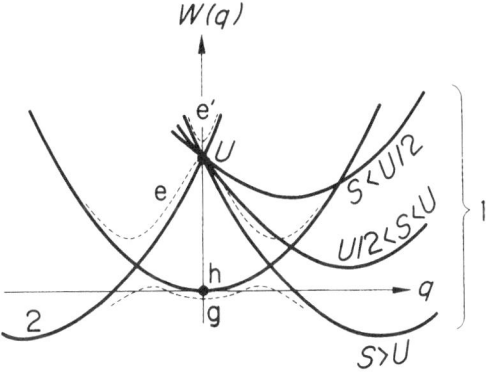

Fig. 4.44. Adiabatic potentials for singlet states of the 2-site 2-electron system, for vanishing (——) and non-vanishing ($\cdots$) transfer energy

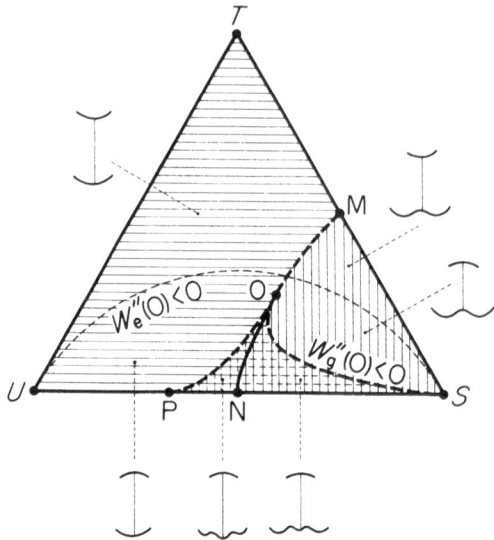

Fig. 4.45. Phase diagram of 2-site 2-electron system on triangular coordinates with transfer energy t, on-site e–e repulsive energy U and lattice relaxation energy $S \equiv 2E_{LR}$ for double occupancy

$E_e = U$, of which the ground state (g) represents the Heitler-London and the Hund-Mulliken states in the limits of $4t \ll U$ and $4t \gg U$, respectively. With vanishing U but finite t and cq, the corresponding eigenenergies are given by $E_{g,e'} = \mp \sqrt{4t^2 + 2c^2q^2}$ and $E_e(q) = 0$. The minimum point of the adiabatic potential $W_g(q) = E_g(q) + q^2/2$ bifurcates from $q = 0$ to $q = \pm \sqrt{2(S^2 - t^2)}/S$ as S exceeds t, namely, the charge transfer instability sets in at $S = t$ as a second-order transition.

Figure 4.45 is the phase diagram [4.160] indicating, over the whole region of (t, U, S) triangular coordinates, the stable (solid hatching) and metastable (broken hatching) configurations of the lowest electronic state, where the horizontal and vertical hatchings represent the symmetric ($q = \langle v \rangle = 0$) and the asymmetric ($q = c\langle v \rangle/\sqrt{2} \neq 0$) states, respectively. The adiabatic potentials versus q for the ground and the lowest excited (e) states are also shown schematically for each region. Symmetry breaking in the ground state takes place as a first-order transition across the solid line NO and as a second-order transition across the broken line OM. The state e is also subject to second-order symmetry breaking as t decreases across the dotted circular arc; even a small S can cause this because of the pseudo Jahn-Teller effect between the states e and e' which have an energy difference of $O(t^2/U)$.

4.6.3 Hückel's $(4n + 2)$ Rule for Ring Systems

It is possible to exactly solve similar systems with not too great N. Instead of describing the individual systems, however, we note the following general features.

Let us start from the case of $U = S = 0$. Then we have the one-electron Bloch states

$$\frac{1}{\sqrt{N}} \sum_{l=1}^{N} e^{ikla} a_{l\sigma}^{\dagger} |0\rangle$$

with the pseudo energy band $E(k) = -2t \cos(ka)$, where the wave number k can take the values given by

$$\frac{Nak}{2\pi} = -\frac{N}{2} + 1, -\frac{N}{2} + 2, \ldots, 0, \ldots, +\frac{N}{2}$$

since the N-th site is a nearest neighbor of the first site in this ring system (a is the nearest neighbor distance). Because of the spin degeneracy and of the orbital degeneracy $E(k) = E(-k)$ except for $k = 0$, the highest occupied level is saturated when $N = 4M - 2$ but only doubly occupied when $N = 4M$ (M is a positive integer). In the latter case, we have orbital degeneracy of the electron states, and hence, instability against the alternating mode $q = (Q_1 - Q_2 + \ldots + Q_{N-1} - Q_N)/\sqrt{N}$ even with an infinitesimal S, due to the Jahn-Teller (J–T) theorem. Thus the Peierls instability exists already for a finite length chain of $N = 4M$. If already U has a small but finite value, the orbital degeneracy is removed since the two uppermost electrons will occupy $\cos(kna)$ and $\sin(kna)$ orbitals ($k = \pi/2a$) one by one to avoid the Coulomb repulsion. We then have only the pseudo J–T effect, and a finite S, enough to cope with the Coulomb splitting, is needed for the destabilization to take place. For this reason, the boundary between the distorted and undistorted region on the $B(= 2|t|)$-U-S triangle reaches the B vertex, as shown in the lower-right corner of Fig. 4.46 which is the exact result for the $N = 4$ system [4.161].

In contrast, the $N = 4M - 2$ systems have neither a J–T nor pseudo J–T effect as long as M is finite. The lattice instability occurs only when S is large enough. This is the reason why the distorted region does not reach the B vertex as shown in the lower-left corner of Fig. 4.46 which is a reproduction of Fig. 4.45. The existence of a finite energy gap for electronic excitation makes these systems relatively stable in comparison with $N = 4M$ and $N =$ odd (described below) systems. This stability is well known as Hückel's $4n + 2$ rule in molecular science [4.162]. The stability of the benzene molecule is an example. As M increases, the gap decreases as $O(B/N)$ and the pseudo J–T effect due to neighboring $|k|$'s becomes important, making $N = 4M - 2$ systems indistinguishable from $N = 4M$ systems as it should.

From the foregoing argument on the "closed shell" structure of $N = 4M - 2$ systems, we see immediately that the odd N systems ($N = 4M - 2 \pm 1$) behave similarly to each other, as one-electron or one-hole systems. The unpaired electron (or hole) has an orbital degeneracy (of sine and cosine type), which is no longer removed by introducing U since there is no opponent for mutual repulsion. This means that the region around the B vertex is subject to the J–T

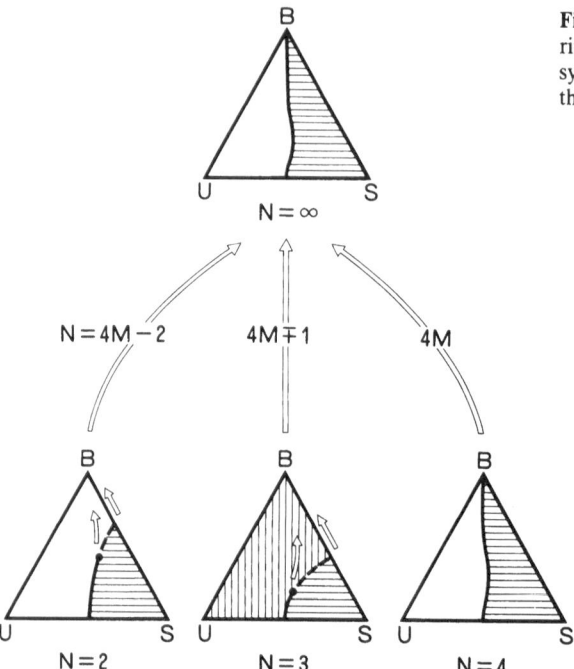

Fig. 4.46. Phase diagrams of ring-shaped N-site N-electron systems, behaving differently for three series of N

distortion. This is borne out by an exact calculation for the $N = 3$ system whose phase diagram [4.163] is shown in the lower center of Fig. 4.46. The lattice instability always takes place except along the U–B edge and the left half of the U–S edge. A bisectric symmetry remains in the distorted three-sites ring in the region of the diagram with vertical hatching, while no symmetry remains in the region with horizontal hatching. In the S-predominant region, two electrons with opposite spins occupy one site (negative U effect), while the third electron is self-trapped on one of the remaining two sites.

Obviously, the phase diagram along the U–S edge with negative U region on its right half is common to any N or even to any dimensionality since there is no interatomic interaction B.

4.6.4 One-Dimensional Hubbard-Peierls System

The foregoing adiabatic argument on the N-site N-electron systems can be extended to the infinitely long chain ($N \to \infty$) on the assumption that the only symmetry breaking distortion realized in the lowest electronic state is the alternating mode

$$q \equiv \sum_{l=1}^{N/2} (Q_{2l-1} - Q_{2l})/\sqrt{N} .$$

The molecular field approximation [4.164, 165] or the RPA, for further consideration of the fluctuation [4.165], have been taken for the electrons. The former exactly bisects the t-U-S triangle, with the first-order transition line, into the spin density wave (SDW) $(U > S)$ and the charge density wave (CDW) $(U < S)$ regions, while the latter has the effect of energetically favoring the SDW state so as to move the border line slightly toward the CDW region for intermediate B values, as shown at the top of Fig. 4.46.

The CDW state with relatively small t can be viewed as a condensate of self-decomposed e–h pairs (see also Fig. 4.43) or equivalently as a condensate of bound (negative U) electron-electron pairs; we have a single band half-filled by electrons, namely, with the same number of electrons and holes. The Wolffram's red salt described in Chap. 9 is a typical example of such a situation.

It is clearly seen from Fig. 4.46 how the different features of the $4M - 2$, $4M \mp 1$, and $4M$ systems approach a common set of features as M increases. An apparent mismatch between the odd number systems and the infinite system is related to the formation of a soliton. We will not give any more detail since it is beyond the scope of this book.

4.6.5 Prospects

In spite of the basic importance of the t-U-S problem in molecular and condensed matter physics, its study is still at a primitive stage. While the Hamiltonian (4.157) represents the simplest conceivable model for the competition of electron-electron and electron-phonon interactions, we have to extend this in various respects in order to describe realistic situations. In the first place, the site-off-diagonal electron-phonon interaction should be considered which plays an even more important role than the site-diagonal one in systems such as polyacetylenes as well as for two-center type self-trapping. Secondly, the intersite electron-electron repulsive potential should be considered in order to reasonably describe the exciton states appearing in the Peierls-Hubbard gap. Inclusion of transfer energy $t_{ll'}$ to more distant sites will give band structures with various possibilities of nesting. Different numbers of sites and electrons should be considered as is realized in systems consisting of donor and acceptor chains, e.g., TTF-TCNQ. Which of the electron-phonon and electron-electron interactions is more dominant has been a matter of much controversy particularly in one-dimensional systems which are most sensitive to these interactions. The instability of the metallic phase against SDW or CDW formation for infinitesimally small U or S is characteristic of one-dimension, while in a three-dimensional system there will be a stable metallic phase in a certain region surrounding the B vertex.

Of course, the individual problems mentioned above have already been considered by a great number of authors in a variety of contexts, and a qualitative argument on the phase diagram in a triangular coordinate space essentially equivalent to ours was presented by *Chakraverty* [4.166]. In our context of excitonic instability in a deformable lattice, we would like to emphasize the importance of the study of excited electronic states of the t-U-S systems and their lattice relaxations, which will manifest themselves in optical absorption and emission processes as well as in resonant secondary radiation processes (Chap. 9). *Global study of the ground and excited state manifold in the multi-dimensional* C.C. *space will provide deeper understanding of the possibilities and impossibilities of various types of electronic and structural states.* In this context, laser-induced phase transitions of a nonthermal type will be one of the exciting topics in the future, although the available data on such observations are ascribed to the mere heating of the lattice following electronic de-excitation.

Finally we would like to emphasize that a thorough study of the t-U-S problem and its extensions will be useful in designing new materials with desired properties and functions, including molecules of various sizes and condensed matter of different dimensionalities. A new age of alchemy based on physical science has indeed started.

5. Excitons in Condensed Rare Gases

Interest in the study of condensed rare gases (He, Ne, Ar, Kr, and Xe) is centered on the physics of the simplest type of solid and liquid on which thorough investigations can be made both experimentally and theoretically. Furthermore, there are specific features in the properties of condensed rare gases. First, the rare-gas atoms have closed electron shells and do not form stable diatomic molecules, so consequently only a small cohesive energy due to the van der Waals' force is available in forming the condensed phase. As shown in Table 5.1, the condensed rare gases can exist only at low temperatures. Secondly, the rare-gas atoms have the largest value of the first ionization energy in the corresponding row of the periodic table. As a result, the energy required for electronic excitation of condensed rare gases is the largest among the condensed systems. These features of *the smallest cohesion as the condensed phase and the largest excitation energy for the electronic system* make the condensed rare gases interesting materials having extreme properties.

However, the above features present several difficulties in experimental studies, especially in the spectroscopic study of electronic excited states in the condensed rare gases. Accordingly, the experimental studies have been delayed in comparison with those on similar insulators such as alkali halides. For example, spectroscopic studies on solid rare gases started in the sixties with the pioneering work by *Baldini* [5.1]. His work, however, was restricted to the valence excitation of Ar and Xe solids with a gas discharge lamp as the light source.

Table 5.1. Melting and critical points of rare gases

	Melting point[a]		Critical point	
	Temperature [K]	Pressure	Temperature [K]	Pressure [atm]
He^3	~ 0.3[b]	29.3 atm	3.31	1.13
He^4	≤ 1[b]	25 atm	5.19	2.25
Ne	24.55	325.2 Torr	44.4	26.2
Ar	83.81	517.15 Torr	150.7	47.9
Kr	115.76	547.5 Torr	209.5	54.5
Xe	161.39	612.2 Torr	289.7	57.7

[a] Except He, the values at the triple point are shown.
[b] Temperature at the lowest pressure of solidification.

An experimental breakthrough occurred in the seventies with the use of synchrotron radiation, the clean and tunable light source unique at higher photon energies above the vuv region. The group at the synchrotron radiation laboratory at
DESY (now HASYLAB) not only extended the study to Ne and Ar solids under
both valence and core-level excitations, but succeeded in reaching higher levels
of sophistication such as photoelectron spectroscopy and high resolution studies
of exciton spectra. These developments are reviewed in several articles, such as
those on the experimental aspect by *Sonntag* [5.2] and the theoretical aspect by
Rössler [5.3].

On the basis of these spectroscopic studies which identified many of the
optical transitions in solid rare gases, recent interest has been directed toward the
elucidation of the relaxation processes, both radiative and nonradiative, from the
electronic excited states. The main emphasis in the present chapter will be on the
current understanding of the nature of the relaxed excited states and the
dynamics of the relaxation processes in condensed rare gases. The recent review
by *Schwentner* et al. [5.4] tends in a similar direction and is concerned with the
relaxation processes in rare gas alloys and metal rare gas mixtures as well as those
in pure rare gases. In this chapter, we concentrate on the microscopic nature of
the condensed phase of pure rare gases and our discussion will be mainly based
on recent developments which have revealed unique features of the relaxed
excited states and especially their relaxation processes.

5.1 Electronic Structure of Condensed Rare Gases

We start from the basic electronic structure of the condensed rare gases. Figure
5.1 shows a schematic energy band structure expected for rare gas solids. The
valence bands are derived from the outermost np electrons of rare gas atoms, and
the lowest conduction band from the excited $(n + 1)s$ electrons. The energy
dispersion in Fig. 5.1 is typical of the bands derived from p and s electrons. Table
5.2 summarizes the energy band parameters obtained experimentally. In the
table, ε_g is obtained as the series limit of the exciton absorption bands
(Sect. 5.3.1), and λ corresponds to the spin-orbit splitting observed in the exciton
spectra. All other parameters E_V, E_A $(= E_V - \varepsilon_g)$, and W_{VB} are obtained from
the energy distribution curves (EDC) of photoelectrons as a function of excitation energy [5.7]. Table 5.2 also includes the results of typical band calculations;
the nonrelativistic calculation by *Kunz* and *Mickish* [5.6] and the relativistic
calculation by *Rössler* [5.8]. Although the systematic trend of energy parameters
such as ε_g and W_{VB} among different rare gases is reproduced in the band calculations, quantitative disagreements are evident between theory and experiment.
The only exceptions are the values of λ which show good agreement between
experiment, theory, and atomic data.

In Table 5.2 the disagreements for E_A between theory and experiment are
especially distinct. Experimental values of the electron affinity E_A are negative in

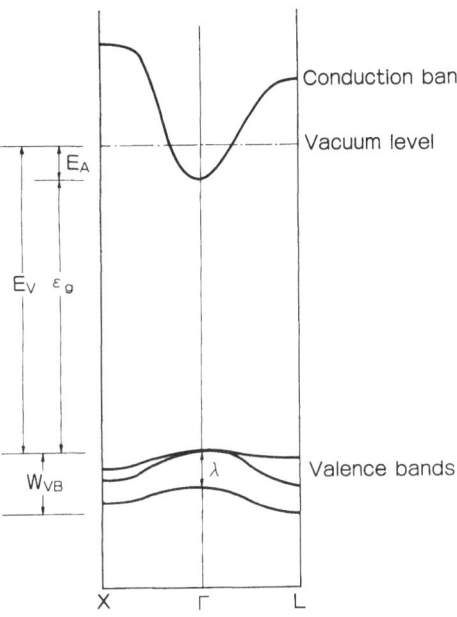

Fig. 5.1. Schematic energy band diagram for a rare gas solid. Only the lowest conduction band and the valence bands are shown. Here ε_g is the band gap energy, E_V the lowest excitation energy to the vacuum level, E_A the electron affinity $(= E_V - \varepsilon_g)$, W_{VB} the total width of the valence bands and λ the spin-orbit splitting

Table 5.2. Band structure parameters of rare gas solids. Experimental values at liquid helium temperature (LHeT) are compared with theoretical values (theo.) and atomic data (atom.). All energies are in eV

	Ne	Ar	Kr	Xe
ε_g [a]	21.50	14.16	11.62	9.32
(theo.)[b]	22.16	15.22	13.57	–
E_V [c]	20.3	13.8	11.9	9.8
(theo.)[b]	21.8	14.1	12.3	–
E_A [c]	−1.3	−0.4	0.3	0.5
(theo.)[b]	−0.36	−1.12	−1.27	–
λ [a]	0.09	0.2	0.64	1.3
(theo.)[d]	–	0.18	0.70	1.37
(atom.)	0.10	0.18	0.67	1.31
W_{VB} [c]	1.3	1.7	2.3	3.0
(theo.)[b]	0.4	1.3	1.6	–
(theo.)[d]	0.4	1.0	1.6	1.8

[a] [5.2, 5], [b] [5.6], [c] [5.7], [d] [5.8].

Ne and Ar, and positive in Kr and Xe while band calculations predict negative electron affinity for Ne, Ar, and Kr. The origin of the discrepancy is apparently the failure of band calculations for the conduction-band width, especially for Ar and Kr solids. On the other hand, the disagreement in the valence-band width W_{VB} is rather unexpected, in view of the successful description of valence band structure in semiconductors and metals [5.9] from both photoelectron spectroscopy and band calculations. Furthermore, one would expect band calculations in rare gas solids to be more successful for the narrow valence bands compared to the wider conduction bands. However, we need more detailed experimental information on the valence band structures. Photoemission experiments on single crystals, such as those on epitaxially grown films of Xe [5.10], will give more convincing evidence of stronger dispersion of valence bands compared to band calculations, as suggested in Table 5.2.

5.2 Charge Carriers in Condensed Rare Gases

Experimental information on the electronic structure of a condensed system can be obtained from transport studies on charge carriers, as will be shown in Chap. 8 for ionic crystals. Transport properties of charge carriers in condensed rare gases have been studied fairly extensively since the 1950s [5.11]. By using various excitation techniques, such as high-energy excitation by electron or α particle bombardment and carrier injection in a strong electric field, the drift mobilities of carriers have been measured for both liquid and solid phases. These experiments presented a unique chance of studying the electronic structure of liquids, because spectroscopic studies on liquids are severely limited by experimental difficulties due to the high vapor pressure of the liquid rare gases. It turned out that free electrons are usually stable carriers in the liquids as well as in the solids, but holes are stabilized as *self-trapped holes*. The properties of these two types of carriers, free and self-trapped, in a series of condensed rare gases will be our main concern in this section.

Table 5.3 summarizes the drift mobility data on condensed rare gases. The mobility values can be grouped into two categories; one with high mobilities of 10^2–10^3 cm^{2}V^{-1}s^{-1} and the other with low mobilities of 10^{-2}–10^{-3} cm^{2}V^{-1}s^{-1}. On the basis of the general theory for drift mobility in condensed systems [5.17], we can estimate that the border line between free carriers and hopping-type carriers corresponds to mobility of 1–10 cm^{2}V^{-1}s^{-1}. The negative carriers having high mobilities in Table 5.3 correspond to free electrons in the conduction band, while the low mobility carriers represent the hopping-type motion of self-trapped holes.

In addition to the high mobility values, the negative charge carriers in solid Ar, Kr, and Xe show a temperature dependence of mobility typical for the acoustic-phonon scattering of free carriers,

$$\mu \propto (\beta/\varrho^2) T^{-3/2} \qquad (5.1)$$

Table 5.3. Summary of the drift mobilities [cm^2/V sec] of carriers with $-e$ and $+e$ charges in condensed rare gases (S: solid, L: liquid). Temperature dependence of the thermal activation type is indicated by $*$. Except for He^4, the data at the triple point are shown for comparison between the solid and liquid

	Temperature [K]	μ_S^-	μ_L^-	μ_S^+	μ_L^+
He^4	4.2	$\sim 10^{-4*}$ [a]	2×10^{-2} [b]	$\sim 10^{-4*}$ [b]	5×10^{-2} [b]
Ne	25	600 [c]	1.6×10^{-3} [c]	$1.05 \times 10^{-2*}$ [c]	$1.6 \times 10^{-3*}$ [c]
Ar	84	1000 [d]	475 [d]	$2.3 \times 10^{-2*}$ [e]	$-$
Kr	116	3600 [d]	1800 [d]	$4 \times 10^{-2*}$ [e]	$-$
Xe	161	4000 [d]	1900 [d]	1.7×10^{-2} [e]	$-$

[a] [5.12], [b] [5.13], [c] [5.14], [d] [5.15], [e] [5.16].

where β is the bulk modulus and ϱ the density. The ratio of μ_S^-/μ_L^- in Table 5.3 is very close to the experimental value of $(\beta_S/\beta_L)(\varrho_L/\varrho_S)^2$ [5.15]. There seems to be no doubt about the free-electron conduction in these cases. The conclusion is in accord with the theoretical expectation [5.17] that the conduction band originating from the s-type atomic orbitals can describe the charge carriers even in an amorphous system such as a liquid.

Compared with Ar, Kr, and Xe, the negative charge carriers in Ne and He behave differently, as shown in Table 5.3. The low values of μ^- in liquid Ne and He^4 and solid He^4 have been ascribed to the formation of *electron bubbles* [5.18]. The electron bubble is an electron localized in a microscopic cavity, and corresponds to a special type of self-trapped electron in a condensed system having a large negative value of electron affinity. As shown in Table 5.2, the electron affinity of solid Ne is -1.3 eV and the measured value for liquid He^4 [5.19] is -1.05 eV. In the case of liquid He^4, extensive studies on carrier mobilities have been made for superliquid He II [5.13], in order to elucidate the nature of elementary excitations in the quantum liquid. The magnitude of μ^- in liquid He II as well as its temperature dependence has been successfully analyzed by the electron-bubble model [5.20]. Furthermore, from an ingenious electric resonance experiment [5.21] on the carriers trapped below the liquid surface, the effective mass of negative carriers is obtained as

$$m_-^* = (76 \pm 2)m_{He^4} \tag{5.2}$$

which corresponds to a 12 Å radius of the electron bubble cavity, which agrees with predictions from mobility analysis. Thus the existence of electron bubbles in liquid He II is clearly confirmed experimentally, but there is another interesting aspect of the electronic structure of the bubble which is not yet fully clarified experimentally. The electron bubble is an electron localized in a square well potential and the energy levels of the localized electron are predicted by *Fowler* and *Dexter* [5.22] for liquid He. Future spectroscopic studies of optical transitions in the system will be interesting. As for the negative carriers in liquid Ne, the evidence of the electron bubble is still insufficient. Further investigations on liquid Ne, such as on the possible trapping of negative carriers below the liquid surface, will be interesting.

The low mobility values of the positive carriers shown in Table 5.3 suggest that all of the positive carriers in condensed rare gases are self-trapped. In the case of liquid He II, the positive carriers have been assigned as holes trapped at a cluster of He atoms called a *snow ball* [5.23]. The effective mass is determined [5.21] as

$$m_+^* = (45 \pm 2)m_{\mathrm{He^4}} \tag{5.3}$$

corresponding to a positive hole trapped in a solid He particle with radius 6 Å. The mobilities and their temperature dependence are well explained by the snow ball model [5.21]. In the other rare gases, all of which have a valence band of p electron character, the self-trapped holes are believed to be the R_2^+ ion (R: Ne, Ar, Kr, and Xe), which is similar to the V_k center in alkali halides. Although stable R_2^+ ions in rare gas solids are predicted theoretically [5.24] and their existence is confirmed experimentally in the gaseous phase [5.25], more direct evidence such as in electron spin resonance experiments is still lacking in rare gas solids. This situation contrasts with that of alkali halides in which the V_k centers were originally discovered by spin resonance experiments. There seems to be an uncertainty about the R_2^+ type of self-trapped holes in all of the rare gas solids. For example, the temperature dependence data of mobilities μ^+ in the solid rare gases have been analyzed in terms of the small-polaron hopping motion of R_2^+ in [5.16], however, it is suggested in [5.26] that the μ^+ in solid Xe can be reasonably well explained by acoustic-phonon scattering of free positive holes subject to polaron mass enhancement.

Summarizing, there are various pieces of experimental evidence which indicate that charge carriers in condensed rare gases interact strongly with surrounding *lattices*, and become self-trapped in extreme cases such as liquid He II. However, most of the information is from mobility data in a limited temperature range. Mobility data for a wider range of temperatures as well as more direct information on self-trapped carriers are highly desirable.

5.3 Excitons and Exciton-Phonon Interactions in Condensed Rare Gases

Following discussion of the basic electronic structure and the charge carrier, we proceed to the microscopic nature of the excited electronic states and their de-excitation processes. Experimental information on the excited states has been obtained primarily by fundamental absorption spectra for optically allowed excitations from the ground state of the system. The de-excitation processes, on the other hand, contain various decay channels which can be radiative or nonradiative; mainly the radiative decay processes have been studied. In this section we start from the present understanding of absorption and luminescence spectra. The absorption spectra correspond to optical excitation of the "free" excited states in the ground-state configuration, and mark the starting point of de-excitation processes which follow. The luminescence spectra reveal the nature of the radiative decay processes and give information on the "relaxed" excitons which are the initial states of the radiative transition.

Typical absorption and luminescence spectra are summarized in Figs. 5.2, 3, for liquid He and solid Ne, Ar, Kr, and Xe. Absorption spectra of the solid rare gases are those obtained using synchrotron radiation at liquid helium temperature (LHeT) in [5.5, 31, 32]. Luminescence spectra of the solid rare gases are those obtained under electron bombardment at LHeT [5.30]. By comparing the absorption with the luminescence spectra in Figs. 5.2 and 5.3, it is evident that the relaxed excitons, in most cases, are characterized by the Stokes shift relative to the exciton absorption and accompanied by lattice relaxation. This is one of the essential features of excitation and de-excitation processes in the condensed rare gases and we are concerned mainly with this aspect in the following sections. The nature of the free excitons and the relaxed excitons will be discussed in

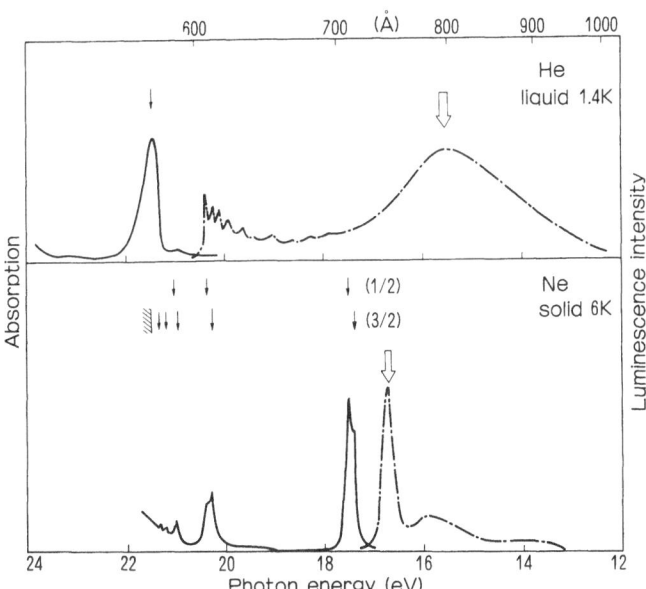

Fig. 5.2. Absorption (——) and luminescence (– · – · –) spectra for liquid He and solid Ne. Thin arrows indicate the optical transitions discussed in the text [$\Gamma(3/2)$ and $\Gamma(1/2)$ excitons for Ne]. Thick arrows correspond to the maxima of luminescence intensity. Absorption data: He [5.27, 28], Ne [5.5]. Luminescence data: He [5.29], Ne [5.30]

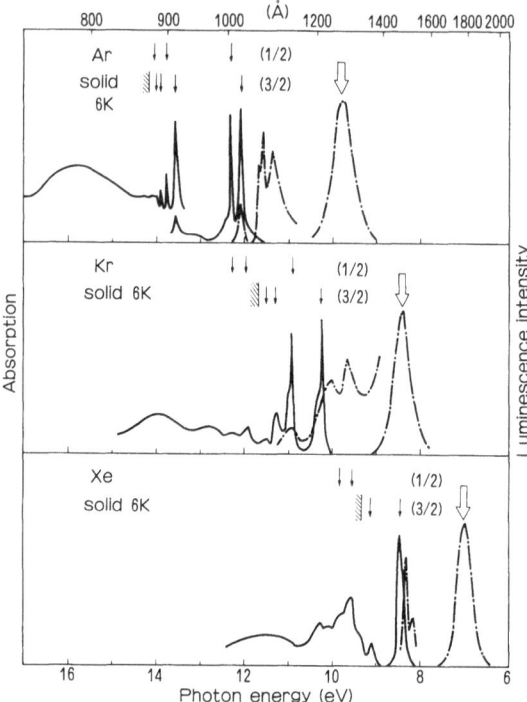

Fig. 5.3. Absorption (——) and luminescence (–·–·–) spectra for solid Ar, Kr, and Xe. Thin arrows indicate the optical transitions [$\Gamma(3/2)$ and $\Gamma(1/2)$ excitons] discussed in the text. Thick arrows correspond to the maxima of luminiescence intensity. Absorption data: Ar [5.31], Kr [5.32], Xe [5.32]. Luminescence data: [5.30]

Sects. 5.3.1 and 5.3.2, respectively. The dynamical aspect of the lattice relaxation process and the de-excitation mechanism from the photoexcited states will be treated in Sects. 5.3.3 and 5.3.4, respectively.

5.3.1 Exciton Absorption Spectra

The fundamental absorption edge of rare gas solids is characterized by a series of absorption bands due to free direct-exciton transitions. Above the exciton series limit, the broad spectrum due to band-to-band transition follows. As far as the fundamental absorption from the valence band is concerned, the exciton series is the best understood part of the spectrum [5.2]. Fig. 5.4 indicates the transition energies of the exciton series in solids against $1/n^2$ according to the Wannier exciton model. As has been well established, the agreement with the $1/n^2$ rule is satisfactory for excitons in a rare gas solid except with the *central cell correction* for the lowest exciton $n = 1$. The exciton parameters obtained on the basis of a fairly good fit with the Wannier model are summarized in Table 5.4. The necessity of the central cell correction may be obvious because the exciton wave function for $n = 1$ is almost completely localized at the rare gas atom as indicated in Fig. 5.4. At the time of its discovery, the existence of Wannier excitons was a surprise for such wide band-gap materials as rare gas solids in which the Frenkel

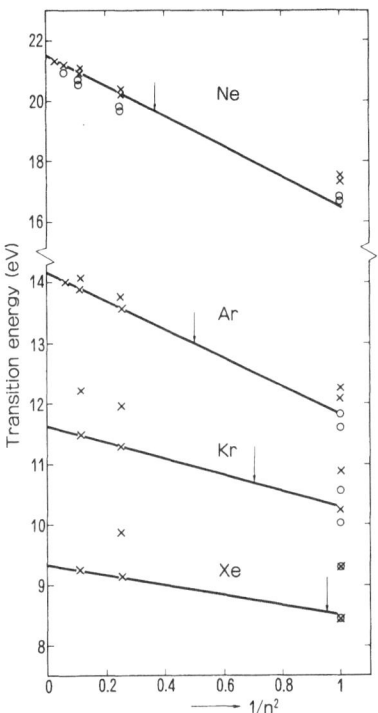

Fig. 5.4. Exciton transition energies at LHeT in rare gas solids ($\times$) vs. $1/n^2$, according to the Wannier model. Both $\Gamma(3/2)$ and $\Gamma(1/2)$ excitons are indicated. Arrows indicate the border lines at which the radius of the Wannier exciton is equal to the nearest neighbor distance in the solid. Corresponding atomic transition energies ($\bigcirc$) are plotted for comparison. Ne [5.5], Ar [5.31], Kr [5.32] and Xe [5.32]

exciton picture was supposed to be more appropriate. However, it is now reasonably agreed that excitons in rare gas solids can be described as an intermediate betwen the Frenkel and Wannier type for an $n = 1$ exciton, and a Wannier type for $n \geq 2$ excitons [5.3].

The magnitude of the central cell correction for an $n = 1$ exciton is characterized by the following parameter

$$\Delta = \frac{E^b_{1s} - E^b_{ex}}{E^b_{ex}} \tag{5.4}$$

which is called the relative hydrogenic defect [5.33]. The absolute value of Δ is much less than unity for excitons in a rare gas solid as shown in Table 5.4. As

Table 5.4. Exciton parameters for $\Gamma(3/2)$ excitons in rare gas solids at LHeT

	Ne	Ar	Kr	Xe
E^b_{ex} [eV]	5.00	2.32	1.4	0.8
Δ in (5.4)	−0.2	−0.1	−0.03	+0.1
ϵ	1.24	1.66	1.86	2.18
m_{ex}/m_e	0.565	0.47	0.356	0.28

pointed out by *Hermanson* and *Phillips* [5.33], the magnitude of the hydrogenic defect is not necessarily correlated with the extension of the exciton wave function nor with the magnitude of the exciton binding energy, but should be correlated with the microscopic nature of the exciton wave function itself. Thus, contrary to the usual criterion for the validity of the effective mass approximation, there should be no surprise in the good agreement of exciton energies in rare gas solids with predictions from the Wannier model and also with atomic transition energies. As indicated in Fig. 5.4, the exciton transition energies are fairly close to the atomic transition energies $np \rightarrow (n + 1)s$ for $n = 1$ excitons in all rare gas solids. Furthermore, in solid Ne, the whole exciton series is identified as very close to the atomic transition energies for $n \geq 2$ also. Thus, at least in the case of solid Ne, there is experimental evidence indicating that the excitons correspond to $2p^6 \rightarrow 2p^5 ns$ $(n \geq 3)$ atomic transitions.

In this context, the evolution of excitons in a less dense system such as a liquid or dense gas is relevant to the nature of excitons in a solid. Due to the experimental difficulties of high vapor pressures, only a few spectroscopic studies have been reported for liquid rare gases. The first experiment was on the reflectance spectra of liquid Xe by *Beaglehole* [5.34] who intended to clarify the correlation between exciton transitions and the band structure, which was a controversy at the time. It turned out that exciton absorption bands in the solid persisted in the liquid accompanied by a shift to lower energy and a broadening with increasing temperature. Further extensive studies on liquid Xe in comparison with the solid have been carried out by *Steinberger* et al. [5.35–37]. The energy shift of exciton spectra in Xe as a function of temperature up to the triple point has been shown to be exactly proportional to the density of the condensed Xe including the liquid. The results are shown in Table 5.5 and should be compared with the corresponding values at LHeT in Tables 5.2 and 5.4. These results can be reasonably understood in terms of the change of electronic band structure with lattice parameter as expressed by the deformation potentials [5.38].

The spectroscopic studies are further extended to liquid Xe with densities ranging from below the triple point value to near the critical point [5.37]. The evolution of the $n = 1$ Γ (3/2) exciton takes place above a number density of 4.9×10^{21} atoms cm^{-3}, and it is estimated that the $n = 1$ exciton appears in a cluster containing at least 10 atoms, corresponding to the cluster radius of 7.1 Å. The estimated values should be compared to the Bohr radius of 3.52 Å [5.36] for

Table 5.5. Exciton parameters of solid and liquid Xe at the triple point [5.35]. The data are to be compared with those at LHeT (Tables 5.2 and 5.4) at the number density of 17.3×10^{21} atoms cm^{-3}

	Temperature [K]	Number density [cm^{-3}]	ε_g [eV]	E_{ex}^b [eV]	ϵ	m_{ex}/m_e
Solid Xe	161.2	15.6×10^{21}	9.272	1.063	2.00	0.31
Liquid Xe	163	14.1×10^{21}	9.22	1.084	1.85	0.27

the $n = 1 \, \Gamma \, (3/2)$ exciton in liquid Xe. The exciton in such a case is certainly a tightly bound state well-confined in the threshold cluster. On the other hand, quasi-free electron conduction is observed in a much less dense liquid with a density of 3×10^{20} atoms cm^{-3} [5.39]. It appears that the conduction electrons with large de Broglie wavelength are not confined in a threshold cluster such as required for exciton formation.

Apart from liquid Xe, only a study on liquid He II has been reported. The spectroscopic study on liquid He was started in order to elucidate the *neutral excitation* in liquid He II which is one of the elementary excitations in the superfluid [5.40]. As far as fundamental absorption spectra are concerned, only the reflectivity data at 1.4 K are available [5.27, 28]. Two maxima are observed in the spectrum and have been assigned as $1s \rightarrow 2p$ for the strong maximum at 21.6 eV and as the $1s \rightarrow 2s$ forbidden transition for the weaker structure at 20.8 eV. The corresponding transition energies in the He atom are 21.2 and 20.6 eV. Due to the vapor pressure difficulty of liquid He, the experimental results are not sufficient to give any detailed information about the exciton states. Further studies will be required at lower temperatures using synchrotron radiation.

This section shows that the free exciton spectra in rare gas solids are fairly well understood. However, there remain several unresolved aspects which may be worth mentioning to stimulate further investigations. Experimentally, most of the studies have been made on evaporated films which are not well characterized and some of the spectroscopic data are subject to ambiguities depending on the sample preparation procedures. Above all, the absolute magnitude of absorption strength remains uncertain. Future studies on well-characterized samples, preferably on good single crystals, will give more quantitative information about the excitons, in areas such as the spin-orbit *vs.* exchange interactions discussed by *Onodera* and *Toyozawa* [5.41] (Sect. 1.4). Another example is the study of absorption spectra for thicker samples at the lower-energy side of the exciton threshold as a function of temperature, which will clear up details of the exciton band shape associated with the exciton-phonon interaction during optical excitation.

5.3.2 Nature of Relaxed Excitons in Condensed Rare Gases

The first identification of the luminescence centers was made for liquid and solid Ar, Kr, and Xe bombarded by α particles [5.42]. The close similarity of the luminescence spectra for condensed and gaseous phases led to the conclusion that the luminescence is due to the radiative decay of the *excimer* which corresponds to the excited state of a diatomic rare gas molecule. Extensive studies have been made of the luminescence spectra of dense gases (at pressures of the order of 100 Torr), with the purpose of developing continuum light sources [5.43] or to elucidate the mechanism of excimer lasers [5.44].

Luminescence spectra of dense gases are very similar among the various rare gases and consist of two bands, called the first and the second continua [5.45].

The second continuum is more intense and located at a lower photon energy. From the comparison with absorption spectra of diatomic molecules, the origin of the second continuum has been assigned as the transition from the lowest vibrational levels, $^1\Sigma_u^+$ and/or $^3\Sigma_u^+$, of excited molecules to the ground state $^1\Sigma_g^+$ of dissociative character. The spectra of the most intense components (indicated by thick arrows in Figs. 5.2 and 5.3) in condensed He, Ar, Kr, and Xe agree very well with those of the second continuum in the gaseous phase. The model proposed for the luminescence center in these cases corresponds to the *self-trapped* exciton (STE) in the form of the excited state of a rare gas molecule; it will be called m-STE (molecular-type STE). The electronic structure of m-STE is very similar to the self-trapped exciton in alkali halides [5.46], however, there is a distinct difference between their luminescence spectra. In alkali halides, the spectra are usually composed of two bands; one is singlet and the other is triplet in nature [5.47]. In rare gas solids, the singlet-triplet energy separation is very small, 30 to 50 meV [5.30], in contrast to the values of 1 to 2 eV in alkali halides.

Contrary to the other rare gases, the luminescence spectra of solid Ne are found to be exceptional [5.48]. The position of the most intense band in Fig. 5.2 is very close to the atomic lines and the fine structure in the spectra resolved by *Fugol'* et al. [5.49] agree more or less exactly with atomic lines of $^1P_1$, $^3P_1$, and $^3P_2 \rightarrow {}^1S_0$ in the free atom. The origin of the luminescence component has been considered as the STE in the form of an excited Ne atom; it will be called a-STE (atomic-type STE). The less intense and broader two luminescence bands observed at lower energies (Fig. 5.2) have been considered as due to the m-STE in solid Ne. This suggests that the coexistence of a- and m-STEs is realized in solid Ne. The detailed nature of the weaker bands, however, is not well understood. The 14 eV band is observed also in gaseous Ne [5.48] and is located near the second continuum of the gas spectra. This may suggest that the band is due to the vibrationally relaxed m-STE [5.50]. However, the thermalization time of the diatomic molecule, especially in the condensed lighter gases such as solid Ne, can be longer than the radiative lifetime of the excited state [5.51]. Under these circumstances, luminescence of the vibrationally unrelaxed m-STE is more probable, as concluded theoretically for solid Ne [5.52]. On this basis, we tentatively assume that the origin of both the two weaker bands is the decay of the unrelaxed m-STE. In liquid He, on the other hand, the evidence of unrelaxed m-STE decay is very convincing for the higher-energy band (19–20.5 eV) which is accompanied by vibrational structures. The corresponding transition in liquid He has been identified as the decay of the $^1\Sigma_u^+$ m-STE at the quasi-bound vibrational levels of $v = 16$–17 [5.29], in contrast to the more intense and broader band with a 15.5 eV peak due to the vibrationally relaxed m-STE.

In addition to the dominant luminescence bands due to a- and m-STEs described above, the free-exciton luminescence resonant with exciton absorption is observed for solid Xe, Kr, and Ar, as shown in Fig. 5.3. As pointed out by *Fugol'* [5.50] and *Zimmerer* [5.30], the intensity of the free-exciton luminescence decreases in the sequence Xe, Kr, and Ar, and is not observable in Ne. According to the theoretical calculation of the tunneling probability of the free exciton to

STE [5.53], it is expected that the tunneling rate increases sensitively with a decrease of exciton band width and results in the reduction of free-exciton luminescence. In view of the decrease of valence band width with a decrease in atomic number (Table 5.2), the observed tendency among rare gas solids is well explained by the increase of the tunneling probability in the sequence Xe, Kr, Ar, and Ne. The situation is very similar to the tendency observed among alkali iodides which has been analyzed successfully in [5.53].

In addition to the luminescence components described above, several weaker components are observed in solid Ar, Kr and Xe. The transition energies of the luminescence bands and their assignments are summarized in Table 5.6. It may be mentioned that the data of these weaker bands, especially in solid Ar and Kr, have been found to be sensitive to the method of preparation of the evaporated films and should be treated with caution. Here again, further studies are required using better-characterized samples.

Up to this point, the study of relaxed excitons has been based on the luminescence spectra. As a second type of investigation, the transient absorption spectroscopy of the relaxed excitons has presented new detailed information on condensed rare gases [5.54], as well as on alkali halides [5.55] and liquid and solid He [5.56]. Transient absorption is concerned with optical transitions from the relaxed excitons to their higher excited states and gives information complementary to the luminescence which corresponds to transitions between the relaxed excitons and the ground state of the system. In the transient absorption study of rare gases by *Suemoto* and *Kanzaki* [5.57], solid samples grown from the melt were studied, in contrast to evaporated films used for most of the other optical studies. Liquid and gaseous samples were also studied using the same experimental apparatus.

Figure 5.5 shows the spectra of solid and liquid Ar excited by an electron pulse from Febetron 706 (maximum electron energy 500 keV and pulse duration 5 ns) [5.57]. The decay time of the transient absorption is found to be exactly the same as that of the triplet component of the m-STE luminescence. This indicates that the initial state of absorption is the same as that of the triplet m-STE

Table 5.6. Summary of the transition energies of luminescence peaks in condensed rare gases shown in Figs. 5.2 and 5.3 and their assignments. The most intense bands indicated by thick arrows in the figures are here indicated by *

	Free exciton [eV]	a-STE [eV]	m-STE (unrelaxed) [eV]	m-STE (relaxed) [eV]
liquid He	–	–	19.4	15.5*
solid Ne	–	16.7*	16.5	–
			14.0	
solid Ar	12.1	11.6	11.37	9.8*
solid Kr	10.1	9.7	~9	8.4*
solid Xe	8.3	–	–	7.1*

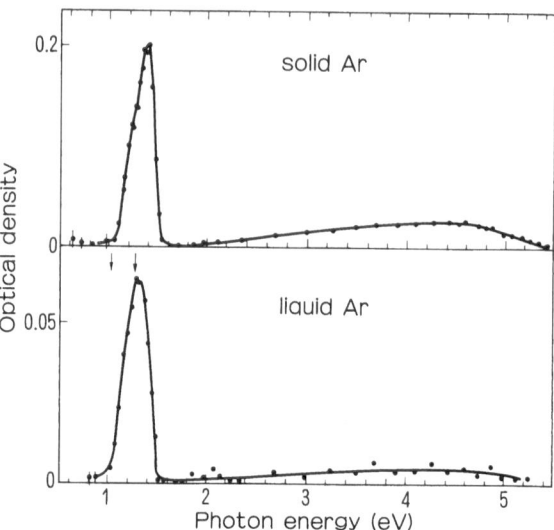

Fig. 5.5. Transient absorption spectra in solid (80 ± 3 K) and liquid (93 ± 3 K) Ar measured 1 μs after electron-pulse excitation [5.57]. The absorption decays with lifetimes of 1.41 μs and 1.11 μs in solid and liquid, respectively. The arrows indicate the position of the absorption peaks of the m-STE in the gaseous phase [5.58]

luminescence. The spectra in Fig. 5.5 consist of two components; one is the sharp band near 1 eV and the other the broad band around 4 eV. The nature of the optical absorption can be understood in terms of the potential energy diagram for the excited Ar_2 molecule. The 1 eV band is a doublet band of $^3\Sigma_u \to {}^3\Sigma_g$ and $\to {}^3\Pi_g$, corresponding to the Rydberg transition of the outer electrons. The 4 eV band is from $^3\Sigma_u$ to $^3\Sigma_g'$, corresponding to the excitation of holes in the Ar_2^+ ion core. The dissociative nature of the final state $^3\Sigma_g'$ is responsible for the broadness of the 4 eV band. The similarity of the absorption spectra of the solid, liquid, and gas, as well as the assignment based on the molecular energy levels establishes the existence of the m-STE in condensed Ne, Ar, and Kr [5.57]. However, there is a small but definite energy shift between the gas, liquid, and solid. The energy shift from gas to liquid or solid is a *blue shift* (toward higher energy) in Ne and Ar, and a *red shift* (toward lower energy) in Kr. The tendency can be explained by the systematic change of electron affinity in rare gases. As shown in Table 5.2, the electron affinity is negative in Ne and Ar, and positive in Kr. The final state of the absorption transition seems to be pushed up or down in energy, according as the electron affinity is negative or positive.

The transient absorption of condense Ne shows another feature different from Ar and Kr. In Ne, there are two kinds of absorption bands with very different decay times. One is located at 1.8 eV due to the m-STE (lifetime $\tau = 3.9$ μs), and the other is at the higher energy side due to the a-STE ($\tau = 560$ μs). The latter spectrum shows a series of sharp absorption bands as shown in Fig. 5.6 which contrasts with the single absorption band due to the $^3\Sigma_u \to {}^3\Pi_g$ transition of the m-STE. The series of bands at higher energy can be understood as transitions from the lowest 3s ($3s_{12}$) level to several 3p multiplets in the excited Ne atom. As also shown in Fig. 5.6, we can successfully fit the solid

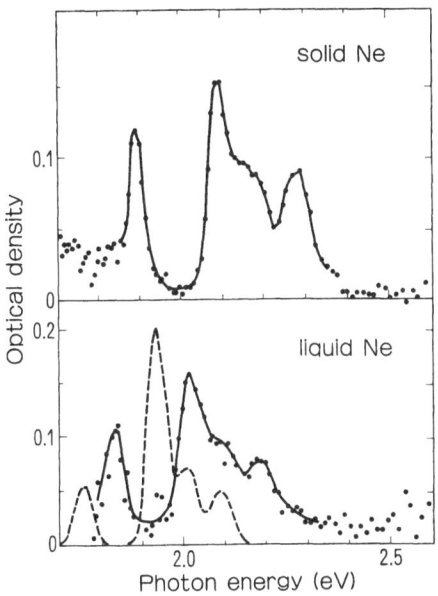

Fig. 5.6. Transient absorption spectra of a-STE in solid (50–150 μs after excitation) and liquid (0–10 μs after excitation) Ne, near the triple point [5.57]. The dashed curve is the calculated spectrum of the $3s_{12}(^3P_2)$ state of the Ne atom, with broadening assumed. The absorption in the solid decays with a lifetime of 560 μs, but that in the liquid decays nonexponentially within several microseconds

and liquid spectra by convoluting the atomic lines assuming a Gaussian broadening width of 0.05 eV and adding a parallel blue shift. These studies established the coexistence of a- and m-STEs in condensed Ne which was suggested by the luminescence spectra. In the calculated potential curve of Ne$_2$ molecule, there

Table 5.7. Parameters for the transient absorption bands of STEs in condensed rare gases [5.57]

	Energy [eV]			Assignment	Lifetime [μsec]	
	solid	liquid	gas		solid	liquid
Ne	2.29	2.18	(2.093)	$3s_{12} \to 3p'_{01},$ $3p'_{12}, 3p'_{11}$		
	2.19	2.11	(2.013)	$3s_{12} \to 3p_{12},$ $3p_{11}$	560	–
	2.10	2.02	(1.943)	$3s_{12} \to 3p_{22},$ $3p_{23}$		
	1.90	1.84	1.763	$3s_{12} \to 3p_{01}$		
	1.76	1.67	1.52–1.54	$^3\Sigma_u \to {}^3\Pi_g$	3.9	2.9
Ar	1.38	1.30	1.25–1.26	$^3\Sigma_u \to {}^3\Pi_g$	1.41 ± 0.05	1.11 ± 0.05
	1.23	1.13	0.95–1.09	$^3\Sigma_u \to {}^3\Sigma_g$		
Kr	1.6					
	1.21	1.18	1.26–1.28	$^3\Sigma_u \to {}^3\Pi_g$	0.09 ± 0.005	0.11 ± 0.005
	1.05					
	(0.93)	(0.9–1.0)	0.90–1.07	$^3\Sigma_u \to {}^3\Sigma_g$		

exists a small energy barrier between $^3\Sigma_u$ and the separated atomic levels [5.59, 60], and this is the origin of the coexistence of STEs in condensed Ne.

The locations and lifetimes of the main absorption bands described above are summarized in Table 5.7 with their assignments. The experimental transition energies of absorption and luminescence due to a- and m-STEs in rare gas solids have been successfully reproduced by the pseudopotential calculation by *Song* and *Lewis* [5.61, 62].

5.3.3 Formation of Self-Trapped Exciton Bubbles in Condensed Neon

In time-resolved absorption spectroscopy of solid Ne under electron pulse excitation, a distinct change of spectra with time was observed for peak energies as well as band shapes. Detailed studies of the kinetics of the phenomena as a function of temperature led to the conclusion that the change of spectra is due to the formation of microscopic cavities around a STE; named the *STE bubble* by *Suemoto* and *Kanzaki* [5.63, 64]. Models for the formation process of STE bubbles in solid Ne are as follows.

In the first step of self-trapping, the STE is formed by deforming the surrounding lattice; called the *primary STE*. The distortion around the primary STE will be the lattice expansion as pointed out by *Song* [5.62]; the outward displacement of the first nearest neighbors is estimated as about 2 a.u. for the primary a-STE in solid Ne [5.65], which corresponds to an expansion of the first nearest neighbor distance of nearly 30%. During the lifetime of a STE, thermal vacancies are captured by the STE whenever they approach the STE during their random motion. Growth of the bubble size continues until it reaches the equilibrium size. The number of vacancies n_e attached in equilibrium can be estimated as 3–7 for an a-STE, and larger for a m-STE.

The idea of formation of microscopic cavities around a STE is not new. It was previously proposed for a- and m-STEs in liquid He [5.66], on the basis of the similarity between the spectra of the STEs and the free atom or molecule. Also, the existence of cavities was concluded for rare gas solids by *Fugol'* [5.50] from the similarity between the luminescence spectra in the solid and gaseous phases. Thus, the formation of bubbles has been suggested for condensed He and Ne, both of which have large negative electron affinity as shown in Table 5.2. The repulsive interaction between STEs and the surrounding atoms is a plausible origin of the bubble formation. On the other hand, the studies by *Suemoto* and *Kanzaki* [5.63, 64] gave more convincing evidence for the formation process of the STE bubbles in solid Ne. In the following, their experimental results will be outlined.

Figure 5.7 shows a typical example of the change of the a-STE absorption spectra in solid Ne. The structure clearly observed at 5 μs is broadened at 20 μs, and then shifted to a lower energy. Considering the blue shift of the transition energies in the sequence gas, liquid, and solid (Table 5.7), it is suggested that the observed energy shift with time corresponds to an increase of the a-STE to Ne

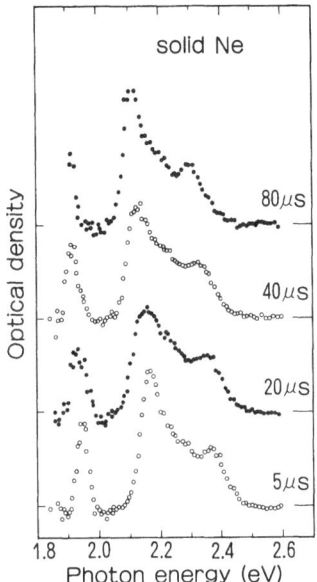

Fig. 5.7. Time-resolved absorption spectra of the a-STE in solid Ne at various delay times after pulse excitation, at 19.0 K [5.63]

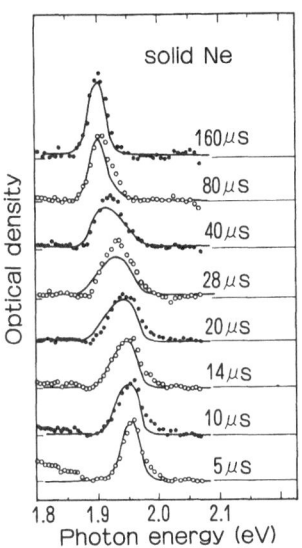

Fig. 5.8. Time-resolved absorption spectra of the $3s_{12} \rightarrow 3p_{01}$ transition of the a-STE in solid Ne at 19.0 K, at various delay times after pulse excitation [5.63]. The curves are the simulation based on the vacancy capture model described in the text

distance with the growth of bubbles. More details of time-resolved absorption spectra are shown in Fig. 5.8 for the absorption band corresponding to the $3s_{12} \rightarrow 3p_{01}$ transition. At the beginning (5 μs delay), there exist only primary a-STEs showing a sharp band. Toward the end (160 μs delay), there are only a-STE bubbles of equilibrium size, which again show a sharp band. For the intermediate stage during bubble growth, there will be a wide distribution of bubble size, which results in broader bands. The curves in Fig. 5.8 are obtained by a simulation based on the vacancy capture model assuming that $n_e = 3$ and the energy shift is proportional to the number of attached vacancies. The fit to the data points is found to be reasonably good for $3 \lesssim n_e \lesssim 7$. Alternatively, the value of n_e can be estimated as 3–4 vacancies, from the gas-to-solid energy shift.

The kinetics of the spectrum change as a function of temperature and between 18 and 23 K can be described as follows. Firstly, both the initial and the final energy positions of the $3s_{12} \rightarrow 3p_{01}$ transition are independent of temperature; 1.957 ± 0.005 eV and 1.902 ± 0.005 eV, respectively. This suggests that the sizes of the primary a-STE and the full-grown bubble are independent of temperature. Secondly, the characteristic relaxation time t_r is very sensitive to temperature. Figure 5.9 shows the plot of log t_r against $1/T$. The negative slope of the data gives an activation energy 41 meV, which agrees with the activation energy of

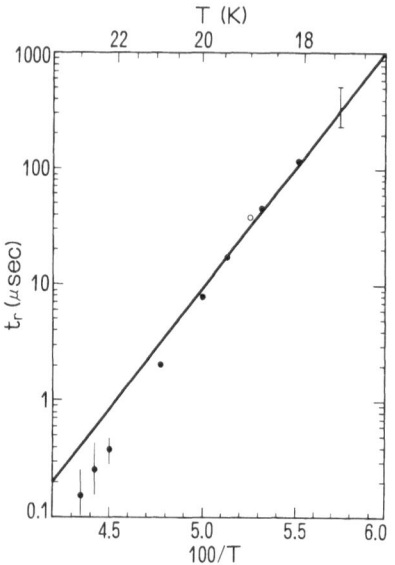

Fig. 5.9. The characteristic relaxation time t_r of the growth of an a-STE bubble in solid Ne against reciprocal temperature [5.63]. The solid line represents (5.5)

self-diffusion in solid Ne, 40.95 ± 0.26 meV obtained by the NMR method [5.67]. The self-diffusion in rare gas solids occurs via migration of thermal vacancies [5.68]. Considering the random motion of vacancies in a fcc lattice and using the same assumptions as in the simulation of Fig. 5.8, the following expression is obtained for t_r:

$$t_r = 4.48 \times 10^{-16} \exp(475/T) \tag{5.5}$$

which corresponds to the solid line in Fig. 5.9. The calculated temperature dependence of t_r shows good agreement with experimental data. Although the agreement of the absolute value of t_r may be fortuitous, this is convincing evidence which supports the model of growth of a-STE bubbles by capturing thermal vacancies in solid Ne.

The formation of m-STE bubbles is also found to proceed similarly to that of a-STE bubbles [5.64]. The temperature dependence of t_r can be fitted by

$$t_r = 8.4 \times 10^{-16} \exp(475/T) \tag{5.6}$$

The same activation energy as (5.5) suggests that m-STE bubbles also grow by capturing thermal vacancies. The preexponential factor in (5.6) is nearly twice as large as in (5.5), suggesting a larger equilibrium bubble size of the m-STE compared to the a-STE.

Because of the variety of higher excited states of STEs identified unambiguously and the precise information on the relaxation of STEs by bubble formation, it seemed worthwhile to investigate the de-excitation processes of the higher excited states of the a-STE in solid Ne. For this line of study, *Suemoto* and

Kanzaki [5.64] used the double excitation by an electron pulse and a laser pulse, which had previously been applied to alkali halides [5.69]. The lowest triplet state $3s_{12}$ of the a-STE produced by an electron pulse is optically excited to the $3p_{23}$ and $3p_{22}$ states by a subsequent laser pulse.

In the experiments shown in Fig. 5.10, the conversion of the optically excited a-STE to the m-STE is observed. Firstly, in Fig. 5.10a at low temperature, the a-STE remains as the primary STE without bubble growth. In curve (a2), without laser excitation, the m-STE has vanished. However, in curve (a3), with intermediate laser irradiation, the a-STE has decreased and the m-STE appears. Conversion occurred from the primary a-STE to the primary m-STE through excited states of the a-STE. In Fig. 5.10b, at a higher temperature, the a-STE bubble grows. In curve (b2), without the laser, the a-STE has grown to its equilibrium size and the m-STE has vanished. On the other hand, in curve (b3), after laser irradiation, the a-STE has decreased and the m-STE appears. The m-STE in (b3) has an intermediate bubble size and subsequently grows after conversion. In this case, the a-STE of equilibrium size is converted to an m-STE of intermediate size. These results suggest: (1) the a-STE is converted to the m-STE in the same bubble through higher excited states and (2) the equilibrium size of the m-STE bubble is larger than that of the a-STE. The formation, the bubble growth, and the a-STE to m-STE conversion processes in solid Ne are illustrated in Fig. 5.11.

When the a-STE is excited by a laser, intense luminescence bands are observed. Figure 5.12 shows the spectra in the solid (a), (b) and liquid (c), in comparison with the electron-excited Ne gas (d). The radiative transitions correspond to those from the lowest $3p$ state $(3p_{01})$ to the $3s$ multiplets in the a-STE. By comparing the luminescence and absorption spectra, the Stokes shift of the $3p_{01}$–$3s_{12}$ transition is estimated as 0.07 eV, 0.04 eV, and 0.05 eV for (a), (b), and (c), respectively, indicating the dependence of the lattice relaxation on the environment of the a-STE. These results by *Suemoto* and *Kanzaki* [5.64] indicate that the de-excitation from the higher $3p$-type excited states of the a-STE in condensed Ne consists of two competing channels: (1) radiative decay to the $3s$ states and (2) conversion to the m-STE. These processes in condensed Ne present a clear-cut example in which lattice rearrangement proceeds through well-defined excited states in the condensed phase.

Belov et al. [5.70] observed the luminescence corresponding to $3p \rightarrow 3s$ atomic transitions in evaporated films of Ne under low-energy electron excitation at 5 K and 2 K. The transition energies they observed are very close to those of the free Ne atom and the line widths are very narrow, less than 0.2 meV. *Belov* et al. [5.70] assigned the luminescence to the $3p \rightarrow 3s$ transitions of the a-STE in solid Ne. Their interpretation is inconsistent with the results of Fig. 5.12 from *Suemoto* and *Kanzaki* [5.64]. The origin of the discrepancies will be discussed in Sect. 5.3.4.

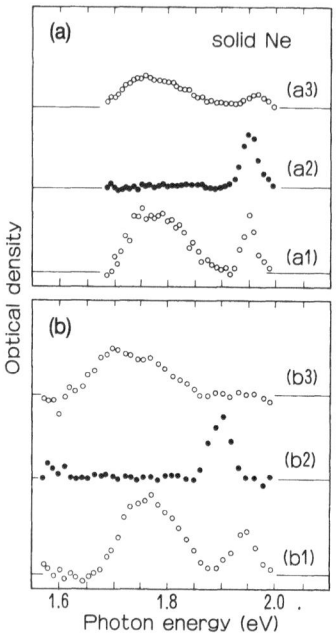

Fig. 5.10a, b. Transient absorption spectra showing the conversion of the a-STE to the m-STE via higher excited levels of the a-STE, in solid Ne [5.64]. (**a**): at 17.0 K. Delay time after electron pulse: (*a1*) 0.2 μs, (*a2*) 19.7 μs, and (*a3*) 19.7 μs with a laser pulse excitation at 17.3 μs. (**b**): at 21.0 K. Delay time after electron pulse: (*b1*) 0.8 μs, (*b2*) 42 μs, and (*b3*) 44 μs with a laser pulse excitation at 43 μs

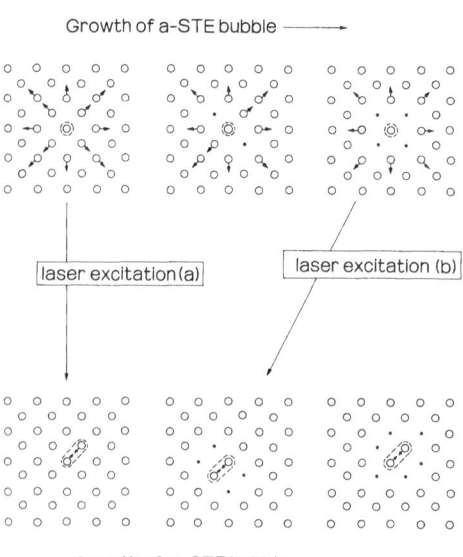

Fig. 5.11. Schematic diagram showing the growth of a- and m-STE bubbles in solid Ne from the primary to the equilibrium size bubble. Laser excitations (**a**) and (**b**) correspond to the conversion processes shown in Fig. 5.10a and b, respectively

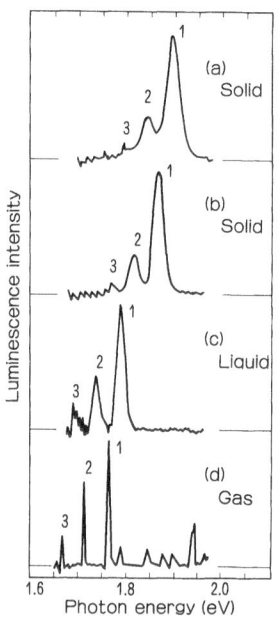

Fig. 5.12. Luminescence spectra from higher excited states of the a-STE in condensed Ne under laser excitation delayed with respect to electron excitation [5.64]. (*a*) primary STE in the solid at 17.0 K, (*b*) STE bubble in the solid at 20.4 K. (*c*) in liquid at 25.3 K. (*d*) Ne gas (2.5 atm) by electron pulse excitation at 45 K. The luminescence peaks 1, 2, and 3 correspond to $3p_{01} \rightarrow 3s_{12}$, $\rightarrow 3s_{11}$, and $\rightarrow 3s'_{00}$, respectively

5.3.4 Relaxation of Free Excitons in Photo-Excited Rare Gas Solids

The relaxation process from photo-excited states in condensed matter is a subject which is not yet completely clarified in spite of its importance. The condensed rare gases are unique materials for the detailed study of the relaxation process. For example, starting from selected photo-excited states which are fairly well understood as given in Sect. 5.3.1, it should be possible to follow the de-excitation process toward the ground state through the metastable relaxed states such as discussed in Sects. 5.3.2, 3. Until recently, however, only the excitation spectra of the m-STE luminescence in solid Ar, Kr, and Xe [5.71] had been reported from this angle. In the following, the recent work by *Inoue* et al. [5.72] on the de-excitation and relaxation processes in photo-excited solid Ne, Ar, and Kr will be outlined.

In solid Ne, the coexistence of a- and m-STEs is firmly confirmed as described in Sects. 5.3.2, 3, and it is interesting to distinguish differences between the formation of the two types of STE from photo-excited states in solid Ne. *Inoue* et al. compared the excitation spectra of yield between the two luminescence components due to a- and m-STEs. Solid Ne films with sufficient thickness are excited at LHeT by monochromatic light using synchrotron radiation from the SOR-RING of the Institute for Solid State Physics, the University of Tokyo. From the luminescence spectra under monochromatic excitation, or conversely, from the excitation spectra for each luminescence component, information is obtained on the relaxation channels from a final state of optical absorption to the various initial states of luminescence. Figure 5.13 shows the comparison between yield spectra for the a- and m-STE luminescence components. The yield spectra for the m-STE luminescence monitored at 16.1 and 15.1 eV in the figure and also at 14 eV (see the luminescence spectrum in Fig. 5.2), are very similar to each other, while the yield for the a-STE luminescence monitored at 16.7 eV has a different spectral shape. The difference between the a- and m-STEs is more clearly displayed by the ratio of yield spectra [a-STE luminescence]/[m-STE luminescence] which is shown in Fig. 5.14a.

The enhancement of the a-STE relative to the m-STE is distinct for excitation at 17 and 19 eV, but the intensity ratio is almost constant at the other excitation energies. The question is, what is the origin of the two peaks, at 17 eV and 19 eV, in the spectrum? The 17 eV peak is located at the low-energy tail of the $n = 1$ exciton transition. One possibility is the exciton absorption at the surface observed at 17.15 eV in solid Ne [5.5]. However, this possibility must be disregarded because of the negative result in the surface coverage experiment and we have to look for another interpretation for the a-STE exhancement at 17 eV.

In the following, we proceed by adopting the atomic or Frenkel-type description of free excitons, partly for the sake of convenience. In atomic terms, the free excitons $\Gamma(3/2)$ and $\Gamma(1/2)$ correspond to the atomic transitions from $^1S_0$ to $^3P_1$ and $^1P_1$, respectively. The absorption peak of the $n = 1$ exciton in solid Ne is a $^1P_1$ exciton at 17.50 eV and the shoulder peak is a $^3P_1$ exciton at 17.36 eV [5.5]. The $^3P_2$ forbidden exciton transition is located at the lower energy. From energy

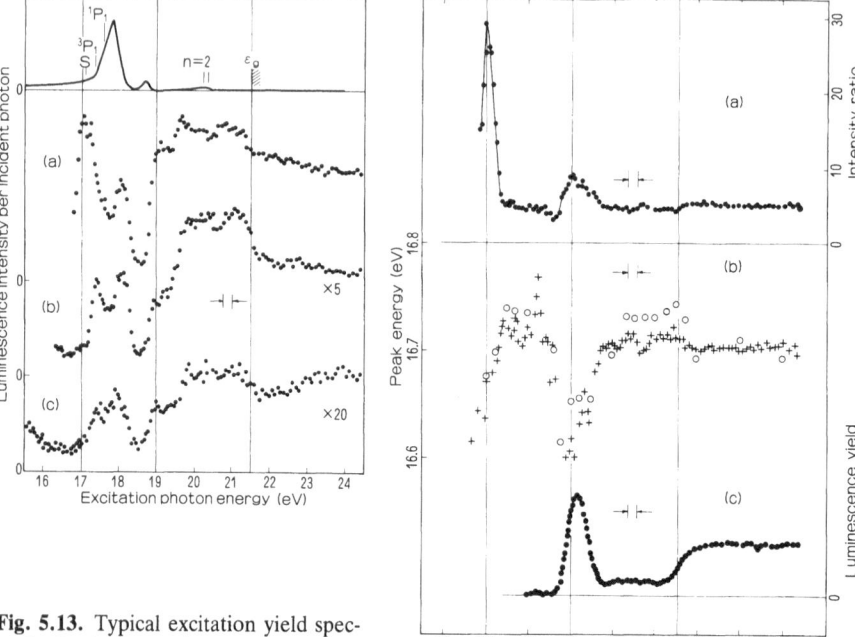

Fig. 5.13. Typical excitation yield spectra for luminescence in solid Ne at LHeT [5.72]. Luminescence intensity is monitored at (a) 16.7, (b) 16.1, and (c) 15.1 eV. In the upper part is shown the reflectance spectrum of the sample with the energy position of the exciton transitions (S is the surface exciton). The reflectance peak at 18.7 eV is due to reflection from the substrate

Fig. 5.14. Excitation spectra of photo-excited solid Ne at LHeT [5.72]: (a) the ratio of luminescence yield spectra [a-STE component]/[m-STE component], (b) the peak energy of the a-STE luminescence spectra and (c) the yield spectrum for the visible luminescence observed in the spectral range from 1.7 to 2.0 eV

considerations, the enhancement of the a-STE formation at 17.0 eV can be assigned as due to the $^3P_2$ exciton transition.

The origin of the a-STE enhancement can be reasonably explained as follows. Generally, there will be a potential barrier for the relaxation of the free exciton to the STE [5.53] (see Sect. 4.4). In the case of relaxation to the m-STE in solid Ne, however, there exists an additional energy barrier of specifically molecular origin. The relativistic calculation of molecular potential [5.59], as well as the luminescence measurements in the gaseous phase [5.73], confirm the existence of potential barriers in the formation of molecules from Ne atoms at the $^3P_2$ and $^3P_1$ levels. Then, in solid Ne, analogous barriers for m-STE formation will cooperate in the relaxation of free excitons, in addition to the deformation of the surrounding lattice which is generally required. The $^3P_2$ or $^3P_1$ free exciton in solid Ne will thus have a much larger barrier for m-STE than a-STE formation, and will relax almost exclusively as the a-STE. On the other hand, the $^1P_1$ free exciton will relax equally to both STEs, because it can relax to m-STE through the potential crossing between molecular levels and no potential barrier is effective for the m-STE formation.

The 19 eV enhancement peak in Fig. 5.14a can be assigned as the $3p$ forbidden exciton transition from comparison with atomic spectra. During the possible de-excitation from $3p$ to $3s$ states in atomic terms, the lower $3p$ levels will decay mainly to the lower $3s$ states, $^3P_2$ or $^3P_1$, having the same angular momentum of the hole, $J = 3/2$. Therefore, the low-energy portion of the $3p$ exciton will decay to the $^3P_2$ or $^3P_1$ exciton first, and subsequently be self-trapped mainly as the a-STE for the same reason as given above for the 17 eV band.

There are two sets of further experimental evidence which support the relaxation process described above. The first is the change of a-STE luminescence spectra with excitation energy, which is shown in Fig. 5.14b. The lower-energy shift of the luminescence peak is observed exactly in association with the a-STE enhancement peaks in Fig. 5.14a. The luminescence spectra due to a-STE consist of luminescence components from $^3P_2$, $^3P_1$, and $^1P_1$ to $^1P_0$ as described in Sect. 5.3.2, among which the $^3P_2$ and $^3P_1$ components are dominant. The $^3P_2$ component is at lower energy than the $^3P_1$ component, and the increase of the $^3P_2$ component will cause the low-energy shift of the peak energy as observed. The amount of shift, 0.1 eV, in the figure coincides with the energy separation between the $^3P_2$ and $^3P_1$ components. The results suggest that the a-STE enhancement is related to the $^3P_2$ free-exciton state for both the 17 eV and 19 eV peaks, and the $^3P_2$ exciton is mainly relaxed to the $^3P_2$ atomic level of the a-STE.

The second result is concerned with the de-excitation of the $3p$ exciton to the lower $3s$ exciton states. Figure 5.14c shows the excitation yield spectrum for the visible luminescence due to atomic $3p \rightarrow 3s$ transitions. The prominent peak at 19.2 eV corresponds to the 19 eV enhancement peak in Fig. 5.14a, and can be assigned as the $3p$ exciton transition. The de-excitation of the $3p$ exciton to the $^3P_2$ and $^3P_1$ excitons proceeds through radiative decay observed as the visible luminescence. In Fig. 5.14c, the excitation region between 20 and 21 eV corresponds to higher excitons $4s$, $5s$, and so on. In this region, the visible luminescence is weak and the de-excitation pathway through the $3p$ exciton is not dominant. Under excitation above 21.5 eV which corresponds to the onset of interband transitions (ε_g), the visible luminescence yield increases again. This is evidence that the self-trapped holes Ne_2^+ recombine dissociatively with free electrons, and produce the $3p$ excitons which subsequently decay to the $^3P_2$ and $^3P_1$ excitons by emitting visible luminescence. A similar process is known to occur in gaseous Ne [5.74].

The visible luminescence observed here is the same as that observed by *Belov* et al. [5.70], and contains narrow lines very close to the $3p \rightarrow 3s$ atomic transitions and the accompanying weak acoustic-phonon sidebands. However, the interpretation by *Belov* et al. [5.70], as a transition in the a-STE, is hardly acceptable in view of the result by *Suemoto* and *Kanzaki* [5.64], as explained in Sect. 5.3.3, while the proposition by *Inoue* et al. [5.72] that the visible luminescence corresponds to the interexciton transition $3p \rightarrow 3s$ is consistent with all of

the experimental information. As far as the transition energies of the luminescence lines are concerned, the energy difference between the $3p$ and ^3P excitons
in the solid assigned above agrees with that of atomic levels within the experimental accuracy of 0.1 eV, and the assignment as the interexciton transition is
consistent with the closeness of the visible luminescence lines to atomic transitions.

In the relaxation processes described above for solid Ne, the $3p$ (19.2 eV) and
the $^3P_2$ (17 eV) excitons play important roles. Both of the exciton transitions are
forbidden in the free atom and are also very weak in the solid. It is to be noticed
that they are not observable in the absorption or reflectance spectra and are
revealed in the excitation yield spectroscopy, for the first time, by using sufficiently thick samples.

The excitation yield spectra for the a- and m-STE luminescence in solid Ar
and Kr are very different from those in solid Ne. Contrary to solid Ne, both of the
luminescence components are excited at the low-energy tail of the allowed exciton absorption in solid Ar and Kr. It is known that the potential curves for Ar_2
and Kr_2 are different from those for Ne_2. There is no potential barrier which
separates the molecular state and the atomic $^3P_2$ level [5.75], and the barrier will
be very small for the $^3P_1$ level if it even exists. On the basis of the different nature
of the molecular potential, the relaxation of excitons in solid Ar and Kr will be
different from that in solid Ne. Firstly, almost all of the $^3P_2$ excitons will be self-
trapped as m-STEs, because of the absence of a potential barrier. Secondly, a
small proportion of the $^3P_1$ excitons can be self-trapped as a-STEs, due to the
small potential barrier. On the other hand, the $^1P_1$ exciton can relax to both the
m-STE and the ^3P exciton. The probability to the a-STE, however, will be much
smaller than the direct excitation of the ^3P exciton. These considerations on the
different relaxation process in solid Ar and Kr compared with solid Ne are also
consistent with the difference in the luminescence spectra themselves. As shown
in Figs. 5.2, 3, the m-STE luminescence is the most intense in solid Ar and Kr, in
contrast to the most intense a-STE luminescence in solid Ne.

Summarizing, the nature of the de-excitation and relaxation processes of the
free excitons is now fairly well understood for solid Ne, Ar, and Kr. Further
studies, however, are required for more quantitative information on the branching ratio among various channels. As for the relaxation from the still higher
excited states including core-level excitations, very little is known at present.
Furthermore, there are other relaxation channels which have not been studied
yet for condensed rare gases. For example, Frenkel defect formation under x-ray
irradiation has been suggested for solid Ar, from the observation of a lattice
parameter change [5.76]. Another example is photon-stimulated desorption
which will certainly be an efficient decay channel of the photo-excited states in
solid rare gases, in view of the electron-stimulated desorption data [5.77]. It is
hoped that future studies will clarify these unresolved and interesting subjects.

6. Exciton-Phonon Processes in Silver Halides

The unique properties of silver halides among simple ionic crystals have been recognized for a long time. Many of the physical properties of silver halides are clearly different from those of alkali halides, e.g., a small band gap energy with a large dielectric constant, and excellent photoconductivity as well as ionic conductivity. Furthermore, it is accepted that these features are closely related to the application of silver halides in photography. Consequently, a large number of investigations have been carried out on silver halides resulting from both basic interest and the need to understand photographic phenomena.

The first review of these researches appeared in the book by *Mott* and *Gurney* in 1940 [6.1], in which the importance of both ionic and electronic processes in photographic sensitivity was pointed out for the first time. The second review by *Seitz* in 1951 [6.2] was concerned with more general properties of silver halides and stimulated further investigations from the standpoint of solid state physics. In subsequent years, an impressive development was achieved in our understanding of the basic properties of silver halides. The progress made in this period has been surveyed in several reviews, such as those by *Brown* [6.3] and *Kanzaki* [6.4].

In the present chapter, we concentrate on the phenomena related to exciton-phonon interactions in the silver halides, mostly AgBr and AgCl on which thorough investigations have been made. The main emphasis will be on the various aspects of the exciton relaxation process in which the exciton-phonon interaction plays a key role.

6.1 Electronic and Lattice Properties of Silver Halides

The first experimental evidence for the unique features of the electronic band structure in silver halides was obtained through the fundamental absorption edge spectra for a wide range of the absorption coefficient measured by *Okamoto* [6.5]. Figure 6.1 shows the spectra of AgBr and AgCl, together with KBr which shows spectra typical of alkali halides. The distinct difference between silver and alkali halides is evident in Fig. 6.1, and reflects the essential difference of band structure between the two. The absorption spectra of silver halides consist of two components with different characters. One is the high-energy part, above 3.9 and 4.6 eV for AgBr and AgCl, respectively. This component is very similar to the

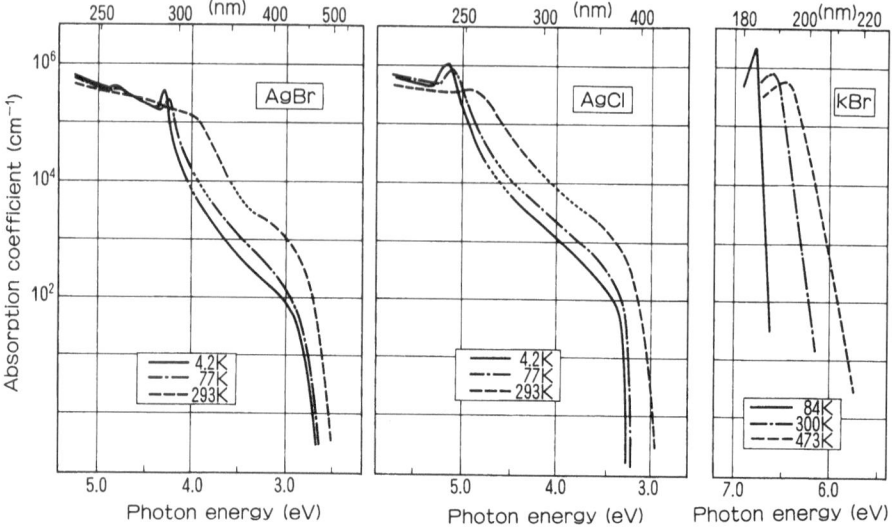

Fig. 6.1. Temperature dependence of fundamental absorption spectra of AgBr [6.4] and AgCl [6.6] in comparison to KBr [6.7]

first exciton peak of alkali halides in spectral shape as well as its temperatue dependence. The other is the low-energy part with a smaller absorption coefficient which extends toward the absorption threshold. This component is a broad band without any peak and its absorption strength increases with increasing temperature. The essential difference in the spectra of silver and alkali halides is the absence of the low-energy component in alkali halides. The correspondence of the absorption peak in the high energy component in silver halides to the first exciton peak in alkali halides is also understandable from the comparison of their transition energies. As shown in Table 6.1, the peak energies of both silver and alkali halides agree fairly well with the empirical formula of *Hilsch* and *Pohl* [6.8] for the exciton transition energy E_{ex},

$$E_{\mathrm{ex}} = E_{\mathrm{A}} - E_{\mathrm{I}} + \alpha_{\mathrm{M}} e^2/a \tag{6.1}$$

Table 6.1. Comparison of experimental absorption peak energy with that calculated by the Hilsch-Pohl formula (6.1). All energies are in eV

	$\alpha_{\mathrm{M}} e^2/a$	E_{A}	E_{I}	E_{ex}(calc.)	E_{ex}(exp.)
AgCl	9.07	3.63	7.57	5.13	5.13[a]
AgBr	8.71	3.38	7.57	4.52	4.29[b]
KBr	7.68	3.38	4.34	6.72	6.79[c]

[a] [6.6]; [b] [6.4]; [c] [6.7]

in which E_A is the electron affinity of the halogen, E_I the ionization energy of the alkali, α_M the Madelung constant, and a the nearest neighbor distance. Apparently, the low-energy shift of transition energies in silver halides in comparison with alkali halides is due to the larger value of the ionization energy of silver compared to an alkali atom.

At the time of discovery of the low-energy tailing of absorption spectra in silver halides [6.5], its origin was ascribed to the partly homopolar binding of silver halides based on the similarity of the spectra to those of silicon and germanium. This conjecture was substantiated by later studies by *Brown* et al. [6.9] who revealed the structures in the absorption threshold of silver halides and identified the indirect excitons unambiguously. Thus, the similarity of the low-energy absorption component in silver halides and silicon or germanium is due to the similarity of the band structure; both of them are indirect band gap materials. The dissimilarity between the silver and alkali halides, on the other hand, is due to the difference of the nature of the band-gap transition; the indirect energy gap in silver halides and the direct gap in alkali halides. The high-energy absorption peak in silver halides corresponds to the direct-exciton transition, which is common to alkali halides. The low-energy absorption component is unique for silver halides and corresponds to the indirect gap transition including the indirect exciton at the absorption threshold.

In the mean time, other experimental information on the band structure was obtained through transport studies on photo-excited charge carriers which will be described in Chap. 8. The transport of electron polarons in silver halides has been studied extensively by *Brown* and his collaborators [6.10, 11]. Their results have definitely shown that the minimum of the conduction band in AgBr and AgCl is located at the Γ point in momentum space which, in combination with the indirect gap concluded from absorption spectra, indicates the maximum of the valence band is located at some other point in momentum space. The band structure calculation on silver halides which emerged at the time by *Bassani* et al. [6.12] and *Scop* [6.13] also proposed the indirect gap, but did not have enough confidence to specify the exact location of the valence-band maximum. Later on, the successful observation of cyclotron resonance of a positive hole in AgBr by *Tamura* and *Masumi* [6.14] established that the valence band maximum of AgBr is located along the [111] direction. As will be shown later in this section, it is now established, both experimentally and theoretically, that the maximum of the valence band is located at the L point in AgBr and AgCl.

Figure 6.2 shows the band structure of AgBr based on the nonrelativistic calculations of [6.13, 15]. The band structure of AgCl [6.12, 13, 15, 16] is qualitatively very similar to AgBr. As concluded in all of the calculations, the lowest conduction band is mostly s-like and nearly spherical in the vicinity of its minimum at Γ_1. On the other hand, the valence bands are derived from the mixing of the np state of the halogen with the $4d$ state of the silver ions, and the degree of p–d mixing or hybridization varies with the location in k-space. For example, the valence bands at the Γ point arise from either a p- or d-state with hardly any p–d mixing; the eigenfunction of Γ_{15} is mainly halogen np with some

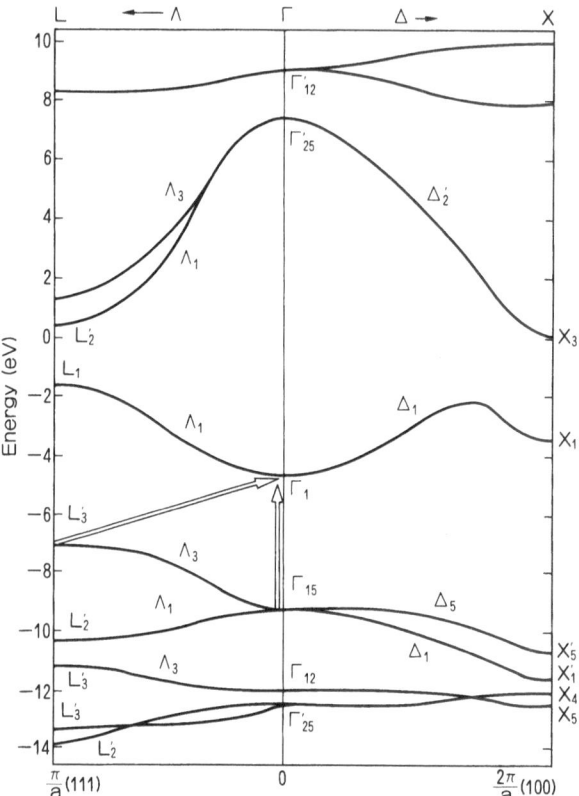

Fig. 6.2. Band structure of AgBr according to non-relativistic calculations of [6.13, 15]. Arrows indicate the indirect ($L'_3 \rightarrow \Gamma_1$) and the direct ($\Gamma_{15} \rightarrow \Gamma_1$) band gap transitions. The notation applies to the case where the origin of the coordinates is located on the halogen ion

Ag^+ $5p$, and those of Γ_{12} and Γ'_{25} are dominantly Ag^+ $4d$. At other points in k-space away from Γ, the valence bands tend to exhibit stronger p–d mixing. For example, the valence band maximum at L'_3 exhibits some of the strongest p–d mixing with almost equal amounts of p- and d-states.

On the basis of these band calculations, we can visualize the origin of the indirect energy gap of silver halides in contrast to alkali halides. The character of the valence bands of silver halides comes from the near degeneracy of the two constituents derived from halogen np and silver $4d$, respectively. The repulsive interaction between the two components results in the "pushing apart" of the valence bands, away from the Γ point. In contrast, the valence bands of alkali halides are essentially formed from the np electrons of halogen ions and the bands from the alkali ions are isolated lower down. This type of distinction of valence bands between alkali halides and other metal halides such as silver, thallium, and copper was first recognized by *Krumhansl* [6.17]. His conjecture correctly predicted not only the strong mixing of cation and halogen levels in these salts, but also the energy positions of cation-derived bands relative to halogen *np* bands which were substantiated by later investigations as described in Chaps. 3 and 7, for copper and thallium halides, respectively, as well as for silver halides.

In addition to the indirect gap nature which is well reproduced in the band calculations, the following experimental results relevant to the band structure are available for silver halides. Firstly, the energy distribution curve (EDC) of photo-electrons excited from valence bands are obtained by x-ray photoelectron spec-troscopic (XPS) [6.18] and ultraviolet photoelectron spectroscopic (UPS) [6.19] methods, and compared with the density of valence states obtained from band calculations [6.16]. The agreement between experiment and theory is fair for the density peaks due to Ag^+ $4d$, except for AgF. Secondly, the energy shift of indirect and direct exciton transitions are studied under high pressure [6.20]. Interestingly, the sign of the pressure shift is different between indirect (toward lower energy) and direct (toward higher energy) exciton transitions. *Fowler* [6.15] calculated the energy band of AgCl at points of high symmetry as a func-tion of lattice parameters and obtained the deformation potential D defined as

$$D = \Delta E/(\Delta V/V) \tag{6.2}$$

where ΔE is the energy shift and $(\Delta V/V)$ the fractional volume change. The theoretical values D for AgCl are $+0.1$ eV and -4.2 eV, for indirect and direct gaps, respectively. The results can be favorably compared with experimental D of $+0.51$ to $+0.65$ eV and -2.77 to -2.70 eV, respectively for the corresponding exciton transitions.

Summarizing, the band structure of silver halides has been well elucidated as far as the band-edge states are concerned. Further experimental investigations will be required for the detailed analysis of the band structure, using more sophis-ticated techniques such as angle-resolved photoelectron spectroscopy of single crystals using synchrotron radiation. Although such experiments have been con-sidered difficult for insulators, their feasibility has already been demonstrated for alkali halides [6.21]. In the case of silver halides, such studies should preferably be done at lower temperatures, in view of a unique temperature dependence of photoelectron emission reported in [6.22].

In the early proposal of the indirect edge for silver halides by *Seitz* [6.2], it was postulated that momentum conservation during the optical transition is realized through scattering by the high density of imperfections. The temperature dependence of absorption edge spectra, such as that obtained by *Okamoto* [6.5], however, clearly showed that the momentum selection rule for the indirect transi-tion is satisfied through the emission and absorption of phonons. On the basis of a detailed study of the absorption edge spectra to be described in Sect. 6.2.1, the energy values of momentum-conserving MC phonons in the indirect exciton transition of silver halides are obtained as follows:

$$\left. \begin{array}{l} 8.0 \pm 0.2 \text{ meV, and} \\ 12.0 \pm 0.4 \text{ meV} \quad \text{for AgBr [6.23] ,} \end{array} \right\} \quad \text{and} \tag{6.3}$$

$$8 \pm 1 \text{ meV, and}$$
$$13 \pm 1 \text{ meV} \quad \text{for AgCl [6.24] .} \qquad (6.4)$$

The later experimental determination of the momentum of the MC phonons by neutron scattering actually led to the identification of the exact position of the valence band maximum in silver halides. In spite of the demand for information on phonons, the inelastic neutron scattering study of silver halide was delayed due to the appreciable absorption of neutrons by silver. The first result came out for AgCl [6.25], and subsequently for AgBr [6.26–28]. The neutron scattering results on AgCl at 78 K [6.25] revealed the following two kinds of phonons at the L point:

$$TA(L) = 8.4 \text{ meV} ,$$
$$\text{and} \quad LA(L) = 12.4 \text{ meV} . \qquad (6.5)$$

The closeness of the L point phonons (6.5) to the MC phonons (6.4) confirmed the location of the valence band maximum in AgCl at the L point, in agreement with the band calculations. From the selection rules of indirect-exciton transitions, the TA(L) and LA(L) phonons had been predicted for MC phonons in AgCl [6.12, 24]. Experimentally, however, the exact assignment of the nature of MC phonons is not definite because of the proximity of the TO(L) and TA(L) phonons in AgCl.

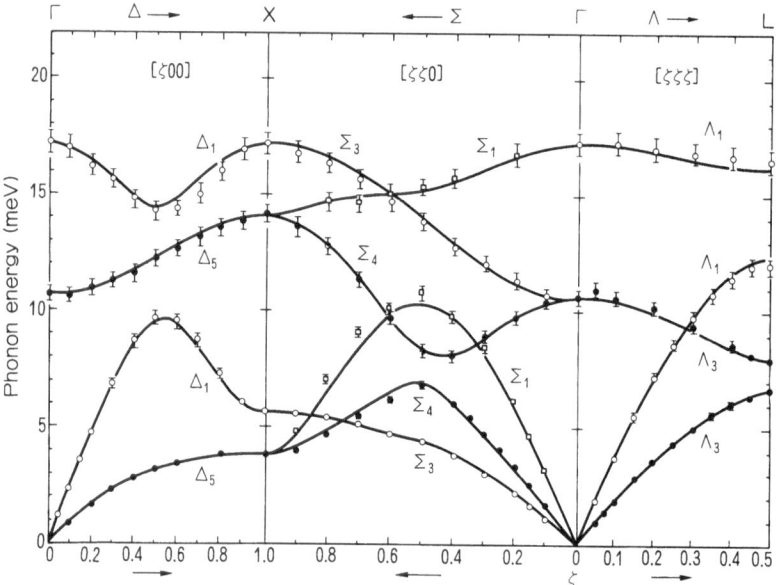

Fig. 6.3. Phonon dispersion relation for AgBr at 4.4 K [6.27]. Solid curves represent an extended-shell-model fit to the data points. Momentum vectors q are given in terms of $\zeta = q(a/2\pi)$ where a is the lattice constant

Figure 6.3 shows the phonon dispersion relations along the three high-symmetry directions of AgBr at 4.4 K. Solid curves represent an extended-shell-model fit which enables us to calculate the phonon density of states. The energies of phonons at the L point which we are most interested in are determined as follows

$$\left. \begin{array}{rcl} LA(L) & = & 12.00 \pm 0.30 \text{ meV} \\ TO(L) & = & 7.99 \pm 0.10 \text{ meV} \\ TA(L) & = & 6.68 \pm 0.10 \text{ meV.} \end{array} \right\} \tag{6.6}$$

Excellent agreement between the MC phonon energies in (6.3) and the TO(L) and LA(L) energies in (6.6) leads to the conclusion that the valence-band maximum in AgBr is located exactly at the L point [6.26].

On the other hand, the assignment of MC phonon modes differs from the previous prediction of TA(L) and LA(L) modes. Closer inspection, however, shows that there is no real disagreement between the two. What actually happens is as follows [6.26, 27]: In the case of phonons at the L point, both the Ag^+ and Br^- ions vibrate independently of each other. Ordinarily the heavier ion will vibrate with the lower frequency, but this natural tendency can be *inverted* under certain circumstances so that the lighter ion vibrates with the lower frequency, which is the case in AgBr. The experimental scattering intensity measurements on TO and TA phonons in AgBr along the [111] direction gave convincing evidence that *character inversion* occurs between TO(L) and TA(L), and the TO(L) phonon corresponds to the mode in which only Ag^+ ions move and Br^- ions stand still. This mode has precisely the symmetry property of the MC phonon required from the optical selection rule of the phonon-assisted indirect-exiton transition derived in [6.12, 24] and the assignment of MC phonons as TO(L) and LA(L) is fully consistent with theoretical expectations.

The TA–TO character inversion of phonons at the L point in AgBr has been attributed to the deformability of the Ag^+ ion in the lattice dynamical model by *Fischer* et al. [6.29], which successfully reproduced the data on silver halides by introducing both a quadrupolar deformabilitiy and a rigid rotation of the Ag^+ shell. These specific phonon features can be correlated to the large value of ionic conductivity of silver halides due to Frenkel defects (silver vacancy and silver interstitial) [6.30]. The ionic conductivity of silver halides is several orders of magnitude larger than that of alkali halides due to Schottky defects (alkali and halogen vacancy). For example, the ionic conductivity of AgBr amounts to $1 \ \Omega^{-1} \ cm^{-1}$ at the relatively low melting point of 420 °C. The value is in the range of superionic conductors and in strong contrast to the corresponding value of 10^{-4}–$10^{-5} \ \Omega^{-1} \ cm^{-1}$ observed for alkali halides. As shown in Table 6.2 which summarizes the formation and migration energies of thermal defects, both of the energies are much smaller in silver halides in comparison to alkali halides.

The unique properties of thermal defects in silver halides, high concentration and large mobility, can be traced back to the extraordinarily large deformability of silver ions in silver halides which is also reflected in the phonon properties.

Table 6.2. Formation and migration energies of Frenkel defects in AgBr and AgCl [6.31] compared to those of Schottky defects in alkali halides. All energies are in eV

	AgBr	AgCl	KCl	NaCl
Formation energy	1.13	1.45	2.1	2.4
Migration energy of cation vacancy	0.32	0.28	0.8	0.8
Migration energy of cation interstitial:				
Collinear interstitialcy motion	0.04	0.01	–	–
Noncollinear interstitialcy motion	0.28	0.10	–	–

This line of reasoning is also in accord with the analysis by *Phillips* [6.32] which suggests that the cohesive energy of silver halides is near the border line between heteropolar and homopolar nature, and the energy difference will be small between configurations with coordination number 6 (NaCl structure) and 4 (wurzite or zinc blende structure). The unique lattice properties of silver halides mentioned above are closely related to the features of the bonding character of the silver halides, which is in strong contrast to the entirely ionic bonding of alkali halides.

6.2 Excitons and Exciton-Phonon Interactions in Silver Halides

Following a discussion of the specific features of the electronic and lattice properties of silver halides, we now proceed to show how these features are reflected in the exciton-related phenomena found in silver halides. On the basis of spectroscopic studies to be described for the absorption and luminescence at low temperature and using the high spectral resolution as far as possible, we now have a fairly complete picture of exciton transitions in silver halides, at least for the lowest exciton states. In addition to the phonon-assisted indirect transitions mentioned in Sect. 6.1, various interesting features of exciton-phonon interactions have been discovered, indicating that the excitons in silver halides are characterized by a borderline nature between strong and weak coupling with phonons. This contrasts the exciton in silver halides to that in alkali halides or rare gas solids which have strong exciton-phonon coupling, on one hand, and to that in homopolar semiconductors with weak coupling, on the other.

In the following, the exciton transitions in pure AgBr and AgCl are described in Sect. 6.2.1, including both absorption and luminescence spectra at low temperature, in Sect. 6.2.2, the exciton transitions in mixed crystals are shown to contribute to the deeper understanding of excitons in pure crystal, and in Sect. 6.2.3, the bound exciton transitions at isoelectronic iodine impurity in AgBr and AgCl are presented.

6.2.1 Exciton Transitions in Pure Crystals

a) Fundamental Absorption Spectra. The overall fundamental absorption spectra of silver halides can be grouped into two categories. One is the optical excitation of the valence bands and the other is that of the deeper core levels. The valence excitation spectrum starts from the indirect-exciton threshold and is followed by the direct-exciton transition. This part of the spectrum was shown in Fig. 6.1 and will be described later in more detail. Above the direct-exciton peaks corresponding to the $\Gamma_{15} \rightarrow \Gamma_1$ direct gap transition in Fig. 6.2, several absorption peaks have been observed since as early as 1956 by *Okamoto* [6.5] and thoroughly studied by *White* et al. [6.33, 34] based on the reflectivity spectrum of bulk crystals at room temperature.

Figure 6.4 shows the absorption spectra obtained by Kramers-Kronig transformation of reflectivity spectra at 4.2 K [6.35]. The energies of absorption peaks and thresholds observed in the valence excitation spectra are tabulated in Table 6.3 along with the assignment suggested for the transitions. Except for the indirect-exciton threshold and the direct-exciton peak, most of the assignments in Table 6.3 are based on the comparison of the transition energies with the results of band calculations. Further experimental information on the symmetry characters, such as that based on piezo-optical and magneto-optical studies, will be required for the more direct assignment of the optical transitions which is indispensable for the elucidation of higher-conduction-band structure as well as that of the valence band.

The spectral structures originating from the valence excitation cease to be observed above 20 eV, and the structures due to core-level excitation start to appear superposed on the broad background of valence excitation. The core-level excitation spectra of silver halides have been studied by *Carrera* and *Brown*

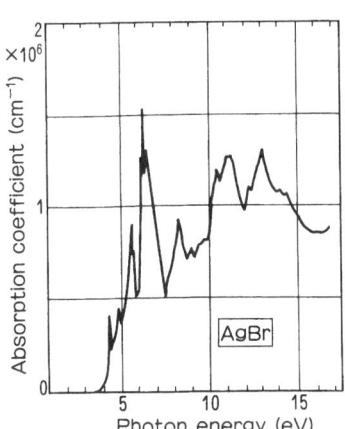
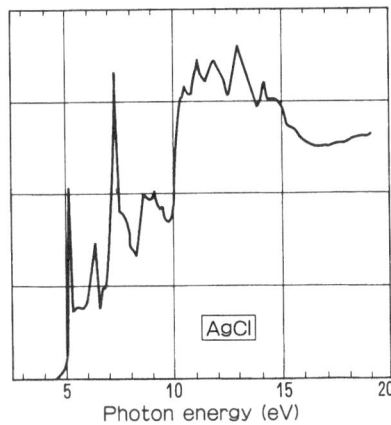

Fig. 6.4. Absorption spectra of silver halides calculated from the reflectivity spectra of bulk crystals at 4.2 K [6.35]

Table 6.3. Main spectral structures observed in the valence excitation spectra of silver halides [6.23, 35] at 4.2 K

	Transition energy [eV]	Assignment
AgBr	2.6922	indirect-exciton threshold (phonon emission) $L_3' \rightarrow \Gamma_1$
	4.280	direct exciton $\Gamma_{15} \rightarrow \Gamma_1$
	4.820	(spin-orbit splitting)
	5.576	direct exciton $L_3' \rightarrow L_1$
	5.736	
	6.025	direct exciton at X
	6.32	
	8.26	$L_3' \rightarrow L_2'$
	10.47	$X_5' \rightarrow X_3$
	11.2	–
	13.1	–
AgCl	3.2560	indirect-exciton threshold (phonon emission) $L_3' \rightarrow \Gamma_1$
	5.13	direct exciton $\Gamma_{15} \rightarrow \Gamma_1$
	6.26	–
	7.30	$L_3' \rightarrow L_1$
	8.58	$L_3' \rightarrow L_2'$
	10.53	–
	11.16	–
	11.89	–
	12.95	$X_5' \rightarrow X_3$
	14.3	–
	22 (broad peak)	–

Table 6.4. Spectral features observed in the core-excitation spectra [6.36]

	Transition energy [eV]	Assignment
AgBr	48	–
	59.5	Ag $4p \rightarrow \Gamma_1$
	72	Br $3d \rightarrow \Gamma_1$
	75.5	Br $3d \rightarrow$ higher conduction band
	185	Br $3d \rightarrow \Gamma_1$
AgCl	49	–
	61	Ag $4p \rightarrow \Gamma_1$
	197.6	threshold of Cl $2p \rightarrow \Gamma_1$
	199.0	Cl $2p \rightarrow \Delta_1 - X_1$
	200.8	(spin-orbit splitting)
	201.7	Cl $2p \rightarrow L_2'$ or L_1
	205.4	Cl $2p \rightarrow X_3$
	210.5	Cl $2p \rightarrow \Gamma_{25}'$
	212.2	Cl $2p \rightarrow \Gamma_{12}$

[6.36] using thin films and synchrotron radiation. Their results are shown in Table 6.4 with the suggested assignments. Although the assignments of the initial core states in the transitions are fairly unambiguous, those of the final conduction-band states are tentatively based on comparisons with the band calculations. Some of the transitions could be assigned as core-exciton transitions, but the experimental basis for core-exciton is not well founded at the present stage [6.36]. In principle, the core-excitation spectra starting from the nearly flat core levels are expected to give explicit information on the final conduction band density of states modulated by the selection rules imposed by the symmetry of the initial and final states. Future experimental improvement is hoped for in reflectance spectroscopy on bulk crystals, possibly by using the second derivative methods which turned out to be very useful in elucidating the core-exciton transitions superposed on the broad background due to valence excitation [6.37].

This section indicates that the present understanding of the fundamental absorption spectra of silver halides is more or less confined to the transitions between the upper part of the valence and the lower part of the conduction bands. As for the higher-energy transitions, not even qualitative discrimination between exciton and band-to-band transitions is conclusive.

b) Direct-exciton Transition. Figures 6.5 and 6.6 show the direct-exciton absorption spectra of AgBr and AgCl, respectively, as a function of temperature. For

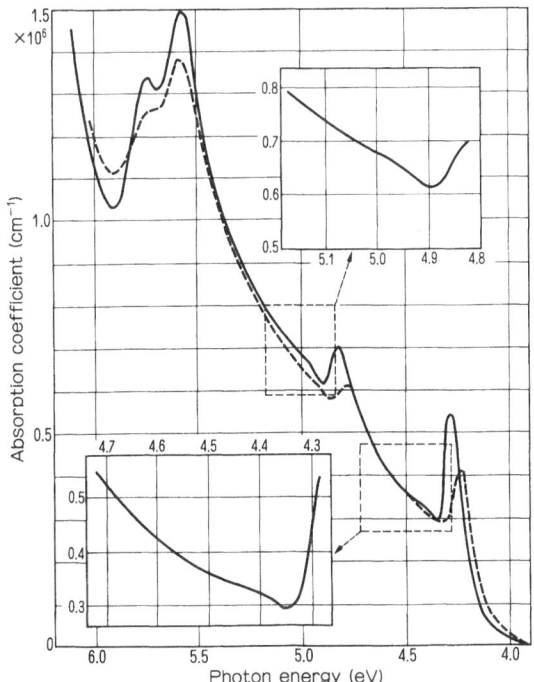

Fig. 6.5. Thin film absorption spectra of AgBr at 4.2 K (——) and 77 K (– – –) [6.38]

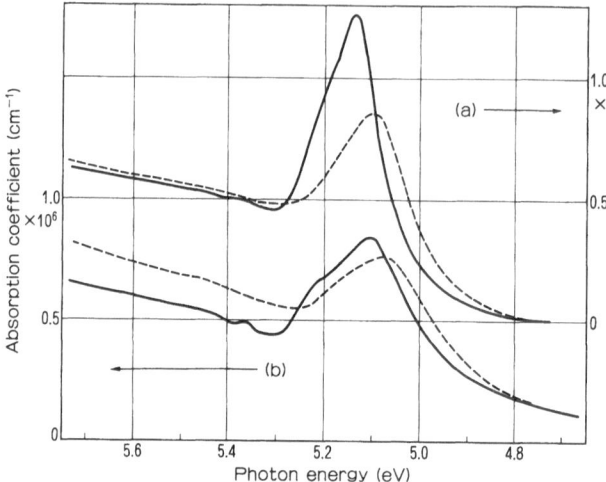

Fig. 6.6. Thin film absorption spectra of AgCl at 4.2 K (—) and 77 K (---) [6.38]. The cases (a) and (b) correspond to the films prepared with different deposition rates

measurement in such a high absorption range, thin evaporated films had to be used and this resulted in a variety of absorption spectra depending on the sample preparation conditions [6.38]. Typically, the half-widths of absorption bands are found to depend on the deposition rate for the same film thickness, and the peak positions depend on the choice of substrate, as well. For example, use of a cleaved alkali halide crystal as the substrate sharpened the absorption bands, and the absorption peaks were shifted to slightly higher energies. The spectra shown in Figs. 6.5 and 6.6a are for the films evaporated on fused silica at room temperature, with the deposition rate optimized for sharp spectrum shape. These results can be considered to represent the intrinsic spectra of silver halide films of good quality. For example, the AgCl spectrum in Fig. 6.6a shows reasonably good agreement with those on carefully prepared films in [6.36].

The lower-energy two peaks of AgBr, 4.280 and 4.820 eV, in Fig. 6.5 have been identified as the spin-orbit-split direct excitons. The energy splitting of 0.54 eV is reasonable in comparison to the minimum splitting (0.432 eV) expected for a hole in the $j = 3/2$, $j = 1/2$ bromine valence bands [6.39], corresponding to the splitting of Γ_{15} in Fig. 6.2 into Γ_8^- and Γ_6^-. The spin-orbit splitting is not apparently clear for the direct-exciton transition of AgCl in Fig. 6.6a, but the absorption spectra of samples with the exceptionally broad line shapes shown in Fig. 6.6b indicate a clear spin-orbit splitting of 0.10 eV, which is evidently similar to the data of *Okamoto* [6.5].

Recently, a reflectivity measurement has been made on AgCl single crystals at 4.2 K, by *Yanagihara* et al. [6.35]. Their results after Kramers-Kronig transformation confirmed that the absorption spectrum of AgCl in Fig. 6.6a reflects the intrinsic spectrum of the bulk crystal. The half-width of 0.175 eV at 4.2 K in Fig. 6.6a, which is to be compared with 0.15 eV in [6.36], can be concluded as partly due to the spin-orbit splitting. After deconvolution of the AgCl spectra of bulk crystals in [6.35], the energy splitting is estimated as 0.10 eV, which is reasonably comparable with an estimated minimum for chlorine of 0.103 eV [6.39]. In the AgCl spectra in Fig. 6.6, on the other hand, the existence of two steps are clearly observed at the higher-energy side of the absorption peak, in contrast to only one "step" in the AgBr spectra in Fig. 6.5. The separation of the two steps in AgCl is approximately equal to 0.10 eV and can be concluded to represent the spin-orbit splitting effect from the analysis of spectrum shape to be described in the following.

Now we proceed to discuss the spectrum shape of the direct exciton transition. In the effective mass theory of excitons, the eigenstates of excitons are, from Sect. 1.2,

$$
\left.
\begin{aligned}
E_n(k) &= \varepsilon_g - \frac{G}{n^2} + \frac{\hbar^2 K^2}{2M} , \\[2mm]
G &= \frac{\mu e^4}{2\,\epsilon_{\text{eff}}\hbar^2}
\end{aligned}
\right\}
\tag{6.7}
$$

where G is the binding energy of the $n = 1$ exciton, μ and M are exciton masses for internal and translational motion, respectively, K the exciton wave vector, and ϵ_{eff} the dielectric constant describing the effective Coulomb interaction in the exciton. In the dipole-allowed transition of direct excitons, the expressions for the absorption coefficient $\alpha(\hbar\omega)$ have been given by *Elliot* [6.40]. The results indicate: a) a series of exciton absorption lines in the region of the forbidden energy gap with absorption strength proportional to

$$
\frac{1}{a_B^3 n^3} , \quad \text{for} \quad n = 1, 2, 3 \ldots
\tag{6.8}
$$

where the exciton radius a_B for the $n = 1$ exciton is given by

$$
a_B = \hbar^2 \epsilon_{\text{eff}}/\mu e^2 ,
\tag{6.9}
$$

b) the continuous absorption above the direct energy gap ε_g, and c) a quasi-continuum region near $\hbar\omega \approx \varepsilon_g$, giving

$$
\alpha(\hbar\omega)_{\text{quasi-cont.}} \propto \frac{1}{a_B^3 G} .
\tag{6.10}
$$

In the case of direct exciton spectra of silver halides such as shown in Figs. 6.5 and 6.6, however, the exciton series are not separately observed, and only the $n = 1$ exciton band along with the "steps" at the higher-energy side are observed. Apparently, the broadening effects caused the amalgamation of the $n \geq 2$ exciton series converging toward the direct energy gap. Spectrum shapes for such circumstances have been analyzed by *Kotani* and *Toyozawa* [6.41]. The $n = 1$

exciton line is now an absorption band, and the quasi-continuum absorption (6.10) starts by making a step at

$$(\hbar\omega - \varepsilon_g)/G = -2\left(\sum_{n=2}^{\infty} \frac{1}{n^3}\right) = -0.404 . \tag{6.11}$$

In this model, the absorption coefficient at the step will be given by,

$$\frac{1}{2G} \text{ [integrated absorption of } n = 1 \text{ exciton band]} \tag{6.12}$$

which is equal to

$$\text{[absorption coefficient at the } n = 1 \text{ exciton peak]} \cdot (\pi W/4G) , \tag{6.13}$$

when the $n = 1$ exciton band exhibits Lorentzian shape with half-width W, as in the present data at lower temperatures. The above expectations are also supported by the following observation: the integrated absorption of the $n = 1$ exciton band as well as the absorption coefficient at the step is found to be constant for various temperatures between 4.2 K and 77 K, within experimental error of a few percent. The result also suggests that the direct-exciton binding energy is independent of temperature in the range studied. The exciton binding energies for direct excitons estimated by (6.13) can reproduce the observed spectra consistently according to (6.11). The direct-exciton parameters thus obtained are shown in Table 6.5, together with those for indirect excitons to be described later.

As for the binding energies, the values for direct excitons are definitely larger than for indirect excitons. This is as expected from the following consideration. The reduced mass of excitons μ will be

$$\frac{1}{\mu} = \frac{1}{m_e^*} + \frac{1}{m_h^*} \tag{6.14}$$

Table 6.5. Exciton parameters for silver halides estimated from absorption spectra at 4.2 K [6.23, 38]. All energies are in eV

	AgBr	AgCl
Direct exciton		
peak energy	4.280	5.13
exciton binding energy	0.16	0.33
direct band gap	4.44	5.46
Indirect exciton		
exciton energy gap	2.684	3.248
exciton binding energy	0.022	(0.04)
indirect band gap	2.706	(3.29)

where m_e^* and m_h^* correspond to the effective mass of electron and hole, respectively, at the band extrema relevant to the transition. In view of the band structure of silver halides, the sign of m_h^* will be different for direct and indirect excitons; negative for Γ_{15} relevant to the direct exciton and positive for L_3' for the indirect exciton. Then, the value of μ will be larger for the direct exciton in comparison to the indirect exciton. Such a difference in the reduced masses, along with the change of ϵ_{eff} due to the different exciton radius as expected from the formula by *Haken* [6.42], will be the origin of the different binding energies between direct and indirect excitons in Table 6.5. However, it should be noticed that the above estimate of exciton binding energies are based on more or less indirect procedures. In the future, more direct information on the effective mass is highly desirable for the valence band extrema at Γ_{15}.

c) **Indirect-Exciton Absorption Spectra**. In contrast to the discrete absorption lines expected for the direct excitons as in (6.8), the absorption spectra for indirect transitions to the exciton states given by (6.7) will show a continuous absorption band to all states with momentum K in the exciton band [6.40]. The spectrum shape of the indirect-exciton absorption will be thus determined by the state-density of the exciton band (Sect. 4.3).

The indirect-exciton absorption in pure silver halides proceeds by the assistance of momentum-conserving MC phonons, as suggested from its temperature dependence shown in Fig. 6.1. The absorption coefficient $\alpha(\hbar\omega)$ in the phonon-assisted indirect transition will consist of two components $\alpha_e(\hbar\omega)$ and $\alpha_a(\hbar\omega)$, due to emission and absorption of MC phonons, respectively. On a general basis, the following expressions are valid for a single MC phonon of energy $\hbar\omega_{MC}$

$$\left.\begin{aligned} \alpha_e(\hbar\omega) &= (n_{MC} + 1) \cdot g(\hbar\omega - E' - \hbar\omega_{MC}) \\ \alpha_a(\hbar\omega) &= n_{MC} \cdot g(\hbar\omega - E' + \hbar\omega_{MC}) \end{aligned}\right\} \tag{6.15}$$

where n_{MC} is the number of available MC phonons, and the function g is a basic shape function which rises from a threshold $(E' \pm \hbar\omega_{MC})$ with a characteristic shape depending on the nature of the transition. The temperature dependence of n_{MC}

$$n_{MC} = [\exp(\hbar\omega_{MC}/k_B T) - 1]^{-1} \tag{6.16}$$

determines the temperature dependence of the absorption spectra.

The phonon absorption component vanishes as $T \to 0$ and the phonon emission component approaches $g(\hbar\omega - E' - \hbar\omega_{MC})$. The basic shape function g in the effective mass approximation for the indirect allowed exciton will be proportional to

$$\frac{1}{a_B^3 n^3}\left[\hbar\omega - \left(\varepsilon_g - \frac{G}{n^2} \pm \hbar\omega_{MC}\right)\right]^{1/2} \tag{6.17}$$

where + and − stand for phonon emission and phonon absorption, respectively. The proportional factor which includes phonon matrix elements is assumed to be constant. Although the expression of the g function for the band-to-band transition is fairly complicated and will not be given here, the overall spectrum shape for indirect absorption in the effective mass approximation will be as follows [6.40]: 1) the absorption starts at the indirect exciton threshold of $(\varepsilon_g - G \pm \hbar\omega_{MC})$ with the spectrum shape given by

$$[\hbar\omega - (\varepsilon_g - G \pm \hbar\omega_{MC})]^{1/2} , \tag{6.18}$$

2) the spectrum near the indirect energy gap of

$$\varepsilon_g \pm \hbar\omega_{MC} \tag{6.19}$$

is proportional to

$$[\hbar\omega - (\varepsilon_g \pm \hbar\omega_{MC})]^{3/2} , \tag{6.20}$$

and 3) the spectrum proportional to

$$[\hbar\omega - (\varepsilon_g \pm \hbar\omega_{MC})]^2 , \tag{6.21}$$

is expected above the indirect energy gap for $[\hbar\omega - (\varepsilon_g \pm \hbar\omega_{MC})] \gg G$.

The absorption edge spectra of AgBr and AgCl as a function of temperature are shown in Figs. 6.7 and 6.8, respectively. On increasing the temperature, the

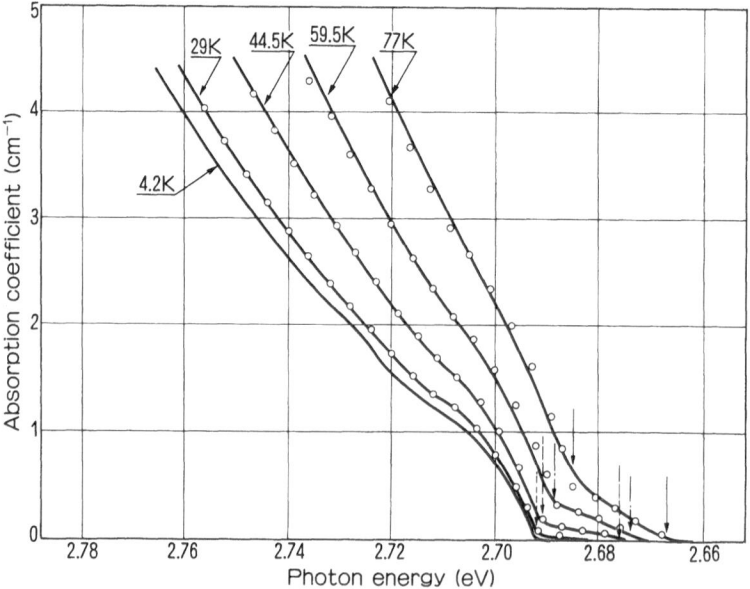

Fig. 6.7. Absorption spectra of a AgBr crystal as a function of temperature [6.38]. Circles are the fit to the temperature dependence using $\hbar\omega_{MC} = 8.0$ meV. Arrows indicate the MC phonon thresholds

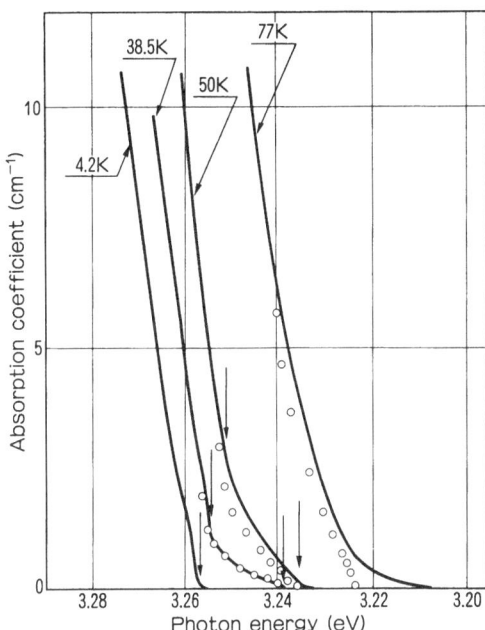

Fig. 6.8. Absorption spectra of a AgCl crystal as a function of temperature [6.38]. Circles are the fit to the temperature dependence using $\hbar\omega_{MC} = 8.0$ meV. Arrows indicate the MC phonon thresholds

absorption strength increases according to (6.16), and the absorption threshold due to the phonon-absorption component appears at the lower-energy side of the phonon-emission component. As shown in the figures, the fit to the data for the temperature dependence of the absorption spectra given by (6.15) and (6.16) is found to be fairly satisfactory, for the values of $\hbar\omega_{MC}$ obtained from absorption thresholds, which are

$$\hbar\omega_{MC} = \tfrac{1}{2}(\alpha_e \text{ theshold} - \alpha_a \text{ threshold}) . \qquad (6.22)$$

The estimated values of $\hbar\omega_{MC}$ dominating the absorption edge are 8.0 ± 0.2 and 8.0 ± 0.8 meV for AgBr and AgCl, respectively. As the result of above analysis, the value of the $n = 1$ exciton energy $(\varepsilon_g - G)$, called the *exciton energy gap*, is obtained as a function of temperature and shown in Fig. 6.9.

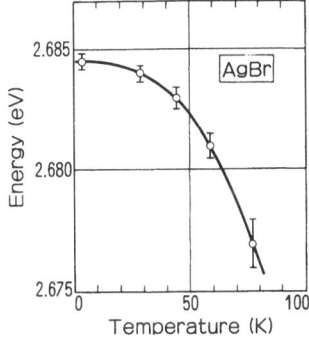

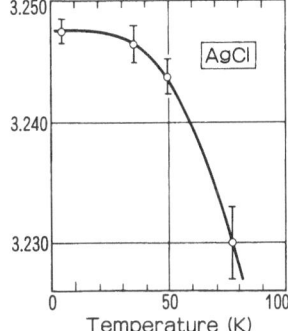

Fig. 6.9. Indirect-exciton energy gap of silver halides vs temperature [6.38]

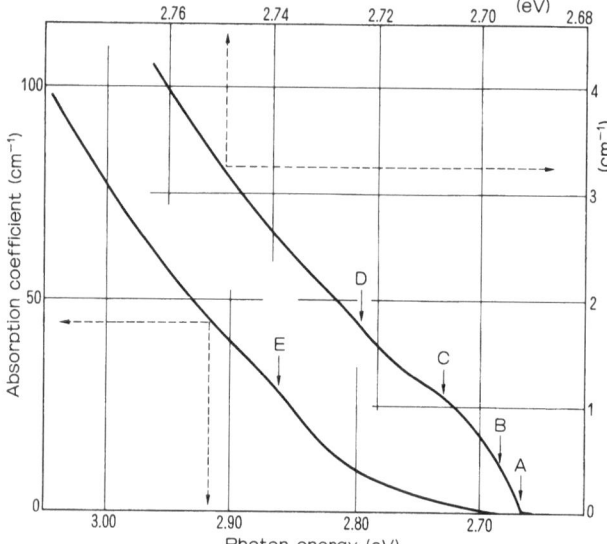

Fig. 6.10. Fine structures in the absorption edge spectra of a AgBr crystal at 4.2 K [6.38]. Energy positions are in Table 6.6

Figures 6.10 and 6.11 show the indirect absorption spectra at 4.2 K, for AgBr and AgCl, respectively. These spectra represent the spectral shape of the basic g function (6.17), corresponding to the phonon emission spectra $\alpha_e(\hbar\omega)$. The energies of the fine structure observed in the spectra are tabulated in Table 6.6, along with the assignments. The first two structures A and B, in both AgBr and AgCl, correspond to the absorption thresholds due to emission of the two kinds of MC

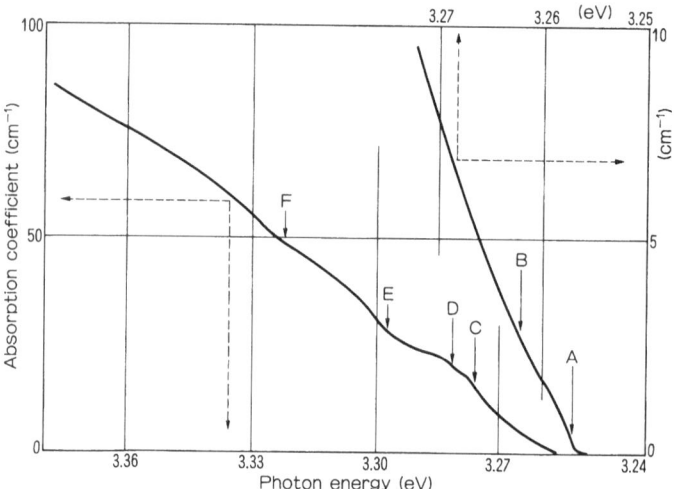

Fig. 6.11. Fine structures in the absorption edge spectra of AgCl crystal at 4.2 K [6.38]. Energy positions are in Table 6.6

Table 6.6. Energies of the fine structure in the indirect transition spectra of silver halides at 4.2 K in Figs. 6.10 and 6.11. All energies are in eV

AgBr	Assignment
A. 2.6922	threshold for indirect exciton with emission of MC phonon TO(L)
B. 2.6962	threshold for indirect exciton with emission of MC phonon LA(L)
C. 2.708	$\approx E_A + LO(\Gamma)$
D. 2.725	$\approx E_A + 2LO(\Gamma)$
E. 2.862	indirect exciton due to valence band maximum at Σ or due to spin-orbit splitting at L

AgCl	Assignment
A. 3.257	threshold for indirect exciton with emission of MC phonon TA(L)
B. 2.262	threshold for indirect exciton with emission of MC phonon LA(L)
C. 3.276	$E_A + 0.8 LO(\Gamma)$
D. 3.281	$E_B + 0.8 LO(\Gamma)$
E. 3.298	$E_C + LO(\Gamma)$
F. 3.322	$E_C + 2LO(\Gamma)$

phonons given in (6.3, 4) assisting the same indirect-exciton transition. The structures observed at the higher-energy side (C in AgBr and C, D in AgCl) are assigned as the one-phonon sidebands of the $n = 1$ exciton transition. The energies of the $LO(\Gamma)$ phonon are 17 and 24 meV, for AgBr and AgCl, respectively [6.25, 27], and the observed energy separation of the phonon sidebands relative to the *zero-phonon* absorption is somewhat less than that of $LO(\Gamma)$. The nature of the present phonon sidebands of the indirect-exciton transition will be similar to the bound and quasi-bound phonon sidebands which were discussed by *Toyozawa* [6.43] for direct-exciton transitions in Sect. 4.3. In the present case, the exciton binding energy is slightly larger than the $LO(\Gamma)$ phonon energy and the situation may correspond to the indirect-exciton–phonon bound state proposed by *Toyozawa* and *Hermanson* [6.44]. As for the higher phonon sidebands (D in AgBr and E, F in AgCl), the indirect-exciton–phonon quasi-bound state may be a more appropriate explanation.

The assignments proposed for the observed structures mentioned above are qualitatively common to AgBr and AgCl. However, there are distinct quantitative differences between the two. As is evident in Figs. 6.10 and 6.11, the increase of absorption coefficient above the thresholds is far steeper in AgCl than in AgBr. The observed difference originates from the following two factors. Firstly, the intensity ratio of two MC components, B component/A component, is about 0.2 in AgBr [6.23] but is nearly twice as large in AgCl. This is due to the difference between AgBr and AgCl of the relative magnitude of the phonon matrix elements for the two MC phonons. Secondly, the strength of the phonon sidebands relative to the zero-phonon band is far larger in AgCl than in AgBr. This clearly suggests stronger exciton-LO-phonon coupling during the indirect-exciton transition in AgCl than in AgBr. The spectrum shape for AgBr in Fig. 6.10, in the spectral region up to about 2.75 eV, can be well fitted simply by (6.17–21) assuming the value of the binding energy

$$G_{ex} = 22 \pm 2 \text{ meV} . \tag{6.23}$$

The binding energy thus estimated is close to that expected from (6.7) and (6.14) in the effective mass theory which is 25.6 meV for AgBr using the polaron mass values $m_e^* = 0.289 \, m_0$, m_h^* (transverse) $= 0.79 \, m_0$, and the static dielectric constant $\epsilon_0 = 10.6$ for ϵ_{eff}. It can be concluded that the absorption edge spectra of AgBr are fairly well described by the effective mass approximation, and the deviation from (6.17) caused by the LO phonon sidebands is not dominant.

A similar attempt to fit the AgCl spectra in Fig. 6.11 resulted in the value of $G_{ex} = 0.13 \pm 0.01$ eV which is unreasonably large compared to the effective mass value of 0.065 eV for AgCl assuming $\mu = m_e^* = 0.43 \, m_0$ and $\epsilon_{eff} = \epsilon_0 = 9.50$. The situation may be that the AgCl spectrum in the region of Fig. 6.11 is dominated by the multi-LO-phonon sidebands rather than the exciton series converging to the continuum. The quantitative difference between the indirect-exciton spectra observed for AgBr and AgCl is a manifestation of far stronger exciton-phonon coupling in AgCl than in AgBr.

The structure E in the AgBr spectra is unique and its origin is qualitatively different from that of the other structures. There are two possibilities; one is that it is the indirect exciton associated with a valence-band maximum at somewhere other than L, possibly at Σ, and the other is that it is due to the spin-orbit splitting of the L_3' valence band. The valence-band maximum L_3' will be split into (L_4^-, L_5^-) and L_6^- states [6.13] by the spin-orbit interaction. The magneto-absorption study of silver halides by *Matsushita* [6.45] concluded unambiguously that the (L_4^-, L_5^-) state corresponds to the lowest indirect exciton dominating the absorption threshold of AgBr and AgCl. The contribution from the L_6^- state may be expected to appear at a higher-energy. The energy separation between the structure E and the absorption threshold A is about 0.17 eV in Table 6.6, and it is interesting to note that the energy separation coincides with the separation of 0.16 eV between the 5.576 and 5.736 eV doublet peaks in Table 6.3, within experimental uncertainties. Although the spin-orbit splitting at L_3' is theoretically estimated as less than 0.1 eV [6.13], it is noticeable that the value is expected to be larger than the spin-orbit splitting for the valence band maximum at Σ [6.13]. At the present stage, the assignment of the indirect exciton of $L_6^- \rightarrow \Gamma_1(\Gamma_6^+)$ may be preferred for the structure E observed in AgBr. More direct information based on magneto-absorption spectroscopy in high magnetic fields will discriminate between the alternative possibilities.

d) Luminescence Spectra of Pure Crystals. In contrast to absorption spectra which give information on the free excited states in the ground-state configuration, the luminescence spectra reveal the optical transitions in the relaxed-state configuration in which various lattice relaxation phenomena will be dominant. Although the luminescence of silver halides has long been studied to obtain information on recombination processes competing with photographic sensitivity, fundamental understanding of the luminescence has been achieved only recently through high resolution spectroscopy studies at liquid helium temperature (LHeT) [6.4].

Photoluminescence spectra of nominally pure AgBr at LHeT consist of three components with different origins [6.23]. They are, (i) a 2.50 eV band with multiphonon structures, (ii) a 2.2 eV broad band and (iii) a high-energy component which is located at the high-energy end of the spectra and consists of a group of narrow luminescence lines. The high-energy component contains an intrinsic luminescence due to decay of the indirect exciton.

The 2.50 eV band is due to the decay of a bound exciton at a residual iodine impurity and will be discussed in Sect. 6.2.3 in detail. The 2.2 eV band is also of an extrinsic origin with a feature of pronounced delayed luminescence and has been considered to be related to hole trapping at silver ion vacancies which are charge-compensating the residual divalent cation impurities [6.4]. In the luminescence of nominally pure AgBr, even for zone-refined samples, the intrinsic luminescence due to free-exciton decay observed in the high-energy luminescence component is not a dominant decay channel and the luminescence bands of extrinsic origin, such as the 2.5 and 2.2 eV bands, are much more intense. This may be understood from the indirect gap nature of AgBr, because the radiative decay of a free exciton assisted by phonons is an inefficient process and the decay of an exciton localized at an impurity or defect proceeds with higher efficiency.

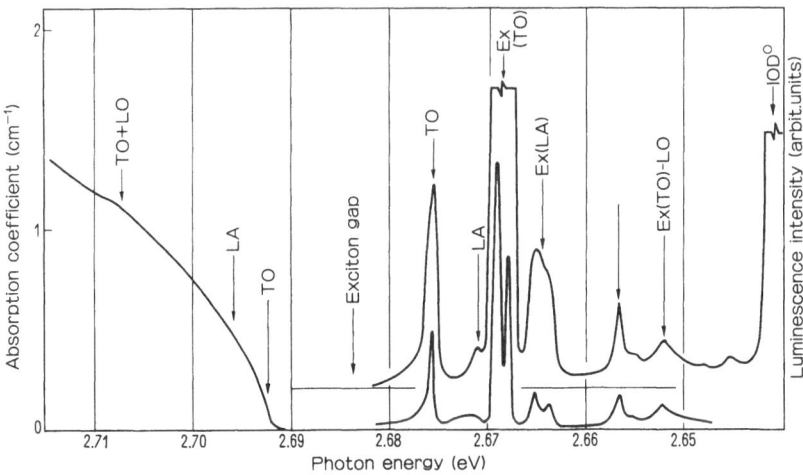

Fig. 6.12. Photoluminescence spectrum (*right*) of pure AgBr at 2 K, together with the absorption spectrum (*left*) [6.23]. Excitation used is the 300–410 nm light from a high-pressure mercury lamp

A typical spectrum of the high-energy luminescence component in AgBr is shown in Fig. 6.12. The two bands, labeled TO and LA, observed at the high-energy end correspond to the radiative decay of indirect excitons assisted by MC phonons, TO(L) and LA(L), respectively. In a similar way to the absorption spectrum given by (6.17), the luminescence spectrum due to the decay, with the emission of MC phonons, of an indirect exciton at a low temperature will be proportional to

$$\left.\begin{array}{l} \varepsilon^{1/2} \cdot e^{-\varepsilon/k_B T_{eff}} \quad \text{where} \\ \varepsilon = \hbar\omega - (\varepsilon_g - G - \hbar\omega_{MC}) \end{array}\right\} \tag{6.24}$$

and the effective temperature T_{eff} is not necessarily equal to the lattice temperature T. The TO and LA bands have intensity profiles of the Maxwell-Boltzmann distribution as expected from (6.24), with half-widths of 0.5 meV at 2 K. The low-energy thresholds are determined by fitting the spectrum shape to (6.24) and estimated as 2.6754 and 2.6712 eV, for TO and LA bands, respectively. Half of the energy interval between the threshold energies of the luminescence bands and those of the corresponding absorption spectra agrees with energies of MC phonons given by (6.3). Furthermore, the intensity ratio between the TO and LA bands agrees with the corresponding ratio observed in the absorption spectra. On the basis of these results, it is now firmly established that the free indirect excitons in AgBr are stable even at 2 K and their decay at low temperatures proceeds through the emission of MC phonons.

Recently, the de-excitation processes from the photo-excited states in the indirect exciton absorption region have been studied in detail by *von der Osten* and his coworkers [6.46] through resonant Raman spectroscopy under selective excitation in the $n = 1$ exciton absorption band. Their results revealed the scattering of photo-excited states in the exciton band by various well-defined phonons through inter- and intraexciton valley scattering mechanisms, which leads finally to the Maxwell-Boltzmann distribution observed in the free-exciton luminescence spectra under higher-energy excitation, such as shown in Fig. 6.12. One noticeable result concerning the free-exciton decay in AgBr is that the experimental value of T_{eff} in (6.24) is much higher than lattice temperature T. Obviously, the lifetime of the free exciton is too short for excitons to reach thermal equilibrium with the lattice. The measured decay time is reported as less than 10^{-7} s [6.46, 47] which is much shorter than the radiative lifetime of 10^{-4} s expected from the transition probability for indirect-exciton absorption.

The luminescence bands, labeled EX bands in Fig. 6.12, are identified as extrinsic in origin being due to residual divalent cation impurities. Spectral features of the EX bands can be understood by assuming that the bands correspond to the radiative decay of *shallow-bound* indirect excitons with assistance of the same MC phonons, TO and LA, as in the free-exciton decay. This is in contrast with the *deep-bound* excitons such as bound excitons at iodine which decay without the assistance of MC phonons, as will be described in Sect. 6.2.2. The energy for *shallow localization* varies with the kind of impurities and this causes the coexistence of two EX bands in the case of Fig. 6.12. The localization energy of these shallow-bound excitons can be estimated as 6 to 7 meV from the energy shift relative to the free-exciton luminescence bands. It is observed that the addition of cadmium ions causes the enhancement of both the EX band in the high-energy luminescence and the 2.1 eV band which has a character very similar to the 2.2 eV band in pure AgBr. A possible explanation may be that the shallow localization of the EX exciton is related to shallow localization of electrons and the deep localization of excitons of 2.1 or 2.2 eV bands is due to deeper localization of holes at charge-compensating vacancies.

In contrast to AgBr in which free indirect excitons are stable, the excitons in AgCl are stabilized by forming self-trapped excitons [6.48]. As will be shown later in Fig. 6.16, the intrinsic luminescence of AgCl is observed as a broad emission band with a 2.5 eV peak which was first observed by *Smith* [6.49] under excitation at LHeT. In the meantime, the self-trapped hole Ag^{2+} was identified in AgCl through an ESR study by *Höhne* and *Stasiw* [6.50]. On the basis of later studies using transient absorption spectroscopy [6.51, 52] and optically detected magnetic resonance (ODMR) [6.53, 54], it is now established that the self-trapped exciton in AgCl is an electron localized around a self-trapped hole which is Ag^{2+} at a substitutional silver site. The self-trapped exciton, as well as the self-trapped hole, is stabilized by the tetragonal distortion of surrounding lattice. The recent study of AgCl by resonant Raman spectroscopy [6.55] also concluded that the self-trapping process is by far the most efficient channel of relaxation of the indirect exciton in AgCl. It is also interesting to note that the observation of weak scattering by two MC phonons resonant with the exciton transition in AgCl [6.46, 55] indicates the non-zero probability of free-exciton luminescence in AgCl; hot luminescence in this case.

Although the two silver halides, AgBr and AgCl, are very similar in their electronic and lattice properties, it is now evident that they exhibit distinct differences in the nature of relaxed excited states; free holes and excitons are stable in AgBr even at the lowest temperature, but both of them are self-trapped in AgCl at low temperatures. These differences observed between AgBr and AgCl suggest that the stability of relaxed states is determined by a critical magnitude of electron- and exciton-phonon coupling strength, as will be further discussed for AgBr–AgCl mixed crystals in Sect. 6.2.2.

6.2.2 Exciton Transitions in Mixed Crystals

On the basis of experimental data on the intrinsic absorption spectra of substitutional binary solid solutions, *Onodera* and *Toyozawa* [6.56] gave a unifying theoretical picture for the two different types of behavior of mixed crystal systems (see also Sect. 4.2). In the first type, called the persistence type, two structures due to each of the constituents persist in the spectra but in the second type, called the amalgamation type, they amalgamate into a single structure. In alkali halides, halogen substitution results in the persistence type and alkali substitution gives the amalgamation type. These features are explained in terms of the ratio.

[energy difference Δ between the energy band centers
of the two constituents]/
[width T of the energy band] .   (6.25)

When the ratio Δ/T is large, the energy band is split into two, each of which presents an individual absorption spectrum, which corresponds to the persistence type. When the ratio is small, the two energy bands unite to give a single absorption spectrum, which corresponds to the amalgamation type.

In the case of silver halides, the first experiment was carried out for the direct exciton in AgBr–AgCl mixed cystals [6.57] which showed amalgamation-type behavior. In the following, the spectra of indirect-exciton transitions are discussed for the mixed crystals of silver halides [6.23, 58]. It will be shown that the study of mixed crystals gives valuable information for deeper understanding of the indirect exciton in silver halides. Most of the results on absorption spectra of the indirect exciton in mixed crystals of silver halides will be shown to belong to the amalgamation type. For such cases, the mixed crystal may be regarded as a perfect crystal, as envisaged in the virtual crystal model proposed originally for interband transitions in the Si–Ge system [6.59]. In indirect-exciton transitions, however, the deviation from the virtual crystal description appears as the disorder-assisted transition in AgBr : Cl which was first noticed by *Joesten* and *Brown* [6.60] in absorption and later substantiated by luminescence spectra [6.23, 58]. A further interesting observation in the mixed crystal is the transition of the stable relaxed state from the free to self-trapped exciton in the AgBr–AgCl system.

a) AgBr–AgCl Mixed Crystal. The AgBr–AgCl mixed crystal forms a perfect solid solution in the entire range of composition x in $AgBr_{1-x}Cl_x$. Figure 6.13 shows a series of indirect absorption edge spectra of AgBr : Cl, corresponding to the AgBr side of the mixed crystal [6.23]. As observed by *Joesten* and *Brown* [6.60], one of the features in the absorption spectra of AgBr : Cl is the existence of indirect-exciton absorption without the assistance of MC phonons, the threshold of which corresponds to the exciton energy gap $(\varepsilon_g - G)$ and is specified as NP (no-phonon) in Fig. 6.13. The absorption shape of the NP component can be described by

$$\alpha(\hbar\omega) \propto (\hbar\omega - E_{NP})^{1/2} .$$ (6.26)

When the extrapolated absorption of the NP absorption according to (6.26) is subtraced from the spectra of AgBr : Cl, the remaining absorption is exactly the same as the absorption spectrum of AgBr, except in the higher-energy region where the indirect band-to-band transition due to the NP component becomes appreciable. The extra absorption due to chlorine increases its magnitude in proportion to chlorine concentration and is found to be independent of temperature between 2 and 77 K. This is consistent with the extra absorption corresponding to the indirect transition without phonon assistance which is allowed by momentum conservation due to scattering of the chlorine impurity during the

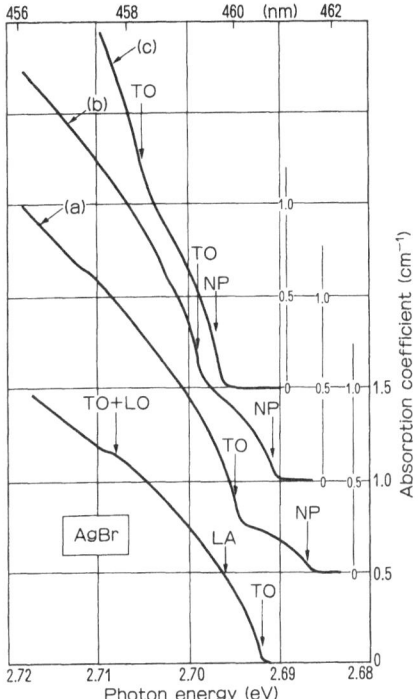

Fig. 6.13. Absorption edge spectra of AgBr : Cl at 2 K together with that for AgBr [6.23]. Chlorine concentrations: (a) 1.01 × 10^{-2}, (b) 2.07 × 10^{-2}, and (c) 4.02 × 10^{-2} mole fractions. Arrows indicate threshold energies of absorption structures

optical transition. Except for the NP component described above, the indirect absorption spectra of AgBr : Cl can be described perfectly in terms of the virtual crystal model; the absorption threshold shifts continuously in proportion to the chlorine concentration and the optical transition proceeds with the assistance of MC phonons of nearly the same energy as with pure AgBr [6.23].

The behavior of absorption spectra for the AgCl side is a little different. Figure 6.14 shows a typical absorption edge spectra of AgCl : Br at the low bromine concentration of below 1 × 10^{-3} mole fraction [6.61]. An extra absorption component appears at the low-energy side of the indirect absorption threshold, 3.2560 eV, of pure AgCl. The extra absorption due to bromine has a weak structure, possibly due to a phonon sideband, at 3.246 eV in Fig. 6.14, but otherwise the spectrum shape of the extra absorption is Gaussian which is similar to that of AgBr : I at temperatures where most of multi-phonon structures are smoothed out [6.61]. The bromine-induced extra absorption can be assigned as the bound-exciton transition at bromine impurities at a low concentration, which is similar in nature to the bound-exciton transiton in AgBr : I to be described later. The different shape of absorption spectra for AgCl : Br and AgBr : I will be due to the difference in the exciton-phonon coupling strength between the two. The bound excitons in both cases are of isoelectronic origin and the dominant exciton-phonon coupling is expected to come from the coupling of bound

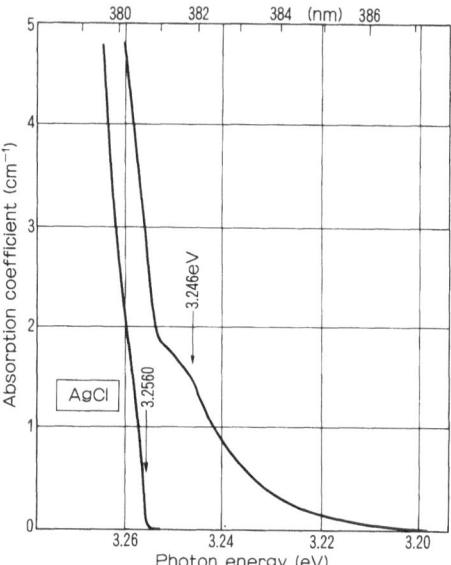

Fig. 6.14. Absorption edge spectra of AgCl : Br (2.12×10^{-4} mole fraction at 2 K together with AgCl [6.61]

holes at the impurity. Therefore, the observed difference will be due to the stronger phonon-coupling of the bound hole in AgCl : Br than AgBr : I.

Above a bromine concentration of 10^{-2} mole fraction in AgCl : Br, however, the bound-exciton transition described above amalgamates with the free-exciton absorption edge and the absorption shape resembles that of pure AgCl shifted parallel in energy as shown later in Fig. 6.16. Summarizing, the indirect-exciton absorption of $AgBr_{1-x}Cl_x$ can be described as the amalgamation or the virtual crystal type, except for the extreme end close to AgCl, corresponding to AgCl : Br below 10^{-2} mole fraction. The situation is in accord with the analysis by *Onodera* and *Toyozawa* [6.56] which predicted the possible occurrence of the persistence type in the limit end of the composition even when the amalgamation type prevails in most of the composition range of the mixed crystal (see Sect. 4.2).

Now we proceed to the luminescence spectra of the AgBr–AgCl mixed crystal. As concluded in Sect. 6.2.1, the free exciton is a stable relaxed state in AgBr, in contrast to AgCl in which the exciton is self-trapped during its relaxation. Then it is expected that a transition in the nature of the stable relaxed exciton will occur at a certain composition of the AgBr–AgCl mixed crystal. Experimental results on a series of $AgBr_{1-x}Cl_x$ [6.48] have shown that this kind of transition actually occurs around $x = 0.45$ rather abruptly against composition and this corresponds to the attainment of a critical value of exciton-phonon coupling strength which varies continuously with composition. Figure 6.15 shows the photoluminescence spectra of $AgBr_{1-x}Cl_x$ for $x \leq 0.4$ along with the absorption edge spectra. The luminescence spectra for the AgBr-rich crystals $x \leq 0.4$ are dominated by the radiative decay of the free exciton in the virtual crystal. The decay proceeds

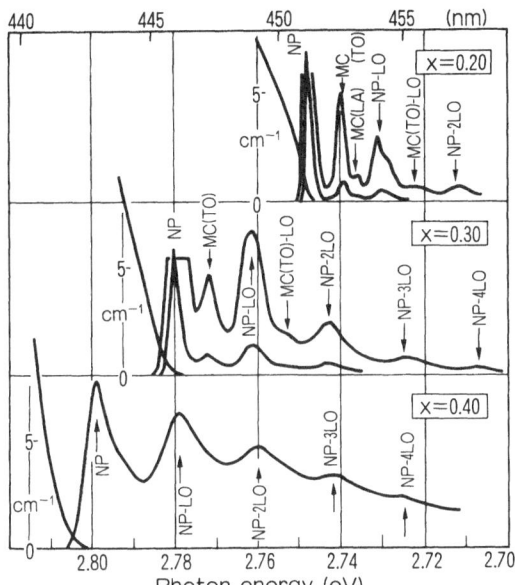

Fig. 6.15. Absorption edge (*left*) and photoluminescence (*right*) spectra at 2 K of $AgBr_{1-x}Cl_x$ for $x \leq 0.4$ [6.48]

through various modes, each of which corresponds to various luminescence bands specified in Fig. 6.15; without the assistance of MC phonons (NP), with the assistance of MC phonons [MC(TO) and MC(LA)], and through multi-phonon processes (MC-LO, NP-LO, NP-2LO, etc.). These radiative processes correspond exactly to those observed in the absorption spectra of AgBr : Cl described above with reference to Fig. 6.13.

On the other hand, the luminescence spectra of AgCl-rich crystals $0.50 < x < 1.0$ shown in Fig. 6.16 are dominated by a single broad band of Gaussian shape, with its peak around 2.5 eV. Similarity of the band shape among the luminescence bands for $0.50 < x < 1.0$ suggests that the origin of the luminescence is essentially the same; radiative decay of the self-trapped exciton. This interpretation is consistent with the observation that the luminescence peak shift between $x = 0.5$ and 1.0 is small (~ 0.06 eV) in comparison with the shift of absorption edge with x. The nature of the self-trapped exciton in $AgBr_{1-x}Cl_x$ for $0.47 \leq x \leq 1$ has been studied recently by an ODMR method by *Yamaga* et al. [6.62] and the self-trapped exciton for $0.47 \leq x \leq 0.7$ near the transition range of concentration has been assigned as a $(AgBr_6)^{5-}$ complex.

In the intermediate range of composition, $0.40 < x < 0.50$ in Fig. 6.16, the luminescence spectra consist of two kinds of luminescence components; one is the higher-energy component characterized by multi-phonon structures similar to those observed for $x \leq 0.40$ in Fig. 6.15, and the other is the lower-energy broad band similar to those for $x > 0.50$ with a peak around 2.56 eV. Variation of the intensity ratio between the two luminescence components against composition x indicates that a change of the main relaxed exciton occurs from free for $x < 0.40$

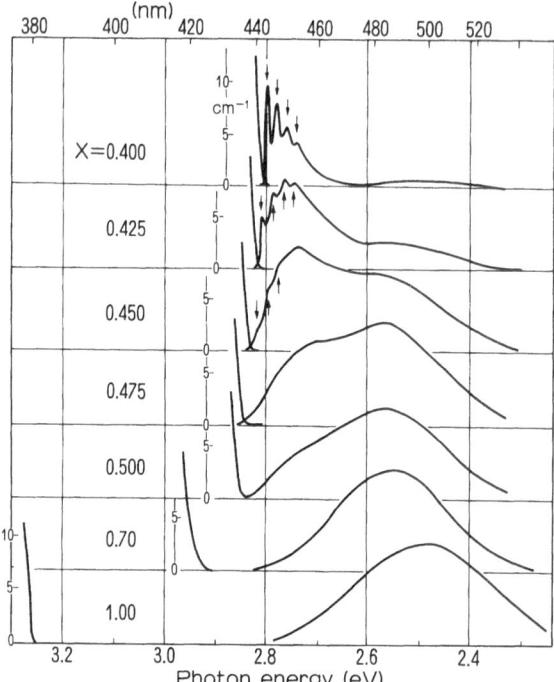

Fig. 6.16. Absorption edge (*left*) and photoluminescence (*right*) spectra at 2 K of $AgBr_{1-x}Cl_x$ for $x \geq 0.4$ [6.48]

to self-trapped for $x > 0.50$. It is concluded that the nature of the relaxed exciton changes discontinuously around $x = 0.45$ from free to self-trapped excitons, in contrast to the unrelaxed excitons which shift continuously in energy by forming a series of virtual crystals. The spectrum features of the two coexisting luminescence components at the composition range $0.4 < x < 0.5$ have recently been analyzed by *Toyozawa* [6.63], in terms of the formation of a *free–self-trapped* resonance state near the critical concentration of $x = 0.45$.

Considering that the self-trapping of an exciton is a result of strong exciton-phonon coupling, the discontinuous change observed in $AgBr_{1-x}Cl_x$ will correspond to the attainment of sufficient exciton-phonon coupling strength at the composition $x \geq 0.45$. According to the theoretical study [6.64] on the Urbach rule for fundamental absorption tail

$$a(\hbar\omega) = a_0 \exp\left(-\sigma\, \frac{\hbar\omega_0 - \hbar\omega}{k_B T}\right), \tag{6.27}$$

the value of the steepness coefficient σ can be correlated to the exciton-phonon coupling strength S (see Sects. 4.2 and 4.4). The theory predicts that σ is proportional to B/S (B is the exciton band width) and the self-trapping of the exciton is expected below a critical value of B/S close to unity. The absorption tail of $AgBr_{1-x}Cl_x$ near room temperature is well described by the Urbach rule, and it is found that σ varies linearly with x between $\sigma_{AgCl} = 0.78$ and $\sigma_{AgBr} = 0.97$ as

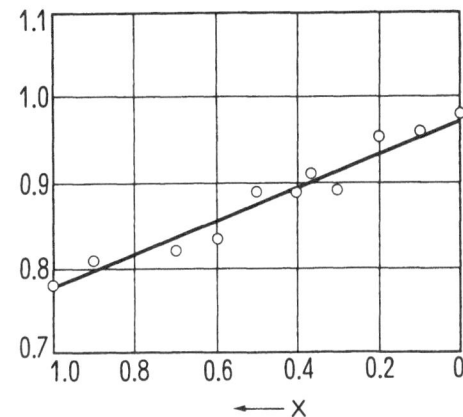

Fig. 6.17. Change of steepness coefficient σ in the Urbach rule for the absorption edge at 293 K against composition x of $AgBr_{1-x}Cl_x$ [6.48]

shown in Fig. 6.17. The stabilization of the self-trapped exciton for $x \geq 0.45$ corresponds to the critical value of $\sigma_{cr.} = 0.89$, consistent with the theoretical predictions. The experimental results on the luminescence spectra of $AgBr_{1-x}Cl_x$ can be considered as the first demonstration that an exciton is self-trapped below a critical value of B/S when its magnitude is varied continuously with composition x. Recently, *Schreiber* and *Toyozawa* [6.65] computed the line shape of the indirect absorption edge in detail and derived an expression for the correlation between the occurrence of exciton self-trapping and the steepness coefficient σ. Their results on the NaCl-type lattice with an indirect absorption edge predict a critical value of $\sigma_{cr.} = 0.88$, in excellent agreement with experimental value of 0.89 obtained above from the luminescence spectra of $AgBr_{1-x}Cl_x$.

b) AgBr : Na and AgBr : Li. Solid solutions are formed for AgBr : Na and AgBr : Li up to a concentration of 10^{-1} mole fraction. The effect of the addition of Na^+ and Li^+ on the absorption edge spectra of AgBr is found to be simply a parallel shift of spectra to higher energies with very little change in spectrum shape [6.58]. In the range of alkali concentration studied, the energy shift is proportional to the concentration, and extra absorption such as the NP component which is observed in AgBr : Cl is hardly observable for AgBr : Na and : Li.

Figure 6.18 shows a typical photoluminescence spectrum of AgBr : alkali for the concentration around 1×10^{-2} mole fraction. The distinct features are as follows [6.58]. (a) The intrinsic luminescence component due to free-exciton decay assisted by MC phonons and its LO phonon sidebands dominate the spectra at 2 K. The energies of MC and LO phonons agree with those observed for pure AgBr. (b) The NP luminescence component due to free-exciton decay without the assistance of MC phonons is observed. The intensity ratio of NP component/MC(TO) component, however, is about 0.01 for alkali concentration of 1×10^{-2} mole fraction and is an order of magnitude smaller than the corresponding value of 0.1 for AgBr : Cl of the same concentration. This is consistent with the NP component not being observed in the absorption spectra of AgBr : alkali in contrast to AgBr : Cl.

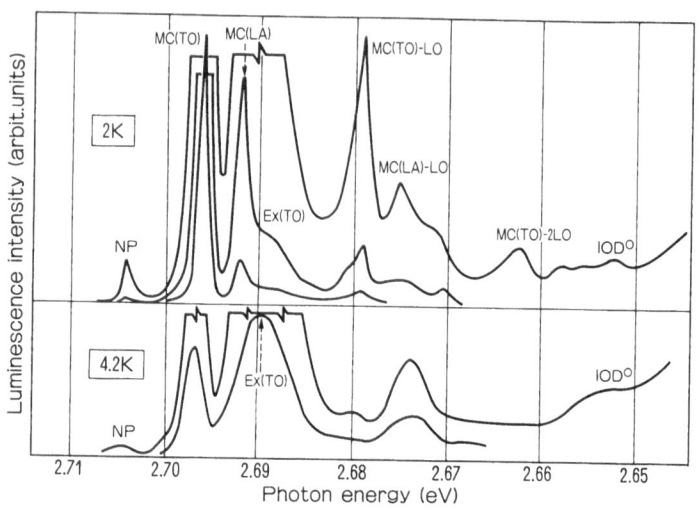

Fig. 6.18. Photoluminescence spectra at 4.2 K and 2 K of AgBr : Na (1.1 × 10⁻² mole fraction) [6.58]

These observations in the absorption and luminescence spectra of AgBr : Na and AgBr : Li lead to the conclusion that the addition of alkali results in the virtual crystal with exciton spectra of the amalgamation type. The situation is qualitatively similar to AgBr : Cl, except the main effect of alkali addition will be the raising of the conduction band in forming the virtual crystal while in AgBr : Cl chlorine addition will mainly lower the valence band. This expectation is substantiated by the shift observed in the transition energy of the iodine luminescence in forming the mixed crystals. As will be described in Sect. 6.2.3, the iodine luminescence is due to the decay of a bound-exciton at residual iodine, and the main origin of exciton localization comes from the binding of holes at iodine. To first order, the transition energy of iodine luminescence will be determined by the energy difference between the level of the hole trapped at iodine and the bottom of the conduction band. The location of the zero-phonon band of iodine luminescence is shifted to higher energy in AgBr : alkali, almost parallel to that of the free-exciton transition, but no shift is observed for chlorine addition in AgBr : Cl [6.23, 58]. These observations are consistent with the expectation that the main effect of alkali and chlorine addition is the change of conduction and valence bands, respectively, of AgBr.

Summarizing, the spectral behavior of mixed crystals of AgBr : Cl and AgBr : Na or Li are of the amalgamation or virtual crystal type. Both cases correspond to a small value of Δ/T in the analysis of [6.56]. On increasing the value of Δ/T, a transition is expected from amalgamation to persistence type. This situation is realized, for AgBr : I and AgCl : I to be described in the following section, which exhibit bound-exciton transitions in both absorption and luminescence spectra.

6.2.3 Bound-Exciton Transitions at an Isoelectronic Iodine Impurity

The solubility limits at room temperature of AgI in AgBr and AgCl are 3.9×10^{-1} and 5×10^{-2} mole fractions, respectively [6.66]. Most practical photographic emulsions contain AgI of order of 10^{-2} mol. In the course of spectroscopic studies on AgBr, *Kanzaki* and *Sakuragi* [6.61] added AgI of 10^{-4} mole fraction to AgBr in order to increase the yield strength for the uniaxial stress experiments at LHeT. They then discovered unique and interesting spectral features in both absorption and luminescence due to a bound-exciton at an iodine impurity. In a sense, they were fortunate in starting from the study of AgBr : I, because the photoluminescence spectrum of AgBr with the highest purity at the time was dominated by the presence of residual iodine of order of 10^{-6} mole fraction and the study of intentionally doped AgBr was essential to identify the origin of the luminescence. In the following, we review the bound-exciton transitions in AgBr : I and AgCl : I [6.61].

a) AgBr : I. Addition of iodine at low concentration which substitutes bromine produces an extra absorption near the indirect-exciton threshold of AgBr as shown in Fig. 6.19. The iodine-induced absorption consists of two components; the lower-energy component below 2.7 eV and the second component at higher

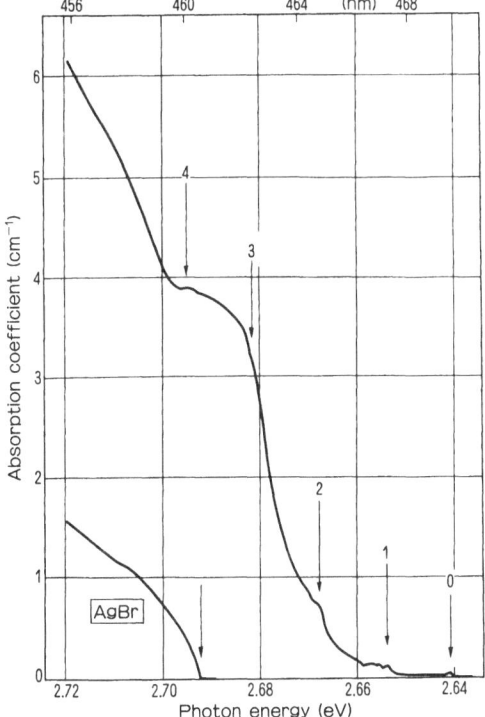

Fig. 6.19. Absorption spectrum of AgBr : I (5.7×10^{-4} mole fraction) at 2 K in comparison with pure AgBr [6.61]

Table 6.7. Energy position of multi-phonon structures in absorption spectrum of AgBr : I $(5.7 \times 10^{-4}$ mole fraction) at 2 K [6.61]. Here n refers to the number of phonons and ΔE is the energy interval

n	$\hbar\omega$ [eV]	ΔE [meV]
0	2.6413	–
1	2.6538	12.5
2	2.6676	13.8
3	2.6814	13.8
4	2.6950	13.6
Average		13.3

energy. The lower-energy component is characterized by a series of multi-phonon structures with nearly equal intervals as shown in Table 6.7. The lowest-energy peak at 2.6413 eV can be assigned as the zero-phonon line from the agreement of its transition energy with the luminescence peak to be described later. Substitution of bromine by isoelectronic iodine will produce an isoelectronic trap for holes due to the difference of electron affinity between the two halogens, and result in a bound exciton by subsequently attracting an electron. The lower-energy component thus corresponds to the bound-exciton transition at iodine. On the other hand, the higher-energy component has a different absorption shape which consists of a superposition of various transitions. The spectral shape of a hump at 2.71 eV suggests that one of the underlying transitions corresponds to unbound-exciton transitions enhanced by the presence of iodine. There will be other transitions occurring in the higher-energy region, such as the creation of an unbound electron and a bound hole at iodine, for example.

The multi-phonon structures in the lower-energy component at low temperature become diffuse for iodine concentrations above 10^{-3} mole fraction and are accompanied by a broad background absorption extending toward lower photon energy. Finally, only a broad absorption tail is observed for iodine of 1×10^{-2} mole fraction even at 2 K. This behaviour of the absorption spectra with increasing iodine concentration corresponds to the change of spectra from the persistence type (bound exciton) at the lowest concentration to the amalgamation type (virtual crystal), as discussed for the change of spectra from AgCl : Br to $AgBr_{1-x}Cl_x$ in Sect. 6.2.2. On the other hand, the energy positions of the multi-phonon series remain constant in the low concentration range of 5×10^{-4} to 2×10^{-3} mole fraction.

The strength of extra absorption due to iodine is found to be nearly independent of temperature between 2 and 77 K, and this is consistent with the transition being allowed by momentum conservation by the iodine impurity in contrast to the phonon-assisted indirect-exciton transition in pure AgBr. In the lowest temperature range, no drastic changes are observed in both absorption and luminescence spectra, except the energy shift of zero-phonon line toward higher photon energy by 0.1–0.2 meV in increasing temperature from 2 to 4.2 K. This type of *blue shift* with temperature between 2 and 4.2 K is also observed for the indirect-exciton threshold of pure AgBr and AgCl in contrast to the *red shift* observed appreciably above 20 K for all of these transitions. At temperatures above 20 K, the multi-phonon structures are smoothed out completely, and the absorption spectra of AgBr : I become very similar to AgCl : Br at 2 K (Fig. 6.14). Finally at 77 K, the absorption tail of AgBr : I and AgCl : Br at lower impurity concentrations follows well the Urbach rule of (6.27), with $\sigma = 0.5$ for both AgBr : I and AgCl : Br. The smaller value of σ compared with those of $\sigma_{AgBr} = 0.97$ and $\sigma_{AgCl} = 0.78$ is similar to the case of $\sigma = 0.51$ for KCl : Br [6.67] which is substantially smaller than $\sigma_{KCl} = 0.80$.

Now we proceed to more details of the iodine-induced absorption for low concentration and at low temperature. In addition to the main multi-phonon series, there is additional fine structure in the absorption spectra as shown in Fig. 6.20, which can be understood as due to transitions creating the bound exciton in the higher excited states.

The optical transition of the main multi-phonon series corresponds to creation of the $n = 1$ bound exciton. On the basis of energy intervals, the following assignments are proposed for the additional fine structure [6.61].

2.6551 eV: zero-phonon transition of the $n = 2$ exciton,

2.6572 eV: zero-phonon transition of higher excitons above $n = 3$ amalgamated to form a quasi continuum as in (6.11), and

2.6699 eV: one-phonon transition of the quasi continuum of excitons above $n = 2$.

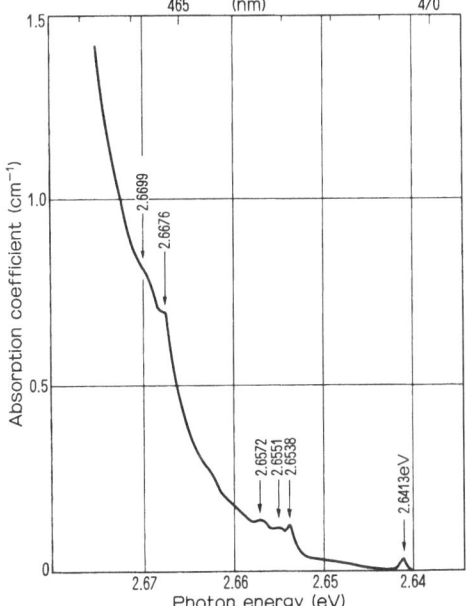

Fig. 6.20. Fine structure in the absorption spectrum of AgBr : I (5.7×10^{-4} mole fraction) at 2 K [6.61]

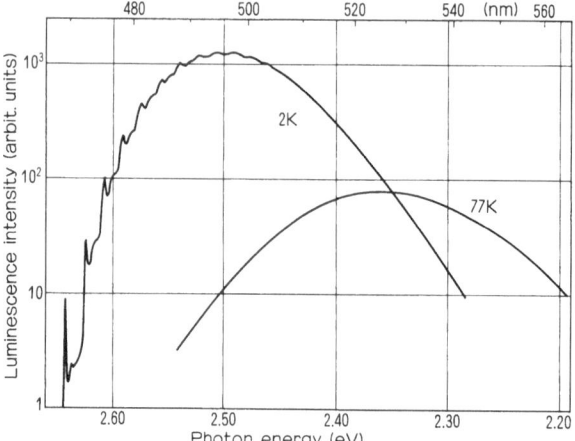

Fig. 6.21. Photolumines-
cence spectra of AgBr : I
$(5.7 \times 10^{-4}$ mole fraction)
at 2 K and 77 K [6.61]

These assignments lead to a value of the ionization energy of the $n = 1$ bound exciton of

$$18.5 \text{ meV} \tag{6.28}$$

which is close to the value of 22 meV estimated for the free exciton of pure AgBr in (6.23).

The luminescence spectra of AgBr : I at low concentration are shown in Fig. 6.21 as a function of temperature. The spectra at 2 K are characterized by a series of equally spaced luminescence lines, which decrease in intensity above 40 K and are finally replaced by a single broad band at 77 K. The same broad luminescence band is also observed for higher concentrations of iodine at 4.2 K, in addition to the series of luminescence lines at higher energy. The energy positions of the line series in Fig. 6.21 do not show any noticeable change on reducing the iodine concentration to of the order of 10^{-6} mole fraction which is the case of nominally pure AgBr. This shows that the series of luminescence lines is associated with an isolated iodine impurity dispersed in the AgBr matrix. On the other hand, the broad band at 4.2 K increases its intensity relative to the line series with increasing iodine concentration. On the basis of linear increase of the intensity ratio (broad band/line series) with iodine concentration, *Tsukakoshi* and *Kanzaki* [6.68] concluded that the origin of broad band was related to the existence of iodine pairs.

The peak energies of the line series in luminescence spectra are tabulated in Table 6.8. The average energy interval of 16.5 meV is very close to the LO(Γ) phonon energy 17.24 meV in AgBr [6.27]. This energy value is in contrast to the smaller energy interval of 13.3 meV in Table 6.7 observed in absorption spectra and this indicates the strong binding of excitons with phonons which leads to the exciton-phonon bound state as first pointed out by *Toyozawa* and *Hermanson* [6.44].

Table 6.8. Energy position of the peaks in the luminescence spectrum of AgBr : I (5.7×10^{-4} mole fraction) at
2 K [6.61]. Here n refers to the number of phonons and ΔE
is the energy interval

n	$\hbar\omega$ [eV]	ΔE [meV]
0	2.6412	–
1	2.6240	17.2
2	2.6075	16.5
3	2.5908	16.7
4	2.5741	16.7
5	2.5574	16.7
6	2.5410	16.4
7	2.5245	16.5
8	2.5082	16.3
9	2.4923	15.9
10	2.4760	16.3
Average		16.5

Details of the luminescence spectrum in the highest-energy region are shown
in Fig. 6.22 together with the absorption spectrum. The energy of the zero-
phonon band, 2.6412 eV, coincides with that of the absorption spectrum,
2.6413 eV, within experimental accuracy, and the multi-phonon series can be
concluded to be the radiative decay of bound excitons at isolated iodine

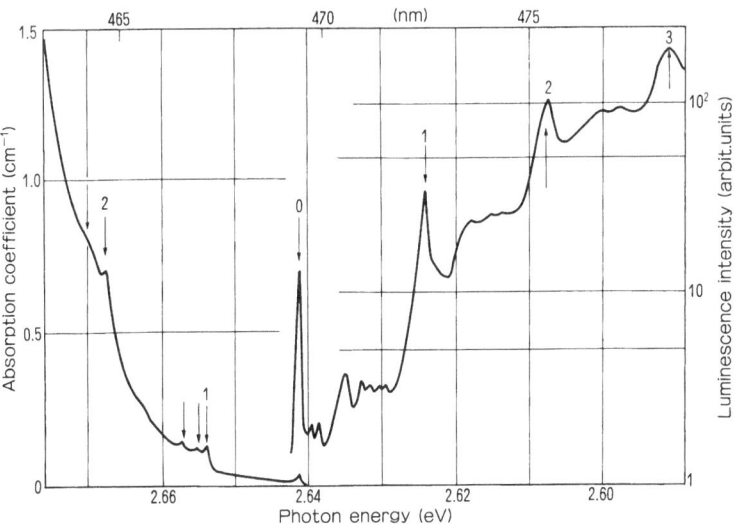

Fig. 6.22. Photoluminescence (*right*) and absorption (*left*) spectra of AgBr : I (5.7×10^{-4} mole
fraction) at 2 K [6.61]. Multi-phonon structures are specified by the number of phonons coupled
with the transition

impurities. The localization energy of the iodine-bound exciton can be estimated as the energy difference between the $n = 1$ free indirect-exciton energy and the zero-phonon line of the iodine-bound-exciton transition, which is

$$2.6845 - 2.6412 = 0.0433 \text{ eV} . \tag{6.29}$$

The origin of the localization energy will be mainly due to binding of the positive hole at iodine, because the binding energy of the electron in an iodine-bound exciton is close to that of a free exciton as indicated in (6.28).

In the luminescence spectra of Fig. 6.22, the half-width of the zero-phonon line is about 0.5 meV, which is close to that of the zero-phonon absorption band. The relative intensity distribution of multi-phonon series in luminescence spectra can be described by the linear coupling theory [6.69] using the value of coupling strength $S = 6.0$, which is close to the value of $S = 5.6$ estimated for the absorption spectra. *Czaja* and *Baldereschi* [6.70] measured the luminescence spectra of AgBr : I with higher resolution in a lower iodine concentration range, and resolved the exchange splitting between $J = 2$ and 1 of the zero-phonon luminescence line. The exchange splitting they observed is 0.1 meV which is only one-third of the splitting observed by *Matsushita* [6.45] for the free exciton in AgBr.

Recently, *Czaja* [6.71] reported concentration-induced variations of luminescence spectra in AgBr : I, such as the shift and broadening of the zero-phonon line with increasing iodine concentration. The luminescence spectrum shown in Fig. 6.22 for iodine concentration of 5.7×10^{-4} mole fraction is certainly influenced by the concentration effect which is most evident in the phonon sidebands, especially in the one-phonon sideband. Figure 6.23 shows the one-phonon sidebands of luminescence spectrum for AgBr : I at a concentration of the order of 10^{-6} mole fraction [6.27]. This spectrum corresponds to the lowest concentration limit in the systematic study of [6.71]. Most of the phonon sidebands in Fig. 6.23 can be ascribed to the singularities observed in the one-phonon density of AgBr in [6.27], indicating that various acoustic phonons contribute to the phonon sidebands in addition to the more strongly coupled optical phonons such as LO(Γ). This can be explained, at least qualitatively, by a coupling of nearly all the phonon modes with the deeply localized hole at the iodine site having a small orbital radius [6.72]. The only exception in Fig. 6.23 is the 2.1 meV sideband which is difficult to assign to a lattice phonon and has been tentatively assigned as a quasi-localized impurity mode due to iodine.

Now let us compare the phonon sidebands of Figs. 6.22 and 6.23 for different iodine concentrations. The distinct effects due to higher concentration in Fig. 6.22 are (a) enhancement of the 6.3 meV structure which is also distinct in the two-phonon sidebands, and (b) the splitting of the 2.1 meV structure. The enhancement of the 6.3 meV band is proposed as due to coupling with the TA(L) phonon in [6.71].

The other distinct effect of higher concentration in AgBr : I is the occurrence of the broad luminescence band first identified by *Tsukakoshi* and *Kanzaki* [6.68]. As mentioned above, this band is related to the iodine pair and can be

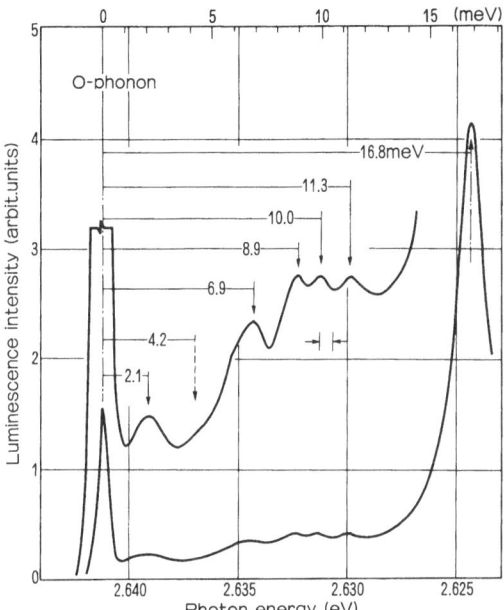

Fig. 6.23. Photoluminescence spectra due to residual iodine impurity (order of 10^{-6} mole fraction) in nominally pure AgBr at 2 K [6.27]. The zero-phonon line is located at 2.6412 eV. The upper curve is drawn with an expanded scale to show details of the sidebands

correlated to the broad absorption background extending toward the low-energy side of the zero-phonon band. Both of the optical transitions, absorption and luminescence, are due to iodine pairs and can be reasonably understood as due to the bound exciton at an iodine-pair which is analogous to a bound exciton at a NN pair in GaP : N [6.73]. It is concluded that the localization energy is enhanced for the exciton bound at an iodine pair and the exciton radius is reduced in comparison with the bound exciton at isolated iodine, because of the stronger binding of the hole at the iodine pair. This is also consistent with the greater thermal stability of the iodine-pair luminescence than that of the isolated iodine as indicated in Fig. 6.21. Due to stronger coupling with phonons in exciton transitions at an iodine pair compared with at isolated iodine, no phonon structures are observed in both absorption and luminescence spectra. The recent proposal by *Czaja* [6.71] for the low-energy luminescence band is based on the radiative decay of a kind of exciton molecule formed by the overlap of two iodine-bound excitons, which is hardly acceptable from the experimental results presented above.

b) AgCl : I. Iodine impurities in AgCl present a distinct absorption band of Gaussian shape with a peak at 3.13 eV [6.74]. This is in contrast to the absorption spectrum of AgBr : I, in which the iodine-induced absorption band is not distinctly separated from the absorption edge of AgBr, as shown in Fig. 6.19. Furthermore, no phonon structures are resolved in the 3.13 eV absorption band of AgCl : I. The luminescence spectrum of AgCl : I consists of a band with a peak at 2.64 eV at 2 K [6.74]. Weak phonon structures have been observed in the high-energy tail of the luminescence band [6.75], which are the peaks at 2.9115 (sharper) and 2.8871 eV (broader) with an interval of 24.4 meV close to the LO(Γ) of 24 meV in AgCl [6.25]. Although the zero-phonon line is not observed in absorption and the assignment is ambiguous, the 2.9115 eV peak may be reasonably considered as the zero-phonon line in luminescence. Then, the value of the localization energy of the iodine-bound exciton will be

$$3.2480 - 2.9115 = 0.337 \text{ eV} \tag{6.30}$$

which is an order of magnitude larger than the value of 0.043 eV in (6.29) for the corresponding value in AgBr : I. The coupling strength S with phonons is nearly 10, which is far larger than 5.6–6.0 in AgBr : I. Irrespective of the zero-phonon assignment, it is certain that the localization energy of the iodine-bound exciton is much larger in AgCl than in AgBr. This can be reasonably understood, in view of the larger electron affinity difference between chlorine and iodine, in comparison with that between bromine and iodine.

One of the noticeable observations in relation to the exciton localization at iodine in silver halides is the following. The maximum of the iodine-induced absorption in AgBr : I shown in Fig. 6.19 is located at 2.695 eV which is slightly above the $n = 1$ indirect-exciton energy of AgBr, 2.684 eV. In terms of the configuration coordinate diagram for exciton transition, this result indicates that the free exciton is more stable than the iodine-bound exciton in the ground state or unrelaxed configuration of the AgBr lattice. In other words, lattice relaxation is essential in stabilizing the exciton bound at iodine in AgBr : I. The situation in AgCl : I is different, and the iodine-bound exciton is already stable in the unrelaxed configuration of the AgCl lattice. These considerations indicate that the nature of exciton localization changes systematically in going from AgBr : I to AgCl : I. The assistance of lattice relaxation is essential in the shallower exciton binding in AgBr : I but not required in AgCl : I. The coupling strength with phonons also increases in the deeper localization in AgCl : I in comparison with the shallower localization in AgBr : I.

In the absorption spectra of AgCl : I, a series of multi-phonon structures are observed in the high-energy tail of the main 3.13 eV absorption band, as shown in Fig. 6.24 [6.61]. The following assignments have been proposed:
3.2075 eV: zero-phonon band of the lowest exciton transition,
3.2254 eV: one-phonon band,
3.2419 eV: zero-phonon band of higher exciton, and
3.2433 eV: two-phonon band.
The energy interval 17.9 meV in the above assignments is to be compared with the LO(Γ) phonon of 24 meV in AgCl. The coupling strength S is estimated as 3 which is much smaller than that of $S \approx 10$ for the 3.13 eV absorption band. The above assignments give the ionization energy of the assumed exciton as

$$34-46 \text{ meV} . \tag{6.31}$$

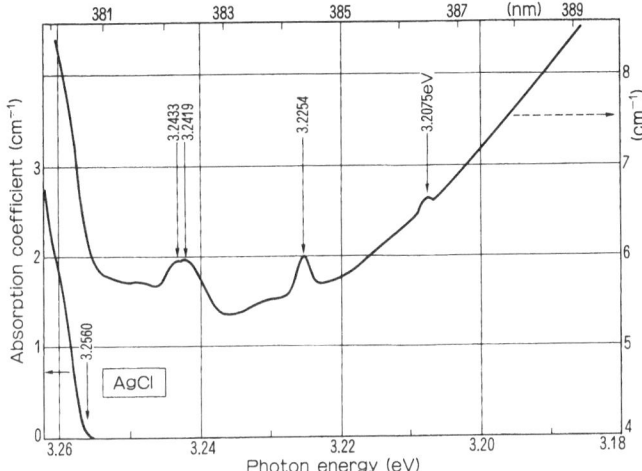

Fig. 6.24. Part of the absorption spectrum of AgCl : I (2.9×10^{-4} mole fraction) at 2 K showing the multiphonon structures superposed on the high-energy tail of the 3.13 eV band, in comparison with pure AgCl [6.61]

In comparison with the corresponding energy of 18.5 meV in (6.28) for the iodine-bound exciton in AgBr : I, the energy range estimated in (6.31) is reasonable in terms of the effective mass approximation [6.61]. However, the exact origin of the multi-phonon series in Fig. 6.24 is not clear at present. One of the possibilities may be that there are two types of bound-exciton states in AgCl : I, one with stronger lattice relaxation which corresponds to a stable bound exciton, and the other with weaker lattice relaxation corresponding to a metastable bound-exciton state.

Summarizing, the various phenomena observed for iodine-bound excitons in silver halides have elucidated the interesting aspects of localization of excitons due to the short-range attractive potential of the isoelectronic halogen impurities. However, more quantitative understanding is required, both theoretically and experimentally, for some of the unresolved problems mentioned above.

6.3 Relaxation Processes of Photo-Excited States in Silver and Alkali Halides

So far in this chapter, we have been mainly concerned with the fundamental aspects of exciton transitions based on spectroscopic studies, and have elucidated various *elementary excitations* which are important in understanding excitonic processes in the silver halides. Let us now look at the problem from another standpoint and consider the de-excitation of photo-excited states on a more general basis.

Starting from the photo-excited states, the nature of which can be assigned from absorption spectra, their de-excitation proceeds through various relaxation channels, radiative and nonradiative. During these processes, all of the elementary entities, such as excitons and the exciton-phonon interactions discussed above, will certainly play important roles, in various combinations. Although the luminescence spectra described in Sect. 6.2 presented important information about the nature of radiative relaxation channels, we need further quantitative knowledge about the *branching ratio* between radiative and nonradiative relaxation channels to understand de-excitation phenomena as a whole. Since the intrinsic optical spectra have been fairly well investigated for various solids, especially in the wide photon energy range available with synchrotron radiation, the study of relaxation following high-energy excitation including core-electron excitation has become an interesting subject. So far, however, only a few studies have been reported on the excitation spectroscopy of intrinsic luminescence in alkali halides [6.76, 77] and rare gas solids [6.78, 79]. Furthermore, the absolute value of luminescence yield has not been measured in any of these studies and the discussion has been only on a qualitative basis.

Recently, *Yanagihara* et al. [6.35] studied the excitation spectra of luminescence quantum yield at low temperatures for silver halides (AgCl, AgBr) as well as for alkali halides (KCl, KBr, NaCl, NaBr) using synchrotron radiation from SOR-RING, and clarified the quantitative aspects of radiative relaxation from photo-excited states, for the first time. In the following the main results of their work will be described.

Typical excitation spectra for luminescence quantum yield are shown in Figs. 6.25, 26, and 27 for AgCl, KBr, and NaBr, respectively. The luminescence yield shown in the figures is the quantum yield per absorbed photon, and has been corrected for the reflectivity, the spectrum of which was measured simultaneously and is shown in the upper part of each figure. Photoexcitation in the spectrum region below 30 eV corresponds mostly to excitation of the valence band, except for the excitation of the potassium $3p$ core exciton (KCl and KBr) [6.80] corresponding to structures around 20 eV which are clearly seen in the reflectivity spectrum of KBr in Fig. 6.26. Under the low-temperature excitation used in the experiments, all of the luminescence observed in AgCl, KBr and NaBr is due to decay of self-trapped excitons. Excitation spectra are taken for both π (triplet) and σ (singlet) luminescence of the STE in the case of KBr.

As shown in the figures, the first of the noticeable features in the yield spectra is the increase of quantum yield with the excitation energy. However, there are two distinct types among the spectral shapes of yield spectra. One is a group represented by AgCl and NaBr, in which the yield increases more or less monotonically through "steps" in increasing excitation energy; called the *S-type*. The following also belong to the S-type: σ luminescence in KBr, π and σ luminescence in NaCl and iodine luminescence in nominally pure AgBr. The other is a group represented by π luminescence in KBr and KCl, in which the yield spectra show "peaks" at the lowest exciton band as well as in the higher energy region; called the *P-type*. Extrinsic luminescence due to sodium-perturbed STEs in KBr : Na [6.81] also belongs to this group.

Generally, it is expected that the luminescence yield will show the S-type stepwise increase corresponding to the occurrence of multiple excitation process-

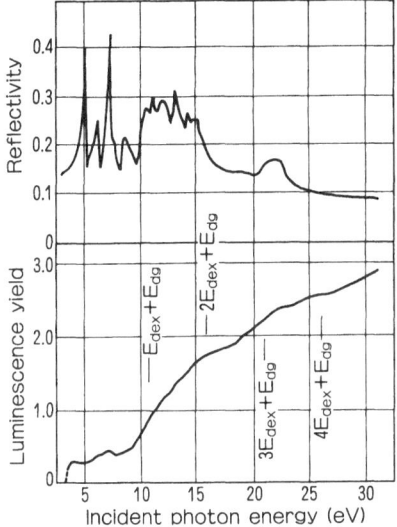

Fig. 6.25

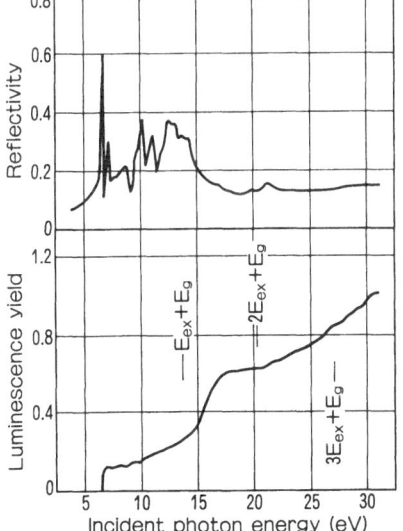

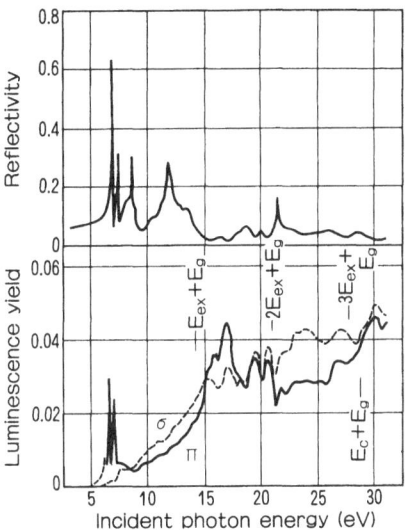

Fig. 6.26

Fig. 6.25. Luminescence yield spectrum (*bottom*), and reflectivity spectrum (*top*) for AgCl at 4.2 K [6.35]

Fig. 6.26. Luminescence yield spectra (*bottom*) for π (——) and σ luminescence (- - -), and reflectivity spectrum (*top*) for KBr at 4.2 K [6.35]

◀ **Fig. 6.27.** Luminescence yield spectrum for π luminescence (*bottom*) and reflectivity spectrum (*top*) for NaBr at 77 K [6.35]

es. In the figures, the possible threshold energies for multiplication processes are indicated as $E_{ex} + E_g$, $2E_{ex} + E_g$, or $E_c + E_g$, where E_{ex} is the lowest exciton energy, E_g the band gap and E_c the core exciton energy (in silver halides, E_{dex} and E_{dg} correspond to the direct exciton and direct band gap, respectively). In the case of solid argon studied by *Möller* et al. [6.78], a series of distinct steps are observed in the luminescence-excitation spectra, and multiple exciton scattering is concluded as the dominant multiplication mechanism. In the present case, however, the location of the steps is ambiguous in the spectra and it is not possible to discriminate which is the dominant scattering mechanism; multiple exciton or multiple electron-hole production.

As for the P-type spectra in some of the alkali halides, it has been known since the work by *Ikezawa* and *Kojima* [6.76] that the yield spectra of π luminescence in potassium halides show a peak at the first exciton band, contrary to σ luminescence which starts from the high-energy side of the first exciton band. The present results indicate the existence of peaks in the yield spectra of π luminescence in KBr and KCl in the high-energy region above 15 eV, in addition to the peak at the first exciton band. The high-energy peaks are less distinct in the yield spectrum of σ luminescence in KBr as shown in Fig. 6.26. *Yanagihara* et al. [6.35] proposed a model for the existence of peaks in the P-type spectra in contrast to the S-type without peaks. Their explanation is based on the repetition of the spectral features near the absorption threshold region again at higher energy near the multiplication threshold.

The second important result from the yield spectra is concerned with the magnitude of quantum yield. As indicated in the figures, the luminescence yield in the lowest exciton region is very low in alkali halides; the values for π luminescence of KCl, KBr, and NaBr are 0.09, 0.03, and 0.13, respectively. It must be concluded that most of the energy of the excited states is dissipated through nonradiative relaxation channels. Since the electron affinity is positive for these materials, photoelectron emission is not expected to be dominant in this energy region. As a candidate for the nonradiative process, the production of color centers may be proposed. It is known that color centers can be formed more easily in KCl and KBr than NaBr, under x-ray irradiation at 4 K [6.82] and this tendency is in parallel with the larger luminescence yield of NaBr than KCl or KBr. Further studies on the yield of color center formation are required under photoexcitation of alkali halides at low temperature. In contrast to alkali halides, the luminescence yield of AgCl is 0.3 to 0.4 in the lowest-valence exciton region, and much larger than the corresponding values of alkali halides studied. Combined with the quantum yield of 0.6 for producing a conduction electron in AgCl at LHeT [6.83], it is concluded that the relaxation of low-energy excited states in AgCl can be described by two dominant channels; the STE luminescence and the production of e–h pairs which are subsequently trapped nonradiatively without appreciable contribution to the luminescence.

Under photoexcitation at higher energies, the luminescence yield of AgCl and NaBr exceeds unity at 12 eV and 28 eV, respectively. The results are quantitative evidence for the occurrence of multiple exciton production through inelastic scattering of the excited electrons. In contrast with the silver halides having a very low yield of 0.1 for photoelectron emission even above 10 eV [6.84], the photoelectron emission yield in KBr and KCl is large, exceeds unity above 14 eV for KBr [6.85] and can be a dominant relaxation channel of the excited states in alkali halides above the photoelectron thresholds. In view of the low value of the luminescence yield of less than 0.1 in this energy region, there is obviously a competitive relation between the luminescence and electron emission processes, as suggested by *Beaumont* et al. [6.77] on the basis of qualitative comparison of spectrum shape between the luminescence and photoelectron emission yield spectra. However, the experimental conditions are entirely different between the present study of bulk crystals at low temperature and the photoelectron emission of evaporated films at room temperature in [6.85]. Further experimental studies are required for more quantitative understanding of the competitive relation between the two kinds of relaxation processes in alkali halides.

The third feature in the luminescence yield spectra is the existence of *dips*, which have been found in alkali halides [6.76] and also in rare gas solids [6.79]. The sharp dips in the yield spectra of KBr and KCl in the present study are located close to the reflection peaks. In previous studies such as in [6.76], the dips are observed in the yield spectra per incident quantum and ascribed mainly to reflectivity losses. The yield in the present experiments, however, is the yield per absorbed quantum and so the dips cannot be ascribed to reflectivity losses. The enhancement of photoelectron emission yield may be a possible origin of the dips in the higher energy region, but this cannot explain the dip observed at the first exciton band. As another origin of the dips, a nonradiative relaxation specific to the surface has been proposed for rare gas solids in [6.79]. In the present experimental results, however, the occurrence of dips is not necessarily correlated with large values of absorption coefficients, as pointed out in [6.35]. Although the exact origin cannot be identified at present, the experimental results in [6.35] suggest that the sharp dips observed at some of the absorption peaks should be correlated with the nature of the excited states themselves, rather than with the magnitude of the absorption coefficient. Interesting observations relevant to this point are the nonexistence of the dips in yield spectra of AgCl and NaBr in Figs. 6.25 and 27 and also for iodine-induced luminescence in AgBr. Further investigations will be necessary especially for the nonradiative relaxation processes including formation of defects, photoelectron emission, and photon-induced desorption.

6.4 Localized Electrons and Holes in Silver Halides

The study of localized electronic states in solids started with that on the color center in alkali halides at Pohl's laboratory at Göttingen in the 1930s [6.86]. This was followed by that on the shallow impurity center in semiconductors after the invention of the transistor in the 1940s [6.87]. These two lines of study represent two extreme examples of localized electronic states with entirely different characters. The orbital radii of localized electrons and holes are very small in the color centers and almost confined to the unit cell of the lattice, but very diffuse in the shallow impurities extending over many lattice distances. For example, although both the F center in alkali halides and the shallow donor in semiconductors correspond to an electron bound in a Coulomb potential, there are striking contrasts between the natures of the two. The F center ground state can be described by the *particle in a box model* which is an electron localized in a three dimensional square well [6.88]. The orbital radius of the F center ground state is limited by the size of the halogen vacancy as evidenced by the Mollwo-Ivey relation for the variation of F band energy among various alkali halides. By contrast, the energy levels of a localized electron in the shallow donor can be well described by the effective-mass approximation [6.87]. The orbital radius of the shallow donor ground state is very large, of the order of 50 Å, because of the large dielectric constant and the small effective mass of the electron. As for the excited electronic states of an F center, however, a strong lattice relaxation results in a diffuse relaxed excited state as evidenced by the large Stokes shift and the invalidity of the Mollwo-Ivey relation for the luminescence energy. This kind of lattice relaxation phenomena is essentially absent in the case of shallow donor.

The experimental study of the localized electrons and holes in pure silver halides has been much delayed in comparison with that in alkali halides. Early studies on the absorption spectra revealed only the excitation-induced band due to colloidal silver [6.89]. It was not until 1969 that the absorption spectrum due to shallow localized electrons in AgBr was discovered by *Brandt* and *Brown* [6.90]. Subsequently, an extensive spectroscopic study was carried out on the localized electrons and holes in silver halides by *Kanzaki* and *Sakuragi* [6.91]. As will be shown in the following, the localized electrons and holes in the silver halides can be understood as of an intermediate nature between color centers and shallow impurities which correspond to the strong and weak coupling limits of the electron-phonon interaction, respectively. This indicates again the borderline character of silver halides between heteropolar and homopolar substances.

In the following, the proposed models for localized centers in silver halides at low temperature are summarized in Sect. 6.4.1. In Sect. 6.4.2, the optical transition of the shallow localized electron center in silver halides is discussed as a typical example of the bound polaron and in Sect. 6.4.3, the photochemical reaction in silver halides at higher temperatures is described in connection with photographic processes.

6.4.1 Nature of Localized Centers in Silver Halides Compared to Color Centers in Alkali Halides

One of the experimental features in the study of localized centers in pure silver halides is that the excitation-induced localized states of atomic dimensions exist only temporarily during irradiation at low temperatures and the stable centers are not observed even after prolonged irradiation [6.51]. At higher temperatures, the localized centers tend to coagulate and form clusters, such as the silver colloids. This is in strong contrast to alkali halides in which stable color centers are efficiently produced by irradiation in a wide range of temperatures. The origin of the difference can be understood as follows. As will be shown, most of the trapped electrons and holes in silver halides at low temperatures have a smaller binding energy than the color centers in alkali halides. In particular, the dominant localized electron at low temperatures is a shallow state having binding energy of 30 to 40 meV, which is in contrast to the F center with binding energy of 2 to 3 eV. Accordingly, the shallow electron center in silver halides has a large orbital radius and is expected to decay by tunneling recombination with localized holes located at distant sites [6.51, 91].

Figure 6.28 shows typical examples of luminescence decay in silver halides at low temperatures. The initial fast decay, called the prompt luminescence, is characterized by an exponential time decay with a lifetime of the order of microseconds [6.68] and is due to radiative decay of the bound exciton; the self-trapped exciton and iodine-bound exciton, in AgCl and AgBr : I, respectively, in Fig. 6.28. The later slow decay, called the delayed luminescence, shows nonexponential time decay extending over 10 ms and is due to the radiative decay of distant e–h pairs; recombination of the shallow electron center with the self-trapped hole and the iodine-bound hole, respectively, in Fig. 6.28. Properties of the delayed luminescence [6.91], such as the nonexponential decay kinetics, the saturation of luminescence intensity under high-intensity excitation, and the transfer of the delayed component to the prompt component by infrared excitation, are entirely different from those of the prompt luminescence and can be

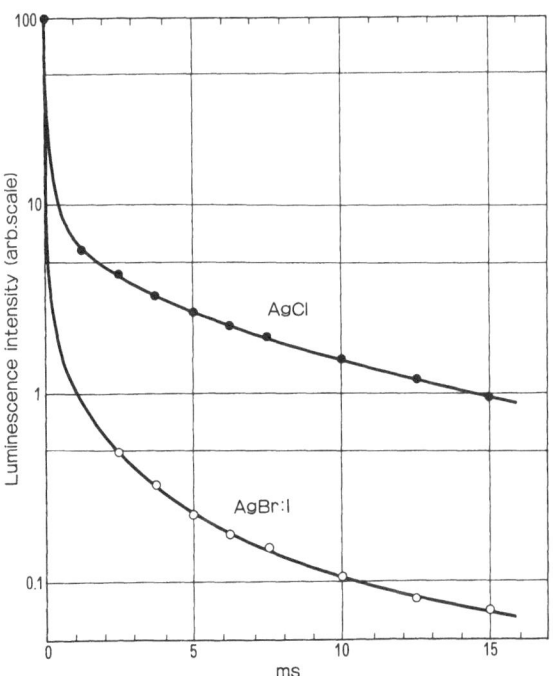

Fig. 6.28. Photoluminescence decay of AgCl and AgBr : I ($\sim 10^{-4}$ mole fraction) after pulse excitation at 2 K [6.4]. Luminescence intensity is normalized at $t = 0$

understood only as a decay of distant pairs having long and variable lifetimes. In this way, most of the trapped electrons and holes produced by excitation decay by recombination and no stable centers remain after the excitation is terminated.

Experimental study of the temporarily-existing localized centers in silver halides at low temperatures required the development of new spectroscopic techniques for the measurement of excitation-induced absorption during irradiation. The first spectroscopic method used is transient absorption spectroscopy which measures a small change of absorption due to the transient localized centers produced by photoexcitation. The method was first applied to the near-infrared transient absorption due to localized hole centers in AgCl, AgCl : Cd, AgCl : I, and $AgCl_{0.7}Br_{0.3}$ [6.51]. Although the transient absorption method can measure directly the absorption spectra of the transient centers, we need additional information on the nature of the optical transition, such as the relaxation of the excited states and the decay kinetics of the center. The experimental methods developed for this purpose are yield spectroscopy for the *enhancement of conductivity and luminescence intensity accompanied by the optical transition of localized centers* [6.4, 91, 92]. The enhancement yield spectroscopy, called modulation spectroscopy for transient centers, has several advantages. The first is the improvement of detection sensitivity which is essential for the study of usually low concentrations of transient centers. Secondly, the enhancement of conductivity indicates that localized electrons, not localized holes, are responsible for the optical transition, because the conduction due to holes in silver halides can be estimated as too small to be observable and the enhancement will be almost exclusively due to electrons. Thirdly, the enhancement of luminescence intensity can detect both electron and hole transitions in contrast to conductivity modulation. It is possible to identify the hole transition when only the luminescence enhancement is observed [6.4, 92]. On the basis of extensive spectroscopic studies using these various techniques, a fairly satisfactory picture is now available for the models of localized electrons and holes in silver halides at low temperatures [6.93].

Figure 6.29 summarizes the models of localized electron centers in silver halides in comparison with those in alkali halides. The basic electron center in silver halides is the intrinsic shallow center; an electron localized in the Coulomb field of a silver interstitial ion as shown in Fig. 6.29. The absorption spectrum of the shallow electron center in AgBr was first observed by *Brandt* and *Brown* [6.90], and the identification of the positive charge in the core was made later by *Sakuragi* and *Kanzaki* [6.94] on the basis of a series of experiments on silver halides containing various impurities at a very low concentration level. The intrinsic electron center in silver halides is in contrast to the F center in alkali halides which is an electron deeply localized at a halogen ion vacancy. The shallow center in silver halides has a small binding energy of 24 and 40 meV for AgBr and AgCl, respectively [6.94], and shows an absorption spectrum with multi-LO-phonon sidebands up to 7 or 8 phonons. The situation is very different from an F center with deep localization and showing a broad spectrum indicative of a strong electron-phonon interaction. The intrinsic shallow center is dominant

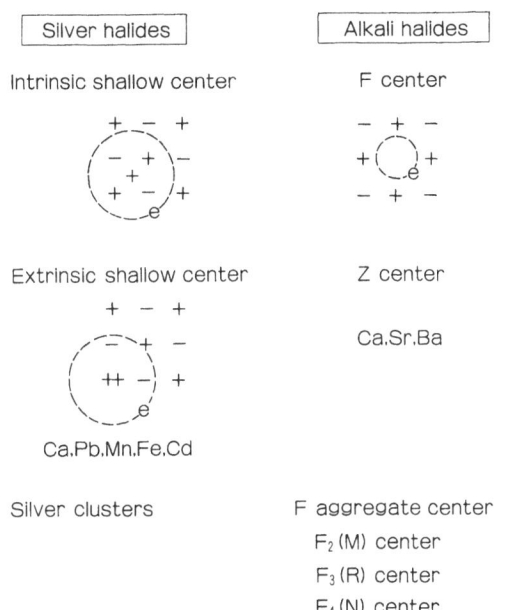

Fig. 6.29. Models of localized electron centers in silver and alkali halides [6.93]

in pure silver halides excited at LHeT and its concentration is increased by the addition of divalent sulfur ions. On the other hand, the extrinsic shallow center shown in Fig. 6.29 is dominant for silver halides containing divalent cation impurities of 10^{-6} mole fraction [6.94]. The extrinsic center consists of an electron localized in the Coulomb field of a substitutional divalent cation and its binding energy is very close to that of the intrinsic shallow center.

Figures 6.30 and 6.31 show the zero-phonon bands of the $1s \to 2p$ transition for the shallow electron centers in AgBr and AgCl, respectively. It is shown that the transition energy of the extrinsic shallow center is shifted depending on the kind of impurity ion. The spectrum shift has been explained quantitatively in terms of the central cell correction to the effective mass state, the magnitude of which is proportional to the effective ionization energy of the impurity [6.94]. Corresponding to the extrinsic center in silver halides, there are Z centers in alkali halides containing substitutional alkaline-earth impurity ions. However, the Z centers have been assigned as F centers perturbed by the divalent impurity–vacancy complex [6.95] and their nature is entirely different.

Models of localized hole centers in silver halides at low temperature are summarized in Fig. 6.32. The nature of hole centers is different between AgCl and AgBr because positive holes are self-trapped in AgCl but not in AgBr. The self-trapped hole (STH) in AgCl is a substitutional Ag^{2+} center as shown in Fig. 6.32 and belongs to the category of atomic-type self-trapping. In contrast, the STH in alkali halides is a halogen molecular ion, called a V_k center. The STH in AgCl was identified from transient absorption spectra [6.51] which are identi-

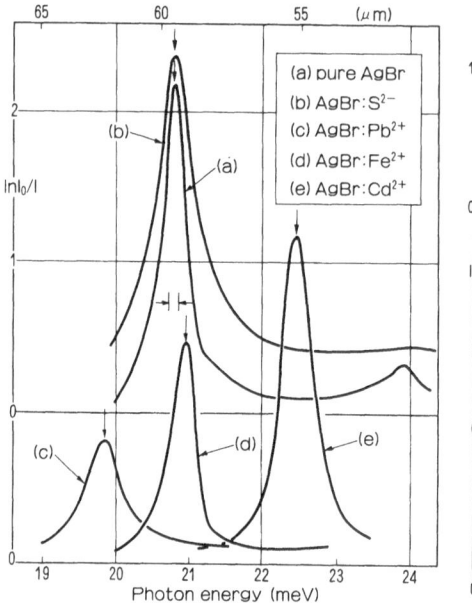

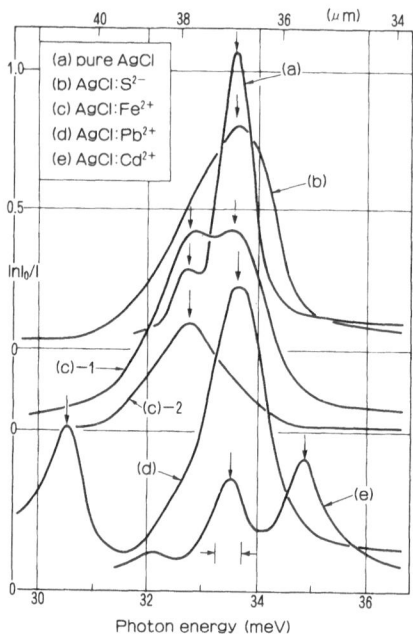

Fig. 6.30. Typical absorption spectra of zero-phonon bands of shallow electron centers during ultraviolet excitation of AgBr at 2 K [6.94]

Fig. 6.31. Typical absorption spectra of zero-phonon bands of shallow electron centers during ultraviolet excitation of AgCl at 2 K [6.94]. The excitation intensity in (c)–2 is reduced to 1/20 of (c)–1

cal with those of stable STHs previously observed after photo-ejection of positive holes from impurities [6.96, 97]. The transient spectroscopy also revealed the localization of holes at defects or impurities. The STH perturbed by a neighboring silver vacancy in Fig. 6.32 shows a spectrum similar to the unperturbed STH but shifted by 0.1 eV toward higher energies [6.51]. The perturbed STH center becomes dominant with the addition of divalent cations to AgCl and also at temperatures above 50 K at which the unperturbed STH becomes thermally unstable. Such a deeper localization due to a perturbing vacancy is also observed for the similar case in alkali halides which is a V_k center perturbed by an alkali vacancy (called a V_F center), shown in Fig. 6.32 [6.95].

Positive holes in AgBr are localized exclusively at impurities or defects. As shown in Sect. 6.2.3, the trapping of a hole at a residual iodine impurity is dominant in nominally pure AgBr at LHeT. Localization energy of the hole trapped at iodine can be estimated as 0.04 eV for isolated iodine and as 0.1 to 0.15 eV for an iodine pair. The deeper localization at an impurity pair compared to a single impurity has been observed commonly for isoelectronic traps in compound semiconductors [6.73]. Positive holes in AgBr can also be trapped at a

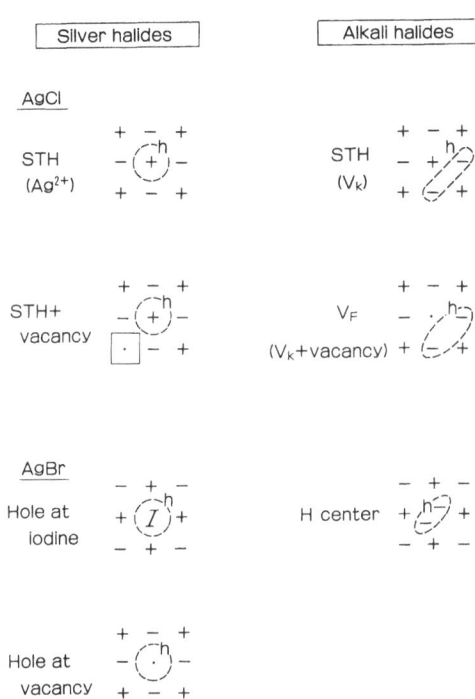

Fig. 6.32. Models of localized hole centers at low temperatures in silver and alkali halides [6.93]

silver ion vacancy as shown in Fig. 6.32 which is a dominant hole center in AgBr when the residual iodine content is substantially reduced. The binding energy of such a hole center has been estimated as 0.4 eV from the luminescence modulation spectra [6.4].

In comparing the localized hole centers in silver and alkali halides, it is evident that silver halides have no center corresponding to the H center in alkali halides. The H and F centers are produced as a pair of Frenkel defect centers in the halogen sublattice of alkali halides. The corresponding defect pair in silver halides is produced in the silver sublattice: the shallow electron center at a silver interstitial ion and the hole center at or near a silver ion vacancy. There is some circumstantial evidence which suggests the production of a pair of Frenkel defect centers during ultraviolet irradiation of silver halides at LHeT [6.93]. However, more direct experimental evidence, such as macroscopic volume expansion during photoexcitation at LHeT, is highly desirable in order to substantiate the defect formation process in silver halides.

Summarizing, the spectroscopic studies by *Kanzaki* [6.4, 93] led to the proposal of models for localized centers in silver halides the nature of which is very different from that of the color centers in alkali halides. Future studies should be directed toward the more detailed characterization of these centers as well as the relaxation mechanism of photoexcited states leading to the formation of localized centers. For example, the interstitial configuration of the intrinsic shallow centers has not been identified yet. As for the relaxation of a metastable free hole in AgCl to the self-trapped state, *Laredo* et al. [6.98] studied the lifetime of the free hole as a function of temperature in the range 4.5 to 30 K and obtained a value of the thermal activation energy of 1.7 meV corresponding to the energy barrier in the self-trapping process. Except for the STH in AgCl, the formation of localized centers can proceed by trapping of photo-excited carriers at defects or impurities. However, the availability of silver interstitial ions at LHeT is one of the unsolved problems. It has been suggested that the origins of the silver interstitials are frozen-in defects in AgBr and defect formation in AgCl [6.4, 93].

6.4.2 Bound Polarons in Silver and Alkali Halides

The shallow binding of an electron localized in a Coulomb field in the silver halides was first predicted by *Simpson* [6.99]. Later on, the problem was studied in detail by *Buimistrov* [6.100] as an example of the bound polaron; the polaron in a Coulomb field. In the following, the absorption spectra of the shallow intrinsic center in silver halides are discussed, including the multi-phonon transition range, and their spectral features are compared with predictions from the bound polaron theory. The interesting features observed for the phonon sidebands in silver halides are then compared with the similar spectra of the relaxed excited state (RES) of the F center in alkali halides. Although the bound polaron has been of wide theoretical interest, most of the studies have been concerned only with the energies of the ground or low-lying excited states. The present examples of multi-phonon absorption spectra are related to the electron-phonon coupling in the higher excited states and require a new development of the bound-polaron theory.

As previously discussed by *Takegahara* and *Kasuya* [6.101], there are various categories of bound-polaron ground states depending on the host material. In the strong coupling limit, the orbital radius of the electron is so small that the electron can hardly feel the phonon polarization. This corresponds to the situation of the F center ground state in which the binding energy is far greater than the LO phonon energy. In the weak coupling limit, on the other hand, the free polaron is only weakly bound at the Coulombic center and its orbital radius is very large. This corresponds to the shallow donors in III-V and II-VI compound semiconductors, in which the binding energy is nearly equal to or less than the LO phonon energy. The present cases of both the intrinsic shallow center and the F center RES belong to an intermediate type in the above classification: the coupling of the electron with the phonon is not weak and the polaron radius is nearly comparable to the orbital radius of the electron in the center.

The energy level scheme relevant to the absorption transition being discussed is shown in Fig. 6.33. The optical absorption of the intrinsic shallow center in AgBr starts from the zero-phonon line of the $1s$ to $2p$ transition and is followed by the multi-LO-phonon sidebands. The energy intervals in the multi-phonon series are found to be smaller than the LO(Γ) phonon energy suggesting the

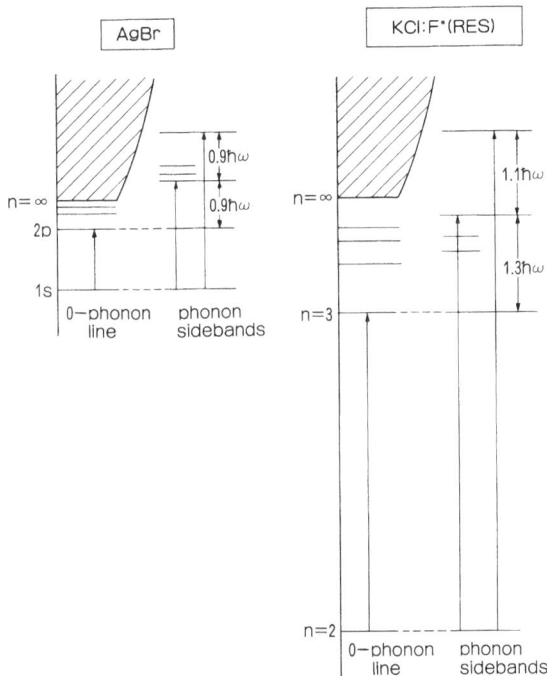

Fig. 6.33. Energy levels relevant to absorption transitions from bound polarons: the intrinsic shallow center in AgBr and the RES of the F center in KCl [6.102]. Energy intervals in multi-phonon transitions are given in units of LO phonon energy $\hbar\omega$

binding of an excited electron with a phonon. The absorption spectra from the F-center RES were first measured by *Kondo* and *Kanzaki* [6.103]. The multiphonon intervals in this case are found to be greater than the LO phonon energy suggesting a difference in the nature of the excited states between the zero- and one-phonon bands.

Figures 6.34 and 6.35 show the transient absorption spectra due to the intrinsic shallow center in the silver halides [6.75]. The sharp zero-phonon lines are located at 20.8 and 33.5 meV in AgBr and AgCl, respectively [6.94]. Table 6.9 summarizes the comparison of the transition energy with theoretical estimates.

Table 6.9. Comparison of the zero-phonon transition energy [meV] of the intrinsic shallow electron center in silver halides with various theoretical estimates of the $1s \rightarrow 2p$ transition energy using the second-order perturbation method

	AgBr	AgCl
Exp.[a]	20.8	33.5
EMA	26	48
Bajaj-Clark[b]	20.8	33.8
Matsuura[c]	31	49

[a] [6.94];   [b] [6.104];   [c] [6.105]

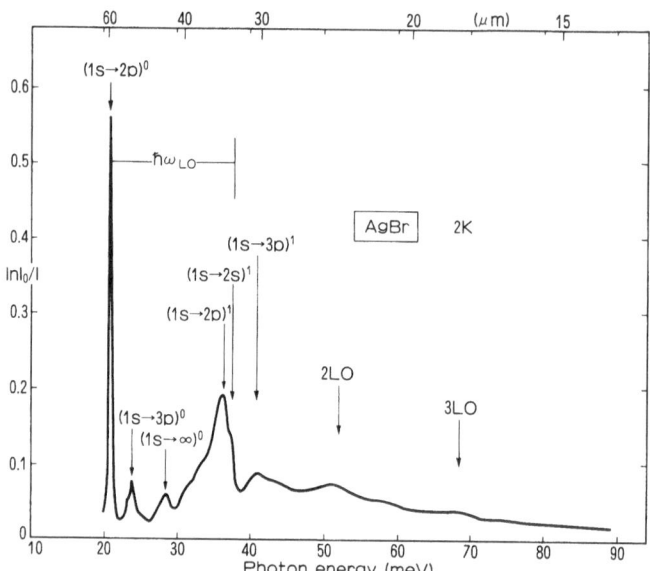

Fig. 6.34. Absorption spectrum of the intrinsic shallow electron center in photo-excited AgBr at 2 K [6.75]. See the text for the notation of the transition assignments

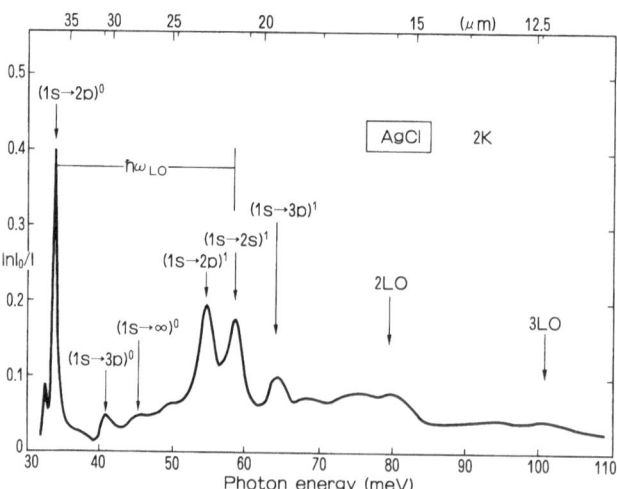

Fig. 6.35. Absorption spectrum of the intrinsic shallow electron center in photo-excited AgCl at 2 K [6.75]. See the text for the notation of the transition assignments

The simple effective mass approximation (EMA) gives the $1s \to 2p$ transition energy as 26 and 48 meV, respectively, using the polaron mass and the static dielectric constant. Considering the nature of the electron center, there is no doubt that the transition corresponds to $1s \to 2p$ of the bound electron polaron.

The results of a bound-polaron calculation by *Bajaj* and *Clark* [6.104] using an approximate perturbation method show a striking agreement with experimental values, but, the recent calculation by *Matsuura* [6.105] using an exact numerical method results in a definite deviation from the experimental energy. One of the origins of the discrepancies will be the central cell correction due to the structure of the interstitial core with surrounding lattice distortion which has not been taken into account in any of the theoretical calculations. More detailed calculations as well as experimental studies are called for in order to settle the controversies.

The assignments of transitions for spectral structures in Figs. 6.34 and 6.35 are based on the unpublished calculation by *Matsuura* using the variational method [6.105]. The zero-phonon line is expressed as $(1s \rightarrow 2p)^0$; the zero-phonon transition of $1s \rightarrow 2p$. The higher transitions are $(1s \rightarrow 3p)^0$ and $(1s \rightarrow \infty)^0$. In the one–LO-phonon region of the spectra, the structures are assigned as one-phonon transitions; $(1s \rightarrow 2p)^1$, $(1s \rightarrow 2s)^1$, and $(1s \rightarrow 3p)^1$. In principle, the usually forbidden optical transition such as $1s \rightarrow 2s$ is expected to be enhanced due to the interaction with phonons in the case of bound polarons. This is in contrast with the zero-phonon transition in which only the allowed transition is observed. Then it is expected that the intensity of the zero-phonon line will be reduced in comparison with the one-phonon and the multi-phonon sidebands. This is actually observed for the spectra of shallow electron centers in silver halides. The electron-phonon coupling strength S is estimated as near unity for the spectra of Figs. 6.34 and 6.35 from the intensity distribution of multi-phonon components in the range of $n \geq 2$. On the other hand, the absorption intensity of the zero-phonon band is reduced to nearly one-tenth of the value extrapolated from the higher members of the multi-phonon series by assuming linear coupling with phonons. These arguments, however, are of a qualitative nature. Quantitative aspects, especially the enhancement of one-phonon absorption, remain still open for future studies [6.105].

Figure 6.36 shows the transient absorption spectra from the F-center RES in potassium halides. After *Swank* and *Brown* [6.106] found an anomalously long decay time of F-center luminescence, extensive studies on the F-center RES were made, both experimentally and theoretically. However, most of the studies were concerned with the optical transitions between the ground state and the first excited states of the F center. The transient absorption study by *Kondo* and *Kanzaki* [6.103] gave spectroscopic information on the higher excited states for the first time by starting from the relaxed first excited state. The common spectral feature in Fig. 6.36 is the strong zero-phonon peak followed by multi-LO-phonon sidebands. The energy separation between the zero- and the one-phonon band corresponds to 1.3, 1.1, and 1.0 in units of LO(Γ) phonon energy for KCl, KBr, and KI, respectively. The electron-phonon coupling strength S is estimated as 0.6 to 0.4 assuming linear coupling. The small value of the S factor is in striking contrast to that of the F band transition which is about 10.

The nature of the F center RES has been concluded as an S-like state composed of nearly degenerate $2s$ and $2p$ electronic states mixed through vibronic

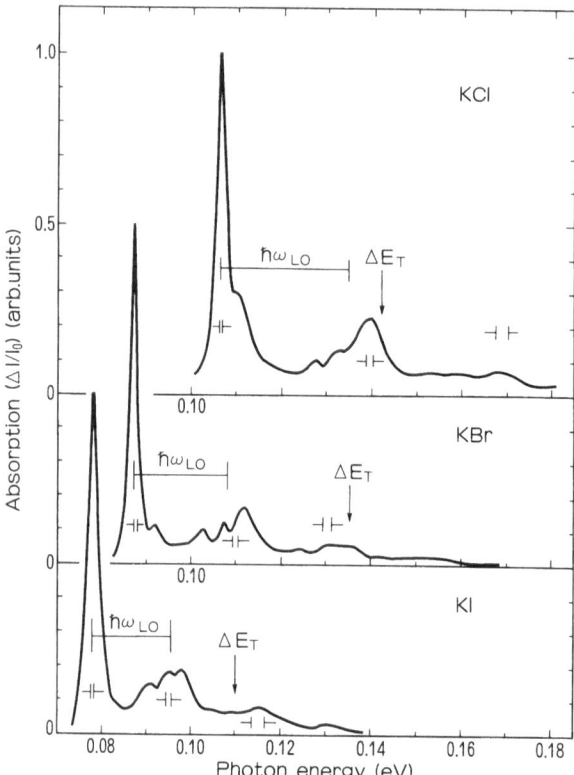

Fig. 6.36. Transient absorption spectra of the F center RES in potassium halides at 10 K. The absorption is normalized at the maximum value. The symbol ΔE_T represents the thermal ionization energy of the RES [6.103]

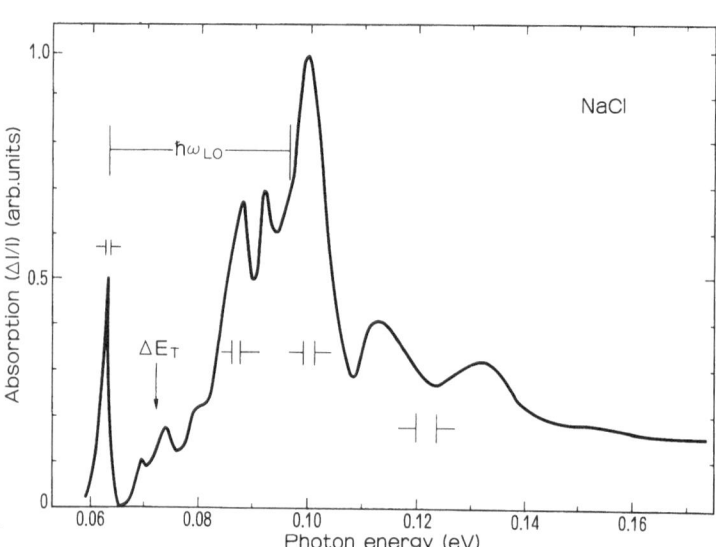

Fig. 6.37. Transient absorption spectrum of the F center RES in NaCl. The symbol ΔE_T represents the thermal ionization energy of the RES [6.102]

interaction [6.107], and the zero-phonon transition can be assigned as RES → 3p. In the one-phonon sideband, on the other hand, phonon enhancement will be expected for some of the forbidden or weak electronic transitions, and the situation will be very similar to the one-phonon band of the shallow center in silver halides.

Kayanuma and *Kondo* [6.108] analyzed the RES spectra in terms of a bound polaron by using a new approach taking advantage of the spherical symmetry of the system. Their results on the RES spectrum in KCl indicate an apparent triplet splitting of the one–LO-phonon sideband in excellent agreement with the observed spectrum. According to their results, three absorption peaks correspond to one-phonon transitions from RES to 3s, 3p, and 3d (in the sequence of energies) each of which is coupled with phonons of a different symmetry character. It is to be mentioned that the numerical parameters adopted by *Kayanuma* and *Kondo* [6.108] are determined not only to fit the spectral shape of one-phonon sideband but also to fit the degree of 2s–2p mixing in the RES derived from the luminescence experiments. Thus we now have a satisfactory picture of the F center RES in KCl which can describe all of the phonon-coupled optical transitions between the ground state, the RES, and the higher excited states of the F center.

The reduction of strength of the zero-phonon band is not appreciable in the RES spectra of potassium halides shown in Fig. 6.36, but is much enhanced for the RES spectra in some of the other alkali halides. A typical example is that of NaCl shown in Fig. 6.37. On the basis of spectroscopic studies on the RES in a series of alkali halides, *Kondo* [6.109] concluded that the reduction of the zero-phonon strength relative to the one-phonon band is pronounced whenever the final state of the one-phonon transition is degenerate with the conduction band. Notice that the position of thermal ionization energy ΔE_T is located above the one-phonon band peak in Fig. 6.36 for potassium halides but well below the one-phonon peak in Fig. 6.37 for NaCl. The shallow center in silver halides in Figs. 6.34 and 6.35 shows an ionization limit below the one-phonon peak, similarly to the RES in NaCl of Fig. 6.37.

The overall situation pointed out by *Kondo* [6.109] is summarized in Fig. 6.38 which was originally invented by *Kondo*. The units of the coordinates of Fig. 6.38 are chosen to arrange various systems in a consistent way. For example, the dotted lines specified as $n = 1$, 2, and 3 indicate the location of effective mass states having an effective mass equal to the bare electron mass. The above conclusion by *Kondo* [6.109] is equivalent to the statement that the reduction of the zero-phonon band is pronounced when the final state of the zero-phonon transition (○) is located above the horizontal line of $E/\hbar\omega_{LO} = 1$ shown in Fig. 6.38. The corresponding cases are the shallow intrinsic center in AgBr and AgCl, and the RES in NaCl, NaF, KF, and RbF. In these cases, various higher electronic states will simultaneously contribute to the one–LO-phonon sideband, in contrast to such a case as RES in KCl in which the contribution from only the $n = 3$ state is dominant [6.107]. However the quantitative analysis of the "anomalous" reduction of the zero-phonon band is not yet successful. Further

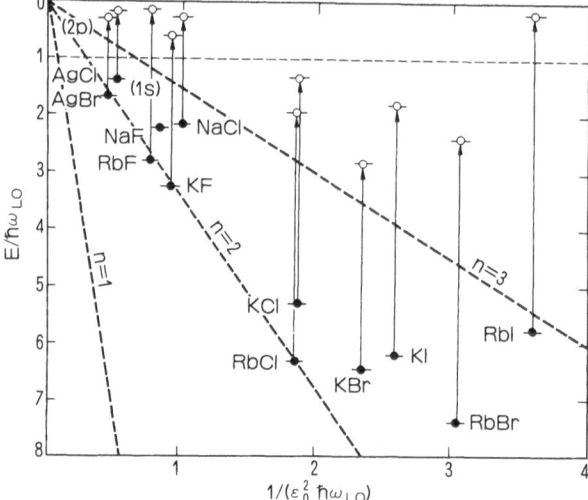

Fig. 6.38. Summary of optical transitions in the shallow electron center in silver halides and the F-center RES in alkali halides [6.102]. The ordinate is the binding energy E of the electronic state in units of LO phonon energy $\hbar\omega_{LO}$. The abscissa is chosen for convenience to arrange various materials using the static dielectric constant ϵ_0 and divided by $\hbar\omega_{LO}$ to match the ordinate

development of the bound-polaron theory is required to substantiate the qualitative picture described above.

6.4.3 Photochemical Reactions in Silver Halides at Higher Temperatures

So far in this chapter, we have been mainly concerned with the photo-excited states and their relaxation in the silver halides mostly at liquid helium temperatures (LHeT). As shown in Sect. 6.3, the relaxation of photo-excited states in silver halides at LHeT proceeds dominantly through two channels. One is the

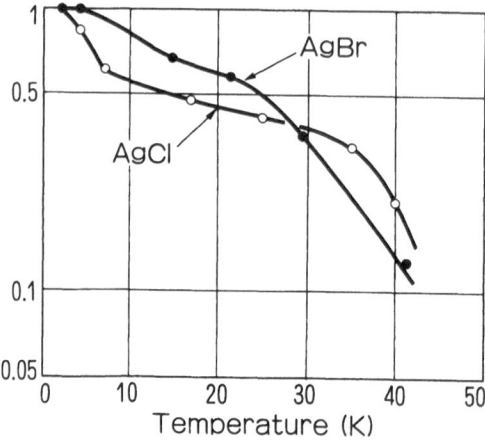

Fig. 6.39. Temperature dependence of the zero-phonon absorption strength due to the intrinsic shallow electron center in silver halides produced under a constant excitation intensity [6.93]. The absorption strength is normalized at 2 K

recombination luminescence due to decay of excitons and the other is the formation of localized electrons and holes which decay subsequently by tunneling recombination as described in Sect. 6.4.1. On increasing the temperature of excitation above LHeT, the luminescence yield starts to decrease around 20 and 30 K in AgBr and AgCl, respectively. Figure 6.39 shows the temperature dependence of the formation yield of intrinsic shallow centers which is more or less parallel to that of the luminescence yield in the range of temperature above 20 K. The change of formation yield below 20 K is specific for shallow centers and its origin has been discussed in [6.93]. These observations can be understood as mainly due to the thermal instability of both shallow electron centers and excitons with increasing temperature. At higher temperatures, above liquid nitrogen temperature (LNT), the excitons are thermally unstable and the e–h recombination proceeds nonradiatively. At the same time, the nature of the dominant localized electrons changes from the shallow electron center to deep electron centers. In the following, the formation of deep-localized electrons will be discussed on the basis of yield spectra of conductivity modulation under photoexcitation at higher temperatures.

Figures 6.40 and 41 show the excitation-induced absorption spectra of AgBr and AgCl, respectively, at various temperatures. At the lowest temperature of 2 K, the induced spectra are entirely due to the intrinsic shallow centers and specified as "A" spectra in the figures. The A spectra at 2 K decay away after photoexcitation is terminated. On increasing the temperature of excitation, the shallow electron centers become unstable as indicated by the decrease of A spectra above 20 K and 30 K for AgBr and AgCl, respectively. The observed thermal quenching of shallow centers is accompanied by the growth of the spectra at higher energies which is due to various deep electron centers. It is found that these deep electron centers do not decay away even after excitation is terminated. This is shown for AgBr at 23 K and 34 K in Fig. 6.40, in which the solid and dashed curves correspond to the spectra during and after irradiation, respectively. The AgBr spectra drawn by dash-dotted curves in Fig. 6.40 are measured at 77 K after irradiation at 77 K or at room temperature. The above results indicate that the optical ionization energies as well as the stabilities of deep electron centers are greater than those of shallow electron centers and increase with the irradiation temperature.

As shown in Figs. 6.40, 41, only the "C" spectra are dominant above LNT. The spectral shape of the C band at the low-energy threshold can be fitted well by the so-called *Fowler curve* [6.110, 111] which is expected to be valid for the excitation spectra of external photoelectron emission from metals in general. The threshold energies obtained by Fowler-curve analysis of the C spectra are close to those expected for the injection of photoelectrons into the conduction band of silver halides from silver metal in contact with silver halides. The results suggest that the deep electron centers responsible for the C spectra correspond to the clusters of several silver atoms embedded in the silver halides. This proposition is further supported by the agreement between the threshold energy of the C band and the threshold energy of the Herschel effect of latent images in photographic

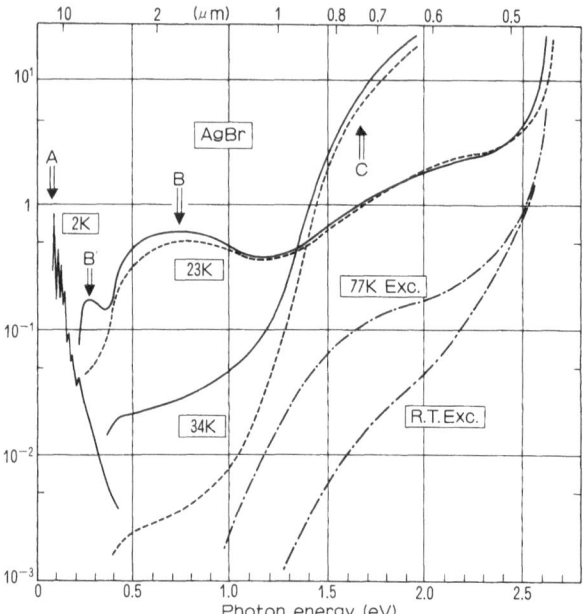

Fig. 6.40. Excitation-induced absorption spectra of AgBr photo-excited at various temperatures, obtained by the conductivity-modulation method [6.4]. Solid and dashed curves are the data during and after ultraviolet irradiation, respectively. Dash-dotted curves are measured at 77 K, after irradiation at 77 K or at room temperature

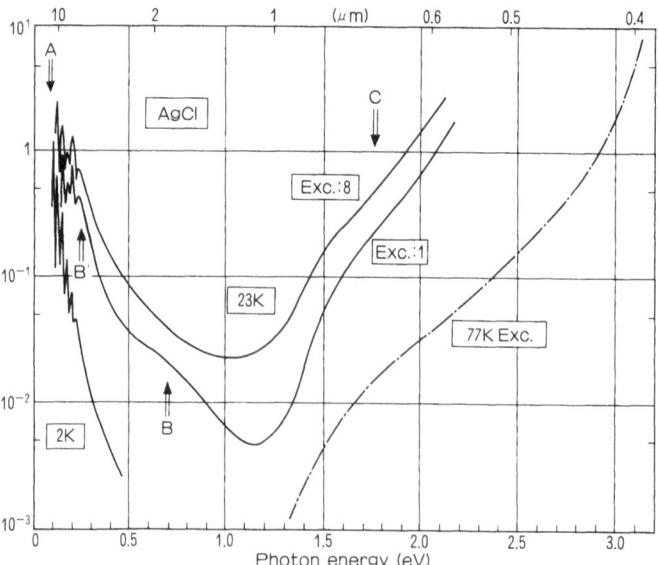

Fig. 6.41. Excitation-induced absorption spectra of AgCl photo-excited at various temperatures, obtained by the conductivity-modulation method [6.4]. Solid curves are the data during ultraviolet irradiation and the dash-dotted curve after irradiation. Two curves for 23 K correspond to the change of excitation intensity from 8 to 1

emulsions which is 1.13 eV for AgBr [6.112]. Furthermore, it is reasonable to assume that the centers responsible for the "B" and "B'" spectra observed in the intermediate range of temperature, 20 to 30 K, are silver clusters with a smaller size than those giving the C spectra. Although the microscopic structures of these deep electron centers are not clear yet, it seems certain that the excitation-induced spectra at higher temperatures are due to the formation of small silver clusters having deeper binding energies than shallow centers. The transformation from the shallow electron centers with atomic dimensions below LHeT to the cluster centers above LNT can be considered as a parallel to the sequence of silver aggregation during the photographic process. Further studies on the structure of the deep electron centers will be very valuable in elucidating the detailed mechanism of latent image formation in photographic emulsions.

Compared with the formation of silver clusters from localized electrons, information about the fate of photo-excited holes was until lately very scarce. The single exception was the study by *Luckey* [6.113] of the evolution of halogens from the surface of photo-excited silver halides near room temperature. Recently *Kanzaki* and *Mori* [6.114, 115] studied photon-stimulated desorption (PSD) from silver and alkali halides in a wider temperature range and as a function of excitation wavelength and excitation intensity. They used a sensitive detection method, which consisted of the phase-sensitive detection of PSD rate by a quadrupole mass spectrometer, and succeeded in obtaining much more detailed information than the results by *Luckey* [6.113].

The main results under low-energy photoexcitation are as follows [6.114, 115]. Firstly, different PSD species were observed. Only the neutral diatomic halogen molecules are emitted from silver halides, in contrast to alkali halides in which the desorption species are neutral alkali and halogen atoms. The difference observed between silver and alkali halides suggests that a halogen atom (a positive hole localized at a surface halide ion) can be efficiently desorbed from alkali halides but not from silver halides. A possible reason for this observation is that the wave function of a localized hole at a halogen atom is sufficiently localized, in alkali halides, for the desorption to occur, but not in silver halides. In silver halides, the formation of the diatomic molecule is required for sufficient localization for efficient desorption. These considerations are consistent with the tendency for the localized states in silver halides to be generally shallower than those in alkali halides.

The second set of interesting results are on the PSD yield spectra in silver halides. Figure 6.42 shows the results at various temperatures. The data points can be well fitted to the theoretical expectation assuming that the photo-excited holes diffuse to the surface with diffusion coefficient D during their lifetime τ. The values of the diffusion length λ

$$\lambda = (2D\tau)^{1/2} \tag{6.32}$$

can be estimated by fitting the data of yield spectra and are summarized in Table 6.10. Combining the values of λ with τ obtained from the study of PSD

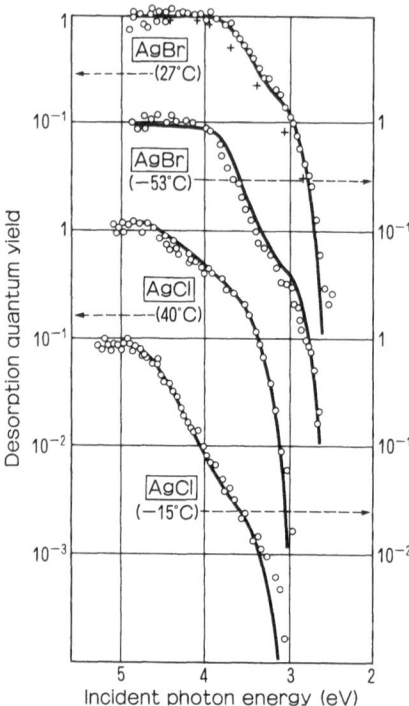

Fig. 6.42. Desorption yield spectra for halogen molecules from silver halides at various temperatures [6.115]. Solid curves are the yield spectra obtained by fitting the data points to the theoretical expression explained in the text. The crosses are the quantum yield estimated from *Luckey*'s data in [6.113]

kinetics, the diffusion coefficient D is estimated. The following values are obtained for AgBr at room temperature

$$(D\tau)^{1/2} = 1 \times 10^{-4} \text{ cm} ,$$
$$\tau = 1 \times 10^{-2} \text{ s} , \quad \text{and}$$
$$D = 1 \times 10^{-6} \text{ cm}^2 \text{s}^{-1} . \tag{6.33}$$

Table 6.10. Diffusion length λ in (6.32) of photoholes in silver halides estimated from the PSD yield spectra in silver halides [6.115]

	Temperature [°C]	λ [10^{-4} cm]
AgBr	27	1.4 $\pm$ 0.5
	−11	1.1 $\pm$ 0.3
	−53	0.45 $\pm$ 0.2
AgCl	40	1.4 $\pm$ 0.5
	20	0.7 $\pm$ 0.3
	−5	0.2 $\pm$ 0.1
	−15	0.14 $\pm$ 0.1

The estimated value of D is much smaller than that of free holes 3×10^{-2} cm s^{-1}, but is fairly close to that of 3×10^{-7} cm^2 s^{-1} obtained by *Malinowski* [6.116] for the diffusion of a neutral hole complex photo-generated in AgBr at room temperature. We can conclude that the photo-excited holes in AgBr diffuse to the surface in the form of a neutral hole–silver vacancy complex having lower mobility than that of free holes. The PSD yield spectra in Fig. 6.42 show that the quantum yield of PSD is near unity in the photon energy range where the absorption coefficient is larger than $(1–2) \times 10^5$ cm^{-1} as seen from the absorption spectra of Fig. 6.1. This suggests that the desorption of halogen molecules from the surface is a very efficient decay channel of photo-excited holes near the surface in the temperature range studied. One interesting future study using the PSD technique will be directed to the surface properties of silver halides including microcrystals used in photographic emulsions [6.117]. The photographic process in silver halides is essentially a photochemical reaction at or near the surface of silver halides. However, the physical properties of silver halide surfaces have not been investigated in any detail and much work still remains to be done.

7. Excitons and Their Interactions with Phonons and External Fields in Thallous Halides

In the preceding chapters, we have discussed excitonic processes in cuprous halides, silver halides, and rare gas solids. We find that they are certainly interesting materials in many respects for the study of the excitonic molecule (EM) and exciton-phonon coupling. In this chapter we introduce other ionic crystals important for the study of excitons, that is thallous halides [7.1].

The extrema of the valence and the conduction band in thallous halides, TlCl, and TlBr, are not at the Γ point. They have many-valley structures both for electrons and holes at a direct gap and are different from other typical ionic and rare gas solids in this respect. Because of this band structure, the direct exciton has a unique and complex structure due to intervalley exchange and Coulomb interactions. The lowest exciton state is an indirect exciton of a novel type. In this exciton, the scattering from the virtual direct-exciton state to the indirect-exciton state is forbidden for any momentum-conserving phonons so that it is a phonon-forbidden indirect exciton which is rarely found in ionic crystals. Owing to this nature, the indirect exciton can be accumulated and so the EM can be generated very easily. Thus thallous halides are nice materials for the study of the EM.

In thallous halides, the electron–LO-phonon coupling is strong and an exciton-polaron pinning effect has been found. This is the only case so far studied, although an electron-polaron pinning effect has been found in III–V compounds. Also large dielectric constants make the binding energies of excitons in thallous halides so small that the interaction energy with an external field available in a laboratory can be made comparable or larger than the exciton binding energy. Thus the effect of a high external field on an exciton can be studied in these materials which is difficult to do in other ionic crystals.

In both TlCl and TlBr, it is found that a self-trapped state of an exciton is a metastable state because the energy of the self-trapped state is higher than that of the free-exciton state. However in their mixed crystal, it is expected that, if the potential of the atom of lower-energy excitation cooperates with the electron-phonon coupling, the exciton self-trapped state at the site of the lower-energy atom will become the lowest excited state in the matrix of the other component atom. Indeed only in a mixed crystal, and not in the pure crystals of the component thallous halide, a self-trapped state is realized and this gives a good example of extrinsic self-trapping.

In the following sections, we shall see several different aspects of excitons which will become apparent through these particular properties of thallous halides. It is noted that in this and in the following chapter, the symbol H will be used for the magnetic field rather than the Hamiltonian as in the preceding chapters. The Hamiltonian will be denoted by $\mathcal{H}$.

7.1 Band Structures and Exciton States of Thallous Halides

7.1.1 Thallous Halides

Thallous halides, TlX, where X is a halogen, are III–VII compounds. In a crystalline form, thallium and halogen atoms are both ionized as Tl^+ and X^-. The Szigeti effective ionic charges e^*/e are 0.87 in TlCl and 0.84 in TlBr at room temperature which are a little larger than the 0.69 and 0.68 of AgCl and AgBr and are comparable with 0.81 of KCl [7.2]. Thus the thallous halides are typical ionic crystals.

Thallous chloride and bromide crystallize in CsCl structure in an ordinary atmosphere. The stable form of thallous iodide at room temperature and atmospheric pressure is orthorhombic (D_{2h}^{17} – Cmcm) and it is transformed into CsCl structure above 170 °C.

In TlCl, it has been observed that the phonons whose wave vectors cover from the midpoint q_1 to the zone boundary q_2 of the [110] TA_2 branch are softened by a temperature decrease [7.3]. The phonons at q_1 and q_2 are necessary modes for shear displacement and their softening results in the transformation from the CsCl structure to the orthorhombic TlI structure.

NaCl structure can be obtained also from the orthorhombic TlI structure by the shear displacement associated with the q_1 and q_2 phonons. Thus the CsCl structure, the orthorhombic TlI structure, and the NaCl structure of thallous halides are intimately connected. One can prepare TlCl and TlBr with an unstable NaCl structure by evaporating them onto a KBr substrate [7.4] while TlI of NaCl and CsCl structure can be prepared by evaporating it onto RbI, CsI, and organic film [7.5, 6].

The thallous halides whose excitonic processes are well investigated are of CsCl structure and so the discussion in this chapter will be based on them. In the following, thallous halides without any indication of crystal structure should be understood as those of CsCl structure.

The advantages of thallous halides for the study of excitonic processes are manifold. Firstly, its crystal structure is simple; the space lattice is simple cubic and so is first Brillouin zone. This greatly reduces the complexity of electronic structure usually encountered in anisotropic materials such as II–VI compounds. Secondly, its strong ionic nature gives a significant interaction between electrons and LO phonons and thus a Fröhlich coupling of intermediate range. One will see its effect in exciton spectra and polaron transport phenomena. Thirdly, the dielectric constant is quite large compared with other ionic crystals. In TlCl and TlBr, the static dielectric constants ϵ_0 are 37.6 and 35.1, respectively, at 1.5 K while in AgCl, AgBr, and KCl they are 9.50, 10.60, and 4.49, respectively, at the same temperature [7.7]. The high frequency dielectric constant ϵ_∞ is 5.10 and 5.41 in TlCl and TlBr. These high dielectric constants should result in a small exciton binding energy. The small binding energy will give a large exciton orbit and thus the effective mass approximation can be applied very successfully in thallous halides. Further the energy change of an exciton by the application of an external field can be made comparable to this small binding energy without much difficulty. Thus a drastic change in exciton structure, which has been rarely studied in ionic crystals, can be realized by the external fields available in a laboratory. Fourthly, we should mention the advantages of thallous halides in the purification of the material. Their high vapor pressure is convenient for purification by distillation and their low melting point, of a little above 400 °C, makes zone refining easy. In practice, thallous halides are synthesized in aqueous solution from chemically purified $TlNO_3$ and HX and they are distilled in a high vacuum several times. Zone melting is performed in an HX atmosphere and single crystals are grown in a high vacuum [7.8, 9]. The concentrations of impurities in pure TlCl and TlBr are less than 10^{-7} mole fraction for metal ions, less than 10^{-6} mole fraction for iodine ions and less than 10^{-4} mole fraction for other isoelectronic halogen ions.

7.1.2 Band Structures

Let us first look at the electronic energy of free-ion states of Tl^+ and X^- embedded in a TlX matrix of CsCl structure under the effect of the Madelung potential. Figure 7.1 is the energy diagram in the case of TlBr [7.10]. Here, in the ground state, the levels are filled up to $(6s_{1/2})^2$ in Tl^+ and $(4p_{3/2})^4$ in Br^-. Thus one can expect the highest valence band to be composed of Tl^+ $6s$ and Br^- $4p$ states. For the lowest conduction band, the lowest empty $6p$ state of Tl^+ would be important. A similar situation holds in TlCl and TlI.

Since the space group of thallous halides of CsCl structure is simple cubic, the first Brillouin zone is also simple cubic. The symmetry point at the zone center is called Γ and those at the zone boundaries in the $\langle 100 \rangle$, $\langle 110 \rangle$ and $\langle 111 \rangle$ directions are called X, M, and R, respectively, in the simple cubic Brillouin zone. the band calculations of TlCl, TlBr, and TlI have been made by several workers [7.5, 10–16] along typical symmetry lines in k-space. In Fig. 7.2, the band structures of TlCl and TlBr calculated by *Overhof* and *Treusch* [7.11] are shown. The symmetry assignments given here are made by taking the origin at a Tl site.

All the calculations show that the highest valence band consists of Tl $6s$ and halogen p orbitals. The maximum in this band is X_6^+ at the X point where Tl $6s$ and halogen p are mixed considerably. The state at the Γ point Γ_8^- is due to a pure halogen $p_{3/2}$ orbital. There are three states closely spaced at the R point. They are the states R_6^+, R_8^+ and R_7^+ which originate from Tl $6s$, halogen $p_{3/2}$, and halogen $p_{1/2}$ respectively. The energy sequence of the three states is still in question and no agreement is found between the workers, perhaps because the energy differences are so small that they are beyond the accuracies of the calculations. The

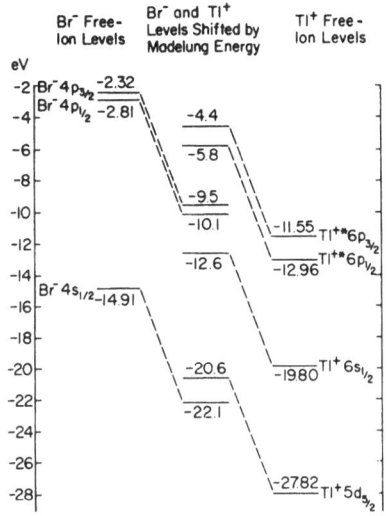

second and third highest valence bands are relatively narrow. At the X point, they are X_7^+ and X_6^+; the former originates from halogen $p_{3/2}$ whereas the latter is made from an admixture of halogen $p_{1/2}$ and $p_{3/2}$ states. The energy difference between the X_7^+ and X_6^+ states is close to the spin-orbit splitting energy of the free halogen p orbital.

The lowest conduction band is mostly from the Tl $6p$ state. Two minima are present. One is at the X point and the other is at the R point. The state at the X point is X_6^- and is a mixture of $6p_{1/2}$ and $6p_{3/2}$ orbitals of thallium. In TlCl, it is suggested that the halogen s-state of more than 10% is mixed with this state [7.12]. At the R point, the lowest conduction-band state is due to Tl $p_{1/2}$ which is mixed with a halogen d-state of about 20%. The second lowest band at the R point is R_8^- and consists of Tl $p_{3/2}$ with some halogen d-state.

From the band calculation, it is difficult to determine whether the X_6^- state or the R_6^- state in the lowest conduction band is higher in energy. It is found by

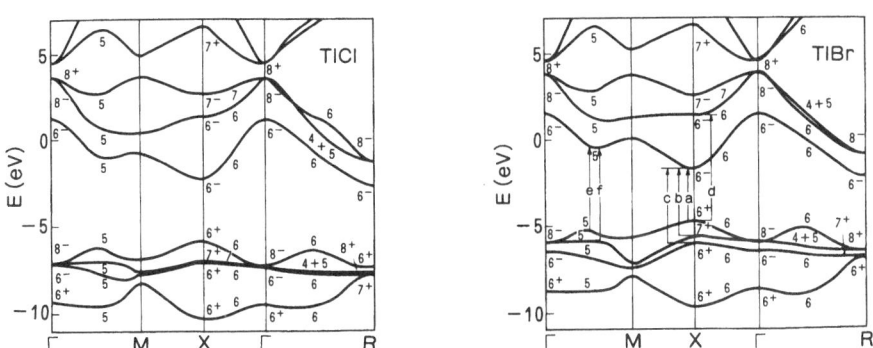

Fig. 7.2. Band structures of TlCl and TlBr of CsCl structure [7.11]

optical and cyclotron resonance experiments, which will be described later, that the R_6^- state is lower than the X_6^- state by a few hundred meV.

The band structures of TlCl, TlBr [7.4], and TlI [7.5] of NaCl structure (fcc space lattice) have been calculated and their differences from those of CsCl structure (sc space lattice) have been discussed [7.17]. The main differences in the band-edge structure between the fcc and sc crystals are as follows: Firstly, the smallest direct gap is at the L point in the fcc crystal whereas in the sc crystal it is at the X point. This is a consequence of the nearest neighbour coordination geometry of the lattice. Secondly, in the sc crystal, the smallest gap is the indirect X–R gap which is a result of the shape of the Brillouin zone. Thirdly, a strong mixing of the Tl 6s and the halogen p is found in the fcc crystal, leading to a larger degree of covalency of the thallium-halogen bond.

7.1.3 Exciton States

In most of wide band gap compound semiconductors and insulators, such a typical II–VI compounds, cuprous halides, alkali halides, and rare gas solids, the valence and the conduction band consist of p-like and s-like orbitals, respectively. In thallous halides, however, the highest valence band is s-like whereas the lowest conduction band is p-like, giving a special electronic structure of the exciton which will not be found in ordinary wide band gap materials.

The smallest direct band gap in thallous halides is at the X point and not at the Γ point, as can be seen in Fig. 7.2. The lowest-energy direct exciton is composed of the X_6^+ state of the valence band the X_6^- state of the conduction band. The smallest indirect gap is between the X_6^+ state of the valence band and the R_6^- state of the conduction band. We call the former direct exciton the $X_6^+ \times X_6^-$ exciton and the latter indirect the $X_6^+ \times R_6^-$ exciton.

$X_6^+ \times X_6^-$ **Direct Excitons.** The electronic state of the $X_6^+ \times X_6^-$ exciton has been investigated in detail both theoretically and experimentally [7.18–20]. There are three inequivalent X points in the *sc* Brillouin zone at the zone boundaries in the [100], [010], and [001] directions. We call these three X valleys the x, y, and z valleys. In each valley, the X_6^- conduction band and the X_6^+ valence band are doubly degenerate. They are in the z valley

$$X_6^- \begin{cases} az^{Tl}|\uparrow\rangle - (\beta/\sqrt{2})(x+iy)^{Tl}|\downarrow\rangle : c_\uparrow^z \\ az^{Tl}|\downarrow\rangle + (\beta/\sqrt{2})(x-iy)^{Tl}|\uparrow\rangle : c_\downarrow^z , \end{cases} \tag{7.1}$$

$$X_6^+ \begin{cases} \gamma s^{Tl}|\uparrow\rangle - i\delta z^x|\uparrow\rangle : v_\uparrow^z \\ \gamma s^{Tl}|\downarrow\rangle + i\delta z^x|\downarrow\rangle : v_\downarrow^z , \end{cases} \tag{7.2}$$

where x, y, and z are the p functions and s is the s function. Spins are indicated as $|\uparrow\rangle$ and $|\downarrow\rangle$. It has been given that $\alpha^2 = 0.85$, $\beta^2 = 0.15$, $\gamma^2 = 0.3$, and $\delta^2 = 0.7$ [7.19]. In the X_6^- conduction-band state, the halogen s-state is mixed to some extent but its amount is small so that it is omitted from (7.1). There is strong mixing of thallium $6s$ and halogen p in the valence band X_6^+. However, if one consider the $X_6^+ \times X_6^-$ exciton state, the halogen p-state can be ignored in the first approximation because the conduction-band X_6^- state is made mostly of thallium p-state so that the oscillator strength of the transition from the halogen p-state in the valence band X_6^+, whose amplitude will be large on the halogen sublattice, to the conduction band X_6^- must be fairly small.

Four exciton states can be constructed from two X_6^- conduction-band states and two X_6^+ valence-band states in the z valley. We denote them as ψ_q^p where p stands for the direction of the valley and q the direction of the polarization of the wave function. They are written, using (7.1, 2), as

$$\psi_x^z = (1/\sqrt{2})(- v_\uparrow^z c_\uparrow^z + v_\downarrow^z c_\downarrow^z) \ ,$$

$$\psi_y^z = (i/\sqrt{2})(v_\uparrow^z c_\uparrow^z + v_\downarrow^z c_\downarrow^z) \ ,$$

$$\psi_z^z = (1/\sqrt{2})(v_\downarrow^z c_\uparrow^z + v_\uparrow^z c_\downarrow^z) \ ,$$

$$\psi_0^z = (1/\sqrt{2})(v_\downarrow^z c_\uparrow^z - v_\uparrow^z c_\downarrow^z) \ . \tag{7.3}$$

Here ψ_0^z is a pure spin triplet state and the others are singlet-triplet mixed states. In the expressions (7.3), the envelope function $F(r) \exp(iK \cdot R)$ is not shown explicitly. In the x and y valleys, similar states can be constructed so that there are twelve states for the $X_6^+ \times X_6^-$ exciton which are degenerate as long as the electron-hole inter- and intravalley exchange interaction and the intervalley Coulomb interaction are not taken into account.

In Sect. 1.4 the effective mass Hamiltonian of the exciton [7.21, 22] has been derived as

$$\mathcal{H}_{vc; v'c'} = \delta_{vv'} E_{cc'}\left(- i\nabla + \frac{K}{2}\right) - \delta_{cc'} E_{vv'}\left(- i\nabla - \frac{K}{2}\right)$$

$$+ \delta_{vv'}\delta_{cc'} V(r) + \Omega\delta(r)J_{vc, v'c'}(K) \ , \tag{7.4}$$

where K is the total wave vector of the exciton and Ω the volume of a unit cell. The first and the second terms are the kinetic term and $V(r)$ is the attractive Coulomb potential. The exciton states expressed by (7.3) are the states which are constructed by taking into account only the kinetic terms and the Coulomb term for an electron and a hole in the same valley. The fourth term is the exchange term between electron and hole and is given by

$$J_{vc, v'c'}(K) = |F(0)|^2 \int\int a_c^*(r_1)a_v(r_1) \frac{e^2}{|r_1 - r_2|} a_{v'}^*(r_2)a_{c'}(r_2)dr_1dr_2$$

$$+ \left(\frac{4\pi}{3\Omega}\right) \frac{3(\mu_{cv} \cdot K)(\mu_{v'c'} \cdot K) - (\mu_{cv} \cdot \mu_{v'c'})K^2}{K^2} \ , \tag{7.5}$$

Table 7.1. Energies of the z-polarized states and the spin triplet state of the $X_6^+ \times X_6^-$ exciton. The energy of the spin triplet state Ψ_0^z is taken as zero

	Ψ_z^z	$(\Psi_z^x + \Psi_z^y)/\sqrt{2}$	$(\Psi_z^x - \Psi_z^y)/\sqrt{2}$	Ψ_0^z
Ψ_z^z	$2\Delta_1$	$2\sqrt{2}\Delta_3$	0	0
$(\Psi_z^x + \Psi_z^y)/\sqrt{2}$	$2\sqrt{2}\Delta_3$	$\Delta_2 + 2\Delta_4$	0	0
$(\Psi_z^x - \Psi_z^y)/\sqrt{2}$	0	0	$\Delta_2 - 2\Delta_4$	0
Ψ_0^z	0	0	0	0

where $a(r)$ is the Wannier function and μ is a transition dipole moment

$$\mu = eF(0)\int a_c^*(r)ra_v(r)dr \ . \tag{7.6}$$

The second term in (7.5) gives a transverse-longitudinal splitting when $K \neq 0$. The 12-fold degeneracy of the exciton state expressed in (7.3) is lifted by introducing the intervalley Coulomb interaction and the inter- as well as intravalley exchange interactions which couple excitons of same wave vector. Since the excitons (7.3) are constructed by an electron in the conduction band and a missing electron in the valence band in the same valley, the total wave vectors of excitons in the x, y, and z valleys are all zero. This gives twelve new states which are split by these interactions. Four of them are shown in Table 7.1. Here ψ_z^z, $(\psi_z^x + \psi_z^y)/\sqrt{2}$ and $(\psi_z^x - \psi_z^y)/\sqrt{2}$ are z-polarized states which interact with z-polarized light by a dipole interaction. The spin triplet state ψ_0^z has no polarization. There are similar corresponding states for x-polarization and for y-polarization and spin triplet states of other valleys. In this table, we take the energy of the spin triplet state as the origin, Δ_1 and Δ_2 are intravalley interaction energies and Δ_3 and Δ_4 are the intervalley interaction energies. They are

$$\Delta_1 = (v_\uparrow^z c_\uparrow^z |g| v_\uparrow^z c_\uparrow^z)\Omega|F(0)|^2 = \alpha^2\gamma^2 M_{AT} \ , \tag{7.7}$$

$$\Delta_2 = (v_\uparrow^z c_\downarrow^z |g| v_\uparrow^z c_\downarrow^z)\Omega|F(0)|^2 = \beta^2\gamma^2 M_{AT} \ , \tag{7.8}$$

$$\Delta_3 = (v_\downarrow^z c_\downarrow^z |g| v_\downarrow^x c_\downarrow^x)\Omega|F(0)|^2 = \frac{\alpha\beta\gamma^2}{2}\left(M_{AT} - \frac{1}{2}M_{CT}^{xz}\right) \ , \tag{7.9}$$

$$\Delta_4 = (v_\uparrow^x c_\downarrow^y |g| v_\uparrow^z c_\uparrow^z)\Omega|F(0)|^2 = \frac{\beta^2\gamma^2}{2}\left(M_{AT} - \frac{1}{2}M_{CT}^{xy}\right) \ , \tag{7.10}$$

where g is $e^2/|r_1 - r_2|$ [7.19, 20]. The exchange energy M_{AT} is given as

$$M_{AT} = \frac{4\pi e^2}{\epsilon}|F(0)|^2 \sum_{G \neq 0} \frac{\langle s^{Tl}|e^{iG \cdot r}|z^{Tl}\rangle\langle z^{Tl}|e^{-iG \cdot r}|s^{Tl}\rangle}{|G|^2} \ , \tag{7.11}$$

where G is a reciprocal wave vector. The $x - y$ intervalley Coulomb energy M_{CT}^{xy} is

$$M_{CT}^{xy} = \frac{4\pi e^2}{\epsilon} |F(0)|^2 \sum_G \frac{\langle s^{Tl}|e^{i(G + K_{xy}) \cdot r}|s^{Tl}\rangle \langle z^{Tl}|e^{-i(G + K_{xy}) \cdot r}|z^{Tl}\rangle}{|G + K_{xy}|^2} \qquad (7.12)$$

where K_{xy} is the wave vector connecting the valleys x and y [7.9]. We note that the off-diagonal intervalley scattering energy $2\sqrt{2}\Delta_3$ has been found to be small [7.19] and the states given in Table 7.1 are almost diagonalized. There are other corresponding x- and y-polarized states with another two spin triplet states. Among the z-polarized states the transitions to the Ψ_z^z and $(\Psi_z^x + \Psi_z^y)/\sqrt{2}$ states are optically dipole allowed and the transitions to the $(\Psi_z^x - \Psi_z^y)/\sqrt{2}$ and Ψ_0^z states are forbidden. A similar selection rule holds for x- and y-polarized light.

b) $X_6^+ \times R_6^-$ Indirect Excitons. The wave vector of the $X_6^+ \times R_6^-$ indirect exciton is at the M point, the zone boundary in the [110] direction. In the sc Brillouin zone, all R points are equivalent so that there is only one R valley. For the X point, there are three inequivalent valleys. This leads to three $X_6^+ \times R_6^-$ exitons whose wave vectors are at inequivalent M points, that is, $110\,(M_1)$, $101\,(M_2)$, and $011\,(M_3)$. The exciton states can be constructed [7.23] by using the basis function given in (7.2) and

$$R_6^- \begin{cases} -z^{Tl}|\uparrow\rangle - (x + iy)^{Tl}|\downarrow\rangle : c_1^z \\ -(x - iy)^{Tl}|\uparrow\rangle + z^{Tl}|\downarrow\rangle : c_2^z . \end{cases} \qquad (7.13)$$

In the expression of the R_6^- state, we use only the Tl $6p$ state but it is mixed with halogen d orbitals to some extent which are omitted here. The resultant excitons are at the M_1 point and their symmetry axes are in the [001] direction. There are four $X_6^+ \times R_6^-$ exciton states in the 110 valley (M_1) and they are

$$\Psi_x^{110} = (1/\sqrt{2})\,(v_\uparrow^z c_1^z - v_\downarrow^z c_2^z) ,$$

$$\Psi_y^{110} = (-i/\sqrt{2})\,(v_\uparrow^z c_1^z + v_\downarrow^z c_2^z) ,$$

$$\Psi_z^{110} = (-1/\sqrt{2})\,(v_\uparrow^z c_2^z + v_\downarrow^z c_1^z) ,$$

$$\Psi_0^{110} = (1/\sqrt{2})\,(v_\uparrow^z c_2^z - v_\downarrow^z c_1^z) . \qquad (7.14)$$

The state Ψ_0^{110} is a pure spin triplet state and the others are singlet-triplet mixed states. Here, 110 and q in Ψ_q^{110} are the position of the valley and the directions of the orbital polarization due to the singlet part. Four states analogous to (7.14) are constructed for both the 101 (M_2) and 011 (M_3) valleys and $\Psi_z^{110} = \Psi_y^{101} = \Psi_x^{001}$, $\Psi_x^{110} = \Psi_z^{101} = \Psi_y^{011}$, $\Psi_y^{110} = \Psi_x^{101} = \Psi_z^{011}$, and $\Psi_0^{110} = \Psi_0^{101} = \Psi_0^{011}$.

The energy of the Ψ_z^{110} state is Δ_1' and those of Ψ_x^{110} and Ψ_y^{110} are Δ_2' which are measured from the triplet Ψ_0^{110} state. The exchange energies Δ_1' and Δ_2' are

$$\Delta_1' = 2(s(1)z(2)|g|s(2)z(1))\Omega|F(0)|^2 ,\tag{7.15}$$

$$\Delta_2' = 2(s(1)x(2)|g|s(2)x(1))\Omega|F(0)|^2 .\tag{7.16}$$

If the valence band X_6^+ and the conduction band R_6^- are made purely by Tl $6s$ and Tl $6p$, respectively, Δ_1' is equal to Δ_2' and the energy of the $X_6^+ \times R_6^-$ exciton splits into two levels. One is ninefold-degenerate and all the exciton states belonging to this level are singlet-triplet mixed states. The other is threefold-degenerate and the exciton states belonging to this level are all pure spin triplet states. If one takes into account the contributions of halogen p orbitals to the X_6^+ state and halogen d orbitals to the R_6^- state, there might be a possibility of finding a difference between Δ_1' and Δ_2', and further splitting of the singlet-triplet mixed states into threefold- and sixfold-degenerate states. It is to be noted that, in contrast to the $X_6^+ \times X_6^-$ direct excitons, intervalley scattering by Coulomb and exchange interactions does not operate in the $X_6^+ \times R_6^-$ exciton because of a change of the total wave vector by the scattering among the M_1, M_2, and M_3 excitons.

7.2 Optical Spectra of Thallous Halides

7.2.1 Absorption and Reflection Spectra in a Wide Energy Range

General features of the fundamental optical absorption spectra of TlCl and TlBr in the energy range from 3 to 6 eV were first clarified by *Zinngrebe* [7.24] and *Tutihasi* [7.25] and later by *Bachrach* and *Brown* [7.26] and others. In the energy range between 195 and 240 eV, $Cl^- L_{2,3}$ absorption has been measured in TlCl [7.27].

Figure 7.3 gives the absorption spectra of TlCl and TlBr at 2 K in the ultraviolet region obtained from the reflectivity spectra, shown in Fig. 7.4, by the

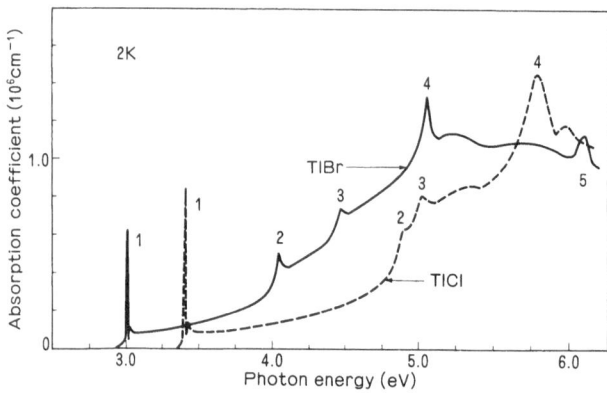

Fig. 7.3. Absorption spectra of TlCl and TlBr at 2 K [7.28, 29]

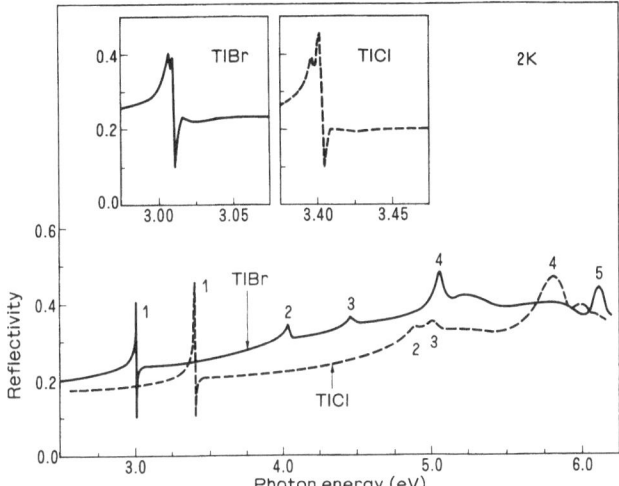

Fig. 7.4. Reflectivity spectra of TlCl and TlBr at 2 K [7.28, 29]

Kramers-Kronig relation [7.28, 29]. Spectra almost identical in shape are obtained by a direct absorption measurement on evaporated thin films. In the absorption spectra in Fig. 7.3, only the structures of high absorption coefficient of the order of 10^5 cm^{-1} and more can be seen and they are associated with dipole-allowed direct transitions. Any structure associated with an indirect transition cannot be identified in this scale of absorption coefficient.

Peak 1 at the low-energy end in both halides is assigned to the direct-exciton absorption at the X point from the X_6^+ highest valence band to the lowest X_6^- conduction band. Peaks 2 and 3 are the direct transitions from the X_7^+ and X_6^+ states in the second and third highest valence bands, respectively, to the lowest X_6^- conduction band. This is because the energy separations between peaks 2 and 3 are 0.10 eV and 0.423 eV in TlCl and TlBr, respectively, whereas the energy differences expected theoretically between the X_7^+ and X_6^- state are 0.09 eV and 0.38 eV [7.11] for TlCl and TlBr, which are close to the spin-orbit splitting energy between $p_{1/2}$ and $p_{3/2}$ states of respective halogens. The intensities of peaks 2 and 3 are comparatively large. This is an indication of the admixture of the halogen s orbital in the X_6^- conduction-band state because both the X_7^+ and the X_6^+ valence-band states are made purely of the halogen p-state.

It is accepted by almost all workers that peak 4 originates from a direct transition at the R point, however, its detailed assignment has not been settled. It is interpreted as the transition $R_6^+ \rightarrow R_6^-$ in TlCl [7.13] and TlBr [7.10, 11], or $R_6^+ \rightarrow R_8^-$ in TlCl [7.11] and $R_6^+ \rightarrow R_6^-$ in TlBr [7.14]. The R_6^- and R_8^- states in the conduction band are basically thallium s-state whereas the R_8^+ state originates from the halogen p-state. Then it would be most probable that peak 4 is due to the $R_6^+ \rightarrow R_6^-$ transition in TlBr and to the $R_6^+ \rightarrow R_6^-$ or the $R_6^- \rightarrow R_8^-$ transition in TlCl, because both the transitions are s–p like on the thallium sublattice, giving a strong absorption as observed in Fig. 7.3. Other broad structures observed in the absorption spectra at high energy are assigned to the transitions at the states of high energy [7.10, 11] but further convincing experimental evidence is necessary.

The absorption spectrum of TlI of CsCl structure is known [7.5]. The structure of the spectrum and the assignment are essentially the same as those given to the other halides mentioned above.

7.2.2 Spectra of $X_6^+ \times X_6^-$ Direct Excitons

The shape and the energy of the absorption peak due to the direct allowed $X_6^+ \times X_6^-$ exciton, which is peak 1 in Fig. 7.3, are quite sensitive to the strain in the crystal. This peak had been believed to be a single peak until *Bachrach* and *Brown* [7.26] found that it splits into a double peak and is accompanied by a further peculiar structure on its high-energy side. They used almost strain-free evaporated thin films of thallous halides which were prepared by an ingenious method developed by them. To do an optical measurement with a thin film at low temperature, one prepares the film first by evaporating onto a substrate at a high temperature to get a sample in a nice crystalline form and then cools it to a low temperature. The strain is usually introduced into the film during the cooling because of a difference in the thermal expansion coefficients of the film and the substrate. Their method is such that a thallous halide sample is evaporated onto a Lucite thin film a few hundred angstroms thick which is placed over a hole drilled through single crystal of the same thallous halide. In this way the sample film can be cooled to a low temperature without strain because the thermal contraction of the thin Lucite film will follow the contraction of the thallous halide crystal so that there is little difference between the contractions of the evaporated sample and the Lucite film substrate.

Figure 7.5 shows how beautifully their method works. Here the absorption spectra of thin TlCl films evaporated onto various substrates, including the Lucite film, are shown [7.30]. We note that the energy of the absorption peak increases by almost 90 Å with the disappearance of the double structure when the thallous halide is evaporated onto a quartz plate where the difference in the thermal contraction is greatest. That the electronic energy of thallous halide is so susceptible to the strain might be due to the easily deformable $6s$ shell of the Tl ion from which the X_6^+ valence-band state is formed. The softness of the $6s$ shell is proposed as the origin of the anomalously large static dielectric constant [7.17].

The most reliable optical spectrum which gives the intrinsic nature in the energy range of the direct exciton in thallous halides is a reflectivity spectrum of a single crystal, because a single crystal is free from the unavoidable residual strain of an evaporated thin film. Fig. 7.6 is the reflectivity spectrum of TlCl [7.28, 29]. The spectra of the absorption coefficient α are calculated for TlCl and TlBr from the reflectivity spectra by the Kramers-Kronig relation [7.28, 29] and are given in Figs. 7.7, 8. Here α is $\omega \epsilon_2/nc$ where ϵ_2 is the imaginary part of the dielectric function, ϵ, and n is the real part of the index of refraction. In these figures, the

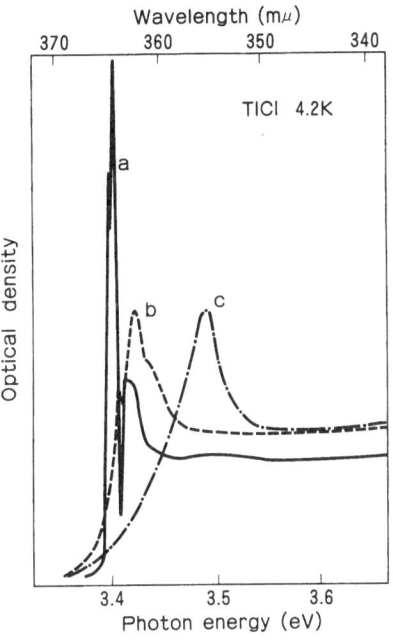

Fig. 7.5. Absorption spectra of TlCl at 4.2 K measured in thin films evaporated on different substrates. (*a*) on Lucite film by Bachrach and Brown's method, (*b*) on a CsBr crystal, (*c*) on a quartz plate [7.30]

optical density spectra measured directly in an evaporated thin film are given in the insets. We note that the Kramers-Kronig relation cannot be applied, in principle, to the transformation of the reflectivity spectrum to the ϵ spectrum for the case of an exciton with spatial dispersion because the reflectivity and the resultant ϵ are only functions of ω whereas ϵ of an exciton with spatial dispersion

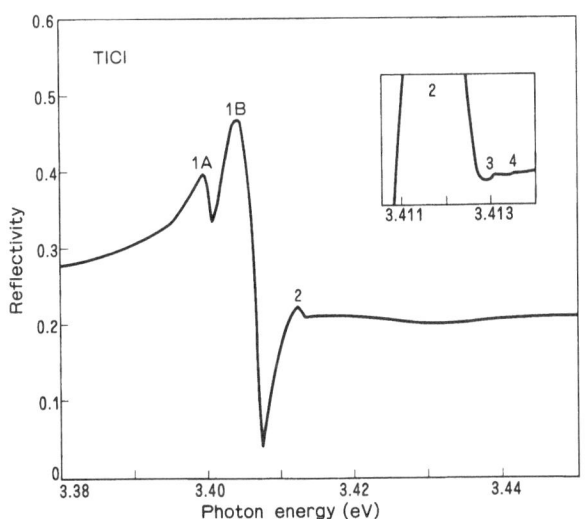

Fig. 7.6. Reflectivity spectrum of TlCl at 4.2 K in the vicinity of the $X_6^+ \times X_6^-$ direct-exciton energy [7.28, 29]

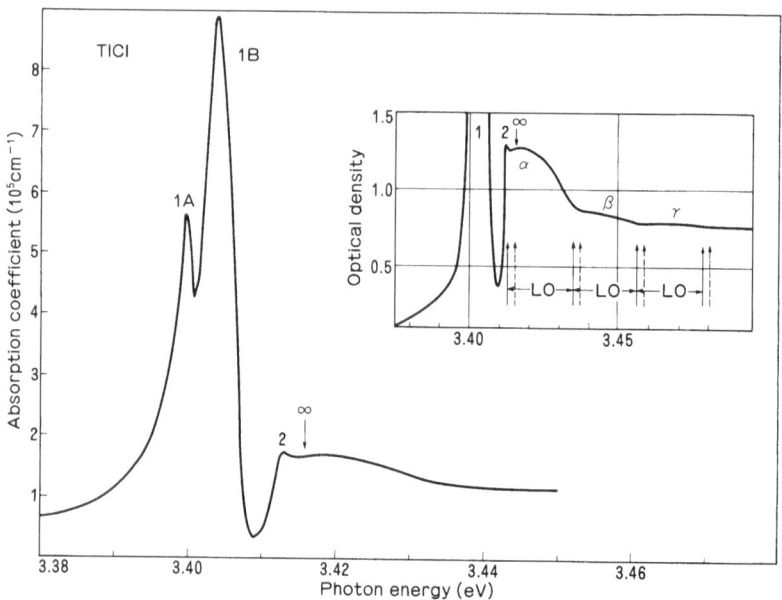

Fig. 7.7. Absorption spectrum of TlCl at 4.2 K in the vicinity of the $X_6^+ \times X_6^-$ direct-exciton energy. It is estimated from the reflectivity spectrum in Fig. 7.6 by the Kramers-Kronig relation [7.28, 29]. In the inset, the absorption spectrum directly measured in an evaporated thin film is shown [7.31]. Here LO is the energy of the LO(Γ) phonon which is 21.7 ± 0.4 meV at 4.2 K

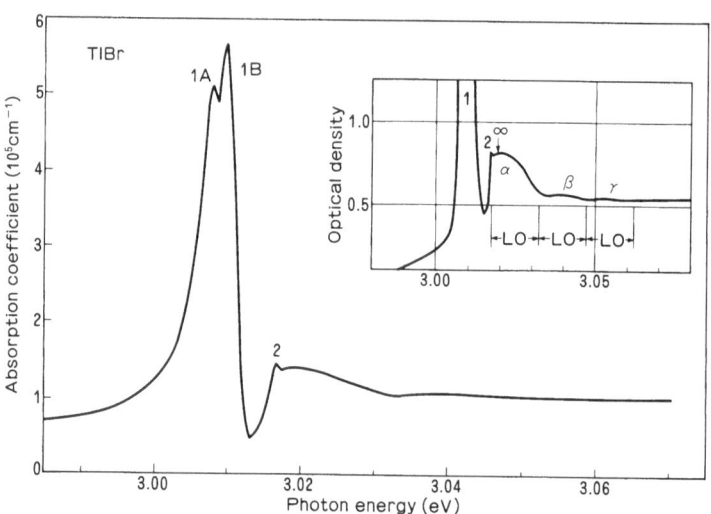

Fig. 7.8. Absorption spectrum of TlBr at 4.2 K in the vicinity of the $X_6^+ \times X_6^-$ direct-exciton energy. It is estimated from the reflectivity spectrum by the Kramers-Kronig relation [7.28, 29]. In the inset, the absorption spectrum directly measured in an evaporated thin film is shown [7.31]. Here LO is the energy of the LO(Γ) phonon which is 14.5 ± 0.1 meV at 4.2 K

must be a function of ω and k. However, the difference between the two dielectric functions $\epsilon(\omega)$ and $\epsilon(\omega, k)$ is sufficiently small in practice that the absorption spectra in Figs. 7.7, 8 can be understood as true absorption spectra within an experimental error. It is found that all the structures in the absorption spectrum of a TlCl single crystal are always several tenths of a milli-electron Volt lower in energy than those measured in evaporated thin films prepared by the method of *Bachrach* and *Brown*. This difference may be understood as due to a residual strain still remaining in the evaporated thin film.

From Figs. 7.7, 8, we find that the structures of the absorption spectra of TlCl and TlBr are essentially the same. The sharp peaks 1A and 1B are found at the lowest energy. They are followed by a small peak 2 and very small structures 3 and 4. The latter two are not observable in Figs. 7.7, 8 but are found in the inset of Fig. 7.6. The shape of the hump α is asymmetric, as if it is truncated by peak 2. The hump α is followed by nearly symmetric broad bands β and γ at higher energies with kinks between them. The peak 1A and 1B are the absorption due to the 1s state of the $X_6^+ \times X_6^-$ exciton. There are two dipole-allowed states for this exciton as is explained in Sect. 7.1.3. They are the Ψ_z^z state and the $(\Psi_z^x + \Psi_z^y)/\sqrt{2}$ state. The oscillator strength of the transition to both states will be proportional to $|\langle\Psi_z^z|p_z|0\rangle|^2$ and $|\langle\Psi_z^x|p_x|0\rangle|^2$, the latter of which is equal to $|\langle\Psi_x^z|p_x|0\rangle|^2$. If one considers that the transition occurs mainly on the Tl sublattice, $|\langle\Psi_z^z|p_z|0\rangle|^2 > |\langle\Psi_x^z|p_x|0\rangle|^2$ because of a larger value of α than β in (7.1). We see in Figs. 7.7 and 7.8 that 1B is of greater intensity than 1A. Thus 1A is due to the $(\Psi_z^x + \Psi_z^y)/\sqrt{2}$ exciton and 1B to the Ψ_z^z exciton. Structures 2, 3, and 4 on the high energy side of 1B are the Rydberg series of the $X_6^+ \times X_6^-$ Wannier excitons and correspond to the 2s, 3s, and 4s envelope states. From their energies, the series limit energy or the band-gap energy ε_g and the binding energies E_{ex}^b of 1A and 1B excitons in TlCl at 4.2 K are estimated and they are tabulated in Table 7.2. For TlBr, the structures 3 and 4 are not resolved and the energy of the 3s and 4s states are estimated from the magneto-optical study given in Sect. 7.5.1. The energy positions of ε_g are shown by ∞ in Figs. 7.7, 8.

The small exciton binding energies of TlCl and TlBr are the consequence of high dielectric constants. They are also interpreted in terms of strong exciton-polaron effects in these substances. Due to the e–h cancellation effect on exciton-phonon coupling [7.32] described in Sects. 4.1.4 and 4.3.3, the polaron self-energy shift is larger for a higher exciton state and for a larger coupling constant. This leads to a large decrease of exciton binding energy when a bare exciton couples with phonons to form an exciton polaron. In fact, as is shown in Fig. 4.18, the binding energy of the bare exciton is reduced by a factor of 5 for the 1s state and of 14 for the 2s state for the case of $\mu/m_0 = 0.2$, $Ry'_\infty/\hbar\omega_{LO} = 5$, $m_e/m_h = 0.5$, and $\alpha = 4$ where Ry'_∞ is the exciton Rydberg energy for ϵ_∞. These values are close to those of TlCl. When one takes $\hbar\omega_{LO}$, the Γ point LO phonon energy, as 21.7 meV and uses the above factor for 2s, one finds that the bare exciton Rydberg energy is 113 meV and the Rydberg energy of the exciton polaron is 8 meV in TlCl which is close to the Rydberg energy of 10 meV estimated from the measured 2s and 4s separation.

Table 7.2. Exciton and related parameters of thallous halides of CsCl structure

		TlCl	TlBr	TlI
$X_6^+ \times X_6^-$ exciton (4.2 K)				
1s state energy [eV]	Ψ_z^z	3.4030	3.0098	2.777_8
	$(\Psi_z^x + \Psi_z^y)/\sqrt{2}$	3.3992	3.0079	
	Ψ_0^z	3.3978		
band gap energy [eV]		3.4144	3.0190	
binding energy of 1s exciton [meV]	Ψ_z^z	11.4	9.2	11 ± 1
	$(\Psi_z^x + \Psi_z^y)/\sqrt{2}$	15.2	11.1	
	Ψ_0^z	16.6		
$X_6^+ \times R_6^-$ exciton (4.2 K)				
1s state energy [eV]		3.225 ± 0.001	2.6444 ± 0.0005	$\sim 2.1(77\ \text{K})$
band-gap energy [eV]		3.248 ± 0.004	2.6634 ± 0.0015	
binding energy of 1s exciton [meV]		23 ± 3	19 ± 1	
Polaron mass (in units of m_0)				
electron at R point		0.56 ± 0.02	0.525 ± 0.03	
hole at X point		$0.58 \pm 0.03(\parallel)$	$0.55\ \pm 0.03(\parallel)$	
		$0.98 \pm 0.04(\perp)$	$0.75\ \pm 0.03(\perp)$	
Fröhlich coupling constant (electron)		2.5	2.7	

We observe no splitting in the 2s and higher excitons. From the matrix elements given in Table 7.1 and (7.7, 10), the origin of the splitting into Ψ_z^z and $\Psi_z^x + \Psi_z^y$ states is mostly the intra- and intervalley exchange interactions which are proportional to $|F(0)|^2$. The magnitude of $|F(0)|^2$ is proportional to the oscillator strength f which varies as n^{-3} for the ns Wannier exciton. Thus the splitting will be very small and will not be observable in the excitons of 2s and higher.

The humps α, β, and γ in Figs. 7.7, 8 are the phonon structures of the exciton. The hump α is a broad peak with a maximum at around 3.418 eV which is higher than the 1A and 1B exciton energy by 0.87 $\hbar\omega_{LO}$ and 0.69 $\hbar\omega_{LO}$. There are no particular structures at the energy of $\hbar\omega_{LO}$ measured from the 1A and 1B exciton energies. Thus the hump α is not a free-phonon sideband of the 1A or 1B exciton. The characteristic feature of the shape of the hump α is found at its low-energy end where it is truncated by the 2s exciton absorption. In Sect. 4.3.3 we found that the absorption line shape of the 1s exciton-phonon quasi-bound state [7.33] is Lorentzian (4.57), with width function $\Gamma_0(E)$ modulated by the background 2s and higher exciton states. The energy of the maximum is in the continuum of the 1s exciton and is less than the 1s exciton energy plus one-LO-phonon energy by the amount of a bound-state self-energy. The line shape of the hump α satisfies all these properties and thus it is the absorption due to the $1s$ exciton–LO-phonon quasi-bound state.

From (4.130) in Sect 4.3.3 we find that the self-energy of the exciton-phonon quasi-bound state will be reduced in higher n exciton states because of the negative contribution of exciton states lower than n. This suggests that there will be no stable exciton-phonon quasi-bound state for all excitons of $2s$ and higher. In the spectrum in Figs. 7.7 and 7.8, we find no structures responsible for the $2s$ and higher exciton-phonon quasi-bound state in the energy region of $\hbar\omega_{LO}$ (21.7 meV) measured from $E_{2s} \sim \mathscr{E}_g$ energy. Instead, we find a broad hump β which begins to rise almost at the energy of $E_{2s} + \hbar\omega_{LO}$. As has been discussed in Sect. 4.3.1, a free one–LO-phonon sideband should rise at a threshold energy, which is equal to the exciton energy plus $\hbar\omega_{LO}$, and have a maximum at a higher energy, because of the exciton dispersion. Since the dip between α and β near the energy of $\hbar\omega_{LO}$ from $2s \sim \mathscr{E}_g$ is broad, the exact threshold energy of the hump β is difficult to determine. It is likely that the slow increase of the hump β begins at the energy of the $2s$ exciton plus $\hbar\omega_{LO}$ and we interpret that the hump β is the one free LO phonon sideband of $2s$ and higher excitons including the continuum. The hump γ is at an energy higher than β by $\hbar\omega_{LO}$ so that it is the two free LO phonon sideband of $2s$ and higher exciton states.

The optical spectra discussed above are one-photon spectra where the one-photon energy is resonant with the energy of the electronic transition. In the two-photon absorption experiment, where the sum of two photon energies are resonant with the electronic transition energy, the energies of two photons are chosen in such a way that the crystal under study does not absorb either photon separately. Usually, the energy of one photon is fixed and the other is varied so that the sum of the two energies sweeps the energy range of a single electronic transition. We consider two beams incident on the crystal. The energy of one beam is $\hbar\omega_1$ and its polarization is e_1, while or the other beam, they are $\hbar\omega_2$ and e_2. The two-photon absorption coefficient α of the transition from the state a to b is proportional to

$$\alpha \propto \left| \sum_i \left(\frac{\langle b|e_1 \cdot p|i\rangle\langle i|e_2 \cdot p|a\rangle}{E_i - E_a - \hbar\omega_2} + \frac{\langle b|e_2 \cdot p|i\rangle\langle i|e_1 \cdot p|a\rangle}{E_i - E_a - \hbar\omega_1} \right) \right|^2 \qquad (7.17)$$

where i is an intermediate state.

One of the features of two-photon absorption which differs from one-photon absorption is the selection rule. In two-photon absorption, the parities of the initial and the final states are the same for a dipole-allowed transition so that it is an even-parity transition, whereas the one-photon transition is an odd-parity transition. Thus the two methods of investigation are complementary and their combination gives a wealth of information about the electronic structure. Another feature is that, by varying the polarizations e of the two beams with respect to one another and to the crystal axes, one can investigate the symmetry of the electronic state.

Two-photon absorption spectra of TlCl in the energy range of the $X_6^+ \times X_6^-$ exciton measured by *Fröhlich* et al. [7.34] are shown in Fig. 7.9. In this figure,

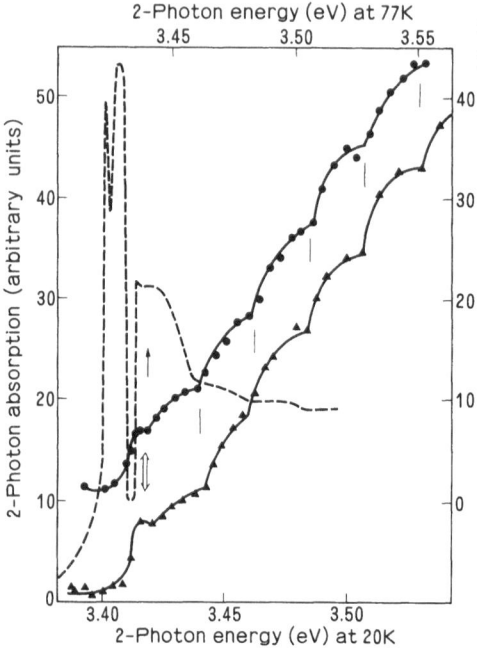

2-Photon energy (eV) at 77K

2-Photon energy (eV) at 20K

2-Photon absorption (arbitrary units)

Fig. 7.9. Two-photon absorption spectra of TlCl in the energy range of the $X_6^+ \times X_6^-$ exciton. Triangles are the measured values at 20 K *(left-hand energy scale)* and the circles are those at 77 K *(right-hand energy scale)*. The dashed line is the one-photon spectrum at 4.2 K taken from Fig. 7.7; its energy scale is shifted by 2 meV to allow comparison with two-photon data [7.34]

two-photon spectra measured at 20 K and 77 K are shown with a one-photon spectrum at 4.7 K which is taken from Fig. 7.7. We note that the strong 1A and 1B absorption peaks in the one-photon spectrum are missing in the two-photon spectra. At the energy of peak 2 of the one-photon spectrum, which is assigned to the 2s-exciton absorption, a small peak is found in the two-photon spectra. Since the overall parity of the 2p state of the $X_6^+ \times X_6^-$ exciton is even, the transition to it from the ground state is dipole allowed in the two-photon absorption, whereas the transition to the 2s state is forbidden. Thus the small peak observed in the two-photon spectra at the energy of the 2s exciton is the absorption of the 2p $X_6^+ \times X_6^-$ exciton because the 2s and 2p states are degenerate in the hydrogenic wave function. Absence of any structure at the energy of the 1s-exciton state in the two-photon spectra is due to the absence of p state for $n = 1$. Thus the assignments given for the absorption peaks in Fig. 7.7 are verified by the two-photon absorption experiment.

At higher energies in the two-photon spectra of Fig. 7.9, there are several dips between the humps. The positions of the dips are multiples of the LO phonon energy measured from the band-gap energy which are indicated by vertical lines. These features are exactly the same as those in the one-photon spectrum. Thus the humps in two-photon spectra are the free-phonon replicas of 2p and higher states of excitons including the continuum but not of the 1s exciton. The number of phonon replicas resolved is five in TlCl and eight in TlBr in the two-photon absorption spectra whereas only two are found in the one-photon spectra.

The phonon emission sidebands in the two-photon absorption spectrum are treated by *Fröhlich* et al. [7.34] in a similar way to resonance Raman scattering. They calculate the phonon-assisted two-photon band-to-band transition in the case of the Fröhlich interaction by taking band states as resonant intermediate states. In the case of phonon scattering in the conduction band, the absorption of the *n*-phonon sidebands $a_n(\omega)$ is

$$a_n(\omega)$$
$$\propto (\hbar\omega_1 + \hbar\omega_2 - \mathscr{E}_g)^{1/2} \cdot \left[\hbar\omega_1 + \hbar\omega_2 - \mathscr{E}_g - n \left(1 + \frac{m_e}{m_h} \right) \hbar\omega_{LO} \right]^{1/2} \quad (7.18)$$

where m_e and m_h are the electron masses in the conduction and the valence band. From this expression, one expects that the one-phonon sideband sets in at the energy $\mathscr{E}_g + [1 + (m_e/m_h)] \hbar\omega_{LO}$ and it is followed by successive higher-order sidebands separated by equal distances of $[1 + (m_e/m_h)] \hbar\omega_{LO}$. In TlCl, m_e/m_h at the X point is 0.8 [7.35] so that the separation between the phonon sidebands should be 1.8 $\hbar\omega_{LO}$, that is 39 meV at 4 K. The measured value is 22.5 meV which is almost equal to 1 $\hbar\omega_{LO}$ of 21.7 meV. This shows that the exciton effect on the phonon sidebands will be quite important, since the exciton one-phonon sideband in two-photon absorption spectrum should rise at the energy 1 $\hbar\omega_{LO}$, if one takes the discrete and continuum states of the exciton as virtual states.

The usefulness of the two-photon spectroscopy when the polarization directions of the two beams are varied with respect to one another and to the crystal axes was first pointed out and analysed by *Inoue* and *Toyozawa* [7.36]. They gave a table of a polarization dependence of two-photon absorption of two linearly polarized beams for the 32 crystal point groups. Later it was extended to include

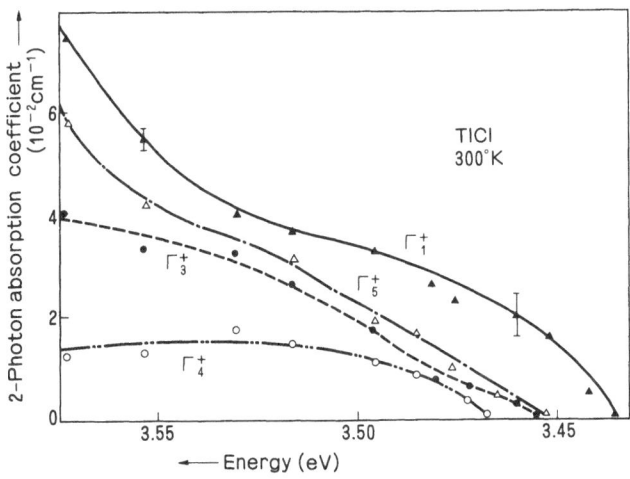

Fig. 7.10. Four contributions of different symmetry to the two photon absorption spectrum of TlCl near the X point band gap [7.38]

transitions between pairs of states including double-valued representations [7.37] and the absorption of circularly polarized light [7.38]. An experimental study of the polarization dependence was made by *Matsuoka* [7.39] in TlCl with linearly polarized light and by *Fröhlich* et al. [7.38] in TlCl with circularly polarized light. The result at 300 K obtained by the latter is shown in Fig. 7.10. Here the curves are the two-photon absorption spectra originating in the excitation of Γ_1^+, Γ_3^+, Γ_4^+, and Γ_5^+ symmetries which are obtained by the appropriate combination of the polarization dependent two-photon spectra with linearly and circularly polarized light. The absorption starts with the spectrum of Γ_1^+ symmetry. About 15 meV higher, Γ_3^+ and Γ_5^+ set in and 15 meV higher still the Γ_4^+ contribution appears. However, what they mean has not been clarified and further study is necessary.

7.2.3 Spectra of $X_6^+ \times R_6^-$ Indirect Exciton

In Figs. 7.7, 8, we observe that peak 1 has a tail which extends to the lower-energy side and ends at a threshold energy of 3.225 eV in TlCl and 2.6485 eV in TlBr at 2 K. The absorption coefficient α rises monotonically at the edge and if one plots $\alpha^{2/3}$ and $\alpha^{1/3}$ against the photon energy $\hbar\omega$, the energy dependence of α becomes apparent. Figure 7.11 gives the plots for TlBr [7.40]. At low energy, $\alpha^{2/3}$ is almost linear to $(\hbar\omega - E_t)$ so that α is proportional to $(\hbar\omega - E_t)^{3/2}$ where E_t is the threshold energy. At higher energy, $\alpha^{1/3}$ is linear against $(\hbar\omega - E_t')$ so that α varies as $(\hbar\omega - E_t')^3$. The fine structure of this spectrum is resolved by wavelength modulation spectroscopy. The spectrum of the wavelength derivative of the absorption coefficient $d\alpha/d\lambda$ [7.41] is shown in Fig. 7.12. We observe two kinds of structure. One is the structure which rises as $(\hbar\omega - \text{const.})^{1/2}$ and the

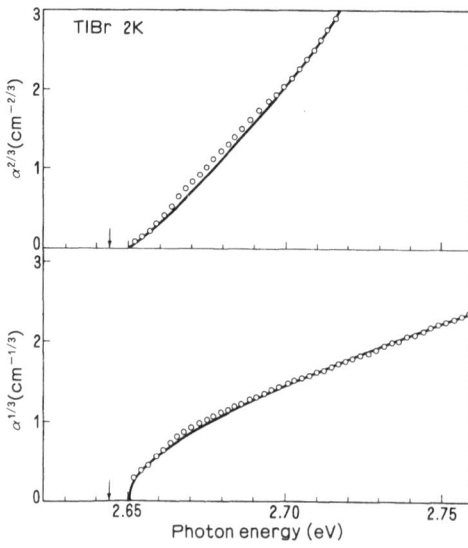

Fig. 7.11. Plots of $\alpha^{1/3}$ and $\alpha^{2/3}$ in TlBr against photon energy at the indirect absorption edge. The arrows indicate the energy of the zero-phonon $1s$ $X_6^+ \times R_6^-$ indirect exciton. The curves are the calculated values of the absorption of the phonon-forbidden indirect exciton [7.40]

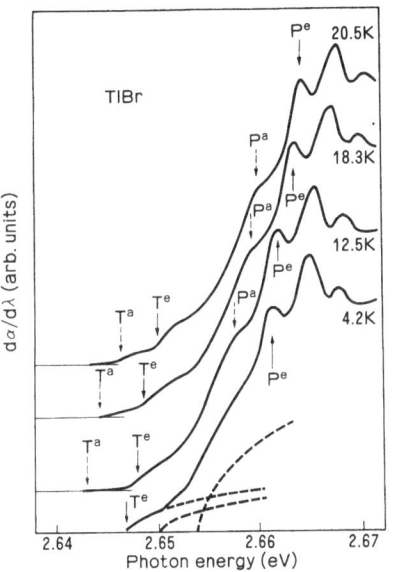

Fig. 7.12. Wavelength-derivative absorption spectra of TlBr at several different temperatures. The curves are vertically displaced relative to each other for visual clarity. Solid arrows T^e and P^e indicate the thresholds of the optical absorptions due to the phonon-forbidden indirect $1s$ and $2s$ exciton, respectively, by the phonon emission process. The thresholds of the absorptions by the phonon absorption process are shown by dashed arrows T^a and P^a. The edge of the spectrum at 4.2 K is decomposed into three square-root structures which originate in the emissions of phonons of three different modes. They are shown by broken lines [7.41]

other is in the form of a peak. At 4.2 K, the spectrum begins at the energy T^e. As the temperature increases a new tail begins to grow at an energy lower than T^e. It rises at the energy T^a.

The low absorption coefficient and the existence of structure suggest that the absorption edge is due to the indirect-exciton absorption edge. If it is so, the indirect transition must be from the X_6^+ state of the valence band to the R_6^- state of the conduction band, as can be seen from the energy bands shown in Fig. 7.2. This transition occurs with the assistance of a zone-boundary M point phonon via the $X_6^+ \times X_6^-$ exciton as a virtual state, since the scattering from the X point to the R point is accompanied by a change of wave vector equal to the M point in the Brillouin zone.

There are four M point phonons in thallous halides. They are the zone-boundary phonons on the TA, LA, TO, and LO branches. Their symmetries are M_4', M_5', M_2', and M_5' around the Tl^+ ion respectively. We note that their parities are all odd [7.42]. Since the scattering from X_6^- to R_6^- is allowed only by an even parity M point phonon, the indirect transition of the $X_6^+ \times R_6^-$ exciton is phonon-forbidden with any M point phonon and the absorption will not be observed. However if the $X_6^+ \times R_6^-$ exciton has some kinetic energy so that its wave vector differs from the M point, the parity changes and scattering of the exciton by phonons from the $X_6^+ \times X_6^-$ state to this state becomes partially allowed. This will give some absorption at energies higher than the M point $X_6^+ \times R_6^-$ exciton. We note that the phonon-forbidden indirect absorption has been found in germanium at the absorption edge [7.43, 44]. In germanium, however, the phonon-allowed indirect absorption also exists at the edge and they are superimposed to give a complex structure. In thallous halides, the edge consists of only a phonon-forbidden exciton transition and they seem to be the best substances for work on this particular type of transition.

The absorption coefficient α_n for the phonon-forbidden indirect exciton of the ns discrete state is [7.40, 41],

$$
\alpha_n = \frac{A}{\hbar\omega} \frac{1}{n^3} \left(\hbar\omega - \varepsilon_g + \frac{Ry}{n^2} \pm \hbar\omega_q \right)^{3/2}
$$

$$
+ \frac{B}{\hbar\omega} \frac{n^2 - 1}{n^5} \left(\hbar\omega - \varepsilon_g + \frac{Ry}{n^2} \pm \hbar\omega_q \right)^{1/2} \tag{7.19}
$$

where $\hbar\omega$ is the incident photon energy, ε_g the indirect band gap energy, Ry the exciton Rydberg energy and n the quantum number. Here $\hbar\omega_q$ is the energy of the momentum-conserving phonon with wave vector q and the plus and minus signs are for the phonon absorption and emission. We take q in the vicinity of K_0 which is the wave vector connecting the minima of the virtual direct exciton and the indirect exciton. The first term gives the contribution of all s excitons and α rises as $(\hbar\omega - E_{ns} \pm \hbar\omega_q)^{3/2}$ where E_{ns} is the ns exciton energy. The second term is the contriubution from $2s$ and higher excitons and gives the absorption rising as $(\hbar\omega - E_{ns} \pm \hbar\omega_q)^{1/2}$. At an energy sufficiently higher than $\mathscr{E}_g$, the absorption varies as

$$
\alpha = \frac{C}{\hbar\omega} (\hbar\omega - \varepsilon_g \pm \hbar\omega_{K_0})^3 \tag{7.20}
$$

where $\hbar\omega_{K_0}$ is the energy of a phonon with a wave vector K_0 [7.40]. We see in Fig. 7.11 that the absorption rises as $(\hbar\omega - E_t)^{3/2}$ at lower energies and $(\hbar\omega - E_t')^3$ at higher energies. This is what we expect from the phonon-forbidden indirect-exciton absorption.

The fine structure of this indirect-exciton absorption can be investigated by measuring a wavelength modulation spectrum. The line shape of a modulation spectrum is obtained by differentiating α_n in (7.19) with respect to ω. Here the lifetime broadening is taken into account phenomenologically by replacing $\hbar\omega$ by $\hbar\omega + i\Gamma$, Γ being the Lorentz broadening parameter [7.45]. For the $1s$ phonon-forbidden exciton absorption

$$
\frac{d\alpha_{1s}}{d\omega} = \frac{3A}{2\sqrt{2}\hbar\omega} \Gamma^{1/2} [(x_1^2 + 1)^{1/2} + x_1]^{1/2} , \tag{7.21}
$$

and for the $2s$ exciton

$$
\frac{d\alpha_{2s}}{d\omega} = \frac{3A}{16\sqrt{2}\hbar\omega} \Gamma^{1/2} [(x_2^2 + 1)^{1/2} + x_2]^{1/2}
$$

$$
+ \frac{7B}{64\sqrt{2}\hbar\omega} \Gamma^{-1/2} \frac{[(x_2^2 + 1)^{1/2} + x_2]^{1/2}}{(x_2^2 + 1)^{1/2}} . \tag{7.22}
$$

where $x_n = [\hbar\omega - \varepsilon_g + (Ry/n^2) \pm \hbar\omega_{K_0}]/\Gamma$ [7.41]. For the $1s$ exciton, the wavelength-derivative absorption spectrum is proportional to $x_1^{1/2}$ which is the square root of the energy measured from the threshold energy, $\varepsilon_g - Ry \mp \hbar\omega_{K_0}$, when the damping is sufficiently small so that $x_1 \gg 1$ in (7.21). For the $2s$ exciton, the derivative spectrum will show a peak near the energy $\varepsilon_g - (Ry/4) \mp \hbar\omega_{K_0}$ with a line shape proportional to $(\hbar\omega - \varepsilon_g + (Ry/4) \pm \hbar\omega_{K_0})^{-1/2}$ due to the second term, overlapped by the square-root increase due to the first term in (7.22) when $x_2 \gg 1$.

In the small energy range of Fig. 7.12, $d\alpha/d\lambda \sim -d\alpha/d\omega$ and the above discussion is applied to the analysis of the spectra. At 4.2 K, we find three square-root type structures, with different threshold energies, at lower energy followed by three peaks at higher energy. Three square-root structures are due to the $1s$ $X_6^+ \times R_6^-$ exciton assisted by the emissions of three different M point phonons. The three peaks are due to the $2s$ excitons with the emissions of M point phonons of three different modes. As the temperature rises, a new square-root structure appears whose threshold energy is T^a, which is lower than the low-temperature threshold T^e. Another new absorption appears at P^a on the lower-energy side of the low-temperature peak P^e. We note that T^a-T^e is almost equal to P^a-P^e. These new structures are due to a phonon absorption process. The middle point between T^e and T^a is the energy of the $1s$ $X_6^+ \times R_6^-$ exciton E_{1s} and is 3.2252 eV in TlCl and 2.6444 eV in TlBr at 4.2 K.

Once the zero-phonon $X_6^+ \times R_6^-$ exciton energy is determined, we can get the M point phonon energies associated with this indirect transition from the energy difference between the threshold energy of the square-root spectrum and E_{1s}. Three phonon modes are found in TlCl and in TlBr whose energies are listed in Table 7.3. In this table the energies of the M point phonons determined by neutron scattering studies [7.3, 46] are listed for both substances. The agreement is satisfactory.

Table 7.3. Phonon energies in thallous halides at 4 K in meV

		TlCl		TlBr	
Γ point	TO	7.5 ± 0.1		5.6 ± 0.04	
	LO	21.7 ± 0.4		14.5 ± 0.1	
		NS	WD	NS	WD
M point	TA	1.42	1.3	2.2	2.4
	LA	5.9	5.7	5.1	
	TO	10.4		6.2	6.1
	LO	11.9	11.8	9.1	9.4

NS: Data from neutron scattering experiments at 80 K for TlCl [7.3] and at 100 K for TlBr [7.46]
WD: Data from wavelength-derivative absorption spectra

The energy separation between the $1s$- and the $2s$-exciton absorption with the same phonon emission gives the binding energy E_{ex}^b of the $X_6^+ \times R_6^-$ exciton. It is 23 ± 3 meV in TlCl and 19 ± 1 meV in TlBr [7.41].

7.2.4 Free-Exciton Emission

Photo-excited excitons relax to the crystal ground state through two channels, namely radiative and nonradiative processes. Light emitted by the radiative process is luminescence. In the nonradiative process, the exciton energy flows into the lattice system by electron-phonon coupling. The exciton loses its kinetic energy by phonon emission in the exciton band and sometimes it is self-trapped accompanied by a local deformation of the lattice. Self-trapping is realized either in an intrinsic lattice or at an impurity site. The self-trapped exciton relaxes to the crystal ground state with, or sometimes without, an emission of light, as was discussed in Sect. 4.5.4.

From electrical transport experiments, which will be described in Sect. 8.3.2, it is found that the lowest states of electrons and holes in pure thallous halides are free states and not self-trapped states. Thus the lowest state of the exciton is believed to be a free state. For this reason, the luminescence expected in pure thallous halides is free-exciton luminescence and a Stokes-shifted luminescence from a self-trapped state will not be observed. This has been confirmed by experiment [7.23], and the luminescence discussed in this section is from the radiative recombination of free excitons.

a) $X_6^+ \times X_6^-$ **Direct Exciton.** Luminescence from the annihilation of the free $X_6^+ \times X_6^-$ direct exciton was found in TlBr by *Grabner* [7.47]. His result is shown in Fig. 7.13. Here the energy and the width of peak A in the emission spectrum almost coincide with those of the $1s$ $X_6^+ \times X_6^-$ exciton peak in the absorption spectrum. Thus peak A is a resonance emission of the free $X_6^+ \times X_6^-$ exciton. Peaks B_1 and B_2 are suggested to be the emissions from the radiative recombination of bound excitons with binding energies of 4 and 7 meV. A similar emission spectrum, but without any structure, is observed at 3.008 eV at 2 K [7.48]. In TlCl, a resonance emission of $X_6^+ \times X_6^-$ is observed at 3.402 eV which is very close to the energy of the 1B exciton peak [7.48].

The resonance emission found at the energy of the direct-exciton absorption is the radiative decay of the $1s$ $X_6^+ \times X_6^-$ exciton at $K \sim 0$ because of the negligible wave vector of the emitted light. It is observed wherever the exciton is excited in the band by light of higher energy than the band minimum. This is because the exciton of higher energy will lose its kinetic energy and will relax to the lowest energy state at $K \sim 0$ before it changes into light by a dipole transition.

The most probable relaxation process of the exciton in its band is the phonon-emission process. The relaxation time of the optical-phonon emission τ_0 is of the order of 10^{-13} s and that of the acoustic-phonon emission τ_a is of the order of 10^{-9} s. Thus the exciton of high kinetic energy will descend toward the bottom of

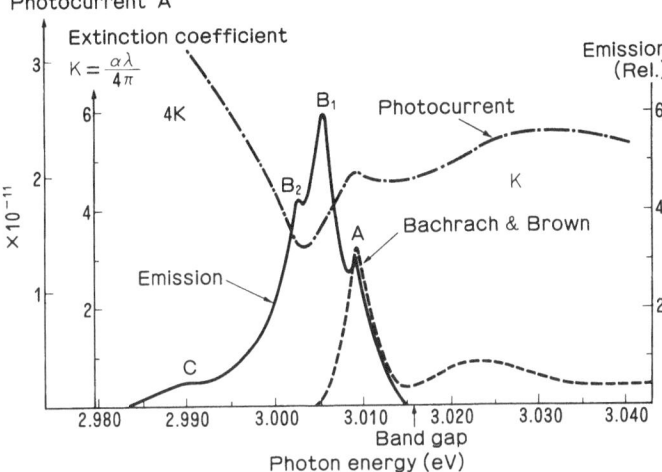

Fig. 7.13. Emission spectrum of TlBr at 4 K in the energy range of the $X_6^+ \times X_6^-$ direct exciton. The absorption spectrum and the excitation spectrum of the photocurrent are also shown [7.47]

the band mainly by the emission of LO phonons with equal energy because the LO phonon is almost dispersionless near $q \sim 0$.

Suppose that the energy of the incident light is just $E_0 + n\hbar\omega_{LO}$ where E_0 is the energy of the exciton at $K = 0$ and n is an integer. Because of the dispersion of the exciton, the incident light creates an exciton of energy $(n - 1)\hbar\omega_{LO}$ with the emission of a LO phonon with an appropriate wave vector to satisfy momentum conservation. The exciton then decays by emitting mainly LO phonons, because $\tau_0 \ll \tau_a$, and reaches the bottom at $K = 0$ where it annihilates and emits light whose energy is equal to the exciton absorption peak energy. Figure 7.14a shows this process.

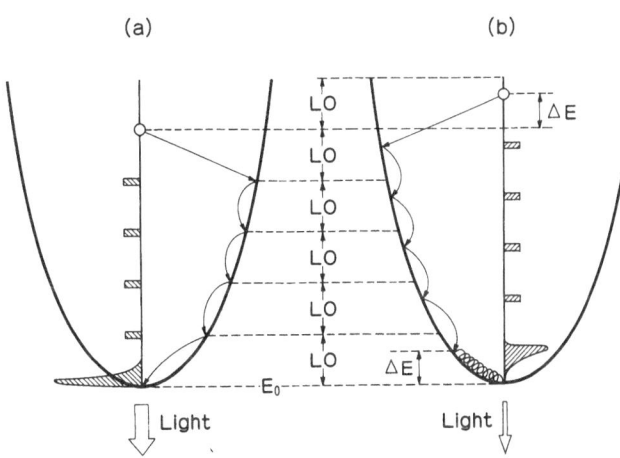

Fig. 7.14. Decay scheme of exciton in the exciton band. Energy of the LO(Γ) phonon is given as LO. Distribution of excitons is shown schematically by hatched area

When the exciting energy is $E_0 + n\hbar\omega_{LO} + \Delta E$, where $\Delta E < \hbar\omega_{LO}$, the exciton will decay towards the minimum by emitting LO phonons until it reaches $E_0 + \Delta E$ as is shown in Fig. 7.14b. Then it can emit only acoustic phonons because $\Delta E < \hbar\omega_{LO}$ and the process is slowed down considerably due to the large τ_a. If the lifetime of the exciton τ_e is shorter than τ_a, due to other nonradiative processes, the number of excitons which reach the $K = 0$ state will be quite small and the intensity of resonance emission will be weak. In these figures, the distributions of the excitons are shown schematically as a function of exciton kinetic energy for both cases, and they are quite different from that in thermal equilibrium whose dependence on the exciton kinetic energy E_{kin} is $E_{kin}^{1/2} \exp(-E_{kin}/k_B T)$. Such nonequilibrium excitons are called hot excitons and a review of this subject is given by *Permogorov* [7.49].

When the resonance emission intensity is measured by varying the energy of exciting light $\hbar\omega$ from E_0 to higher energies, a periodic change of the luminescence yield spectrum or excitation spectrum, with a period of $\hbar\omega_{LO}$, will be observed. The first observation of such excitation spectra due to hot excitons was made by *Goto* and *Ueta* [7.50] in CuBr and its details were described in Sect. 3.2.4. In Fig. 7.15, excitation spectra of the resonance emission of the $X_6^+ \times X_6^-$ excitons in TlCl and TlBr at 1.8 K obtained by *Shimizu* et al. [7.51] are shown. Periodic oscillation is clearly observed. The separations between adjacent maxima in the spectra are almost equal and are 14.5 ± 0.7 meV in TlBr and 22 ± 1 meV in TlCl, which equal the energies of the Γ point LO phonons $\hbar\omega_{LO}$ in the respective substances. The integer N labels the peaks starting from 0 at the energy E_0, the $K = 0$ exciton energy. Thus the peaks are found at the exciting

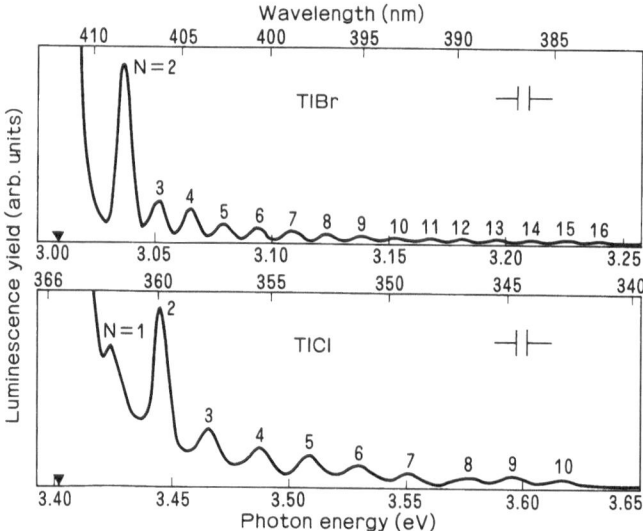

Fig. 7.15. Excitation spectra of the $X_6^+ \times X_6^-$ direct exciton resonance emissions in TlBr and TlCl at 1.8 K. Solid triangles indicate the exciton energies of respective halides [7.51]

energies of $E_0 + Nh\omega_{LO}$. This is just what is expected from theoretical considerations. They estimate that the lifetime of the $X_6^+ \times X_6^-$ direct exciton is of the order of 10^{-11} s which fulfils the condition, $\tau_a > \tau_e > \tau_0$, necessary for observing a hot-exciton oscillatory spectrum.

In normal pure semiconductors, the lifetime of the direct exciton is comparable to or even longer than the relaxation time of acoustic-phonon scattering, and to get an oscillatory hot-exciton spectrum it is necessary to reduce the exciton lifetime by intentional doping or some other means. The short lifetimes of excitons in the band in TlCl and TlBr will be due to the presence of the indirect $X_6^+ \times R_6^-$ exciton state at lower energy to which the $X_6^+ \times X_6^-$ exciton is scattered. Although phonon scattering is forbidden from the Γ point state of the $X_6^+ \times X_6^-$ exciton to the M point state of the $X_6^+ \times R_6^-$ exciton, scattering between states having kinetic energy is partially allowed, as was discussed in Sect. 7.2.3, and this makes the lifetime of the $X_6^+ \times X_6^-$ exciton short but not short enough to quench the resonant emission due to a complete decay before reaching the $K = 0$ state.

We note that the $N = 1$ line is missing or fairly weak in TlBr and TlCl in Fig. 7.15. For excitation by light with energy $E_0 + h\omega_{LO}$, the wave vector of the phonon which is emitted in the scattering from the light branch to the exciton branch must be almost zero. For the $q \sim 0$ LO phonon, the exciton–LO-phonon coupling is quite small because of the charge cancellation discussed in Sect. 4.1.4. Thus the resonance emission will be difficult to observe by excitation at $E_0 + h\omega_{LO}$ as is seen in Fig. 7.15. Quantitative calculation of the intensity variation of the excitation spectrum has been made by *Planel* et al. [7.52].

b) $X_6^+ \times R_6^-$ Indirect Exciton. Edge emission due to the annihilation of $X_6^+ \times R_6^-$ excitons assisted by phonon emission should be observed at the energy lower than the exciton energy E_{ex} by the M point phonon energy $h\omega_M$. This can be understood from Fig. 7.16. Here, at low temperatures, the $X_6^+ \times R_6^-$ exciton is created by the scattering of the $X_6^+ \times X_6^-$ direct allowed exciton (which is excited

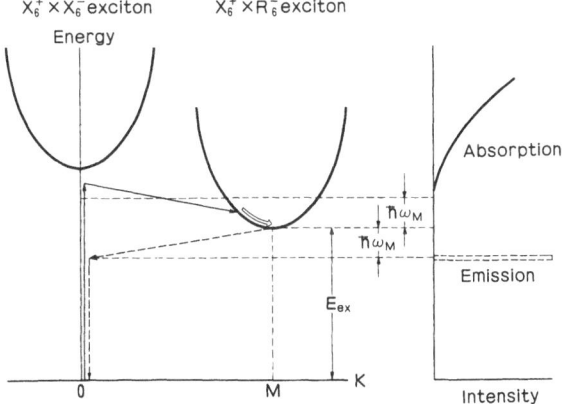

Fig. 7.16. Absorption and emission of the indirect $X_6^+ \times R_6^-$ exciton. The energy of the associated M point phonon is $h\omega_M$

by light virtually or really) into the $X_6^+ \times R_6^-$ exciton band by the emission of appropriate momentum-conserving phonons. It then decays to its minimum energy state at the M point from which it returns to the ground state through an indirect process by the emission of a photon with energy $E_{ex} - \hbar\omega_M$ and a momentum-conserving M point phonon of energy $\hbar\omega_M$. Thus the edge emission at the energy of $E_{ex} - \hbar\omega_M$ will be observed as a sharp peak of width about $k_B T$.

In thallous halides, the indirect transition of the $X_6^+ \times R_6^-$ exciton is phonon forbidden so that the probability of edge emission is very small. Thus if there are any impurities and imperfections in the crystal, even a very small amount, the $X_6^+ \times R_6^-$ excitons will decay more efficiently through these channels than through the M point phonon-assisted indirect recombination process. Thus the edge emission will be observed only in highly purified and perfect crystals. An example is TlCl containing iodine of 10^{-6} mole fraction which gives an impurity level in the band gap. In this crystal edge emission is comletely quenched and, instead, only a broad emission band due to a localized state at the iodine site is observed [7.23].

Figure 7.17 is the emission spectrum in the vicinity of the indirect $X_6^+ \times R_6^-$ absorption edge in TlBr measured at 2 K [7.23]. In this figure, the absorption spectrum due to the indirect exciton is also shown. The energy of the $1s\, X_6^+ \times R_6^-$ exciton with zero kinetic energy E_{1s}, which is determined as 2.6444 ± 0.0007 eV from the wavelength-modulated absorption spectra at low and high temperatures [7.41], is indicated. We find that E_{1s} is almost at the midpoint between the absorption edge and the emissions a, b, and c. The energy separation between E_{1s} and a, b, and c are 4.1 ± 0.7, 4.7 ± 0.7, and 5.2 ± 0.7 meV, respectively, which are almost equal to 5.1 meV, the M point LA phonon in TlBr [7.46]. At this temperature, $k_B T$ is 0.2 meV and is comparable to their line widths. Thus they are the emissions due to the annihilation of free $X_6^+ \times R_6^-$ excitons with the assistance of a momentum conserving M point LA phonon. From the temperature dependences of the intensities of a, b, and c, the ratio of the oscillator strength of a, b, and c is found to be $5:8:1$ assuming that a quasi thermal equilibrium is established for the distribution between the three levels. The origin of the three emission peaks is examined by referring to the electronic states of the

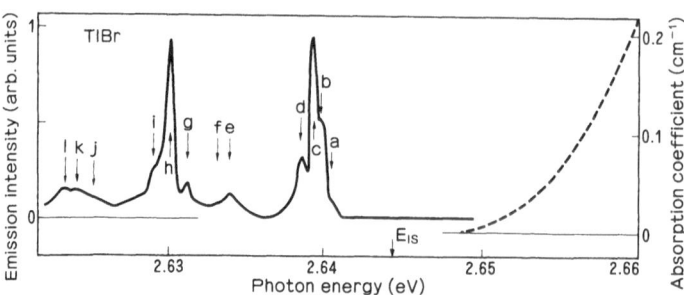

Fig. 7.17. Emission spectrum (——) and absorption spectrum (---) of TlBr at the indirect edge at 2 K [7.23]

$X_6^+ \times R_6^-$ exciton discussed in Sect. 7.1.3. The emission c whose energy is the lowest and oscillator strength is the smallest among the three, is the emission from the annihilation of the spin-triplet state, Ψ_0 in (7.14), of the $X_6^+ \times R_6^-$ exciton. This transition will be weakly allowed because it is an indirect transition via the singlet-triplet mixed state of the $X_6^+ \times X_6^-$ direct exciton as a virtual state. The emissions a and b are due to the recombinations of the $\Psi_x(\Psi_y)$ and Ψ_z states. These three states are degenerate if the X_6^+ and R_6^- states originate purely in Tl $6s$ and $6p$ respectively, but will split into Ψ_z and $\Psi_x(\Psi_y)$ if the contributions of halogen p and d are taken into account. Thus the peaks a and b are assigned to the annihilations of Ψ_z and $\Psi_x(\Psi_y)$ respectively because the oscillator strength of b is almost twice as large as a. The energy splitting of the $1s$ $X_6^+ \times R_6^-$ exciton is not observed in the absorption spectrum and the wavelength-modulation spectrum perhaps because of the monotonically increasing line shape of each split component.

The emission line d in Fig. 7.17 grows in proportion to the square of exciting light intensity and will be discussed in terms of the emission from the excitonic molecule in Sect. 7.2.5. The origins of other low-energy lines are not known. Some might be multi-phonon emission lines and others might be bound-exciton emissions.

In TlCl, a quite similar emission spectrum is found at the $X_6^+ \times R_6^-$ exciton absorption edge [7.23]. The emission peaks a, b, and c are observed and are assigned to the free indirect-exciton emission with an M point LA phonon emission just as in TlBr. These three emissions are followed by the emission d which originates from the decay of excitonic molecules.

In TlBr, an oscillation with a period of twice the Γ point LO phonon energy is observed in the excitation spectrum of the emission due to the $X_6^+ \times R_6^-$ indirect exciton [7.48]. This is in contrast to the oscillation usually found in the excitation spectrum where the period is $\hbar\omega_{LO}$.

This $2\hbar\omega_{LO}$ oscillation cannot be explained by the usual cascade decay process in a single exciton band with a successive one–LO-phonon emission as is the case of the $X_6^+ \times X_6^-$ direct exciton shown in Fig. 7.14. *Shimizu* and *Koda* [7.53] explain the origin by a particular property of the $X_6^+ \times R_6^-$ exciton. In TlBr, the energy separation between the $1s$ and the $2p$ state of the $X_6^+ \times R_6^-$ exciton is 14.3 meV which is very close to the energy of the Γ point LO phonon energy of 14.5 meV. Fig. 7.18 shows the situation. There are two processes for the decay of the exciton in the $1s$ band when it is excited to an energy $2\hbar\omega_{LO}$ above the bottom. One is successive one–LO-phonon emission in the $1s$ band, and in this case an oscillation with a period of $1\hbar\omega_{LO}$ is observed in the excitation spectrum. The other is a two-phonon process in which an exciton in the $1s$ band emits one LO phonon to be scattered virtually to the $2p$ exciton state and then scattered back to the $1s$ state at the bottom. In this case, no one–LO-phonon oscillation will be observed because the $2p$ state is optically forbidden. Usually this process will give a negligible contribution to the emission because it is a second-order process. However this is not the case for the $X_6^+ \times R_6^-$ exciton in TlBr. The wave vector of the LO phonon associated with this scattering is very small and is close

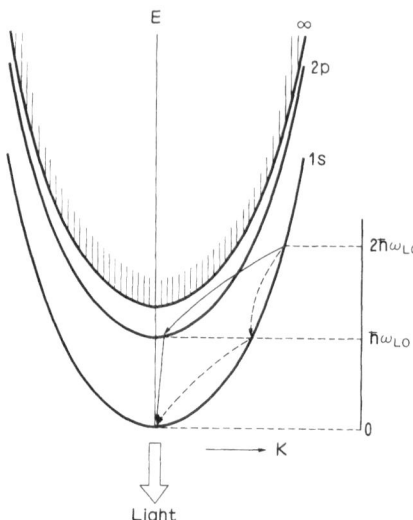

Fig. 7.18. Decay of an exciton in the exciton band through one-phonon (– – –) and two-phonon (——) cascade processes [7.53]

to zero. For the LO phonon of vanishing wave vector, exciton-phonon scattering by the Fröhlich interaction is forbidden between $1s$–$1s$ exciton states whereas it is allowed between $1s$–$2p$ exciton states [7.54]. Thus the scattering matrix element within the $1s$ band is smaller than that between $1s$–$2p$. Further, and perhaps more important, $\hbar\omega_{LO}$ is nearly equal in magnitude to the energy separation of $1s$–$2p$. This leads to a small energy denominator of this second-order process via the $2p$ state. *Shimizu* and *Koda* calculate the transition probabilities for the one–LO-phonon process in the $1s$ band and for the two–LO-phonon process via the $2p$ state and find that the latter is sixteen times larger than the former in TlBr.

7.2.5 Excitonic Molecules of $X_6^+ \times R_6^-$ Excitons

The $X_6^+ \times R_6^-$ excitons in thallous halides are interesting from the point of view of a high density of excitation. They are phonon-forbidden excitons and once they are formed, their decay to the ground state is slow. This leads to high density accumulation of excitons being easily attainable and high-density excitation phenomena such as the formation of an excitonic molecule (EM) can be studied without using a high-power excitation source. The EM of a phonon-forbidden exciton has scarcely been investigated and thallous halides give nice opportunities to investigate this novel high-density excited state.

The highest valence-band state X_6^+ is threefold star-degenerate and has $\langle 100 \rangle$ many-valley structure. The mass ratios of electron and hole, σ, of the $X_6^+ \times R_6^-$ exciton are $0.97\,(\|)$–$0.57\,(\perp)$ for TlCl and $0.96\,(\|)$–$0.71\,(\perp)$ for TlBr. They are close to unity and the EMs in thallous halide are like a positronium molecule rather than a hydrogen molecule. Therefore it is interesting to find whether or not an EM is stable against two free spin-triplet excitons because of the reason discussed in Sect. 2.2.2, 3.

A notable feature of this EM will reside in the M point momentum of the $X_6^+ \times R_6^-$ excitons from which the molecule is formed. Because of this, there are two kinds of EM. One is the molecule whose total wave vector is at the M point and the other is that at the Γ point. They are formed by the following combinations of the M point $X_6^+ \times R_6^-$ excitons:

$$\text{exciton}\,(M_1) + \text{exciton}\,(M_2) \rightarrow \text{molecule}\,(M_3) \ , \tag{7.23}$$

$$\text{exciton}\,(M_1) + \text{exciton}\,(M_1) \rightarrow \text{molecule}\,(\Gamma) \ , \tag{7.24}$$

where M_1, M_2, and M_3 are three inequivalent M points such as $110, 011$, and 101 in the simple cubic Brillouin zone. Fig. 7.19 shows the electron and hole states of the M and Γ molecule in the Brillouin zone. The radiative decay of the M molecule leaving one exciton will proceed in the following two ways:

1) A direct transition which leaves an M exciton ($X_6^+ \times R_6^-$) in the same M valley as that of the molecule. This occurs by the Coulomb scattering of electron and hole without any assistance from phonon emission. For example, in the case of the M molecule at 101 in Fig. 7.19, the decay of the molecule is achieved by the process in which the electron at 111 is scattered to 001 and annihilates with the hole there, emitting a photon, whereas the hole at 100 is scattered to 010 to form an M exciton in the 101 valley. The wave vector changes of the two scatterings are $\bar{1}\bar{1}0$ and $\bar{1}10$ which are equivalent to each other and the sum is zero. In this case, we expect an emission at the energy $E_{ex}(X_6^+ \times R_6^-) - E_{mol}^b$, where $E_{ex}(X_6^+ \times R_6^-)$ is the lowest energy of the $X_6^+ \times R_6^-$ exciton and E_{mol}^b is the binding energy of the EM.

2) A decay to the ground state which emits a photon leaving an M exciton in a different M valley from that of the M molecule. This is just the reverse process of (7.23) where one M exciton in the molecule changes to a photon with the assistance of an M point phonon emission through an indirect process. The energy of the photon is $E_{ex}(X_6^+ \times R_6^-) - E_{mol}^b - \hbar\omega_M$.

There are two decay processes for the Γ-molecule:

1) A direct process which involves two electrons at R being scattered to X to form two Γ point excitons whose electron and hole are in the same X valley, as

M molecule at 101
from 110 exciton and
011 exciton

Γ molecule at 000
from two 110 excitons

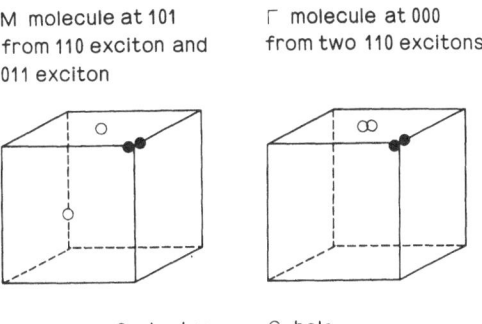

• electron   ○ hole

Fig. 7.19. Electrons and holes of the M EM and Γ EM of $X_6^+ \times R_6^-$ excitons in the Brillouin zone

can be understood from Fig. 7.19. In this scattering, the change of the total wave vector is zero and no phonon assistance but a Coulomb interaction is necessary. This is just the change of the Γ molecule to two Γ excitons, one of which annihilates to become a photon. The Γ exciton associated with the X point direct transition in thallous halides is the $X_6^+ \times X_6^-$ exciton and thus the energy of emitting light must be $2E_{ex}(X_6^+ \times R_6^-) - E_{mol}^b - E_{ex}(X_6^+ \times X_6^-)$ to satisfy energy conservation. It is to be noted that this energy is different from the energy of the emission from the M molecule by an amount almost equal to the energy difference between the direct and indirect excitons.

2) A decay which leads to the M point exciton and a photon. Light is emitted through the indirect recombination process of an electron at R and a hole at X by the assistance of an M point phonon. This process is just the same process as that of the second case of the M molecule. The energy of the emitted light is $E_{ex}(X_6^+ \times R_6^-) - E_{mol}^b - \hbar\omega_M$.

In Fig. 7.20, the energies of the emissions from the M and Γ molecules as well as from the $X_6^+ \times R_6^-$ exciton at low temperatures, where only phonon emission and not phonon absorption takes place, are summarized schematically. The absorptions of the $X_6^+ \times R_6^-$ exciton and the $X_6^+ \times X_6^-$ exciton are also shown.

Figure 7.21 shows the luminescence spectra of TlCl in the vicinity of the indirect absorption edge at various excitation intensities [7.55]. With low intensity excitation, the structures a, b, and c are resolved which are the emissions due to the radiative decay of the $X_6^+ \times R_6^-$ indirect exciton from three energy states, with the assistance of the M point LA phonon. In Sect. 7.2.4, it was shown that they are same as the emissions a, b, and c in TlBr given in Fig. 7.17. When the excitation intensity I_0 increases in this low excitation range, their emission intensites increase proportional to I_0. The emission d found at a lower energy than c increases as $I_0^{1.8}$. At a high intensity of excitation, its growth becomes linear with

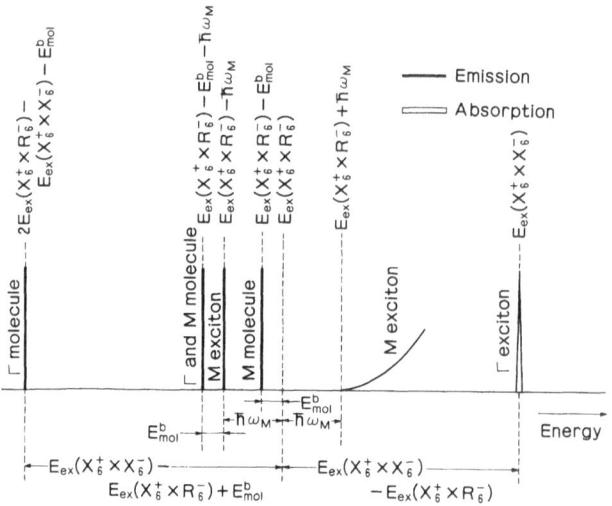

Fig. 7.20. Energies of the emissions from the M indirect exciton, M excitonic molecule and Γ excitonic molecule. The absorptions of the M indirect exciton and the Γ direct exciton are also shown

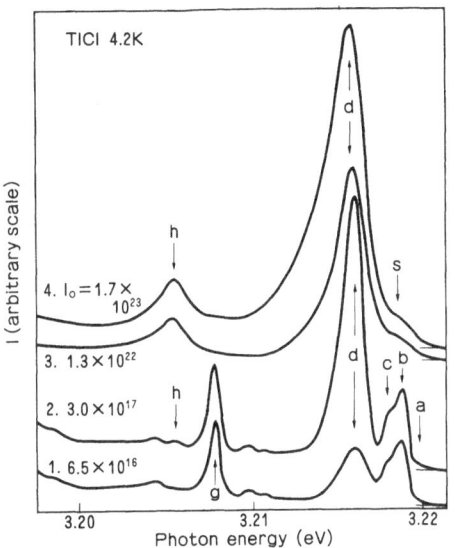

Fig. 7.21. Edge emission spectra of TlCl at 4.2 K by the excitation of various intensities. Curves 1 and 2 are obtained by mercury lamp excitation and the curves 3 and 4 by nitrogen laser excitation. The relative magnitudes of the intensities between the curves are arbitrary. The number given in each curve is the number of incident photons per cm² s used for the excitation [7.55]

I_0. From the analysis based on the rate equation [7.55], this dependence is just what we expect for the two-exciton collision process. The line shape of emission d is asymmetric and has a tail on the low-energy side, characteristic of the decay of an EM. The line shape of the emission from the decay of the EM of the indirect exciton to a phonon-forbidden indirect exciton and a photon is calculated [7.55] by modifying CHO's theory [7.56]. Fig. 7.22 shows the line shapes calculated by the theory and the emission d at 4.2 K. The abscissa is the photon energy ex-

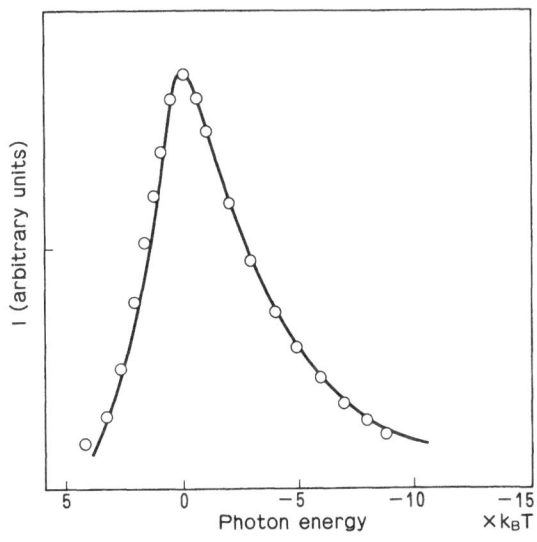

Fig. 7.22. Line shape of the emission from the M EM of the phonon-forbidden indirect $X_6^+ \times R_6^-$ excitons. The energy in the abscissa is measured in units of $k_B T$. Details are given in the text [7.55]

pressed in units of $k_B T$ where k_B is Boltzmann's constant. Its zero is taken as the energy $E_0 = E_{ex}(K_0) - E_{mol}^b - \hbar\omega_{K_0}$ where K_0 is the wave vector of the indirect exciton at the minimum energy. The line shape is temperature dependent because of the distribution of the EM in its band and the curve in the figure is for 4.9 K. The good agreement between the theory and the experiment suggests that the emission d is due to the decay of the M or Γ EM, of temperature of about 5 K, to the $X_6^+ \times R_6^-$ exciton, a photon and an M point phonon.

In Fig. 7.22, we find that the peak energy is nearly equal to E_0. Since the energy of the peak of the emission by the radiative recombination of the phonon-forbidden indirect $X_6^+ \times R_6^-$ exciton is $E_{ex}(X_6^+ \times R_6^-) - \hbar\omega_M + (3/2)k_B T$ and its low-energy threshold is at $E_{ex}(X_6^+ \times R_6^-) - \hbar\omega_M$, the energy separation between the peak of the emission from the exciton and that from its molecule must be $E_{mol}^b + (3/2)k_B T$.

The lowest energy state of the $X_6^+ \times R_6^-$ exciton is a spin-triplet state. Twice the energy of the triplet exciton is the energy of the lowest dissociated state of the EM from which the binding energy of the molecule E_{mol}^b is measured. The emission c is from the triplet state and its separation from the emission d must be $E_{mol}^b + (3/2)k_B T$. It is 2.0 ± 0.3 meV at 4.2 K for TlCl which gives E_{mol}^b as 1.5 ± 0.3 meV [7.55]. A similar emission spectrum from the decay of the EM is measured in TlBr and its E_{mol}^b is found to be 0.7 ± 0.4 meV [7.55]. A theoretical estimate of the binding energy of the EM can be made using the theory of *Akimoto* and *Hanamura* [7.57, 58]. As was discussed in Sect. 2.2.3 the theoretical binding energies are small due to the almost equal masses of the electron and hole and are 0.9 ± 0.2 meV for TlCl and 0.7 meV for TlBr. They are close to the experimental values. However, as is pointed out by *Chung* et al. [7.59], the electron-hole exchange energy, which can be obtained from the free-exciton data, is of the same order of magnitude as E_{mol}^b. They calculate the binding energy by including the electron-hole exchange interaction and find that the EM is unstable in thallous halides. This result is in contrast to the experimental result and suggests that some additional stabilization energy must be taken into account. The luminescence at the energy of $E_{ex}(X_6^+ \times R_6^-) - E_{mol}^b$ due to the decay of the M molecule through a direct process is not found, which is perhaps due to a small transition probability.

The emission due to the decay of the Γ molecule through a direct process to a photon and the $X_6^+ \times X_6^-$ direct exciton is found both in TlCl and TlBr. It is observed just at the energy of $2E_{ex}(X_6^+ \times R_6^-) - E_{mol}^b - E_{ex}(X_6^+ \times X_6^-)$ expected for this process. Figure 7.23 is the spectrum for TlCl [7.60]. Very weak emission peaks α, β, and γ are the emissions from the Γ molecule and their peak energies are 3.0542, 3.0525, and 3.0513 eV for α, β, and γ, respectively. These weak emissions can be observed only in very pure crystals since any impurities will mask these weak emissions by their strong luminescences. Because the binding energy of the Γ molecule is not known, the energies $2E_{ex}(X_6^+ \times R_6^-) - E_{ex}(X_6^+ \times X_6^-)$ are shown by full vertical lines in the figure. Here three energies in TlCl (3.0474, 3.0512, and 3.0526 eV) correspond to the decays leaving the $X_6^+ \times X_6^-$ excitons in the Ψ_z^z, $(1/\sqrt{2})(\Psi_z^x + \Psi_z^y)$, and Ψ_0^z states. A similar emis-

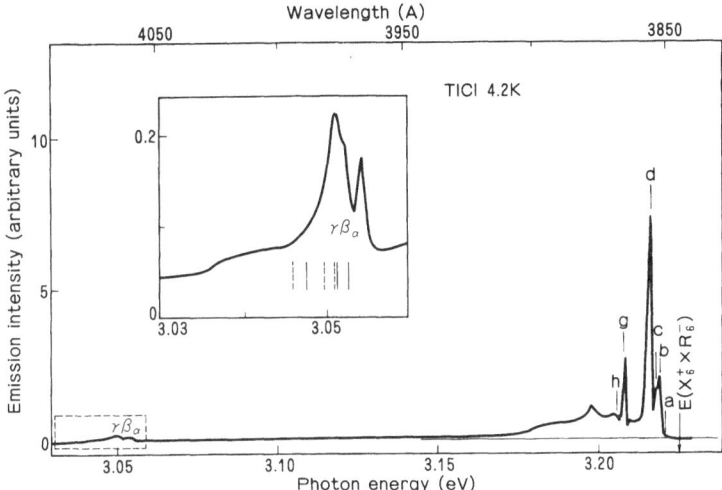

Fig. 7.23. Emission spectrum of TlCl at 4.2 K in the energy range below the indirect absorption edge. The energy $E(X_6^+ \times R_6^-)$ is the energy of the zero-phonon $X_6^+ \times R_6^-$ indirect exciton. In the inset, the enlarged spectrum near the emissions α, β, and γ is shown. They are the emissions from the decay of the Γ EM leaving the $X_6^+ \times X_6^-$ direct exciton. Energies of $2E_{ex}(X_6^+ \times R_6^-) - E_{ex}(X_6^+ \times X_6^-)$ are shown by full vertical lines and those of $2E_{ex}(X_6^+ \times R_6^-) - E_{ex}(X_6^+ \times X_6^-) - E_M^b$ by broken vertical lines [7.60]

sion is observed in TlBr. If one takes into account E_{mol}^b and assumes that it is the same in both the M molecule and the Γ molecule, the theoretical energies must be reduced by E_{mol}^b which is 1.5 meV in TlCl. Their energy positions are indicated by vertical broken lines in the figure.

We observe that the energies of the emissions α, β, and γ in TlCl are fairly close to the theoretical energies. Thus it is concluded that the emissions α, β, and γ in TlCl are due to the direct radiative decay of the Γ molecule of the indirect $X_6^+ \times R_6^-$ excitons leaving the direct $X_6^+ \times X_6^-$ exciton. This result means that an indirect exciton can be converted into a direct exciton through the interaction of two indirect excitons. It seems that the larger peaks β and γ are the emissions leaving the $X_6^+ \times X_6^-$ excitons of $\Psi_z^{z'}$ and $(\Psi_z^x + \Psi_z^y)/\sqrt{2}$ states, but the detailed structure of the spectra is not clarified yet. It might reflect the structure of the Γ molecule as well as that of the $X_6^+ \times X_6^-$ exciton and further theoretical and experimental investigations are needed.

7.3 Resonant Raman Scattering by Excitons in Thallous Halides

7.3.1 LO Phonon Scattering Resonant to a Direct Exciton

In Sects. 7.2.4, 5, the emissions due to the annihilation of photo-excited excitons and EMs were discussed. There the excitons and the molecules are in quasi thermal equilibrium with the lattice. The emissions are called ordinary luminescence and the excited state responsible for the emission is the lowest state of the excitation band, to which the initial photo-excited state relaxes by interaction with the lattice. This is a first-order optical process and is realized when the lifetime of the excited state is sufficiently long compared to the relaxation time to enable quasi thermal equilibrium to be reached. Thus the cold luminescence does not retain the memory of the excitation and its energy does not depend on the excitation energy.

When the life time of the exciton is shorter than or comparable to the thermal relaxation time, the process is no longer adequately described by a separate, independent, absorption and emission. Instead, the process should be treated as a second-order optical process where the absorption and emission take place successively, with strong correlation between them. When the energy of the incident photon is much less than the exciton energy, the exciton acts only as a virtual intermediate state and the process is called scattering. In the case of an incident photon whose energy is nearly resonant to the energy of an exciton of strong absorption, the exciton state acts as a real, as well as a virtual, intermediate state and the process is discussed in terms of an absorption followed by an emission or light scattering under a resonant condition.

A typical spectrum of the Raman scattering of a thallous halide in resonance with the $X_6^+ \times X_6^-$ direct exciton, which was measured by *Stolz* and *von der Osten* in TlBr [7.61], is shown in Fig. 7.24. The abscissa is the energy of the Raman

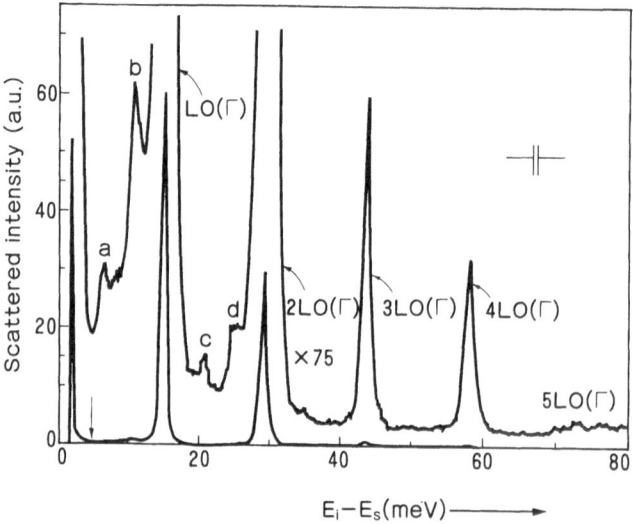

Fig. 7.24. Resonant Raman spectrum of TlBr at 1.8 K for incident photon energy of 3.0158 eV. The arrow on the abscissa indicates the 1s energy of the $X_6^+ \times X_6^-$ direct exciton. The Raman lines a–d correspond to TO(Γ), 2LA(M), TO(Γ) + LO(Γ), and 2LA(M) + LO(Γ) emissions, respectively [7.61]

shift. Here the energy of the incident photon is slightly higher than the $X_6^+ \times X_6^-$ 1s direct-exciton energy whose position for the Ψ_z^z state is indicated by the arrow on the abscissa. We observe a series of lines shifted by integer multiples of the LO(Γ) phonon energy from the exciting energy E_i.

There are two origins of the appearance of the LO overtones; the hot luminescence effect and the resonant Raman effect. The spectra of both processes have many apparently common features but they can be distinguished. In the hot luminescence, the exciton is really excited and is scattered by phonons through real states to cascade down in its band conserving the energy and wave vector of the system. In this condition, the 1LO line is very weak and is almost absent in the spectrum [7.49].

By contrast, in the resonant Raman scattering, the 1LO line is found for the following reasons. The scattering cross section R_c of the first-order Raman effect near the resonance is

$$R_c \propto \left| \sum_{ab} \frac{\langle f | \mathcal{H}_{eR} | b \rangle \langle b | \mathcal{H}_{eL} | a \rangle \langle a | \mathcal{H}_{eR} | i \rangle}{(\omega_s - \omega_b)(\omega_i - \omega_a)} \right|^2, \tag{7.25}$$

where ω_i and ω_s are the incident and the scattered photon frequencies and $\mathcal{H}_{eR}$ and $\mathcal{H}_{eL}$ are the exciton-radiation and the exciton-lattice interactions. The initial and the final states and the two intermediate states are $|i\rangle$, $|f\rangle$, and $|a\rangle$ and $|b\rangle$ respectively. We take the Fröhlich coupling as the exciton-lattice interaction which is the case of LO phonon scattering. The Hamiltonian of the coupling between an exciton and a phonon of wave vector q, which operates on the exciton envelope state, can be extracted from (4.31, 32). For small q it is expanded as

$$\mathcal{H}_F \propto \frac{1}{q} \left[iq \cdot r + \frac{1}{2} \frac{m_e - m_h}{m_e + m_h} (q \cdot r)^2 + \ldots \right]. \tag{7.26}$$

The first term is linear in r and is a q-independent dipole term. The second term is a q-dependent term second-order in r.

In crystals which contain a center of inversion, such as thallous halides, the overall exciton states in $|a\rangle$ and $|b\rangle$ are both odd-parity states and their envelope functions are s-state for the case of optically allowed transitions because of a dipole interaction of $\mathcal{H}_{eR}$. The first term $iq \cdot r/q$ in $\mathcal{H}_F$ connects only different-parity states and thus the 1LO Raman scattering is forbidden in crystals having a center of inversion.

However in resonant Raman scattering, the 1LO and other higher odd-number LO phonon lines are excited, for two reasons. One is the electric-field-induced mechanism [7.62]. When an electric field is formed by the surface space charge it destroys the spherical symmetry of the envelope functions of $|a\rangle$ and $|b\rangle$, due to the polarization of the exciton state, and the matrix element of the q-independent $iq \cdot r/q$ term will not vanish. Thus the surface space charge will induce normally forbidden 1LO and other odd-number Raman scattering in the energy range where strong absorption occurs. This effect will become appreciable in semiconductors of large carrier concentration and in insulators of high photo-conductivity.

The other mechanism is the wave-vector-induced mechanism which originates in the q-dependent $(q \cdot r)^2/q$ term in $\mathscr{H}_F$ [7.63]. This term connects two dipole-allowed exciton states having the same parities. Thus 1LO and other odd-number LO phonon Raman scattering will be observed. Its intensity will usually be quite small compared to the allowed 2LO Raman scattering in a crystal having an inversion symmetry. However it has been shown that the forbidden 1LO line will be enhanced considerably as the energy of the incident photon comes close to the exciton energy levels of $|a\rangle$ and $|b\rangle$ and can become comparable to the allowed lines near resonance for large Wannier excitons like those in thallous halides [7.63].

We thus interpret that the series of the LO lines is due mostly to the resonant Raman effect. Besides these LO lines, several weak Raman peaks are observed in Fig. 7.24 and they will be discussed later.

In Fig. 7.25, the Raman cross sections R_c of 1LO and 2LO scattering in TlBr are shown for the backward scattering geometry as a function of incident photon energy. In the backward scattering, the directions of the incident and scattered beams differ 180° and this geometry is usually used for strongly absorbing materials. The cross section R_c of the backward scattering is obtained from the intensity I_1 of the incident photon with energy $\hbar\omega_1$ and the intensity I_2 of the scattered photon with energy $\hbar\omega_2$, both of which are detected at the surface, by the following relation [7.64]:

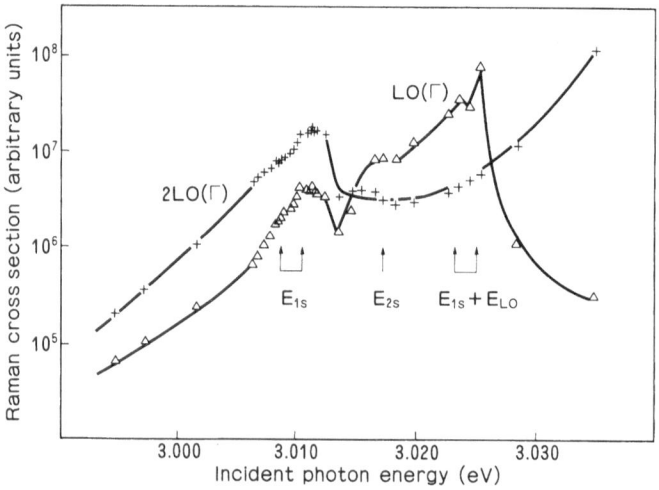

Fig. 7.25. Experimental Raman cross sections R_c of 1LO and 2LO scatterings in TlBr. The energies of the 1s and 2s direct $X_6^+ \times X_6^-$ excitons are shown by E_{1s} and E_{2s}. The energy of the 1LO phonon state of the 1s exciton is also indicated [7.61]

$$I_2 = \frac{R_c I_1}{R_c + \alpha_1 + \alpha_2} \{1 - \exp[-(R_c + \alpha_1 + \alpha_2)L]\} \ . \tag{7.27}$$

Here α_1 and α_2 are the absorption coefficients of the material at ω_1 and ω_2 and are related to κ_1 and κ_2, the imaginary parts of the complex refractive indices at ω_1 and ω_2, by $\alpha_1 = 2\omega_1 \kappa_1/c$ and $\alpha_2 = 2\omega_2 \kappa_2/c$.

We find that both the 1LO scattering and the 2LO scattering have peaks at the 1s exciton energy E_{1s} and at the energy of the 2s exciton E_{2s}, shoulders are found for both the scatterings. This shows that when the energy of the incident light comes close to the energies E_{1s} and E_{2s}, resonance occurs because of the term $(\omega_i - \omega_a)$ in the denominator of R_c. A further resonance occurs at the energy $E_{1s} + 1LO$ of the incident light in the 1LO Raman scattering. In this case, the energy of scattered light is equal to E_{1s} and it is the resonance of the scattered light with the 1s exciton as can be understood from the term $(\omega_s - \omega_b)$ in R_c. The spectra of the 1LO and 2LO Raman cross sections in TlCl have also been measured as a function of incident photon energy [7.65]. The resonance is similar to the case of TlBr but further resonances at the energies of incident light of $E_{1SA} + 1LO$, $E_{1SB} + 1LO$, $E_{2S} + 1LO$, $E_{1SA} + 2LO$, and $E_{1SB} + 2LO$ are observed.

To describe the resonant Raman scattering, two pictures for the exciton-photon-phonon system are usually used. One is the exciton picture where the crystal ground state is excited by interaction with a photon to a direct-exciton state, which is then scattered by phonons through the exciton-phonon interaction and returns to the ground state leaving a photon and phonons. This is a third-order perturbation process and the exciton and the photon behave as if they are independent particles. The decay of the exciton is taken into account as a damping constant in the energy denominator. The other is the polariton picture where the polariton represents the exciton-photon system in the crystal and its interaction with phonons determines the Raman intensity. The damping is expressed in terms of a polariton decay.

To calculate the Raman scattering intensity as a function of ω_i by the two pictures, values of several parameters must be known. They are the energies, the L–T splitting energies and the damping constants Γ of excitons. They can be estimated from the spectra of the real part ϵ_1 and the imaginary part ϵ_2 of the dielectric function of the excitons. The ϵ_1 and ϵ_2 spectra will be obtained from the reflectivity spectrum by the Kramers-Kronig relation but this is true only for the case of a dispersionless oscillator. For the case of an exciton with a spatial dispersion, the Kramers-Kronig relation does not give the true ϵ_1 and ϵ_2 spectra in principle because the resultant spectra will contain the effect of spatial dispersion other than ϵ_1 and ϵ_2, although they may be not very different from the true spectra. Thus the ϵ spectra transformed from the reflectivity spectrum are only an approximation in the case of an exciton in thallous halides. The parameter values used in the following calculation are those estimated from the ϵ spectra transformed from the reflectivity spectrum, and thus they are approximate. The approximate values of Γ are 3.6 meV and 3.2 meV for 1SA and 1SB excitons in

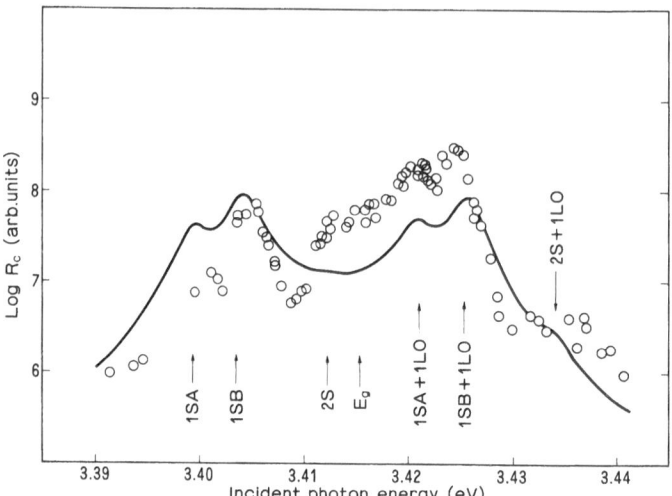

Fig. 7.26. Experimental Raman cross sections R_c of 1LO scattering in TlCl and the theoretical curve calculated from the exciton picture. Arrows indicate the energies of excitons, their one-phonon states, and the band gap [7.65]

TlCl. The L–T splitting energies are estimated as 2.1 meV for 1SA and 2.8 meV for 1SB.

To calculate the energy dependence of the 1LO Raman scattering cross section R_c from the exciton picture *Bendow* and *Birman* [Ref. 7.66, Eq. (3.9)] is followed. In the calculation for TlCl, three discrete exciton states, 1SA, 1SB, and 2s of the $X_6^+ \times X_6^-$ exciton, are taken into account. Exciton damping is introduced by adding $i\Gamma$ to the energy denominator. The result is shown in Fig. 7.26 by a full line, together with experimental values [7.65]. In general, the agreement between the experiment and the theory is good. In this figure, the theoretical cross section is smaller than the experimental value in the energy range from 3.41 to 3.43 eV. The difference is due to the contribution of the resonance to the continuum which is not taken into account in the calculation. In Fig. 7.27, the Raman cross section of the 2LO scattering in TlCl is shown in the region of the $X_6^+ \times X_6^-$ exciton as well as in the energy range distant from it on the low-energy side. The theoretical curve is calculated by the exciton picture similarly to that for 1LO scattering with the approximation of taking only the $X_6^+ \times X_6^-$ exciton as the intermediate state, because the absorption and so the transition matrix element of the phonon-forbidden indirect $X_6^+ \times R_6^-$ exciton which exists in this energy range is small. We find again that the exciton approximation describes fairly well the resonance behavior of the Raman scattering to the direct exciton in a wide energy range.

In interpretations using the polariton picture, the Raman efficiency R_e is used to represent the Raman scattering intensity, rather than the Raman cross section.

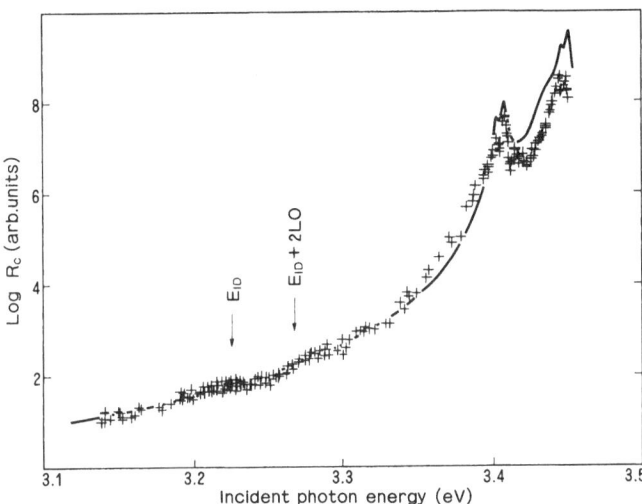

Fig. 7.27. Experimental Raman cross sections R_c of 2LO scattering in TlCl and the theoretical curve calculated from the exciton picture. Here E_{ID} is the $1s$ energy of the indirect $X_6^+ \times R_6^-$ exciton and $E_{ID} + 2LO$ is the energy of its two–LO-phonon state [7.65]

It is expressed as [7.67]

$$R_e \propto \frac{\omega_s^2}{v_g v_e} T_i T_s A_{is}^2 L \ . \tag{7.28}$$

Here v_g is the scattered polariton group velocity, v_e the energy transfer velocity of the incoming polariton, ω_s the scattered polariton frequency, T_i the transmissivity of the incident light to the incoming polariton, T_s the transmissivity of the scattered polariton to the outgoing light. Also, A_{is} is the exciton strength of the polaritons during the scattering of a phonon and L is the effective Raman active depth which is given in a rough approximation by the reciprocal of the sum of the imaginary parts of the incoming and scattered polariton wave vectors. The damping constant is contained in L, v_e, T_i, T_s, and A_{is}. We note that the Raman efficiency is better than the Raman cross section for describing the efficiency of the process in the case of the polariton picture because the effect of the additonal boundary condition (ABC) is included in the Raman efficiency through the transmissivity terms. The experimental result which should be compared with the theoretical R_e is the uncorrected, observed Raman intensity and not that after correction by the absorption coefficient, reflectivity, and refractive index as was done in Figs. 7.25–27 for the case of the Raman cross section. This is because these contributions are already taken into account in the effective Raman active depth and the transmissivities in R_e.

In Fig. 7.28, the uncorrected, experimental, 1LO Raman intensities and the theoretical R_e for TlCl are plotted as functions of the incident photon energy

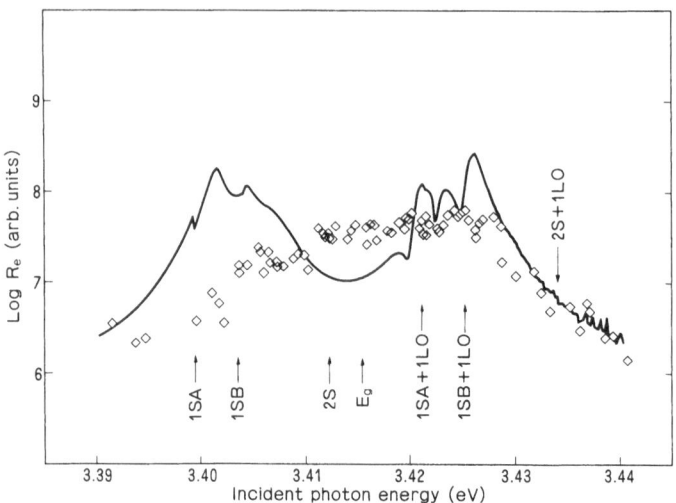

Fig. 7.28. Experimental Raman efficiency R_e of 1LO scattering in TlCl and the theoretical curve calculated by the polariton picture. Arrows indicate the energies of excitons, their one phonon states and the band gap [7.65]

[7.65]. Here the magnitude of the damping constant taken in the calculation is the approximate value used in the calculation by the exciton model. With this damping constant, a difference in model for the ABC has little effect on the shape of the spectrum. We find that the agreement between the experiment and the theory is poor and the exciton picture seems to be better at describing the resonant Raman scattering in TlCl than the polariton picture.

The reason why the exciton picture is better than the polariton picture in explaining the resonance Raman scattering in TlCl is not clear. However we note that the recent study of *Matsushita* et al. [7.67] is suggestive. They show that there is a critical damping constant Γ_c and the usual polariton dispersion relation is significant only when $\Gamma < \Gamma_c$. They give $\Gamma_c = 2E_{ex}(4\pi a E_{ex}/Mc^2)^{1/2}$, where E_{ex} is the transverse exciton energy, M the total exciton mass, and a the polarizability of the exciton. When $\Gamma > \Gamma_c$, the upper and lower polariton branches of the usual dispersion become a photon-like branch and an exciton-like branch and, as a result, an exciton and a photon propagate in a crystal as if they hardly mix with each other.

The exact values of Γ and Γ_c are not known for the $X_6^+ \times X_6^-$ exciton in TlCl because of the inapplicability of the Kramers-Kronig relation to absorption by an exciton with spatial dispersion. As before, we use the approximate values of Γ and Γ_c which are estimated from the approximate ϵ_1 and ϵ_2 spectra obtained from the reflectivity spectrum by the Kramers-Kronig relation. These values show that there is a possibility that $\Gamma > \Gamma_c$ holds in TlCl and this might be the reason for the success of the exciton model for resonant Raman scattering in TlCl. It should be pointed out that a direct experimental determination of the polariton dispersion curve in thallous halides would be an interesting way to examine this speculation.

7.3.2 Intervalley Scattering of a Direct Exciton

In the resonant Raman spectrum of TlBr given in Fig. 7.24, weak Raman lines a, b, c, and d are observed as well as the strong $LO(\Gamma)$ overtones. From the energies of the phonons in TlBr [7.46], they are assigned as $TO(\Gamma)$, $2LA(M)$, $TO(\Gamma) + LO(\Gamma)$, and $2LA(M) + LO(\Gamma)$ lines, respectively. When the energy of the incident light is varied in the vicinity of the $X_6^+ \times X_6^-$ exciton energy, more Raman lines, $2TA(M)$, $TA(M) + LA(M)$, $2TA(M) + LO(\Gamma)$, and $LO(\Gamma) + TO(\Gamma)$, are observed as is seen in Fig. 7.29 [7.61]. Their growths show a behavior resonant to the $X_6^+ \times X_6^-$ exciton, so they are intimately related to it.

One of the characteristic features of these Raman lines is the association of the M point phonons and they always appear in the two phonon scattering. The direct $X_6^+ \times X_6^-$ exciton, whose wave vector is at Γ, is three-fold degenerate because of the three inequivalent X valleys. In the scattering of one direct $X_6^+ \times X_6^-$ exciton to another degenerate direct-exciton state, two M point phonons of the same wave vector must be scattered, one for an electron and the other for a hole. By this process, the direct $X_6^+ \times X_6^-$ exciton at Γ is first scattered to the indirect $X_6^+ \times R_6^-$ exciton at M by one M phonon and then scattered back to another direct $X_6^+ \times X_6^-$ exciton at Γ by another similar M phonon. Thus the appearance of the 2M phonon scattering is a direct indication of the intervalley scattering of the $X_6^+ \times X_6^-$ exciton by phonons. This would suggest that, in the calculation of the $X_6^+ \times X_6^-$ exciton state made in Sect. 7.1.3, some contribution of phonon scattering to the intervalley scattering must be taken into account other than the Coulomb and exchange interactions, although it is predicted to be small [7.19].

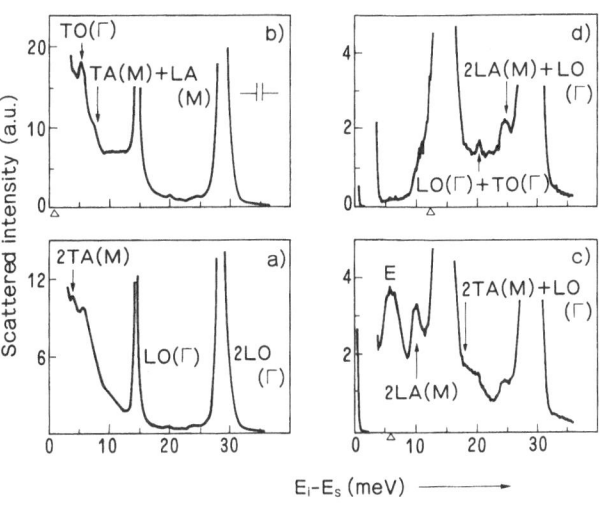

Fig. 7.29a–d. Resonant Raman spectra of TlBr at 1.8 K for various incident photon energies E_i near the 1s $X_6^+ \times X_6^-$ direct-exciton energy which is shown by open triangles on the absissa.
(a) $E_i = 3.0071$ eV,
(b) $E_i = 3.0109$ eV,
(c) $E_i = 3.0165$ eV,
(d) $E_i = 3.0225$ eV
[7.61]

7.4 Excitons and Induced Self-Trapping in Mixed Crystals of Thallous Halides

7.4.1 Exciton States in a Mixed Crystal

It is known that TlCl and TlBr make a stable solid solution maintaining a CsCl structure throughout the whole range of composition. The change of the lattice constant with composition nearly satisfies Vegard's law. The change of the absorption spectrum with composition is shown in Fig. 7.30 in the energy range of the direct transition [7.28] and at the $X_6^+ \times R_6^-$ indirect absorption edge in Fig. 7.31 [7.68].

In the spectra in Fig. 7.30, peak 1 is the $X_6^+ \times X_6^-$ exciton associated with the direct band gap transition at the X point and peak 4 is the direct absorption at the R point. Peaks 2 and 3 are the transitions from the second- and third-highest valence bands at the X point, which originate in the spin-orbit split halogen p states, to the lowest conduction-band state X_6^-. These assignments were discussed in Sect. 7.2.1.

We find that peaks 1 and 4 vary their energies continuously from TlCl to TlBr and maintain their shapes in the whole composition range although they are broadened by the alloying. The absorptions 2 and 3 behave differently. They

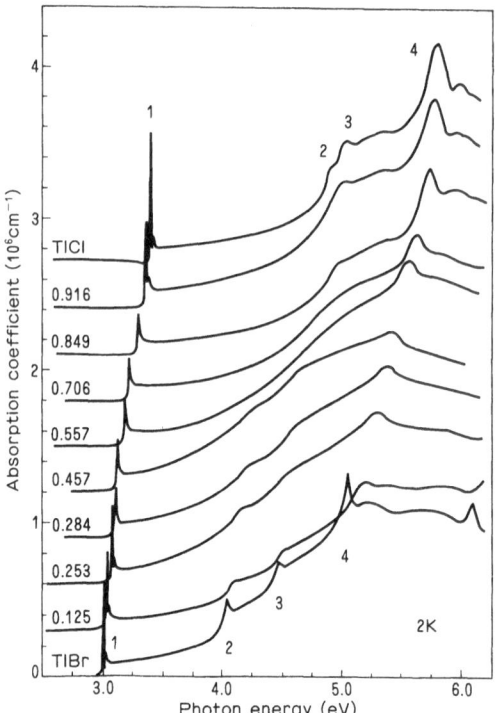

Fig. 7.30. Absorption spectra of TlCl–TlBr mixed crystals at 2 K. The number shown on the left of each spectrum is the concentration of TlCl in mole fractions. The vertical axis is graduated in 3×10^5 cm^{-1} intervals [7.28]

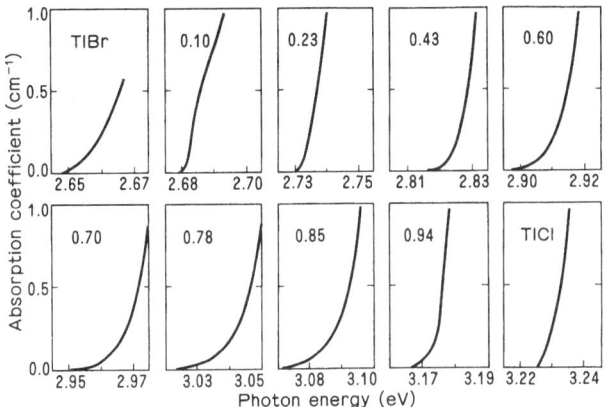

Fig. 7.31. Absorption spectra of TlCl–TlBr mixed crystals near the indirect absorption edge at 2 K. The concentration of TlCl in mole fractions is shown in each figure [7.68]

become broad and weak as bromine ions in TlBr are replaced by chlorine ions and are finally unobservable at a composition of about 0.5. Peaks 2 and 3 in pure TlCl behave in a similar way as bromine ions are mixed in and there is no continuation to the peaks 2 and 3 on the TlBr side.

The energy of the edge of the indirect $X_6^+ \times X_6^-$ exciton absorption shifts continuously from TlBr to TlCl as is found in Fig. 7.31. However the shape changes with the composition.

The energy shifts of the peaks and the edge are plotted as a function of the composition in Fig. 7.32. Here 0 is the indirect edge and the other numbers are the numbers of the peaks in Fig. 7.30. The edge 0 and peak 1 shift with curves which dip below the straight line of a linear shift and similar phenomena are found in alkali-substituted alkali halides [7.69] and in semiconductor alloys [7.70]. Peak 4 also shifts continuously but with a curve above the linear-shift line. Peaks 2 and 3 do not show a continuous shift from TlBr to TlCl and behave independently on the TlCl and TlBr sides.

At a glance, one may think that the continuous shifts of 0, 1, and 4 and the discontinuous changes of 2 and 3 can be explained by the atomic orbital states responsible for the bands relating to the respective transitions. The electronic states associated with the transitions 0, 1, and 4 are X_6^+ and R_6^+ in the highest valence band and X_6^- and R_6^- in the lowest conduction band. The atomic orbitals responsible for X_6^+ and R_6^+ are Tl $6s$ and those for R_6^- and X_6^- are Tl $6p$ and thus all three transitions occur on the Tl sublattice. Since the effect of the substitution of Br$^-$ for Cl$^-$ on the electronic states on the Tl sublattice will be a change of potential, the energies of the transitions 0, 1, and 4 will vary continuously as the composition changes. The transitions 2 and 3 are from the second- and the third-highest valence bands X_7^+ and X_6^+ to the conduction band X_6^-. Here both the X_7^+ and X_6^+ originate in the halogen p state while the X_6^- originates in the Tl $6s$ state. Thus 2 and 3 are the inter-sublattice transition from halogen to thallium and they will persist when changing halogen ions.

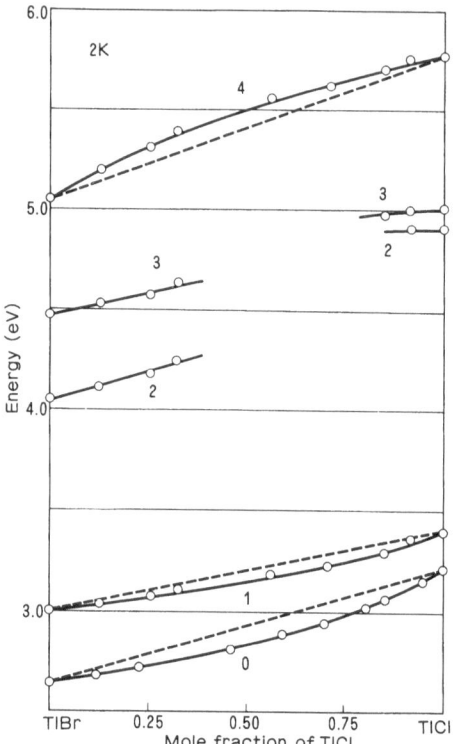

Fig. 7.32. Energies of the absorption peaks 1, 2, 3, and 4 in Fig. 7.30 and of the indirect absorption edge (0) in Fig. 7.31 are plotted against the concentration of TlCl in TlCl–TlBr mixed crystal. The broken lines are the linear dependences of the energy on the concentration [7.28, 29]

Similar persistence is found in the exciton in KCl–KBr mixed crystals [7.71] where the excitation is from the halogen p band to the potassium s band so that the above interpretation is applicable. However, in alkali-substituted mixed crystals of alkali halides, such as KCl–RbCl [7.72], the energy of the exciton due to the transition from the chlorine p band to the alkali s band varies continuously with a change of the composition. In this case, the excitons in KCl and RbCl should persist if the above interpretation is applied but they do not. This shows that the interpretation using the atomic picture is inadequate.

These phenomena are explained by the theory of the amalgamation and persistence of the band in a mixed crystal developed by *Onodera* and *Toyozawa* [7.73] which was given in Sect. 4.2.3. According to the theory, the parameter which determines whether the band is of the amalgamation or persistence type is Δ/T where Δ is the difference of the atomic energies of the atoms A and B, from which the mixed crystal AB is formed, and T the band width of the crystals A and B. From Fig. 7.2, the T of the lowest conduction band are about 3 eV in TlCl and TlBr. For the highest valence band, it is about 1.5 eV in both crystals. Since the atomic functions consisting of the two bands are almost those of thallium in both TlCl and TlBr, Δ can be taken as almost zero for the lowest conduction band and the highest valence band in a TlCl–TlBr mixed crystal. Thus for both the bands

$\Delta/T \gg 0.5$ holds and they are amalgamated in the mixed crystal. This is why the absorptions 0, 1, and 4 continue from TlCl to TlBr in Fig. 7.32.

The second- and the third-highest valence bands associated with the absorptions 2 and 3 are very narrow and their T are of the order of 0.3 eV. We measure the magnitudes of Δ for both valence bands by the difference of the electron affinities between free chlorine and bromine atoms, which is 0.2 eV. Then $\Delta/T \gtrsim 0.5$ for both the bands and they will persist in a TlCl–TlBr mixed crystal. Thus although the conduction band associated with the absorptions 2 and 3 is the lowest band, which is amalgamated, the valence bands persist in the mixed crystal, giving the discontinuities of the absorptions 2 and 3 in Fig. 7.32.

We observe in Fig. 7.32 that the curves 0 and 1 bend downward whereas the curve 4 bends upward and they deviate from a linear shift. The downward bending is observed quite commonly in amalgamated bands in mixed crystals. It is due to the dependency of the transfer energy upon the type of atoms and the second-order perturbation of the potential fluctuation which were discussed in Sect. 4.2.3. The upward bending found in curve 4 is unusual and is found in hardly any other mixed systems. Its origin is suggested to be a possible saddle shape of the highest valence band at the R point but further investigation is needed.

With the mixing of TlCl and TlBr, no particular absorption band is observed below the absorption edge even in the case of a low concentration of bromine ions and so there is no electronic state originating in the minority ions in the band gap as a final state of the optical transition. This is because $\Delta/T < 0.25$ for the $X_6^+ \times R_6^-$ exciton. However, we observe a change in the shape of the absorption edge of the indirect exciton as the composition is varied, particularly at both extremes of composition, although the gross appearance of the energy shift is smooth over the whole range of composition.

The changes of line shape of the indirect edge are shown for TlBr containing a small amount of Cl$^-$ in Fig. 7.33 and for TlCl containing a small amount of Br$^-$ in Fig. 7.34 [7.74]. We observe that when a very small amount of Cl$^-$ or Br$^-$ ion is mixed in TlBr or TlCl, respectively, the shape of the absorption edge changes from an $E^{3/2}$ shape, due to the phonon-forbidden indirect-exciton absorption, in the pure crystal to an $E^{1/2}$ shape in both cases. This is the absorption due to the impurity-assisted zero-phonon indirect-exciton excitation where the assistance of a phonon is not necessary because of the breaking of the translational symmetry by the presence of impurities. The absorption edge is the energy of the amalgamated $X_6^+ \times R_6^-$ exciton. If this energy is extrapolated to zero concentration of added impurity, the energies of the $X_6^+ \times R_6^-$ excitons in pure TlCl and TlBr are determined and they are found to coincide fairly well with the energies determined from the temperature dependence of the absorption edge, discussed in Sect. 7.2.3, and from the edge emission spectrum of doped crystals, in which a zero-phonon emission of the $X_6^+ \times R_6^-$ exciton is observed [7.23].

A difference is found in the behavior of the indirect absorption edge for the dilute mixed crystals of TlCl(Br) and TlBr(Cl), where the parenthesis indicates the minority ion. This is seen in Figs. 7.33 and 7.34. A long tail appears in

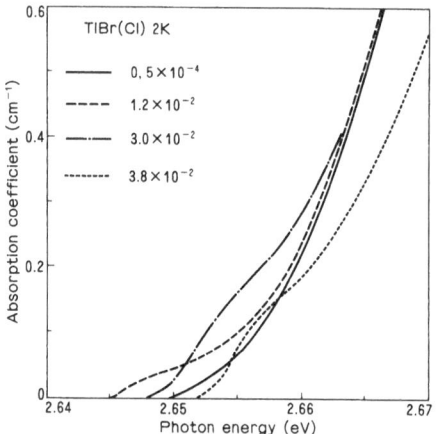

Fig. 7.33. Absorption spectra near the indirect absorption edge at 2 K in dilute mixed crystals TlBr(Cl) of four different TlCl concentrations which are given in mole fractions in the figure [7.74]

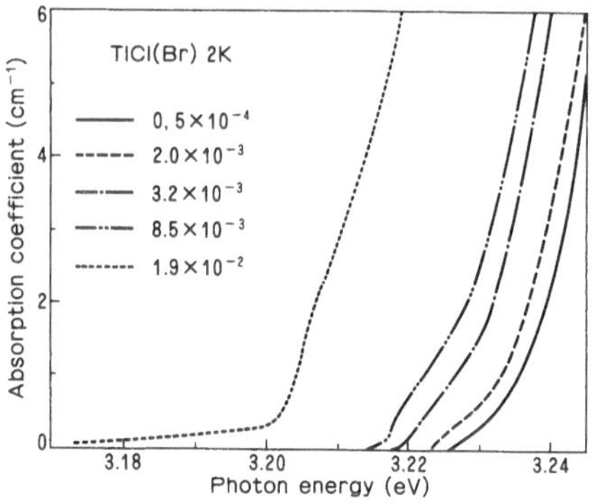

Fig. 7.34. Absorption spectra near the indirect absorption edge at 2 K in dilute mixed crystals TlCl(Br) of five different TlBr concentrations which are given in mole fractions in the figure [7.74]

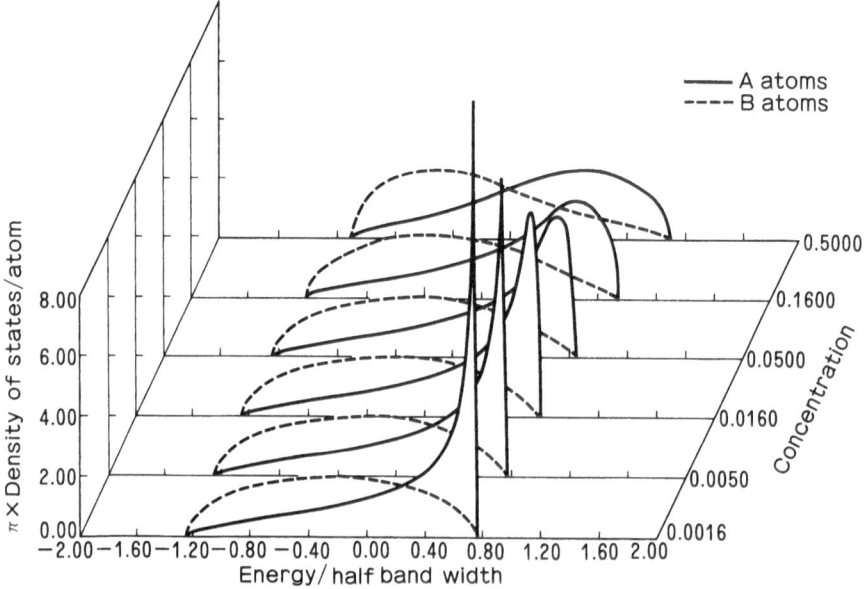

Fig. 7.35. Component densities of states ϱ_A (———) of atoms A from the minority sites and ϱ_B (----) of atoms B from the majority sites in a mixed crystal AB where $E_A > E_B$ and $\Delta/T = 0.25$ [7.75]

TlCl(Br) at a Br^- concentration as low as 1.9×10^{-2} mole fraction whereas no such tail is found in TlBr(Cl) at a Cl^- concentration as high as 3.8×10^{-2} mole fraction. This tendency can be seen also in Fig. 7.31 over a much wider range of concentration where the edge maintains the $E^{1/2}$ shape on the TlBr side up to a Cl^- concentration of 0.23 mole fraction whereas on the TlCl side the tail begins to appear at a low concentration of Br^- and continues to grow as its concentration increases.

These phenomena are understood, by the theory of *Velicky* et al. [7.75], as consequences of the local densities of the exciton states of TlCl and TlBr in the amalgamated band. Suppose that in a mixed crystal AB, where the concentration of A is x and B is $y = 1 - x$, an amalgamated band is formed. If E_A, the atomic level of A, is higher than E_B, the atomic level of B, the average density of states of the A component, ϱ_A, is concentrated at the high-energy edge of the amalgamated single band and the average density of states of the B component, ϱ_B, is distributed over the band almost uniformly when $x \ll y$. In the opposite situation $x \gg y$, ϱ_B is concentrated at the low-energy edge of the amalgamated band and ϱ_A is distributed over the band almost uniformly. When $x \sim y$, both ϱ_A and ϱ_B are distributed over the band but in the high-energy region of the band $\varrho_A > \varrho_B$ and in the low-energy region of the band $\varrho_B > \varrho_A$. The total density of states is $x\varrho_A(E) + y\varrho_B(E)$. In Fig. 7.35 are shown the spectra of ϱ_A and ϱ_B for various compositions in the case of $\Delta/T = 0.25$ [7.75].

The spectra shown in Figs. 7.31, 33, 34 are the absorptions at the low-energy edge which reflect the electronic state at the bottom of the amalgamated band. In the mixed crystal of TlCl–TlBr, $E_{TlCl} > E_{TlBr}$. Thus in TlCl(Br) the $X_6^+ \times R_6^-$ exciton state of TlBr is concentrated at the absorption edge and its intensity $(1 - x)\varrho_{TlBr}$ grows as the concentration of TlBr increases where x is the mole fraction of TlCl in the mixed crystal. The appearance of the tail in Fig. 7.31 in the spectrum of $1 - x = 1.9 \times 10^{-2}$ can be understood as the result of the concentration of the TlBr state at the edge. The model used in the theoretical calculation given in Sect. 4.2.3 is based on a single-site coherent potential approximation (CPA) and clusters of the component atoms are not taken into account. Thus the edge of the amalgamated band originates from a single-atom electronic state in this calculation. Since the electronic energy at a cluster of atoms is lower than that at a single atom because of the kinetic energy, the exciton energy at a TlBr cluster will be lower than that at a single TlBr in a TlCl matrix. Thus the absorption due to TlBr clusters will appear at the energy below the CPA band edge and its shape will be in a form of tail because of the various sizes and shapes of clusters. In TlBr(Cl), ϱ_{TlCl} is concentrated at the upper edge of the band so its influence will not be significant at the absorption edge. It contributes to the lower band edge only by disturbing a translational symmetry and induces a zero-phonon indirect transition with an $E^{1/2}$ line shape. This is what we see in Figs. 7.31 and 33.

7.4.2 Self-Trapping Induced by Alloying

It is known from photocarrier transport experiments [7.76–79] that electrons and holes are mobile and the stable photo-excited carrier states are free polaron states in TlCl and TlBr. The lowest energy state excitons, whether they are direct excitons or indirect excitons, are also mobile because only the edge emissions from the excitons in the band are observed and no Stokes-shifted emissions from immobile self-trapped excitons are found.

When a small amount of isoelectronic halogen ion impurity of lower excitation energy is added, such as Br^- or I^- to TlCl and I^- to TlBr, an intense and broad emission band appears at a low energy in place of the edge emission, which is observed only in pure a halide. Its shape is Gaussian which shows that it originates from a localized state [7.68]. Its energy is 2.66 eV in TlCl(I), 2.19_5 eV in TlBr(I) and 2.35 eV in TlCl(Br) at 4.2 K.

As will be explained in Sect. 8.3.2, it is known from the anomalous decrease of mobility of holes at high temperatures [7.9] that there is a metastable self-trapped state of holes in thallous halides whose energy is higher by a few hundred meV than the free hole polaron state [7.80]. Since a free electron will be bound to a self-trapped hole to form a self-trapped exciton, a metastable self-trapped exciton state will be present at energies above a free exciton state in thallous halides.

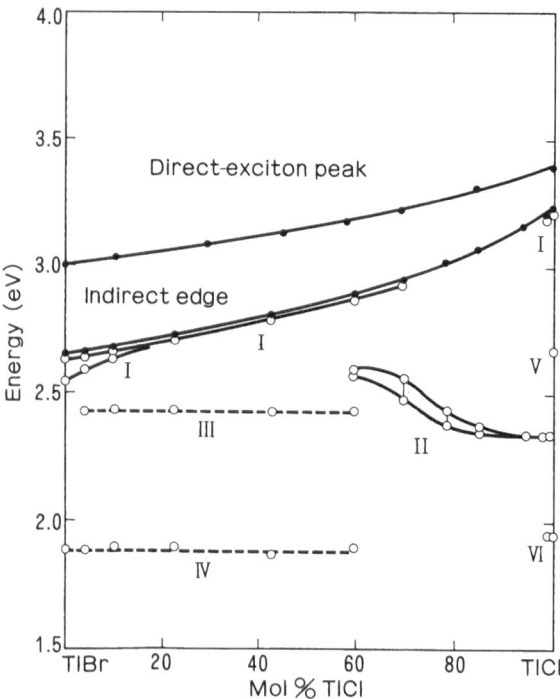

Fig. 7.36. Energies of the absorption peak of the direct exciton and the absorption edge of the indirect exciton at 2 K (•••) and of the emission peaks at 4.2 K (*open circles*) as a function of the composition of the TlCl–TlBr mixed crystal. Emissions I and II are of intrinsic origin and III, IV, and V are of extrinsic origin [7.68]

From the discussion in Sect. 4.4.5, it is readily conceivable that a stable self-trapped exciton will be formed in halogen doped thallous halides, such as TlCl(Br), as a result of the cooperation of exciton-phonon coupling and the isoelectronic impurity potential. Thus the Stokes-shifted emissions found in TlCl(Br), TlCl(I), and TlBr(I) are emissions from the self-trapped excitons induced by alloying.

The edge emission from the free exciton state and the emission from the self-trapped exciton state induced by alloying have been studied over the whole range of composition [7.68]. Figure 7.36 shows the changes of energies of the absorptions and emissions as a function of the composition of the TlCl–TlBr mixed crystal. Here the dots give the peak energies of the $X_6^+ \times X_6^-$ direct exciton absorption and the edge energy of the $X_6^+ \times R_6^-$ indirect exciton absorption, showing both the exciton states are amalgamated. The open circles are the energies of the emission peaks. Emission I is the edge emission associated with the indirect exciton and is observable only in a narrow range of Br$^-$ concentration on the TlCl side, that is from 0 to 2%, whereas it is observed in a wide range of Cl$^-$ concentration on the TlBr side from 0 to 70%. Emission II is the emission from the self-trapped state at the Br$^-$ sites in TlCl and is observed in the range of Br$^-$ concentration from a very small amount, perhaps less than 0.1%, to 40%. In pure TlCl and TlCl containing Br$^-$ of more than 40% mole fraction, emission II is not detectable. From 30% to 40% of Br$^-$ concentration, emission II decreases and emission I grows in the manner shown in Fig. 7.37, as if they exchange their intensities.

Emission II has two important features. One is the shift of its energy to higher energies with increasing Br$^-$ concentration in TlCl. The other is the shift of the peak energy to lower energies with increasing excitation energy at a fixed concen-

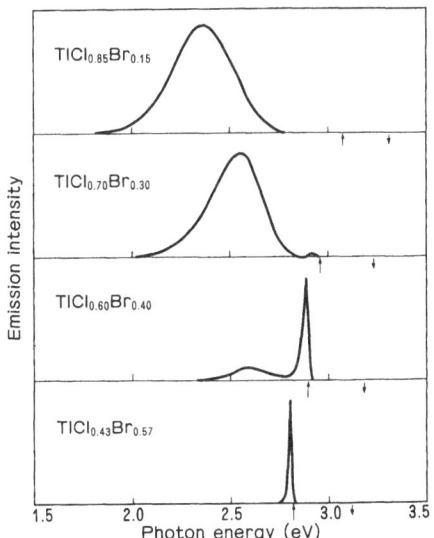

Fig. 7.37. Change with composition of the emission spectrum for the TlCl–TlBr mixed crystal at 4.2 K. Upward and downward arrows indicate the energies of the indirect-exciton absorption edge and the direct-exciton absorption peak, respectively. The relative intensity of the spectra arbitrary [7.68]

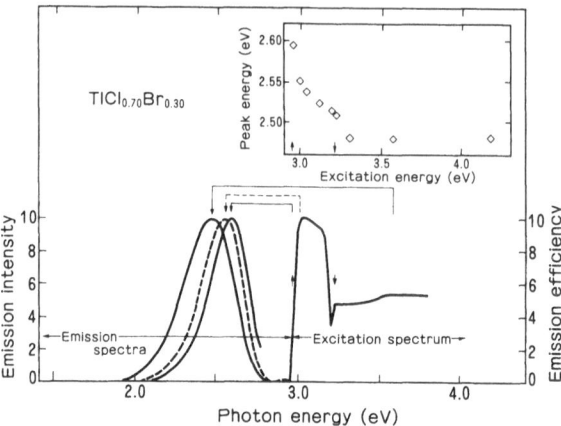

Fig. 7.38. Emission spectra and their excitation spectrum in $TlCl_{0.7}Br_{0.3}$ at 4.2 K. Upward and downward arrows indicate the energies of the indirect exciton absorption edge and the direct exciton absorption peak. The energies of excitations and the corresponding emission spectra are connected by lines. Variation of the peak energy of the emission with excitation energy is shown in the inset [7.68]

tration of Br^-. The latter shift is shown in Fig. 7.38 for $TlCl_{0.7}Br_{0.3}$ in which are given the spectra of emission II for excitation at three different energies, two being in the indirect-exciton absorption region and one in the direct-exciton absorption region. In the inset of the figure, the energy of the emission peak is plotted against the excitation energy. We find that the peak energy decreases as the excitation energy increases in the indirect-exciton absorption region and is unchanged when the excitation is in the direct-exciton absorption region. The range of this shift is shown by the vertical lines in the curve of emission II in Fig. 7.36. It is almost zero at a low concentration of Br^- and increases with an increase of the Br^- concentration until 30%. It then decreases and at 40% of Br^- emission II shows little shift and becomes unobservable.

The above mentioned features of the edge emission I and the self-trapped exciton emission II are interpreted in terms of the self-trapping of excitons induced by alloying and by taking into account the clustering of TlBr. Fig. 7.39 is the configuration coordinate diagram of the exciton-phonon system in a mixed crystal [7.68]. The abscissa Q is the amount of lattice distortion which is normalized in such a way that the ground states of all kinds of clusters look the same. The lattice at Q_0 is the undistorted lattice whose electronic state is a virtual crystal state. The lower curve is the adiabatic potential of the ground state and the upper two curves are those of an exciton at TlBr clusters of fifferent sizes. The states denoted by S_{large} and S_{small} are the self-trapped states of excitons at a large and a small cluster, respectively, and F is the free-exciton state in a virtual crystal state of a mixed crystal.

When the Br^- concentration is very small, Br^- will be distributed in a single-ion form in the TlCl lattice. By absorbing light of energy greater than E_F, excitons will be formed in a virtual crystal band because the transition frequency is so much higher than the phonon frequency that the lattice does not follow during the transition. The excitons will then be localized at the bromine sites to become self-trapped excitons with the cooperation of a single bromine ion potential and exciton-phonon coupling, as was discussed in Sect. 4.4.5. Light with energy E_{small}

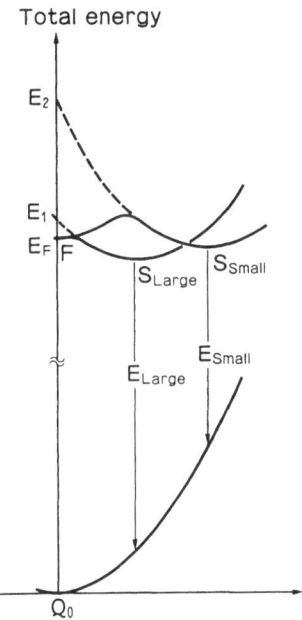

Total energy

Fig. 7.39. Configuration coordinate diagram of an exciton in the TlCl–TlBr mixed crystal. The abscissa is the amount of lattice distortion which is normalized so that the ground stats of all kinds of cluster look the same. State F corresponds to a free virtual crystal state. Here S_{large} and S_{small} are the self-trapped states of exciton at a large and a small cluster, respectively, and E_{large} and E_{small} are the energies of the emissions from these states. The energies E_F, E_1, and E_2 are of excitons of a virtual crystal state, undistorted localized states at a small, and a large cluster, respectively [7.68]

will be emitted as a luminescence whose energy is much less than the absorption energy. This is emission II in a crystal of low Br$^-$ concentration.

When the bromine concentration increases, clusters of TlBr will be formed and their sizes and numbers increase as the concentration increases. In Fig. 7.39 the adiabatic potential curve for a large cluster is also shown. It is found that the amount of lattice distortion of a self-trapped state at a large cluster is smaller than that at a small cluster [7.81]. Thus the energy of the emission from an exciton self-trapped at a large cluster E_{large} is larger than that at a small cluster E_{small}. This is the reason for the shift of emission II to higher energies with an increase of the TlBr concentration in TlCl–TlBr mixed crystal observed in Fig. 7.36.

The shift of emission II with a change of the excitation energy shown in Fig. 7.38 is explained by the same picture. The broken lines in Fig. 7.39 are the extrapolations of the adiabatic potential curves from the side of the self-trapped state to the undistorted virtual crystal state. They intersect the ordinate at the energies E_1 and E_2. The states at E_2 and E_1 are excitons localized at a small and a large cluster, respectively, without any accompanying lattice distortion. These states are embedded in the extended state of the amalgamated band in a rigid lattice. If it is assumed that when the crystal is excited by light of low energy E_1, some portion of the exciton states formed adiabatically relaxes directly to the self-trapped state S_{large} without decaying down to the band bottom of the amalgamated band, then the emission with high energy, E_{large}, will be observed with such low energy excitation. For the same reason, an emission of lower energy E_{small} will be observed for excitation at the higher energy E_2. We find that when the exciting energy exceeds the direct exciton energy, there is no further shift of the emission energy. The exciton excited into the direct-exciton band will first decay to the bottom of its band regardless of its initial energy and then scatter to the indirect-exciton band from which it relaxes to the self-trapped state. The initial state for decaying to the self-trapped state in the indirect-exciton band is thus independent of the energy of the excitation in the direct-exciton band and this makes the energy of the emission peak independent of the excitation energy in the direct-exciton absorption region. It is noted that the decay of an exciton with nonzero kinetic energy directly to a self-trapped state in a crystal of strong electron-phonon coupling, which is assumed here, is an interesting problem which must be examined in a future study.

When the bromine concentration exceeds 30% in the TlCl–TlBr mixed crystal, the edge emission I appears in place of the emission band II. At a concentration of bromine higher than 40%, emission II is hardly observed. This indicates that the lowest excited state is changed from the self-trapped localized state induced by alloying to the amalgamated extended state in an undeformed virtual crystal, in this composition range. At this transition the energy of the emission will jump from E_{large} to E_F.

The origin of this abrupt transition will not be simple because of the many kinds and complicated distribution of clusters in the mixed crystal of this composition. However the results of the Monte Carlo calculation on site-percolation are quite suggestive. The probability that randomly distributed impurity ions at the lattice points form percolated clusters has been calculated for several kinds of lattice structures [7.82]. For the case of the simple cubic lattice, to which the TlCl–TlBr system belongs, the probability of finding an impurity ion in a percolated cluster becomes finite at an impurity concentration of 0.30 mole fraction and increases sharply to about 90% at 0.40 mole fraction. This concentration range is just the same as that of the transition from emission II to emission I. This shows that the self-trapped state induced by alloying is no longer stable with the percolation of the TlBr cluster and the extended state in the amalgamated band is the stable electronic state at the percolation.

7.5 Excitons in Thallous Halides in External Fields

7.5.1 Magnetic Field

The effect of a magnetic field on the excitons is the sum of the effects on the motions of all the individual electrons in the conduction and valence bands which interact with each other by Coulomb and exchange forces. The effect can be divided into two parts in the first approximation. One is the effect on the electron-hole (e–h) pair functions which gives a mixing between them. By the mixing, a change of fine structure of the e–h state belonging to one exciton envelope state will occur. The other effect is the change of the envelope state. The relative motion of electron and hole will be modified by a magnetic field.

The e–h state of the $X_6^+ \times X_6^-$ direct exciton consists of four states in each valley, as given in (7.3), which are mixed by the intervalley Coulomb and intra- and intervalley exchange interactions (7.7–10). The mixings are the results of the V and J terms of the Hamiltonian (7.4). In a similar way to these Coulombic interactions, the magnetic field further mixes the e–h states in the same valley. The magnetic field also deforms the hydrogenic envelope state which is the eigenstate of the first and second terms plus the intravalley Coulomb interaction in the V term of the Hamiltonian (7.4).

The mixing of the e–h states of the $X_6^+ \times X_6^-$ exciton in the missing electron scheme is written for the z valley as

$$\langle v'c' | \mathcal{H}_{\text{Zeeman}} | vc \rangle = \frac{\mu_B}{2} \delta_{vv'} \langle c' | g_\parallel^c \sigma_z H_z + g_\perp^c (\sigma_x H_x + \sigma_y H_y) | c \rangle$$

$$\tag{7.29}$$

$$- \frac{\mu_B}{2} \delta_{cc'} \langle v | g_\parallel^v \sigma_z H_z + g_\perp^v (\sigma_x H_x + \sigma_y H_y) | v' \rangle ,$$

where g^c and g^v are the effective g values of the electrons in the conduction and the valence band, respectively, and $\parallel$ and $\perp$ denote the components of the aniso-tropic g value parallel and perpendicular to the z direction. In (7.29) $\mathcal{H}_{\text{Zeeman}}$ is $\mu_B \sigma g H/2$.

In Table 7.4, the matrix elements are given where the energy origin is taken as the energy of the spin-triplet state Ψ_0^z in zero magnetic field. As will be found in (a), where H is applied in the z direction, the z-polarized dipole-allowed states Ψ_z^z and $(\Psi_z^x + \Psi_z^y)/\sqrt{2}$ mix with the spin triplet states Ψ_0^z and $(\Psi_0^x + \Psi_0^y)/\sqrt{2}$, respectively, in the same valley. Thus the spin-triplet state Ψ_0 becomes observ-able by z-polarized light.

Table 7.4. Off-diagonal matrix elements for excitons in a magnetic field applied in the z-direction. Energy is expressed in units of $\mu_B H/2$

a)

	Ψ_0^z	$(\Psi_0^x + \Psi_0^y)/\sqrt{2}$	$(\Psi_0^x - \Psi_0^y)/\sqrt{2}$
Ψ_z^z	$g_\parallel^c - g_\parallel^v$		
$(\Psi_z^x + \Psi_z^y)/\sqrt{2}$		$g_\perp^c + g_\perp^v$	
$(\Psi_z^x - \Psi_z^y)/\sqrt{2}$			$g_\perp^c + g_\perp^v$

b)

	Ψ_x^x	$(\Psi_x^y + \Psi_x^z)/\sqrt{2}$	$(\Psi_x^y - \Psi_x^z)/\sqrt{2}$
Ψ_y^y		$(-i/\sqrt{2})(g_\perp^c - g_\perp^v)$	$(-i/\sqrt{2})(g_\perp^c - g_\perp^v)$
$(\Psi_y^x + \Psi_y^z)/\sqrt{2}$	$(-i/\sqrt{2})(g_\perp^c - g_\perp^v)$	$(-i/2)(g_\parallel^c + g_\parallel^v)$	$(i/2)(g_\parallel^c + g_\parallel^v)$
$(\Psi_y^x - \Psi_y^z)/\sqrt{2}$	$(-i/\sqrt{2})(g_\perp^c - g_\perp^v)$	$(i/2)(g_\parallel^c + g_\parallel^v)$	$(-i/2)(g_\parallel^c + g_\parallel^v)$

The matrix (b) shows that, with the same magnetic field in the z direction, the y-polarized optically active Ψ_y^y is mixed with the x-polarized $(\Psi_x^y + \Psi_x^z)/\sqrt{2}$ and $(\Psi_x^y - \Psi_x^z)/\sqrt{2}$, and also the y-polarized $(\Psi_y^x + \Psi_y^z)\sqrt{2}$ is mixed with Ψ_x^x, $(\Psi_x^y + \Psi_x^z)/\sqrt{2}$, and $(\Psi_x^y - \Psi_x^z)/\sqrt{2}$. Thus when we measure the absorption with light whose direction is in x and polarization is in y, the Ψ_x^x and $(\Psi_x^y + \Psi_x^z)/\sqrt{2}$ excitons with longitudinal energies of $2\Delta_1^L$ and $\Delta_2^L + 2\Delta_4^L$, respectively, will be excited. The reason for the longitudinal state is that both the excitons are polarized in x and the direction of light and so the directions of the wave vectors of the excitons are also in x. In this geometry, the optically forbidden $(\Psi_x^y - \Psi_x^z)/\sqrt{2}$ exciton will also become observable. Its energy is $\Delta_2 - 2\Delta_4$ because of the absence of the longitudinal-transverse splitting.

Figure 7.40 shows the absorption spectra of the $1s\ X_6^+ \times X_6^-$ direct exciton in TlCl in various magnetic fields [7.20]. The full lines are the spectra for $H \parallel [110]$, $\kappa \parallel [1\bar{1}0]$, and $E \parallel [001]$ where κ and E are the wave vector and the electric field of light. This case is similar to the case of $H \parallel z$, $E \parallel y$, and $\kappa \parallel x$ discussed above and the observable states in the magnetic field are $(\Psi_x^x - \Psi_y^y)/\sqrt{2}$, $(\Psi_x^y + \Psi_x^z - \Psi_y^z - \Psi_y^x)/2$, and $(\Psi_x^y - \Psi_x^z + \Psi_y^z - \Psi_y^x)/2$ whose energies are $2\Delta_1^L$, $\Delta_2^L + 2\Delta_4^L$, and $\Delta_2 - 2\Delta_4$, respectively. With the application of a magnetic field, a new peak A′ appears between 1A and 1B, and it is believed to originate from the longitudinal $(\Psi_x^y + \Psi_x^z - \Psi_y^z - \Psi_y^x)/2$ exciton because the intensities of other possibly observable excitons will be small. The spectra for $H \parallel E \parallel [110]$ and

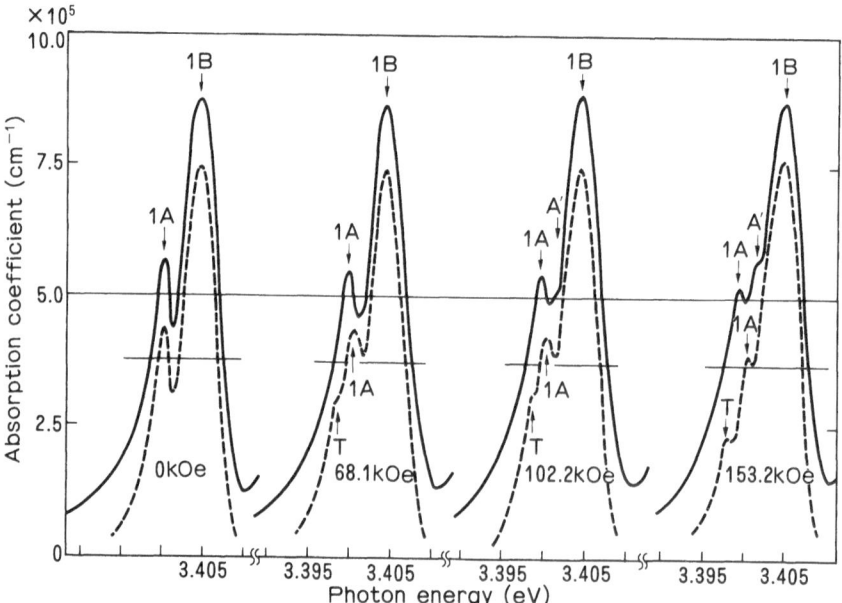

Fig. 7.40. Absorption spectra of the $1s\ X_6^+ \times X_6^-$ exciton in TlCl in magnetic fields at 4.2 K. (——): $H \parallel [110]$, $\kappa \parallel [1\bar{1}0]$ and $E \parallel [001]$. (---): $H \parallel E \parallel [110]$, and $\kappa \parallel [1\bar{1}0]$ [7.20]

$\kappa \| [1\bar{1}0]$ are shown by broken lines. The absorption T, at a lower energy than 1A, grows with the magnetic field. This is the absorption of the spin-triplet Ψ_0^z and $(\Psi_0^x + \Psi_0^y)/\sqrt{2}$ states. We note that all these results support the exciton and the band structures of thallous halides discussed in Sects. 7.1.2 and 7.1.3. Similar effects of magnetic field will be observed, in principle, in 2s and higher direct-exciton absorption but in practice this is not possible because their very small intensities make the A' and T absorptions below the noise level.

The overall envelope function of the exciton in a magnetic field, Φ, is determined by the effective mass equation

$$\left\{ E_c\left[-i\nabla_{r_e} - \frac{e}{ch}A(r_e)\right] - E_v\left[-i\nabla_{r_h} + \frac{e}{ch}A(r_h)\right] - \frac{e^2}{\epsilon|r_e - r_h|} \right\}\Phi = E\Phi$$

(7.30)

Here we take only the diagonal Coulomb term for the e–h interaction and other terms are neglected because they determine the fine structure. In (7.30) E_c and E_v are the conduction and valence band energies and we take their dispersions as of the standard form with the effective masses m_e and m_h. The vector potential A is given by $A(r_i) = (H \times r_i)/2$. Equation (7.30) is reduced by a canonical transformation to the following equation, which is expressed in terms of the e–h relative coordinate $r = r_e - r_h$ [Ref. 7.21, p. 79].

$$\left[-\frac{\hbar^2}{2\mu}\nabla_r^2 - \frac{e^2}{\epsilon r} + \frac{ie\hbar}{c}\left(\frac{1}{m_e} - \frac{1}{m_h}\right)A(r) \cdot \nabla_r + \frac{e^2}{2c^2\mu}A^2(r) \right.$$

$$\left. - \frac{2e\hbar}{c(m_e + m_h)}A(r) \cdot K \right]F(r)$$

$$= \left[E - \mathscr{E}_g - \frac{\hbar^2 K^2}{2(m_e + m_h)} \right]F(r) .$$

(7.31)

Here μ is the reduced mass such that $\mu^{-1} = m_e^{-1} + m_h^{-1}$ and $\mathscr{E}_g$ is the band-gap energy. The envelope function $F(r)$ is related to Φ by $\Phi = F(r)\exp\{i[K - (e/\hbar c)A(r)] \cdot R\}$ where R is the center-of-mass coordinate and is $(m_e r_e + m_h r_h)/(m_e + m_h)$. In the absence of a magnetic field, the first and the second terms of (7.31) give the usual hydrogenic wave function for $F(r)$.

As a measure of the strength of the magnetic field, we introduce γ which is equal to $\hbar\omega_c/2E_{ex}^b$. Here $\hbar\omega_c$ is the cyclotron energy, $e\hbar H/\mu c$, and E_{ex}^b the exciton binding energy, $\mu e^4/2\hbar^2\epsilon^2 n^2$. For the $X_6^+ \times X_6^-$ exciton in TlCl, $\gamma = 1$ can be achieved with a magnetic field of 500, 120, and 60 kOe for the 1s, 2s, and 3s states, respectively. Thus, by using thallous halides, the low magnetic field effect can be studied in low n excitons and the high magnetic field effect in high n excitons with the magnetic fields available in a laboratory.

When $\gamma \ll 1$, the $-e^2/\epsilon r$ term is much larger than the other magnetic terms. In this case, the magnetic effect can be treated as a perturbation to the hydrogenic wave function $F(r)$ which is the eigenstate of $-(\hbar^2 \nabla_r^2/2\mu) - (e^2/\epsilon r)$. Then the third term represents a linear Zeeman effect and the fourth term a diamagnetic effect of the hydrogen atom. The fifth term is the magneto-Stark effect [7.83] which is caused by the motion of an exciton with nonzero velocity in a magnetic field and will not be discussed here because no experiment has been done on this effect in thallous halides.

The linear Zeeman term in (7.31) can be written as $-(e/2c)(1/m_e - 1/m_h)\boldsymbol{H} \cdot \boldsymbol{L}$ where $\boldsymbol{L}$ is the angular momentum of the relative motion of the electron and hole. For the case of a direct allowed exciton like the $X_6^+ \times X_6^-$ exciton in thallous halides, the linear Zeeman effect due to orbital motion will not be observed in the optical spectrum because the optically observable state is an s state whose $\boldsymbol{L}$ is zero. In the case of a dipole-forbidden exciton, like the yellow series exciton in Cu_2O, the splitting due to the linear Zeeman effect is observed [7.84] because the envelope is a p state.

The diamagnetic term can be written as $e^2 H^2 (x^2 + y^2)/8c^2\mu$ when $\boldsymbol{H}$ is in the z direction. It has two effects on the hydrogenic wave function. One is the energy shift ΔE due to the diagonal matrix element of the $|nlm\rangle$ state. For the s state it is

$$\Delta E = \frac{5\hbar^4 H^2 \epsilon^2}{24\mu^3 e^2 c^2} n^4 \left(1 + \frac{1}{5n^2}\right), \tag{7.32}$$

which is proportional to H^2 and n^4. The other is the mixing of two states which differ by ± 2 in l. This is because the diamagnetic term can be expressed by $e^2 H^2 r^2 \sin^2 \theta/8c^2\mu$ where θ is the angle between $\boldsymbol{H}$ and r. Here $\sin^2 \theta$ can be expressed by the spherical harmonics of the order 0 and 2 and the hydrogenic wave function contains a spherical harmonic function as an angular part. Thus the optically inactive d state of the exciton will become observable due to the mixing of the active s state with the application of a magnetic field.

The spectral change of the $X_6^+ \times X_6^-$ exciton of $2s$ and higher states by the magnetic field was measured by *Kurita* and *Kobayashi* [7.31] and was extended further by *Nakahara* and *Fujii* [7.35, 85]. Figure 7.41 shows the spectra measured in an evaporated thin film of TlBr [7.31]. Here the absorption intensity is given by the transmittance. We find strong oscillations which grow with the field in the energy regions of the humps α, β, and γ of Fig. 7.8. The large arrows in the figure are the energy position of the dip between the humps α and β at which the one-free-phonon sidebands of $2s$ and higher excitons begin to rise. The energies of the peaks of the oscillation measured in the reflectivity spectra of single crystals and in the absorption spectra of evaporated thin films are plotted against magnetic field in Fig. 7.42 [7.85]. The numbers given in the region of the hump α are the quantum numbers n of the zero-phonon ns $X_6^+ \times X_6^-$ excitons. The absorptions $2'$ and $3'$ are the absorptions which appear only in a high magnetic field. Their intensities are strong in the absorption spectrum of an evaporated thin film but weak in the reflectivity spectra of single crystals. The reason has not been clarified.

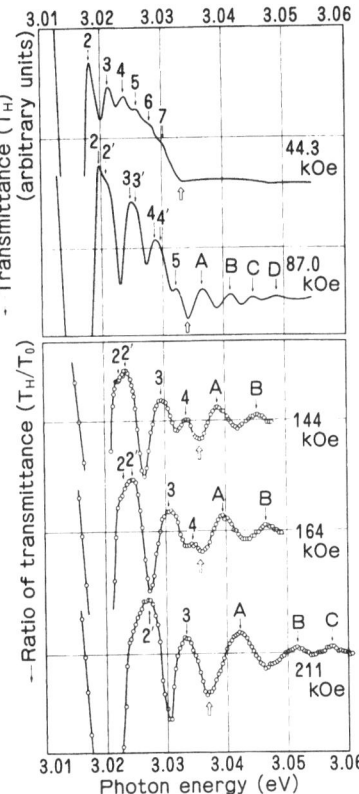

Fig. 7.41. Change of the absorption spectrum of TlBr with magnetic field at 4.2 K in the energy range of the $2s$ and higher states of the $X_6^+ \times X_6^-$ excitons and their phonon sidebands. Here, T_H and T_0 are the transmittances with and without magnetic field, respectively. The arrows indicate the energy of the kink between the humps α and β [7.31]

In Fig. 7.42, the curves of zero-phonon excitons show positive curvature at low field and become linear at high field as can be seen from curves 2 and 3. It is found that the energy shift at the bending region is proportional to H^2, showing that it is due to the diamagnetic effect on the hydrogenic state [7.31]. The critical field strength for the transition from the quadratic shift to the linear shift is smaller for a higher n exciton. This is because the γ value is larger for a higher n exciton at a given magnetic field and thus the low field approximation becomes invalid at a lower field for a higher n exciton.

If one extrapolates curve 2′ to lower fields, its energy coincides with the energy of 3 at zero field. This suggests that the state responsible for 2′ is a state which is degenerate with $3s$ in zero field. This is the $3d$ state of exciton which is optically inactive at zero field but becomes active through the mixing of the allowed $2s$ state by the diamagnetic interaction. The origin of the absorption 3′ is the same.

When γ is increased by increasing the magnetic field, the shift becomes linear. This suggests that the transition relates to an interband Landau-level transition in a high magnetic field. In the high field limit, we take $-e^2/\epsilon r$ as a perturbation in

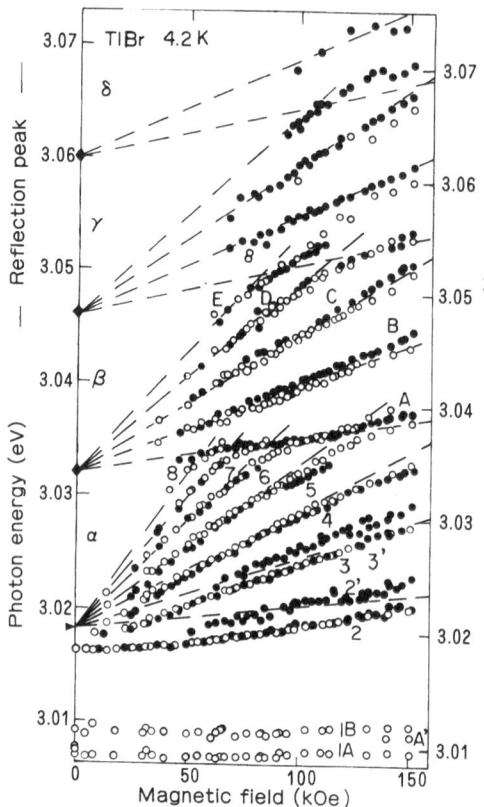

Fig. 7.42. Energies of the reflection peaks measured for a bulk crystal of TlBr (*open circles*) and those of the absorption peaks in an evaporated thin film of TlBr (•••) as a function of magnetic field. The relative positions of the ordinates for the reflection and absorption peaks are shifted against each other in such a way that the observed energies of the peaks are in the same position. The triangle on the ordinate is the energy of the ionization limit of the $X_6^+ \times X_6^-$ exciton. The rhombs are multiples of the LO phonon energy measured from the ionization limit of the exciton. Broken lines represent the interband Landau-level transition energies [7.85]

(7.31). The unperturbed state is a pair state of conduction- and valence-band Landau levels and its energy is [7.86]

$$E_{NMk} = \frac{e\hbar H}{c}\left[\frac{1}{\mu}\left(N + \frac{1}{2}\right) - \frac{M}{m_h}\right] + \frac{\hbar^2 k^2}{2\mu}, \qquad (7.33)$$

where N is the quantum number of the Landau level, and M the z component of the angular momentum of the e–h relative motion. Applying $-e^2/\epsilon r$ as a perturbation and taking an adiabatic approximation, the energy of the e–h bound state v associated with the N, M Landau level is [7.86, 87],

$$E_{NMv}^{\pm} = E_{NM} - \frac{R_y}{(v + \delta_v^{\pm})^2}, \qquad (7.34)$$

where v is the quantum number of the one-dimensional hydrogen atom [7.87]. The quantum defect $\delta_v^{\pm}$ takes + and – for the even and odd parity states, respectively.

Between the low and high field limits, there exists a range of intermediate strength of magnetic field. In this range, there is no convincing theory to descibe the system. It is the classic problem of a hydrogen atom in a magnetic field of intermediate strength but its investigation is still at the very beginning. It is not even clear, which level in the high field limit connects with which level in zero field. This is known as the problem of level correspondence. Two theoretical attempts to answer this problem have been made.

One is the theory based on the conjecture that the number of nodal surfaces of the wave function will be conserved in any strength of magnetic field. It was proposed by *Kleiner* [7.88] and was used by *Elliott* and *Loudon* [7.86]. Later the theory was modified by *Shinada* et al. [7.89, 90] by applying the conservation rule to the total number of nodal surfaces and allowing mixing between levels of same total number. The result is shown in Fig. 7.43. Here the level correspondence is $1s$–(000^+), $2s$–(001^+), $3s$–(002^+), $3d_0$–(100^+), $4d_0$–(101^+) and so on. We note that $1s$, $2s$, and $3s$ will not cross the lowest Landau level at any magnetic field strength. The $3d_0$ level will become optically active by the mixing of $2s$ in a magnetic field and will cross the lowest Landau level in a high magnetic field.

The other theory is based on the noncrossing rule given by *Boyle* and *Howard* [7.91] which states that any states with the same parity and same orbital angular momentum along the magnetic field cannot be degenerate at any strength of magnetic field, thus two such levels cannot cross each other. This is because the Hamiltonian is invariant under inversion and under rotation about the axis along the magnetic field. The theory was extended by *Baldereschi* and *Bassani* [7.92] and *Lee* et al. [7.93]. According to them, the $M = 0$ even-parity exciton states at $H = 0$, that is $1s$, $2s$, $3d_0$, $4d_0$, $4s$ and so on, should not cross the edge of the lowest Landau level at any magnetic field strength from zero to infinity. They also propose that the discrete states associated with high Landau levels at high fields have to merge into the exciton continuum for vanishing magnetic field.

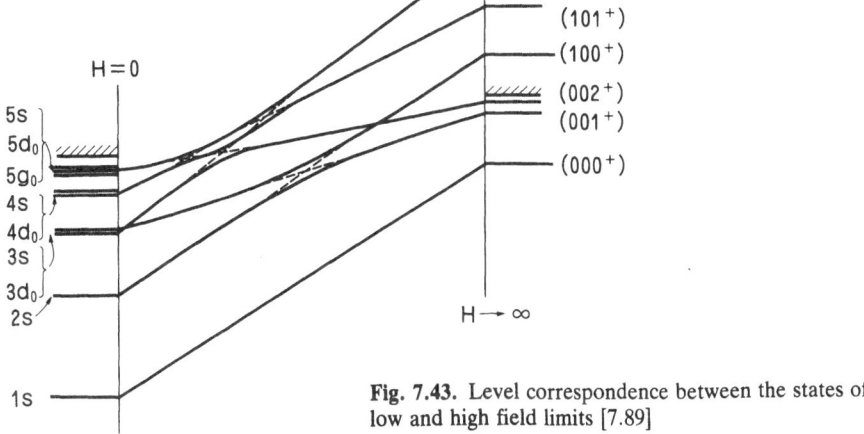

Fig. 7.43. Level correspondence between the states of low and high field limits [7.89]

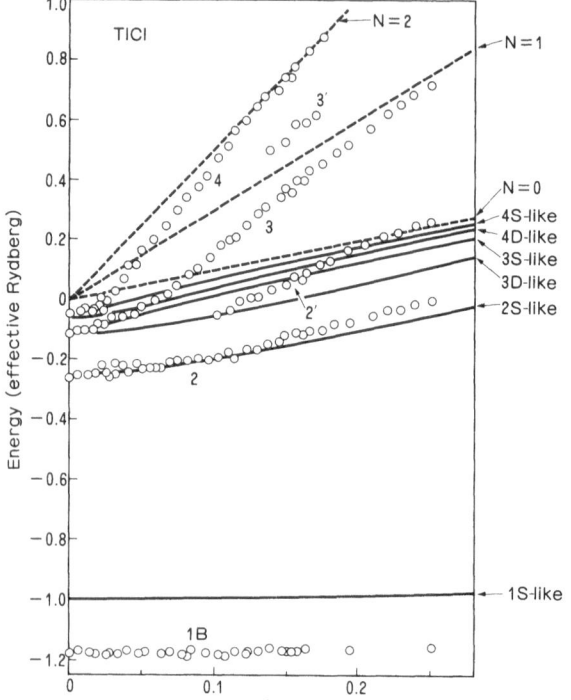

Fig. 7.44. Energies of the $X_6^+ \times X_6^-$ exciton in TlCl against magnetic field. Energy is measured in units of the effective Rydberg Ry (= 10.0 meV) and magnetic field in units of $\gamma = \hbar\omega_c/2Ry$. The solid lines are the theoretical energies given by *Lee* et al. [7.93] and the broken lines are the interband Landau-level transition energies [7.94]

In Fig. 7.44, the energies of the absorption peaks 1B, 2, 3, and 4 of TlCl, which correspond to the $1s$, $2s$, $3s$, and $4s$ excitons at zero magnetic field, are plotted, in units of effective Rydberg energy, as a function of γ, which is a measure of magnetic field [7.94]. The broken lines are the Landau level energies (7.33) of $M = 0$ and $k = 0$. The full lines are the theoretical curves based on the noncrossing rule taken from the work by *Lee* et al. We find that the experimental curves of the $1s$ and $2s$ excitons are below the $N = 0$ Landau level over the whole range of magnetic field studied. This supports both the theories, which predict that $1s$ and $2s$ excitons at zero field tend to the (000^+) and (001^+) bound states associated with the first Landau level at high fields. The $3s$ and $4s$ states and further higher s states, which are not shown in this figure, cross the $N = 0$ Landau level. These experimental results contradict the prediction derived from the theory based on the noncrossing rule. The crossing of the $3s$ state also violates the prediction by the theory of nodal surface conservation. The same evidence is found in TlBr [7.85], as can be seen in Fig. 7.42. The absorption peak 2' in TlCl, which is the $3d_0$ state of the exciton, seems to cross the $N = 0$ Landau level in Fig. 7.44. In a much higher field, it is found to cross definitely [7.31]. In TlBr, the same crossing is observed, as can be seen in Fig. 7.42. This crossing is in agreement with the theory by *Shinada* et al. [7.89, 90] which gives $3d_0$–(100^+) level correspondence.

The origin of the failure of the theories to explain the experimental results has not been clarified. In Figs. 7.42, 44, we find some regularity exists in the energy shift. The energy of the ns exciton higher than $n = 3$ seems to approach asymptotically the Landau-level energy of $\varepsilon_g + [(n - 2) + (1/2)]\hbar\omega_c$ at high field. This result is suggestive but again the reason for it has not been explained. It is necessary to do further experimental work in a more precise manner, because the energy splitting between mixed levels near the Landau edge will be quite small, which will cause great difficulties in tracing the experimental points as a function of magnetic field.

In Fig. 7.41, it is observed in TlBr that peaks 4 and 5 first grow and shift to higher energies with increasing magnetic field. However, when they come close to the threshold energy of the hump β, which is the one-phonon sideband of $2s$ and higher states, their intensities decrease and they finally disappear [7.31]. It can be seen in Fig. 7.42 that their energy shifts deviate from the linear relation at high magnetic fields and show a tendency to saturation. This behavior is also observed in other high-n exciton absorptions. Entirely similar features are found in TlCl [7.31]. This is the pinning of the zero-phonon exciton at the threshold of the one-phonon exciton state and is just the same as the electron-polaron pinning effect found in the interband magneto-absorption and cyclotron resonance experiments in InSb [7.95, 96]. In the case of thallous halides, it is an exciton-polaron pinning effect.

In Fig. 7.42, the energies of peaks which newly appear in the absorption and reflection spectra, with increasing magnetic field, are plotted in the energy range of the one-, two-, and three-phonon sidebands, that is β, γ, and δ in Fig. 7.8. The energy of the band gap ε_g is indicated by a filled triangle and $\varepsilon_g + n\hbar\omega_{LO}$, where n is 1, 2, and 3, by filled rhombs on the ordinate. We note that the shifts of the one-phonon peaks A, B, C, D, and E are almost identical to the shifts of the zero-phonon peaks 2, 3, 4, 5, and 6. This suggests that they are the phonon replicas of the excitons of $2s$, $3s$ and so on. However, if one extrapolates their energies to zero field, they seem to come to the same energy $\varepsilon_g + \hbar\omega_c$. In the zero phonon region, by contrast, the structures 1, 2, 3, and so on start from energies of different discrete exciton states. Thus the structures B, C, D, and so on could be the phonon sidebands of the e–h pair Landau states instead of the phonon sidebands of the discrete exciton states. Since we do not have any knowledge about the absorption line shape of the one-phonon state of the exciton in a magnetic field from which we can know the meaning of the peak maximum, the detailed analysis has been left for future study.

The pinning behavior in Fig. 7.42 is analysed by *Nakahara* and *Fujii* [7.85] by the calculation of the interband Landau transition under LO phonon coupling. They found that the effective Fröhlich coupling constant of the e–h system in TlBr is about 0.1, which is smaller than those of the free electron and hole by a factor of more than twenty. This shows that the interaction of an LO phonon with an exciton in thallous halides is very much weaker than the interaction with an electron alone because in an exciton the effective charge is small due to the charge cancellation effect discussed in Sect. 4.1.4. In InSb, the Fröhlich coupling constant deduced from the degree of pinning observed in the cyclotron resonance experiment is close to the value of the free electron, because the exciton effect is not involved in this case.

The exciton states which couple each other dominantly at the exciton polaron pinning are of different parity. The first-order term in the Fröhlich coupling in (7.26) is proportional to r. Thus the optically allowed zero-phonon s exciton will be most strongly pinned by the one–LO-phonon state of the p exciton. It is not clear at this moment whether the character of the one-phonon absorption A is $2s$ $2p_0$, or a free e–h pair. We find in Fig. 7.42 that the zero-phonon exciton absorptions penetrate little into the one-phonon region. This is in contrast to the case of InSb and is due to a strong Fröhlich coupling in thallous halides, by which the zero-phonon exciton state is unstable against the one-phonon exciton state in this energy region.

7.5.2 Electric Field

Taking the valence and the conduction band of the standard form, the effective mass equation for an exciton in an electric field $\mathscr{E}$ is reduced to

$$
\left(-\frac{\hbar^2}{2\mu} \nabla_r^2 - \frac{e^2}{\epsilon r} - e\mathscr{E} \cdot r \right) F(r)
$$

$$
= \left(E_{vK} - E_0 - \varepsilon_g - \frac{\hbar^2 K^2}{2(m_e + m_h)} \right) F(r) . \tag{7.35}
$$

$F(r)$ is the hydrogenic wave function in the absence of the external electric field. For $K = 0$, this is just the equation for a hydrogen atom in an electric field, for which there is no known analytic solution, as for the case of the hydrogen atom in a magnetic field.

An approximate solution can be obtained in the low field limit where the potential energy of the electric field is much less than the Coulomb binding energy. In this case, the electric field can be treated as a perturbation to the hydrogenic wave function. This is just the Stark effect in the hydrogen atom. As a measure of the field strength, we take $\mathscr{E}/\mathscr{E}_I$ where $\mathscr{E}_I$ is the field strength for the ionization of exciton and is equal to Ry/ea_B where Ry is the exciton effective Rydberg energy and a_B the exciton Bohr radius. When $\mathscr{E}$ is equal to $\mathscr{E}_I$, the potential energy across the exciton radius is equal to the exciton binding energy so that the exciton will be ionized by this field. The hydrogenic picture for an exciton becomes meaningless at this field strength. For the $1s$ $X_6^+ \times X_6^-$ excitons, $\mathscr{E}_I$ is 3.1×10^4 V cm^{-1} in TlCl and 2.5×10^4 V cm^{-1} in TlBr, and these magnitudes are easily accessible.

The exciton absorption in this high field has been investigated theoretically by *Blossey* [7.97]. Figure 7.45 shows the effect of an electric field on the bound exciton state. The value of $F(0)^2$, which is proportional to the absorption intensity, is plotted against the energy measured from the band-gap energy ε_g for

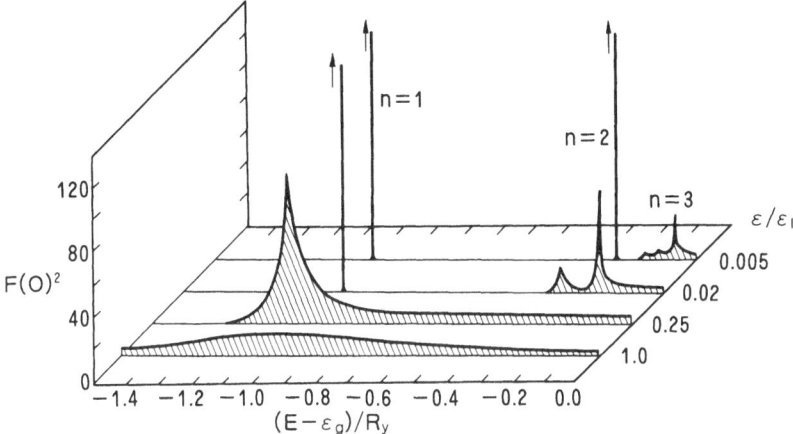

Fig. 7.45. The effect of an electric field on exciton levels for $\mathcal{E}/\mathcal{E}_\mathrm{I} = 0.005, 0.02, 0.25,$ and 1.0 [7.97]

various electric field strength $\mathcal{E}/\mathcal{E}_\mathrm{I}$. For $\mathcal{E}/\mathcal{E}_\mathrm{I} = 0.005$, the electric field has little effect on the $1s$ and $2s$ hydrogenic levels whereas the $n = 3$ and higher levels are greatly affected. The $n = 3$ level is split into three broadened levels by the first-order Stark effect, which are mixtures of the $3s$, $3p_0$, and $3d_0$ hydrogenic states. In this field, all other higher levels are smeared into a continuum. On increasing the field from $\mathcal{E}/\mathcal{E}_\mathrm{I} = 0.005$ to 0.02, the $n = 3$ level is smeared into the continuum. The $n = 2$ level splits into two levels which are mixtures of $2s$ and $2p_0$ states. Increasing the field further from 0.02 to 0.25, the $n = 2$ levels are smeared into the continuum and the $n = 1$ level is broadened. It is shifted to a lower energy by the quadratic Stark effect. For $\mathcal{E}/\mathcal{E}_\mathrm{I} = 1$, no bound exciton states are distinguishable in the spectrum and the exciton is completely ionized by the electric field. The broadening of the exciton level by the electric field is a result of the distribution of the wave function in the potential well due to the Coulomb field plus electric field. The electric field reduces the Coulomb potential well in one direction so that the change of an electron escaping from a hole is improved in this direction. This makes the level more susceptible to ionization and results in the broadening.

In investigating the effect of an electric field on excitons by an optical experiment, one always encounters a difficulty. This is space charge polarization by photoconducting carriers, which reduces the electric field inside the crystal considerably so that the true electric field strength acting on the exciton cannot be estimated. In particular thallous halides are very good photoconductors and a quantitative analysis of the experimental result is difficult.

Experimental investigations of the effect of an electric field on the $X_6^+ \times X_6^-$ exciton in thallous halides are made by means of electro-modulation spectroscopy [7.98, 99]. Figure 7.46 is the electro-transmission spectra of TlCl by *McClel-*

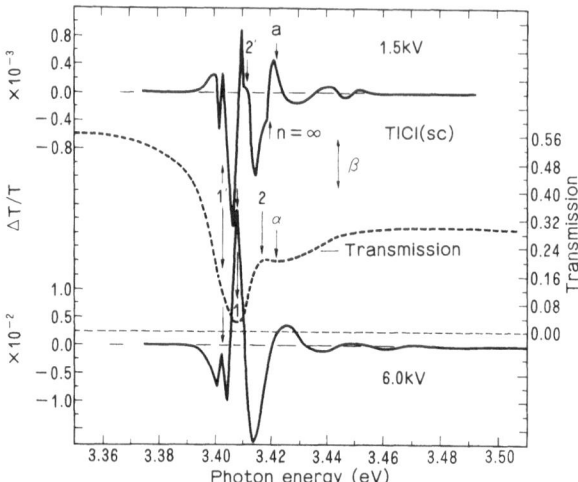

Fig. 7.46. Electron-modulated transmission spectra (——) at two applied voltages at 5–6 K and the transmission spectrum (··) of TlCl [7.99]

land et al. [7.99]. Here the electro-modulated transmissions $\Delta T/T$ are plotted for two applied fields. The energies of the 1A and 1B absorptions are indicated by 1′ and 1 and the band gap energy by $n = \infty$. From the optical transmission spectrum, which is also given in this figure, it is found that the sample is rather strained because the splitting of the $1s$ exciton absorption into 1A and 1B is not resolved. This and the uncertainty of the field strength make quantitative analysis difficult and not much information is obtainable. However, general features of the spectrum can be obtained from this figure. With increasing field, the structures due to the bound exciton states below the band-gap energy are broadened. This is what we expect from the theory. Further, the oscillation grows with the field in the energy range of the exciton continuum. This is the Franz-Keldysh type oscillation which is expected in this energy region from the theory [7.97].

The effect of an electric field on an exciton is a fundamental and classic problem just like the effect of a magnetic field on an exciton. However, there is a difference between them from the experimental point of new. In the case of a magnetic field, the exciton wave function shrinks and the binding energy increases with the field. This gives a larger absorption intensity and a larger level splitting of the exciton which make experimental detection easier at higher fields. However, in the case of an electric field, the energy level of the exciton is broadened by the field. This makes the experimental detection and analysis more difficult at higher fields. A detailed line-shape analysis of the broad absorption band will be important in this case and a precise and quantitative study is needed. Experimental investigations on the exciton in electric fields which are quantitative and precise enough for theoretical analysis are quite rare. Thallous halides are nice substances to study from this point of view because the exciton states are sharp and are understood in some detail and also $\mathscr{E}_\mathrm{I}$ are small which makes the study of high field effects possible in comparatively weak applied fields.

7.5.3 Uniaxial Stress Field

When an uniaxial stress is applied to a crystal, changes of symmetry and lattice parameters are brought about. These then change the electronic states through deformation-potential coupling, and lead to mixing, splitting and shifts in energy of the levels.

In thallous halides, the effect of uniaxial stress has been studied mainly for the $1s$ state of the $X_6^+ \times X_6^-$ exciton by optical measurements. Figure 7.47 is the change of the ϵ_2 spectrum of the $X_6^+ \times X_6^-$ exciton in TlCl with the application of stress P in the [001] direction. This is the result by *Mohler* et al. [7.18] who used incident light of two polarizations, $E \parallel P$ and $E \perp P$, where E is the electric field of the light. At zero stress, the two peaks 1A and 1B, which are the absorptions of the $(\Psi_z^x + \Psi_z^y)/\sqrt{2}$ and Ψ_z^z excitons, are seen similarly to those in Fig. 7.7. They are isotropic. By applying the stress in the z direction and measuring the spectrum with light polarized in z, that is $E \parallel P$, it is found that the oscillator strength is transferred from 1B to 1A and they are both shifted to lower energies by the stress. For light with $E \perp P$, 1B remains almost unchanged whereas 1A decreases in intensity as if it is transferred to the absorption C which appears and grows between 1A and 1B with increasing stress.

The matrix elements of the z-polarized $X_6^+ \times X_6^-$ exciton (which is active to z-polarized light) under stress P applied in the z direction are given in Table 7.5a. We omit the spin triplet Ψ_0^z because a stress does not mix different spin states.

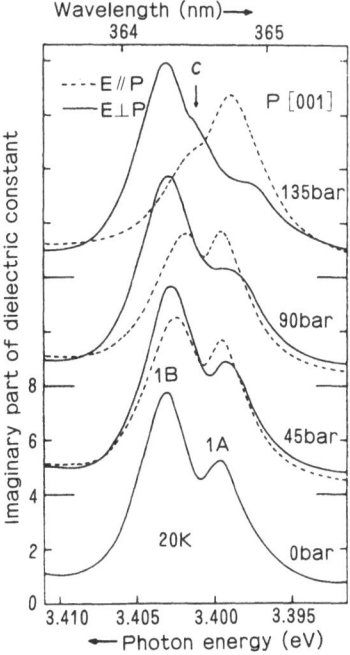

Fig. 7.47. Imaginary part of the dielectric function of uniaxially stressed TlCl at 20 K. The ordinate scale is shifted for different pressures. Here P is the stress and E the electric field of light [7.18]

Table 7.5. Matrix elements for excitons under uniaxial stress applied in the z direction

a) $P \parallel z$, z-polarized states

	Ψ_z^z	$(\Psi_z^x + \Psi_z^y)/\sqrt{2}$	$(\Psi_z^x - \Psi_z^y)/\sqrt{2}$
Ψ_z^z	$2\Delta_1 + \delta + \gamma$	$2\sqrt{2}\Delta_3$	
$(\Psi_z^x + \Psi_z^y)/\sqrt{2}$	$2\sqrt{2}\Delta_3$	$\Delta_2 + 2\Delta_4 - \frac{\delta}{2} + \gamma$	
$(\Psi_z^x - \Psi_z^y)/\sqrt{2}$			$\Delta_2 - 2\Delta_4 - \frac{\delta}{2} + \gamma$

b) $P \parallel z$, x-polarized states

	Ψ_x^x	$(\Psi_x^y + \Psi_x^z)/\sqrt{2}$	$(\Psi_x^y - \Psi_x^z)/\sqrt{2}$
Ψ_x^x	$2\Delta_1 - \frac{\delta}{2} + \gamma$	$2\sqrt{2}\Delta_3$	
$(\Psi_x^y + \Psi_x^z)/\sqrt{2}$	$2\sqrt{2}\Delta_3$	$\Delta_2 + 2\Delta_4 + \frac{\delta}{4} + \gamma$	$-\frac{3}{4}\delta$
$(\Psi_x^y - \Psi_x^z)/\sqrt{2}$		$-\frac{3}{4}\delta$	$\Delta_2 - 2\Delta_4 + \frac{\delta}{4} + \gamma$

Here γ is the hydrostatic deformation energy shift and is $D_1^1(s_{11} + 2s_{12})P/\sqrt{3}$ where D and s are the deformation potential and the elastic compliance constant. The energy shift for tetragonal deformation δ is $D_1^3\sqrt{2}(s_{11} - s_{12})P/\sqrt{3}$. We find that the stress induces energy shifts of the zero-stress states. Thus the absorptions 1A and 1B shift as observed in Fig. 7.47. The stress-induced exchange of the oscillator strength between 1A and 1B is due to the nondiagonal matrix element $2\sqrt{2}\Delta_3$.

In the case of $E \perp P$, the matrix elements are given in Table 7.5b. Here we take the polarization of light, and so the polarization of the exciton, in the x direction. We find that, other than the energy shifts of the dipole-allowed Ψ_x^x and $(\Psi_x^y + \Psi_x^z)/\sqrt{2}$ excitons, mixing of the $(\Psi_x^y + \Psi_x^z)/\sqrt{2}$ exciton to the dipole-forbidden $(\Psi_x^y - \Psi_x^z)/\sqrt{2}$ exciton is induced by the stress. This results in the appearance of the absorption of the $(\Psi_x^y - \Psi_x^z)/\sqrt{2}$ exciton at the expense of the absorption of the $(\Psi_x^y + \Psi_x^z)/\sqrt{2}$ exciton. This is what we see in Fig. 7.47 as the growth of the new absorption C and the decrease of the absorption 1A.

Quantitative study of the effect of stress on the $X_6^+ \times X_6^-$ exciton has been extended by *Fujita* et al. [7.101] by means of a piezoreflectance measurement and a line-shape analysis. They found that $\delta/\gamma = 1.7 \pm 0.3$, $D_1^3 = 1.6 \pm 0.3$ eV, and $D_1^1 = 4.16$ eV, for TlCl and 0.5 ± 0.3 eV, 0.4 ± 0.2 eV, and 4.05 eV, respectively, for TlBr.

8. Photocarrier Motion in Ionic Crystals

When an ionic crystal is excited by light at the energy of a discrete exciton state, excitons are formed. Since the total charge of an exciton is zero, it does not carry charges as an electric current in an externally applied electric field. If the energy of the incident light is higher than the ionization threshold energy of the exciton, equal numbers of free electrons and holes can be formed as a result of the ionization of excitons. The excited electrons and holes finally annihilate and the crystal returns to its ground state. Before the annihilation, electrons and holes decay to the band minima within a time of 10^{-12}–10^{-13} s and slow electrons and slow holes are formed. Since they are in an unbound state, they drift independently in an electric field and form electric currents. They are called photocarriers. In a crystal in which a self-trapped state is the lowest-energy state of the photocarriers, they are self-trapped and become immobile in a short time of the order of 10^{-12} s. In this case, little current is carried by photocarriers within their lifetimes and the material is nonphotoconductive with regard to the respective carriers.

In the absence of a stable self-trapped state for photocarriers, they move freely for a comparatively long time and the material is photoconductive. The lifetime of the photocarriers is determined by their capture by traps, decay to an exciton state and direct recombination of an electron and hole. Trapping is the most important factor for the determination of the carrier lifetime and so the intensity of the photocurrent at low temperatures, and thus a large photoresponse is expected in highly pure and perfect crystals. As we saw in the preceding chapters, high purity samples of ionic crystals are available and transport phenomena of photocarriers have been studied quite extensively.

Free electrons and free holes in ionic crystals are in polaron states. As was discussed in Sect. 4.1.2, a polaron is a quasi particle consisting of a free carrier and an associated phonon cloud; a result of the electron-phonon interaction. In the usual case, the associated phonons are longitudinal optical (LO) phonons and the polarons in ionic crystals having inversion symmetry, such as alkali, silver, and thallous halides, are of this type. In a crystal without inversion symmetry, such as cadmium sulfide, another interaction contributes to the polaron. It is an interaction with acoustic phonons through their piezoelectric field and the resultant state is called a piezoelectric polaron.

In the investigation of polarons, the galvanomagnetic effect and cyclotron resonance have been studied most extensively because they are related to the carrier mass and scattering. In these cases, the strength of the externally applied electric field is important because of the nature of the polaron band which levels off at a one-phonon energy from the bottom of the polaron band. The transport properties of cold polarons, the polarons at the bottom of the band, have been well explored by low field mobility and cyclotron resonance experiments as well as by theories. Polarons in a strong electric field whose kinetic energies are high in the polaron band are called hot polarons and their particular properties have become apparent.

In this chapter, the experimental method, which differs from the usual method and is essential for the investigation of photocarrier transport phenomena in ionic crystals, is discussed first. Secondly, experiments of spectral dependence, mobility, and cyclotron resonance are presented for the low field case and they are discussed with the aid of polaron transport theories. Other related phenomena such as the spin-dependent current and photo-magneto-current are also given. The problem of hot polarons and the importance of the polaron band are discussed in the last part.

Alkali, silver and thallous halides are taken as the substances on which to base our discussion because they are the substances which have been investigated most extensively from the standpoint of polaron transport. In addition, cadmium sulfide is presented as an example of piezoelectric scattering.

8.1 Photocurrent and Measurement

8.1.1 Photocurrent

In semiconductors, there are freely moving electrons which carry an electric current. If the semiconductor is in thermal equilibrium, the number of carriers per unit volume n is determined by the temperature. The current density J in a uniform electric field E is $ne\mu E$ where μ is the carrier mobility and e the electronic charge.

In insulating materials such as ionic crystals, there are no carriers in thermal equilibrium. However if they are placed in a field of radiation whose energy is high enough to create an electron-hole (e–h) pair, carriers are formed. They decay through various channels, among which trapping is the most important decay process. If the trap depth is deep enough at a given temperature, the carriers become immobilized by the trapping. Thus in most cases in ionic crystals, the trapping lifetime τ_t determines the photocarrier lifetime τ with which the photocarriers decay as $\exp(-t/\tau)$.

When a crystal is illuminated by light continuously, the number of carriers in a steady state is $n_0\tau$ where n_0 is the number of carrier excited in unit time and unit volume. The steady-state current density J is then

$$J = n_0 \tau e \mu_d E .\tag{8.1}$$

Here μ_d is a drift mobility and is equal to the distance drifted by a carrier in a unit electric field divided by the time required. In some cases, particularly when shallow traps are present, the drift mobility is less than the microscopic mobility μ which is the mobility determined solely by scatterings in the band. Using a schubweg w which is equal to $\mu_d \tau E$, the current density is written as $n_0 e w$. The schubweg is sometimes called the range and the schubweg in a unit field $\mu_d \tau$ is called the unit range.

8.1.2 Blocking Electrode Method and Response

In the measurement of electric conduction in semiconductors and metals, electrodes are placed on both ends of the sample to apply a voltage and to inject and draw a current. In this case, ohmic contacts between the sample and the electrodes are required to maintain a charge neutrality inside the sample. If ohmic contact is not established, a space charge is developed and the electric field is no longer uniform inside the crystal. In an extreme case, the electric field is concentrated at the contacts and no field is present in almost the entire crystal. In this situation, any quantitative measurements are impossible.

In ionic crystals, it is always found that ohmic contact is not established for photo-excited carriers and a buildup of space charge by the photocurrent is unavoidable. To overcome this difficulty, a blocking electrode method has been used. It is believed that only by this method can experimental data quantitative enough for analysis be obtained.

In the blocking electrode method, two electrodes are completely blocked from a sample by inserting thin insulating materials between the sample and the electrodes, thus establishing a well-defined condition for the contact. The number of photocarriers generated by a light pulse is kept low enough to avoid a measurable contribution of polarization due to carrier drift in the externally applied field.

By using blocking electrodes, the electric current or motion of charges is measured by the time dependent charge induced on the electrodes. Fig. 8.1 shows the principle of the measurement.

Suppose that a crystal is sandwiched between two electrodes A and B both of which are completely insulated from the crystal so that A and B just play the role of capacitor plates between which a dielectric is inserted. First the switch S is closed and an electric field E develops in the crystal whose strength is V/l when the thickness of the insulating film is small enough. After S is opened, the crystal is illuminated through the transparent electrode B by a short light pulse whose energy is chosen in such a way that the light is absorbed only at the surface of the crystal. The electrons and holes created at the surface will drift along the electric field E but in this case only electrons can drift into the crystal because of the polarity of V. If the total charge of the photo-excited electrons is $-q$, the charge Q induced on the electrode after moving the electrons with a drift velocity v a distance d in time t is qd/l. The potential drop ΔV between the two terminals of

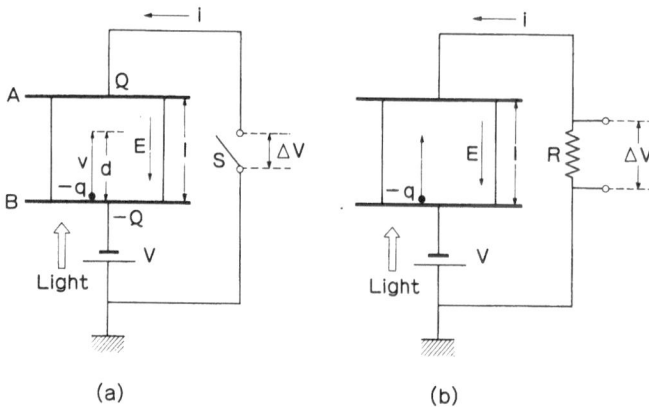

Fig. 8.1a, b. Blocking electrodes, external circuits and charge drifts

(a)   (b)

the switch is then $-qd/lC$ where C is the capacitance between the electrodes when the crystal is in the dark. We note that this argument is correct only when $\Delta V \ll V$, that is, the space charge polarization due to the carrier drift is negligible compared with the externally applied field. When the electrons reach the opposite side of the crystal at time t_t, their drifts cease because their motions are blocked by the blocking film on electrode A. The magnitude of ΔV becomes a constant value of $-q/C$ after t_t. In Fig. 8.2a, ΔV is shown as a function of time. Here the crystal is illuminated by a light pulse with infinitesimal width at time zero.

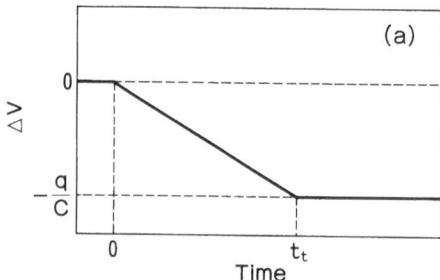

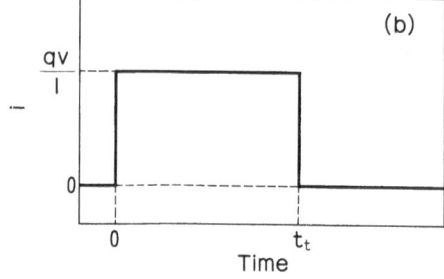

Fig. 8.2a, b. Response observed by the blocking electrode method. (a) switch S in Fig. 8.1 is opened. (b) S is closed

If the crystal is illuminated with the switch S closed in Fig. 8.1a, there is no potential drop across the switch because it is shorted. In this case, a current i flows in the outer circuit following the drift of the charge $-q$ inside of the crystal. This amount is qv/l as long as the electrons are drifting in the crystal and becomes zero after t_t. This is seen in Fig. 8.2b.

In the actual measurement, the switch is replaced by a resistance R as shown in Fig. 8.1b and the potential drop ΔV across R is measured. If the value of R is large enough so that $CR \gg t_t$, the waveform of ΔV is quite similar to that of Fig. 8.2a. For the case of small R where $CR \ll t_t$, ΔV is almost equal to qvR/l and the waveform is close to that of Fig. 8.2b.

In the above discussion, we assume that the absorption coefficient α and the carrier life time τ are almost infinite. When τ is finite, the number of carriers, and so the magnitude of q, decays during the drift. This gives a rounding off of the rise of the ΔV curve before t_t and a reduction of its saturated value after t_t. For the curve of i, this gives an exponential decay of i with time. The effect of the finite α on q is through the spatial distribution of carriers at $t = 0$. The saturated value of ΔV at t_t is reduced from q/C, because of the distribution, since it is proportional to the distance which is traversed by the carriers. The spatial distribution causes another contribution to Q. Since electrons and holes are both excited by light, holes can now drift in the crystal along the electric field in the reverse direction to the direction of light in Fig. 8.1. This gives a contribution to q or ΔV from the hole drift in the reverse direction, in addition to the contribution from the electron drift in the forward direction. The charge induced after t_t by the forward motion, Q_f, for finite τ and α is [8.1]

$$Q_f = e\eta N_0(1 - R) \frac{w}{l} \left(1 - \frac{1}{1 - \alpha w} e^{-\alpha l} + \frac{\alpha w}{1 - \alpha w} e^{-l/w}\right), \tag{8.2}$$

and that induced by the reverse motion, Q_r, is

$$Q_r = e\eta N_0(1 - R) \frac{w}{l} \left\{\frac{1}{1 + \alpha w} + \frac{\alpha w}{1 + \alpha w} \exp\left[-\left(\alpha + \frac{1}{w}\right)l\right] - e^{\alpha l}\right\}. \tag{8.3}$$

Here η is the quantum yield of the photocarrier, N_0 the total number of incoming photons at the surface, and R the reflectivity. In these equations, the schubweg w is used rather than τ by using the relations $x = \mu_d E t$ and $w = \mu_d \tau E$ which give $n = n_0 \exp(-t/T) = n_0 \exp(-x/w)$. In the geometry of Fig. 8.1, w is taken as the electron schubweg w_e in (8.2) and as the hole schubweg w_h in (8.3). The total induced charge Q is the sum of Q_f and Q_r. For more detailed expressions for Q and i, the reader may refer to [8.2]. We note that, since the photoresponse given here is by a single pulse of light, it is a transient photoconductivity and not the steady-state photoconductivity which is usually measured in semiconductors.

8.1.3 Spectral Dependence of Photoconductivity

It can be seen from (8.2) and (8.3) that the photoresponse Q/eN_0 depends on the absorption coefficient α. Furthermore, if the crystal has a surface layer in which the schubweg is smaller than that in the bulk because of a large number of traps due to disorder in the surface layer, the spectral response is greatly modified at the wavelength of high absorption coefficient.

Let us call the polarity of the electric field given in Fig. 8.1a, with respect to the incoming light direction, an electron polarity, and the opposite polarity of the electric field, a hole polarity. If only electrons are mobile and holes are not, the one-carrier photoresponse in electron polarity is Q_f of (8.2). In the hole polarity, the photoresponse by electron drift is Q_r of (8.3). In the case of two-carrier conduction where electrons and holes are both mobile with their schubwegs w_e and w_h, respectively, the photoresponse in the electron polarity is the sum of Q_f of electrons and Q_r of holes. In the hole polarity it is the sum of Q_f of holes and Q_r of electrons. Thus by measuring the spectral photoresponses in the electron and hole polarities, one can distinguish between one-carrier and two-carrier photoconductions. The schubwegs w_e and w_h can also be estimated from the responses in both the polarities in one- [8.1] and in two-carrier [8.3] cases.

Figure 8.3 is the spectral photoresponse measured by *van Heyningen* and *Brown* [8.1] in AgCl whose surfaces had been carefully treated and thus the crystal was free from the surface effect. It is known that, in AgCl, electrons are mobile whereas holes are immobile due to self-trapping and thus AgCl is an example of a one-carrier case. The absorption rises at about 387 µm with an indirect-exciton excitation, at this temperature, and has a peak at about 245 µm

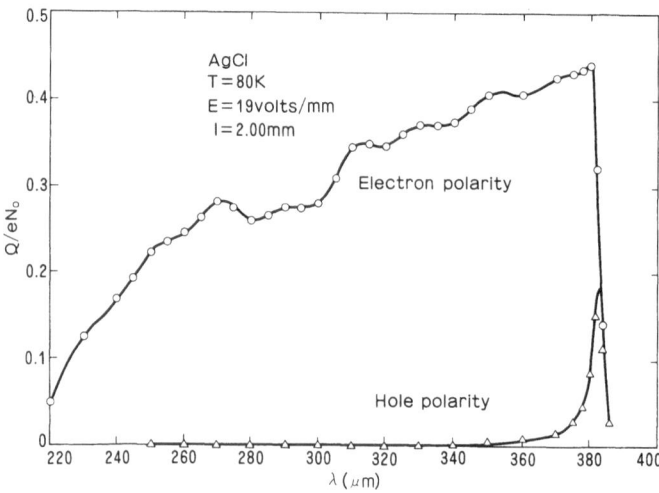

Fig. 8.3. Wavelength dependences of photoresponses in electron polarity (*circles*) and in hole polarity (*triangles*) at 80 K in AgCl in which only electrons are mobile [8.1]

with a direct-exciton excitation with $\alpha \sim 10^5$ cm^{-1}. In the electron polarity, the normalized photoresponse Q/eN_0 rises at the absorption edge because of the increase of the total number of excited electrons in the crystal. With excitation at high energy where $\alpha l \gg 1$ is satisfied, the number of excited electrons is almost independent of α. In this energy region,the photoresponse becomes flat against the excitation energy, particularly when $w_e \ll l$. This behavior is given by (8.2) and Q/eN_0 in the electron polarity in Fig. 8.3 is just what is expected. The gradual decrease with energy increase is believed to be due to $\eta(1 - R)$. The response in the hole polarity in Fig. 8.3 is explained entirely in terms of an electron drift toward the illuminated anode expressed by Q_r/eN of (8.3).

Figure 8.4 is the spectral photoresponse of a two-carrier case, observed in TlCl, where the open circles are the photoresponses in electron polarity and the filled ones are those in hole polarity [8.3]. The absorption begins at 384 μm with the onset of the indirect-exciton transition and peaks at 364 μm with the direct-exciton absorption. We find that the spectra in both the polarities are quite similar to the electron polarity spectrum in Fig. 8.3 in the indirect-absorption region because of strong contributions of forward responses of electrons and of holes in the electron and hole polarities, respectively. At higher energy than the direct-exciton absorption, where α is of the order of 10^6 cm^{-1}, the electron polarity response is practically absent in this particular sample because of the presence of defects in a surface layer a few hundred angstroms thick which are effective for trapping at this temperature which therefore give a very small w_e. The traps are not effective for holes at this temperature and Q/eN_0 in the hole polarity is still high due to hole drift.

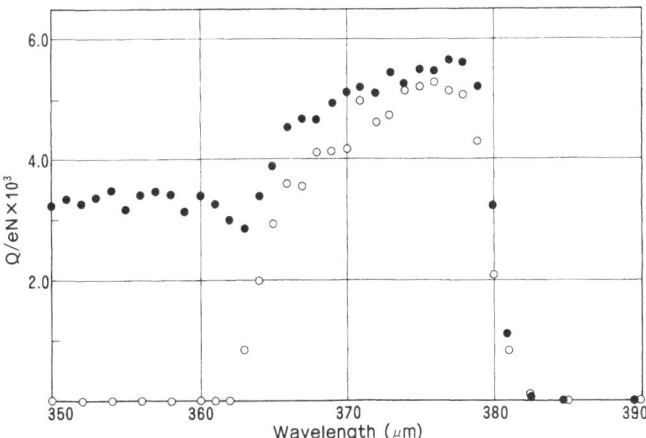

Fig. 8.4. Wavelength dependences of photoresponses in electron polarity (*open circles*) and in hole polarity (*filled circles*) at 17 K in TlCl in which both electrons and holes are mobile [8.3]

8.2 Measurements of Carrier Mobility and Cyclotron Resonance in Insulating Photoconductors

Among the parameters which determine the photoconductivity $\sigma = ne\mu$, the most interesting one is the mobility μ because it reflects the intrinsic nature of a crystal such as carrier mass and scattering probability, whereas n is a function of externally controllable parameters such as the exciting light intensity and the time taken for carriers to be trapped by defects.

Three methods are available for a direct measurement of the mobility of photocarriers in insulating material using the blocking electrode method. One is a time of flight method and the other two are based on galvanomagnetic effects, that is, the Hall effect and the magnetoconductive effect. The former method is easier to perform than the latter two, however, in some conditions the result of the former method is influenced by the effect of temporary trapping by shallow traps and thus it is usually used for measurements above liquid nitrogen temperature. The mobilities obtained by the latter two methods are not influenced by the temporary trapping because a galvanomagnetic effect is the effect of a magnetic field on only the carriers in motion in a band and not those in traps. The mobility at low temperatures has been investigated mainly by the latter two methods.

8.2.1 Carrier Mobility

Before beginning the discussion on the experimental method for the measurement of the mobility of photocarriers, we will discuss the motion of carriers in an electric field E and a magnetic field H from which the mobility is determined.

Let us assume that the band is isotropic and single valley and that the electronic energy ε is expressed by $\hbar^2 k^2/2\, m$. We assume also that the scattering relaxation time τ_s for carrier can be defined and its relation to ε is

$$\tau_s = \tau_c \left(\frac{\varepsilon}{k_B T}\right)^p = \tau_c x^p . \tag{8.4}$$

Here p is a scattering index and takes the value $-1/2$ for deformation potential scattering, 0 for scattering by polar LO phonons and by neutral impurities, $1/2$ for scattering by the piezoelectric field due to acoustic phonons, and $3/2$ for ionized-impurity scattering. When the electric field and the magnetic field are sufficiently small so that the carrier motion is linear in E and $\omega_c \tau_s \ll 1$ is satisfied, where ω_c is the cyclotron angular frequency eH/mc, and further, the carrier distribution is expressed by the Boltzmann distribution function $\exp(-x)$, the current density J is expressed from the Boltzmann equation as [8.4],

$$J = \sigma(H)E + \alpha(H)E \times H + \gamma(H)(H \cdot E)H . \tag{8.5}$$

Here

$$\sigma(H) = \frac{ne^2}{m} \left\langle \frac{\tau}{1 + \omega_c^2\tau^2} \right\rangle , \tag{8.6}$$

$$a(H) = \frac{ne^2}{m} \frac{\omega_c}{H} \left\langle \frac{\tau^2}{1 + \omega_c^2\tau^2} \right\rangle , \tag{8.7}$$

$$\gamma(H) = \frac{ne^2}{m} \frac{\omega_c^2}{H^2} \left\langle \frac{\tau^3}{1 + \omega_c^2\tau^2} \right\rangle . \tag{8.8}$$

The brackets indicate the average over the energy distribution of electrons of the form

$$\langle g \rangle = \frac{1}{\Gamma(5/2)} \int_0^\infty g x^{3/2} e^{-x} dx .$$

where Γ is the gamma function.

In the absence of a magnetic field, (8.5) reduces to $J = \sigma(0)E = ne^2 \langle \tau_s \rangle E/m$ where J is in the direction of E and the mobility μ, which is equal to J/neE, is

$$\mu = \frac{e}{m} \langle \tau_s \rangle . \tag{8.9}$$

This mobility is determined from the motion of carriers in an electric field alone and is called a drift mobility or a conductivity mobility.

In the presence of a magnetic field in the z direction with an electric field in the x direction, the current density has a component in y as well as in x because of the Lorentz force. Then $J_x = \sigma(H)E$, $J_y = -a(H)HE$, and $J_z = 0$. The Hall angle J_y/J_x is

$$\frac{J_y}{J_x} = -a(H)H/\sigma(H) , \quad \text{where} \tag{8.10}$$

$$\frac{a(H)}{\sigma(H)} = \frac{e}{mc} \left(\left\langle \frac{\tau_s^2}{1 + \omega_c^2\tau_s^2} \right\rangle \Big/ \left\langle \frac{\tau_s}{1 + \omega_c^2\tau_s^2} \right\rangle \right) . \tag{8.11}$$

In a vanishingly small magnetic field

$$\frac{a(0)}{\sigma(0)} = \frac{e}{mc} \frac{\langle \tau_s^2 \rangle}{\langle \tau_s \rangle} = \frac{1}{c}\mu_H(0) . \tag{8.12}$$

The mobility $\mu_H(0)$ has the same dimensions as that of (8.9) and is called a Hall mobility at zero field. In general, the Hall mobility μ_H is defined as

$$\mu_H = \frac{ca(H)}{\sigma(H)} . \tag{8.13}$$

The Hall angle $\tan\theta = J_y/J_x$ is

$$\frac{J_y}{J_x} = -\frac{\mu_H H}{c} \,.$$
(8.14)

The magnetoresistance mobility μ_M is related to the magnetoresistance $M(H)$ which is

$$M(H) = \frac{\varrho(H) - \varrho(0)}{\varrho(0)} = \frac{J_E(0)}{J_E(H)} - 1 \,,$$
(8.15)

where J_E is the transient current density in the direction of the applied electric field. The ratio $J_E(H)/J_E(0)$ is called a magnetoconductivity in the transient condition where there is no Hall field resulting from polarization due to carrier drift in the direction perpendicular to E.

The magnetoconductive effect in the transient condition has been analysed by *Tippins* and *Brown* [8.5]. When E is applied as $(E_x, 0, 0)$ and H as $(H \sin\theta, 0, H \cos\theta)$, the magnetoconductivity $J_x(H)/J_x(0)$ is given from (8.5) as [8.5]

$$\frac{J_x(H)}{J_x(0)} = \frac{\sigma(H) + \gamma(H)H^2 \sin^2\theta}{\sigma(0)} \,.$$
(8.16)

In the transverse case where $\theta = 0$, it reduces to

$$\left(\frac{J_x(H)}{J_x(0)}\right)_{\mathrm{T}} = \frac{\sigma(H)}{\sigma(0)} \,,$$
(8.17)

and for the longitudinal case where $\theta = 90°$, it is

$$\left(\frac{J_x(H)}{J_x(0)}\right)_{\mathrm{L}} = 1 \,.$$
(8.18)

Here $\sigma(H) + \gamma(H)H^2 = \sigma(0)$ is used to get (8.18). The transverse magnetoresistance $M_{\mathrm{T}}(H)$ is then

$$M_{\mathrm{T}}(H) = \frac{\sigma(0)}{\sigma(H)} - 1 = \frac{\gamma(H)}{\sigma(H)}H^2 \,.$$
(8.19)

We define the magnetoresistance mobility μ_M as

$$\mu_M = c \left(\frac{\gamma(H)}{\sigma(H)}\right)^{1/2} \,,$$
(8.20)

which reduces to $(e/m)(\langle\tau^3\rangle/\langle\tau\rangle)^{1/2}$ in a vanishingly small magnetic field so that it has the dimensions of a mobility. The transverse magnetoresistance is written as

$$M_{\mathrm{T}}(H) = \left(\frac{\mu_M H}{c}\right)^2 \,.$$
(8.21)

8.2.2 Drift Mobility Measurement

For the measurement of the drift mobility (8.9), a time of flight method is commonly used, which is just the method given in Fig. 8.1b. When the resistance R is large enough to satisfy $CR \gg t_t$, the wave form of the photoresponse after the illumination of a short pulse of strongly absorbed light at $t = 0$ is that shown in Fig. 8.2a. From this wave form, the transit time t_t of carriers across the crystal thickness l is determined. The drift mobility μ_d is then calculated from $\mu_d = l^2/t_t V$. In the case of $CR \ll t_t$, which is more frequently used than the above case, the wave form shown in Fig. 8.2b is observed and t_t and μ_d are determined. An important condition in this method is that the carrier lifetime τ must be longer than t_t or the schubweg must be longer than l. This condition is achieved only by using a pure and perfect sample. We note that by the time of flight method it is possible to measure the electron drift mobility and the hole drift mobility separately in two-carrier photoconductive materials by using light of high absorption coefficient and by changing the polarity of the applied electric field.

8.2.3 Hall and Magnetoresistance Mobility Measurements

In the blocking electrode method a transient current is measured, and so the Hall angle and magnetoresistance, from which a Hall and magnetoresistance mobility are deduced, must be measured by a somewhat different method from the ordinary steady-state method usually applied to semiconductors and metals.

Let us examine first the ordinary method which can be applied only to the case in which an ohmic contact is established between the electrodes and a crystal. Figure 8.5 illustrates this case. When a voltage is applied between two electrodes, an electric field E is developed in the x direction and a current I flows,

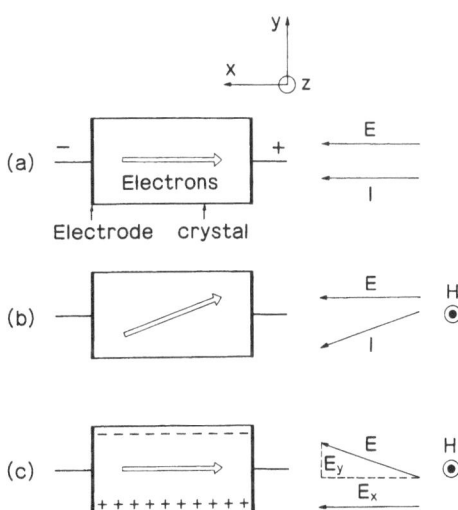

Fig. 8.5a–c. Electric field, electric current, magnetic field, and space charge in an ordinary Hall effect measurement

as shown in (a), if there are mobile electrons in a crystal. If the crystal is isotropic, I is always parallel to E and no space charge builds up in any part of the crystal because of the ohmic contact. With the application of the magnetic field H in the z direction the current rotates due to the Lorentz force as in (b). As time elapses, space charge builds up at both the surfaces of the crystal perpendicular to y due to the y component of I. Finally the current flows again in the x direction with the combined field E of the externally applied field E_x in x and the newly developed space charge field E_y in y as shown in (c). The time required to reach the steady state (c) is very short and is less than a nano second in common semiconductors. By measuring E_y, the Hall field, and E_x, the externally applied field, the angle E_x/E_y can be obtained. From (8.5), the angle is the Hall angle and is

$$\frac{E_y}{E_x} = \frac{\alpha(H)}{\sigma(H)} H = \frac{\mu_\mathrm{H} H}{c} .$$ (8.22)

Thus μ_H is calculated from E_y/E_x. The transverse magnetoresistance measured in the steady state (c) is

$$M_\mathrm{T}(H) = \frac{\sigma(0)}{\sigma(H) + [\alpha(H)^2/\sigma(H)]H^2} - 1 .$$ (8.23)

When $\omega_c \tau \ll 1$ in low magnetic fields, M_T reduces to $\omega_c^2(\langle \tau^3 \rangle \langle \tau \rangle - \langle \tau^2 \rangle^2)/\langle \tau \rangle^2$.

If the ohmic contact between the electrodes and a crystal is not established, as in the case of insulating photoconductors, the situation is different from what we saw in Fig. 8.5. Because of the nonohmic contact, a space charge builds up in the x direction in (a) and E_x is reduced from the magnitude estimated from the applied voltage. The Hall field E_y is also reduced and when the contact is completely blocking, E_x and E_y will become zero in the steady state. Further, another difficulty exists because of a high resistance of insulating photoconductors. The time required to build up the Hall field is CR and is of the order of several tens of seconds or more in a typical case. This presents a difficulty for Hall and magnetoresistance measurements because of the instability of high impedance instruments. For these reasons, a quantitative measurement of a galvanomagnetic effect in an insulating photoconductor cannot be made by the ordinary steady-state method.

The only method in which the above difficulties are overcome is the transient method with blocking electrodes discussed in Sect. 8.1.2. In this method a small number of carriers excited by a light pulse is used to avoid the effect of space charge polarization. In the measurement of a transient Hall angle, a transient Hall current, which is the y component of I in (b), becomes the subject of the measurement instead of a steady-state Hall voltage. In the case of a magnetoresistance, the x component of a transient I in (b) instead of the total current in the steady state (c) is measured.

The measurement of Hall angle by the transient current with blocking electrodes was initiated by *Redfield* [8.6] and it was modified by the application of a fast pulse technique by *Brown* and *Kobayashi* [8.7, 8]. Later a considerable

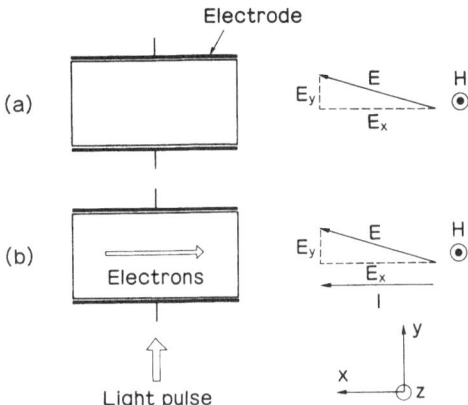

Fig. 8.6a, b. Electric field, electric current, and magnetic field in the Redfield method

improvement in sensitivity was achieved by *Borders* and *Hodby* [8.9] by using elaborate circuitry.

Figure 8.6 shows the principle of the Redfield method. A crystal is placed between two blocking electrodes and, by some means, an electric field E is applied to the crystal in the x-y plane, as shown in (a), so that its direction can be rotated in the plane. A magnetic field H is applied in the z direction. Then the crystal is illuminated by a light pulse to excite photocarriers. The blocking electrodes will respond to the y component of the transient current I in the manner described in Sect. 8.1.2. If the direction of E is adjusted by rotation in the x-y plane in such a way that no transient current is detected in the y direction, so that I is exactly in the x direction as is shown in (b), E_y/E_x is just the Hall angle and the Hall mobility is obtained from $\mu_H = (c/H)(E_y/E_x)$.

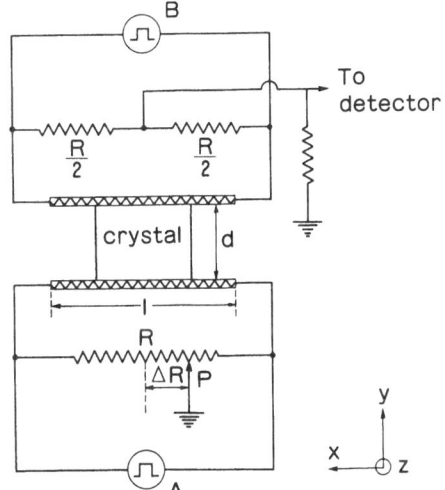

Fig. 8.7. External circuit of the Redfield method

The key to this method is how to apply and rotate E in the x-y plane. This problem was solved by *Redfield* [8.6]. Figure 8.7 is the basic circuit of the Redfield method after modification for the convenience of practical use [8.7, 8, 10]. The upper and lower electrodes are the resistance plates which are transparent to the exciting light. When the crystal is sufficiently small compared to the electrodes, an electric field V/l in the x direction is produced in the crystal by applying synchronized voltage pulses of the same height V, from the pulse generators A and B, to the lower and the upper electrodes, with the potentiometer terminal P at the center of the uniform resistance R. When P is displaced from the center position of R by ΔR, a uniform electric field $\Delta V/d$ in the y direction is developed where $\Delta V = V\Delta R/R$. In this way, the electric field E can be rotated from the x direction by θ where $\tan \theta = E_y/E_x$, $E_x = V/l$, and $E_y = \Delta V/d$. Thus by changing ΔR, the electric field E can be rotated in any direction in the crystal. If no signal due to the transient current is observed by the detector at a certain θ when the crystal is illuminated by a light pulse at a fundamental absorption, then the Hall mobility is calculated from this E_y/E_x.

The measurement of magnetoresistance by the blocking electrode method is essentially the same as that of the zero-field photoresponse discussed in Sect. 8.1.2. The only difference is the application of a magnetic field. The magnetoresistance is obtained from the photoresponse Q at zero field and at H. When $\alpha w \ll 1$, which is the usual case at low temperatures, Q is proportional to $e\eta N_0 w$ as is known from (8.2) and (8.3). Since w is equal to $\mu\tau E$, Q is proportional to $e\eta N_0\mu E$ or the current density J, if τ is independent of H. Thus the magnetoresistance $M(H)$ is $[Q(0)/Q(H)] - 1$ from (8.15). It is noted that the magnetoresistance measured by the blocking electrode method is the transient magnetoresistance (8.19) and not the steady-state magnetoresistance (8.23).

8.2.4 Detection of Cyclotron Resonance

The blocking electrode method with the detection of a transient photocurrent has been applied to a cyclotron resonance measurement of photocarriers in insulating materials. It was done by *Mikkor* et al. [8.11] and is called an electron heating or cross modulation cyclotron resonance technique. It was improved by *Borders* and *Hodby* [8.9] by using elaborate circuitry to achieve an exceedingly high sensitivity.

A dc electric field E is applied to a crystal in the z direction by blocking electrodes and the variable static magnetic field H is applied in the same direction. A microwave field of frequency ω is applied in the x-y plane and the transient photocurrent excited by the illumination of absorbing light is measured as a function of the magnetic field. If the microwave field is not present, no change of photocurrent is observed with a variation of the magnetic field because the transient longitudinal magnetoresistance is zero for a spherical energy band [8.5]. In the presence of the microwave field, the carriers absorb microwave power and so their average temperature rises as the cyclotron resonance condition, $\omega = eH/m^*c$, is approached by changing the magnetic field [8.12]. As a result, the mobility changes due to a hot electron effect, which will be discussed in Sect. 8.5, leading to a change of the transient photocurrent at the resonance. From a magnetic field at a resonance peak in a photocurrent spectrum against magnetic field, a polaron mass m^* can be estimated. This method is very powerful for cyclotron resonance experiments in insulating photoconductors where it is often difficult to apply the usual cyclotron resonance technique.

8.3 Polaron and Mobility

8.3.1 Polaron Masses and Coupling Constants

To measure a cyclotron resonance for the determination of a polaron mass, the condition $\omega_c\tau(= \mu H/c) > 1$ must be satisfied. If a resonance occurs at several kilo-oersted, which corresponds to a microwave with a wavelength of several millimeters and a effective mass ratio of carriers of near unity, the carrier mobility must be higher than $10^4 \, \mathrm{cm^2 \, V^{-1} \, s^{-1}}$ in order to get a measurable signal in a cyclotron resonance experiment. Mobilities of photocarriers of this order or even higher have been observed and cyclotron resonance experiments have been performed successfully on highly purified ionic crystals. However, because of the high mobility of carriers, the dc and microwave electric fields must be kept sufficiently low in an experiment to avoid the hot polaron effect in which the

Table 8.1. Polaron masses m^*, Fröhlich coupling constants α, and bare band masses m in ionic crystals. Here m_0 is the free electron mass. The electron heating method and the ordinary method of cyclotron resonance experiment are denoted by H and O, respectively

substance	carrier	polaron mass			α	m/m_0
		m^*/m_0	method	ref.		
KCl	e	0.922 ± 0.04	H	[8.13]	3.5	0.43
KBr	e	0.700 ± 0.03	H	[8.13]	3.1	0.37
KI	e	0.536 ± 0.03	H	[8.14]	2.5	0.33
RbCl	e	1.03 ± 0.10	H	[8.13]	3.8	0.43
RbI	e	0.72 ± 0.07	H	[8.13]	3.2	0.37
AgCl	e	0.43 ± 0.02	H, O	[8.13], [8.15]	1.86	0.30
AgBr	e	0.288 ± 0.005	H, O	[8.13], [8.15]	1.60	0.215
AgBr	h($\parallel$ 111)	1.71 ± 0.06	O	[8.16]	2.8	1.25
	h($\perp$ 111)	0.79 ± 0.01	O	[8.16]		0.52
TlCl	e	0.56 ± 0.02	H, O	[8.17], [8.18]	2.5	0.35
TlCl	h($\parallel$ 100)	0.58 ± 0.03	H	[8.17]		
	h($\perp$ 100)	0.98 ± 0.04	H	[8.17]		
TlBr	e	0.52 ± 0.03	H, O	[8.17], [8.18]	2.7	0.31
TlBr	h($\parallel$ 100)	0.55 ± 0.03	H	[8.17]		
	h($\perp$ 100)	0.74 ± 0.03	H	[8.17]		

measured cyclotron mass will differ from the slow polaron mass due to the nature of the polaron band, which will be discussed in Sect. 8.5.1.

In Table 8.1, the polaron masses m^*, the Fröhlich coupling constants α and the bare band masses m of several ionic crystals are given. The polaron masses are obtained from cyclotron resonance experiments by the electron heating method or by the ordinary method. The electron masses of all alkali halides and silver halides are at the Γ point and those of thallous halides at the R point. The anisotropic hole masses in silver bromide are those at the L point of the fcc Brillouin zone and in thallous halides at the X point of the sc Brillouin zone. The coupling constant and the bare band mass are calculated from the polaron mass by the following expression given by *Langreth* [8.19]:

$$m^* = (1 - 0.0008\alpha^2)/[1 - (\alpha/6) + 0.0034\alpha^2] \tag{8.24}$$

and the coupling constant given by the definition (4.28). We find that the coupling constants in all the materials listed in the table are larger than unity and the couplings are intermediate. The coupling is strongest in alkali halides, weakest in the silver halides and the thallous halides are in between.

8.3.2 Polaron Mobilities

The drift mobility of an electron in AgCl measured by the time of flight method [8.20] is shown as a function of temperature in Fig. 8.8. The measured mobilities on curve 1 are the intrinsic drift mobilities and are reproducible from sample to sample in most cases. Sometimes the measured mobility is different from curve 1 at low temperatures; the mobilities on curve 2 are examples of this. It shows that

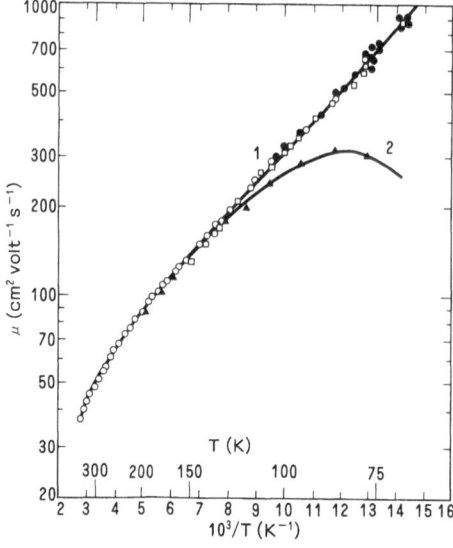

Fig. 8.8. Electron drift mobility in AgCl as a function of reciprocal temperature [8.20]

the mobility coincides with the intrinsic mobility at high temperatures but as the temperature decreases, it becomes smaller than the intrinsic value and bends downwards. The amount of the decrease depends on the sample and is structure sensitive.

A multiple trapping of electrons by shallow traps in a crystal is responsible for the decrease of the drift mobility at low temperature. In the time of flight method, a drift mobility is computed from the measured transit time t_t of a sheet of electrons moving in an applied electric field. If there are shallow traps in the crystal, the electrons will spend a fraction of their time in the conduction band and the rest of it in shallow traps. This gives a longer time for t_t than the intrinsic t_t which is due only to a drift of electrons in the band. Thus the presence of shallow traps will cause a decrease of mobility. This effect is stronger at lower temperatures because of the longer time the electrons spend in shallow traps.

Assume that the conduction electrons are distributed classically between the states near the bottom of the conduction band and the states in shallow traps of depth E and uniform concentration N_t. Let τ_t be the mean trapping time for electrons in the conduction band and τ_g the average time an electron remains in a shallow trap. The apparent drift mobility μ_d is then equal to $\mu[1 + (\tau_g/\tau_t)]^{-1}$ where μ is the intrinsic drift mobility determined by the electron scattering in the band. The ratio τ_g/τ_t can be estimated by assuming that the rate of trapping is equal to the rate of release from the traps. The drift mobility is then [8.21]

$$\mu_d = \mu[1 + (g_1/g_0)(N_t/N_c)\exp(E/k_BT)]^{-1} , \tag{8.25}$$

where N_c is the effective density of states near to bottom of the conduction band and g_1 and g_0 are the statistical weights of the full and empty traps, respectively. Curve 2 in Fig. 8.8 is the theoretical curve for μ_d with E being 0.02 eV and N_t being 5.7×10^{14} cm^{-3}. It represents the experimental results very well.

In thallous halides, photo-excited electrons and holes are both mobile and the electron mobility and the hole mobility can be measured separately by the time of flight method. Figure 8.9 is the result. We find that the multiple trapping effect is effective in determining both the electron mobility and the hole mobility in some crystals [8.22, 23]. It is found that the temperature dependences of the electron mobility and the hole mobility are very different from each other at high temperatures. The reason will be discussed in a later part of this section.

Figure 8.10 shows the Hall mobility of photoelectrons in AgBr in a small electric field measured by *Burnham* et al. [8.4]. It is observed that the mobility rises exponentially with a decrease of temperature in the relatively high temperature region and saturates below 15 K. This temperature dependency is analysed by the combination of three electron scattering processes; by optical phonons, acoustic phonons, and impurities. For the optical phonon scattering, the scattering time τ_0 must be proportional to $\exp(\Theta/T) - 1$ by the polaron mobility theory which will be discussed later. The scattering time τ_a for scattering by acoustic phonons through a deformation potential coupling must be proportional to $T^{-3/2}$. The scattering time τ_i for scattering by ionized impurities depends on tempera-

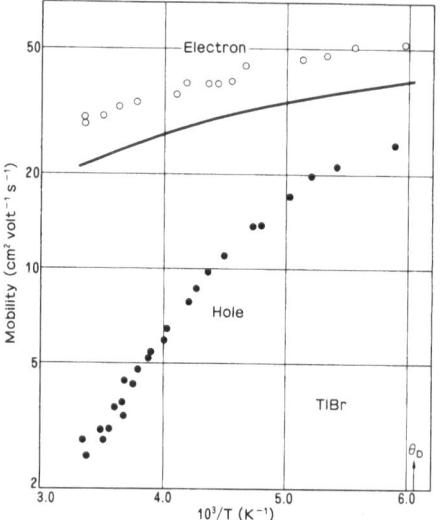

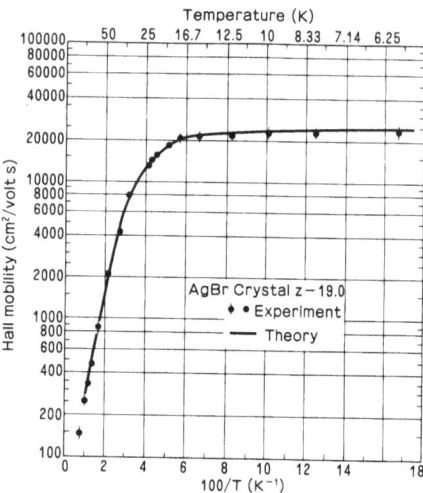

Fig. 8.9. Electron and hole drift mobilities in TlBr as functions of reciprocal temperature. Here θ_D is the Debye temperature of the LO phonon [8.22]

Fig. 8.10. Hall mobility of electrons in AgBr as a function of reciprocal temperature [8.4]

ture as $AT^{3/2}/[B \ln (1 + CT^2)]$. The Hall mobility $\mu_H = (e/m)(\langle \tau_s^2 \rangle / \langle \tau_s \rangle)$, where $1/\tau_s = (1/\tau_0) + (1/\tau_a) + (1/\tau_i)$, is drawn in Fig. 8.10 and it represents fairly well the experimental values. In the relatively high-temperature region, where the mobility increases steeply with lowering temperature, the mobility is determined mainly by the optical phonon scattering. In the intermediate temperature range between 30 and 10 K, the acoustic phonon scattering contributes mostly. At 6 K, the impurity scattering is the dominant process in determining the mobility. It is noted that Θ determined experimentally from the higher-temperature mobilities in Fig. 8.10 is 195 K which is just the Debye temperature of the LO phonon in AgBr. The steep exponential rise of the low field mobility with lowering temperature in this relatively high-temperature region has been found in many other ionic crystals such as AgCl [8.8, 24], TlCl [8.25], TlBr [8.26], KCl [8.27, 28], KBr [8.10], KI [8.10, 28], NaCl [8.10, 28], Rb halides [8.28], Cs halides [8.28], alkaline-earth fluorides [8.29], and CdS [8.30]. In many cases, Θ is close to the Debye temperature of the LO phonon of the respective material.

In the case of two-carrier photoconduction, where electrons and holes are both mobile, the transient photoresponse Q due to the motions of carriers in the y direction due to the Hall effect is the sum of the photoresponses due to the electron motion and the hole motion in the geometry of Fig. 8.7. If it is assumed that the schubwegs of both carriers are much smaller than the crystal dimensions, the Hall mobility to be measured from the Hall angle by this transient method is [8.31]

$$\mu_H(\text{meas.}) = \mu_{He}\left(1 - \left|\frac{\mu_{Hh}}{\mu_{He}}\right| \left|\frac{w_{0h}}{w_{0e}}\right|\right)\left(1 + \frac{w_{0h}}{w_{0e}}\right)^{-1}. \tag{8.26}$$

Here μ_{He} and μ_{Hh} are the Hall mobilities (8.13) of electrons and holes and w_{0e} and w_{0h} are the unit ranges $\mu_d T$ of electrons and holes, respectively. When $w_{0e} \gg w_{0h}$, $\mu_H(\text{meas.})$ becomes μ_{He} and in the opposite case $-\mu_{Hh}$. The sign of the measured mobility or the Hall angle in a two-carrier case depends strongly on the relative magnitudes of the unit ranges of electron and hole. Since the unit range is proportional to the carrier lifetime τ, the measured Hall mobility will change its sign at a temperature where the lifetime of one type of carrier becomes short due to trapping by deep traps which becomes effective at this temperature.

The Hall mobility estimated from the measured Hall angle in the case of two-carrier photoconduction in TlCl is shown in Fig. 8.11 [8.25]. Open circles are the mobilities corresponding to a negative carrier sign and the filled circles to a positive carrier sign. The Hall angle, and so the Hall mobility, change sign at about 30 K. In this figure, the drift mobilities of electrons measured by the time of flight technique [8.23] are shown by crosses. Good agreement between the

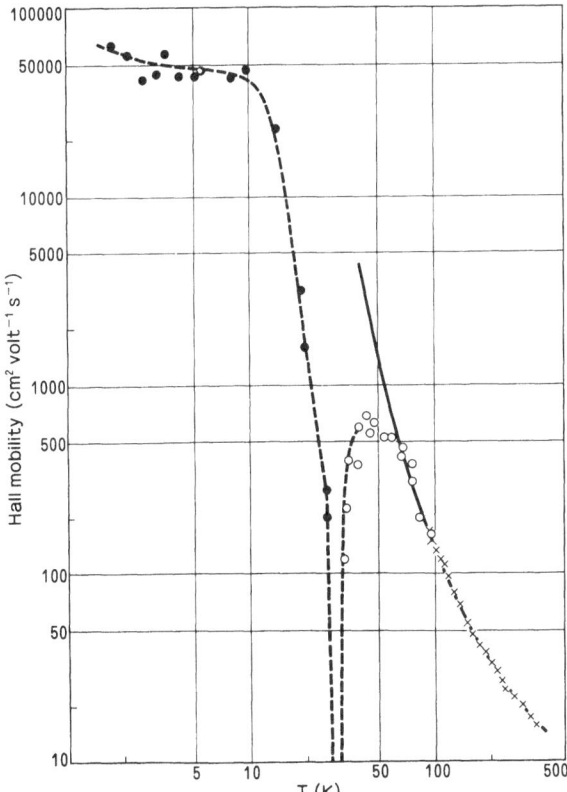

Fig. 8.11. Hall mobility as a function of reciprocal temperature in TlCl. The mobilities of negative and positive carrier signs are distinguished by open circles and filled circles, respectively. The drift mobilities measured by the time of flight method are plotted by crosses [8.25]

Hall mobility of negative carrier sign and the electron drift mobility around 100 K shows that the measured Hall mobility is almost purely an electron mobility and $w_{0e} \gg w_{0h}$ is satisfied at this temperature. As the temperature decreases, electron trapping becomes effective, leading to a sharp drop in the mobility of negative carrier sign because of a large decrease of w_{0e} compared with w_{0h}. Finally, w_{0e} becomes comparable to w_{0h} at around 30 K where the transition from the electron sign to the hole sign in the Hall effect is observed in this particular sample. Below this temperature, the sign of the Hall angle is always of positive carrier type and the mobility increases sharply and saturates below 10 K at about $5 \times 10^4 \text{ cm}^2 \text{V}^{-1} \text{s}^{-1}$ where $w_{0e} \ll w_{0h}$ is realized. The mobility above 70 K in Fig. 8.11 is essentially a one-carrier mobility due to electrons and below 10 K due to holes.

The exponential increases of the mobilities with a decrease of temperature with a characteristic temperature equal to the Debye temperature of the LO phonon, which are found in many ionic crystals at rather high temperatures, are an indication of the importance of the electron–LO-phonon coupling in this temperature range. Since the lowest mobile state of an electron–LO-phonon system in ionic crystals is a large polaron state, the analysis of the mobility must be made by the polaron mobility theory.

A number of theories on the mobility of slow polarons have been developed. At low temperatures, well below the Debye temperature of the LO phonon, and for a relatively small coupling constant, the polaron mobility μ_p of slow electrons may be written as

$$\mu_p = \frac{e}{2am^* \omega \overline{N}} f(\alpha) , \tag{8.27}$$

where α is the Fröhlich coupling constant, m^* the polaron mass, and ω the frequency of the LO phonon. The average number of LO phonons $\overline{N}$ is equal to $[\exp(\hbar\omega/k_B T) - 1]^{-1}$ and $f(\alpha)$ is a slowly varying function of α given by Low and Pines [8.32]. The mobility expression (8.27) was given by Langreth [8.33] who derived it from the mobility formula given by Low and Pines [8.32] with some correction. It can be applied perhaps to the case of intermediate coupling where α is not greater than 6. Equation (8.27) shows that the mobility is proportional to $1/\overline{N}$, that is, the polaron mean free time is inversely proportional to the number of LO phonons present. At low temperatures and in small electric fields, the mean free time of the slow polaron is determined mainly by the phonon absorption process, since the phonon emission process of a slow polaron will occur only after the absorption of LO phonon by resonance scattering. At low temperatures, the phonon absorption process is very much slower than the phonon emission process so that the former determines the scattering rate of the polarons. At low enough temperatures, $1/\overline{N}$ becomes $\exp(\hbar\omega/k_B T)$ or $\exp(\Theta/T)$ where Θ is $\hbar\omega/k_B$, the Debye temperature of the LO phonon, and the mobility is proportional to it. This temperature dependence is just what we see in Figs. 8.8, 10, 11 and in many other ionic crystals.

The polaron mobility theory at finite temperature has been worked out by *Osaka* [8.34] who uses the Boltzmann equation of the Feynman polaron of *Kadanoff* [8.35] and by *Langreth* [8.19] who uses the Kadanoff-Baym equation. These theories are expected to be applicable in the temperature range $T < \Theta/2$. It is found that these theories are not good enough to describe the experimental data near and above the temperature $\Theta/2$.

These theories are based on the assumption that the relaxation time can be defined for the scattering of polarons, but this is not always clear at high temperatures, particularly when the temperature approaches Θ. To overcome this difficulty, *Thornber* and *Feynman* [8.36] calculate the expectation value of the velocity of electrons, which are interacting with LO phonons in a given electric field, by the path integral method of *Feynman* [8.37]. The mobility obtained is believed to be applicable without the limitations of low field and low temperature. Their expression is in a rather complicated form and the reader should refer their original paper for details.

If the polaron mass and the LO phonon energy are known, the mobility can be calculated from the theories without any adjustable parameters. Since in many ionic crystals the polaron masses are accurately known from cyclotron resonance experiments and the LO phonon energies from neutron and Raman scattering experiments or from the TO phonon energies with the Lyddane-Sachs-Teller relation, a comparison between the mobility theories and the experiments can be made in a quantitative sense. Figure 8.12 is an example which is given by *Brown*

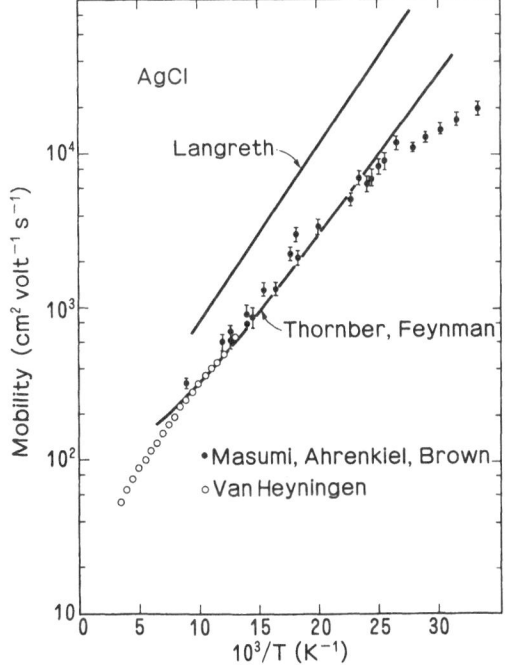

Fig. 8.12. Comparison between experimental and theoretical mobilities for AgCl. The measured Hall mobilities and the drift mobilities are shown by filled and open circles. The theoretical curves shown by the solid lines are calculated for $m/m_0 = 0.303$ and $\Theta = 284$ K [8.38]

[8.38] for AgCl. The Hall mobilities measured by *Masumi* et al. [8.24] and the drift mobilities from *van Heyningen* [8.20] are plotted in the figure. Two theoretical curves from the expressions of *Langreth* [8.19] and of *Thornber* and *Feynman* [8.36] are shown. It is found that the latter curve represents the experimental results fairly well. The deviation at low temperatures is due to the contributions of acoustic-phonon scattering and impurity scattering which become dominant at these temperatures. At high temperatures the theory is not good enough to reproduce the experiment. Appreciable deviation begins at about 130 K which is almost half the Debye temperature of the LO phonon (283 K). The Langreth value is found to be higher by a factor of 3–4 but its temperature dependency is as good as that of Thornber and Feynman's value. These tendencies of the mobility and its temperature dependence are found in KCl and KBr [8.38]. It seems that Thornber and Feynman's expression describes the mobility of the polaron of intermediate coupling better than other theories in the temperature range below $\Theta/2$.

In the high-temperature region, the mobility decreases more steeply with increasing temperature than the finite temperature polaron theory, as is seen in Fig. 8.12. At present, there are no polaron theories in the continuum approximation which are good enough to describe the measured high-temperature mobility of polarons of intermediate coupling. This deviation has been found in electron mobilities in silver halides [8.38], in thallous halides [8.22, 23] and particularly in alkali halides [8.28] in the strongest fashion. Figure 8.13 shows the high temperature electron mobilities in cesium halides. The broken lines are the mobilities calculated by the weak coupling theory of *Howarth* and *Sondheimer* [8.39] which

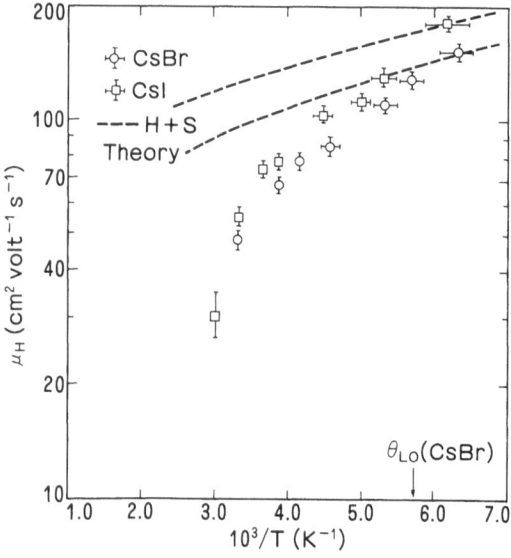

Fig. 8.13. Temperature dependences of the Hall mobilities of electrons in cesium halides at high temperatures. The Debye temperature of the LO phonon is indicated by θ_{LO}. The temperature dependences of mobilities predicted by the theory of Howarth and Sondheimer are given by broken lines [8.28]

is not strictly a polaron mobility theory but is a solution of the Boltzmann equation for bare band electrons with emission and absorption of LO phonons and is applicable at high temperatures. Similar sharp decreases in mobility with temperature are also found in KCl, KBr, and rubidium halides [8.28]. An explanation of this decrease is suggested by *Seager* and *Emin* [8.28] in terms of a band narrowing effect originally given by *Holstein* [8.40] for small polarons in a discrete lattice. The band narrowing with a temperature increase will make the carrier mass temperature dependent, leading to a smaller mobility than Howarth and Sondheimer's mobility at high temperatures. They also suggested that the band narrowing effect will become less pronounced as the polaron size is increased from that of a small polaron. The stronger decrease of electron mobility in alkali halides than in silver and thallous halides may be explained by this effect because of the smaller polaron radii of the former.

A strong decrease in mobility with rising temperature has been also found in the high-temperature hole mobility in TlBr [8.22] which is shown in Fig. 8.9. Similarly to the case of alkali halides, the electron mobility in TlBr at high temperatures is lower than that expected from the usual polaron mobility theory, but the difference is not as large as that found in alkali halides. The decrease of the hole mobility is very much larger than that of the electron mobility. The solid line in Fig. 8.9 is the expected free-polaron mobility of holes in TlBr which is estimated from the measured electron mobility by considering the ratio of am^* of electron to hole in (8.27) which is 0.74, when the hole polaron mass used is the average of the anisotropic hole polaron masses. It is noted that the measured hole mobility is only one tenth of the estimated mobility at the highest temperature. It is unlikely that the strong decrease of hole mobility with temperature is due to band narrowing as for the electron polaron in alkali halides. This is because the Fröhlich coupling constant of holes in TlBr is 2.9, which is estimated from the mean value of the anisotropic hole-polaron mass, and the radius of the Feynman polaron of the hole is almost ten times larger than the lattice spacing in TlBr. Thus the small polaron picture, and so the band narrowing concept, will not explain the strong decrease of the hole mobility. Also the large difference in the temperature dependence between the electron mobility and the hole mobility cannot be understood from the band narrowing theory because the coupling constant of holes in TlBr is only a little larger than that of electrons which is 2.7.

The strong decrease of hole polaron mobility at high temperatures has been explained by *Toyozawa* and *Sumi* [8.41]. They have suggested that there is a metastable self-trapped state of holes whose energy is higher by ε than that of a mobile large polaron state in TlBr, corresponding to the F(S) region of Fig. 4.21 (see also Sect. 4.4.7). The adiabatic potential energy curve for this case is shown in Fig. 4.1b. At low temperatures most of the electrons are in the mobile large-polaron state. As the temperature rises, the number of electrons immobilized by trapping in the metastable self-trapped states will increase. The fraction $f(T)$ of mobile free holes is

$$f(T) = [1 + \gamma A T^{-3/2} \exp(-\varepsilon/k_{\mathrm{B}}T)]^{-1} , \qquad (8.28)$$

where $A = (2\pi\hbar^2/m^*k_B)^{3/2}/v_0$, $\gamma = (g'/g)(v/v')$, and m^* is the hole-polaron mass. The volume of the unit cell is v_0 and the degeneracies of the free and self-trapped states are g and g'. The ratio (v/v') represents the entropy factor coming from the frequency changes of lattice vibrations around the self-trapped hole. The drift mobility of holes to be measured will then be $f(T)\mu$, where μ is the mobility of large polarons shown by the solid line in Fig. 8.9. The calculated mobility is found to fit to the measured mobility fairly nicely with $\varepsilon = 138$ meV. Thus the drastic decrease of the high-temperature hole mobility in TlBr is due to the distribution in the metastable self-trapped states.

The steep decrease of hole mobility with an increase of temperature at high temperatures has also been found in AgBr [8.42, 43], in a somewhat narrower temperature range, and again this decrease is explained by the mechanism of a metastable self-trapped state. Even the strong decrease of the electron mobility in alkali halides with a temperature increase at high temperatures, explained by the band narrowing mechanism, could be interpreted by this mechanism. We believe that it is quite important to investigate high-temperature electron transport phenomena in an ionic crystal, although an experimental difficulty exists due to ionic conduction, because they will give information on the localization of electrons in a highly excited phonon field.

In the case of a small coupling constant, the mobility of electrons under an interaction with LO phonons can be analysed by the theory of *Howarth* and *Sondheimer* [8.39]. Cadmium sulfide is a weak coupling material whose Fröhlich coupling constant is about 0.4. Its crystal structure is wurtzite with the c-axis perpendicular to the hexagonal plane. Figure 8.14 gives the low field Hall and magnetoresistance mobilities of photoelectrons in insulating CdS of high purity

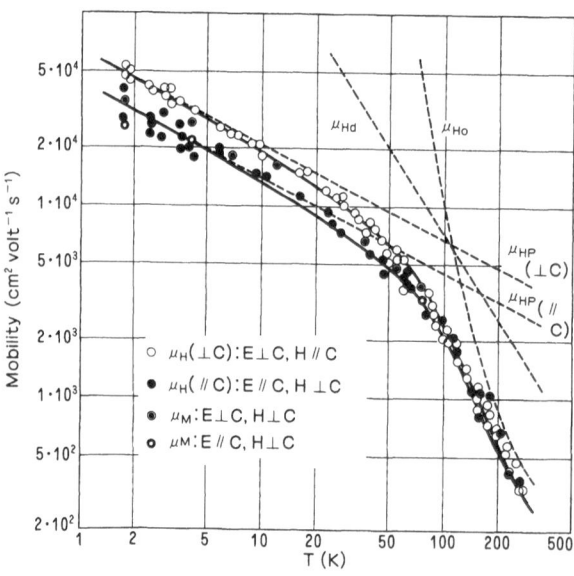

Fig. 8.14. Electron Hall mobilities in CdS. The open circles are $\mu_H (\perp c)$ and the filled circles are $\mu_H (\|c)$. The full lines are theoretical mobilities calculated by the combination of optical phonon scattering μ_{Ho}, deformation potential scattering μ_{Hd}, and piezoelectric scattering μ_{Hp}. Magnetoresistance mobilities are also given both for $\perp c$ and $\|c$ [8.44]

measured by the blocking electrode method [8.44, 45]. Since CdS is an aniso-tropic material, two Hall mobilities are given. One is the mobility in the direction perpendicular to c, $\mu_H(\perp c)$, which can be measured by $E \perp c$, $H \parallel c$, and the other is the mobility in the direction parallel to c, $\mu_H(\parallel c)$, measured with $E \parallel c$, $H \perp c$. Two magnetoresistance mobilities for $E \perp c$ and $E \parallel c$ are also shown. The mobilities in both the cases change exponentially with temperature T above 150 K and are proportional to $T^{-1/2}$ below 25 K. The magnitudes of both the mobilities are essentially the same above 50 K whereas $\mu_H(\perp c)$ is always higher than $\mu_H(\parallel c)$ at low temperatures.

Since CdS has no inversion symmetry and is piezoelectric, electrons are scat-tered by the piezoelectric field associated with the acoustic lattice vibration as well as by the electric field of the LO lattice vibration and by the deformation potential of the acoustic lattice vibration. The broken lines are the corresponding three mobilities. The first one is the mobility determined by the LO phonon scattering, μ_{H0}, and is calculated by the *Howarth* and *Sondheimer* expression which is proportional to $\exp(\Theta/T) - 1$. The second one is the piezoelectric scat-tering mobility μ_{Hp} which is given by *Hutson* [8.46] as $A(m_0/m)^{3/2}$ $(300/T)^{1/2}$ cm^2 V^{-1} s^{-1} where A is a constant, which is anisotropic and determined by the piezoelectric and dielectric constants, and m/m_0 is the effective mass ratio of the electron. The two μ_{Hp} in Fig. 8.14 are the anisotropic piezoelectric scatter-ing mobilities. The third broken line is the mobility determined by the deforma-tion-potential scattering μ_{Hd}. The solid lines in the figure are the theoretical mobilities μ_H calculated by the combination of three mobilities, $(1/\mu_H) = (1/\mu_{H0}) + (1/\mu_{Hp}) + (1/\mu_{Hd})$. They represent the experimental values quite well in a very wide temperature range. It is found that LO phonon scattering is dominant at high temperatures and piezoelectric scattering at low temperatures. The tem-perature Θ for the curve of μ_{H0} is 440 K which is just the Debye temperature of the LO phonon in CdS. It is noted that a simple LO phonon scattering theory can describe the high temperature mobility in CdS quite well, in contrast with alkali halides which are materials of large coupling constants.

8.4 Magnetoconductivity

8.4.1 Spin-Dependent Magnetoconductivity

According to (8.18), the longitudinal magnetoconductivity $J_x(H)/J_x(O)$ mea-sured in the transient condition in a low magnetic field is unity for the electrons in an isotropic energy band. Experimentally this has been confirmed in some ionic crystals such as AgBr [8.5].

In alkali halides containing F centers, it has been found by *Hodby* et al. [8.47] that the electron longitudinal magnetoconductivity increases from unity at zero magnetic field with increasing magnetic field, although the conduction band is spherical in these materials. Figure 8.15 shows how the longitudinal magnetocon-

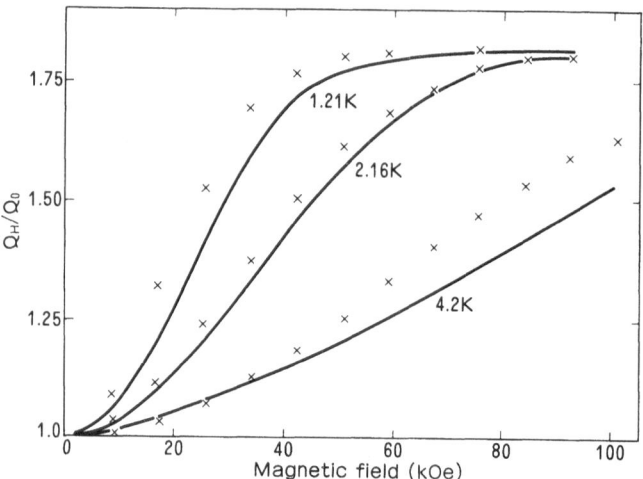

Fig. 8.15. Longitudinal magnetoconductivity of electrons in KCl doped with 10^{16} F centers cm^{-3} at three different temperatures. The full curves are experimental results and crosses are theoretical ones. The electric field is 87 Vcm^{-1} [8.47]

ductivity $J(H)/J(0)$, which is equal to the transient photoresponse ratio $Q(H)/Q(0)$, in KCl containing 10^{16} F centers cm^{-3} increases with the magnetic field and how it depends on temperature.

This behavior is explained in terms of a spin orientation effect of photoelectrons and F centers [8.47]. An F center is an electron trapped at a negative-ion vacancy and is paramagnetic with a spin of 1/2 while a conduction electron also has a spin of 1/2. A cross section of scattering of an electron by an F center with parallel spins, S_t, which is the triplet scattering cross section, is different from that with antiparallel spins, S_s, which is the singlet scattering cross section. By applying a magnetic field, the numbers of up-spins and down-spins of electrons and of F centers become different and this results in a change in the number of singlet and triplet scatterings, leading to a change of current density or to a magnetoconductivity effect.

Let us define a spin polarization P by

$$P = (n\uparrow - n\downarrow)/(n\uparrow + n\downarrow) \ . \tag{8.29}$$

Here $n\uparrow$ and $n\downarrow$ are the numbers of electrons or F centers of spin-up and spin-down, respectively. The spin polarization of the electrons is denoted by P_E and that of the F centers by P_F. According to Boltzmann statistics, the polarization in thermal equilibrium at temperature T and in magnetic field H is given by

$$P_i = \tanh\left(g_i \beta H/2k_B T\right) \ , \tag{8.30}$$

where β is the Bohr magneton. The g values for the conduction electron g_E and for the F center g_F are both 2.0. In a specimen containing N_F F centers and N_0 nonmagnetic scattering centers per cm^3 with nonmagnetic scattering cross section S_0, *Hodby* et al. [8.47] give the total scattering cross section S after *Honig* [8.48] as

$$S = N_F[\tfrac{1}{4}(1 - P_E P_F)S_s + \tfrac{1}{4}(3 + P_E P_F)S_t] + N_0 S_0 . \qquad (8.31)$$

The total cross section is a function of H and T through P_E and P_F by (8.30). In zero magnetic field where $P_E = P_F = 0$, $S(0)$ is $N_F(S_s + 3S_t)/4 + N_0 S_0$ and in a high enough field, where both the conduction electrons and the F centers are fully polarized so that $P_E = P_F = 1$, $S_{\text{full}}(H)$ is $N_F S_t + N_0 S_0$. If the scattering mechanism is independent of magnetic field, the longitudinal magnetoconductivity is expected to change with the magnetic field as

$$\frac{J(H)}{J(0)} = \frac{Q(H)}{Q(0)} = \frac{S(0)}{S(H)} . \qquad (8.32)$$

In the case of Fig. 8.15, photoelectrons are excited from F centers. If the reorientation of the spin of an electron during a photoexcitation and in its lifetime is negligible, and perhaps this is the case in KCl [8.49], (8.32) can be calculated by taking $P_E = P_F$ for electrons excited from F centers. The results of the theoretical calculations together with experimental values for three temperatures are shown in Fig. 8.15. The agreement between experiment and theory is good and shows that spin-dependent scattering operates in alkali halides containing paramagnetic F centers.

We note that the above discussion is based on the change of the scattering of electrons by F centers but it is not the only explanation. A change of the cross section for the trapping of paramagnetic electrons by paramagnetic F centers will also explain the experimental result of Fig. 8.15 by the same means as employed in the scattering. Similar spin-dependent magnetoconductivities have been found in KBr and KI doped with F centers [8.47].

Spin resonance experiments on F centers can be made by the measurement of longitudinal magnetoconductivity in the presence of microwave radiation. Figure 8.16 is the longitudinal magnetoconductivity $Q(H)$ of KCl doped with 5×10^{15} F centers under 138 GHz radiation at 4.2 K plotted against magnetic

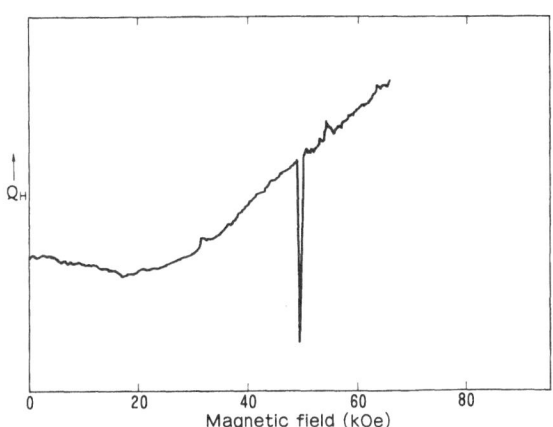

Fig. 8.16. Longitudinal magnetoconductivity of electrons in KCl doped with 5×10^{15} F centers cm^{-3} at 4.2 K in the presence of microwave power at 138 GHz. The electric fields is 11.1 Vcm^{-1}. The spin resonance effect of F centers is seen near 50 kOe [8.47]

field [8.47]. At about 50 kOe, a sharp drop in $Q(H)$ is found. At this field, a full saturation of the spin resonance of F centers, where the polarization of the F centers becomes zero with the resonant microwave field, is expected. This gives a reduction of $Q(H)$ to $Q(0)$ at this field and a resonance in $Q(H)$ as is seen in the figure. It proves that electron spin resonance experiments on paramagnetic centers in insulating photoconductors can be made by measuring a transient longitudinal magnetoconductivity by the blocking electrode method.

8.4.2 Photomagnetocurrent

When an insulating photoconductor is illuminated by light for which it has a high absorption coefficient, carriers are generated in a very narrow region near the surface with a large gradient of carrier density. Due to this gradient, carriers will diffuse into the crystal in the direction of the gradient without an application of an electric field. This effect was studied by *Makita* and *Kobayashi* [8.50] in externally applied magnetic and electric fields.

A crystal is sandwiched between blocking electrodes and the coordinates are taken as shown in Fig. 8.17. A y-z surface of the crystal is illuminated by light in a fundamental absorption region where the absorption coefficient is 10^6 cm^{-1}. As a result, electrons and holes are generated within a layer a few hundred angstroms thick near the surface. When there are no externally applied magnetic and electric fields, electrons and holes diffuse into the crystal along the concentration gradient which is in the x direction as shown in (a) in the figure. The diffusion currents for electrons J_e^D and for holes J_h^D are

$$J_e^D(x) = eD_e \frac{dn}{dx} , \qquad J_h^D(x) = -eD_h \frac{dn}{dx} , \tag{8.33}$$

where D_e and D_h are the diffusion constants of both carriers. The carrier density n is a function of x and is equal to the number of absorbed light quanta per unit volume at x, $\alpha' N_0 (1 - R) \exp(-\alpha' x)$, times the quantum yield η. Here α' is the

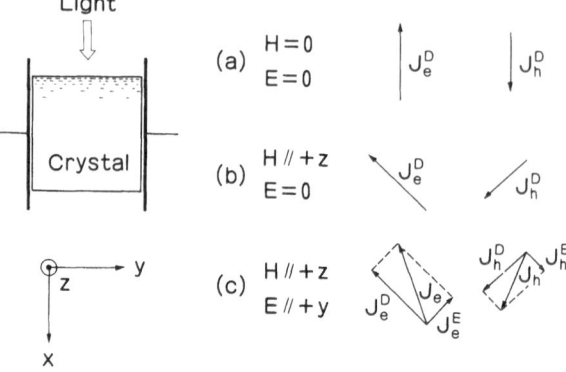

Fig. 8.17. Schematic diagram of the experimental arrangement for the photomagnetocurrent effect. The electric field E and the magnetic field H are applied in the y and z directions, respectively. Photocarriers are excited by light in the x direction. The superscripts D and E stand for a diffusion and an electric field. The subscripts e and h indicate electron and hole, respectively

absorption coefficient, R the reflectivity and N_0 the number of light quanta at the surface. The distance x is measured from the illuminated surface. By using Einstein's relation, $D = \mu k_B T/e$, (8.33) is

$$J_e^D(x) = -n(x)e\mu_e E^* ,$$ (8.34a)

$$J_h^D(x) = +n(x)e\mu_h E^* ,$$ (8.34b)

$$E^* = \frac{k_B T a'}{e} ,$$ (8.34c)

where E^* is an effective electric field. In this case no photoresponse will be observed at the electrodes with illumination by light because there is no current component in the y direction.

When a magnetic field is applied in the z direction, the diffusion current rotates because of the Lorentz force as shown in Fig. 8.17b and the y component of the current density J_y^D appears which induces a photoresponse through the electrodes. The currents J_y^D for electrons and holes are obtained from (8.5) and they are

$$J_{ey}^D(x) = -a_e(H)HE_x^* = -a_e(H)H\frac{k_B T a'}{e} ,$$ (8.35a)

$$J_{hy}^D(x) = -a_h(H)HE_x^* = -a_h(H)H\frac{k_B T a'}{e} .$$ (8.35b)

where $a(H)$ is a function of x through n in (8.7). The photoresponse Q_y^D measured in the geometry in Fig. 8.17 is obtained after integrating (8.35) over the volume and summing the electron and hole responses. It is

$$Q_y^D(H) = G\left[a_{0e}(H)H + \left(\frac{w_{0h}}{w_{0e}}\right)\left(\frac{\Gamma(p_h + 5/2)}{\Gamma(p_e + 5/2)}\right)^2 a_{0h}(H)H\right]$$

$$\times \frac{k_B T}{e}a'(1 - e^{-a'l}) .$$ (8.36)

Here $a_0(H) = (N_0/n)a(H)$, $G = \tau_e SlN_0\eta(1 - R)$ and τ_e is the trapping lifetime of electrons. In this expression S and l are the surface area and the thickness of the specimen, w_0 the unit range, and p the scattering index. This effect is essentially the same as the Kikoin-Noskov effect or the photomagnetoelectric effect [8.51]. The Kikoin-Noskov effect is the effect in a steady state condition when the buildup of the Hall field is completed, whereas the photomagnetocurrent effect discussed here is that in the absence of the Hall field.

If the electric field E is further applied in the y direction by the electrodes, as is shown in Fig. 8.17c, the current observed in the y direction is the sum of the Hall current due to the diffusion current discussed above and the transverse magnetoconductive current due to the externally applied electric field. The latter is $\sigma(H)E$ which is known from (8.5). It is noted that when the applied electric field is weak compared to the effective field of the diffusion current, as is the case of Fig. 8.17c, the current in the y direction flows in the direction opposite to the applied electric field.

In Fig. 8.18, the photoresponse $Q_y(H)$ in TlBr at 1.8 K, which is proportional to J_y in Fig. 8.17, is plotted against a transverse magnetic field in z with and without an electric field in y. The absorption coefficient of the exciting light at 390 nm is 1.5×10^5 cm^{-1}. When there is no electric field applied from outside, that is $E_y = 0$, the measured photoresponse a is found to be an odd function of H which is just what is expected from (8.35) and (8.36). The shape of the curve is a result of the magnetic field dependence of $\alpha(H)$ given by (8.7). When an electric field E_y of 0.57 Vcm^{-1} is applied, the curve shifts to a positive magnetic field. The magnetic field at which curve b crosses $Q_y(H) = 0$ is the field where the strength of the applied electric field is such that the Hall currents due to the diffusion of electrons and holes are balanced by the transverse magnetoconductive currents of both carriers driven by the electric field. The difference between curves b and a is curve c and must be the transverse magnetoconductivity $\sigma(H)E$. From (8.6), $\sigma(H)$ is an even function of H and decreases as H increases. The line shape of the curve c is just what we expect for $\sigma(H)$.

The diffusion current depends on the absorption coefficient through E^* in (8.34). As a result, the photomagnetocurrent J_y depends on the wavelength of the exciting light. Figure 8.19 shows the photoresponse $Q_y(H)/Q_y(0)$ for illumination by light of three different wavelengths. Here a is at 390 nm where $\alpha' = 1.5 \times 10^5$ cm^{-1}. At this wavelength, the transition is interband. Curve b is at 412 nm which is just at the $n = 1$ direct-exciton peak where

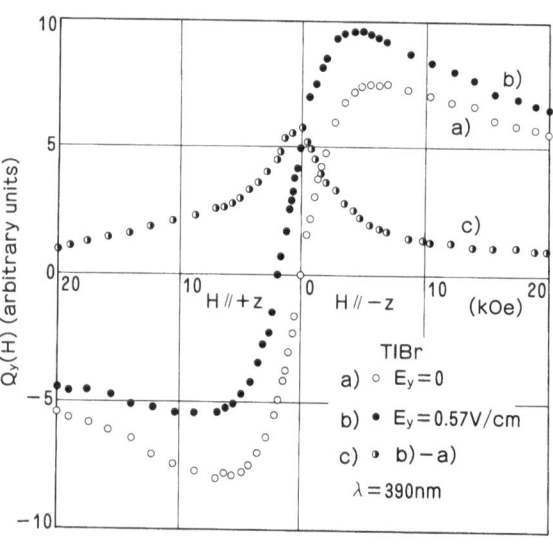

Fig. 8.18. Dependence of the photoresponse Q_y in TlBr on a transverse magnetic field in the z direction. Carriers are excited by light in the x direction at a fixed wavelength of 390 nm, where $\alpha' = 1.5 \times 10^5$ cm^{-1}. (a) is the photoresponse without an applied electric field externally and (b) is with an electric field applied in the y direction. (c) is the difference between (b) and (a) [8.50]

$\alpha' = 5 \times 10^5$ cm^{-1} and c is at 464 nm where $\alpha' = 1.6$ cm^{-1} and light is absorbed almost uniformly. The applied electric field is 12.5 Vcm^{-1}. Curve a is for the same conditions as curve b in Fig. 8.18, the only difference being the magnitude of the applied electric field. Curve c is very different from a. This is because the light is absorbed almost uniformly so that no density gradient of carriers is present which results in the absence of the diffusion current. Thus $J_y(H)$ in case c must be almost equal to the magnetoconductivity $\sigma(H)H$ in the applied electric field. The shape of curve c in the figure is just what we expect and is similar to that of curve c in Fig. 8.18. The absorption coefficient of the exciting light for the case of b is larger than the case of a. As a result, E^* or J_y^D in b must be larger than those in a. However in Fig. 8.19, we find that the opposite is the case. This is because in b the excitation is at the exciton absorption, but at the band to band absorption in a, and so the net free-carrier concentration is smaller in b than in a, since the free carriers are produced only by thermal dissociation of excitons in the case of b. This leads to a smaller concentration gradient of carriers and so a smaller diffusion current, giving a smaller $Q_y(H)/Q_y(0)$ in b. Thus a study of the photomagnetocurrent effect would give important information on the thermal dissociation of excitons, which has been rarely investigated.

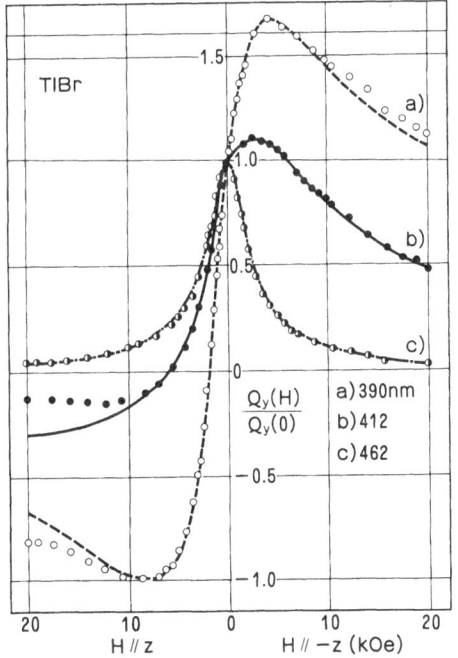

Fig. 8.19. Dependence of photoresponse Q_y in TlBr on a transverse magnetic field in the z direction. Carriers are excited by light in the x direction at three different wavelengths. The electric field applied parallel to y is 12.5 Vcm^{-1}. (a) is at 390 nm where $\alpha' = 1.5 \times 10^5$ cm^{-1}. (b) at 412 nm at the exciton absorption where $\alpha' = 5 \times 10^5$ cm^{-1} and (c) at 464 nm where $\alpha' = 1.6$ cm^{-1}. The lines are theoretical values [8.50]

8.5 Polarons with High Energy

8.5.1 Nonparabolicity of the Polaron Energy Spectrum

In the measurements of cyclotron resonance by the electron heating technique, *Hodby* found that the polaron masses in potassium and silver halides estimated from resonant magnetic fields depend on the electric field applied parallel to the magnetic field [8.52]. He attributed this effect to the variation of the effective mass of the polaron with energy. When the applied electric field is sufficiently high, the translational energy of a polaron will become a significant fraction of the energy of a LO phonon and the measurement will give the effective mass of a hot polaron. The mass of a hot polaron is known to be greater than that of a cold polaron having a low kinetic energy.

This can be understood from Fig. 8.20 where the polaron energy is shown as a function of its momentum in the case of $\alpha = 1$ [8.53]. When the polaron energy is close to $\hbar\omega_{LO}$, the LO phonon energy, a strong admixture of the polaron and its one free LO phonon state will occur by the Fröhlich interaction. As a result, the zero free LO phonon polaron state is pinned at the LO phonon energy by the one free LO phonon state as is shown in the figure. Thus the polaron effective mass increases as the energy increases and this explains *Hodby*'s results.

Similar evidence was found by *Tamura* and *Masumi* [8.54] who measured cyclotron resonance in AgBr by an ordinary method. In this case, the excitation of polarons to a high-energy state is made by the microwave electric field. Their results are shown in Fig. 8.21. We observe that as the microwave power is

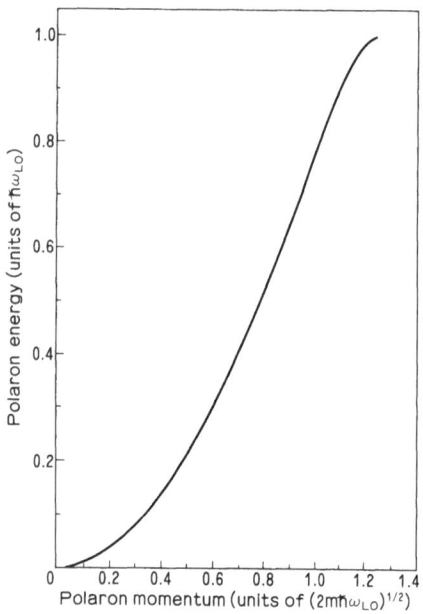

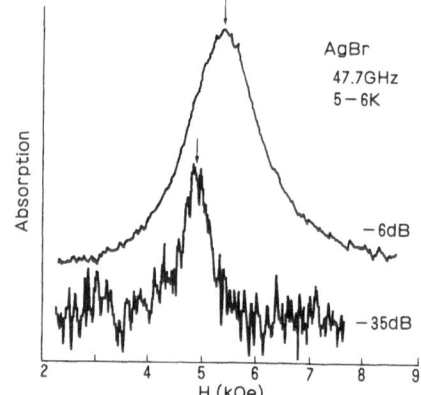

Fig. 8.21. Traces of cyclotron resonance absorption in AgBr at two different levels of microwave power at 47.7 GHz [8.54]

◀ **Fig. 8.20.** Polaron energy versus momentum curve for $\alpha = 1$ [8.53]

increased, the resonance line is shifted, from the cold polaron mass of 0.29 m_0 to a heavier mass, and is broadened. A shift of more than 10% is observed with a field of 20 Vcm^{-1}. At this field, the mean energy of hot polarons at the resonance is estimated to be about half the LO phonon energy (17.4 meV or 150 K).

Heating of the polarons is also achieved by raising the lattice temperature. A shift of a cyclotron resonance peak to a higher magnetic field or higher mass by an increase of temperature is observed by *Tamura* and *Masumi* [8.15] and by *Baxter* et al. [8.55] in silver halides.

A more direct method of investigating the nonparabolicity of the polaron band by cyclotron resonance is by changing the frequency of the electromagnetic wave and measuring the resonant magnetic field. This is because the energy corresponding to the frequency is equal to the polaron Landau level splitting at the resonance and thus a higher energy state of a polaron can be studied by using a higher frequency of the radiation.

We consider a free polaron without an interaction with free LO phonons in a magnetic field as an unperturbed state. Its energy eigenvalues are $(N + 1/2)\hbar\omega_c$ where $\omega_c = eH/m^*c$. As the magnetic field is increased, the $N = 1$ level will become degenerate with the $N = 0$ plus 1 LO phonon state. In the presence of electron–LO-phonon coupling, both the states mix and splitting will occur. This is a polaron pinning effect and is similar to the exciton-polaron pinning discussed in Sect. 7.5.1. The polaron pinning has been worked out in detail by *Larsen* [8.56] who uses the theory of *Haga* [8.57] on the polaron excitation energy. Figure 8.22 shows the pinning schematically. We find that the transition energy from a zero-phonon $N = 0$, $k_z = 0$ Landau level to a zero-phonon $N = 1$, $k_z = 0$ level, which is the cyclotron resonance energy, is linearly related to the magnetic field at low magnetic fields, giving a constant polaron effective mass. As the field increases, the transition energy deviates from the linear relationship and gives a heavier mass than the low field mass, due to the polaron pinning effect.

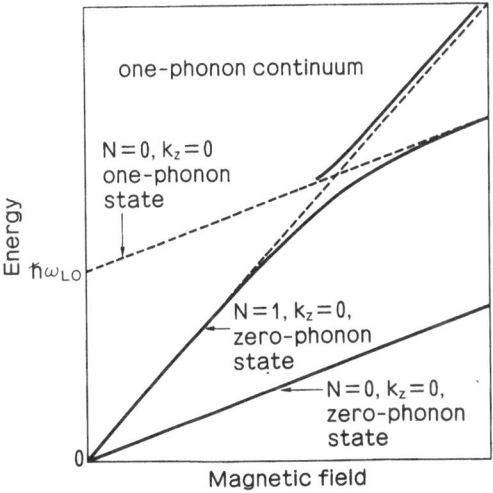

Fig. 8.22. Polaron Landau level in a magnetic field

The change of the polaron mass by polaron pinning has been studied extensively by cyclotron resonance experiments in InSb [8.58] in which the Fröhlich coupling constant is only 0.02. Also CdTe, a material of higher coupling constant, has been studied [8.59] but its α is 0.4 and it is in a weak coupling range.

A material of intermediate coupling, AgBr, whose α is 1.6, has been studied by *Hodby* et al. [8.60] using a cross modulation technique. In AgBr, it is expected that the $N = 1$ Landau level will cross the one phonon $N = 0$ Landau level at a magnetic field of about 430 kOe assuming no interaction between the zero- and one-phonon states. They found that the cyclotron mass obtained from the resonance at 890.76 GHz is larger by about 5.2% than that measured at 135.88 GHz. This amount is found to be in good agreement with the theory by *Larsen* [8.56] and gives direct evidence of the nonparabolicity of the polaron energy spectrum in an ionic crystal.

8.5.2 Hot-Polaron Transport Phenomena

In the preceding section, we saw that the polaron mass increases with the polaron energy. This gives a reduction of the polaron mobility on heating the polaron to a high-energy state. Beside the mass change, there is another, more drastic, effect of heating polarons which leads to the reduction of mobility. It is an increase of scattering.

When electrons are accelerated by a sufficiently high electric field, the distribution of electrons is altered from the equilibrium distribution determined by a lattice temperature. Electrons in such a nonequilibrium distribution are called hot electrons. The first evidence of the hot electron was given by *Ryder* and *Shockley* [8.61]. They measured a current against an applied voltage in Ge and found that Ohm's law holds up to a certain critical electric field, above which the current is approximately proportional to the square root of the field. They explained this in terms of hot electrons which suffer from acoustic-phonon scattering. The magnitude of the velocity of electrons at this critical field was found to be of the order of the velocity of sound in the crystal.

Transport phenomena of hot electrons in ionic crystals, in which electrons are subjected to a strong Fröhlich interaction to form polarons, are other attractive problems of hot carriers because, if the energy of a hot polaron reaches the LO phonon energy, a strong LO phonon emission will occur and a drastic change of current will be observed.

The first observation of the hot polaron effect in an ionic crystal was made by *Masumi* in AgCl [8.62] who found a square root dependence of a transient current on electric field and a reduction of mobility with increasing electric field in a strong electric field. Similar phenomena have been found in alkali halides [8.63, 64, 65] and CdS [8.66, 67].

Detailed investigations have been made most thoroughly in silver halides and we will discuss mainly the work on these materials. Figure 8.23 shows the photoresponse Q_x of electrons in AgBr, by open circles, against the applied electric

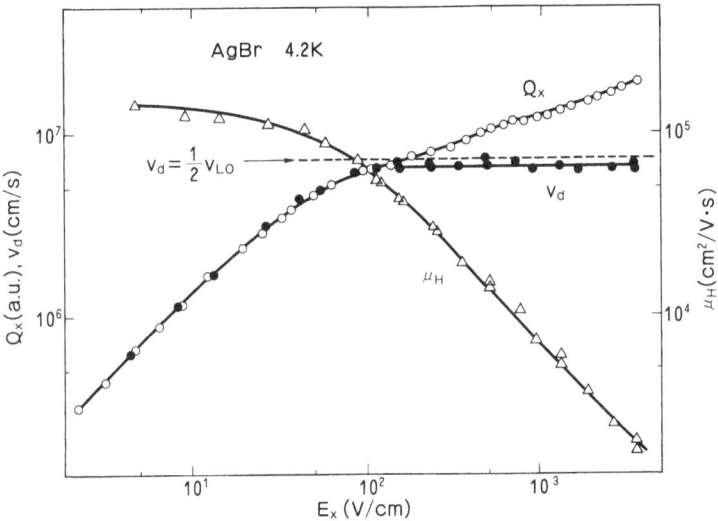

Fig. 8.23. The photoresponse Q_x, the Hall mobility μ_H, and the drift velocity $v_d = \mu_H E_x$ as a function of externally applied electric field E_x. Here v_{LO} is the velocity of the polaron with kinetic energy $\hbar\omega_{LO}$ [8.68]

field $E = (E_x, 0, 0)$. It was measured by *Komiyama* et al. [8.68] by a transient method with blocking electrodes. Here Q_x is proportional to $N_0 ew = N_0 e\mu\tau_t E_x = J_x$ when the schubweg w is sufficiently small compared to the dimensions of the specimen. We find that Ohm's law holds in the low field region and Q_x changes almost as $E_x^{0.5}$ in the high field region. They also measured the Hall current Q_y by further applying a small transverse magnetic field $H = (0, 0, H_z)$ using a special geometry of the electrodes. From Q_y, they obtained the Hall mobilities $\mu_H = -(c/H)(J_y/J_x) = -(c/H)(Q_y/Q_x)$ which are shown in the same figure as open triangles. The Hall mobility decreases as the field increases at high fields. The drift velocities $v_d = \mu_H E$ are calculated and are also plotted in the figure, by filled circles. It shows that the change of the drift velocity against E_x represents fairly well the change of Q_x up to 1×10^2 Vcm^{-1}. Above this field, v_d becomes saturated whereas Q_x increases almost as $E^{0.5}$. The origin of the departure of the Q_x–E_x curve from the v_d–E_x curve has been explained by *Komiyama* et al. [8.69] who show that the increase of Q_x in high fields is caused by the increase of the trapping time τ_t with the field E_x. We note that the current subjected only to scattering is proportional to v_d but not Q_x.

Above 1×10^2 Vcm^{-1} in Fig. 8.23, v_d becomes saturated at nearly $v_{LO}/2$ where v_{LO} is defined by $m^* v_{LO}^2/2 = \hbar\omega_{LO}$ and m^* is the cold polaron mass. *Shockley* [8.70] has suggested that, when the applied electric field E is large enough that an electron in an ionic crystal is accelerated from $v = 0$ to v_{LO} (at which the electron kinetic energy is $\hbar\omega_{LO}$) in so short a time that it is not scattered by acoustic phonons, the electron emits optical phonons by a strong

Fröhlich interaction, loses all its energy and returns to $v = 0$. This process repeats with a period of $m^* v_{LO}/eE$. The average velocity of the electrons in this periodic motion is $v_{LO}/2$. If a larger and larger field is applied to accelerate the electron, the energy flows more and more into the LO phonon system merely increasing the frequency of the repeated motion and thus the mean velocity of the electrons is not altered from $v_{LO}/2$. This is just what we saw in Fig. 8.23. In k-space this repeating motion gives an oscillating trajectory between 0 and $mv_{LO}/\hbar$ in the direction of E and is called a cyclic streaming motion [8.71], in sharp contrast to the diffusion picture for the carrier motion. The distribution and velocity of electrons in the streaming motion without a magnetic field has been calculated by *Kurosawa* and *Maeda* [8.72].

Further extensive studies of the streaming motion of photocarriers in crossed electric and magnetic fields have been made both theoretically and experimentally. *Maeda* and *Kurosawa* [8.73] proposed that there must be a peculiar distribution of hot carriers subjected to LO phonon scattering in crossed electric and magnetic fields of appropriate strengths; that is, a population inversion. Figure 8.24 is a k_x–k_y plane of k-space for an isotropic conduction band. A surface of a constant energy at $\varepsilon = \hbar \omega_{LO}$ is shown by the broken circle. This corresponds to the upper edge of the polaron band. Suppose that there is an electric field in the x direction which is large enough to accelerate electrons at $v = 0$ ($\varepsilon = 0$) to v_{LO} ($\varepsilon = \hbar \omega_{LO}$). The electrons will make a streaming motion between $k = 0$ and $k_{LO} = \sqrt{2} \, m \omega_{LO}/\hbar$ as is indicated by line a in the figure. Photocarriers excited at any energy in the exciton ionization continuum will reach the same final steady state. This is because the excited electrons will at first be distributed widely in k-space but they will be accelerated by the electric field to k_{LO} and then scattered to $k = 0$ from where they will make a streaming motion.

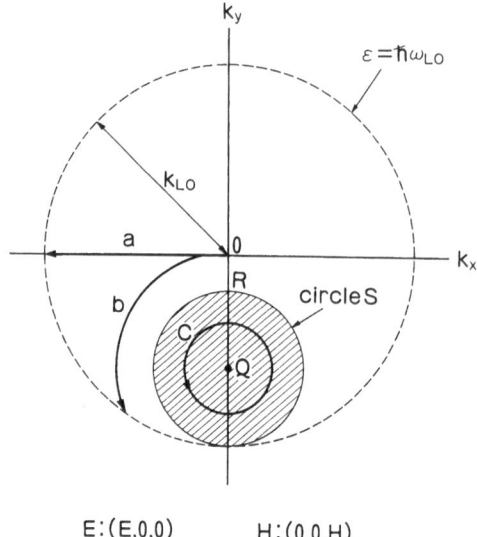

E: (E,0,0)   H: (0,0,H)

Fig. 8.24. The streaming motion and the accumulation of electron polarons in the k_x–k_y plane in k-space

When there is a transverse magnetic field H in the z direction besides E in x, the electrons will make a circular motion with its center Q at $(0, -m^*cE/\hbar H, 0)$ in k-space.

Maeda and *Kurosawa* [8.73] have suggested that if the point Q is inside the $\varepsilon = \hbar\omega_{LO}$ surface at which k is k_{LO}, the carrier motion is strongly dependent on where the carrier is in k-space. When $m^*cE/\hbar H$ is larger than $\sqrt{m^*\omega_{LO}/2\hbar}$, that is, the distance from $k = 0$ to Q is larger than $k_{LO}/2$ and this is the case of Fig. 8.24, the trajectory of the motion of an electron starting from $k = 0$ will follow the line b, hit the surface $\varepsilon = \hbar\omega_{LO}$ and be scattered back to $k = 0$ by the emission of a LO phonon. Any electrons on the outside of the circle S, whose center is at Q and which is tangential to the $\varepsilon = \hbar\omega_{LO}$ surface, will behave in a similar way. As a result, such electrons will make a streaming motion. However, the electrons initially inside the circle S will make circular motion, drawn as curve c, with a frequency eH/m^*c. The circular motion will continue for quite a long time because of the absence of the LO phonon emission which occurs at the $\varepsilon = \hbar\omega_{LO}$ surface. Thus the carriers are long-lived in S and short-lived on its outside, leading to an accumulation of carriers within S. When E and H are such that the distance of Q from the origin is smaller than $k_{LO}/2$, the accumulation is more effective since no carriers are outside S. In three-dimensional k-space, the carriers are accumulated in a volume having a spindle shape whose axis is parallel to k_z and whose cross section in the k_x–k_y plane is S in Fig. 8.24.

Experimental investigations of the streaming and the accumulation of photo-carriers in electric and magnetic fields have been made in AgCl and AgBr by *Komiyama* et al. [8.69, 74, 75]. In Fig. 8.25, the Hall angle tan θ which is equal to

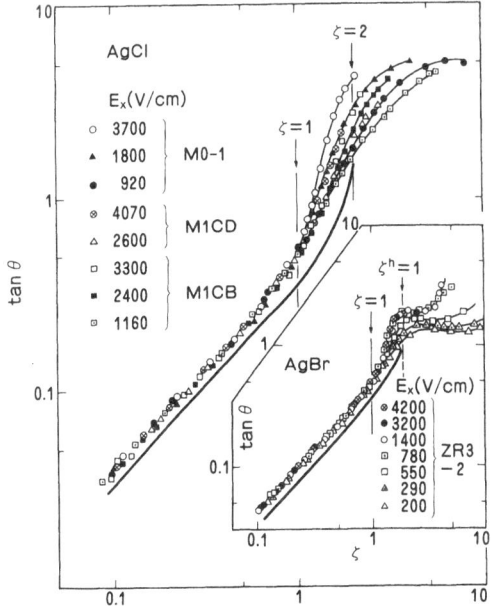

Fig. 8.25. Tangent of the Hall angle as a function of ζ for AgCl where ζ is v_{LO}/v_y and v_y is cE/H. The solid line is the theoretical value for circular streaming motion [8.69]

J_y/J_x in transient conditions is plotted against ζ in large electric fields. Here ζ is v_{LO}/v_y where v_{LO} is $\sqrt{2\hbar\omega_{LO}/m}$ for electrons. The velocity in the y direction is v_y and is equal to cE/H. The position of Q in k-space, k_Q, in Fig. 8.24 is then $mv_y/\hbar$ and v_{LO}/v_y is equal to k_{LO}/k_Q which is a measure of the position of Q relative to the $\varepsilon = \hbar\omega_{LO}$ surface. When $\zeta < 1$, the point Q is outside the $\varepsilon = \hbar\omega_{LO}$ sphere and only the streaming motion is expected for all carriers in a large electric field. When $\zeta > 1$ on the other hand, accumulation as well as streaming are expected and the former sets in at $\zeta = 1$. Above $\zeta = 2$ all carriers will be accumulated in the vicinity of k_Q. The electric field in Fig. 8.25 is large enough that v_d is saturated in each sample in the absence of a magnetic field. In the region $\zeta < 1$, all the data points from different samples and in different electric fields are found to fall on a single line with a slope of $\zeta^{1.0}$. In the range $\zeta > 1$, the curves begin to depart from the linear relation and increase steeply with ζ. This shows that a different mechanism sets in at about $\zeta = 1$. *Komiyama* et al. [8.69] calculated the Hall angle due to electrons in a circularly streaming motion as a function of ζ and the result is shown by the solid line in the figure. The good agreement gives evidence of the streaming motion of hot polarons.

The abrupt rise of the curves found in the range $1 < \zeta < 2$ is interpreted in terms of circular streaming and carrier accumulation. In this range, the electrons accumulated in the region of S in Fig. 8.24 make a large contribution to the Hall current J_y but little to J_x. As a result, the Hall angle becomes larger than that expected from the streaming electrons alone. This is the reason for the abrupt rise of the Hall angle in this range. In the range over $\zeta = 2$ in AgCl, all the streamings cease and all the electrons are distributed in the cyclotron orbit within S. The average drift velocity becomes nearly equal to $(0, v_y, 0)$ so that the Hall angle becomes quite large, being consistent with the experiment shown in Fig. 8.25. In the case of AgBr in which photoconduction is two carrier, the streaming of holes also contributes to the Hall angle. In Fig. 8.25, ζ for holes is shown as ζ^h. The bend over of the curve for AgBr above $\zeta^h = 1$ is ascribed to the accumulation of holes. From this evidence, it is almost certain that the high-energy polarons in ionic crystals have a streaming motion and they are accumulated at a certain place in k-space to give a population inversion in appropriate high electric and magnetic fields. This picture has been confirmed further by other magnetoconductive experiments [8.69] and a cyclotron resonance experiment [8.75].

9. Excitons and Phonon Couplings in Quasi-One-Dimensional Crystals

A quasi-one-dimensional electron lattice system is not simply a highly anisotropic three-dimensional (3-d) system when it is viewed from the standpoint of an excited electronic state and its phonon coupling. In 3-d crystals, one finds almost always an excited electronic state in the form of a Wannier or a Frenkel exciton. The former case is found in typical ionic crystals and semiconductors and the latter in molecular crystals. In quasi-one-dimensional crystals composed of a bundle of linear chains, the most frequently found excited electronic state is a charge transfer exciton. In the charge transfer exciton, an electron in a ground state in one atom is excited to a different atom at a neighbouring site and the radius of the exciton is of this order. In 3-d crystals, a charge transfer exciton is rare. Perhaps the $1s$ exciton in alkali halides is an example, but its characteristic property does not appear as clearly as in one-dimensional (1-d) crystals.

A 1-d crystal in its ground state at the absolute zero of temperature is an insulator. This is because a 1-d metal is always unstable against electron-phonon coupling or the electron-electron interaction and transforms into a Peierls insulator or a Hubbard insulator, respectively, by opening an energy gap. With electron-phonon coupling, a metal becomes a charge density wave (CDW) state and with electron-electron repulsion a spin density wave (SDW) state, as was discussed in Sect. 4.6.4. The optical excitation from these ground states in a 1-d crystal will reflect the CDW or SDW states. For example, a charge transfer excitation in a Peierls insulator is the reverse process of the formation of the CDW state. Thus the excitons in a 1-d crystal are strongly related to the instability of the 1-d system.

The electron-phonon interaction in a 1-d system has a unique nature and it gives several attractive properties for the excited electronic state. As was discussed in Sects. 4.1.1–3, the short-range deformation-potential coupling in a 1-d system (like the long range Fröhlich coupling in a 3-d system) is "limp", while in a 3-d system it is "stiff", see (4.133). In a 3-d crystal, free excitons are self-trapped after passing through a potential barrier only when the electron-phonon coupling is strong enough. In a 1-d crystal, excitons are self-trapped without a barrier by the presence of electron-phonon coupling of any strength and thus they are always localized. In some cases the self-trapped states, which are sometimes called small polarons, transform into solitons so that 1-d crystals are nice systems for studying solitons in solid materials.

To study these unique properties of the excited state in a 1-d system, one must find crystals which are one dimensional. Strictly speaking, there is no such crystal and real crystals are always three dimensional. Thus one must use crystals in which atoms are bonded strongly only in one direction to form chains and the bonding between chains is sufficiently weak. These are called quasi-one-dimensional (quasi-1-d) crystals. The success of the study of the properties of the 1-d electron-phonon system in quasi-1-d crystals depends upon how clearly one can isolate 1-d characteristics from other 3-d properties. This depends on the materials used and thus a search for a suitable material is the key to the study of the 1-d system.

There are a number of chain compounds and the only ones studied so far, from the point of view of quasi-1-d crystals, are some organic and organo-metallic compounds having chains or stacks of molecules extending in one direction. Although a variety of properties have been disclosed already in known quasi-1-d materials, more and more new phenomena will be found in other unexplored materials. In this section, we will discuss the characteristic properties of optical excitation of quasi-1-d crystals by introducing a few typical examples.

9.1 Halogen-Bridged Mixed-Valence Chain Compounds

The structure of a single chain in a halogen-bridged mixed-valence chain crystal is shown in Fig. 9.1. Here M is platinum of palladium and X is a halogen. The valence of M is either $M^{2+}(d^8)$ or $M^{4+}(d^6)$ and they alternate along the chain. The halogen ion X^- on the chain is located not at the center between two metal ions but is closer to M^{4+} than to M^{2+}. Both M^{2+} and M^{4+} are coordinated by the same ligand molecules L by which the backbone metal-halogen chain is separated distantly enough from other chains to give a fairly good 1-d character to the backbone chain. The molecules L are usually alkyl amines. This structure is ordinarily interpreted in terms of a charge transfer complex of two radicals $(M^{2+}L)$ and $(M^{4+}X_2^-L)$ but it can be understood in the following way as well.

Let us imagine a metallic chain of $-M^{3+}-X^--M^{3+}-X^--$. In this chain, the halogens must be positioned at the middle point between the metal ions and the band consisting of d_{z^2} of metal ions will be half-filled. However, as was discussed in Sects. 4.6.3, 4, this structure is unstable against the Peierls transition in the 1-d

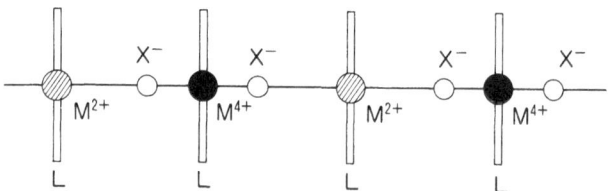

Fig. 9.1. Model for the halogen-bridged mixed-valence chain compound. Here M and X are the metal and halogen atoms and L is a ligand molecule

system. The $2k_F$ Peierls transition occurs which is accompanied by a localization of charges on alternate metal ions to form M^{2+} and M^{4+} and a displacement of halogen ions from the middle point towards the more positive M^{4+} to double the original periodicity of the lattice. The resultant structure is the ground state of the halogen-bridged mixed-valence chain shown in Fig. 9.1.

This structure can also be understood by a picture of charge transfer excitons which are coupled with phonons in the 1-d system. We take the $-M^{3+}-X^--M^{3+}-X^-$- chain in which d_{z^2} electrons are localized at metal ions to form an antiferromagnetic insulator or a Hubbard insulator. We excite electrons from alternate M^{3+} ions to their neighbouring M^{3+} ions thus forming an array of charge transfer excitons $-Pt^{4+}-X^--Pt^{2+}-X^-$-. This excited state is unstable against a lattice distortion in a 1-d system for electron-phonon coupling of any strength, as was discussed in Sects. 4.4.1-3, and will be subject to stabilization by self-trapping of excitons accompanied by a displacement of the halogen ions towards the M^{4+} ions to form the structure of Fig. 9.1. Thus the structure in the figure is understood as an array of self-trapped charge transfer excitons whose energy is lower than the original antiferromagnetic state.

The optical excitation by the charge transfer of an electron from M^{2+} to M^{4+} in Fig. 9.1 is just the reverse process of the above-mentioned processes. Accordingly, a displacement of the halogen ions towards the middle point between the metal ions is expected for an optical charge transfer excitation, as a result of a strong exciton-phonon coupling. In the excitation, the metal ions will not move as drastically as the halogens because of the ligands whose masses and mutual interactions will prevent the intermetal-ion distance changing. This allows us to deal with the electronic system and the lattice system separately, in the first approximation, and the electronic potential on the metal site is determined almost solely by the position of the halogen ion. It is noted that the polarization of the charge transfer excitation and the displacement of the halogen ion are both in the chain direction so that the exciton and its phonon coupling will have a 1-d character.

Among the halogen-bridged mixed-valence chain crystals, Wolffram's red salt (WRS), $[Pt^{2+}(ea)_4][Pt^{4+}(ea)_4Cl_2]Cl_4 \cdot 4H_2O$, where ea is ethylamine $C_2H_5NH_2$, is the material which has been studied most extensively with regard to optical excitation and relaxation. In WRS, the metal ions are platinum ions and the halogen ions are chlorine ions. Both Pt^{2+} and Pt^{4+} are coordinated tetragonally by four ethylamines in an equivalent fashion. The structure of the chain is shown in Fig. 9.2.

The absorption spectrum of WRS at low temperature was measured first by *Day* [9.1]. Fig. 9.3 shows the spectra at low and high temperatures [9.2]. The spectrum is anisotropic and a strong dichroism is observed for $E \parallel z$ and $E \perp z$ where E is the electric field of light and z is the direction of the chain. The absorptions higher than 3.0 eV are due to intramolecular transitions within the component radicals $[Pt^{2+}(ea)_4]$ and $[Pt^{4+}(ea)_4Cl_2]$. The absorption below 3.0 eV is observed only in WRS and not in the component radicals. Particularly at the absorption edge at about 2.3 eV, the spectrum is highly anisotropic and light is

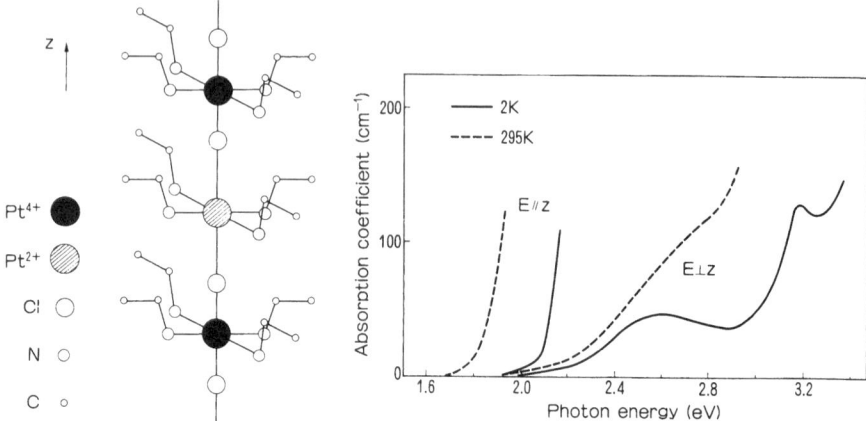

Fig. 9.2. Structure of the chain in the Wolffram's red salt crystal

Fig. 9.3. Absorption spectra of WRS after some modification [9.2]

absorbed only for $E \parallel z$. The z-polarized absorption at the band edge has been assigned to a charge transfer excitation from Pt^{2+} $5d_{z^2}$ to Pt^{4+} $5d_{z^2}$ by *Clark* et al. [9.4] who found that the Raman scattering intensity of the breathing mode of Cl^- along the chain around Pt ion resonates at the absorption edge at room temperature.

More extensive optical work on WRS, from the point of view of a reverse Peierls transition and self-trapping in a quasi-1-d system, was done by *Tanino* and *Kobayashi* [9.2, 3]. Figure 9.4 shows the emission spectra of WRS excited at 2.410 eV where the absorption is due to the charge transfer excitation from Pt^{2+} to Pt^{4+}. For cases of polarization, $E_i \parallel z \parallel E_s$, $E_i \parallel z \perp E_s$, $E_i \perp z \parallel E_s$ and $E_i \perp z \perp E_s$, have been investigated where E_i and E_s are the electric fields of the incident and scattered light. For $E_i \parallel z \parallel E_s$, a series of Raman lines with almost equal separa-

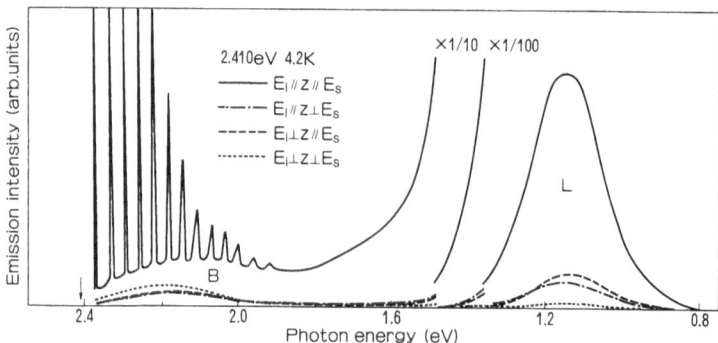

Fig. 9.4. Emission spectra of WRS. The electric field of light is E and the direction of the chain is z [9.2]

tions of 38.1 meV are observed. Those are the fundamental and the overtones of the vibration of Cl^- along the chain and are observed only in $E_i \| z \| E_s$ and not in the other cases. They resonate at the absorption edge at 2.3 eV where the absorption curve for $E \| z$ shows a sharp rise.

The Raman lines are accompanied by a weak and continuous background emission B below them. The background emission always appears as long as the excitation is made at the energy of charge transfer absorption. It is polarized considerably in the z-direction. This emission B has characteristic features which are not found in ordinary Raman scatterings and luminescences. It rises at the energy of excitation and moves with it. The position of its broad maximum depends on the excitation energy and its tail extends toward lower energies to continue to a large luminescence band L which peaks at 1.15 eV. It has no particular structure at the absorption edge.

The large luminescence band L at 1.15 eV has a Gaussian-like shape with a slight asymmetry. It is highly polarized in z. Its intensity is very much stronger than the background emission B, by a factor of 10^3.

Similar spectra of absorption and emission have been observed in other halogen-bridged platinum mixed-valence compounds such as $[Pt(en)_2]$ $[Pt(en)_2Cl_2](ClO_4)_4$ and $[Pt(en)_2][Pt(en)_2I_2](ClO_4)_4$ where en is ethylene diamine [9.5]. Thus the spectra in Fig. 9.3 and 9.4 seem to be common in this class of substance.

The processes of the excitation and relaxation of the charge transfer exciton in WRS are shown schematically in Fig. 9.5. The ground state of WRS is shown in

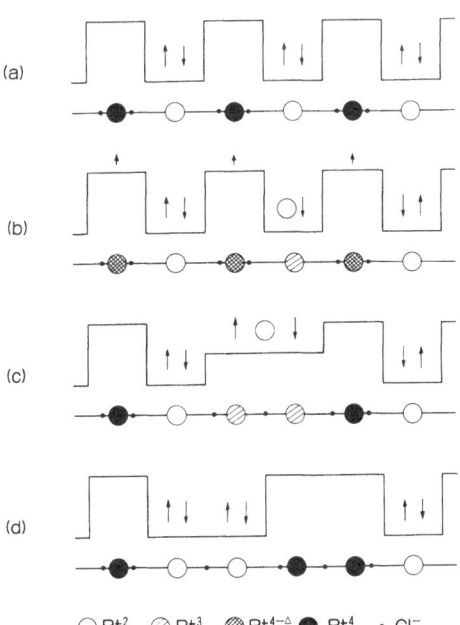

(a)

(b)

(c)

(d)

○ Pt^2 ◎ Pt^3 ◉ $Pt^{4-\triangle}$ ● Pt^4 · Cl^-

Fig. 9.5a–d. Schematic representation of the excitation and relaxation of the charge transfer exciton in WRS. Detailed description is given in the text

(a). Potential energy curves are drawn for an electron at the platinum sites by supposing that they are determined simply by the position of Cl$^-$ in the chain. In the ground state, two extra electrons of $5d_{z^2}$ are localized at each Pt^{2+} site by the Peierls instability. When WRS is irradiated by light with $E \parallel z$ at the energy of the charge transfer absorption, an electron is excited from the Pt^{2+} $5d_{z^2}$ state to the Pt^{4+} $5d_{z^2}$ state. It is believed that the ground-state valence delocalization is small in WRS [9.1] and so the hole left at the Pt^{2+} site will be localized. The electron excited to the Pt^{4+} $5d_{z^2}$ state will spread more or less over the chain. Since the frequency of the excitation is very much higher than that of the Cl$^-$ vibration, Cl$^-$ ions will stay at the ground-state position during the excitation. This is shown in (b). After the excitation, the charge transfer excited electron is self-trapped without a potential barrier and is localized at the hole site as shown in (c) because of the 1-d charactor of the chain. The lattice distortion associated with the self-trapping is a displacement of the Cl$^-$ ion towards the middle point between the platinums along the same chain direction as the polarization of the charge transfer exciton. In this self-trapped state, the exciton state is close to $-$Pt^{3+}–Cl$^-$–Pt^{3+}– which is equivalent to a metallic or an antiferromagnetic domain in the chain. By the recombination of electron and hole in (c), light is emitted and the system returns to the ground state (a) by changing the position of halogens. The state (c) could also be transformed into the configuration (d). This state is essentially a kink-antikink pair soliton.

A 1-d metal is always unstable against an electron-phonon interaction or an electron-electron interaction and it becomes a Peierls insulator or a Hubbard insulator, respectively, as was discussed before. In real materials, these two interactions coexist and their relative importance determines the electronic state and lattice structure. Systematic work on many electrons with finite transfer energy T which are subjected to electron-phonon coupling S and intra- and intersite electron-electron repulsions U and V in the 1-d system has been done by Nasu [9.6–8].

The adiabatic potential energy curves for the ground and the excited state of the electron-phonon system which is believed to be close to the case of WRS is shown in Fig. 9.6 [9.3]. Here the adiabatic potential energy is measured in units of S. The abscissa is the distortion of the lattice which represents the position of Cl$^-$ relative to the platinum ions. Here 0 is the configuration of the WRS ground state, -1 is that with Cl$^-$ at the middle point between platinums and -2 is that with Cl$^-$ at a position symmetric to the ground-state position beyond the middle point. It can be seen that there is no potential barrier between the undistorted and the self-trapped state of the excited state because it is a 1-d system. The adiabatic potential energy curve is rather sensitive to T and Fig. 9.6 is the case of large T. When T is reduced, another minimum appears in the energy surface of the excited state at around $Q \sim -1.5$. This figure tells us that in WRS a charge transfer exciton excited at $Q = 0$ relaxes to a localized self-trapped state at $Q \sim -0.5$ where Cl$^-$ is displaced to, but not exactly at, the middle point. It emits a luminescence whose energy is about half the energy of the absorption edge. This is just what has been found in WRS. It then returns to the ground state and

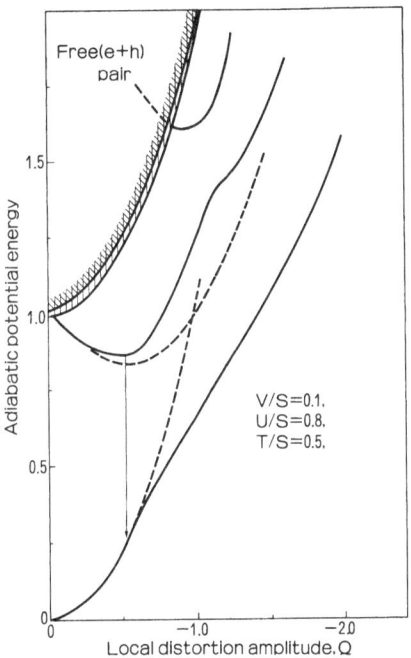

Fig. 9.6. Adiabatic potential energy curves for the ground and excited states of the electron-phonon system which may represent those in WRS. Detailed description is given in the text [9.8]

the Cl$^-$ ions move back to the ground-state position. This process is essentially the same as the process of (a) $\rightarrow$ (b) $\rightarrow$ (c) $\rightarrow$ (a) in Fig. 9.5. The soliton state (d) in Fig. 9.5 which corresponds to $Q \sim -2.0$ will not be formed in WRS if the parameters used in Fig. 9.6 are correct for WRS.

The emission spectrum given in Fig. 9.4 can be interpreted by this picture with some modifications. Figure 9.7 is a diagram which represents a possible process for the relaxation of an excited state in WRS [9.2]. Figure 9.7a shows the energy bands of WRS. The valence band is flat because the Pt^{2+} d_{z^2} state is almost

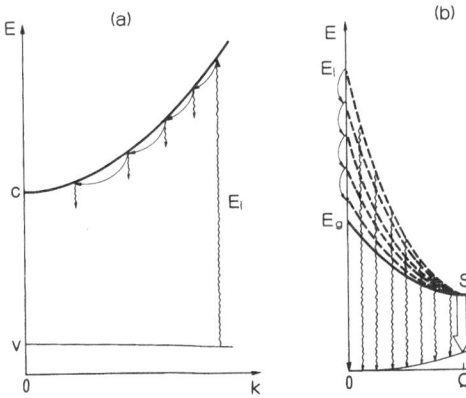

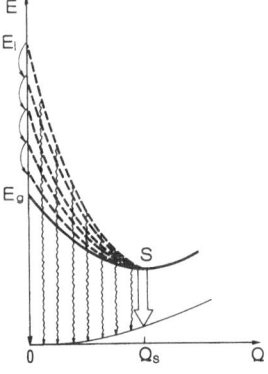

Fig. 9.7a, b. Schematic representation of the relaxation process of an excited state in WRS [9.2]

localized. The electron-hole Coulomb energy is included in the energy of the delocalized conduction band which is made of Pt^{4+} d_{z^2}. When light with energy E_i is absorbed, electrons in the valence band are excited to the conduction band. Fig. 9.5b corresponds to this excitation. The excited electrons then cascade down towards the bottom of the band by emitting phonons corresponding to the breathing vibration of Cl^- around platinum along the chain. After each emission of a phonon, light is emitted to give the multiple Raman lines in Fig. 9.4. Also during the cascading, some of the excited electrons relax to the self-trapped states assisted by displacements of Cl^- ions. This is shown in the configuration coordinate diagram (b). Here Q represents a position of Cl^- as in Fig. 9.6. All the states shown in (a) are on the $Q = 0$ ordinate in (b). Energy E in (a) is the electronic energy whereas in (b) it is the total energy of the system including the electrons and lattice. It is suggested that, during the decay to the self-trapped state S, some electrons and holes recombine to give rise to a weak emission due to the fast decay. As can be seen from (b), this emission begins to rise at the energy of excitation and extends to lower energies as far as the energy at which the emission from the self-trapped state sets in. The broad and weak emission B observed under the Raman lines is this emission. The excited states are finally accumulated at the self-trapped state S, which is the same as the state shown in Fig. 9.5c, and give a strong luminescence L at the low-energy end of the emission spectrum as is observed in Fig. 9.4.

There are questions which must be examined. In a 3-d perfect crystal, it is believed that the relaxation to a self-trapped state occurs for the electrons with practically zero kinetic energy. In the picture shown in Fig. 9.7, it is assumed that self-trapping occurs for electrons with finite kinetic energy in order to explain the rise of the broad emission band B at the exciting energy. Whether this type of relaxation is possible in a 1-d system or not is a future problem which should be clarified. Further, there is a weak absorption band polarized in the chain direction at a lower energy than the absorption edge. It is a characteristic absorption of this class of compound. However, the Raman effect due to Cl^- vibration along the chain is not resonant with this band and the emission from the self-trapped state cannot be excited by irradiation at this band. Its origin is not known but it must be related to the quasi-1-d electronic structure of the chain because it is not found in the component radicals.

9.2 Polyacetylene

Another Peierls insulator which has been studied extensively by optical means is polyacetylene, $(CH)_x$. It is the simplest linear conjugated polymer. Two isomers are known, that is, a *trans*-polyacetylene and a *cis*-polyacetylene. Their chain structures are shown in Fig. 9.8. In a *trans* form, two structures (a) and (b) are topologically degenerate. In a *cis* form, a *cis-transoid* (c) and a *trans-cisoid* (d) are not degenerate and the configuration (c) has a lower energy than (d). Of the *trans*

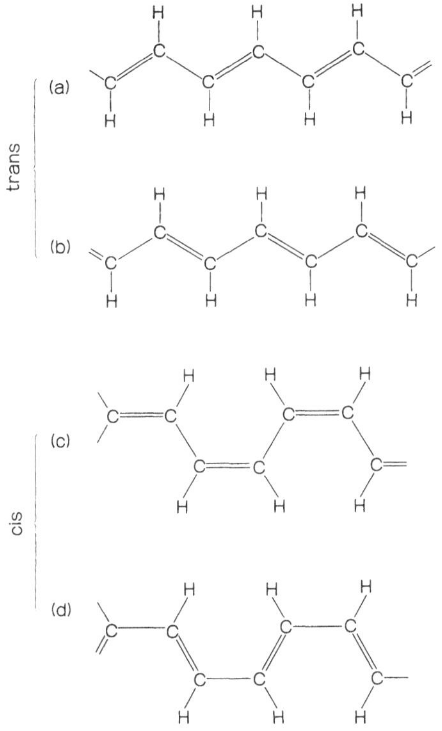

Fig. 9.8. Structures of the chains of poly-acetylene

and *cis* forms, the *trans* form is more stable. These chains form a solid. The shortest interchain carbon atom spacing is about 3.6 A in a *cis* compound [9.9].

In polyacetylene, three out of four carbon valence electrons form σ bonds by sp^2 hybridization by which two neighbouring carbon atoms on the backbone chain and a hydrogen atom are linked. The remaining $2p_z$ valence electron is in a π orbital whose extension is perpendicular to the chain axis. The σ bonds form a filled valence band. The π bond will form a half-filled metallic band if the bond distances between the carbon atoms in the chain are all alike. However this π electron state is unstable against the $2k_F$ Peierls transition, as is the case of WRS discussed in Sect. 9.1, and the structure is transformed into an alternating bond configuration as shown in Fig. 9.8. Here two π electrons are localized at the double bond where the C–C distance becomes shorter than the single σ bond distance, giving a doubling of the periodicity and an opening up of the band gap at the Fermi level to become an insulator. The C–C and C=C bond lengths are known to be 1.46 and 1.37 Å in the *cis* isomer [9.9] and 1.44 and 1.36 Å in the *trans* isomer [9.10]. The lower branch of the Peierls split band is filled by electrons and its $k = 0$ state is equivalent to the bonding combination of carbon $2p_z$ orbitals. The upper split band is vacant and its $k = 0$ state corresponds to the antibonding combination of $2p_z$ orbitals. The electronic structure of the alternating bond *trans*-polyacetylene has been investigated theoretically by *Su* et al.

[9.11] by considering the changes of the energies of the electron transfer and the lattice distortion with a change of bond length. Although in this work an electron-electron Coulomb interaction is not included, the qualitative nature of the electronic state of $(CH)_x$ has been clarified. The nature of the soliton and of the small polaron or self-trapped state in polyacetylene have become apparent from this work. We note that, in this case, the change of the transfer energy with the lattice distortion is site-off-diagonal, in contrast to the case of WRS discussed in the preceding section where the transfer energy is assumed not to change with the displacement of Cl^-, in the first approximation.

Optical absorption spectra of *trans*- and *cis*-polyacetylene have been measured by *Fincher*, Jr. et al. [9.12, 13] at room temperature and these are shown in Fig. 9.9. The spectra were measured on films of *trans*-$(CH)_x$ and *cis*-$(CH)_x$ both being unoriented. It has been found that the absorption in the *trans* isomer is polarized in the chain direction. The spectra of the *trans* and *cis* isomers are similar in shape but some vibronic structures due to a stretching mode are found in the *cis* isomer. The absorption maximum of the *cis* isomer is blue-shifted by about 0.3 eV from that of the *trans* isomer. The absorption coefficient at the maximum is of the order of 10^5 cm^{-1} for both the isomers. This value is comparable to those of direct allowed exciton absorptions in normal 3-d insulators which

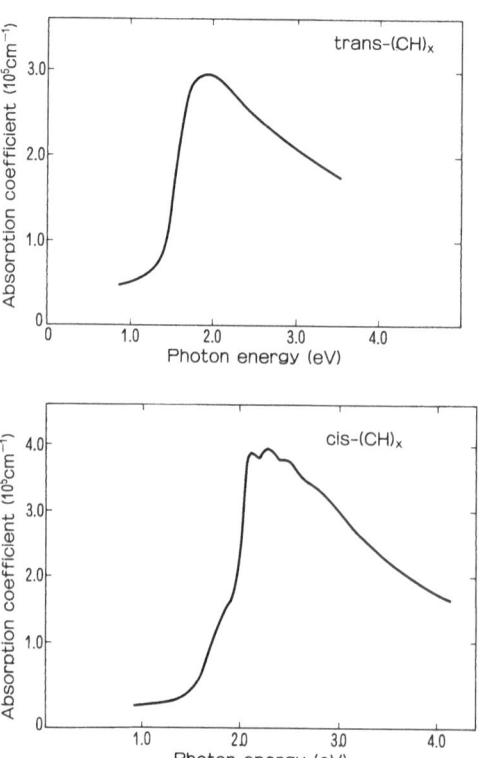

Fig. 9.9. Absorption spectra of *trans*- and *cis*-$(CH)_x$ [9.13]

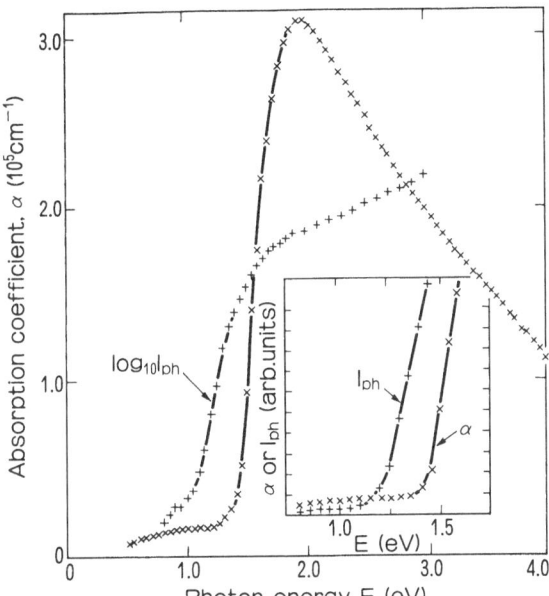

Fig. 9.10. Photoconductivity (I_{ph}) and absorption spectrum (α) for *trans*-(CH)$_x$ at room temperature [9.16]

are in the range 10^5–10^6 cm^{-1}. The absorption shows a characteristic line shape. It is noted that an absorption with a low absorption coefficient is present on the low-energy side of the strong absorption, extending to about 0.5 eV.

From the line shape, *Fincher*, Jr. et al. [9.13] have suggested that the absorptions in Fig. 9.9 are due to a band-to-band transition between the Peierls split π bands. The line shape is the reflection of the 1-d joint density of the band states whose singularity at the edge is quenched by the interchain coupling. In their work, they discussed the possibility of the contribution of an electron-hole interaction or an exciton state to the optical absorption and suggested that, according to the measured oscillator strength, an interband transition, rather than an exciton transition, is responsible for the absorption.

By the measurement of spectral dependence of photoconductivity, *Matsui* and *Nakamura* [9.14] showed that the photocurrent rises at about 1.0 eV. Later *Tani* et al. [9.15] measured the spectral dependences of a photovoltaic response as well as a photocurrent at room temperature in *trans*-(CH)$_x$ and found that the threshold energy is at 1.48 eV, which is at the tail of the 2.0 eV absorption band. This fact again suggests that the one-electron band gap in *trans*-(CH)$_x$ is at the absorption tail so that the absorption is due to the band-to-band transition. The study of photoconductivity was further extended by *Lauchlan* et al. at room temperature [9.16]. Their result for *trans*-(CH)$_x$ is shown in Fig. 9.10. It is found that the threshold of the photocurrent is near 1 eV which is well below the onset of the strong absorption, which is also shown in the figure. The photoconductivity increases as the energy of the exciting light increases, and becomes almost flat near the absorption peak. This gives further support to the interband transition at

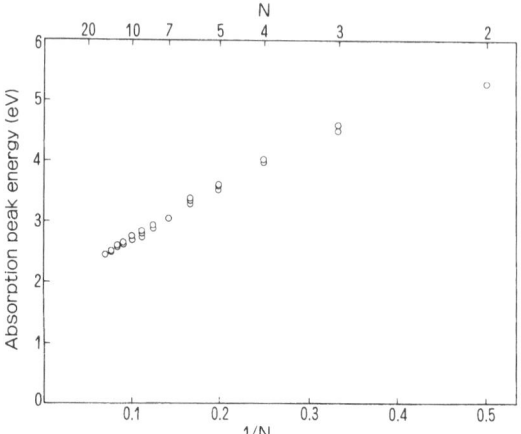

Fig. 9.11. Transition energies from the ground state to the ^1B_u excited state in finite polyene molecules of $2N$ conjugated carbon atoms as a function of $1/N$. Data are taken from [9.18]

the absorption edge. The band calculation by *Grant* and *Batra* [9.17] shows that the band-gap energy of *trans*-$(CH)_x$ is 0.8–2.3 eV although the value is quite sensitive to the degree of bond alternation. It is noted that in *cis*-$(CH)_x$, the photoconductivity is unmeasurable and its upper limit is more than three orders of magnitude smaller than that of *trans*-$(CH)_x$ [9.16].

The exciton energy associated with the transition across the Peierls gap in polyacetylene can be estimated from the measured energies of the transitions from the ground state to the ^1B_u state in finite polyenes of $2N$ conjugated carbon atoms by extrapolation to the limit $N = \infty$. Here the final state of the transition corresponds to an exciton where a hole is in a filled bonding-like π band and an electron in a vacant antibonding-like π band (π^* band). Figure 9.11 shows the transition energies of several polyenes taken from the paper by *Suzuki* and *Mizuhashi* [9.18] plotted against $1/N$. By extrapolation, the energy of the excitation seems to come to about 2 eV at $N = \infty$ which should correspond to the exciton energy in polyacetylene. A theoretical study of the excitation in polyene was made by *Duke* et al. [9.19] and they obtained 2.3 eV as the infinite polyene value or the singlet exciton energy in polyacetylene by the theoretical extrapolation of the ^1B_u energy to $N = \infty$. These values are close to the peak energy 2.0 eV of the absorption in *trans*-polyacetylene in Fig. 9.9 and suggest that the absorption in polyacetylene is due to the exciton, in contrast to the band-to-band transition mentioned above. If the exciton picture is taken for the 2.0 eV absorption, the interband transition energy is likely to be about 5.5 eV because the binding energy of the π-π^* exciton is estimated to be about 3.5 eV [9.13].

The origin of the absorption has not been clarified, the discussion has not been settled and further study is necessary. As for the exciton in a quasi-1-d system, an energy state of the 1-d hydrogen atom which would be a prototype of the Wannier exciton in 1-d lattice is well known [9.20]. In real quasi-1-d crystals, however, nothing is known about the energy and the line shape of exciton absorption. The effects of a discontinuity of the lattice, of the characteristic strong electron-phonon coupling and of the localization of an electronic state in a 1-d system will give exciton absorption a unique line shape which could be different from that in a 3-d system. It would be quite desirable to study this problem theoretically to clarify the origin of the absorption in polyacetylene. Further, reexaminations of photoconductivity and other photo–transport phenomena in $(CH)_x$ will be needed, with measurements made at low temperatures on single crystals to avoid the effect of thermal dissociation of electrons bound to holes and imperfections.

As means of relaxation of the photo-excited state in polyacetylene, a resonance Raman scattering and luminescence have been measured by *Lichtmann* et al. [9.21] and *Lauchlan* et al. [9.16]. In the *cis* isomer, multiple Raman lines of the backbone stretching mode and a broad and weak luminescence were both observed in the same spectral region. The luminescence is centered at 1.9 eV, which is just below the steep increase in the absorption spectrum, and can be excited by irradiation at an energy higher than 2.05 eV. In the *trans* isomer, no indication of luminescence near the absorption edge and no long series of Raman overtone lines were found. The observation of the band edge luminescence in the *cis* isomer but not in the *trans* isomer and the observation of photoconductivity in the *trans* isomer but not in the *cis* isomer were interpreted by *Lauchlan* et al. [9.16] in terms of photo-generated solitons. According to their interpretation, in a *trans* isomer, charged soliton-antisoliton pairs are formed by the relaxation of photo-excited electron-hole pairs because of the presence of the doubly degenerate ground states (a) and (b) shown in Fig. 9.8. For this reason, a band-edge luminescence is not expected in *trans* isomers. Photoconductivity is observed in the *trans* isomer as charged solitons are carriers and so contribute to the photoconductivity. In the *cis* isomer, nondegenerate ground states lead to a difficulty in the formation of solitons and give an absence of photoconductivity and an observation of a recombination luminescence.

In *trans* and *cis* isomers, *Orenstein* and *Baker* [9.22] found a transient optical absorption induced by irradiation at 2.2 eV, which is the energy high enough to excite electrons, at an energy less than the energy of the edge absorption. The spectra of the transient absorptions are shown in Fig. 9.12. A dominant feature at low temperatures is a narrow peak at 1.35 eV in *trans*-$(CH)_x$ and 1.55 eV in *cis*-$(CH)_x$. In the *trans* isomer, there is an additional absorption at 0.5 eV. The high-energy absorptions in *trans* and *cis* isomers are assigned to the transition from the lower self-trapped state, or small polaron state, associated with the valence band to the upper self-trapped state associated with the conduction band. Here the self-trapped state, which is equivalent to a pair of overlapped soliton kinks, is a metastable intermediate state and it decays to a separated kink-antikink pair. The low-energy absorption observed only in the *trans* isomer was suggested to result from a self-trapped state of a single carrier.

Recently, a broad Stokes-shifted luminescence was found in the *cis* isomer by *Imhoff* and *Fitchen* [9.23] and by *Yoshino* et al. [9.24] at 1.4 eV at 77 K, which moves toward lower energy with increasing *trans* isomer concentration. This luminescence can be observed even in the *trans* isomer, with its peak at about

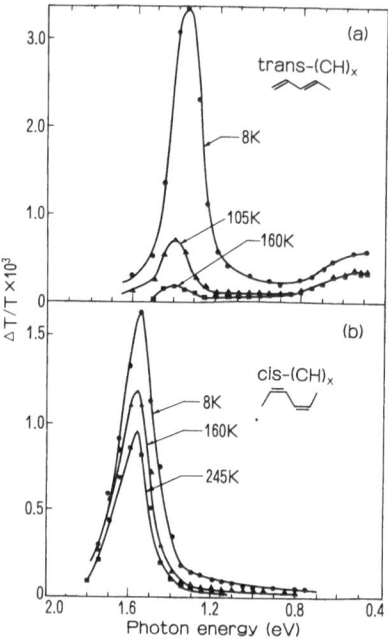

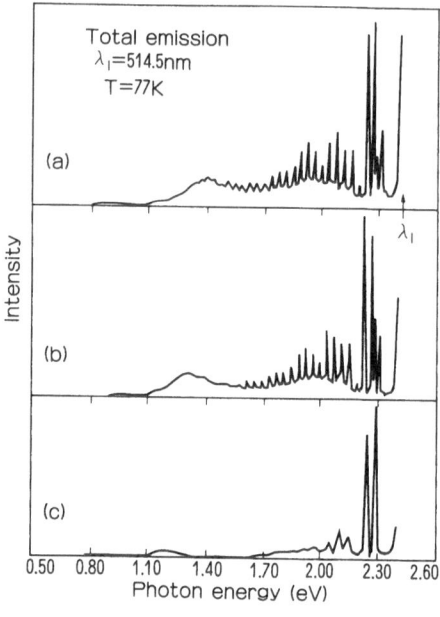

Fig. 9.12a, b. Spectra of transient optical absorption $\Delta T/T$ in (a) *trans*-(CH)$_x$ and (b) *cis*-(CH)$_x$ induced by excitation at 2.2 eV. Here T is the total transmitted signal and ΔT the difference in transmission caused by the excitation [9.22]

Fig. 9.13a–c. Emission spectra of polyacetylene for three different compositions of *trans* and *cis* isomers at 77 K excited at 2.41 eV. The *cis* fractions are about (a) 90%, (b) 80%, and (c) 0 [9.23]

1.2 eV, although its intensity is weak. Figure 9.13 shows the emission spectra of (CH)$_x$ with different *cis* concentrations [9.23]. The emission at 1.2 eV in the pure *trans* isomer was observed also by *Yoshino* et al. [9.25]. We note that the emission spectra in Fig. 9.13 are quite similar to the spectrum of WRS shown in Fig. 9.4. Both consist of a series of Raman lines and a broad and weak background emission band beneath the Raman lines, which is followed by a Stokes-shifted luminescence band at the low-energy end. This would suggest the possibility of interpreting the relaxation process in polyacetylene in terms of a mechanism similar to that given for WRS. The small intensity of the Stokes-shifted luminescence in polyacetylene compared with that in WRS would be due to the short lifetime of the metastable self-trapped state due to a fast decay to a more stable relaxed state such as a soliton state in the *trans* isomer.

At the end of this section, the absorption spectrum of highly doped *trans*-(CH)$_x$ measured by *Suzuki* et al. [9.26] must be mentioned. Figure 9.14 gives the spectra. Here the absorption spectra of undoped and AsF$_5$-doped *trans*-(CH)$_x$ films are shown. With the doping, an absorption appears at a lower energy than the 2.0 eV absorption and grows with the increase of AsF$_5$ concentration. This

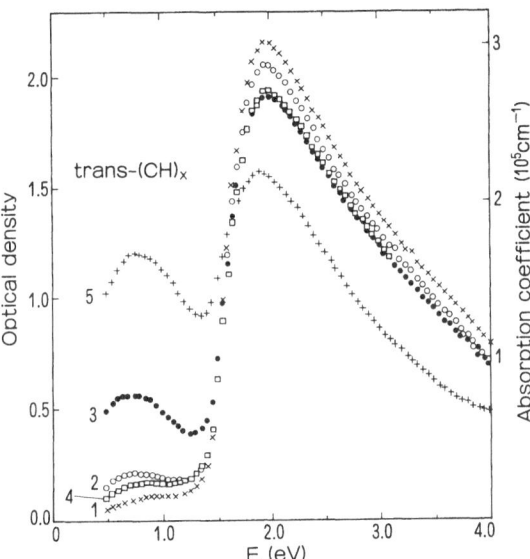

Fig. 9.14. Absorption spectra of *trans*-(CH)$_x$: curve 1, undoped; curve 2, ~0.01% AsF$_5$; curve 3, ~0.1% AsF$_5$; curve 4, compensated with NH$_3$; and curve 5, ~0.5% AsF$_5$ [9.26]

band is believed to be associated with the excitation of an electron from a valence band to a positively charged soliton level, which is formed by the doping with the acceptor AsF$_5$, resulting in the formation of a neutral soliton.

9.3 Mixed Stacked Donor-Acceptor Charge Transfer Complexes

Another class of quasi-1-d compounds is mixed stacked donor-acceptor charge transfer organic complexes, in which donor molecules (D) and acceptor molecules (A) are stacked alternately along a certain direction.

An example is a crystal of tetrathiafulvalene(TTF)-chloranil. The cystal is made by an alternate face-to-face stacking of donor TTF and acceptor chloranil molecules, the structures of which are shown in Fig. 9.15. The potential for electrons alternates along the stack with a period of –D–A– similar to the case of WRS where the potential varies with a period of –Pt^{2+}–Cl$^-$–Pt^{4+}– as was shown in Fig. 9.5a. These compounds are either neutral (D^{0}A^0) or ionic (D$^+$A$^-$) in the

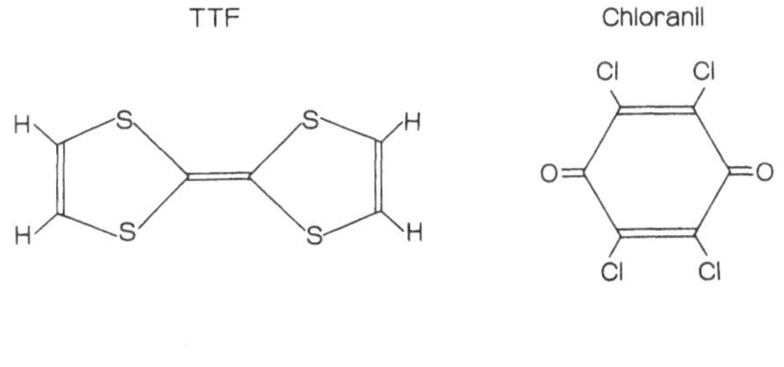

Fig. 9.15. Molecular structures of TTF and chloranil

ground state. In the neutral state, the charge transfer between D and A is vanishingly small and in the ionic state it is almost complete. Whether the neutral state or the ionic state is the stable state depends on the total energy per DA pair, E_{DA}. If we ignore the transfer energy between neighbouring molecules because it is small compared with other energies involved in DA complex, E_{DA} is

$$E_{DA} = (I - A) - \frac{\alpha e^2}{R} . \tag{9.1}$$

Here I and A are the ionization potential of the donor and the electron affinity of the acceptor, respectively. The distance between the nearest D and A molecules is R, and α is the Madelung constant where $\alpha e^2/R$ is the Madelung energy gained when all D and A molecules in the lattice are ionized. If $(I - A) > \alpha e^2/R$ the material is neutral and if $(I - A) < \alpha e^2/R$ it is ionic [9.27]. An inclusion of a transfer energy gives a hybridization of the neutral and ionized states, resulting in a partly ionized state $(D^{\varrho+}A^{\varrho-})$ where $0 < \varrho < 1$. A theoretical investigation of the neutral-ionic problem has been made by *Soos* et al. [9.28-30].

In the neutral complex, the energy of an electronic excitation of an electron from a donor to a neighboring acceptor along the stack by a charge transfer $(D^0A^0 \rightarrow D^+A^-)$ is [9.27]

$$E_{CT}^I = (I - A) - \frac{e^2}{R} . \tag{9.2}$$

Here the Madelung constant α in (9.1) has dropped out because the excitation is between only the nearest neighbours in the neutral chain. Since $\alpha > 1$, $E_{CT}^N > E_{DA}$ holds. For the ionic complex, a charge transfer excitation is a trans-

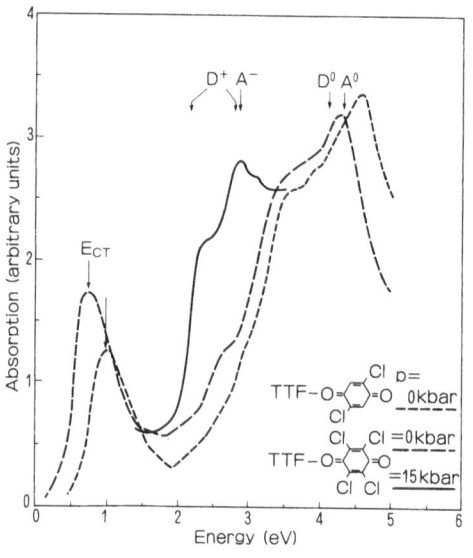

Fig. 9.16. Absorption spectra of TTF-dichlorobenzoquinone at the applied pressure of 0 kbar and of TTF-chloranil at the applied pressures of 0 and 15 kbar at room temperature [9.27]

fer of an electron from a charged acceptor back onto a neighbouring charged donor $(D^+A^- \rightarrow D^0A^0)$. The energy of the transition is [9.27]

$$E_{CT}^I = (2\alpha - 1) \frac{e^2}{R} - (I - A) .$$ (9.3)

Classification of the mixed stacked DA complexes taking account of these relations was made by *Torrance* et al. [9.27]. They determined the magnitude of E_{CT} and the form of the material from optical data. An example is shown in Fig. 9.16. The lines with shorter and larger dashes are the spectra of TTF-dichlorobenzoquinone and TTF-chloranil, respectively, which are both in a neutral form at atmospheric pressure and room temperature. The strong absorptions in the 3–5 eV range are characteristic of the intramolecular absorptions in a neutral molecule. The solid line is the spectrum of TTF-chloranil at the applied pressure of 15 kbar, at which pressure the complex is ionic. The absorptions in the region 1.5–3 eV are due to the intramolecular transitions in the ionic molecule. Thus the spectra can be used for the identification of neutral and ionic natures. The absorption band found below 1.5 eV is the charge transfer band which shifts to lower energies as the electron affinity of the acceptor increases, as is expected from (9.2).

In Fig. 9.17, the charge transfer excitation energies E_{CT}, given by (9.2) and (9.3) for neutral and ionic complexes, are given, against $(I - A)$, by the V-shape straight line [9.27]. The vertical broken line is the neutral-ionic boundary where $I - A = ae^2/R$ is satisfied. In this figure, E_{CT} values measured in a number of compounds listed in Table 9.1 are plotted. As a measure of $I - A$, a redox potential ΔE_{redox}, which is the difference between the oxidation potential of the donor and the reduction potential of the acceptor in solution, is used because of the

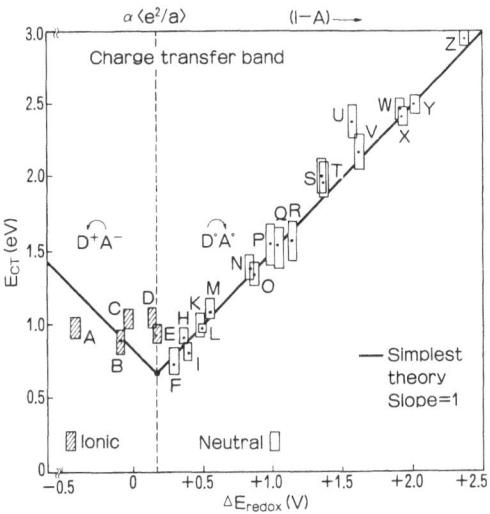

Fig. 9.17. Theoretical charge transfer excitation energies E_{CT} for neutral and ionized donor-acceptor complexes plotted against $(I - A)$ are represented by the V-shape straight line. The vertical broken line is the neutral-ionic boundary. The values of E_{CT} measured in several complexes listed in Table 9.1 are plotted as a function of a redox potential ΔE_{redox} which is a measure of $(I - A)$ [9.27]

Table 9.1. Compounds in Fig. 9.17. TMPD stands for tetramethylphenylenediamine; TCNQ, tetracyanoquinodimethane; TMDAP, tetramethyldiaminopyrene; TTF, tetrathiafulvalene; DEDMTSeF, diethyldimethyltetraselenafulvalene; DDQ, dichlorodicyano-p-benzoquinone; TCNE, tetracyanoethylene; PMDA, pyromelliticdianhydride [9.26]

Symbol	Compound	N/I
A	TMPD-tetrafluoroTCNQ	I
B	dimethylphenazine-TCNQ	I
C	TMPD-TCNQ	I
D	TMPD-chloranil	I
E	TMDAP-TCNQ	N
F	TTF-chloranil	N
G	TTF-fluoranil	N
H	DibenzeneTTF-TCNQ	N
I	DEDMTSeF-diethylTCNQ	N
J	TMDAP-fluoranil	N
K	TTF-dichlorobenzoquinone	N
L	perylene-tetrafluoroTCNQ	N
M	perylene-DDQ	N
N	perylene-TCNE	N
O	perylene-TCNQ	N
P	TTF-dinitrobenzene	N
Q	perylene-chloranil	N
R	pyrene-TCNE	N
S	pyrene-chloranil	N
T	anthracene-chloranil	N
U	hexamethylbenezene-chloranil	N
V	naphthalene-TCNE	N
X	anthracene-PMDA	N
Y	anthracene-tetracyanobenzene	N
Z	phenanthrene-PMDA	N

relation $I - A = \Delta E_{redox} + \Delta G$ where ΔG is a constant representing the solvation energy. The values for neutral compounds are shown as open boxes and those for ionic compounds as hatched boxes. Fairly good agreement can be seen between the simple theory and the experiment.

The compounds found near the neutral-ionic boundary are interesting. Because they are near the boundary, there is the possibility of inducing a change from one phase to the other by applying a perturbation externally. One such perturbation is pressure. If pressure is applied to a neutral compound, the spacing between molecules decreases, the Madelung energy increases and hence the boundary in Fig. 9.17 shifts towards the right, thus causing a neutral compound to change to an ionic compound. The change of the spectrum of TTF-chloranil, which is a substance positioned fairly closely to the boundary, by the application of pressure at room temperature is shown in Fig. 9.16 and it is just what we expect from the above discussion. A similar transition is found in several other compounds [9.27].

Besides pressure, temperature will promote the transition between neutral and ionic compounds. By lowering the temperature, the transition from neutral to ionic is driven due to the increase of Madelung energy caused by thermal contraction. Indeed, *Torrance* et al. [9.31] found a phase transition from neutral to ionic in a TTF-chloranil crystal. By lowering temperature, the transition was observed to begin at about 84 K and extend to about 50 K. The absorption spectra at 300 K and 11 K are very similar to the spectra at the applied pressures of 0 kbar and 15 kbar in Fig. 9.16, respectively. They suggest that the transition is not first order and that there is an intermediate region in which both the neutral and the ionic stacks coexist.

However, *Tokura* et al. [9.32] found that the transition occurs in a very narrow temperature range at $T_c = 84.4 \pm 0.5$ K. They used single crystals and measured reflection spectra with lights polarized parallel and perpendicular to the a-axis which is the stack axis. The charge transfer band is strongly polarized in the a direction because of 1-d overlap along the a-axis. It is absent in the spectrum perpendicularly polarized in a. The intramolecular transition bands in TTF are predominantly polarized perpendicularly to the a-axis since the transition dipoles lie on the TTF molecular plane. Figure 9.18 gives the spectra. In the spectra, A_1, A_2, and A_3 are due to the charge transfer excitations. The structures B, C, D, and E originate from the intramolecular transitions in TTF molecules. The peaks B, C, and D are due to a charged molecule TTF^+ and E to a neutral molecule TTF^0. The spectra exhibit a sudden change at about 84 K. The change is drastic in the TTF intramolecular transition band. The band E from neutral TTF^0 grows above 84 K. The presence of the TTF^+ intramolecular bands B and

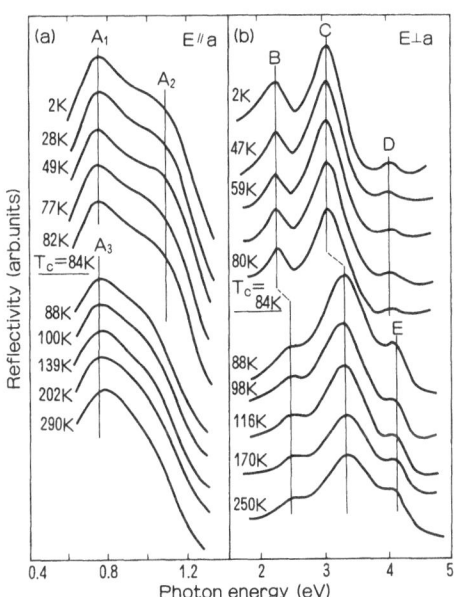

Fig. 9.18. Temperature changes of (**a**) the charge transfer excitation band and (**b**) the intramolecular excitation bands in TTF-chloranil in the vicinity of the neutral-to-ionic phase transition temperature of 84 K [9.32]

Anthracene   PMDA

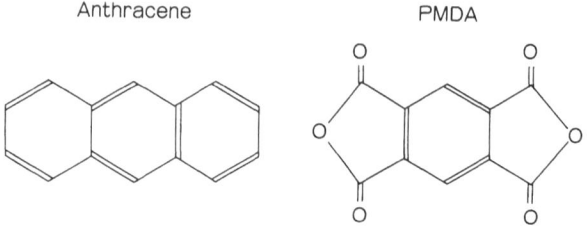

Fig. 9.19. Molecular structures of anthracene and PMDA

C in the neutral phase after discontinuous energy changes at 84 K is ascribed to the partially charge-transferred ground state of the neutral phase which is caused by the hybridization of a perfectly neutral state with an ionic state due to a transfer between TTF and the chloranil molecules.

A charge transfer excitation and its relaxation in a mixed stacked donor-acceptor complex has been studied optically in anthracene-pyromellitic acid dianhydride (A-PMDA) which is a highly neutral complex in the ground state as can be seen from Fig. 9.17 and Table 9.1. It is a crystal in which donor anthracene and acceptor PMDA molecules are stacked alternately face-to-face in the c direction. The component molecules are shown in Fig. 9.19.

Haarer [9.33, 34] made a detailed study of singlet charge transfer absorption and reflection spectra and an emession spectrum in A–PMDA at low temperatures using light polarized nearly perpendicularly to the c-axis. He found an absorption and an emission of zero-phonon origin at the same energy. Narrow

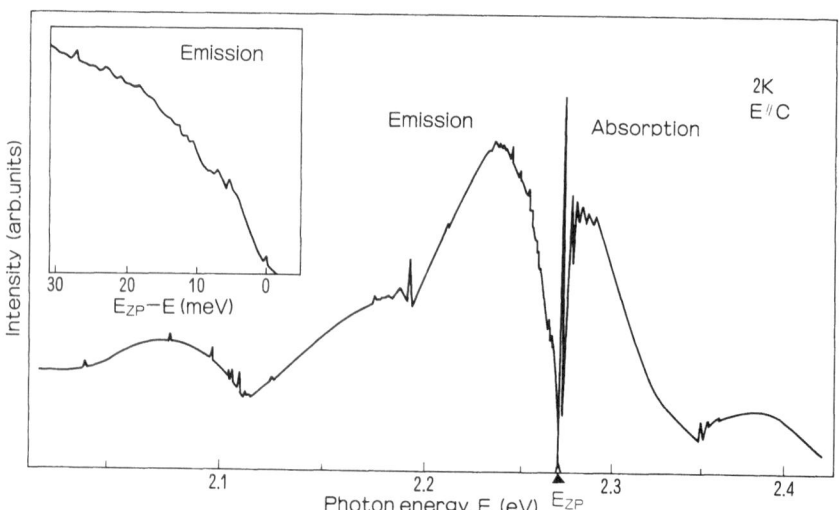

Fig. 9.20. The absorption spectrum measured by light polarized in the stack axis and the emission spectrum of Anthracene-PMDA at 2 K. Here E_{zp} is the energy of a zero-phonon absorption. The inset is the enlarged spectrum of the emission near E_{zp} in which the zero-phonon emission line is clearly observed [9.36]

phonon sidebands corresponding to the excitation of intramolecular vibrational modes of 78.6 meV were observed in the absorption and emission spectra with some fine structure. Further study was carried out by *Brillante* and *Philpott* [9.35] who measured the absorption spectrum with high resolution using light polarized parallel to the c-axis. It was found that several librational modes were associated with the vibronic lines. The absorption and emission spectra measured recently by *Oowaki* et al. [9.36] are shown in Fig. 9.20. These structures suggest a strong electron-phonon coupling and are analysed in terms of a charge transfer vibronic transition within the framework of a configuration coordinate model [9.34, 35].

Haarer [9.34] tried to determine the band width of the charge transfer exciton in A-PMDA by using mixed crystals of A–PMDA and deuterated anthracene-PMDA. The principle of the determination is based on the discussion given in Sect. 4.2.3. If the difference in excitation energy between A–PMDA and deuterated A-PMDA is large in comparison with the exciton band width, one expects a persistence type of spectrum in which two absorptions due to both components are observed separately on alloying. If, on the other hand, the deuteration shift is smaller than the exciton band width, one expects an amalgamation type in which only one spectral feature is observed which shifts with the composition. Thus one can estimate a limit of the band width by alloying. The range of the exciton band width thus estimated was 1 meV or larger. This work was extended by *Tokura* and *Koda* [9.37] who determined the band width as about 12 meV by considering a nonlinear energy shift of an amalgamated exciton band discussed in Sect. 4.2.3 and examining the vibronic structures in luminescence spectra.

The broad structures underneath the sharp phonon sidebands are different in the absorption and the emission spectra. The origin of the difference is suggested by *Brillante* and *Philpott* [9.35] in terms of a self-trapped charge transfer exciton or a charge transfer exciplex as the initial state of the emission.

9.4 Segregated Stacked Donor-Acceptor Charge Transfer Complexes

When molecules of a single kind are lined up periodically to form a chain and electrons with a density of one electron per molecule are supplied from donors, the chain will be a half-filled metal. However, it is unstable and will become an insulator by an electron-phonon interaction through the Peierls transition as in the cases of WRS and $(CH)_x$. When one considers an electron-electron Coulomb interaction, the metal will also become an insulator without the assistance of the electron-phonon interaction. If the Coulomb energy of the electron-electron repulsion is greater than the transfer energy of electron between sites, the electrons are no longer delocalized but are localized with one electron per site in an antiferromagnetic order to form an insulator, which is called a Hubbard insulator [9.38]. In the Hubbard insulator, the metallic band splits into a filled lower

Hubbard band and an empty upper Hubbard band. The gap between them is called the Hubbard gap. Whether a material is a Peierls insulator, a Hubbard insulator, or a metal depends on the energies of the electron-phonon coupling, electron-electron repulsion and electron transfer. This was discussed in Sect. 4.6.4 and by *Nasu* [9.6–8].

A quasi-1-d Hubbard insulator is a simple system because the potentials for an electron are same at all sites, in contrast to the cases of halogen-bridged mixed valence compounds, polyacetylene, and mixed stacked donor-acceptor complexes where the potential varies from site to site periodically.

There are many compounds which belong to this class. Typical examples are the 7,7,8,8-tetracyanoquinodimethane (TCNQ) anion radical salts. The donors are alkali or transition metals, ammonium, and some organic radicals. The Donors and TCNQ are positively and negatively charged, respectively, by the transfer of electrons from the former to the latter. In these compounds, TCNQ molecules, whose structure is shown in Fig. 9.21a, are stacked face-to-face to form a quasi-1-d column as shown in Fig. 9.21b. The salts can be classified into two classes. One consists of salts in which the ratio of the donor to the acceptor is $1:1$, such as K-TCNQ. The other type is the salts whose stoichiometry is not $1:1$, and a typical example is Cs_2-$(TCNQ)_3$.

Absorption spectra were measured first by *Iida* [9.39, 40] for $1:1$, $1:2$, and $2:3$ salts. The investigation was extended further by *Hiroma* et al. [9.41],

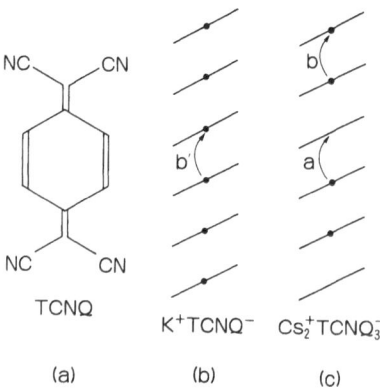

Fig. 9.21. Molecular structure of TCNQ is given in (**a**). The face-to-face stacking of TCNQ column and the electron distribution in K-TCNQ and Cs_2-$TCNQ_3$ are shown in (**b**) and (**c**)

Fig. 9.22. Absorption spectra of TCNQ ▶ charge transfer salts [9.43]

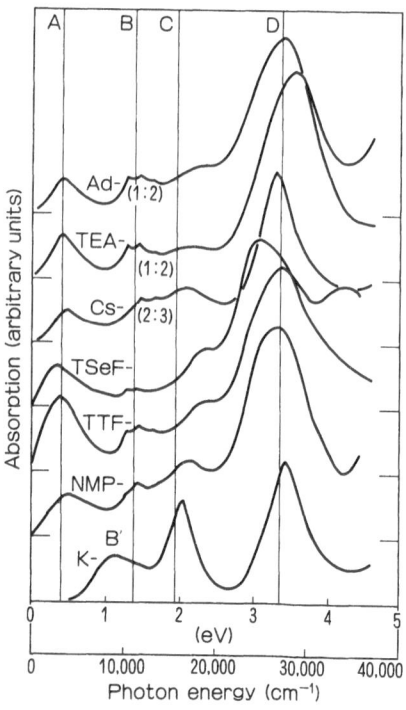

Oohashi and *Sakata* [9.42] and *Torrance* et al. [9.43]. The most complete set of spectra of single crystals of TCNQ salts has been given by *Tanaka* et al. [9.44]. The absorption spectra measured by *Torrance* et al. [9.43] are shown in Fig. 9.22 for a number of TCNQ salts. Here K-TCNQ is a 1:1 insulating salt and Ad(acridinium)- and TEA(triethylammonium)-TCNQ are 1:2 salts. The NMP(N-methylphenazinium)-, TTF(tetrathiafulvalene)- and TSeF(selenium analogue of TTF)-salts are 1:1 salts but they are metallic due to incomplete charge transfer so that the average electron density is less than one per TTF molecule. The spectra are similar to each other except for that of K-TCNQ. In K-TCNQ there are three absorption peaks while in the others there are four. This difference in the spectra was interpreted by *Torrance* et al. [9.43]. The absorptions C and D are common to all the materials and are due to the intramolecular excitations of the $TCNQ^-$ ion [9.39]. The lowest-energy absorption B' in K-TCNQ is a charge transfer excitation between charged $TCNQ^-$ ions [9.39, 41], as is shown by b' in Fig. 9.21b. It is equivalent to $(TCNQ^-)(TCNQ^-) \rightarrow (TCNQ^0)(TCNQ^{2-})$ or to the transition from the lower Hubbard band to the upper Hubbard band. In insulating Cs_2-$(TCNQ)_3$, electrons are localized in the manner shown in Fig. 9.21c. Here pairs of $TCNQ^-$ ions are separated by a $TCNQ^0$ neutral molecule so that it is a mixed-valence compound. In this compound two absorptions A and B are found at low energy in the spectrum instead of a single peak B' in K-TCNQ. This is because there must be two charge transfer excitations in Cs_2-$(TCNQ)_3$ as shown by (a) and (b) in the figure. The energy of the transition b should be close to that of the transition b' in K-TCNQ and the absorption B is due to the transition b. The energy of the charge transfer a from a charged $TCNQ^-$ to a neutral $TCNQ^0$ must be very much smaller than that of *b* because of the absence of a Coulomb repulsion energy at a $TCNQ^0$ site. The absorption A corresponds to the transition a [9.39].

In Ad-$(TCNQ)_2$ and TEA-$(TCNQ)_2$, the possibility of observing the absorption B must be very low and only the absorption A will be observed if the electrons on the TCNQ stack are completely localized on alternate TCNQ molecules so that two electrons are never in neighbouring TCNQ sites. The spectra in Fig. 9.22 show the absorption B, in contrast to the above expectation. Unlike Cs_2-$(TCNQ)_3$, these two compounds are conducting and electrons are not localized in the TCNQ molecules. Thus they are like a half-filled metal and are similar to an incomplete charge transfer complex. We find in the figure that similar spectra are also found in 1:1 metallic compounds TSeF-TCNQ, TTF-TCNQ, and NMP-TCNQ in all of which charge transfers are known to be incomplete. Thus the observation of B in these compounds is related to the delocalized nature of the electrons in incomplete charge transfer complexes. The origin has been suggested by *Torrance* et al. [9.43] as follows: Statistically speaking, there can be a number of pairs of electrons on neighbouring TCNQ sites in an incomplete charge transfer complex, giving a charge transfer absorption band B due to the excitation of an electron from a $TCNQ^-$ site to a neighbouring $TCNQ^-$ site. In terms of the band picture, it is the excitation of an electron in the partially filled lower Hubbard band to the vacant upper Hubbard band. The absorption A

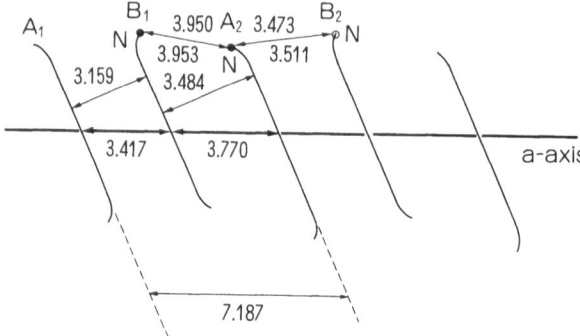

Fig. 9.23. Structure of the TCNQ stack in Rb-TCNQ at −160 °C [9.45]

is due to the transition of an electron from $TCNQ^-$ to a statistically available $TCNQ^0$. In a more exact way, it is the transition within the partially filled lower Hubbard band.

It should be remarked in the final part of this section that the preceding discussion is based on the stack of equally spaced TCNQ molecules which gives rise to an antiferromagnetic ordering to form the usual Hubbard insulator in the ground state. However, this is only true at high temperatures. Alkali metal-TCNQ salts are known to make a structural change, with decreasing temperature, from the equally spaced, uniform, high temperature structure. The transition temperatures are 346, 395, 376, 217, and 299 K in the Na, K, Rb, Cs and NH_4 salts, respectively. Below the transition temperature, the structure changes and becomes a regular array of singlet spin pairs by the dimerization of $TCNQ^-$ accompanied by a dimerized lattice distortion so that the 1-d unit cell contains two molecules [9.45–48]. An example of the structure of the low-temperature phase is shown in Fig. 9.23 for Rb-TCNQ [9.45]. In Na-TCNQ, an equal inter-planar spacing of 3.385 Å between TCNQ radicals above 346 K changes to alter-nating interplanar distances of 3.21 and 3.49 Å [9.46, 47]. In K-TCNQ, an equal spacing of 3.476 Å changes to 3.237 and 3.567 Å below 395 K [9.48]. At these transition temperatures, a magnetic transition is observed [9.49, 50]. This transition has been explained by *Lépine* et al. [9.51] in terms of a spin-Peierls instability [9.52, 53] which is expected for a linear antiferromagnetic chain interacting with 3-d phonons. Since the instability originates from the dependency of the exchange integral on the instantaneous positions of the magnetic ions, the spin-Peierls transition is expected in alkali-TCNQ salts because of their large thermal expansion coefficients of the lattice in the stacking direction. The study of optical excitation and relaxation in relation to the spin-Peierls transition in alkali-TCNQ salts would be a quite interesting field of investigation.

References

Chapter 1

1.1 L. D. Landau, E. M. Lifshitz: *Statistical Physics*, English transl. by J. B. Sykes, M. J. Kersley (Pergamon, Oxford 1968)

1.2 P. W. Anderson: *Basic Notions of Condensed Matter Physics*, (Benjamin Cummings, Menlo Park USA 1984)

1.3 D. Pines: *Elementary Excitations in Solids* (Benjamin, New York 1968)

1.4 P. W. Anderson: *Concepts in Solids* (Benjamin, New York 1963)

1.5 G. H. Wannier: Phys. Rev. **52**, 191 (1937)

1.6 N. F. Mott, R. W. Gurney: *Electronic Processes in Ionic Crystals* (Clarendon, Oxford 1940)

1.7 W. Kohn, J. M. Luttinger: Phys. Rev. **97**, 1721 (1955); ibid. **98**, 915 (1955)

1.8 J. Frenkel: Phys. Rev. **37**, 17, 1276 (1931)

1.9 R. S. Knox: *Theory of Excitons*, Solid State Physics, Suppl. Vol. 5, ed. by F. Seitz, D. Turnbull (Academic, New York 1963)

1.10 K. Cho (ed.): *Excitons*, Topics Curr. Phys., Vol. 14 (Springer, Berlin, Heidelberg 1979)

1.11 H. Haken, W. Schottky: Z. Phys. Chem. (Frankfurt/M) **16**, 218 (1958)

1.12 Y. Toyozawa: Prog. Theor. Phys. **12**, 421 (1954)

1.13 T. Onodera, Y. Toyozawa: J. Phys. Soc. Jpn. **22**, 833 (1967)

1.14 A. S. Davydov: *Theory of Molecular Excitons* (Plenum, New York 1971)

1.15 M. Pope, C. E. Swenberg: *Electronic Processes in Organic Crystals* (Clarendon, Oxford 1982) p. 73;
D. Haarer, M. R. Philpott: "Excitons and Polarons in Organic Weak Charge Transfer Crystals", in *Spectroscopy and Excitation Dynamics of Condensed Molecular Systems*, ed. by V. M. Agranovich, R. M. Hochstrasser (North-Holland, Amsterdam 1983) Chap. 2, p. 27

1.16 R. J. Elliott: Phys. Rev. **108**, 1384 (1957)

1.17 Y. Toyozawa, M. Inoue, T. Inui, M. Okazaki, E. Hanamura: J. Phys. Soc. Jpn. **22**, 1337, 1349 (1967);
Y. Toyozawa: J. Lumin. **24/25**, 23 (1981)

1.18 I. Egri: J. Phys. C **12**, 1843 (1979)

1.19 V. M. Agranovich, V. L. Ginzburg: *Crystal Optics with Spatial Dispersion, and Excitons*, 2nd ed. (Springer, Berlin, Heidelberg 1984)

1.20 J. L. Birman: "Electrodynamic and Non-Local Optical Effects Mediated by Exciton Polaritons", in *Excitons*, ed. by E. I. Rashba, M. D. Sturge (North-Holland, Amsterdam 1982) Chap. 2, p. 27

1.21 M. Born, K. Huang: *Dynamical Theory of Crystal Lattices* (Oxford University Press, Oxford 1954) Chaps. 8, 10

1.22 J. J. Hopfield: Phys. Rev. **112**, 155 (1958)

1.23 D. L. Dexter, R. S. Knox: *Excitons* (Wiley, New York 1965)

1.24 E. I. Rashba, M. D. Sturge (eds.): *Excitons* (North-Holland, Amsterdam 1982)

1.25 H. Haken, S. Nikitine (eds.): *Excitons at High Density*, Springer Tracts Mod. Phys., Vol. 43 (Springer, Berlin, Heidelberg 1974)

1.26 E. A. Silinsh: *Organic Molecular Crystals*, Solid-State Sci., Vol. 16 (Springer, Berlin, Heidelberg 1980)

1.27   V. M. Kenkre, P. Reineker: *Exciton Dynamics in Molecular Crystals and Aggregates*, Springer Tracts Mod. Phys., Vol. 94 (Springer, Berlin, Heidelberg 1982)
1.28   P. Reineker, H. Haken, H. C. Wolf (eds.): *Organic Molecular Aggregates*, Solid-State Sci., Vol. 49 (Springer, Berlin, Heidelberg 1983)
1.29   E. I. Rashba, E. F. Sheka, V. I. Broude: *Spectroscopy of Molecular Excitons*, Chem. Phys., Vol. 16 (Springer, Berlin, Heidelberg 1985)

Chapter 2

2.1   O. Akimoto, E. Hanamura: J. Phys. Soc. Jpn. **33**, 1537 (1972)
2.2   W. F. Brinkman, T. M. Rice, B. Bell: Phys. Rev. B**8**, 1570 (1973)
2.3   E. Hanamura: J. Lumin. **12/13**, 119 (1976)
2.4   E. Hanamura: Solid State Commun. **12**, 951 (1973)
2.5   W. F. Brinkman, T. M. Rice, P. W. Andersosn, S. T. Chui: Phys. Rev. Lett. **28**, 961 (1972)
2.6   R. P. Groff, P. Avakian, R. E. Merrifield: Phys. Rev. B**1**, 815 (1970)
2.7   T. Inui: Proc. Phys. Math. Soc. Jpn. **20**, 770 (1938)
2.8   E. A. Hylleraas, A. Ore: Phys. Rev. **71**, 493 (1947)
2.9   R. K. Wehner: Solid State Commun. **7**, 457 (1969)
2.10   J. Adamowski, S. Bednarek, M. Suffcznski: Solid State Commun. **9**, 2037 (1971)
2.11   F. Bassani, J. J. Forney, A. Quattropani: Phys. Status Solidi (b) **65**, 591 (1974)
2.12   E. Hanamura: J. Phys. Soc. Jpn. **39**, 1506 (1975)
2.13   M. Certier, C. Wecker, S. Nikitine: J. Phys. Chem. Solids **30**, 2135 (1961); W. Staude: Phys. Lett. **29 A**, 228 (1969)
2.14   J. S. Wang, C. Kittel: Phys. Lett. **42 A**, 189 (1972)
2.15   Ya. E. Pokrovskii, K. I. Svistunova: JETP Lett. **9**, 261 (1969)
2.16   E. Hanamura: In Proc. 10th Int. Conf. Phys. Semicond., Cambridge USA (1970), ed. by S. P. Keller, J. C. Hensel, F. Stern (USAEC Div. of Technical Information, Oak Ridge 1970) p. 487
2.17   W. F. Brinkman, T. M. Rice: Phys. Rev. B**7**, 1508 (1973)
2.18   M. Combescot, P. Nozières: J. Phys. C**5**, 2369 (1972)
2.19   M. Inoue, E. Hanamura: J. Phys. Soc. Jpn. **35**, 643 (1973)
2.20   P. Vashishta, P. Bhattacharyya, K. S. Singwi: Phys. Rev. B**10**, 5108 (1974); P. Vashishta, S. G. Das, K. S. Singwi: Phys. Rev. B**10**, 911 (1974)
2.21   G. Beni, T. M. Rice: Phys. Rev. Lett. **37**, 874 (1976)
2.22   L. V. Keldysh, A. P. Silin: Sov. Phys.-JETP **42**, 535 (1976)
2.23   V. D. Kulakovskii, V. B. Timofeev, V. M. Edelstein: Sov. Phys.-JETP **47**, 193 (1978)
2.24   K. Kobayashi: Festkörperprobleme **XVI**, 117 (1976)
2.25   J. Nakahara, K. Kobayashi: J. Phys. Soc. Jpn. **40**, 180 (1976)
2.26   R. Z. Bachrach, F. C. Brown: Phys. Rev. Lett. **21**, 685 (1968); Phys. Rev. B**1**, 818 (1970)
2.27   S. G. Chung, G. D. Sanders, Y. C. Chang: Solid State Commun. **45**, 237 (1983)
2.28   J. Nakahara, K. Kobayashi: J. Phys. Soc. Jpn. **40**, 189 (1976)
2.29   B. L. Joesten, F. C. Brown: Phys. Rev. **148**, 919 (1966)
2.30   A. B. Kunz: Phys. Rev. B**26**, 2070 (1982)
2.31   P. M. Scop: Phys. Rev. **139**, A 934 (1965); F. Bassani, R. S. Knox, W. B. Fowler: Phys. Rev. **137**, A1217 (1965)
2.32   H. Tamura, K. Masumi: Solid State Commun. **12**, 1183 (1973)
2.33   I. Pelant, A. Mysyrowicz, C. Benoît à la Guillaume: Phys. Rev. Lett. **37**, 1708 (1976)
2.34   I. Pelant, J. Hála, L. Parma, K. Vacek: Solid State Commun. **36**, 729 (1980)
2.35   T. Baba, T. Masumi: Nuovo Cimento **39 B**, 609 (1977)
2.36   D. Hulin, A. Mysyrowicz, M. Combescot, I. Pelant, C. Benoît à la Guillaume: Phys. Rev. Lett. **39**, 1169 (1977)
2.37   S. Shionoya, H. Saito, E. Hanamura, O. Akimoto: Solid State Commun. **12**, 223 (1973)
2.38   O. Akimoto: J. Phys. Soc. Jpn. **35**, 973 (1973)
2.39   D. B. Tran Thoai: Z. Phys. B**26**, 115 (1977)

2.40 V. G. Lysenko, V. I. Revenko, T. G. Tratas, V. B. Timofeev: Sov. Phys.-JETP **41**, 163 (1975)
2.41 R. F. Lehemy, J. Shah: Phys. Rev. Lett. **37**, 871 (1976)
2.42 J. C. Merle, M. Robino: Opt. Commun. **14**, 240 (1975)
2.43 F. Bassani, M. Rovere: Solid State Commun. **19**, 887 (1976)
2.44 A. Mysyrowicz, J. B. Grun, A. Bivas, R. Levy, S. Nikitine: Phys. Lett. **26A**, 615 (1968)
2.45 H. Souma, T. Goto, T. Ohta, M. Ueta: J. Phys. Soc. Jpn. **29**, 697 (1970)
2.46 E. I. Rashba: "Gigantic Oscillator Strengths Inherent in Exciton Complexes", in *Excitons at High Density*, ed. by H. Haken, S. Nikitine, Springer Tracts Mod. Phys., Vol. 73, (Springer, Berlin, Heidelberg 1975) p. 150
2.47 E. Hanamura: J. Phys. Soc. Jpn. **39**, 1516 (1975)
2.48 S. Suga, T. Koda: Phys. Status Solidi (b) **66**, 255 (1974)
2.49 E. Ostertag, R. Levy, J. B. Grun: Phys. Status Solidi (b) **69**, 629 (1975)
2.50 Y. Segawa, S. Namba: Solid State Commun. **17**, 489 (1975)
2.51 M. Inoue, E. Hanamura: J. Phys. Soc. Jpn. **41**, 771 (1976)
2.52 H. Risken: *The Fokker-Planck Equation*, Springer Series in Synergetics, Vol. 18 (Springer, Berlin, Heidelberg 1984)
2.53 G. M. Gale, A. Mysyrowicz: Phys. Rev. Lett. **32**, 727 (1974)
2.54 Y. Nozue, T. Itoh and M. Ueta: J. Phys. Soc. Jpn. **44**, 1305 (1978)
2.55 E. Doni, R. Girlanda, G. Pastori Parravicini: Solid State Commun. **17**, 189 (1975)
2.56 Vu Duy Phach, A. Bivas, B. Hönerlage, J. B. Grun: Phys. Status Solidi (b) **84**, 731 (1977)
2.57 N. Nakata, N. Nagasawa, Y. Doi, M. Ueta: J. Phys. Soc. Jpn. **38**, 903 (1975)
2.58 Vu Duy Phach, R. Levy: Solid State Commun. **29**, 247 (1979)
2.59 T. Itoh, S. Watanabe, M. Ueta: J. Phys. Soc. Jpn. **48**, 542 (1980)
2.60 Y. Toyozawa: Prog. Theor. Phys. **27**, 89 (1962)
2.61 E. Hanamura, T. Takagahara: J. Phys. Soc. Jpn. **47**, 410 (1979)
2.62 N. Nagasawa, T. Mita, M. Ueta: J. Phys. Soc. Jpn. **41**, 929 (1976)
2.63 T. Mita, M. Ueta: Solid State Commun. **27**, 1463 (1978)
2.64 Y. Segawa, Y. Aoyagi, O. Nakagawa, K. Azuma, S. Namba: Solid State Commun. **27**, 785 (1978)
2.65 Y. Toyozawa: Solid State Commun. **28**, 533 (1978)
2.66 M. Inoue, E. Hanamura: J. Phys. Soc. Jpn. **41**, 1273 (1976)
2.67 T. Itoh, T. Suzuki: J. Phys. Soc. Jpn. **45**, 1939 (1978)
2.68 F. Henneberger, K. Henneberger, J. Voigt: Phys. Status Solidi (b) **83**, 439 (1977)
2.69 T. Mita, K. Sôtome, M. Ueta: Solid State Commun. **33**, 1135 (1980); J. Phys. Soc. Jpn. **48**, 496 (1980)
2.70 W. R. Heller, A. Marcus: Phys. Rev. **84**, 809 (1951)
2.71 Y. Onodera: J. Phys. Soc. Jpn. **49**, 1845 (1980)
2.72 Y. Masumoto, S. Shionoya: J. Phys. Soc. Jpn. **49**, 2236 (1980); Solid State Commun. **38**, 865 (1981)
2.73 A. Maruani, J. L. Oudar, E. Batifol, D. S. Chemla: Phys. Rev. Lett. **41**, 1372 (1978)
2.74 A. Maruani, D. S. Chemla: Phys. Rev. **23**, 841 (1981)
2.75 G. Mizutani, N. Nagasawa: J. Phys. Soc. Jpn. **52**, 2251 (1983)
2.76 Vu Duy Phach, A. Bivas, B. Hönerlage, J. B. Grun: Phys. Status Solidi (b) **86**, 159 (1978)
2.77 Y. Nozue, N. Miyahara, S. Takagi, M. Ueta: Solid State Commun. **38**, 1199 (1981)
2.78 Y. Nozue: J. Phys. Soc. Jpn. **51**, 1840 (1982)
2.79 R. März, S. Schmitt-Rink, H. Haug: Z. Phys. B**40**, 9 (1980)
2.80 V. May, K. Henneberger, F. Henneberger: Phys. Status Solidi (b) **94**, 611 (1979)
2.81 G. Kurtze, W. Maier, G. Blattner: Z. Phys. B**39**, 95 (1980)
2.82 M. Kuwata, T. Mita, N. Nagasawa: Solid State Commun. **40**, 911 (1981)
2.83 M. Kuwata, N. Nagasawa: J. Phys. Soc. Jpn. **51**, 2591 (1982)
2.84 M. Kuwata, N. Nagasawa: Solid State Commun. **45**, 937 (1983)
2.85 M. Nakayama: Solid State Commun. **45**, 821 (1983)
2.86 T. Takagahara: Solid State Commun. **47**, 345 (1983)
2.87 E. Hanamura: Solid State Commun. **51**, 697 (1984)

2.88 T. Itoh, T. Katohno, T. Kirihara, M. Ueta: J. Phys. Soc. Jpn. **53**, 854 (1984)
2.89 T. Itoh: In Proc. 9th Int. Conf. Raman Spectroscopy, ed. by M. Tsuboi (1984) p. 42
2.90 T. Tokihiro, E. Hanamura: Solid State Commun. **52**, 771 (1984)
2.91 J. B. Grun, B. Hönerlage, R. Levy: Solid State Commun. **46**, 51 (1983)
2.92 E. Hanamura: Solid State Commun. **38**, 939 (1981)
2.93 S. W. Koch, H. Haug: Phys. Rev. Lett. **46**, 450 (1981)
2.94 H. M. Gibbs, S. L. McCall, T. N. C. Venkatesan: Optics News, **5**, No. 3, 6 (1979)
2.95 S. D. Smith, D. A. B. Miller: J. Phys. Soc. Jpn. **49**, Suppl. A597 (1980)
2.96 N. Peyghambarian, H. M. Gibbs, D. A. Weinberger, M. C. Rushford, D. Sarid: Phys. Rev. Lett. **51**, 1692 (1983)
2.97 B. Hönerlage, J. Y. Bigot, R. Levy: In *Optical Bistability*, ed. by H. Gibbs, S. McCall, C. Bowden (Plenum, New York 1984) p. 253
2.98 T. Itoh, T. Katohno, M. Ueta: J. Phys. Soc. Jpn. **53**, 844 (1984)
2.99 N. Nagasawa, N. Nakata, Y. Doi, M. Ueta: J. Phys. Soc. Jpn. **39**, 987 (1975)
2.100 L. L. Chase, N. Peyghambarian, G. Grynberg, A. Mysyrowicz: Opt. Commun. **28**, 189 (1979)
2.101 N. Peyghambarian, L. L. Chase, A. Mysyrowicz: Opt. Commun. **42**, 51 (1982)
2.102 Y. Aoyagi, Y. Segawa, S. Namba: Phys. Rev. B **25**, 1453 (1982)
2.103 N. Nagasawa, T. Mita, M. Ueta: J. Phys. Soc. Jpn. **41**, 929 (1976)
2.104 R. Levy, C. Klingshirn, E. Ostertag, Vu Duy Phach, J. B. Grun: Phys. Status Solidi (b)**77**, 381 (1976)
2.105 E. Ostertag, A. Bivas, J. B. Grun: Phys. Status Solidi (b) **84**, 673 (1977)
2.106 B. Hönerlage, Vu Duy Phach, J. B. Grun: Phys. Status Solidi (b) **88**, 545 (1978)
2.107 M. Ojima, Y. Oka, T. Kushida, S. Shionoya: Solid State Commun. **24**, 845 (1977)
2.108 N. Peyghambarian, L. L. Chase, A. Mysyrowicz: Phys. Rev. B **27**, 2325 (1983)
2.109 L. L. Chase, N. Peyghambarian, G. Grynberg, A. Mysyrowicz: Phys. Rev. Lett. **42**, 1231 (1979)
2.110 P. Peyghambarian, L. L. Chase, A. Mysyrowicz: Opt. Commun. **41**, 178 (1982)
2.111 V. B. Timofeev: "Free Many Particle Electron-Hole Complexes in an Indirect Gap Semidonductor", in *Excitons*, ed. by E. I. Rashba, M. D. Sturge (North-Holland, Amsterdam 1982) Chap. 9
2.112 J. B. Grun, B. Hönerlage, R. Levy: "Biexcitons in CuCl and Related Systems", in *Excitons,* ed. by E. I. Rashba, M. D. Sturge (North-Holland, Amsterdam 1982) Chap. 11
2.113 E. Hanamura: "Excitonic Molecules", in *Optical Properties of Solids – New Development,* ed. by B. O. Seraphin (North-Holland, Amsterdam 1976) Chap. 3
2.114 C. Benoît à la Guillaume, M. Voos: "Condensation of Excitons into Electron-Hole Drops in Semiconductors", in *Optical Properties of Solids – New Development*, ed. by B. O. Seraphin (North-Holland, Amsterdam 1976) Chap. 4
2.115 E. Hanamura, H. Haug: Phys. Rep. **33**C, 209 (1977)
2.116 C. Klingshirn, H. Haug: Phys. Rep. **70**, 315 (1981)

Chapter 3

3.1   S. Nikitine: Prog. Semicond. **6**, 235 (1962)
3.2   M. Cardona: Phys. Rev. **129**, 69 (1963)
3.3   K. Shindo, A. Morita, H. Kamimura: J. Phys. Soc. Jpn. **20**, 2054 (1965)
3.4   K. S. Song: J. Phys. Chem. Solids **28**, 2003 (1967)
3.5   S. Kono, T. Ishii, T. Sagawa, T. Kobayashi: Phys. Rev. B**8** 795 (1973)
3.6   A. Goldmann, J. Tejeda, N. J. Shevchik, M. Cardona: Phys. Rev. B**10**, 4388 (1974)
3.7   Y. Kato, T. Goto, T. Fujii, M. Ueta: J. Phys. Soc. Jpn. **36**, 175 (1974)
3.8   T. Mita, K. Sôtome, M. Ueta: Solid State Commun. **33**, 1135 (1980)
3.9   Y. Nozue: J. Phys. Soc. Jpn. **51**, 1840 (1982)
3.10 S. Suga, K. Cho, Y. Niji, J. C. Merle, T. Sauder: Phys. Rev. B**22**, 4931 (1980)

3.11 S. Nikitine: "Properties of Biexcitons", in *Excitons at High Density*, ed. By H. Haken, S. Nikitine, Springer Tracts Mod. Phys., Vol. 73 (Springer, Berlin, Heidelberg 1975) p. 22

3.12 Y. Nozue, N. Miyahara, S. Takagi, M. Ueta: Solid State Commun. **38**, 1199 (1981); J. Lumin. **24/25**, 429 (1981)

3.13 H. J. Mattaush, Ch. Uihlein: Phys. Status Solidi (b) **96**, 189 (1979)

3.14 T. Nanba, K. Hachisu, M. Ikezawa: J. Phys. Soc. Jpn. **50**, 1579 (1981)

3.15 Y. Kato, C. I. Yu, T. Goto: J. Phys. Soc. Jpn. **28**, 104 (1970)

3.16 M. Certier, C. Wecker, S. Nikitine: Phys. Lett. 28**A**, 307 (1968)

3.17 W. Staude: Phys. Lett. 29**A**, 228 (1969)

3.18 T. Koda, T. Mitani, T. Murahashi: Phys. Rev. Lett. **25**, 1495 (1970)

3.19 S. Sakoda, Y. Onodera: J. Phys. Chem. Solids **32**, 1365 (1971)

3.20 E. Mohler: Phys. Status Solidi **38**, 81 (1970)

3.21 T. Goto, T. Takahashi, M. Ueta: J. Phys. Soc. Jpn. **24**, 314 (1968)

3.22 M. Ueta, T. Goto: J. Phys. Soc. Jpn. **20**, 401 (1965)

3.23 T. Goto, M. Ueta: J. Phys. Soc. Jpn. **22**, 488 (1967)

3.24 E. F. Gross, S. Permogorov, B. Razbirin: J. Phys. Chem. Solids **27**, 1647 (1966)

3.25 T. Goto, H. Souma, M. Ueta: J. Lumin. **1/2**, 231 (1970)

3.26 Y. Toyozawa: Prog. Theor. Phys., Suppl. **12**, 111 (1959)

3.27 M. Certier, C. Wecker, S. Nikitine: J. Phys. Chem. Solids **30**, 2135 (1969)

3.28 T. Anzai, T. Goto, M. Ueta: J. Phys. Soc. Jpn. **38**, 774 (1975)

3.29 M. Ueta, T. Goto, T. Yashiro: J. Phys. Soc. Jpn. **20**, 1022 (1965)

3.30 T. Takahashi, T. Goto: J. Phys. Soc. Jpn. **25**, 461 (1968)

3.31 Y. Toyozawa: J. Phys. Chem. Solids **25**, 59 (1964)

3.32 T. Tomiki: J. Phys. Soc. Jpn. **22**, 463 (1967)

3.33 R. S. Knox, N. Inchauspe: Phys. Rev. **116**, 1093 (1959)

3.34 Y. Onodera, Y. Toyozawa: J. Phys. Soc. Jpn. **22**, 833 (1967)

3.35 C. I. Yu, T. Goto, M. Ueta: J. Phys. Soc. Jpn. **34**, 693 (1973)

3.36 S. Imai, S. Namba: Appl. Phys. Lett. **19**, 41 (1971)

3.37 S. Nikitine, J. Ringeissen, M. Certier, C. Wecker, S. Lewonezuk, J. C. Merle, C. Jung: In Proc. 10th Int. Conf. Phys. Semicond., Cambridge USA (1970), ed. by S. P. Keller, J. C. Hensel, F. Stem (USAEC Div. of Technical Information, Oak Ridge 1970) p. 196

3.38 K. S. Song: J. Phys. (Paris), **28**, Colloq. C3–43 (1967)

3.39 B. Hönerlage, C. Kringshirn, J. B. Grun: Phys. Status Solidi (b) **78**, 599 (1976)

3.40 L. V. Keldysh, A. N. Kozlov: Sov. Phys.-JETP **27**, 521 (1968)

3.41 E. Hanamura: J. Phys. Soc. Jpn. **29**, 50 (1970)

3.42 M. A. Lampert: Phys. Rev. Lett. **1**, 450 (1958)

3.43 S. A. Moskalenko: Opt. Spectrosc. (USSR) **5**, 147 (1958)

3.44 J. Haynes: Phys. Rev. Lett. **4**, 361 (1960); D. G. Thomas, J. J. Hopfield: Phys. Rev. **128**, 2135 (1962)

3.45 R. R. Sharma: Phys. Rev. Lett. **4**, 361 (1960); R. K. Wehner: Solid State Commun. **7**, 2 (1969)

3.46 O. Akimoto, E. Hanamura: Solid State Commun. **10**, 253 (1972)

3.47 T. Goto, T. Ishii, H. Souma, M. Ueta: J. Phys. Soc. Jpn. **24**, 656 (1968)

3.48 H. Souma, T. Goto, T. Ohta, M. Ueta: J. Phys. Soc. Jpn. **29**, 697 (1970)

3.49 W. T. Huang: Phys. Status Solidi (b) **60**, 309 (1973)

3.50 S. Suga, T. Koda: Phys. Status Solidi (b) **61**, 291 (1974)

3.51 N. Nagasawa, S. Koizumi, T. Mita, M. Ueta: J. Lumin. **12/13**, 587 (1976)

3.52 E. Hanamura: J. Phys. Soc. Jpn. **39**, 1506, 1516 (1975)

3.53 E. Hanamura: In *Luminescence of Crystals, Molecules and Solutions,* ed. by F. Williams (Plenum, New York 1973) p. 121

3.54 E. Hanamura: Solid State Commun. **12**, 951 (1973)

3.55 G. M. Gale, A. Mysyrowicz: In Proc. 12th Int. Conf. Phys. Semicond., Stuttgart (1974), ed. by M. H. Pilkuhn (Teubner, Stuttgart 1974) p. 133

3.56 M. Inoue, E. Hanamura: J. Phys. Soc. Jpn. **41**, 1273 (1976)

504   References

3.57   N. Peyghambarian, L. L. Chase, A. Mysyrowicz: Phys. Rev. B27, 2325 (1983)
3.58   N. Nagasawa, T. Mita, M. Ueta: J. Phys. Soc. Jpn. 41, 929 (1976)
3.59   M. Ueta, T. Mita, T. Itoh: Solid State Commun. 32, 43 (1979)
3.60   Vu Duy Phach, A. Bivas, B. Hönerlage, J. B. Grun: Phys. Status Solidi (b) 84, 731 (1977)
3.61   N. Nagasawa, T. Mita, M. Ueta: J. Phys. Soc. Jpn. 45, 713 (1978)
3.62   Y. Nasu, S. Koizumi, N. Nagasawa, M. Ueta: J. Phys. Soc. Jpn. 41, 751 (1976)
3.63   Vu Duy Phach, R. Levy: Solid State Commun. 29, 247 (1979)
3.64   Y. Nozue, T. Itoh, M. Ueta: J. Phys. Soc. Jpn. 44, 1305 (1978)
3.65   M. Itoh, Y. Nozue, T. Itoh, M. Ueta, S. Satoh, K. Igaki: J. Lumin. 18/19, 568 (1979)
3.66   H. Schrey, C. Kringshirn: Phys. Status Solidi (b) 93, 679 (1979)
3.67   T. Ishihara, T. Mita, N. Nagasawa: Solid State Commun. 44, 33 (1982)
3.68   Q. H. Yu, R. Levy, B. Hönerlage: J. Lumin. 24/25, 417 (1981)
3.69   J. J. Hopfield: Phys. Rev. 112, 1555 (1958)
3.70   T. Itoh, T. Suzuki, M. Ueta: J. Phys. Soc. Jpn. 42, 1069 (1977)
3.71   T. Mita, T. Sôtome, M. Ueta: J. Phys. Soc. Jpn. 48, 496 (1980)
3.72   T. Itoh, T. Suzuki: J. Phys. Soc. Jpn. 45, 1939 (1978)
3.73   F. Henneberger, K. Henneberger, J. Voigt: Phys. Status Solidi (b) 83, 439 (1977)
3.74   R. Loudon: *The Quantum Theory of Light* (Oxford University Press, Oxford 1973) p. 297
3.75   C. Kempf, G. Schmieder, G. Kurtze, C. Kringshirn: Phys. Status Solidi (b) 107, 297
        (1981);
        J. B. Grun, B. Hönerlage, R. Levy: Solid State Commun. 46, 51 (1983)
3.76   M. Kuwata, T. Mita, N. Nagasawa: Solid State Commun. 40, 911 (1981)
3.77   T. Itoh, T. Katohno: J. Phys. Soc. Jpn. 51, 707 (1982)
3.78   T. Itoh, T. Kirihara: In Proc. 17th Int. Conf. Phys. Semicond., San Francisco (1984)
        p. 1259
3.79   T. Itoh, S. Watanabe, M. Ueta: J. Phys. Soc. Jpn. 48, 542 (1980)
3.80   M. Inoue, E. Hanamura: J. Phys. Soc. Jpn. 41, 971 (1976)
3.81   Y. Toyozawa: Prog. Theor. Phys. 27, 89 (1962)
3.82   Y. Toyozawa: Prog. Theor. Phys. 20, 53 (1958)
3.83   K. Sôtome, Y. Nozue, M. Ueta: Solid State Commun. 36, 555 (1980)
3.84   R. Levy, C. Kringshirn, E. Ostertag, Vu Duy Phach, J. B. Grun: Phys. Status Solidi (b)
        77, 381 (1976)
3.85   Y. Segawa, Y. Aoyagi, O. Nakagawa, K. Azuma, S. Namba: Solid State Commun. 27,
        785 (1978)
3.86   Y. Masumoto, S. Shionoya, Y. Tanaka, Solid State Commun. 27, 1117 (1978)
3.87   E. Hanamura, T. Takagahara: J. Phys. Soc. Jpn. 47, 410 (1979)
3.88   T. Mita, M. Ueta: Solid State Commun. 27, 1463 (1978)
3.89   Y. Toyozawa: Solid State Commun. 28, 533 (1978)
3.90   T. Mita, K. Sôtome, M. Ueta: J. Phys. Soc. Jpn. 50, 134 (1981)
3.91   T. Suzuki, T. Itoh: J. Phys. Soc. Jpn. 47, 1246 (1979)
3.92   Y. Masumoto, S. Shionoya: Solid State Commun. 41, 147 (1982)
3.93   S. Schmitt-Rink, H. Haug: Phys. Status Solidi (b) 108, 377 (1981)
3.94   M. Kuwata, T. Mita, N. Nagasawa: J. Phys. Soc. Jpn. 50, 2467 (1981); Solid State
        Commun. 40, 911 (1981)
3.95   L. L. Chase, N. Peyghambarian, G. Grynberg, A. Mysyrowicz: Opt. Commun. 28, 189
        (1979)
3.96   T. Itoh, T. Katohno, M. Ueta: J. Phys. Soc. Jpn. 53, 844 (1984)
3.97   Y. Aoyagi, Y. Segawa, S. Namba: Phys. Rev. B25, 1453 (1982)
3.98   T. Itoh, T. Katohno, T. Kirihara, M. Ueta: J. Phys. Soc. Jpn. 53, 854 (1984)
3.99   W. R. Heller, A. Marcus: Phys. Rev. 84, 809 (1951)
3.100   Y. Onodera: J. Phys. Soc. Jpn. 49, 1845 (1980)
3.101   T. Mita: Doctoral Thesis, Tohoku University (1980)
3.102   S. I. Pekar: Sov. Phys.-JETP 6, 785 (1958);
        R. Zeyher, J. L. Birman, W. Brenig: Phys. Rev. B6, 4613 (1972)
3.103   K. Cho: Phys. Rev. B14, 4463 (1976)

3.104 G. Dresselhaus: Phys. Rev. **100**, 580 (1955)
3.105 S. Suga, K. Cho, M. Bettini: Phys. Rev. B **13**, 943 (1976)
3.106 W. Dreybrodt, K. Cho, S. Suga, F. Willmann, Y. Niji: Phys. Rev. B **21**, 4692 (1980)
3.107 M. Luttinger: Phys. Rev. **102**, 1030 (1956)
3.108 A comprehensive review by K. Cho: "Internal Structure of Excitons", in *Excitons* ed. by
       K. Cho, Topics in Current Phys., Vol. 14 (Springer, Berlin, Heidelberg 1979) Chap. 2
3.109 E. O. Kane: Phys. Rev. B **11**, 3850 (1975)
3.110 E. H. Turner, I. P. Kaminov, C. Schwab: Phys. Rev. B **9**, 2524 (1974)
3.111 B. Hönerlage, U. Rossler, Vu Duy Phach, A. Bivas, J. B. Grun: Phys. Rev. B **22**, 797
       (1980)
3.112 Y. Oka, D. P. Vu: In Proc. 15th Int. Conf. Phys. Semicond., Kyoto (1980), J. Phys. Soc.
       Jpn. **49**, Suppl. A, 547 (1980);
       Vu Duy Phach, Y. Oka, M. Cardona: Phys. Rev. B **24**, 121 (1981)
3.113 K. Cho, M. Yamane: Solid State Commun. **40**, 121 (1981)
3.114 R. Levy, B. Hönerlage, J. B. Grun: Phys. Rev. Lett. **44**, 1355 (1980);
       B. Hönerlage, R. Levy, J. B. Grun: Phys. Rev. B **24**, 3211 (1981)
3.115 Y. Nozue, M. Itoh, K. Cho: J. Phys. Soc. Jpn. **50**, 889 (1981)
3.116 I. A. Karp, S. A. Moskalenko: Sov. Phys.-Semicond. **8**, 183 (1974)
3.117 W. Ekardt, M. I. Sheboul: Phys. Status Solidi (b) **74**, 523 (1976)

Chapter 4

4.1   J. Bardeen, W. Shockley: Phys. Rev. **80**, 72 (1950)
4.2   H. Fröhlich: Adv. Phys. **3**, 325 (1954)
4.3   A. R. Hutson: Phys. Rev. Lett. **4**, 505 (1960);
      G. D. Mahan, J. J. Hopfield: Phys. Rev. Lett. **12**, 241 (1964)
4.4   F. Seitz: Phys. Rev. **73**, 549 (1948)
4.5   F. J. Blatt: "Theory of Mobility of Electrons in Solids", in *Solid State Physics,* Vol. 4, ed.
      by F. Seitz, D. Turnbull (Academic, New York 1957) p. 199
4.6   C. G. Kuper, G. D. Whitfield (eds.): *Polarons and Excitons*, (Oliver and Boyd, Edin-
      burgh 1963)
4.7   J. T. Devreese (ed.): *Polarons in Ionic Crystals and Polar Semiconductors*, (North-
      Holland, Amsterdam 1972)
4.8   T. D. Lee, F. Low, D. Pines: Phys. Rev. **90**, 297 (1953);
      M. Gurari: Philos. Mag. **44**, 329 (1953)
4.9   S. I. Pekar: Zh. Eksp. Teor. Fiz. **19**, 796 (1949)
4.10 R. P. Feynman: Phys. Rev. **97**, 660 (1955)
4.11 A. I. Ansel'm, Yu. A. Firsov: Zh. Eksp. Teor. Fiz. **30**, 719 (1956) [English transl.: Sov.
      Phys.-JETP **1**, 139 (1955)]
4.12 Y. Toyozawa: Prog. Theor. Phys. **20**, 53 (1958)
4.13 J. Dillinger, Č. Kŏnák, V. Prosser, J. Sak, M. Zvára: Phys. Status Solidi **29**, 707 (1968)
4.14 W. Y. Liang, A. D. Yoffe: Phys. Rev. Lett. **20**, 59 (1968);
      R. C. Whited, W. C. Walker: Phys. Rev. Lett. 22, 1428 (1969);
      W. C. Walker, D. M. Roessler, E. Loh: Phys. Rev. Lett. **20**, 847 (1968)
4.15 S. A. Permogorov: Phys. Status Solidi (b) **68**, 9 (1975)
4.16 H. Nishimura: J. Phys. Soc. Jpn. **52**, 3233 (1983)
4.17 M. Matsuura, H. Büttner: J. Phys. Soc. Jpn. **49**, Suppl. A, 413 (1980)
4.18 H. Haken: Fortschr. Phys. (Dtsch. Phys. Ges.) **6**, 271 (1958)
4.19 Y. Toyozawa: "Exciton-Lattice Interaction – Fluctuation, Relaxation and Defects For-
      mation", in *Vacuum Ultraviolet Radiation Physics*, ed. by E. E. Koch, R. Haensel,
      C. Kunz (Pergamon-Vieweg, Braunschweig 1974) p. 317
4.20 Y. Toyozawa: "Localization and Delocalization of an Exciton in the Phonon Field", in
      Organic Molecular Aggregates, ed. by P. Reineker, H. Haken, H. C. Wolf, Springer Ser.
      Solid-State Sci., Vol. 49 (Springer, Berlin, Heidelberg 1983) p. 90

4.21 L. Van Hove: Physica **21**, 901 (1955)
4.22 Y. Toyozawa: Prog. Theor. Phys. **27**, 89 (1962)
4.23 H. Sumi: J. Phys. Soc. Jpn. **32**, 616 (1972)
4.24 H. Sumi, Y. Toyozawa: J. Phys. Soc. Jpn. **31**, 342 (1971)
4.25 M. Schreiber, Y. Toyozawa: J. Phys. Soc. Jpn. **51**, 1528, 1537, 1544 (1982); ibid. **52**, 318 (1983); ibid. **53**, 864 (1984)
4.26 T. Tomiki: J. Phys. Soc. Jpn. **22**, 463 (1967);
      T. Miyata: J. Phys. Soc. Jpn. **31**, 529 (1971);
      T. Tomiki, T. Miyata, H. Tsukamoto: J. Phys. Soc. Jpn. **35**, 495 (1973)
4.27 D. M. Burland, U. Konzelmann, R. M. Macfarlane: J. Chem. Phys. **67**, 1926 (1977);
      D. M. Burland, D. E. Cooper, M. D. Fayer, C. R. Gochanour: Chem. Phys. Lett. **52**, 279 (1977)
4.28 Y. Onodera, Y. Toyozawa: J. Phys. Soc. Jpn. **24**, 341 (1968)
4.29 B. Velicky, S. Kirkpatrick, H. Ehrenreich: Phys. Rev. **175**, 747 (1968)
4.30 P. Soven: Phys. Rev. **156**, 809 (1967);
      D. W. Tylor: Phys. Rev. **156**, 1017 (1967)
4.31 F. Yonezawa, K. Morigaki: Prog. Theor. Phys., Suppl. **53**, 1 (1973)
4.32 E. I. Rashba, G. E. Gurgenishvilli: Sov. Phys.-Solid State **4**, 759 (1962)
4.33 Y. Nakai, T. Murata, K. Nakamura: Jpn. J. Appl. Phys. **4**, Suppl. I, 616 (1965);
      K. Nakamura, Y. Nakai: J. Phys. Soc. Jpn. **23**, 455 (1967);
      K. Nakamura: J. Phys. Soc. Jpn. **22**, 511 (1967)
4.34 T. Murata, Y. Nakai: J. Phys. Soc. Jpn. **23**, 904 (1967);
      N. Nagasawa, N. Nakagawa, Y. Nakai: J. Phys. Soc. Jpn. **24**, 1403 (1968)
4.35 G. Baldini: Phys. Rev. A **137**, 508 (1965);
      T. Nanba, N. Miura, N. Nagasawa: J. Phys. Soc. Jpn. **36**, 158 (1974);
      T. Nanba, N. Nagasawa, M. Ueta: J. Phys. Soc. Jpn. **37**, 1031 (1974)
4.36 F. Urbach: Phys. Rev. **92**, 1324 (1953)
4.37 M. V. Kurik: Phys. Status Solidi (a) **8**, 9 (1971)
4.38 D. L. Dexter: Nuovo Cimento, Suppl. **7**, 245 (1958)
4.39 Y. Toyozawa: Prog. Theor. Phys. **22**, 445 (1959)
4.40 H. Mahr: Phys. Rev. **125**, 1510 (1962)
4.41 Y. Toyozawa: Tech. Rep. ISSP, Ser. A, No. 119 (1964)
4.42 P. W. Anderson: Phys. Rev. **109**, 1492 (1958)
4.43 J. D. Dow, D. Redfield: Phys. Rev. B**5**, 594 (1972)
4.44 E. Mohler, B. Thomas: Phys. Rev. Lett. **4**, 543 (1980)
4.45 J. Ihm, J. C. Phillips: Phys. Rev. B**27**, 7803 (1983)
4.46 D. B. Fitchen: "Zero-Phonon Transitions", in *Physics of Color Centers*, ed. by W. B. Fowler (Academic, New York 1968) p. 293
4.47 Y. Toyozawa: J. Lumin. **1/2**, 732 (1970)
4.48 K. K. Rebane, V. G. Fedoseyev, V. V. Hizhnyakov: In Proc. 9th Int. Conf. Phys. Semicond., Moscow (1968), ed. by S. M. Ryvkin (Nauka, Leningrad 1968) p. 430
4.49 Y. Toyozawa: "Vibration Induced Structures in the Absorption Spectra of Localized Electrons in Solids", in *Dynamical Processes in Solid State Optics*, ed. by R. Kubo, H. Kamimura (Benjamin, New York 1967) p. 90
4.50 Y. Toyozawa: J. Phys. Chem. Solids **25**, 59 (1964)
4.51 U. Fano: Phys. Rev. **124**, 1866 (1961);
      U. Fano, W. Cooper: Phys. Rev. A**137**, 1364 (1965)
4.52 S. Nikitine, J. B. Grun, M. Sieskind: J. Phys. Chem. Solids **17**, 292 (1961)
4.53 T. Miyata: J. Phys. Soc. Jpn. **31**, 529 (1971)
4.54 T. P. McLean: "The Absorption Edge Spectrum of Semiconductors", in *Progress of Semiconductors,* ed. by A. F. Gibson, F. A. Kröger, R. E. Burgess, Vol. 5 (Wiley, New York 1961) p. 53
4.55 F. C. Brown, T. Masumi, H. H. Tippins: J. Phys. Chem. Solids **22**, 101 (1961)
4.56 M. Matsuura, H. Büttner: Phys. Rev. B**21**, 679 (1980)
4.57 J. Pollmann, H. Büttner: Phys. Rev. B**16**, 4480 (1977)

4.58   G. Baldini, B. Bosacchi: Phys. Rev. Lett. **22**, 190 (1969);
       G. Baldini, A. Bosacchi, B. Bosacchi: Phys. Rev. Lett. **23**, 846 (1969)
4.59   Y. Kato, T. Goto, T. Fujii, M. Ueta: J. Phys. Soc. Jpn. **36**, 175 (1974)
4.60   R. Z. Bachrach, F. C. Brown: Phys. Rev. Lett. **21**, 685 (1968);
       Phys. Rev. B**1**, 818 (1970)
4.61   S. Kurita, K. Kobayashi: J. Phys. Soc. Jpn. **30**, 1645 (1971)
4.62   Y. Toyozawa, J. Hermanson: Phys. Rev. Lett. **21**, 1637 (1968)
4.63   Y. Toyozawa: In Proc. 3rd Int. Conf. Photoconductivity, Stanford (1969), ed. by
       E. M. Pell (Pergamon, Oxford 1971) p. 151
4.64   H. Kanzaki, S. Sakuragi: J. Phys. Soc. Jpn. **27**, 109 (1969)
4.65   E. I. Rashba: Sov. Phys.-JETP **23**, 708 (1966)
4.66   Y. Toyozawa, Y. Shinozuka: J. Phys. Soc. Jpn. **48**, 472 (1980)
4.67   Y. Toyozawa: "Self-Trapping of an Electron by the Acoustical Mode of Lattice Vibra-
       tion", in *Polarons and Excitons*, ed. by C. G. Kuper, G. D. Whitfield (Oliver and Boyd,
       Edinburgh 1962) p. 221
4.68   K. S. Song: J. Phys. Soc. Jpn. **26**, 1131 (1969)
4.69   M. Umehara: J. Phys. Soc. Jpn. **47** 852 (1979);
       Private communication
4.70   T. Higashimura, Y. Nakaoka, T. Iida: J. Phys. C**17**, 4127 (1984)
4.71   H. Sumi: J. Phys. Soc. Jpn. **53**, 3498, 3511 (1984)
4.72   Y. Toyozawa, A. Sumi: In Proc. 12th Intern. Conf. Phys. Semicond. Stuttgart (1974), ed.
       by M. H. Pilkuhn (Teubner, Stuttgart 1974) p. 179
4.73   D. Emin, T. Holstein: Phys. Rev. Lett. **36**, 323 (1976)
4.74   E. I. Rashba: Opt. Spektrosk. **2**, 75, 88 (1957);
       Izv. Akad. Nauk SSSR, Ser. Fiz. **40**, 1793 (1976) [English transl.: Bull. Acad. Sci. USSR,
       Phys. Ser. **40**, 20 (1976)]
4.75   E. I. Rashba: "Self-Trapping of Excitons", in *Excitons,* ed. by E. I. Rashba, M. D. Sturge
       (North-Holland, Amsterdam 1982) p. 543
4.76   Y. Toyozawa: Prog. Theor. Phys. **26**, 29 (1961)
4.77   B. Pertzsch, U. Rössler: Solid State Commun. **37**, 931 (1981)
4.78   K. Cho, Y. Toyozawa: J. Phys. Soc. Jpn. **30**, 1555 (1971)
4.79   H. Sumi: J. Phys. Soc. Jpn. **36**, 770 (1974)
4.80   H. Sumi: J. Phys. Soc. Jpn. **38**, 825 (1975)
4.81   A. Sumi, Y. Toyozawa: J. Phys. Soc. Jpn. **35**, 137 (1973)
4.82   Y. Shinozuka, Y. Toyozawa: J. Phys. Soc. Jpn. **46**, 505 (1979)
4.83   Y. Toyozawa: Solid-State Electron. **21**, 1313 (1978)
4.84   Y. Toyozawa: Physica **116B**, 7 (1983)
4.85   P. W. Anderson: Phys. Rev. Lett. **34**, 953 (1975)
4.86   H. Hiramoto, Y. Toyozawa: J. Phys. Soc. Jpn. **54**, 245 (1985)
4.87   A. Sumi: J. Phys. Soc. Jpn. **43**, 1286 (1977)
4.88   W. Känzig: Phys. Rev. **99**, 1890 (1955);
       T. O. Woodruff, W. Känzig: J. Phys. Chem. Solids **5**, 268 (1958)
4.89   M. N. Kabler: "Hole Centers in Halide Lattices", in *Point Defects in Solids*, ed. by J. H.
       Crawford, Jr., L. M. Slifkin, Vol. 1 (Plenum, New York 1972) p. 327
4.90   J. Jortner, L. Meyer, S. A. Rice, E. G. Wilson: J. Chem. Phys. **42**, 4250 (1965)
4.91   I. Ya. Fugol': Adv. Phys. **27**, 1 (1978)
4.92   M. Pope, C. E. Swenberg: *Electronic Processes in Organic Crystals* (Clarendon, Oxford
       1982) p. 48;
       J. B. Birks: "Photophysics of Aromatic Molecules – A Postscript", in *Organic Molecular
       Photophysics*, ed. by J. B. Birks, Vol. 2 (Wiley, New York 1975) p. 409
4.93   A. Matsui, M. Iemura, H. Nishimura: J. Lumin. **24/25**, 445 (1981);
       H. Nishimura, A. Matsui, M. Iemura: J. Phys. Soc. Jpn. **51**, 1341 (1982)
4.94   Ch. B. Lushchik: "Free and Self-Trapped Excitons in Alkali Halides: Spectra and
       Dynamics", in *Excitons*, ed. by E. I. Rashba, M. D. Sturge (North-Holland, Amsterdam
       1982) p. 505

508   References

4.95 T. Hayashi, T. Ohata, S. Koshino: J. Phys. Soc. Jpn. **42**, 1647 (1977)
4.96 H. Nishimura, C. Ohhigashi, Y. Tanaka, M. Tomura: J. Phys. Soc. Jpn. **43**, 157 (1977);
      H. Nishimura, H. Miyazaki, Y. Tanaka, K. Uchida, M. Tomura: J. Phys. Soc. Jpn. **47**,
      1829 (1979)
4.97 T. V. Khiem, A. Nouhailhat: J. Phys. Soc. Jpn. **50**, 127 (1981)
4.98 H. Kanzaki, S. Sakuragi, K. Sakamoto: Solid State Commun. **9**, 999 (1971)
4.99 R. C. Hanson: J. Phys. Chem. **66**, 2376 (1962)
4.100 T. Kawai, K. Kobayashi, M. Kurita, Y. Makita: J. Phys. Soc. Jpn. **30**, 1101 (1971)
4.101 J. Nakahara, K. Kobayashi: J. Phys. Soc. Jpn. **40**, 180 (1976);
      K. Takahei, K. Kobayashi: J. Phys. Soc. Jpn. **44**, 1850 (1978)
4.102 T. Hyodo, J. Kasai, Y. Takakusa: J. Phys. Soc. Jpn. **49**, 2248 (1980);
      J. Kasai, T. Hyodo, K. Fujiwara: J. Phys. Soc. Jpn. **52**, 3671 (1983)
4.103 Y. Suzuki, H. Ohtani, S. Takagi, M. Hirai: J. Phys. Soc. Jpn. **50**, 3537 (1981)
4.104 T. Suemoto, H. Kanzaki: J. Phys. Soc. Jpn. **50**, 3664 (1981)
4.105 T. Higashimura, H. Nakatani, M. Itoh, K. Kan'no, Y. Nakai: J. Phys. Soc. Jpn. **53**, 1878
      (1984). This paper contains references to earlier studies.
4.106 M. Höhne, M. Stasiw: Phys. Status Solidi **28**, 247 (1968); M. Fukui, Y. Hayashi, H.
      Yoshioka, H. Kanzaki, S. Sakuragi: J. Phys. Soc. Jpn. **30**, 1510 (1971)
4.107 W. Hayes, I. B. Owen, P. J. Walker: J. Phys. C**10**, 1751 (1977)
4.108 Y. Kondo, M. Hirai, M. Ueta: J. Phy. Soc. Jpn. **33**, 151 (1972)
4.109 R. T. Williams, J. N. Bradford, W. L. Faust: Phys. Rev. B**18**, 7038 (1978)
4.110 N. Itoh: Adv. Phys. **31**, 491 (1982)
4.111 Y. Toyozawa: J. Phys. Soc. Jpn. **44**, 482 (1978)
4.112 R. B. Murray, F. J. Keller: Phys. Rev. **153**, 993 (1967);
      M. Ikezawa, T. Kojima: J. Phys. Soc. Jpn. **27**, 1551 (1969);
      H. Lamatsch, J. Rossel, E. Saurer: Phys. Status Solidi (b) **48**, 311 (1971)
4.113 N. Itoh: Private communication
4.114 J. M. Langer: J. Phys. Soc. Jpn. **49**, Suppl. A, 207 (1980)
4.115 G. D. Watkins: J. Phys. Soc. Jpn. **18**, Suppl. I III, 22 (1963)
4.116 D. V. Lang: J. Phys. Soc. Jpn. **49**, Suppl. A, 221 (1980)
4.117 C. H. Henry: "Large Lattice Relaxation Processes in Semiconductors", in *Relaxation of
      Elementary Excitations*, ed. by R. Kubo, E. Hanamura, Springer Ser. Solid-State Sci.,
      Vol. 18 (Springer-Verlag, Berlin, Heidelberg 1980) p. 19
4.118 D. V. Lang, R. A. Logan, M. Jaros: Phys. Rev. B**19**, 1015 (1979)
4.119 G. D. Watkins, J. R. Troxell: Phys. Rev. Lett. **44**, 593 (1980)
4.120 R. D. Harris, J. L. Newton, G. D. Watkins: Phys. Rev. Lett. **48**, 1271 (1982)
4.121 H. Fritzsche: J. Phys. Soc. Jpn. **49**, Suppl. A, 39 (1980)
4.122 S. Lakkis, C. Schlenker, B. K. Chakraverty, R. Buder, M. Marezio: Phys. Rev. B**14**,
      1429 (1976)
4.123 O. F. Schirmer, E. Salje: J. Phys. C**13**, L1067 (1980)
4.124 O. F. Schirmer: Z. Phys. B**24**, 235 (1976)
4.125 P. T. Landsberg: Phys. Status Solidi **41**, 457 (1970); J. Lumin. **7**, 3 (1973)
4.126 M. Lax: Phys. Rev. **119**, 1502 (1960)
4.127 A. M. Stoneham: Rep. Prog. Phys. **44**, 1251 (1981)
4.128 Y. Toyozawa: Prog. Theor. Phys., Suppl. **12**, 11 (1959)
4.129 H. Sumi: J. Phys. Soc. Jpn. **41**, 526 (1976)
4.130 T. Goto, T. Takahashi, M. Ueta: J. Phys. Soc. Jpn. **24**, 314 (1968)
4.131 P. Wiesner, U. Heim: Phys. Rev. B**11**, 3071 (1975)
4.132 Y. Segawa, Y. Aoyagi, S. Namba: Solid State Commun. **39**, 535 (1981)
4.133 Y. Masumoto, S. Shionoya: J. Phys. Soc. Jpn. **51**, 181 (1982)
4.134 Y. Toyozawa: J. Phys. Soc. Jpn. **41**, 400 (1976);
      Y. Toyozawa, A. Kotani, A. Sumi: J. Phys. Soc. Jpn. **42**, 1495 (1977)
4.135 D. L. Huber: Phys. Rev. **170**, 418 (1968)
4.136 T. Takagahara, E. Hanamura, R. Kubo: J. Phys. Soc. Jpn. **43**, 802, 811, 1522 (1977);
      ibid. **44**, 728, 742 (1978)

4.137 W. Heitler: *The Quantum Theory of Radiation*, 3rd ed. (Clarendon, Oxford 1954) p. 196
4.138 V. Hizhnyakov, I. Tehver: Phys. Status Solidi **21**, 755 (1967); ibid. **39**, 67 (1970)
4.139 H. Kurita, O. Sakai, A. Kotani: J. Phys. Soc. Jpn. **49**, 1920 (1980)
4.140 R. G. Ulbrich, C. Weisbuch: *Festkörperprobleme* **XVIII**, 217 (1978)
4.141 P. Y. Yu: Phys. Rev. Lett. **30**, 283 (1973)
4.142 T. Karasawa, K. Miyata, T. Komatsu, Y. Kaifu: J. Phys. Soc. Jpn. **52**, 2592 (1983)
4.143 Y. Shinozuka: J. Phys. Soc. Jpn. **51**, 2852 (1982)
4.144 M. Stavola, M. Levinson, J. L. Benton, L. C. Kimerling: Phys. Rev. B**30**, 832 (1984)
4.145 H. Sumi: Phys. Rev. Lett. **47**, 1333 (1981); Physica **116**B, 3944 (1983)
4.146 D. L. Dexter, C. C. Klick, G. A. Russel: Phys. Rev. **100**, 630 (1955)
4.147 V. V. Hizhnyakov, I. Tehver: Sov. Phys.-JETP **42**, 305 (1975)
4.148 L. C. Kimerling: Solid-State Electron. **21**, 1391 (1978)
4.149 Y. Toyozawa, M. Inoue: J. Phys. Soc. Jpn. **21**, 1663 (1966)
4.150 H. Sumi: Phys. Rev. B**29**, 4616 (1984)
4.151 S. Iordanskii, E. Rashba: Sov. Phys.-JETP **47**, 975 (1978)
4.152 K. Nasu, Y. Toyozawa: J. Phys. Soc. Jpn. **50**, 235 (1981)
4.153 Y. Masumoto, Y. Unuma, S. Shionoya: J. Phys. Soc. Jpn. **51**, 3915 (1982);
       Y. Unuma, Y. Masumoto, S. Shionoya, H. Nishimura: J. Phys. Soc. Jpn. **52**, 4277 (1983)
4.154 R. Roick, R. Gaethke, G. Zimmerer, P. Gurtler: Solid State Commun. **47**, 333 (1983)
4.155 A. Sumi: J. Phys. Soc. Jpn. **47**, 1538 (1979)
4.156 R. Foster: *Organic Charge Transfer Complexes* (Academic, New York 1969)
4.157 J. B. Torrance, J. E. Vazquez, J. J. Mayerlo, V. Y. Lee: Phys. Rev. Lett. **46**, 253 (1981)
4.158 Y. Tokura, T. Koda, M. Mitani, G. Saito: Solid State Commun. **43**, 757 (1982)
4.159 R. E. Peierls: *Quantum Theory of Solids* (Clarendon, Oxford 1955) p. 108
4.160 Y. Toyozawa: J. Phys. Soc. Jpn. **50**, 1861 (1981)
4.161 J. Takimoto, Y. Toyozawa: J. Phys. Soc. Jpn. **52**, 4331 (1983)
4.162 See, for example, L. Salem: *The Molecular Orbital Theory of Conjugated Systems* (Benjamin, New York 1966) p. 110
4.163 M. Schreiber, Y. Toyozawa: Unpublished
4.164 I. Egri: Solid State Commun. **22**, 281 (1971)
4.165 K. Nasu, Y. Toyozawa: J. Phys. Soc. Jpn. **51**, 2098, 3111 (1982)
4.166 B. K. Chakraverty: Nature (London) **287**, 393 (1980)

Chapter 5

5.1   G. Baldini: Phys. Rev. **128**, 1562 (1962); ibid. **137**, A508 (1965)
5.2   B. Sonntag: "Dielectric and Optical Properties", in *Rare Gas Solids*, ed. by M. L. Klein, J. A. Venables, Vol. II (Academic, London 1976) Chap. 17, pp. 1022–1117
5.3   U. Rössler: "Band Structure and Excitons", in *Rare Gas Solids*, ed. by M. L. Klein, J. A. Venables, Vol. I (Academic, London 1976) Chap. 4, pp. 506–557
5.4   N. Schwentner, E. E. Koch, J. Jortner: "Electronic Excitations in Condensed Rare Gases", to be published in *Springer Tracts Mod. Phys.*, Vol. 107 (Springer, Berlin, Heidelberg 1985)
5.5   V. Saile, E. E. Koch: Phys. Rev. B**20**, 784 (1979)
5.6   A. B. Kunz, D. J. Mickish: Phys. Rev. B**8**, 779 (1973)
5.7   N. Schwentner, F.-J. Himpsel, V. Saile, M. Skibowski, W. Steinmann, E. E. Koch: Phys. Rev. Lett. **34**, 528 (1975)
5.8   U. Rössler: Phys. Status Solidi (b) **42**, 345 (1970)
5.9   F.-J. Himpsel: Appl. Opt. **19**, 3964 (1980)
5.10 K. Horn, A. M. Bradshaw: Solid State Commun. **30**, 545 (1979)
5.11 W. E. Spear, P. G. Le Comber: "Electronic Transport Properties", in *Rare Gas Solids*, ed. by M. L. Klein, J. A. Venables, Vol. II (Academic, London 1976) Chap. 18, pp. 1119–1149
5.12 S. A. Sai Halasz, A. J. Dahm: Phys. Rev. Lett. **28**, 1244 (1972)

5.13 K. W. Schwarz: Phys. Rev. A6, 837 (1972)
5.14 R. J. Loveland, P. G. Comber, W. E. Spear: Phys. Lett. 39A, 225 (1972)
5.15 L. S. Miller, S. Howe, W. E. Spear: Phys. Rev. 166, 871 (1968)
5.16 P. G. Le Comber, R. J. Loveland, W. E. Spear: Phys. Rev. B11, 3124 (1975)
5.17 N. F. Mott, E. A. Davis: *Electronic Processes in Non-Crystalline Materials* (Clarendon, Oxford 1971) pp. 77–79
5.18 B. E. Springett, M. H. Cohen, J. Jortner: Phys. Rev. 159, 183 (1967) and references therein
5.19 M. A. Woolf, G. W. Rayfield: Phys. Rev. Lett. 15, 235 (1965)
5.20 K. W. Schwarz: Phys. Rev. A6, 1958 (1972)
5.21 J. Poitrenaud, F. I. B. Williams: Phys. Rev. Lett. 29, 1230 (1972)
5.22 W. B. Fowler, D. L. Dexter: Phys. Rev. 176, 337 (1968)
5.23 K. R. Atkins: Phys. Rev. 116, 1339 (1959)
5.24 S. D. Druger, R. S. Knox: J. Chem. Phys. 50, 3143 (1969)
5.25 O. Tüxen: Z. Phys. 103, 463 (1936)
5.26 A. Sumi, Y. Toyozawa: J. Phys. Soc. Jpn. 35, 137 (1973)
5.27 C. M. Surko, G. J. Dick, F. Reif, W. C. Walker: Phys. Rev. Lett. 23, 842 (1969)
5.28 A. A. Lucas, J. P. Vigneron, S. E. Donnelly, J. C. Rife: Phys. Rev. B28, 2485 (1983)
5.29 M. Stockton, J. W. Keto, W. A. Fitzsimmons: Phys. Rev. A5, 372 (1972)
5.30 G. Zimmerer: J. Lumin., 18/19, 875 (1979)
5.31 R. Haensel, G. Keitel, E. E. Koch, M. Skibowski, P. Schreiber: Phys. Rev. Lett. 23, 116 (1969)
5.32 R. Haensel, G. Keitel, E. E. Koch, M. Skibowski, P. Schreiber: Opt. Commun. 2, 59 (1970)
5.33 J. Hermanson, J. C. Phillips: Phys. Rev. 150, 652 (1966);
      J. Hermanson: Phys. Rev. 150, 660 (1966)
5.34 D. Beaglehole: Phys. Rev. Lett. 15, 551 (1965)
5.35 I. T. Steinberger, U. Asaf: Phys. Rev. B8, 914 (1973)
5.36 U. Asaf, I. T. Steinberger: Phys. Rev. B10, 4464 (1974)
5.37 P. Laporte, I. T. Steinberger: Phys. Rev. A15, 2538 (1977)
5.38 J. Bardeen, W. Shockley: Phys. Rev. 80, 72 (1950)
5.39 R. Reininger, U. Asaf, I. T. Steinberger, V. Saile, P. Laporte: Phys. Rev. B28, 3193 (1983)
5.40 C. M. Surko, F. Reif: Phys. Rev. Lett. 20, 582 (1968); Phys. Rev. 175, 229 (1968)
5.41 Y. Onodera, Y. Toyozawa: J. Phys. Soc. Jpn. 22, 833 (1967)
5.42 J. Jortner, L. Meyer, S. A. Rice, E. G. Wilson: J. Chem. Phys. 42, 4250 (1965)
5.43 R. E. Haffman, L. C. Larrabee, Y. Tanaka: Appl. Opt. 4, 1581 (1965)
5.44 Y. Tanaka, W. C. Walker, K. Yoshino: J. Chem. Phys. 70, 380 (1979)
5.45 Y. Tanaka: J. Opt. Soc. Am. 45, 710 (1955)
5.46 M. N. Kabler: Phys. Rev. 136, A 1296 (1964)
5.47 M. N. Kabler, D. A. Patterson: Phys. Rev. Lett. 19, 652 (1967)
5.48 R. E. Packard, F. Reif, C. M. Surko: Phys. Rev. Lett. 25, 1435 (1970)
5.49 I. Ya. Fugol', E. V. Savchenko, A. G. Belov: JETP Lett. 16, 172 (1972)
5.50 I. Ya. Fugol': Adv. Phys. 27, 1 (1978)
5.51 M. Martin: J. Chem. Phys. 54, 3289 (1971)
5.52 V. Yakhot, M. Berkowitz, R. B. Gerber: Chem. Phys. 10, 61 (1975)
5.53 K. Nasu, Y. Toyozawa: J. Phys. Soc. Jpn. 50, 235 (1981)
5.54 H. Kanzaki, T. Suemoto: Semicond. Insul. 5, 345 (1983)
5.55 R. T. Williams, M. N. Kabler: Phys. Rev. B9, 1897 (1974)
5.56 J. W. Keto, F. J. Soley, M. Stockton, W. A. Fitzsimmons: Phys. Rev. A10, 872 (1974);
      F. J. Soley, R. K. Leach, W. A. Fitzsimmons: Phys. Lett. 55A, 49 (1975)
5.57 T. Suemoto, H. Kanzaki: J. Phys. Soc. Jpn. 46, 1554 (1979)
5.58 S. Arai, T. Oka, M. Kogoma, M. Imamura: J. Chem. Phys. 68, 4595 (1978)
5.59 J. S. Cohen. B. Schneider: J. Chem. Phys. 61, 3230 (1974)
5.60 S. Iwata: Chem. Phys. 37, 251 (1979)

5.61   K. S. Song, L. T. Lewis: Phys. Rev. B**19**, 5349 (1979)
5.62   K. S. Song: Semicond. Insul. **5**, 213 (1983)
5.63   T. Suemoto, H. Kanzaki: J. Phys. Soc. Jpn. **49**, 1039 (1980)
5.64   T. Suemoto, H. Kanzaki: J. Phys. Soc. Jpn. **50**, 3664 (1981)
5.65   C. H. Leung, L. Emery, K. S. Song: Phys. Rev. B**28**, 3474 (1983)
5.66   A. P. Hickman, W. Streets, N. F. Lane: Phys. Rev. B**12**, 3705 (1975)
5.67   B. E. Sirovich, R. E. Norberg: Phys. Rev. B**15**, 5107 (1977)
5.68   A. V. Chadwick, H. R. Glyde: "Point Defects and Diffusion", in *Rare Gas Solids*, ed. by
       M. L. Klein, J. A. Venables, Vol. II (Academic, London 1976) Chap. 19, pp. 1151–1229
5.69   R. T. Williams: Phys. Rev. Lett. **36**, 529 (1976)
5.70   A. G. Belov, V. N. Svischer, I. Ya. Fugol', E. M. Yurtaeva: Soviet Phys.-JETP Lett. **30**,
       114 (1979)
5.71   Ch. Ackermann, R. Brodmann, G. Zimmerer, R. Haensel, U. Hahn: J. Lumin. **12/13**,
       315 (1976)
5.72   K. Inoue, H. Sakamoto, H. Kanzaki: Solid State Commun. **44**, 1007 (1982); ibid. **49**, 191
       (1984); J. Phys. Soc. Jpn. **53**, 819 (1984)
5.73   Y. Tanaka, W. C. Walker: J. Chem. Phys. **74**, 2760 (1981)
5.74   T. R. Connor, M. A. Biodi: Phys. Rev. **140**, A778 (1965);
       L. Frommhold, M. A. Biodi: Phys. Rev. **185**, 224 (1969)
5.75   Y. Matsuura, K. Fukuda: J. Phys. Soc. Jpn. **50**, 933 (1981)
5.76   R. Balzer, E.-J. Giersberg: Phys. Status Solidi (a) **57**, K141 (1980)
5.77   P. Børgesen, J. Schou, H. Sørensen, C. Claussen: Appl. Phys. A**29**, 57 (1982)

Chapter 6

6.1   N. F. Mott, R. W. Gurney: *Electronic Processes in Ionic Crystals,* (Oxford University
       Press, Oxford 1940) Chap. VII
6.2   F. Seitz: Rev. Mod. Phys. **23**, 328 (1951)
6.3   F. C. Brown: "The Photographic Process", in *Treatise on Solid State Chemistry,* ed. by
       B. Hannay, Vol. 4 (Plenum, New York 1973) Chap. 10, p. 333
6.4   H. Kanzaki: Photogr. Sci. Eng. **24**, 219 (1980)
6.5   Y. Okamoto: Nachr. Akad. Wiss. Göttingen, Math.-Phys. Kl., 2A: Math.-Phys.-Chem.
       Abt. **14**, 275 (1956)
6.6   H. Kanzaki: Unpublished
6.7   W. Martienssen: J. Phys. Chem. Solids **2**, 257 (1957);
       T. Tomiki, T. Miyata, H. Tsukamoto: Z. Naturforsch. **29a**, 145 (1974)
6.8   R. Hilsch, R. W. Pohl: Z. Phys. **57**, 145 (1929); ibid. **59**, 812 (1930)
6.9   F. C. Brown, T. Masumi, H. H. Tippins: J. Phys. Chem. Solids **22**, 101 (1961)
6.10   F. C. Brown: "Conduction by Polarons in Ionic Crystals", in *Point Defects in Solids*, ed.
       by J. H. Crawford Jr., L. M. Slifkin, Vol. 1 (Plenum, New York 1972) Chap. 8,
       pp. 491–549
6.11   P. G. Harper, J. W. Hodby, R. F. Stradling: Rep. Prog. Phys. **36**, 1 (1973)
6.12   F. Bassani, R. S. Knox, W. B. Fowler: Phys. Rev. **137**, A1217 (1965)
6.13   P. M. Scop: Phys. Rev. **139**, A934 (1965)
6.14   H. Tamura, T. Masumi: Solid State Commun. **12** 1183 (1973)
6.15   W. B. Fowler: Phys. Status Solidi (b) **52**, 591 (1972)
6.16   A. B. Kunz: Phys. Rev. B**26**, 2070 (1982)
6.17   J. A. Krumhansl: "Photoexcitation in Ionic Crystals" in *Photoconductivity Conference,*
       ed. by R. G. Breckenridge, B. R. Russel, E. E. Hahn (Wiley, New York 1956)
       pp. 450–462
6.18   M. G. Mason: Phys. Rev. B**11**, 5094 (1975)
6.19   J. Tejeda, N. J. Shevchik, W. Braun, A. Goldmann, M. Cardona: Phys. Rev. B**12**, 1557
       (1975)
6.20   R. B. Aust: Phys. Rev. **170**, 784 (1968); A. D. Brothers, D. W. Lynch: Phys. Rev. **180**,
       911 (1969)

6.21 F.-J. Himpsel, W. Steinmann: Phys. Rev. Lett. **35**, 1025 (1975); Phys. Rev. B**17**, 2537 (1978)

6.22 R. S. Bauer, W. E. Spicer: Phys. Rev. Lett. **25**, 1283 (1970);
R. S. Bauer, S. F. Lin, W. E. Spicer: Phys. Rev. B**14**, 4527 (1976)

6.23 H. Kanzaki, S. Sakuragi: J. Phys. Soc. Jpn. **29**, 924 (1970)

6.24 K. L. Shaklee, J. E. Rowe: In Proc. 3rd Int. Conf. Photoconductivity, Stanford (1969), ed. by E. M. Pell (Pergamon, Oxford 1971) p. 157

6.25 P. R. Vijayaraghavan, R. M. Nicklow, H. G. Smith, M. K. Wilkinson: Phys. Rev. B**1**, 4819 (1970)

6.26 H. Kanzaki, S. Sakuragi, S. Hoshino, G. Shirane, Y. Fujii: Solid State Commun. **15**, 1547 (1974)

6.27 Y. Fujii, S. Hoshino, S. Sakuragi, H. Kanzaki, L. W. Lynn, G. Shirane: Phys. Rev. B**15**, 358 (1977)

6.28 W. von der Osten, B. Dorner: Solid State Commun. **16**, 431 (1975);
B. Dorner, W. von der Osten, W. Bührer: J. Phys. C**9**, 723 (1976)

6.29 K. Fischer, H. Bilz, R. Haberkorn, W. Weber: Phys. Status Solidi (b) **54**, 285 (1972)

6.30 W. G. Kleppman, H. Bilz: Commun. Phys. **1**, 105 (1976)

6.31 R. J. Friauf: J. Phys. (Paris) **138**, 1077 (1977)

6.32 J. C. Phillips: Rev. Mod. Phys. **42**, 317 (1970)

6.33 J. J. White, III, J. W. Straley: J. Opt. Soc. Am. **58**, 759 (1968);
R. S. Bauer, W. E. Spicer, J. J. White, III: J. Opt. Soc. Am. **64**, 830 (1974)

6.34 J. J. White, III: J. Opt. Soc. Am. **62**, 212 (1972)

6.35 M. Yanagihara, Y. Kondo, H. Kanzaki: J. Phys. Soc. Jpn. **52**, 4397 (1983);
M. Yanagihara: Unpublished

6.36 N. J. Carrera, F. C. Brown: Phys. Rev. B**4**, 3651 (1971)

6.37 S. M. Kelso, D. E. Aspnes, C. G. Olson, D. W. Lynch, D. Finn: Phys. Rev. Lett. **45**, 1032 (1980)

6.38 H. Kanzaki, K. Kido: Unpublished

6.39 R. S. Knox, N. Inchauspé: Phys. Rev. **116**, 1093 (1959)

6.40 R. J. Elliott: Phys. Rev. **108**, 1384 (1957)

6.41 A. Kotani, Y. Toyozawa: "Theoretical Aspects of Inner-Level Spectroscopy", in *Synchrotron Radiation,* ed. by C. Kunz, Topics Curr. Phys., Vol. 10 (Springer, Berlin, Heidelberg 1979) Chap. 4, pp. 169–229

6.42 H. Haken: Fortschr. Phys. (Dtsch. Phys. Ges.) **6**, 271 (1958)

6.43 Y. Toyozawa: In Proc. 3rd Int. Conf. Photoconductivity Stanford (1969), ed. by E. M. Pell (Pergamon, Oxford 1971) p. 151

6.44 Y. Toyozawa, J. Hermanson: Phys. Rev. Lett. **21**, 1637 (1968)

6.45 M. Matsushita: J. Phys. Soc. Jpn. **35**, 1688 (1973)

6.46 W. von der Osten: "Excitons and Exciton Relaxation in Silver Halides", in *Polarons and Excitons in Polar Semiconductors and Ionic Crystals,* ed. by J. T. Devreese, F. Peters (Plenum, New York, in cooperation with NATO Scientific Affairs Division 1984) pp. 293–342

6.47 T. Masumi: "Dynamical and Nonlinear Profiles of Polarons and Excitons in Pure and Ultrapure AgCl, AgBr and AgCl$_x$Br$_{1-x}$", in *Polarons and Excitons in Polar Semiconductors and Ionic Crystals*, ed. by J. T. Devreese, F. Peters (Plenum, New York, in cooperation with NATO Scientific Affairs Division 1984) pp. 99–164

6.48 H. Kanzaki, S. Sakuragi, K. Sakamoto: Solid State Commun. **9**, 999 (1971)

6.49 G. C. Smith: Phys. Rev. **140**, A221 (1965)

6.50 M. Höhne, M. Stasiw: Phys. Status Solidi **28**, 247 (1968)

6.51 H. Kanzaki, S. Sakuragi: Solid State Commun. **9**, 1667 (1971)

6.52 C. L. Marquart, R. T. Williams, M. N. Kabler: Solid State Commun. **9**, 2285 (1971)

6.53 W. Hayes, I. B. Owen, P. J. Walker: J. Phys. C**10**, 1751 (1977)

6.54 A. P. Marchetti, D. S. Tinti: Phys. Rev. B**24**, 7361 (1981) and references therein

6.55 K. Nakamura, J. Windscheif, W. von der Osten: Solid State Commun. **39**, 381 (1981)

6.56 T. Onodera, Y. Toyozawa: J. Phys. Soc. Jpn. **24**, 341 (1968) and references therein

6.57 E. Taglauer, W. Waidelich: Z. Phys. **169**, 90 (1962)
6.58 H. Kanzaki, S. Sakuragi: J. Phys. Soc. Jpn. **29**, 936 (1970)
6.59 R. Parmenter: Phys. Rev. **97**, 587 (1955)
6.60 B. L. Joesten, F. C. Brown: Phys. Rev. **148**, 919 (1966)
6.61 H. Kanzaki, S. Sakuragi: J. Phys. Soc. Jpn. **27**, 109 (1969)
6.62 M. Yamaga, N. Sugimoto, H. Yoshioka: J. Phys. Soc. Jpn. **52**, 3637 (1983)
6.63 Y. Toyozawa: "Localization and Delocalization of an Exciton in the Phonon Field", in *Organic Molecular Aggregates*, ed. by P. Reineker, H. Haken, H. C. Wolf, Springer Ser. Solid-State Sci., Vol. 49 (Springer, Berlin, Heidelberg 1983) p. 90
6.64 K. Cho, Y. Toyozawa: J. Phys. Soc. Jpn. **30** 1555 (1971);
      H. Sumi, Y. Toyozawa: J. Phys. Soc. Jpn. **31**, 342 (1971)
6.65 M. Schreiber, Y. Toyozawa: J. Phys. Soc. Jpn. **52**, 318 (1983)
6.66 H. Frieser, G. Haase, E. Klein (eds.): *Physikalische und Chemische Eigenschaften der Silverhalogenide und des Silbers*, Vol. 1 of *Die Grundlagen der Photographischen Prozesse mit Silverhalogeniden* (Akademische, Frankfurt am Main 1968) pp. 72–75
6.67 K. Kobayashi, T. Tomiki: J. Phys. Chem. Solids **22**, 73 (1961)
6.68 M. Tsukakoshi, H. Kanzaki: J. Phys. Soc. Jpn. **30**, 1423 (1971)
6.69 D. B. Fitchen: "Zero-Phonon Transitions", in *Physics of Color Centers*, ed. by W. B. Fowler (Academic, New York 1968) Chap. 6, pp. 293–350
6.70 W. Czaja, A. Baldereschi: J. Phys. C**12**, 405 (1979)
6.71 W. Czaja: J. Phys. C**16**, 3197 (1983)
6.72 Y. Toyozawa: "Vibration Induced Structures in the Absorption Spectra of Localized Electrons in Solids" in *Dynamical Processes in Solid State Optics*, ed. by R. Kubo, H. Kamimura (Benjamin, New York 1967) pp. 90–115
6.73 D. G. Thomas, J. J. Hopfield: Phys. Rev. **150**, 680 (1965)
6.74 F. Moser, R. K. Ahrenkiel, S. L. Lyu: Phys. Rev. **161**, 897 (1967)
6.75 H. Kanzaki, S. Sakuragi: Unpublished
6.76 M. Ikezawa, T. Kojima: J. Phys. Soc. Jpn. **27**, 1551 (1969)
6.77 J. H. Beaumont, A. J. Bourdillon, M. N. Kabler: J. Phys. C**9**, 2961 (1976)
6.78 H. Möller, R. Brodmann, G. Zimmerer, U. Hahn: Solid State Commun. **20**, 401 (1976)
6.79 C. Ackermann, R. Brodmann, U. Hahn, A. Suzuki, G. Zimmerer: Phys. Status Solidi (b) **74**, 579 (1976)
6.80 G. Sprüssel, M. Skibowski, V. Saile: Solid State Commun. **32**, 1091 (1979)
6.81 K. Toyoda, K. Nakamura, Y. Nakai: J. Phys. Soc. Jpn. **41**, 1981 (1976)
6.82 M. Ikezawa, K. Shirahata, T. Kojima: Sci. Rep. Tohoku Univ., Ser. 1 **12**, 45 (1969)
6.83 R. S. Van Heyningen, F. C. Brown: Phys. Rev. **111**, 462 (1958)
6.84 R. S. Bauer, W. E. Spicer: Phys. Rev. B**14**, 4539 (1976)
6.85 H. Sugawara, T. Sasaki: J. Phys. Soc. Jpn. **46** 132 (1979)
6.86 R. W. Pohl: Proc. Phys. Soc., London **49** (extra part), 3 (1937)
6.87 W. Kohn: "Shallow Impurity States in Silicon and Germanium" in *Solid State Physics*, ed. by F. Seitz, D. Turnbull, Vol. 5 (Academic, New York 1957) pp. 257–320
6.88 W. B. Fowler: "Electronic States and Optical Transitions of Color Centers" in *Physics of Color Centers*, ed. by W. B. Fowler, (Academic, New York 1968) Chap. 2, pp. 54–179
6.89 R. Hilsch, R. W. Pohl: Z. Phys. **64**, 606 (1930)
6.90 R. C. Brandt, F. C. Brown: Phys. Rev. **181**, 1241 (1969)
6.91 H. Kanzaki, S. Sakuragi: Photogr. Sci. Eng. **17**, 69 (1973) and references therein
6.92 H. Kanzaki: J. Photogr. Sci. **32**, 117 (1984)
6.93 H. Kanzaki: Semicond. Insul. **5**, 517 (1983)
6.94 S. Sakuragi, H. Kanzaki: Phys. Rev. Lett. **38**, 1302 (1977)
6.95 H. Pick: "Structure of Trapped Electron and Trapped Hole Centers in Alkali Halides 'Color Centers'" in *Optical Properties of Solids*, ed. by F. Abeles (North-Holland, Amsterdam 1972) Chap. 9, pp. 653–754
6.96 F. Moser, R. S. Van Heyningen, S. Lyu: Solid State Commun. **7**, 1609 (1969)
6.97 W. Ulrichi: Phys. Status Solidi **40**, 557 (1970)
6.98 E. Laredo, L. G. Rowan, L. Slifkin: Phys. Rev. Lett. **47**, 384 (1981)

6.99   J. H. Simpson: Proc. Roy. Soc., London **A197**, 269 (1949)

6.100   V. M. Buimistrov: Sov. Phys. Solid State **5**, 970 (1963)

6.101   K. Takegahara, T. Kasuya: J. Phys. Soc. Jpn. **39**, 1292 (1969)

6.102   H. Kanzaki: Semicond. Insul. **3**, 285 (1978)

6.103   Y. Kondo, H. Kanzaki: Phys. Rev. Lett. **34**, 664 (1975)

6.104   K. K. Bajaj, T. D. Clark: Phys. Status Solidi (b) **52**, 195 (1972)

6.105   M. Matsuura: J. Phys. Soc. Jpn. **53**, 284 (1984); Private communication

6.106   R. K. Swank, F. C. Brown: Phys. Rev. **130**, 34 (1963)

6.107   Y. Kayanuma, Y. Toyozawa: J. Phys. Soc. Jpn. **40**, 355 (1976);
        Y. Kayanuma: J. Phys. Soc. Jpn. **40**, 363 (1976)

6.108   Y. Kayanuma, Y. Kondo: Solid State Commun. **24**, 447 (1977); J. Phys. Soc. Jpn. **45**, 528 (1978)

6.109   Y. Kondo: Unpublished

6.110   R. H. Fowler: Phys. Rev. **38**, 45 (1931)

6.111   M. A. Gilleo: Phys. Rev. **91**, 534 (1953)

6.112   R. E. Bacon: "Latent Image Effects Leading to Reversal or Desensitization", in *The Theory of Photographic Processes*, ed. by T. H. James, (Macmillan, New York 1977) Chap. 7, pp. 182–193

6.113   G. W. Luckey: J. Phys. Chem. **57**, 791 (1953); J. Chem. Phys. **23**, 882 (1955)

6.114   H. Kanzaki, T. Mori: Semicond. Insul. **5**, 401 (1983)

6.115   H. Kanzaki, T. Mori: Phys. Rev. B **29**, 3573 (1984)

6.116   J. Malinowski: Contemp. Phys. **8**, 285 (1967); J. Photogr. Sci. **16**, 57 (1968)

6.117   H. Kanzaki, T. Mori: "Photon Stimulated Desorption from Silver Halide Surface", in *The Physics of Latent Image Formation in Silver Halides*, ed. by A. Baldereschi, W. Czaja, E. Tosatti, M. Tosi (World Scientific, Singapore 1984) pp. 137–150

Chapter 7

7.1   For a review of work on thallous halides up to 1976 see K. Kobayashi: Festkörperprobleme **XVI**, 117 (1976)

7.2   R. P. Lowndes, D. H. Martin: Proc. R. Soc. London, Ser. **A308**, 473 (1969)

7.3   Y. Fujii, T. Sakuma, J. Nakahara, S. Hoshino, K. Kobayashi, A. Fujii: J. Phys. Soc. Jpn. **44**, 1237 (1978)

7.4   K. Heidrich, W. Staude, J. Treusch, H. Overhof: Phys. Rev. Lett. **33**, 1220 (1974)

7.5   K. Heidrich, W. Staude, J. Treusch, H. Overhof: Solid State Commun. **16**, 1043 (1975)

7.6   A. Fujii, K. Takiyama, J. Nakahara, K. Kobayashi: J. Phys. Soc. Jpn. **42**, 525 (1977)

7.7   R. P. Lowndes: Phys. Lett. **21**, 26 (1966)

7.8   K. Kobayashi, T. Kawai, M. Kanada: J. Phys. Soc. Jpn. **23**, 305 (1967)

7.9   T. Kawai, K. Kobayashi, M. Kurita, Y. Makita: J. Phys. Soc. Jpn. **30**, 1101 (1971)

7.10   J. Overton, J. P. Hernandez: Phys. Rev. **B7**, 778 (1973)

7.11   H. Overhof, J. Treusch: Solid State Commun. **9**, 53 (1971)

7.12   S. Asano: Unpublished

7.13   M. Inoue, M. Okazaki: J. Phys. Soc. Jpn. **30**, 582 (1971); ibid. **31**, 1313 (1971)

7.14   J. P. Van Dyke, G. A. Samara: Phys. Rev. **B11**, 4935 (1975)

7.15   W. Schafer, M. Schreiber: Solid State Commun. **32**, 591 (1979)

7.16   M. Schreiber, W. Schafer: Phys. Rev. **B21**, 3571 (1980)

7.17   J. Treusch: Phys. Rev. Lett. **34**, 1343 (1975)

7.18   E. Mohler, G. Schlögl, J. Treusch: Phys. Rev. Lett. **27**, 424 (1971)

7.19   Ch. Uihlein, J. Treusch: Solid State Commun. **17**, 685 (1975)

7.20   S. Kurita, K. Kobayashi, Y. Onodora: Prog. Theor. Phys., Suppl. **57**, 10 (1975)

7.21   See for example, R. S. Knox: *Theory of Excitons,* Solid State Physics, Suppl. Vol. 5, ed. by F. Seitz, D. Turnbull (Academic, New York 1963)

7.22   Y. Onodera, Y. Toyozawa: J. Phys. Soc. Jpn. **22**, 833 (1967)

7.23   J. Nakahara, K. Kobayashi: J. Phys. Soc. Jpn. **40**, 180 (1976)

7.24   H. Zinngrebe: Z. Phys. **154**, 495 (1959)

7.25 S. Tutihasi: J. Phys. Chem. Solids **12**, 344 (1960)
7.26 R. Z. Bachrach, F. C. Brown: Phys. Rev. Lett. **21**, 685 (1968); Phys. Rev. B**1**, 818 (1970)
7.27 S. Sato, M. Watanabe, Y. Iguchi, S. Nakai, Y. Nakamura, T. Sagawa: J. Phys. Soc. Jpn. **33**, 1638 (1972)
7.28 K. Takahei, K. Kobayashi: J. Phys. Soc. Jpn. **43**, 891 (1977)
7.29 K. Takahei: Ph. D. Thesis, University of Tokyo (1976)
7.30 S. Kurita: Bussei **10**, 194 (1969)
7.31 S. Kurita, K. Kobayashi: J. Phys. Soc. Jpn. **30**, 1645 (1971)
7.32 M. Matsuura, H. Butter: Phys. Rev. B**21**, 679 (1980)
7.33 Y. Toyozawa: In Proc. 3rd. Int. Conf. Photoconductivity, Stanford (1969), ed. by E. M. Pell (Pergamon, Oxford 1971) p. 151
7.34 D. Fröhlich, J. Treusch, W. Kottler: Phys. Rev. Lett. **29**, 1603 (1972)
7.35 J. Nakahara: Solid State Commun. **29**, 115 (1979)
7.36 M. Inoue, Y. Toyozawa: J. Phys. Soc. Jpn. **20**, 363 (1965)
7.37 T. R. Bader, A. Gold: Phys. Rev. **171**, 997 (1968)
7.38 D. Fröhlich, B. Staginnus, S. Thurn: Phys. Status Solidi **40**, 287 (1970)
7.39 M. Matsuoka: J. Phys. Soc. Jpn. **23**, 1028 (1967)
7.40 J. Nakahara, K. Kobayashi, A. Fujii: J. Phys. Soc. Jpn. **37**, 1312 (1974)
7.41 A. Fujii, J. Nakahara, K. Kobayashi, Y. Fujii: J. Phys. Soc. Jpn. **46**, 1218 (1979)
7.42 T. Inui, Y. Tanabe, Y. Onodera: *Group Theory and Its Applications in Physics,* (Syokabo, Tokyo 1976) p. 415
7.43 G. G. MacFarlane, T. P. McLean, J. E. Quarrington, V. Roberts: Phys. Rev. **108**, 1377 (1957)
7.44 T. Nishio, M. Takeda, Y. Hamakawa: J. Phys. Soc. Jpn. **37**, 1016 (1974)
7.45 B. Batz: In *Semiconductors and Semimetals,* ed. by R. K. Willandsen and A. C. Bear, Vol. 9 (Academic, New York 1972) p. 315
7.46 E. R. Cowley, A. Okazaki: Proc. R. Soc. London, Ser. A**300**, 45 (1967)
7.47 L. Grabner: Phys. Rev. B**4**, 1335 (1971)
7.48 R. Shimizu, T. Koda, T. Murahashi: J. Phys.Soc. Jpn. **36**, 161 (1974)
7.49 S. Permogorov: Phys. Status Solidi (b) **68**, 9 (1975)
7.50 T. Goto, M. Ueta: J. Phys. Soc. Jpn. **22**, 488 (1967)
7.51 R. Shimizu, T. Koda, Y. Kaneko: Solid State Commun. **21**, 811 (1977)
7.52 R. Planel, A. Bonnot, C. B. a la Guillaume: Phys. Status Solidi (b) **58**, 251 (1973)
7.53 R. Shimizu, T. Koda: J. Phys. Soc. Jpn. **37**, 1468 (1974)
7.54 S. Permogorov: "Optical Emission due to Exciton Scattering by LO Phonons in Semiconductors", in *Excitons*, ed. by E. I. Rashba and M. D. Sturge (North-Holland, Amsterdam 1982) p. 178
7.55 J. Nakahara, K. Kobayashi: J. Phys. Soc. Jpn. **40**, 189 (1976)
7.56 K. Cho: Opt. Commun. **8**, 412 (1973)
7.57 O. Akimoto, E. Hanamura: J. Phys. Soc. Jpn. **33**, 1537 (1972)
7.58 O. Akimoto: J. Phys. Soc. Jpn. **35**, 973 (1973)
7.59 S. G. Chung, G. D. Sanders, Y. Chang: Solid State Commun. **45**, 237 (1983)
7.60 J. Nakahara, K. Kobayashi, M. Seki: Solid State Commun. **18**, 245 (1976)
7.61 H. Stolz, W. von der Osten: Solid State Commun. **33**, 319 (1980)
7.62 A. Pinczuk, E. Burstein: In Proc. 10th Int. Conf. Phys. Semicond., Cambridge USA (1970), ed. by S. P. Keller, J. C. Hensel, F. Stern (USAEC Div. of Technical Information, Oak Ridge 1970) p. 727
7.63 R. M. Martin: Phys. Rev. B**4**, 3676 (1971)
7.64 R. Loudon: J. Phys. (Paris) **26**, 677 (1965)
7.65 H. Takenaka, K. Kobayashi, K. Takiyama, J. Nakahara, T. Fujita: J. Phys. Soc. Jpn. **52**, 4377 (1983)
7.66 B. Bendow, J. L. Birman: Phys. Rev. B**4**, 569 (1971)
7.67 M. Matsushita, J. Wicksted, H. Z. Cummins: Phys. Rev. B**29**, 3362 (1984)
7.68 K. Takahei, K. Kobayashi: J. Phys. Soc. Jpn. **44**, 1850 (1978)
7.69 Y. Nakai, T. Murata, K. Nakamura: J. Appl. Phys. **4**, Suppl. I, 616 (1965)

7.70 J. C. Woolley, K. W. Blazey: J. Phys. Chem. Sol. **25**, 713 (1964)
7.71 T. Murata, Y. Nakai: J. Phys. Soc. Jpn. **23**, 904 (1967)
7.72 Y. Nakai, T. Murata, K. Nakamura: J. Phys. Soc. Jpn. **18**, 1481 (1963)
7.73 Y. Onodera, Y. Toyozawa: J. Phys. Soc. Jpn. **24**, 341 (1968)
7.74 J. Nakahara, K. Kobayashi, A. Fujii: J. Phys. Soc. Jpn. **37**, 1319 (1974)
7.75 B. Velicky, S. Kirkpatrick, H. Ehrenreich: Phys. Rev. **175**, 747 (1968)
7.76 T. Kawai, K. Kobayashi, H. Fujita: J. Phys. Soc. Jpn. **21**, 453 (1966)
7.77 Y. Makita, K. Kobayashi, M. Kanada, K. Kawai: J. Phys. Soc. Jpn. **25**, 816 (1968)
7.78 Y. Makita, K. Kobayashi: J. Phys. Soc. Jpn. **32**, 1262 (1972)
7.79 J. W. Hodby, G. T. Jenkin, K. Kobayashi, H. Tamura: Solid State Commun. **19**, 1017 (1972)
7.80 Y. Toyozawa, A. Sumi: In Proc. 12th Int. Conf. Phys. Semicond., Stuttgart (1974), ed. by M. H. Pilkuhn (Teubner, Stuttgart 1974) p. 179
7.81 Y. Shinozuka, Y. Toyozawa: J. Phys. Soc. Jpn. **46**, 505 (1979)
7.82 H. L. Frisch, J. M. Hammersley, D. J. A. Welsh: Phys. Rev. **126**, 949 (1962)
7.83 D. G. Thomas, J. J. Hopfield: Phys. Rev. **124**, 657 (1961)
7.84 A. G. Zhilich, J. Halpern, B. P. Zakharchenya: Phys. Rev. **188**, 1294 (1969)
7.85 J. Nakahara, A. Fujii: J. Phys. Soc. Jpn. **48**, 1184 (1980)
7.86 R. J. Elliott, R. Loudon: J. Phys. Chem. Solids **15**, 196 (1960)
7.87 R. Loudon: Am. J. Phys. **27**, 649 (1959)
7.88 W. H. Kleiner: Lincoln Lab. Progr. Rep., Feb. (1958)
7.89 M. Shinada, O. Akimoto, H. Hasegawa, K. Tanaka: J. Phys. Soc. Jpn. **28**, 975 (1970)
7.90 K. Tanaka, M. Shinada: J. Phys. Soc. Jpn. **34**, 108 (1973)
7.91 W. S. Boyle, R. E. Howard: J. Phys. Chem. Solids **19**, 181 (1961)
7.92 A. Baldereschi, F. Bassani: In Proc. 10th Int. Conf. Phys. Semicond., Cambridge USA (1970), ed. by S. P. Keller, J. C. Hensel, F. Stern (USAEC Div. of Technical Information, Oak Ridge 1970) p. 19
7.93 N. Lee, D. M. Larsen, B. Lax: J. Phys. Chem. Solids **34**, 1059 (1973)
7.94 K. Kobayashi: J. Magn. & Magn. Mater. **11**, 84 (1979)
7.95 D. M. Larsen, E. J. Johnson: In Proc. 8th Int. Conf. Phys. Semicond., Kyoto (1966), J. Phys. Soc. Jpn. **21** Suppl., 86 (1966)
7.96 E. J. Johnson, D. M. Larsen: Phys. Rev. Lett. **16**, 655 (1966)
7.97 D. F. Blossey: Phys. Rev. B**2**, 3976 (1970)
7.98 R. Shimizu, T. Koda: J. Phys. Soc. Jpn. **38**, 1550 (1975)
7.99 J. F. McClelland, D. W. Lynch: Phys. Rev. B**19**, 3244 (1979)
7.100 M. Fujita, N. Ohno, K. Nakamura: J. Phys. Soc.. Jpn. **44**, 1868 (1978)

Chapter 8

8.1 R. S. Van Heyningen, F. C. Brown: Phys. Rev. **111**, 462 (1958)
8.2 K. Kobayashi: "Photo-Carrier Transport Phenomena" in *Experimental Methods in Physics* ed. by S. Iida, K. Ôno, H. Kanzaki, H. Kumagai, S. Sawada vol. 7 (Asakura-shoten, Tokyo 1967) p. 197 (in Japanese)
8.3 T. Kawai, K. Kobayashi, H. Fujita: J. Phys. Soc. Jpn. **21**, 453 (1966)
8.4 D. C. Burnham, F. C. Brown, R. S. Knox: Phys. Rev. **119**, 1560 (1960)
8.5 H. H. Tippins, F. C. Brown: Phys. Rev. **129**, 2554 (1963)
8.6 A. G. Redfield: Phys. Rev. **94**, 526 (1954)
8.7 F. C. Brown: Phys. **92**, 502 (1953)
8.8 K. Kobayashi, F. C. Brown: Phys. Rev. **113**, 507 (1959)
8.9 J. A. Borders, J. W. Hodby: Rev. Sci. Instrum. **39**, 722 (1968)
8.10 R. K. Ahrenkiel, F. C. Brown: Phys. Rev. **136**, A223 (1964)
8.11 M. Mikkor, K. Kanazawa, F. C. Brown: Phys. Rev. Lett. **15**, 489 (1965); Phys. Rev. **162**, 848 (1967)
8.12 E. Hanamura, T. Inui, Y. Toyozawa: J. Phys. Soc. Jpn. **17**, 666 (1962)
8.13 J. W. Hodby: J. Phys. C**4**, L8 (1971)

8.14   G. T. Jenkin, J. W. Hodby: J. Phys. C4, L89 (1971)
8.15   H. Tamura, T. Masumi: J. Phys. Soc. Jpn. 30, 897 (1971)
8.16   H. Tamura, T. Masumi: Solid State Commun. 12, 1183 (1973)
8.17   J. W. Hodby, G. T. Jenkin, K. Kobayashi, H. Tamura: Solid State Commun. 10, 1017 (1972)
8.18   H. Tamura, T. Masumi, K. Kobayashi: In Proc. 3rd Int. Conf. Photoconductivity, Stanford (1969), ed. by E. M. Pell (Pergamon, Oxford 1971) p. 183
8.19   D. C. Langreth: Phys. Rev. 159, 717 (1967)
8.20   R. Van Heyningen: Phys. Rev. 128, 2112 (1962)
8.21   F. C. Brown, K. Kobayashi: J. Phys. Chem. Solids 8, 300 (1959)
8.22   T. Kawai, K. Kobayashi, M. Kurita, Y. Makita: J. Phys. Soc. Jpn. 30, 1101 (1971)
8.23   K. Kobayashi, T. Kawai, M. Kanada: J. Phys. Soc. Jpn. 23, 305 (1967)
8.24   T. Masumi, R. K. Ahrenkiel, F. C. Brown: Phys. Status Solidi 11, 163 (1965)
8.25   Y. Makita, K. Kobayashi, M. Kanada, K. Kawai: J. Phys. Soc. Jpn. 25, 816 (1968)
8.26   Y. Makita, K. Kobayashi: J. Phys. Soc. Jpn. 32, 1262 (1972)
8.27   F. C. Brown, N. Inchauspe: Phys. Rev. 121, 1303 (1961)
8.28   C. H. Seager, D. Emin: Phys. Rev. B2, 3421 (1970)
8.29   C. H. Seager: Phys. Rev. B3, 3497 (1971)
8.30   H. Fujita, K. Kobayashi, T. Kawai, K. Shiga: J. Phys. Soc. Jpn. 20, 109 (1965)
8.31   D. C. Burnham: Ph.D. thesis, University of Illinois (1959) p. 43
8.32   F. E. Low, D. Pines: Phys. Rev. 98, 414 (1955)
8.33   D. C. Langreth: Phys. Rev. 137, A 760 (1965)
8.34   Y. Osaka: J. Phys. Soc. Jpn. 21, 423 (1966)
8.35   L. P. Kadanoff: Phys. Rev. 130, 1364 (1962)
8.36   K. K. Thornber, R. P. Feynman: Phys. Rev. B1, 4099 (1970)
8.37   R. P. Feynman: Phys. Rev. 97, 660 (1955)
8.38   F. C. Brown: "Conduction by Polarons in Ionic Crystals", in Point Defects in Solids, ed. by J. H. Crawford, Jr., L. M. Slifkin, vol. 1 (Plenum, New York 1972) p. 491
8.39   D. Howarth, E. Sondheimer: Proc. R. Soc. London, Ser. A 219, 53 (1953)
8.40   T. Holstein: Ann. Phys. (N.Y.)8, 325 (1959)
8.41   Y. Toyozawa, A. Sumi: In Proc. 12th Int. Conf. Phys. Semicond., Stuttgart (1974), ed. by M. H. Pilkuhn (Teubner, Stuttgart 1974) p. 179
8.42   R. C. Hanson: J. Phys. Chem. 66, 2376 (1962)
8.43   R. K. Ahrenkiel, R. S. Van Heyningen: Phys. Rev. 144, 576 (1966)
8.44   H. Fujita, K. Kobayashi, T. Kawai, K. Shiga: J. Phys. Soc. Jpn. 20, 109 (1965)
8.45   K. Kobayashi: In II–VI Semiconducting Compounds 1967 Int. Conf., ed. by D. G. Thomas Benjamin, New York 1967) p. 755
8.46   A. R. Hutson: J. Appl. Phys. Suppl. 32, 2287 (1961)
8.47   J. W. Hodby, J. A. Borders, F. C. Brown: J. Phys. C. 3, 335 (1970)
8.48   A. Honig: Phys. Rev. Lett. 17, 186 (1966)
8.49   D. Schmidt, V. Zimmermann: Phys. Lett. 27A, 459 (1968)
8.50   Y. Makita, K. Kobayashi: J. Phys. Soc. Jpn. 38, 435 (1975)
8.51   I. K. Kikoin, M. M. Noskov: Phys. Z. Sowjetunion 5, 586 (1934); ibid. 6, 478 (1934)
8.52   J. W. Hodby: Solid State Commun. 7, 811 (1969)
8.53   D. M. Larsen: Phys. Rev. 144, 697 (1966)
8.54   H. Tamura, T. Masumi: J. Phys. Soc. Jpn. 30, 1763 (1971)
8.55   J. E. Baxter, G. Ascarelli, S. Rodriguez: Phys. Rev. Lett. 27, 100 (1971)
8.56   D. M. Larsen: "Polaron Energy Levels in Magnetic and Coulomb Fields" in Polarons in Ionic Crystals and Polar Semiconductors, ed. by J. T. Devreese (North Holland, Amsterdam 1972) p. 237
8.57   E. Haga: Prog. Theor. Phys. 13, 555 (1955)
8.58   For example, D. H. Dickey, E. J. Johnson, D. M. Larsen: Phys. Rev. Lett. 18, 599 (1967)
8.59   J. Waldman, D. M. Larsen, P. E. Tannenwald: Phys. Rev. Lett. 23, 1033 (1969)
8.60   J. W. Hodby, J. G. Crowder, C. C. Bradley: J. Phys. C 7, 3033 (1974)
8.61   E. J. Ryder, W. Shockley: Phys. Rev. 81, 139 (1951)

8.62   T. Masumi: Phys. Rev. **129**, 2564 (1963); ibid. **159**, 761 (1967)
8.63   F. Nakazawa, H. Kanzaki: J. Phys. Soc. Jpn. **20**, 465 (1965)
8.64   M. Mikkor, F. C. Brown: Phys. Rev. **162**, 841 (1967)
8.65   S. B. Bolte, F. C. Brown: In Proc. 3rd. Int. Conf. Photoconductivity, Stanford (1969),
        ed. by E. M. Pell (Pergamon, Oxford 1971) p. 139
8.66   H. Fujita, K. Kobayashi, K. Takano: J. Phys. Soc. Jpn. **21**, 2569 (1966)
8.67   M. Onuki, K. Shiga: In Proc. Int. Conf. Phys. Semicond., Kyoto (1966), J. Phys. Soc.
        Jpn. **21**, Suppl., 427 (1966)
8.68   S. Komiyama, T. Masumi, K. Kajita: In Proc. 13th Int. Conf. Phys. Semicond., Rome
        (1976) ed. by F. G. Fumi, p. 1222
8.69   S. Komiyama, T. Masumi, K. Kajita: Phys. Rev. B**20**, 5195 (1979)
8.70   W. Shockley: Bell Syst. Tech. J. **30**, 990 (1951)
8.71   W. E. Pinson, R. Bray: Phys. Rev. **136**, A1449 (1964)
8.72   T. Kurosawa, H. Maeda: J. Phys. Soc. Jpn. **31**, 668 (1971)
8.73   H. Maeda, T. Kurosawa: In Proc. 11th Int. Conf. Phys. Semicond., Warsaw (1972), ed.
        by M. Miacşek (PWN-Polish Scientific Publisher, Warsaw 1972) p. 602
8.74   S. Komiyama, T. Masumi, K. Kajita: Solid State Commun. **31**, 447 (1979)
8.75   S. Komiyama, T. Masumi: Solid State Commun. **26**, 381 (1978)

Chapter 9

9.1   P. Day: "Mixed Valence Chemistry and Metal Chain Compounds", in *Low-Dimensional
        Cooperative Phenomena,* ed. by H. J. Kaller (Plenum, New York 1974) p. 191
9.2   H. Tanino, K. Kobayashi: J. Phys. Soc. Jpn. **52**, 1446 (1983)
9.3   H. Tanino, K. Kobayashi: In Proc. 15th Int. Conf. Ohys. of Semicond., Kyoto (1980), J.
        Phys. Soc. Jpn. **49** Suppl. A, 695 (1980)
9.4   R. J. H. Clark, M. L. Franks, W. R. Trumble: Chem. Phys. Lett. **41**, 287 (1976)
9.5   M. Tanaka, S. Kurita, T. Kojima, Y. Yamada: Chem. Phys. **91**, 257 (1984)
9.6   K. Nasu: J. Phys. Soc. Jpn. **52**, 3865 (1983)
9.7   K. Nasu: J. Phys. Soc. Jpn. **53**, 302 (1984)
9.8   K. Nasu: J. Phys. Soc. Jpn. **53**, 427 (1984)
9.9   R. H. Baughman, S. L. Hsu, G. P. Pelz, A. J. Signorelli: J. Chem. Phys. **68**, 5405 (1978)
9.10   C. S. Yannoni, T. C. Clarke: Phys. Rev. Lett. **51**, 1191 (1983)
9.11   W. P. Su, J. R. Schrieffer, A. J. Heeger: Phys. Rev. Lett. **42**, 1698 (1979); Phys. Rev.
        B**22**, 2099 (1980)
9.12   C. R. Fincher, Jr., D. L. Peebles, A. J. Heeger, M. A. Dry, Y. Matsumura, A. G.
        MacDiarmid, H. Shirakawa, S. Ikeda: Solid State Commun. **27**, 489 (1978)
9.13   C. R. Fincher, Jr., M. Ozaki, M. Tanaka, D. Peebles, L. Lauchlan, A. Heeger, A. G.
        MacDiarmid: Phys. Rev. B**20**, 1589 (1979)
9.14   A. Masui, K. Nakamura: Jpn. J. Appl. Phys. **6**, 1468 (1967)
9.15   T. Tani, P. M. Grant, W. D. Gill, G. B. Street, T. C. Clarke: Solid State Commun. **33**,
        499 (1980)
9.16   L. Lauchlan, S. Etemad, T.-C. Chung, A. J. Heeger, A. G. MacDiarmid: Phys. Rev.
        B**24**, 3701 (1981)
9.17   P. M. Grant, I. P. Batra: Solid State Commun. **29**, 225 (1979)
9.18   H. Suzuki, S. Mizuhashi: J. Phys. Soc. Jpn. **19**, 724 (1964)
9.19   C. B. Duke, A. Paton, W. R. Salaneck, H. R. Thomas, E. W. Plummer, A. J. Heeger,
        A.G. MacDiarmid: Chem. Phys. Lett. **59**, 146 (1978)
9.20   R. Loudon: Am. J. Phys. **27**, 649 (1959)
9.21   L. S. Lichtmann, A. Sarhangi, D. B. Fitchen: Solid State Commun. **36**, 869 (1980)
9.22   J. Orenstein, G. L. Baker: Phys. Rev. Lett. **49**, 1043 (1982)
9.23   E. A. Imhoff, D. B. Fitchen: Solid State Commun. **44**, 329 (1982)
9.24   K. Yoshino, S. Hayashi, T. Sakai, Y. Inuishi, H. Kato, Y. Watanabe: Jpn. J. Appl. phys.
        21, L653 (1982)

9.25 K. Yoshino, S. Hayashi, Y. Inuishi, K. Hattori, Y. Watababe: Solid State Commun. **46**, 583 (1983)

9.26 N. Suzuki, M. Ozaki, S. Etemad, A. J. Heeger, A. G. MacDiarmid: Phys. Rev. Lett. **45**, 1209 (1980)

9.27 J. B. Torrance, J. E. Vazquez, J. J. Mayerle, V. Y. Lee: Phys. Rev. Lett. **46**, 253 (1981)

9.28 P. J. Strebel, Z. G. Soos: J. Chem. Phys. **53**, 4077 (1970)

9.29 Z. G. Soos, S. Mazumdar: Phys. Rev. B **18**, 1991 (1978)

9.30 Z. G. Soos: Chem. Phys. Lett. **63**, 179 (1979)

9.31 J. B. Torrance, A. Girlando, J. J. Mayerle, J. I. Crowley, V. Y. Lee, P. Batail: Phys. Rev. Lett. **47**, 1747 (1981)

9.32 Y. Tokura, T. Koda, T. Mitani, G. Saito: Solid State Commun. **43**, 757 (1982)

9.33 D. Haarer: Chem. Phys. Lett. **27**, 91 (1974)

9.34 D. Haarer: J. Chem. Phys. **67**, 4076 (1977)

9.35 A. Brillante, M. R. Philpott: J. Chem. Phys. **72**, 4019 (1980)

9.36 Y. Oowaki, Y. Tokura, T. Koda: Unpublished. We are grateful to these authors for supplying us with the unpublished figure used as Fig. 9.20

9.37 Y. Tokura, T. Koda: Solid State Commun. **40**, 299 (1981)

9.38 J. Hubbard: Proc. R. Soc. London, Ser. A **276**, 238 (1963); ibid. **281**, 401 (1964)

9.39 Y. Iida: Bull. Chem. Soc. Jpn. **42**, 71 (1969)

9.40 Y. Iida: Bull. Chem. Soc. Jpn. **42**, 637 (1969)

9.41 S. Hiroma, H. Kuroda, H. Akamatu: Bull. Chem. Soc. Jpn. **44**, 9 (1971)

9.42 Y. Oohashi, T. Sakata: Bull. Chem. Soc. Jpn. **46**, 3330 (1973)

9.43 J. B. Torrance, B. A. Scott, F. B. Kaufman: Solid State Commun. **17**, 1369 (1975)

9.44 J. Tanaka, M. Tanaka, T. Kawai, T. Takabe, O. Maki: Bull. Chem. Soc. Jpn. **49**, 2358 (1976)

9.45 A. Hoekstra, T. Spoelder, A. Vos: Acta Crystallogr., Sect. B **28**, 14 (1972)

9.46 M. Konno, Y. Saito: Acta Crystallogr., Sect. B **30**, 1294 (1974)

9.47 M. Konno, Y. Saito: Acta Crystallogr. Sect. B **31**, 2007 (1975)

9.48 M. Konno, T. Ishii, Y. Saito: Acta Crystallogr., Sect. B **33**, 763 (1977)

9.49 J. G. Vegter, T. Hibma, J. Kommandeur: Chem. Phys. Lett. **3**, 427 (1969)

9.50 J. G. Vegter, J. Kommandeur: Mol. Cryst. Liq. Cryst. **30**, 11 (1975)

9.51 Y. Lépine, A. Caillé, V. Larochelle: Phys. Rev. B **18**, 3585 (1978)

9.52 G. Beni, P. Pincus: J. Chem. Phys. **57**, 3531 (1972)

9.53 E. Pyette: Phys. Rev. B **10**, 4637 (1974)

Subject Index

N. Schwentner, E.-E. Koch, J. Jortner

Electronic Excitations in Condensed Rare Gases

1985. VIII, 239 pages. (Springer Tracts in Modern Physics, Volume 107). ISBN 3-540-15382-9

Contents: Introduction. – Experimental Aspects. – Electronic Structure of Valence and Conduction Bands and Excitonic States. – Electronic Excitations in Rare-Gas Liquids. – Metal Rare-Gas Mixtures. – Excited-State Dynamics. – Electron Transport and Electron-Hole Pair Creation Processes. – Concluding Remarks. – References. – Subject Index.

W. Press

Single-Particle Rotations in Molecular Crystals

1981. 53 figures. IX, 129 pages. (Springer Tracts in Modern Physics, Volume 92). ISBN 3-540-10897-1

Contents: Introduction. – Interaction and Rotational Potentials. – Neutron Scattering. – Stochastic Rotational Motion. – Rotational Excitations at Low Temperatures I. Principles. – Rotational Excitations at Low Temperatures II. Examples. – Rotational Excitations at Low Temperatures III. Special Features. – Appendix: Calculation of Transition Matrix Elements. – List of Symbols. – References. – Subject Index.

Excitons

Editor: **K. Cho**
1979. 118 figures, 8 tables. XI, 274 pages. (Topics in Current Physics, Volume 14). ISBN 3-540-09567-5

Contents: *K. Cho:* Introduction. – *K. Cho:* Internal Structure of Excitons. – *P. J. Dean, D. C. Herbert:* Bound Excitons in Semiconductors. – *B. Fischer, J. Lagois:* Surface Exciton Polaritons. – *P. Y. Yu:* Study of Excitons and Exciton-Phonon Interactions by Resonant Raman and Brillouin Spectroscopies.

Springer-Verlag
Berlin Heidelberg
New York Tokyo

Springer

V. L. Broude, E. I. Rashba, E. F. Sheka

Spectroscopy of Molecular Excitons

1985. 135 figures. XI, 271 pages. (Springer Series in Chemical
Physics, Volume 16). ISBN 3-540-12409-8

Contents: Experimental Background. – Exciton Spectra of
Perfect Crystals. – Exciton Spectra of Doped Crystals. – Exciton
Spectra of Mixed Crystals. – Band-to-Band Transition Spectra. –
Vibronic Spectra of Molecular Crystals. – Conclusion. –
Appendix A. – Appendix B. – References. – Subject Index.

Inert Gases

Potentials, Dynamics, and Energy Transfer in Doped Crystals
Editor: **M. L. Klein**
1984. 89 figures. XI, 266 pages. (Springer Series in Chemical
Physics, Volume 34). ISBN 3-540-13128-0

Contents: *M. L. Klein:* Argon and Its Companions. – *R. A. Aziz:*
Interatomic Potentials for Rare-Gases: Pure and Mixed Inter-
actions. – *S. S. Cohen, M. L. Klein:* Dynamics of Impure Rare-Gas
Crystals. – *H. Dubost:* Spectroscopy of Vibrational and Rotational
Levels of Diatomic Molecules in Rare-Gas Crystals. – Subject
Index.

B. Y. Zel'Dovich, N. F. Pilipetsky, V. V. Shkunov

Principles of Phase Conjugation

1985. 70 figures. X, 250 pages. (Springer Series in Optical
Sciences, Volume 42). ISBN 3-540-13458-1

Contents: Introduction to Optical Phase Conjugation. – Physics
of Stimulated Scattering. – Properties of Speckle-Inhomogeneous
Fields. – OPC by Backward Stimulated Scattering. – Specific
Features of OPC-SS. – OPC in Four-Wave Mixing. – Nonlinear
Mechanisms for FWM. – Other Methods of OPC. – References.
– Subject Index.

A. A. Kaminskii

Laser Crystals

Their Physics and Properties
Translation edited by H. F. Ivey
1981. 89 figures, 56 tables. XIV, 456 pages. (Springer Series in
Optical Sciences, Volume 14). ISBN 3-540-09576-4

"... the book is excellent, being very readable and well balanced
in its choice of subject matter... This book serves as an excellent
introduction to the physics of laser crystals and as such will
strongly appeal to physicists and other researchers in this field.
As a work of reference it will prove of great value to both laser
scientists and engineers.". *Optica Acta*

Springer-Verlag
Berlin Heidelberg
New York Tokyo